The IGBT Device

Elsevier
Radarweg 29, PO Box 211, 1000 AE Amsterdam, Netherlands
The Boulevard, Langford Lane, Kidlington, Oxford OX5 1GB, United Kingdom
50 Hampshire Street, 5th Floor, Cambridge, MA 02139, United States

ISBN: 978-0-323-99912-0

For information on all Elsevier publications
visit our website at https://www.elsevier.com/books-and-journals

Publisher: Matthew Deans
Acquisitions Editor: Stephen Jones
Editorial Project Manager: Maria Elaine D. Desamero
Production Project Manager: Prem Kumar Kaliamoorthi
Cover Designer: Greg Harris

Typeset by STRAIVE, India

I dedicate this book to my wife, Pratima, for her support during my work on the development of the insulated gate bipolar transistor (IGBT) while working at the General Electric Company.

I also dedicate this book to the thousands of engineers from around the world who worked on improving the performance of the IGBT over the last 40 years and applying these devices to serve the needs of all sectors of our worldwide economy.

Contents

About the Author

Jay Baliga is internationally recognized for his leadership in the area of power semiconductor devices. In addition to more than 700 publications in international journals and conference digests, he has authored and edited 26 books (*Power Transistors*, IEEE Press 1984; *Epitaxial Silicon Technology*, Academic Press 1986; *Modern Power Devices*, John Wiley 1987; *High Voltage Integrated Circuits*, IEEE Press 1988; *Solution Manual: Modern Power Devices*, John Wiley 1988; *Proceedings of the 3rd Int. Symposium on Power Devices and ICs*, IEEE Press 1991; *Modern Power Devices*, Krieger Publishing Co. 1992; *Power Electronics Technology and Applications I*, IEEE Press 1992; *Proceedings of the 5th Int. Symposium on Power Devices and ICs*, IEEE Press 1993; *Power Semiconductor Devices*, PWS Publishing Company 1995; *Solution Manual: Power Semiconductor Devices*, PWS Publishing Company 1996; *Cryogenic Operation of Power Devices*, Kluwer Press 1998; *Power Electronics Technology and Applications II*, IEEE Press 1998; *Silicon RF Power MOSFETs*, World Scientific Publishing Company 2005; *Silicon Carbide Power Devices*, World Scientific Publishing Company 2006; *Fundamentals of Power Semiconductor Devices*, Springer Science, 2008; *Solution Manual: Fundamentals of Power Semiconductor Devices*, Springer Science, 2008; *Advanced Power Rectifier Concepts*, Springer Science, 2009; *Advanced Power MOSFET Concepts*, Springer Science, 2010; *Advanced High Voltage Power Device Concepts*, Springer Science, 2011; Chinese Edition of *Fundamentals of Power Semiconductor Devices*, Springer Science 2012; *The IGBT Device: Physics, Design, and Applications*, Elsevier Press 2015; Chinese Edition of *Advanced High Voltage Power Devices Concepts*, China Machine Press 2015; *Gallium Nitride and Silicon Carbide Power Devices*, World Scientific Publishing Company 2016; *Fundamentals of Power Semiconductor Devices, Second Edition*, Springer Science 2018; *Wide Bandgap Semiconductor Power Devices*, Elsevier 2019).

In addition, he has contributed chapters to another 24 books. He holds 122 US patents in the solid-state area. In 1995, one of his inventions was selected for the B.F. Goodrich Collegiate Inventors Award presented at the Inventors Hall of Fame.

Prof. Baliga obtained his bachelor of technology degree in 1969 from the Indian Institute of Technology (IIT), Madras, India. He was the recipient of the Philips India Medal and the Special Merit Medal (as valedictorian) at IIT, Madras. He obtained his masters and Ph.D. degrees from Rensselaer Polytechnic Institute (RPI), in Troy, New York, in 1971 and 1974, respectively. His thesis involved gallium arsenide diffusion mechanisms and pioneering work on the growth of InAs and GaInAs layers using organometallic CVD techniques. This technology is now widely used to manufacture light-emitting diodes, lasers, radio frequency devices, and compound semiconductor-based ICs. At RPI, he was the recipient of the IBM Fellowship in 1972 and the Allen B. Dumont Prize in 1974.

From 1974 to 1988, Baliga performed research and directed a group of 40 scientists at the General Electric Research and Development Center in Schenectady,

New York, in the area of power semiconductor devices and high-voltage integrated circuits. During this time, he pioneered the concept of combining metal oxide semiconductor (MOS) and bipolar physics to create a new family of discrete devices. He is the inventor of the insulated gate bipolar transistor (IGBT), which is now in production by many international semiconductor companies. This invention is widely used around the globe for air conditioning, home appliance (washing machines, refrigerators, mixers, etc.) control, factory automation (robotics), medical systems (CAT scanners, uninterruptible power supplies), and electric streetcars/bullet trains as well as for the drive train in electric and hybrid electric cars. IGBT-based motor control improves efficiency by more than 40%. The IGBT was essential in the deployment of compact fluorescent lamps to replace incandescent lamps, improving efficiency by 75%. Since two-thirds of the electricity in the world is used to run motors and 20% of the electricity in the world is used for lighting, the availability of IGBTs has produced a cumulative electrical energy savings of more than 100,000 TWh. In addition, the IGBT enabled the introduction of the electronic ignition system for running spark plugs in the internal combustion engine of gasoline-powered cars and trucks. The resulting 10% improvement in fuel efficiency has saved consumers 1.8 trillion gallons of gasoline during the last 30 years. The cumulative impact of these electrical energy and gasoline savings is a cost savings of more than $32 trillion for consumers, and a reduction in carbon dioxide emissions from coal-fired power plants by more than 150 trillion pounds. For this achievement, he has been labeled the "man with the smallest carbon footprint on earth." Most recently, the IGBT has enabled the creation of very compact, lightweight, and inexpensive defibrillators used to resuscitate cardiac arrest victims. Deployment of these portable defibrillators in fire trucks, ambulances, buildings, and aircraft is projected by the American Medical Association (AMA) to save 100,000 lives per year in the United States. In recognition of his work on the IGBT, *Scientific American* magazine named him one of the eight heroes of the semiconductor revolution in their 1997 special issue commemorating the solid-state century.

Dr. Baliga is also the originator of the concept of merging Schottky and pn junction physics to create a new family of JBS power rectifiers that is commercially available from various companies. This concept, originally implemented for silicon devices, has become an essential concept for the commercialization of silicon carbide high-voltage Schottky rectifiers since 2005.

In 1979, Dr. Baliga developed a theoretical analysis resulting in Baliga's Figure of Merit (BFOM), which relates the resistance within power rectifiers and FETs to the basic semiconductor properties. He predicted that the performance of Schottky power rectifiers and power MOSFETs could be enhanced by several orders of magnitude by replacing silicon with other materials such as gallium arsenide and silicon carbide. This is forming the basis of a new generation of power devices in the 21st century. These devices have been commercialized since 2010.

In August 1988, Dr. Baliga joined the faculty of the Department of Electrical and Computer Engineering at North Carolina State University (NCSU), Raleigh, North Carolina, as a full professor. In 1991, he established an international center called the Power Semiconductor Research Center (PSRC) there for research in the area of power semiconductor devices and high-voltage integrated circuits, and served as its founding director until 2001. His research interests include the modeling of novel device concepts, device fabrication technology, and the investigation of the impact of new materials, such as GaAs and silicon carbide, on power devices. The first high-performance SiC Schottky rectifiers and power MOSFETs were demonstrated at PSRC in the 1990s, resulting in the release of products by many companies during the last 10 years.

In 2008, Prof. Baliga was a key member of an NCSU team—partnered with four other universities—that was granted an Engineering Research Center from the National Science Foundation, named FREEDM Center, for the development of microgrids that allow the integration of renewable energy sources. Throughout this 10-year program, he was responsible for the fundamental sciences platform and the development of power devices from wide-bandgap semiconductors for utility applications.

In recognition of his contributions to NCSU, he was given the highest university faculty rank of Distinguished University Professor of Electrical Engineering in 1997. He was made the Progress Energy Distinguished University Professor of Electrical Engineering in 2018.

Prof. Baliga has received numerous awards in recognition for his contributions to semiconductor devices. These include two IR 100 awards (1983, 1984), the Dushman and Coolidge Awards at GE (1983), and being selected among the 100 Brightest Young Scientists in America by Science Digest Magazine (1984). He was elected Fellow of the IEEE in 1983 at the age of 35 for his contributions to power semiconductor devices. In 1984, he was given the Applied Sciences Award by the sitar maestro Ravi Shankar at the Third Convention of Asians in North America. He received the 1991 IEEE William E. Newell Award, the highest honor given by the Power Electronics Society, followed by the 1993 IEEE Morris E. Liebman Award for his contributions to the emerging Smart Power Technology. In 1992, he was the first recipient of the BSS Society's Pride of India Award. At 45, he was elected as foreign affiliate to the prestigious National Academy of Engineering, and was one of only four citizens of India to have the honor at that time (converted to a regular member in 2000 after taking US citizenship). In 1998, the University of North Carolina system selected him for the O. Max Gardner Award, which recognizes the faculty member among the 16 constituent universities who has made the greatest contribution to the welfare of the human race. In December 1998, he received the IEEE J.J. Ebers Award, the highest recognition given by the IEEE Electron Devices Society, for his technical contributions to the solid-state area. In June 1999, he was honored at the Whitehall Palace in London with the IEEE Lamme Medal, one of the highest forms of recognition given by the IEEE Board of Governors, for his contributions to the development of an apparatus/technology of benefit to society. In April 2000, he was honored by his alma mater as a distinguished alumnus. In November 2000, he received the R.J. Reynolds Tobacco Company Award for Excellence in Teaching, Research, and Extension for his contributions to the College of Engineering at NCSU. In 2011, Dr. Baliga was selected to receive the Alexander Quarles Holladay

Medal for Excellence, which recognizes members of the NCSU faculty who over their careers have made outstanding contributions to the university through their research, teaching, and extension services.

In 1999, Prof. Baliga founded a company, Giant Semiconductor Corporation, with seed investment from Centennial Venture Partners, to acquire an exclusive license for his patented technology from NCSU with the goal of bringing his inventions to the marketplace. A company, Micro-Ohm Corporation, subsequently formed by him in 1999, has been successful in licensing the GD-TMBS power rectifier technology to several major semiconductor companies for worldwide distribution. These devices have application in power supplies, battery chargers, and automotive electronics. One manufacturer states that this is the most successful new rectifier product in the last 25 years. In June 2000, Baliga founded another company, Silicon Wireless Corporation, to commercialize a novel superlinear silicon RF transistor that he invented for application in cellular base stations, and grew it to 41 employees. This company (renamed Silicon Semiconductor Corporation) received an investment of $10 million from Fairchild Semiconductor Corporation in December 2000 to codevelop and market this technology. Based upon his additional inventions, this company also produced a new generation of power MOSFETs for delivering power to microprocessors in notebooks and servers. This technology was licensed by his company to Linear Technologies Corporation with transfer of the know-how and the manufacturing process. Voltage regulator modules using these transistors are currently available on the market for powering microprocessor and graphics chips in laptops and servers. NCSU conferred the Risley Entrepreneur of the Year Award to Baliga in 2013 to recognize his entrepreneurial leadership.

In 2010, Dr. Baliga was inducted into the Engineering Design Magazine's Engineering Hall of Fame for his invention, development, and commercialization of the IGBT, joining well-known luminaries (e.g., Edison, Tesla, and Marconi) in the electrical engineering field. The award announcement states: "While working at General Electric in the late 1970s, Baliga conceived the idea of a functional integration of MOS technology and bipolar physics that directly led to the IGBT's development... it remains undeniable that Baliga's vision and leadership played a critical role in moving the IGBT from a paper-based concept to a viable product with many practical applications."

President Obama personally presented Dr. B. Jayant Baliga with the National Medal of Technology and Innovation, the highest form of recognition given by the US government to an engineer, in a ceremony at the White House on October 21, 2011. Dr. Baliga's award citation reads: "For development and commercialization of the Insulated Gate Bipolar Transistor and other power semiconductor devices that are extensively used in transportation, lighting, medicine, defense, and renewable energy generation systems."

In October 2012, Governor Beverly Purdue presented Dr. Baliga with the North Carolina Award for Science. This is the highest award given by the state and the governor to a civilian. On October 4, 2013, he was inducted into the Rensselaer Alumni Hall of Fame by Rensselaer Polytechnic Institute President Shirley Jackson. The ceremony included unveiling his portrait etched on a window in Thomsen Hall in the Darrin Communications Center.

On August 23, 2014, Dr. Baliga received the IEEE Medal of Honor "For the invention, implementation, and commercialization of power semiconductor devices with widespread benefits to society" in a ceremony held in Amsterdam, The Netherlands. This, IEEE's highest award, has been given since 1917 to recognize great achievements in the field of electrical engineering.

In 2015, Dr. Baliga was honored with the Global Energy Prize, considered the Nobel Prize for energy, by the Russian Federation at a ceremony in St. Petersburg. His citation states that the award is for the "For invention, implementation, and commercialization of the Insulated Gate Bipolar Transistor—one of the most important innovations for the control and distribution of energy."

In 2016, Prof. Baliga was inducted into the National Inventors Hall of Fame as the sole inventor of the insulated gate bipolar transistor. This invention was recognized, with a display at a museum in Washington, DC, among one of the fewer than 600 patents chosen from more than 10 million patents issued by the US Patent and Technology Office.

In 2016, he was asked to be the chief guest at the Indian Institute of Technology—Madras convocation, where he was conferred a Doctor of Science (Honoris Causa) degree. In 2017, he was inducted into the National Academy of Inventors as an outstanding academic inventor.

Foreword

When I joined the General Electric Company Research and Development (R&D) Center in 1970, the world of power electronics was a completely different environment than exists today. There, I worked under the tutelage of some of the legends of power electronics such as Bill McMurray, Bernie Bedford, Fred Turnbull, and many others. This was still during the era of thyristor technology, dating from 1956, in which the inverters of that time called upon the use of resonant circuits to turn off the thyristors. The McMurray–Bedford inverter, the McMurray inverter, and the Verhoef inverter were just some of the developments during this period. Because of the numerous modes of operation (leading or lagging power factor, no load or full load, low or high frequency, etc.), analyses of these circuits were at the same time both a nightmare and a joy to those who loved to solve differential equations!

While some solutions for self-switching were being attempted at the time (e.g., FETs—small power only, Darlington BJTs—heavy monstrous bricks), they seemed to be going nowhere. This era ended in the late 1970s when it was announced by corporate management that a new switching device had been developed at the R&D Center by Jay Baliga, which removed the need for the turnoff resonant circuit and achieved turnoff with an FET-style gate pulse. By the mid-1980s, some of the remaining problems with this device had been solved (latch-up and secondary breakdown) and a new era of power electronics was set in motion. In the last 30 years, we have seen continuous expansion of insulated gate bipolar transistor (IGBT) technology, with the aid of 6000 V devices, even into high-voltage DC applications. To think that a turnoff device could ever make inroads into that last bastion of thyristor technology, that is, HVDC power conversion, would make one's head spin in 1970. We have all benefitted greatly from Prof. Baliga and his contemporaries working in the field of solid-state device technology. In this book, we learn from the master.

Thomas A. Lipo
University of Wisconsin, Madison, WI, United States

Foreword

Preface to the Second Edition

It has been 5 years since *The IGBT Device* book was published. It was well received and won the prestigious PROSE Award as the best book in engineering and technology published in 2015. Semiconductor technology makes rapid advances due to demands from applications. This tenet is certainly applicable to the case of insulated gate bipolar transistor (IGBT) devices. This has motivated the preparation of a second edition of this book. Modern publication methods also make color illustrations affordable. Color illustrations have therefore been extensively used in the new edition to enhance the reader experience and convey the information more effectively. My intention was to add new published technical information to all sections of the book to make it up to date. It was also important to revise the written materials on the device's social impact because this has grown enormously during the last 5 years.

Before preparation of the second edition, the publisher asked some reviewers for suggestions on the proposal. The reviewers suggested the addition of new IGBT device concepts, such as reverse conducting IGBTs and fine-mesh IGBTs, which were already planned. Many new IGBT structural concepts have therefore been added to the book. The reviewers also suggested adding more information on device fabrication. This has been done in the new edition. The information on packaging and gate drivers has been expanded based on reviewer suggestions. Many new references were added to each chapter when preparing the second edition. This brings the book up to date. A reviewer suggested that more in-depth analysis be provided for all the power circuits described in the book. This suggestion was not practical due to limitations on the size of the book and not appropriate because the book was intended to focus on the IGBT rather than on power electronics.

During the past 5 years, important new technology trends have emerged. The first is a growing role of energy storage to complement various applications, such as renewable energy generation and data centers. Consequently, a new chapter on energy storage has been added to the book. Another new trend was the rapid acceleration of renewable energy deployment. Chapter 15 on renewable energy was therefore expanded with more recent information. In addition, the replacement of gasoline power automobiles with electric vehicles has become a globally recognized imperative to reduce greenhouse gas emissions. The role of the IGBT in enabling electric vehicles was already extensively discussed in the original book. This information has been updated in the new edition. More importantly, the future environmental benefits of these developments have been discussed in the chapter on social impact.

One of the most important changes during the past years is the growth of the impact of IGBT in many sectors of the economy. The 1990–2015 time frame saw a growth in IGBT-enabled technologies from a small scale at the beginning of the 1990s to a major impact by 2015. This high impact has continued during the past 5 years, making the already significant cumulative effect much larger in magnitude. The cost savings enjoyed by consumers has grown from \$15 trillion to \$33.7 trillion while the reduction of carbon dioxide emissions has increased from 100 trillion pounds to 180 trillion pounds. This impact will continue to grow in the future.

Silicon carbide- and gallium arsenide-based power devices finally became commercialized in 2015 after a 30-year period of investment in materials and device development. In principle, these devices could displace the silicon IGBT in various applications due to superior on-state and switching losses. However, it is now recognized that this displacement will occur in the near term only in niche areas due to the high cost of this new technology and the lack of confidence in its reliability on the part of end users. Consequently, the strong position of the silicon IGBT in all sectors of the economy is unlikely to change for a significant length of time. This makes the new edition of this book valuable to all engineers, technologists, economists, and environmentalists for the foreseeable future.

B. Jayant Baliga

Preface to the First Edition

In 1977, I submitted a patent disclosure while working for the General Electric (GE) Company on vertical metal oxide semiconductor (MOS)-gated thyristors that contain the basic insulated gate bipolar transistor (IGBT) structure. After a V-groove process for making the structure was developed, the fabrication of the devices began in November 1978 with completion in July 1979. In addition to the latched-up thyristor mode of operation, my measurements clearly showed the IGBT mode of operation. In response to the need for an improved power switch for adjustable speed drives for heat pump applications by GE, I prepared a patent disclosure in September 1980 that described all the characteristics for the IGBT that we now take for granted. It was immediately apparent that this new device would have a widespread impact on the company's products in the small appliance, large appliance, medical, factory automation, and lighting business units. Due to its impact across the company, my proposal received the attention of Chairman Jack Welch, who supported its commercialization. I was fortunately able to deliver a 600 V, 10 A IGBT within a year by fabricating the device in the existing power MOSFET production facility with my chip and process design, including suppression of the latch-up of the parasitic thyristor. I simultaneously developed an electron irradiation lifetime control process with a unique annealing step that healed the damage created by electron irradiation to the gate oxide. This allowed production of IGBTs optimized for a wide range of switching frequencies and applications. The availability of these IGBTs spurred power electronics designers to rapidly apply them to a large variety of products at GE. GE eventually announced the commercially available IGBT product in 1983. This prompted worldwide interest in manufacturing the device, leading to products from other companies after 1985.

A few years ago, the new head of the Department of Electrical and Computer Engineering at North Carolina State University suggested that I prepare a report about the impact of my work on the IGBT to post on our website. The outcome of my effort was a 140-page document with more than 300 references titled *The IGBT Compendium: Applications and Social Impact*. GE recognized the impact of the IGBT on most of the company's product divisions immediately after my invention of the device. I was personally involved with the design of IGBTs suitable for GE's adjustable speed drives for Trane and Carrier air conditioners (heat pumps), for GE's early efforts on creating more efficient lighting products, and a variety of small and large appliance controls. However, preparing the report on the IGBT after a time span of 30 years was a voyage of discovery. It was apparent that the IGBT had now penetrated literally every sector of the economy and enhanced the comfort, convenience, and health for billions of people around the globe.

Improving the efficiency of power management and delivery is in the very nature of power electronics. It is well recognized that power semiconductor devices play a dominant role in achieving this outcome. However, the impact of the improvements in efficiency on power savings had not been quantified using a rigorous methodology.

Without this metric, it was also not possible to evaluate the environmental consequences of this technology. Because two-thirds of the electricity in the world is used to run motors, I decided to quantify the power savings derived from IGBT-based adjustable speed drives for motors. In addition, since one-fifth of the electricity in the world is used for lighting, I decided to quantify the impact of compact fluorescent lamps (CFLs) because IGBTs are used in the electronic ballast. The third sector of the economy that has benefitted from IGBTs is the transportation sector. It became quickly apparent that electronic ignition systems, enabled in the late 1980s with the availability of the IGBT to control the spark plugs of internal combustion engines in cars and trucks, had enhanced fuel efficiency. With the huge quantities of gasoline consumed around the world, it became important to quantify the impact of this innovation. With just these three applications of the IGBT, I determined that society had saved more than 100,000 terawatt hours in electricity consumption (equivalent to not building 11,000 coal-fired power plants) and more than 1.8 trillion gallons of gasoline. This not only saved worldwide consumers more than $32 trillion but also reduced carbon dioxide emissions by more than 150 trillion pounds from 1990 to 2020.

In 2012, I was encouraged by colleagues to consider writing a book on the IGBT with the above report as a foundation. My reaction was a proposal for creating a comprehensive book on the IGBT that first includes the device operation, chip design, fabrication technology, packaging, and gate drive circuits. It then provides an extensive discussion of its applications in all sectors of the economy with elaboration of the circuit topologies used in each case and the optimized IGBT device structures developed by the power semiconductor industry for each application. I was very pleased that the editors at Elsevier found my proposal compelling. The reaction from reviewers of my IGBT book proposal was also very positive, with the suggestion that I include a discussion of how I invented, developed, and successfully commercialized the IGBT in the early 1980s.

That book (the first edition) was the result of 2 years of effort to create a single resource for the reader regarding the operation and design of the IGBT as well as its social impact. It was immensely gratifying that the first edition of the book was recognized with the Professional and Scholarly Excellence Award as the best book published in the electronics area in 2015. This second edition has been prepared after a period of 7 years to capture the advancements in IGBT technology since the first edition was published. In addition, many papers have been published during these years on IGBT applications in all sectors of the economy; these applications are included in the new edition to bring it up to date. Most importantly, a new chapter on energy storage has been added in the second edition. This technology is of great importance in renewable energy generation and for server farms used in cloud computing and data storage. It is worth pointing out that many of the figures have been redrawn in color in the second edition to make the book more appealing and to make it easier to assimilate the information. I hope this second edition will be received as favorably as the first.

Chapter 1 provides a high-level perspective of the applications of the IGBT and its power ratings. It includes a discussion of the history behind the conception of the device and its commercialization. Chapter 2 describes various IGBT structures that have evolved over the years. The very first IGBT I developed at GE in 1981 was a 600 V symmetric blocking device, followed soon after with the 600 V asymmetric blocking devices. The power semiconductor manufacturers focused their attention on the asymmetric structure for motor drive applications over the next 30 years. More recently, there has been interest in the symmetric blocking IGBT for current source inverters and matrix converters. The first IGBTs made use of the planar gate structure, but significant improvement in the trade-off curve between on-state voltage drop and switching losses was later achieved using trench gate devices. The transparent emitter IGBT structure had an important role in scaling the voltage ratings of the IGBT to allow application to traction drives.

Chapter 3 provides a description of the physics of operation of the IGBT structure to allow its design using analytical models. The symmetric, asymmetric, and transparent emitter structures are systematically analyzed in terms of the blocking characteristics, the on-state voltage drop, and the power loss trade-off curve. Silicon carbide IGBTs are included here for completeness, although no commercial devices are as yet available.

The excellent ruggedness of the IGBT with a wide safe operating area has been one of its prime features from an application standpoint. Chapter 4 provides analytical models for designing the safe operating area of the IGBT. It includes device cell innovations that have been responsible for preventing latch-up of the internal parasitic thyristor, which was considered a show stopper when I originally proposed the IGBT.

Chapter 5 provides a practical description of the layout of the active area for the IGBT chip and its edge termination. Techniques for overcurrent, overvoltage, and overtemperature protection are described here. Lifetime control processes that enable adjustment of the switching speed of the IGBT without damaging its gate oxide are described as well.

Chapter 6 describes packaging technology for both discrete IGBTs and for IGBTs packaged in modules. The power module designs range from low- to high-power levels. In Chapter 7, various gate drive circuits are provided that enable controlling the reverse recovery of the fly-back diodes and switching losses in the IGBT itself. Chapter 8 provides models employed for simulation of IGBTs in power circuits.

In Chapters 9–19, the applications of IGBTs in various sectors of the economy are reviewed. These chapters demonstrate the breadth of the impact of this singular innovation on society. In each chapter, the circuit topology, such as hard switching versus resonant switching, and the IGBT specifications that ensure efficient operation in these circuits, are described. The optimization of the IGBT structure by the device manufacturers to reduce power losses in each case is then provided.

In the transportation sector discussed in Chapter 9, at the individual consumer level, the IGBT is essential for operating the internal combustion engine in gasoline-powered automobiles, and for driving the electric motors in electric vehicles and hybrid electric vehicles. The IGBT is essential for mass transit systems, ranging from electric buses and trams to high-speed rail networks around the world.

With the growth in ratings for the IGBTs, they have even penetrated propulsion systems for large ships and enabled the all-electric aircraft.

The discussion of the industrial sector in Chapter 10 includes adjustable speed drives for motor control, factory automation systems, robotics, welding, induction heating, milling and drilling, mining, and paper, textile, and metal mills. Chapter 11 on the lighting sector provides extensive coverage of various circuits used for this high-volume application. In addition, the use of IGBTs in strobe lights for the flash in cameras and for powering xenon arc lamps in automobiles and movie projectors is described.

Chapter 12 deals with the consumer section where the IGBT has been utilized for a large variety of appliances. Among large appliances commonly used in our homes are the air conditioner, the refrigerator, the washing machine, the microwave oven, the induction cooktop, and the dishwasher. Among small tabletop appliances that are essential conveniences for food preparation in kitchens are portable induction cooktops, rice cookers, blenders, mixers, juice makers, and mixers. In addition, the IGBT is an essential component in older generation television sets with cathode ray tubes and in modern plasma TV sets.

Society has greatly benefitted from IGBTs in the medical sector as well for improving medical diagnosis and life-saving events during cardiac arrest. Chapter 13 describes the use of IGBTs in the power supply for x-ray machines, CT scanners, MRI scanners, and ultrasound machines to produce high-quality images for medical diagnosis and treatment of physical trauma. The automatic external (portable) defibrillators could not have been deployed in low-cost, lightweight, laptop-size units without the availability of the IGBT. The creation of this device saves more than 100,000 lives in the United States each year and many more around the world.

The IGBT was at first reluctantly adopted in the United States by the defense sector, as described in Chapter 14, but now occupies an essential role in equipment deployed by all the armed forces. The navy utilizes it in power distribution systems in warships, aircraft carriers, and nuclear submarines. The Army is developing electric vehicles that depend on IGBTs for their inverters and the Air Force makes use of IGBTs to replace hydraulic systems with more reliable, lightweight electrical actuators.

The mitigation of global warming due to increased carbon in the atmosphere from fossil fuel (carbon and natural gas)-powered electricity generating plants requires increased deployment of solar and wind power generation capacity. These renewable resources all utilize the IGBT in the inverter for delivering well-regulated AC power to the transmission grid. Chapter 15 describes the power electronics technology not only for these renewable resources but also for hydroelectric power, wave power, tidal power, and geothermal power.

Chapter 16 describes the penetration of the IGBT into the power transmission sector. This has occurred more recently after the power ratings of IGBT modules were enhanced by semiconductor suppliers to handle megawatt power levels. IGBT-based static VAR compensators and static synchronous compensators

(STATCOM) have been employed for AC power transmission grids. The HVDC-lite concept using IGBTs is gaining dominance in new power transmission deployments.

The IGBT has even benefitted the financial sector of the economy, as discussed in Chapter 17. With the advent of computer-based high-speed transactions in the banking, credit card, and investment sectors, any interruption of power can lead to a loss of millions of dollars in revenue each hour. IGBT-based uninterruptable power supplies have become essential equipment in protecting data centers against not only power interruption but voltage sags, voltage swells, and other power quality issues.

The addition of energy storage capability has been gaining increasing importance. It is necessary to supplement energy generation from renewable resources due to their intermittent nature. It is essential for the success of cloud computing and information storage to ensure reliable operation, an expectation of millions of subscribers to these modern conveniences. The role of the IGBT in energy storage deployment is described in Chapter 18.

Chapter 19 has been written to capture the myriad of other IGBT applications that do not fit into the sectors of the economy discussed in earlier chapters. These applications include smart homes; printing and copying machines; airport security machines; particle accelerators, including the CERN Large Hadron collider used for the discovery of the Higgs boson; food and water sterilization; water desalination; roller coasters; and the NASA space shuttle and the International Space Station.

The social impact of the IGBT is covered in Chapter 20. Here, three case studies are described: adjustable speed motor drives, CFL lamp ballasts, and electronic ignition systems. These three applications enabled by IGBTs have tremendously improved efficiency.

My intention is to create a book on the IGBT that provides not only a comprehensive description of its operation and design but also its breadth of application across all sectors of the economy as well as to quantify its social impact. This book should be of interest to all power semiconductor and power electronics engineers. In addition, it should be of interest to social scientists who are interested in the impact of technology on society.

Jayant Baliga

CHAPTER

Introduction

1

Today, the insulated-gate bipolar transistor (IGBT) is pervasively used in power electronic systems and their applications to improve the comfort and quality of life for billions of people around the world. The impact of the IGBT on society can be measured by asking the question: " What would happen if IGBTs were removed from all the applications they serve today?" The answer is quite revealing:

1. Our gasoline-powered cars would stop running because the electronic ignition systems would no longer function.
2. Our hybrid and electric cars would stop running because the inverters used to deliver power from the batteries to the motors would no longer function.
3. Our electric mass transit systems would come to a standstill because the inverters used to deliver power from the electric grid to the motors would no longer function.
4. Our air-conditioning systems in homes and offices would stop working because the inverters used to deliver power from the utility company to the heat pumps and compressors would no longer function.
5. Our refrigerators and vending machines would no longer function, making the delivery and storage of perishable products impossible.
6. Our factories would come to a grinding halt because the controllers used to run the robots would cease to function.
7. Our new low-energy compact fluorescent bulbs would stop functioning, significantly increasing power consumption for lighting due to reverting back to incandescent bulbs.
8. Our portable defibrillators recently deployed in emergency vehicles, airplanes, and office buildings would no longer be operational, putting more than 100,000 people at risk of death from cardiac failure every year.
9. Our new solar- and wind-based renewable energy resources would not be able to deliver power to the grid because the inverters would stop functioning.
10. Our uninterruptible power supplies would no longer work, jeopardizing financial transactions conducted by banks and investment firms.
11. Our modern power transmission systems would become less efficient, resulting in greater carbon emissions from fossil fuels.

The IGBT Device. https://doi.org/10.1016/B978-0-323-99912-0.00009-X

In other words, the quality of life in our society would be greatly impaired if the IGBT was no longer available. The IGBT has become an embedded technology that enriches the lives of billions of people around the globe by providing comfort in the home, food preservation, industrial manufacturing, transportation, and even medical assistance.

In September 2005, when celebrating their 30th anniversary of covering trends in power semiconductor technology, Power Electronics Technology magazine published a review article [1] with a milestone chart. In this chart, the first significant event highlighted is the invention of the bipolar transistor by Brattain, Bardeen, and Shockley in 1947, for which they received the Nobel Prize in 1956. The next important milestone is the invention of the integrated circuit by Jack Kilby, who received a Nobel Prize in 2000. The conception of the integrated circuit is also credited to Robert Noyce, who received the National Medal of Technology and Innovation in 1987. During the 1950s, power thyristors were also commercially introduced for high-power applications. Major manufacturers for these bipolar devices were General Electric (GE) and Westinghouse Corporation. According to the milestone chart, the next major innovation in power devices was the introduction of the power metal-oxide semiconductor field-effect transistor (MOSFET) by Siliconix in 1975 and International Rectifier in 1978. At this time, the power semiconductor industry was bifurcated into two tracks, with one group of companies producing bipolar power devices and a separate group of companies producing power MOSFET devices. At that time, the manufacturing of these devices was considered to be incompatible because the MOS devices required know-how in the control of semiconductor surface properties while the bipolar devices relied on control of minority carriers within the bulk regions of semiconductors.

The milestone chart states that I invented the IGBT in 1979–80. This was accomplished by proposing the functional integration of MOS and bipolar physics within the same monolithic structure. In December 2010, I was inducted into the Electronic Design Engineering Hall of Fame for the invention, development, and commercialization of the IGBT. The award announcement states [2]: "*While working at General Electric in the late 1970s, Baliga conceived the idea of a functional integration of MOS technology and bipolar physics that directly led to the IGBT's development … it remains undeniable that Baliga's vision and leadership played a critical role in moving the IGBT from a paper-based concept to a viable product with many practical applications.*" On October 21, 2011, President Obama presented the National Medal of Technology and Innovation to me at the White House in recognition of my development and commercialization of the IGBT and other power semiconductor devices. This recognition put the spotlight on IGBTs and acknowledged the impact of power electronics on our society. In 2015, I was awarded the Global Energy Prize due to the huge impact of the IGBT on the efficient generation, management, and distribution of electrical energy. This was followed by my induction into the National Inventors Hall of Fame as the sole inventor of the IGBT in 2016. This was a fitting recognition of the IGBT, as it was listed among just 500 inventions out of 10 million US patents that had been issued at that time.

1.1 IGBT Applications Spectrum

IGBTs are required for applications that operate over a broad spectrum of current and voltage levels, as shown in Fig. 1.1. Their characteristics are ideal for applications with operating voltages above 200 V. Typical examples are lamp ballasts, consumer appliances that utilize motors, and electric vehicle drives. Other examples are high-power motor control in steel mills and for traction (electric trains). They are now being utilized even in power transmission and distribution systems. The on-resistance of conventional silicon power MOSFET structures is too large to serve these applications. Consequently, these applications utilize silicon IGBTs today. Silicon carbide (SiC) IGBTs offer very promising characteristics for applications that require blocking voltages of above 10–15 kV for use in smart grid applications [3].

It is worth noting that the current ratings for IGBTs increase with increasing voltage rating for these applications, with the exception of the smart grid. In the case of silicon IGBTs, this issue is tackled by resorting to multichip press-pack modules. The smart grid application is unique in requiring very high-voltage devices with low current ratings. These applications can be served by using SiC-based IGBTs despite the lower chip manufacturing yield and higher cost of SiC wafers. These SiC IGBTs can operate at higher frequencies than their silicon counterparts, resulting in a smaller size for the magnetic elements used in the power circuits.

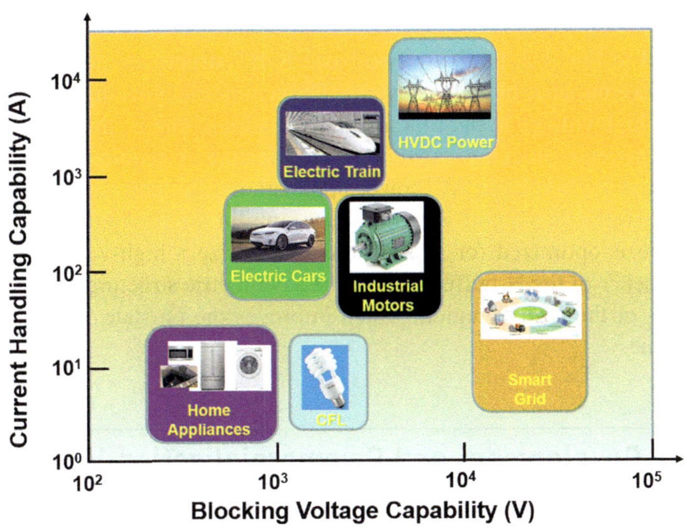

FIGURE 1.1

Application spectrum for insulated gate bipolar transistors (IGBTs).

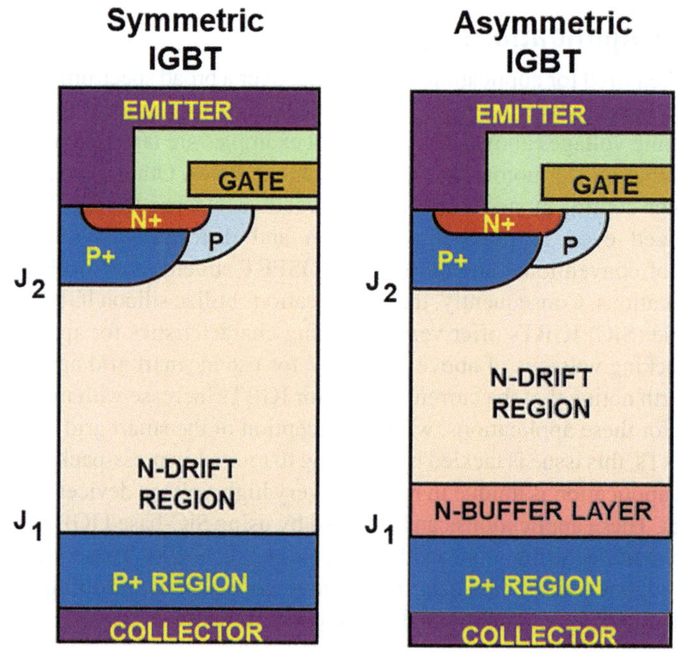

FIGURE 1.2

Basic insulated gate bipolar transistor device structures.

1.2 Basic IGBT Device Structures

As illustrated in Fig. 1.2, there are two basic IGBT structures, namely symmetric blocking and asymmetric blocking devices. The symmetric blocking structure allows supporting a high voltage in the first and third quadrant, that is, the device has high forward and reverse blocking capability. This feature is required for any power devices used in high-voltage AC power applications. In contrast, the asymmetric blocking structure can support a high voltage only in the forward blocking mode. This structure is optimized for applications that utilize a high-voltage DC power bus. The presence of the N-buffer layer in the asymmetric structure allows reducing the thickness of the N-drift region, which improves the on-state voltage drop and switching time.

1.3 IGBT Development and Commercialization History

The first semiconductor power devices were bipolar transistors that evolved out of the invention of the junction transistor in 1947. A thick, lightly doped drift region was added in the power bipolar transistor, as illustrated in Fig. 1.3, to support high voltages. The high blocking voltage capability also required a relatively large base

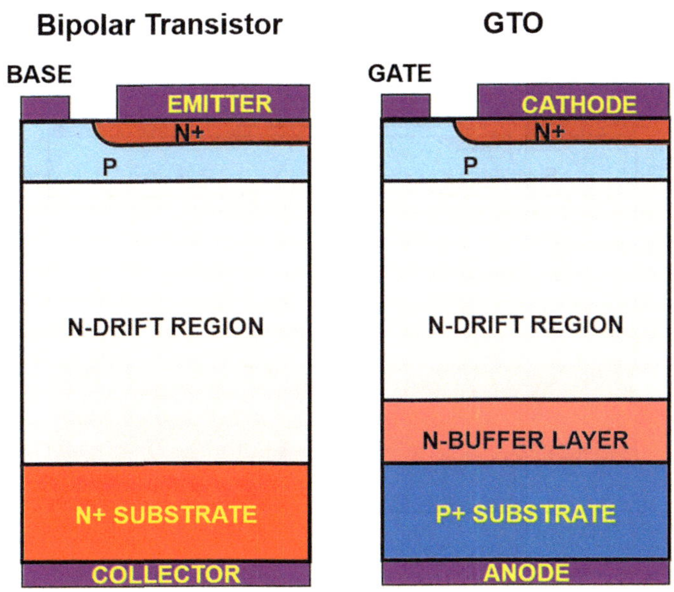

FIGURE 1.3

Bipolar power devices.

width that degraded the current gain. High-level injection effects produced further reduction of the current gain [4]. The large base drive current required for the bipolar transistor restricted its voltage rating to below 600 V.

High-power applications that needed devices capable of supporting more than 1000 V and controlling large currents were served in the 1970s by the development of thyristor or four-layer semiconductor structures. The gate turnoff thyristor (GTO) shown in Fig. 1.3 became popular for motor drives in traction applications for street cars and electric trains. The thyristor regenerative action within these four-layer switches allowed manufacturing high-current devices with low on-state voltage drop. However, the GTO required very large gate drive currents to achieve unity gain turnoff. The complex gate drive and snubber circuits for GTOs increased power losses while adding cost and size to the system.

During the 1970s, a concerted effort was made to create power MOSFETs after the successful development of complementary metal oxide semiconductor (CMOS) technology. The double-diffused metal oxide semiconductor (DMOS) structure shown in Fig. 1.4 was the most commercially successful power device. The MOS gate structure for the device made it a voltage-controlled device with essentially no current flow in the gate circuit during the steady-state on-mode and blocking mode. The gate drive current required for switching the devices was also modest due to the relatively low operating frequencies of power circuits at that time. Although the power MOSFET displaced the power bipolar transistor due to its

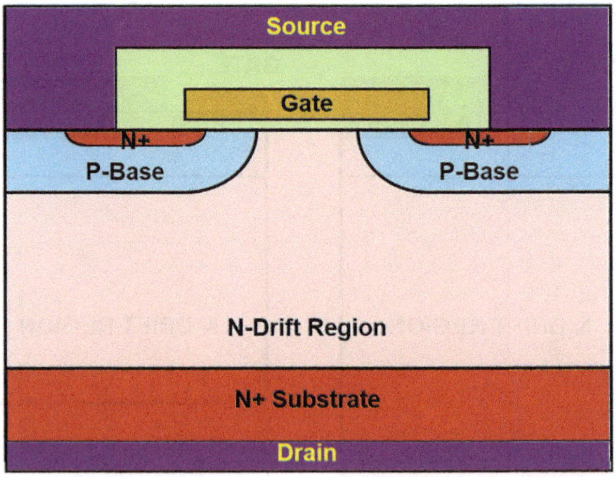

FIGURE 1.4

Power MOSFET structure.

superior gate drive, switching speed, and ruggedness for lower voltage (<200 V) applications, a severe increase in its resistance became an impediment to scaling up its blocking voltage.

In the late 1970s, it was clear that a high-performance power device technology was badly needed for medium- and high-power applications. In early 1977, I began to explore the use of MOS-gated structures to control a four-layer semiconductor device. My analysis indicated that, unlike the conventional thyristor structure, the gate drive current requirement for the thyristor could be decoupled from the [d*V*/d*t*] and the [d*I*/d*t*] capability by using an MOS gate structure. On July 26, 1977, I submitted a patent disclosure at GE on MOS gating of thyristors [5]. This disclosure describes a vertical four-layer structure with a V-groove region in which the MOS gate structure is prepared. The structure described in my patent disclosure is the V-MOS cross-section shown in Fig. 1.5, which is also the IGBT structure. The fabrication of this device structure required setting up and optimizing a potassium hydroxide (KOH)-based silicon etching process to form the (truncated) V-groove region. Once this process was established, the fabrication of the MOS-gated thyristors commenced under my supervision on November 9, 1978 [6], with Margaret Lazeri as my lead process technician.

The processing was successfully completed, leading to the first experimental results obtained on July 30, 1979. These results clearly showed the IGBT mode with MOS gate-bias-controlled current saturation as well as the latch-up of the four-layer thyristor at larger current levels [7]. I discovered that some of the fabricated devices exhibited the expected enhancement mode operation, leading to a snapback in the output characteristics due to latch-up of the thyristor structure. However, other devices displayed depletion mode characteristics, demonstrating for the first time

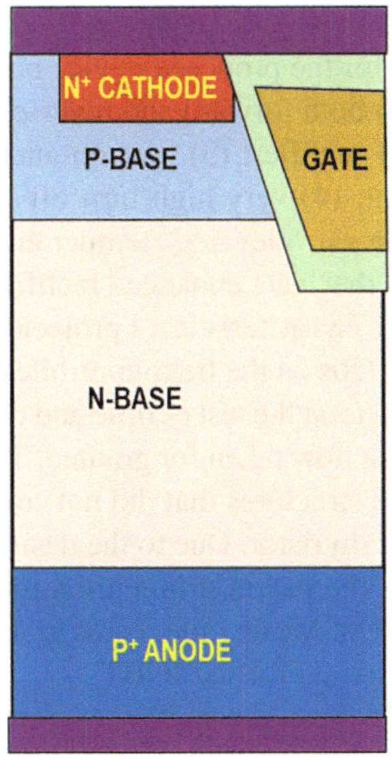

FIGURE 1.5

V-groove insulated gate bipolar transistor/thyristor structure.

that current can flow in a four-layer vertical semiconductor device without latch-up of the thyristor. This represents the first observation of the IGBT mode of operation in a vertical semiconductor device. It demonstrated that current flow through the four-layer structure could be turned off by the application of a negative gate bias to the MOS gate electrode-definitively demonstrating nonlatch-up current transport in a vertical four-layer semiconductor device for the first time. Consequently, the conception and experimental demonstration of the IGBT at GE can be directly attributed to my patent disclosure on July 26, 1977.

After an internal GE review cycle, on August 28, 1979, the above results were submitted for publication to *Electronics Letters*. The results were accepted and appeared in print in this journal on September 27, 1979 [8]. The article contains photographs of the output characteristics of the devices. In particular, Fig. 1.2 in the article shows IGBT-like characteristics, namely a diode-line behavior in the on-state and MOS gate-controlled current saturation capability, with no snapback due to latch-up of the four-layer semiconductor structure. The paper states: "*The structure can be seen to be similar to that of V-groove (MOSFET) devices but with the drain region of the (MOSFET) being replaced by a p-type anode region.*"

In early September 1980, Tom Brock, the vice president of a new GE product division set up to create high-efficiency adjustable-speed motor drives for air conditioning, visited the GE Research Center and described his frustrations with using the available Darlington bipolar transistors for this application. He challenged my group to provide a better device technology. I submitted a patent disclosure on

September 29, 1980, in response to this challenge [9]. In my disclosure, I projected the following characteristics for the proposed device based on my analytical modeling of the IGBT structure: (1) both forward and reverse blocking capability; (2) forward drop similar to a P-i-N rectifier; (3) turn on and turn off using a small gate voltage with low gate current; (4) very high turn off gain; (5) high dV/dt and dI/dt capability; (6) operation at elevated temperatures; and (7) tolerance to radiation. I named the device the "gate enhanced rectifier" (GERECT) to emphasize its P-i-N rectifier-like on-state characteristics. I projected these characteristics based on my previous work in the 1970s on the field-controlled thyristor. In retrospect, this prescient description has withstood the test of time and encapsulated the performance attributes for the IGBT that are now taken for granted. The GE patent application for the IGBT included innovative structures that did not contain an N^+ emitter region to avoid formation of a parasitic thyristor. Due to the desire on the part of GE to obtain the broadest possible claims, the patent prosecution took nearly 10 years, with the patent issued in November 1990. Meanwhile, I was granted many other patents filed by GE on structural enhancements for the IGBT.

My GERECT proposal in September 1980 was met with skepticism by colleagues at GE. They first pointed out that previous efforts at MOS gating of four-layer structures showed latch-up of the thyristors at low current levels [8,10]. They also pointed out that my proposed IGBT structure consisted of an n-channel MOSFET driving a wide-base P-N-P bipolar transistor. Prevailing wisdom based on decades of work on power bipolar transistors recommended using a narrow base N-P-N structure to get good current gain. Based on this, my critics said that my proposed device could be expected to operate at a low on-state current density (below 20 A/cm^2) with a high on-state voltage drop, as observed in recent publications [10]. My projections of an on-state current density of 100–200 A/cm^2 for my proposed device based on the P-i-N diode model that I had proposed were considered unrealistic.

However, in October 1980, numerical simulations of my proposed IGBT structures by Mike Adler confirmed the results of my analytical modeling of its capability, leading to the opportunity to describe my idea to Brock. During this presentation, I made it clear that this innovation could impact other GE divisions, such as lighting, small appliances, and major appliances. Brock was very impressed by the potential for this device and proceeded to brief GE Chairman Jack Welch about the impact of this idea on multiple businesses within the company, prompting his visit to Schenectady to meet with me in November 1980. My presentation was very favorably received by Welch, giving me strong support within GE to commercialize and apply this innovation to GE businesses. However, Welch wanted to maximize the impact of this innovation within GE products and placed an embargo on releasing any information about the IGBT for several years. This delayed external recognition of my early work on the IGBT.

Because the potential impact of the IGBT to GE businesses was recognized by Welch, the preparation and filing of the patent application for my disclosure RD-13,112 was undertaken at the corporate level rather than by the GE Research Center. The patent application was filed on December 2, 1980, and issued only in 1990 after vigorous prosecution at the patent office [11]. With this strategy, GE obtained patent protection for the basic IGBT concept with broad claims until 2009. It is worth pointing out that this patent includes an IGBT structure without the parasitic thyristor where I proposed a tunneling current from the emitter contact metal to the inversion channel. The technology for making such devices did not exist in 1980 but has now been demonstrated for MOSFETs in integrated circuit (IC) applications. Although my patent for the basic IGBT concept was not issued until 1990, a dozen derivative patents were also filed by GE, and many IGBT patents were issued to me in the 1980s [12].

The prevalent practice in the GE Research Center was to first build an innovation in Schenectady and then transfer it into production after it had been proven. This procedure typically required at least 3 years. A major hurdle with this approach was the need to set up a manufacturing capability for the product after the research and development effort. In general, it is very expensive to create a new manufacturing line and the initial product yields can be poor due to wide variability in device characteristics. To avoid these problems, as the acknowledged inventor and principal developer of the device at GE [13], I decided it would be prudent to engineer my IGBT structure to allow its fabrication in an existing power MOSFET product line in California. In 1981, I flew to California to describe my IGBT idea to Nathan Zommer, who was responsible for the power MOSFET product line. After assimilating the power MOSFET production process, I procured the unique IGBT starting material; designed a process flow for the IGBT that included one extra step—a deep P^+ region not included in the power MOSFET process at that time; then designed the cell topology and a multiple floating field ring edge termination. After creating a mask set for the fabrication of IGBTs with layout assistance from Peter Gray, I supervised its production by Zommer in California. The very first IGBT wafers out of the manufacturing line that I tested produced functional chips with a blocking voltage of 600 V at a high yield. The total time between my conception of the IGBT and my successfully producing functional devices on a production line was less than 10 months. This remarkable feat provided a large supply of devices that I could tailor with different switching speeds by using an electron irradiation process that I had already proposed and proven using power MOSFETs. These devices were supplied to numerous engineers at GE working on novel lamps, small appliances such as steam irons and toasters, major appliances such as refrigerators and washing machines, and of course for the variable-speed motor drives.

In parallel with the development of the IGBT, my group at GE undertook the development of the high-voltage integrated circuit (HVIC) that could drive the IGBT in an H-bridge topology. The combination of the IGBT and HVIC produced a unique product with an order of magnitude improvement in size, weight, and manufacturability. The first commercial "Smart Switch" 5 hp. motor drive product announcement was made on October 3, 1983, fulfilling the promise made to Tom Brock in 1981.

Toward the end of 1982, the GE Semiconductor Products Division decided to break the embargo and announce the IGBT as a product. This decision allowed my publication of the first IGBT paper at the IEEE Electron Devices Meeting in

December 1982 [14]. The IGBT product (D94FQ4, R4 with 18 A, 400 V, and 500 V rating) was announced in June 1983, resulting in Electronic Products declaring it "Product of the Year." The IGBT product was introduced by me and a GE Marketing Manager, Marvin Smith, in a trade publication, Electronic Design News, in September 1983 [15] and reported at the IEEE Industrial Applications Society Meeting in October 1983 [16]. I also promoted the IGBT in Europe at the Drives/Motors/Controls Conference held in Harrogate, United Kingdom [17]. In December 1983, I described the unique flexibility of trading-off on-state and switching losses in IGBTs by using electron irradiation [18]. This paper demonstrated that the IGBT could be optimized for a wide variety of applications. During this period at the end of 1983, GE requested my presentation of the work on IGBTs to numerous visiting delegations from outside organizations. This included a visit from a Japanese delegation with corporate representatives from many companies, including the Fuji Electric Company. My presentation to the Japanese delegation of the very promising features of the IGBT and its proven widespread utilization within GE applications prompted vigorous activity in Japan to commercialize the device [19]. During subsequent years, a systematic effort was made at the GE Research Center under my supervision to characterize the high-temperature operating capability of the IGBT [20], develop complementary (p-channel) devices [21], and scale up the voltage rating [22]. The demonstration of the excellent performance characteristics of the IGBT during these formative years played an important role in generating worldwide interest in the commercialization of the device and subsequent exploitation of the IGBT in myriad applications.

It is worth mentioning that there were many skeptics regarding the viability of the IGBT as a power device when I first proposed the idea at GE. They argued that the device would not be useful due to destructive latch-up of the parasitic thyristor within the IGBT structure. I was able to overcome this problem by the addition of a deep P^+ region [23], as shown in Fig. 1.2. The skeptics then argued that even if the structure was functional, it would have very little utility because of its slow switching speed. The known lifetime control processes of heavy metal diffusion and particle bombardment had been reported to severely degrade the gate structure of CMOS devices and power MOSFETs. Fortunately, I was able to discover an annealing process following electron irradiation that removed the damage in the gate oxide of the IGBT structure while leaving di-vacancy-induced traps within the bulk to reduce the switching time [24]. This allowed tailoring the characteristics of the IGBT after complete fabrication of the devices to optimize their performance in applications with operating frequencies from 50 to 10,000 Hz. Without these breakthroughs, the barriers to commercialization of the IGBT would not have been overcome, thus vindicating the skeptics, and the IGBT may have never been developed into a commercial product.

After the IGBT concept was proven to exhibit excellent electrical characteristics, it has sometimes been denigrated as simply a power MOSFET with the drain replaced with a P^+ region. Although an accurate literal description of the structure, this viewpoint does not take into account the perceived shortcomings of such a structure when first proposed. It was understood from the beginning that the IGBT operates as a P-N-P bipolar transistor with base drive current provided via the integrated MOSFET structure. The prevailing wisdom in the early 1980s from the development of power bipolar transistors was that P-N-P transistors are far inferior to N-P-N transistors due to the much greater mobility for electrons in silicon. The prevailing design philosophy for these power bipolar transistors was to make the base region as narrow as possible to obtain a larger current gain and to support the high voltage in the collector region. In stark contrast to the best practices at that time, the proposed IGBT structure contained a wide-base P-N-P transistor in which the voltage was supported in the base region. This led skeptics to conclude that the IGBT structure would work at lower on-state current density than the existing power bipolar transistors and power MOSFETs. They also concluded that the required base drive current could not be supplied sufficiently by the integrated MOSFET structure because of the large perceived series resistance of its drift region. Based on these arguments, it was postulated that devices based on on-state current flow using a thyristor with MOS gate-controlled turnoff capability were preferable to the IGBT [25,26] because these devices would operate at a high on-state current density. To make matters worse, the power device funding agencies in the United States, namely the Department of Defense and the Electric Power Research Institute, decided to only support the development of the MOS-controlled thyristor (MCT) [27] in the 1980s. Even after officials invested tens of millions of dollars in the MCT over a span of more than a decade, the MCT has been completely eclipsed by the IGBT. This occurred because my demonstration of the ease of manufacturing IGBTs, suppressing the parasitic thyristor latch-up, and tailoring its switching speed for numerous applications encouraged the worldwide power semiconductor industry to embrace the technology, driving up the ratings of IGBT products while driving down their cost. Once the IGBT was adopted in an expanding range of applications, it became impossible to displace the device, as alternatives failed to deliver all the features of this device.

Once the IGBT was announced as an excellent product by GE in 1983, there has been an effort by other researchers to take credit for the IGBT, according to Wikipedia [28]. The Wikipedia article states that the IGBT mode was first proposed by Yamagami in a Japanese patent JP 47-021739B issued to Mitsubishi Electric Company in 1972. One of the figures in the patent illustrates a four-layer semiconductor structure with an MOS gate electrode, which is the basis for a single claim that describes this structure but not its electrical characteristics. The structure shown in the patent application is unlike all manufactured IGBTs today. It is also obvious that the existence of the IGBT mode was not recognized by Yamagami, and that he certainly did not appreciate or champion its utility because his employer Mitsubishi failed to commercialize the IGBT until 1987 [29], well after I had demonstrated excellent device performance and commercialized the IGBT at GE in 1983.

A lateral, high-voltage, MOS-gated triac structure was reported by Plummer and Scharf [10] in February 1980 with the work supported by a patent application [30]. This device was created by merging two lateral double-diffused metal-oxide semiconductor field-effect transistors (DMOSFETs) with a single gate electrode. In the paper, it was demonstrated that the structure has a DMOSFET mode where the current flows along the surface through both channels, a bipolar mode "*with a wide-base p-n-p transistor operating in parallel with the surface devices*," and an on-state mode with the thyristor operating with regenerative action. The paper also describes a lateral insulated gate SCR structure in which "*the thyristor function is still directly controlled by the MOS gate*." The electrical characteristics provided in the publication for this structure show a snapback to the on-state, indicating that the device proposed by the authors operates with regenerative action in the on-state. The authors also state that this concept is "*directly adaptable to discrete-power devices*" and illustrates a vertical "equivalent" structure in the paper. Many years later, Plummer added claim language to cover the IGBT mode of operation by using a reexamination certificate filed in October 1994 [31] after the IGBT became established as a product. To my knowledge, neither Scharf nor Plummer pursued the commercialization of the ideas in their paper. The electrical characteristics demonstrated for MOS-gated SCRs in their work indicate very high on-state voltage drop before latch-up of the thyristor at relatively low current levels. These results discouraged any effort to pursue development of an IGBT structure by the worldwide power semiconductor industry because no products were announced outside of GE until 1985.

On March 25, 1980, Becke and Wheatley from Radio Corporation of America (RCA) filed a patent application titled "*A vertical power MOSFET with an anode region*," which was issued on December 14, 1982 [32]. It is remarkable that this patent was allowed without citing and taking into account close prior art [6] published in the literature 6 months before the filing date of March 25, 1980, which not only discussed the structure but even contained a sentence with the name of the patent! The single independent claim in this patent states: "*…and no thyristor action occurs under any device operating conditions*." It is worth pointing out that all commercial IGBT devices made since the 1980s undergo thyristor action under an appropriate rise in temperature and collector current level. Consequently, the claims in this patent do not describe the IGBTs that have been manufactured and used in applications over the past 30 years. According to Wheatley [33], he "*would provide the direction while Becke would handle the time-consuming yet essential 'dogwork'*." However, Becke passed away soon after the filing of this patent and the development of the conductivity modulated field effect transistor (COMFET) at RCA Laboratories was reported by Russel et al., in 1983 [34] with device characteristics resembling the IGBT. It is interesting that the paper states that the four-layer structure of the COMFET device may latch-up at high anode current levels, contradicting the claims of the RCA patent cited in the paper. To my knowledge, RCA did not follow through with a product release or work on applications for the device. The acquisition of RCA by GE in 1986 made this eventually unnecessary.

In Europe, a group at Siemens also pursued the development of MOS-gated thyristors [35,36]. The authors of this work cited the prior reports by Scharf and Plummer [28] and by Baliga [8] in their publications. There is no recognition of the promising IGBT mode of operation disclosed in these publications.

The Wikipedia article on the IGBT also gives credit to Nakagawa at Toshiba for "complete suppression" of the parasitic thyristor resulting in a "non-latch-up" IGBT in 1984 [37]. Nakagawa's innovation was to periodically eliminate the source region to provide a hole-bypass path. This idea does improve the latch-up current density at the expense of a higher on-state voltage drop, but it does not completely eliminate the latch-up of the parasitic thyristor in the IGBT structure, as has been conclusively demonstrated by numerical simulations [4]. Many other methods to improve the latch-up current density of the IGBT have been documented in the literature as well [4]. The only approach for "complete suppression" of the parasitic thyristor in the IGBT is to replace the N^+ emitter region with a metal emitter region, as proposed in my first patent application [7]. To my knowledge, this idea has yet to be successfully exploited to make truly latch-up-free IGBT devices. Products based on Nakagawa's IGBT structure were first commercialized by Toshiba in 1985. Modern IGBTs utilize high-resolution lithography to create small device cell structures that are resistant to thyristor latch-up.

1.4 Scaling of Power Ratings

Symmetric (equal forward and reverse) blocking 600 V IGBT devices with substantial current ratings were first reported in the literature in 1982 in recognition of their potential applications in DC and AC circuits [14]. Soon thereafter, it was demonstrated that the asymmetrical IGBT structure, with high forward blocking capability and low reverse blocking capability, has superior on-state and switching speed capability [38]. One of the important attributes of the IGBT has been the relative ease of increasing its blocking voltage and current ratings. From inception, it was recognized at GE that the voltage rating of the IGBT could be easily scaled up due to its diode-like on-state characteristics. The ability to scale the voltage rating of the IGBT structure with only a modest increase in the on-state voltage drop was demonstrated soon after its invention [39]. Experimentally measured data on symmetric blocking devices with voltage ratings of 300, 600, and 1200 V [22] were published in 1985, providing a strong impetus to continually scale up the voltage rating of IGBT structures, as shown in Fig. 1.6. The information used for this figure is based on data published on IGBT devices at the IEEE International Electron Devices Meeting until 1990 and the IEEE International Symposium on Power Semiconductor Devices and ICs beyond that time frame. Silicon IGBTs capable of supporting more than 3000 V were developed within a short span of 10 years. Interest in using the IGBT for traction (electric locomotive drives) applications in Europe and Japan led to scaling up the voltage ratings to 6500 V by 2001. This is the limit of capability for silicon IGBTs. More recently, there has been interest in the development of IGBTs from SiC material to achieve even higher voltage ratings for utility applications [40]. SiC IGBTs with 9 kV blocking voltage capability were reported in 2007 [41], followed by 13 kV devices in 2009 [42], 16 kV devices in 2015 [43], 20 kV devices in 2016 [44,45], and 27 kV devices in 2015 [46]. A press-pack module has been more recently reported for 15 kV SiC IGBTs to achieve a current of 10 A with an on-state voltage drop of 5.7 V [47]. There is limited activity in these devices due to the narrow

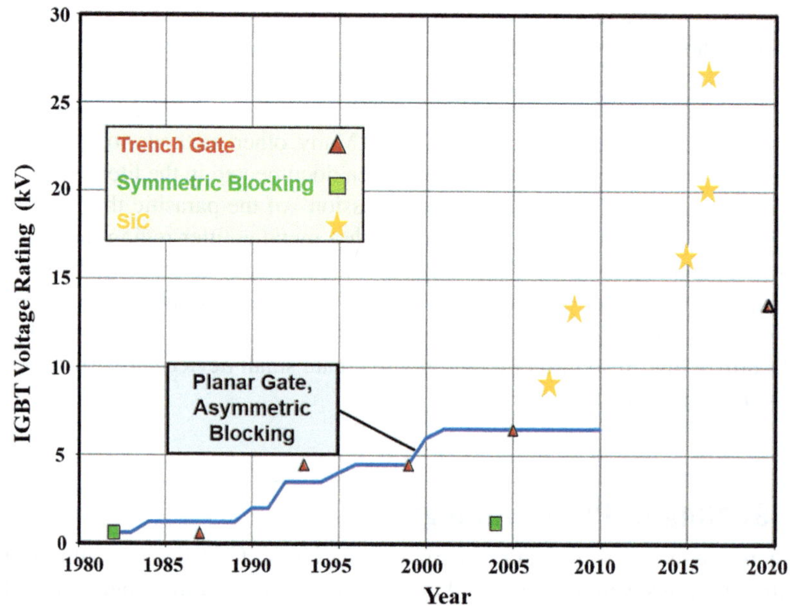

FIGURE 1.6

Growth in insulated gate bipolar transistor voltage ratings.

scope of their application and the long time horizon for their acceptance in the utility industry.

IGBTs with the trench-gate architecture were first demonstrated in 1987 [48] to reduce the on-state voltage drop. The enhanced carrier concentration in the vicinity of the trench-gate region [49] has stimulated the development of devices with voltage and current handling capability similar to those of planar devices. The growth in the voltage ratings of the trench-gate IGBTs is also charted in Fig. 1.6. The voltage rating increased to 6.5 kV and then stagnated for 20 years. It was increased to 13 kV in 2020, as indicated in the figure [50]. This figure includes the evolution of IGBT structures with reverse blocking capability. Although first reported in 1982 [14], reverse blocking devices lagged behind the asymmetric blocking structure due to the higher degree of interest in DC circuit applications and difficulties with the passivation of reverse blocking junctions. Recently, interest in the reverse blocking structure has been motived by applications in matrix (AC-to-AC) converters [51–54].

The IGBT structure operates with the high-level injection of minority carriers into the drift region to reduce its resistance to current flow. The reduced resistance of the drift region enables IGBTs to operate at much higher current density than power MOSFETs with comparable blocking voltages. The high on-state current density for the IGBT structure has allowed rapid scaling up of its current handling capability. The growth in the current handling capability for the IGBT devices is charted in Fig. 1.7 using the same reference sources used for the voltage ratings. Beginning

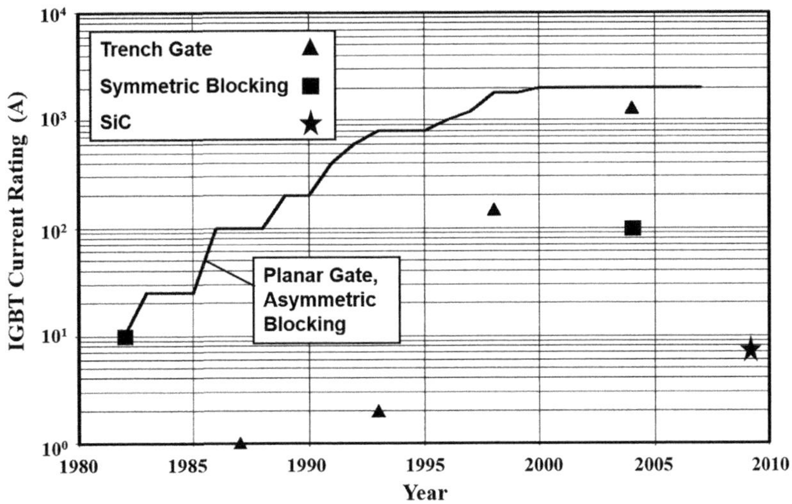

FIGURE 1.7

Growth in current-handling capability for insulated gate bipolar transistors (IGBTs).

with the first IGBTs reported in 1982 with a modest current rating of 10 A, the current handling capability was quickly scaled up to 25 A for use in adjustable-speed motor drives for air-conditioning applications. Since then, steady growth in the current-handling capability of IGBTs has been achieved due to interest in Europe and Japan for traction applications. The current rating can be increased either by scaling the die size or by connecting chips in parallel. Early work on IGBTs demonstrated that the devices could be easily paralleled due to a mild positive temperature coefficient for the on-state voltage drop [20]. Special packages, called power modules, have been developed with multiple IGBT chips that are connected in parallel by using wire bonding to scale up the current rating, as shown in Fig. 1.7. Press-pack IGBTs have now become available with the capability of switching 2000 A. These devices have utilized the planar gate structure due its larger die yield and ruggedness. The current-handling capability of trench-gate IGBT devices has generally lagged behind in time when compared with planar gate devices, but trench-gate IGBTs have now been developed that are capable of handling 1300 A. The current-handling capability of the IGBT modules exceeds 1000 A, making them suitable for applications such as electric trains, including the Shinkansen bullet train in Japan. These modules can be paralleled to handle even higher current levels. The current rating for the recently developed SiC IGBTs is below 10 A [47]. This will have to be scaled up to at least 100 A for use in utility applications in the future.

The tremendously rapid growth in the power-handling capability of IGBT devices is documented in Fig. 1.8. The data in this figure were gathered from the previously cited reference sources for the current and voltage ratings of IGBTs, augmented by review papers that have reported these trends [4,55]. The first IGBTs

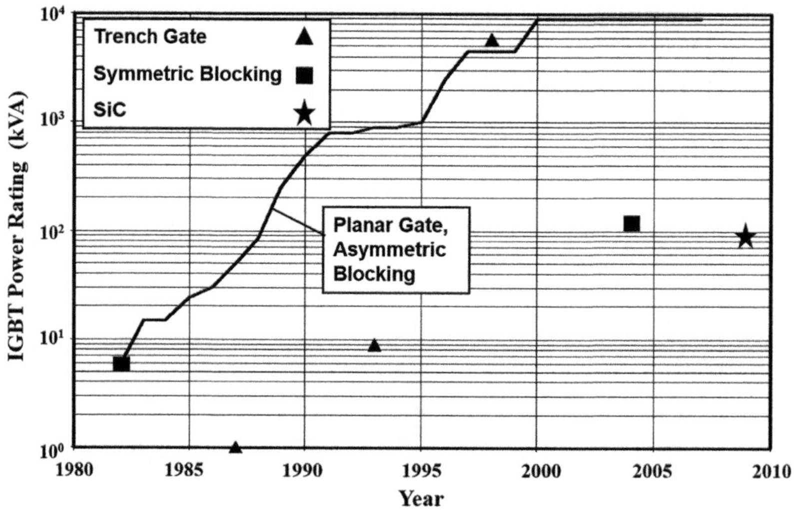

FIGURE 1.8

Growth in power-handling capability for insulated gate bipolar transistors (IGBTs).

reported in 1982 could control 6 kW of power. The rapid growth in the voltage and current ratings has allowed a huge increase in the power-handling capability of IGBTs, expanding their applications. A single IGBT module can control more than 9 MW of power today. This is satisfactory for most applications, making further increases in the power-handling capability for individual modules unnecessary. The IGBT structure has not only completely displaced the bipolar power transistor in medium-power applications, but has also displaced the GTO in high-power applications. Its application spectrum now includes cruise ships and renewable power generation and distribution networks.

1.5 Summary

The applications of IGBTs have been reviewed in this chapter. These devices have become dominant components for applications with operating voltages above 300 V. IGBTs are extensively utilized in the consumer, industrial, lighting, transportation, and renewable energy generation sectors of the economy. The attractive electrical characteristics of the IGBT and the simplicity of its control circuit, together with the ease of manufacturing and tailoring of its switching speed, have prompted worldwide interest in creating products. All the major power semiconductor companies now have extensive IGBT product lines. The high volume of IGBT sales for many of the applications, such as motor drives, automotive electronics, and lighting, has resulted in the design and production of specially tailored device structures, which are discussed in subsequent chapters of this book.

The history of innovations at GE that led to the commercialization of the IGBT has been described in this chapter. It demonstrates that the invention of the device was only one small step required for making the IGBT the well-recognized innovation it is today. The visionary leadership at the top of the company with the mission to make every product division the leader in its area provided a strong impetus to adopt new technologies like the IGBT. My flawless development of an IGBT product in the existing power MOSFET manufacturing line with meticulous attention to detail in the mask and fabrication process design was critical in its success. My successful suppression of the parasitic thyristor latch-up problem in the IGBT and the creation of an electron irradiation lifetime control process for an MOS-gated device enabled IGBTs to be conveniently tailored for a broad range of applications. The broad portfolio of products made by GE for the consumer, industrial, and lighting sectors provided a unique environment for widespread utilization and impact of the IGBT technology.

References

[1] S. Grossman, Advances in discrete semiconductors march on, in: Power Electronics Technology, September 2005, pp. 52–56.

[2] B.J. Baliga, EDN Engineering Hall of Fame, December 2010.

[3] J. Wang, et al., Smart grid technologies: development of 15-kV SiC IGBTs and their impact on utility applications, IEEE Ind. Electron. Mag. 3 (2009) 16–23.

[4] B.J. Baliga, Fundamentals of Power Semiconductor Devices, second ed., Springer-Science, New York, 2019.

[5] B.J. Baliga, MOS Gate Structure for Thyristors, GE Patent Disclosure Letter RD-10,243, submitted July 26, 1977.

[6] B.J. Baliga, V-Groove MOS Gated SCR Process Ticket, GE Corporate Research Center, Schenectady, NY, November 1978.

[7] B.J. Baliga, MOS Gate Structure for Thyristors—Reduction to Practice, Patent Disclosure RD-10,243, July 30, 1979.

[8] B.J. Baliga, Enhancement and depletion mode vertical channel MOS-gated thyristors, Electron. Lett. 15 (1979) 645–647.

[9] B.J. Baliga, Gate Enhanced Rectifier (GERECT), GE Patent Disclosure Letter RD-13,112, September 29, 1980.

[10] J.D. Plummer, B.W. Scharf, Insulated gate planar thyristors, IEEE Trans. Electron Devices 27 (February 1980) 380–394.

[11] B.J. Baliga, Gate Enhanced Rectifier, U.S. Patent 4,969,028, filed December 2, 1980, issued November 6, 1990.

[12] B.J. Baliga, Twelve IGBT Related U.S. Patents Issued to GE: (a) #4443931, issued April 14, 1984; (b) #4618872, issued October 21, 1986; (c) #4620211, issued October 28, 1986; (d) #4717679, issued January 5, 1988; (e) #4782379, issued November 1, 1998; (f) #4801986, issued January 31, 1989; (g) #4823176, issued April 18, 1989; (h) #4883767, issued November 28, 1989; (i) #4901127, issued February 13, 1990; (j) #4933740, issued June 12, 1990; (k) #4980740, issued December 25, 1990; (l) #4994871, issued February 19, 1991.

[13] GE Corporate Research and Development Post, No. 936, September 28, 1983 (as one example among many).

[14] B.J. Baliga, et al., The insulated gate rectifier: a new power switching device, in: IEEE International Electron Devices Meeting, Abstract 10.6, 1982, pp. 264–267.

[15] B.J. Baliga, M. Smith, Modulated conductivity devices reduce switching losses, Electron. Des. News 28 (1983) 153–162.

[16] B.J. Baliga, et al., The insulated gate transistor (IGT)—a new power switching device, in: IEEE Industrial Applications Society Meeting, October 1983, pp. 794–803.

[17] B.J. Baliga, The new generation of MOS power devices, in: Invited Paper, Drives/Motors/Controls Conference, October 1983, pp. 139–141.

[18] B.J. Baliga, Fast switching insulated gate transistors, IEEE Electron Device Lett. 4 (1983) 452–454.

[19] Y. Uchida, Fuji Electric Company Board of Directors (Private Communication).

[20] B.J. Baliga, Temperature behavior of insulated gate transistor characteristics, Solid State Electron. 28 (1985) 289–297.

[21] M.F. Chang, et al., Comparison of N and P channel IGBTs, in: IEEE International Electron Devices Meeting, Abstract 10.6, December 1984, pp. 278–281.

[22] T.P. Chow, B.J. Baliga, Comparison of 300-, 600-, and 1200-V n-channel insulated gate transistors, IEEE Electron Device Lett. 6 (April 1985) 161–163.

[23] B.J. Baliga, M.S. Adler, Method of fabrication of a semiconductor device with a base region having a deep portion, U.S. Patent 4,443,931, filed June 28, 1982, issued April 24, 1984.

[24] B.J. Baliga, J.P. Walden, Improving the reverse recovery of power MOSFET integral diodes by electron irradiation, Solid State Electron. 26 (1983) 1133–1141.

[25] V.A.K. Temple, MOS-controlled thyristors, in: IEEE International Electron Devices Meeting, Abstract 10.7, December 1984, pp. 282–285.

[26] M. Soisiek, H. Strack, MOS-GTO—a turn-off thyristor with MOS-controlled emitter shorts, in: IEEE International Electron Devices Meeting, Abstract 6.5, December 1985, pp. 158–161.

[27] N.G. Hingorani, et al., Research coordination for power semiconductor technology, Proc. IEEE 77 (1989) 1376–1389.

[28] Insulated Gate Bipolar Transistor. en.wikipedia.org/wiki/Insulated_gate_bipolar_transistor.

[29] T. Takahashi, Y. Yoshiura, The 6th generation IGBT & thin wafer diode for new power modules, Mitsubishi Electr. Mag. Adv. 135 (June) (2011) 5–7.

[30] J.D. Plummer, Monolithic Semiconductor Switching Device, U.S. Patent 4,199,774, filed September 18, 1978, issued April 22, 1980.

[31] J.D. Plummer, Monolithic Semiconductor Switching Device, Reexamination Certificate B1 33,209, filed October 17, 1994, issued September 12, 1995.

[32] H.W. Becke, C.F. Wheatley, Power MOSFET With an Anode Region, U.S. Patent 4,364,073, filed March 25, 1980, issued December 14, 1982.

[33] H. Becke, F. Wheatley, EDN Engineering Hall of Fame, December 2010.

[34] J.P. Russel, A.M. Goodman, L.A. Goodman, J.M. Neilson, The COMFET—a new high conductance MOS-gated device, IEEE Electron Device Lett. 4 (1983) 63–65.

[35] J. Tihanyi, Functional integration of power MOS and bipolar devices, in: IEEE International Electron Devices Meeting, Abstract 4.2, December 1980, pp. 75–78.

[36] L. Leipold, et al., A FET-controlled thyristor in SIPMOS technology, in: IEEE International Electron Devices Meeting, Abstract 4.3, December 1980, pp. 79–82.

[37] A. Nakagawa, et al., Non-latch-up 1200-V, 75-A, bipolar-mode MOSFET with large SOA, in: IEEE International Electron Devices Meeting, Abstract 16.8, December 1984, pp. 860–861.

[38] B.J. Baliga, et al., The insulated gate transistor: a new three-terminal MOS-controlled bipolar power devices, IEEE Trans. Electron Devices 31 (1984) 821–828.

[39] B.J. Baliga, Evolution of MOS-bipolar power semiconductor technology, Proc. IEEE 74 (1988) 409–418.

[40] J.W. Palmour, et al., SiC power devices for smart grid systems, in: IEEE International Power Electronics Conference, 2010, pp. 1006–1013.

[41] Q. Zhang, et al., 9-kV SiC IGBTs, Mater. Sci. Forum 556–557 (2007) 771.

[42] M.K. Das, et al., A 13-kV 4H-SiC n-channel IGBT with low Rdiff, on and fast switching, Mater. Sci. Forum 600–603 (2009) 1183.

[43] K. Fukuda, et al., Development of ultra-high-voltage SiC devices, IEEE Trans. Electron Devices 62 (1984) 396–404.

[44] A. Koyama, et al., 20 KV-class ultra-high voltage 4H-SiC n-IE-IGBTs, in: Materials Science Forum, vol. 1004, Trans Tech Publications, 2020, pp. 899–904.

[45] M. Hinojosa, A. Ogunniyi, High voltage, step-down converter design using 20-kV silicon carbide IGBTs, in: 2016 IEEE International Power Modulator and High Voltage Conference (IPMHVC), 2016, pp. 520–524, https:/doi.org/10.1109/IPMHVC.2016.8012883.

[46] E. van Brunt, et al., 27 KV, 20 A 4H-SiC n-IGBTs, in: Materials Science Forum, vol. 821–823, Trans Tech Publication, 2015, pp. 847–850.

[47] Y. Du, et al., 15kV Press-Pack SiC IGBT, in: IEEE Workshop on Wide Bandgap Power Devices and Applications in Asia, 2021, pp. 162–166.

[48] H.R. Chang, et al., Insulated gate bipolar transistor (IGBT) with a trench gate structure, in: IEEE International Electron Devices Meeting, Abstract 29.5, 1987, pp. 674–677.

[49] M. Kitagawa, et al., A 4500V injection enhanced insulated gate bipolar transistor (IEGT) operating in a mode similar to a thyristor, in: IEEE International Electron Devices Meeting, Abstract 28.3.1, 1993, pp. 679–682.

[50] A. Itoh, I. Omura, 13kV UHV-IGBT: feasibility study, in: IEEE International Symposium on Power Semiconductor Devices and ICs, 2020, pp. 50–53.

[51] M. Takei, Y. Harada, K. Ueno, 600V-IGBT with reverse blocking capability, in: IEEE International Symposium on Power Semiconductor Devices and ICs, Abstract 11.1, 2001, pp. 413–416.

[52] T. Naito, et al., 1200V reverse blocking IGBT with low loss for matrix converter, in: IEEE International Symposium on Power Semiconductor Devices and ICs, Abstract 3.2, 2004, pp. 125–128.

[53] H. Takahashi, M. Kaneda, T. Minato, 1200V class reverse blocking IGBT for AC matrix converter, in: IEEE International Symposium on Power Semiconductor Devices and ICs, Abstract 3.1, 2004, pp. 121–124.

[54] N. Tokuda, M. Kaneda, T. Minato, An ultra-small isolation area for 600V class reverse blocking IGBT with deep trench isolation process, in: IEEE International Symposium on Power Semiconductor Devices and ICs, Abstract 3.3, 2004, pp. 129–132.

[55] H. Shigekane, H. Kirihata, Y. Uchida, Developments in modern high power semiconductor devices, in: IEEE International Symposium on Power Semiconductor Devices and ICs, 1993, pp. 16–21.

CHAPTER

IGBT Structure and Operation

2

The basic structures for the insulated gate bipolar transistor (IGBT) were shown in Fig. 1.2 in the previous chapter. The fundamental static and dynamic electrical characteristics for IGBTs that were derived from the structures first introduced in the 1980s have not changed during the last three decades. However, many modifications of the structures have been demonstrated to improve the IGBT performance. One of the most important alterations to the structure has been replacement of the planar-gate architecture with the trench-gate architecture. Other variants to the structure include the transparent emitter, shorted collector, or reverse-conducting and charge-coupled IGBT designs. Lateral IGBTs suitable for monolithic integration have also been developed for implementation in junction isolation or dielectric isolation technologies to create power integrated circuits. Novel IGBT structures have been introduced recently for improving the trade-off between on-state voltage drop and switching losses. This chapter describes these IGBT structures.

2.1 Symmetric D-MOS Structure

The structure for the symmetric blocking, double-diffused, IGBT structure is shown on the left side of Fig. 2.1. It consists of four alternating layers of semiconductor material arranged in a vertical stack. For the n-channel IGBT shown in the figure, the N-drift region has the largest thickness to enable the IGBT to support high voltages. The lower junction J_1 is usually formed across the entire device cell. In contrast, the upper junction J_2 is formed by a planar diffusion process. Junction J_2 is comprised of two portions. One portion, referred to as the P-base region, is formed by using the gate material as a defining edge. The second portion, referred to as the deep P^+ region, is formed by using a mask to define its diffusion edge from the central portion of the opening in the gate material. In addition, the IGBT structure includes an N^+ emitter region, which is defined on one side by the edge of the gate electrode and the other side by using a mask during fabrication. To achieve the IGBT mode of operation for the four-layer device shown in the figure, it is imperative to short-circuit the N^+ emitter region to the P-base region at its edge located away from the gate electrode, as shown in the figure. The fabrication process for the double-diffused metal oxide semiconductor (D-MOS) IGBT structure will be described in detail in another chapter of this book.

The IGBT Device. https://doi.org/10.1016/B978-0-323-99912-0.00004-0

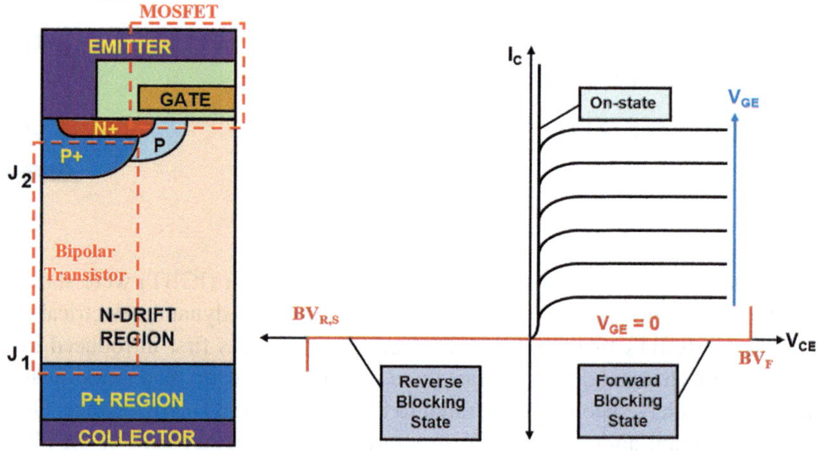

FIGURE 2.1

Symmetric insulated gate bipolar transistor D-MOS structure. Metal oxide semiconductor field effect transistor.

The symmetric IGBT structure is capable of supporting a high voltage when the collector is biased either positive or negative as long as the gate electrode is shorted to the emitter electrode by the control circuit. Under the zero-voltage gate bias operation, there is no path for current to flow from the N^+ emitter region to the collector region. When a negative bias is applied to the collector terminal, junction J_1 becomes reverse biased while junction J_2 becomes forward biased. A depletion layer extends from junction J_1 in the N-drift region toward junction J_2. A high voltage can be supported in the depletion region for negative collector biases, as shown in the output characteristics on the right side of the figure. The maximum blocking voltage capability ($BV_{R,S}$) for the symmetric blocking IGBT is indicated in the figure by the onset of current flow. The maximum reverse blocking capability is determined by open-base P-N-P transistor breakdown, as discussed in the next chapter. The first IGBT reported in the literature had 600 V reverse blocking capability [1]. In the symmetric IGBT structure, the depletion region from junction J_1 cannot extend all the way through the N-drift region. It has therefore been referred to as the non-punch-through IGBT structure.

When a positive bias is applied to the collector terminal, junction J_2 becomes reverse biased while junction J_1 becomes forward biased. A depletion layer extends from junction J_2 in the N-drift region toward junction J_1. A high voltage can be supported in the depletion region for positive collector biases, as shown on the right side of the figure. The maximum blocking voltage capability (BV_F) for the symmetric blocking IGBT is indicated in the figure by the onset of current flow. Because the maximum forward blocking capability is determined by the same open-base P-N-P transistor breakdown as during reverse blocking, the symmetric IGBT structure has essentially equal forward and reverse blocking capability.

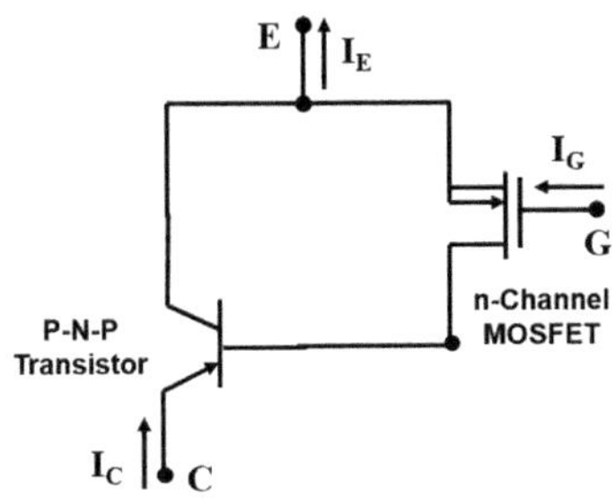

FIGURE 2.2

Equivalent circuit for the insulated gate bipolar transistor structure.

The IGBT structure can carry high currents for positive collector voltages when a positive bias is applied by the control circuit to the gate electrode. The gate bias turns on the n-channel metal-oxide semiconductor field-effect transistor (MOSFET) shown in the figure by creating an electron inversion layer over the P-base region. Electrons are transported via the channel into the N-drift region and serve as the base drive current for the P-N-P bipolar transistor shown in the figure. The IGBT structure operates as a wide-base P-N-P transistor driven by a monolithically integrated n-channel MOSFET structure with an equivalent circuit shown in Fig. 2.2. Because the base drive current for the P-N-P transistor is derived from the collector circuit, current flow in the IGBT exhibits a diode-like on-state characteristic, as shown on the right side of Fig. 2.1, with exponential growth in collector current with increasing collector voltage. In the on-state mode of operation, the collector junction J_1 injects a high concentration of holes into the N-drift region, modulating its conductivity. The holes are collected at junction J_2. This allows the IGBT to carry a high on-state current density with a low on-state voltage drop. One of the chief advantages of the IGBT structure that has enabled the development of high-voltage devices with high current-handling capability is the conductivity modulation of the drift region.

The IGBT structure can be switched from the on-state to the forward blocking state by reducing the gate bias to zero using the control circuit. When the gate bias falls below the threshold voltage, the channel of the MOSFET is removed, cutting off the supply of base drive current to the P-N-P transistor. However, this does not produce an immediate reduction of the collector current. The injection of a high concentration of holes and electrons into the drift region of the IGBT in the on-state produces a large amount of stored charge that must be removed during the turn-off switching transient. Symmetric blocking IGBT devices have a relatively long turn-off time with a collector current tail that produces losses during each switching cycle. For this reason, the development of symmetric blocking IGBTs was abandoned in the 1980s in favor of the development of the asymmetric blocking structure discussed in the next section. There has recently been a revival of interest in the development of reverse blocking IGBTs for cyclo- or matrix-converter applications.

2.2 Asymmetric D-MOS Structure

The asymmetric IGBT structure is shown on the left side of Fig. 2.3. The asymmetric IGBT structure operates in a similar manner to that described for the symmetric blocking structure in the previous section. The key difference from the symmetric structure is the addition of an N-buffer layer in the asymmetric structure. The doping concentration of the N-buffer layer is much larger ($\sim 10^{17}$ cm^{-3}) than that of the drift region ($<10^{14}$ cm^{-3}). The drift region doping concentration for the asymmetric structure is also made much smaller than that for the symmetric structure. This produces a nearly uniform electric field distribution in the drift region for the asymmetric structure when a positive collector bias is applied compared with a triangular electric field distribution for the symmetric structure. A comparison of the electric field profiles for the symmetric and asymmetric cases is shown in Fig. 2.4. It can be seen that the altered electric field profile allows the asymmetric structure to support the same forward blocking voltage (area under the electric field plot) with a much thinner drift region. A narrower drift region thickness allows on-state current flow with a lower on-state voltage drop. In addition, the asymmetric IGBT structure has a superior turn-off time because of the faster recombination of free carriers in the N-buffer layer. Most of the development effort on IGBTs has been focused on the asymmetric structure because of interest in using the device for motor control from a DC link. In the asymmetric IGBT structure, the depletion region from junction J_2 extends all the way through the N-drift region and is stopped by the buffer layer. It has therefore been referred to as the punch-through or field-stop IGBT structure as well.

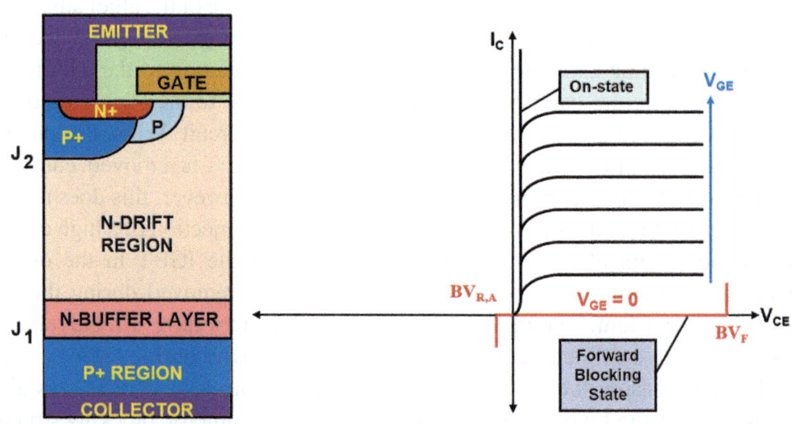

FIGURE 2.3

Asymmetric insulated gate bipolar transistor D-MOS structure.

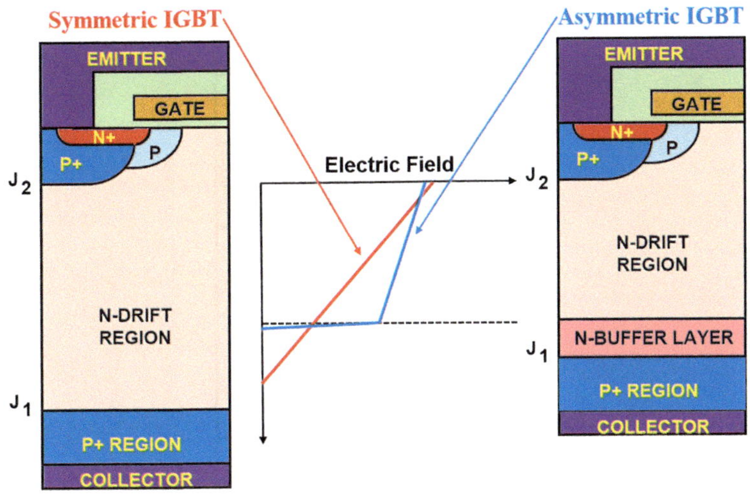

FIGURE 2.4

Comparison of electric field profiles for the symmetric and asymmetric blocking IGBT structures.

2.3 Trench-Gate IGBT Structure

The trench-gate asymmetric IGBT structure is illustrated in Fig. 2.5. In contrast with the D-MOS gate structure, the gate region is formed by etching a trench into the upper surface and locating the gate inside the trench with a gate oxide for isolation from the semiconductor. The trench-gate structure is also referred to as the U-shaped metal oxide semiconductor (U-MOS) gate structure due to the shape of the trenches etched in the semiconductor. In the trench-gate IGBT structure, the device supports voltage at the reverse biased junction J_2 when a positive collector bias is applied. To turn on the device, a channel is formed in the P-base region on the vertical surface of the trench. This allows electrons to be transported from the N^+ emitter to the N-drift region and serve as the base drive current for the P-N-P transistor in the IGBT structure.

Although the basic physics of operation of the trench-gate IGBT structure are the same as that for the D-MOS IGBT structure, a lower on-state voltage drop can be obtained with the trench-gate structure due to the smaller cell pitch and elimination of the junction field effect transistor (JFET) region [2]. The progress in the development of IGBT devices with larger voltage blocking and current handling capability is traced in Figs. 1.6 and 1.7 in the introductory chapter. Trench-gate devices with power handing capability of 5 MW have been reported, as shown in Fig. 1.8. Trench-gate structures with deep trench regions have been called injection-enhanced IGBTs (IEGTs) due to the improved minority carrier profiles leading to lower on-state voltage drop [3].

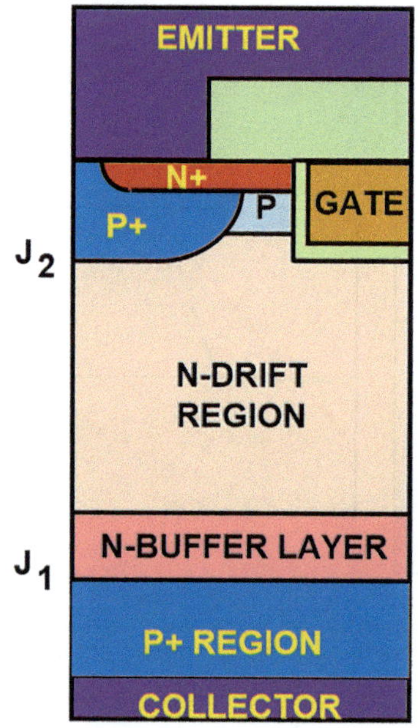

FIGURE 2.5

Trench-gate asymmetric insulated gate bipolar transistor structure.

2.4 Transparent Emitter IGBT Structure

The transparent emitter IGBT structure [4] is shown in Fig. 2.6 together with the doping profiles for the various regions. The doping profiles for the layers formed on the emitter side of the IGBT structure have high surface concentration for the deep P^+ region and the N^+ emitter regions. The surface concentration for the P-base region is adjusted to achieve the desired threshold voltage, as discussed in a later chapter on

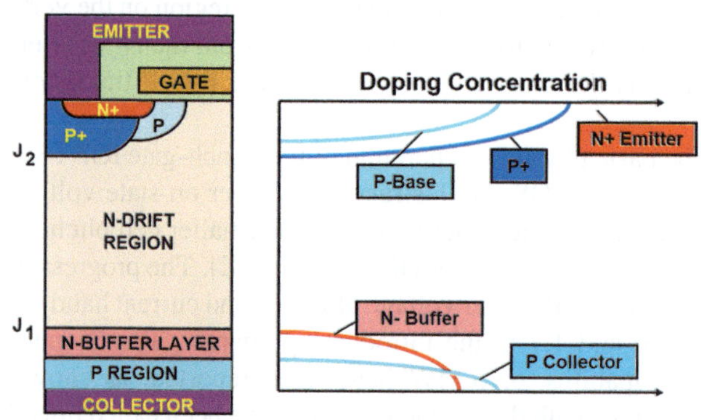

FIGURE 2.6

Transparent emitter asymmetric insulated gate bipolar transistor structure.

device fabrication. These profiles are similar to those used in the symmetric and asymmetric IGBT structures previously shown in Figs. 2.1 and 2.3.

The doping profile on the collector side of the IGBT structure is different for the transparent emitter structure. In the previous structures, the P^+ collector region has a high doping concentration and large thickness. This results in a high injection efficiency at junction J_1 to promote strong injection of holes into the N-drift region. In contrast to this, the surface concentration and thickness for the P collector region for the transparent emitter structure are made much smaller to degrade the injection efficiency at junction J_1. The low injection efficiency at junction J_1 has been found to reduce the hole concentration in the drift region at the collector side of the N-drift region. This results in a faster turn-off time for the IGBT, as discussed in the next chapter, due to not only the smaller amount of stored charge in the drift region but also enhanced recombination due to the injection of electrons from the drift region into the P collector region.

2.5 Novel IGBT Structures

As described earlier, the on-state and switching behavior of the IGBT structure depends upon the free carrier distribution in the on-state. Novel IGBT structures have been proposed to improve the trade-off between the on-state and turn-off losses by tailoring the injected carrier distribution during the on-state. One such structure is shown in Fig. 2.7, where the collector P^+ region is shorted to the N-drift region. Because this approach creates an antiparallel diode within the IGBT structure, it

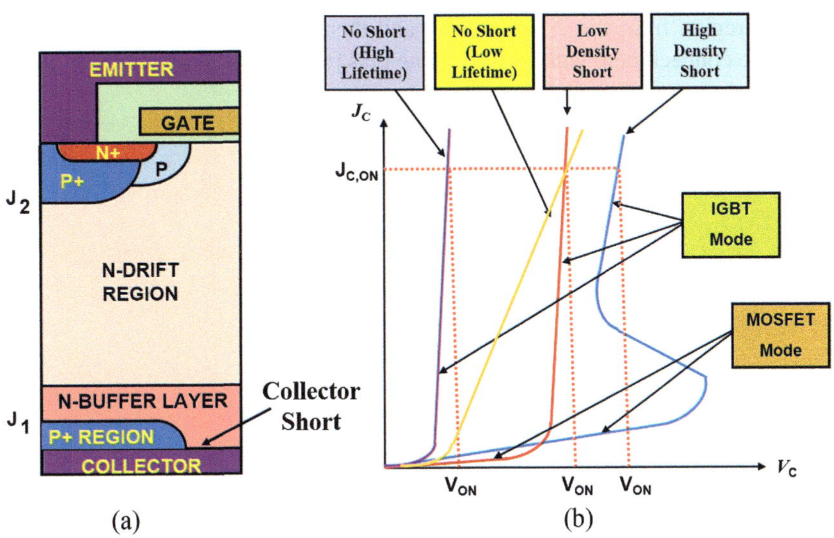

FIGURE 2.7

Reverse conducting IGBT: (a) structure with collector short; (b) on-state characteristic.

is most appropriate to also utilize the asymmetric structure with a buffer layer. The injected hole concentration in the N-drift region is reduced due to the presence of the shorts. The turn-off time is reduced due to the reduced hole concentration and the removal of electrons from the drift region directly through the short. These device structures are also referred to as reverse-conduction (RC) IGBTs.

The on-state i-v characteristics of the RC-IGBT structure have been found to be sensitive to the design of the size of the collector shorting area, as shown in Fig. 2.7. At smaller collector current levels, the RC-IGBT with a high short density behaves like a power MOSFET with linear characteristics, as shown by the blue line. The MOSFET current shunts the collector junction J_1, suppressing its injection until large collector voltages. Once the junction J_1 begins to inject holes into the drift region, conductivity modulation reduces its resistance, producing a snapback behavior. This is undesirable behavior for an IGBT structure. With a low short density, the RC-IGBT structure exhibits the characteristic shown in red without snapback. The on-state voltage drop for this case is close to that for an IGBT after lifetime control to reduce the switching loss to match that of the RC-IGBT. The RC-IGBT creates a structure with low switching losses and an integrated reverse conducting diode.

Another novel IGBT structure utilizes the charge coupling concept [5], as shown in Fig. 2.8. Here, alternating columns of P-type and N-type drift regions are formed to create the region that supports the forward blocking voltage. This approach has been referred to as the superjunction (SJ) concept. The resulting two-dimensional charge coupling effect makes the electric field more uniform along the vertical direction, allowing a reduction of the thickness of the drift region. This has been projected to improve the turn-off losses by 50% for the same on-state voltage drop as the conventional asymmetric IGBT structure. A variation on the SJ concept is to separate the P-column region from the P-base region and then use a second trench-gate MOS structure to connect them with a second gate bias [6]. The P-column floats with zero

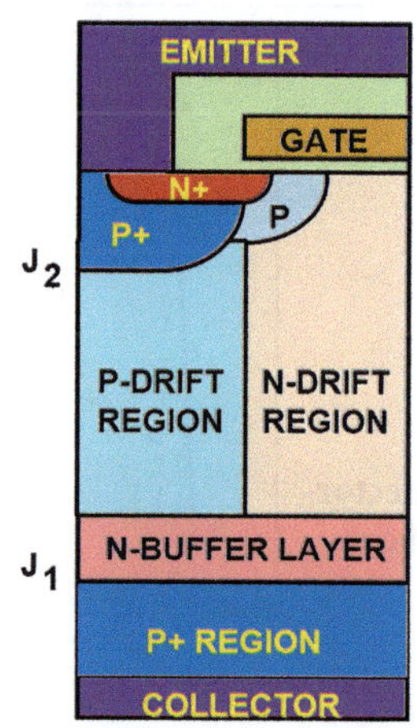

FIGURE 2.8

IGBT with charge-coupled drift region.

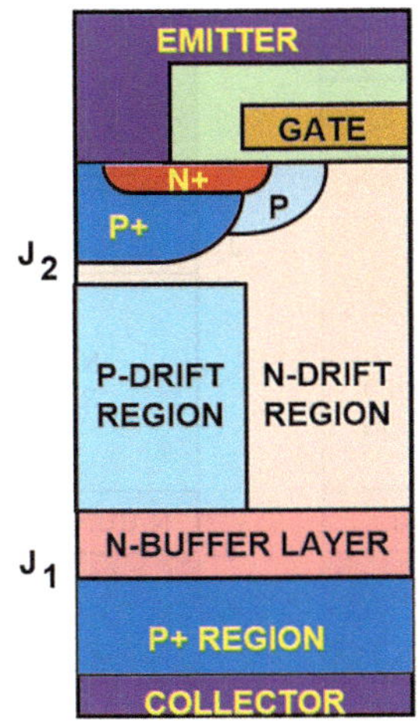

FIGURE 2.9

IGBT with floating P-type charge-coupled drift region.

gate bias, improving the conductivity modulation in the on-state while the turn-off losses are reduced by connecting it to the P-base region with a negative gate bias to the second trench-gate electrode. Although this structure has not been validated by its fabrication, numerical simulations show a superior trade-off between on-state and switching power losses.

A 650 V SJ IGBT structure has been experimentally demonstrated with only 45 μm wafer thickness [7]. The device structure illustrated in Fig. 2.9 was formed using trench etch and refill technology with the P-column disconnected from the P-base region. A superior trade-off curve between on-state voltage drop and switching energy loss was experimentally confirmed.

The on-state voltage drop of the IGBT structure is larger than that for a P-i-N rectifier despite the high-level injection of minority carriers into the drift region. This is due to the reduced free carrier concentration near the upper junction J_2, which is reverse biased in the on-state mode of operation for the IGBT. Two approaches to reducing the on-state voltage drop of the IGBT have been demonstrated by enhancing the concentration of carriers near junction J_2. In the first method, shown in Fig. 2.10(a), the trench-gate region is made much deeper than the P-base region and the mesa containing the junctions is made very narrow in size. This type of structure has been called the injection-enhanced IGBT structure or sometimes the IEGT [8].

Another approach to enhancing the free carrier concentration near junction J_2 is to dope the N-drift region to a higher level in the vicinity of the junction, shown as the current enhancement layer (CEL) in Fig. 2.10(b). The resulting larger doping concentration in the JFET region improves electron current flow from the MOS channel. The larger doping near junction J_2 also reduces the voltage drop in the drift region. Because this modification to the IGBT structure produces a larger collector current

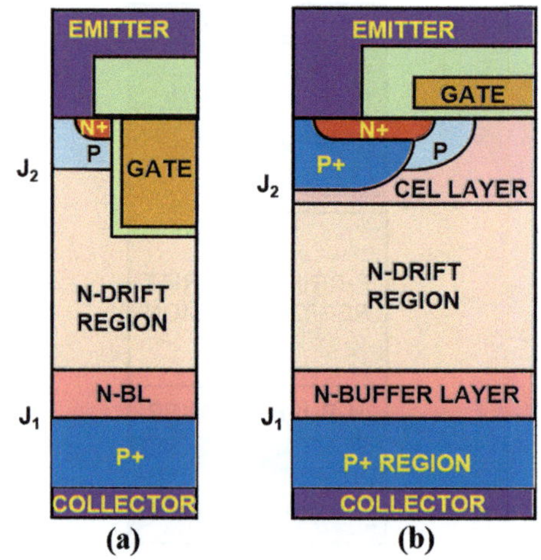

FIGURE 2.10

Novel insulated gate bipolar transistor structures. CEL, current enhancement layer. (a) Deep trench gate IGBT structure; (b) IGBT with current enhancement layer.

flow in the on-state, the more highly doped layer has been called the CEL. It is commonly utilized in the development of silicon carbide IGBT structures [9]. One difficulty with this approach is that the forward blocking voltage can be reduced if the CEL doping concentration is too large.

A charge-coupling concept using a deep trench region with a source connected electrode was proposed [10] and has been widely commercialized for silicon power MOSFETs with blocking voltages up to 150 V. This approach allows greatly increasing the doping concentration in the drift region while maintaining a high breakdown voltage. The idea has been applied to IGBTs to reduce the resistance of the region near the emitter side where conductivity modulation is suppressed by the reverse bias on junction J_2. The trench-shielded planar-gate (TSPG) IGBT, illustrated in Fig. 2.11, is a three-dimensional structure that produces charge coupling with the trench-based electrode orthogonal to the planar gate. It reduced the on-state voltage drop while increasing short-circuit withstand capability by a factor of two.

2.6 Lateral IGBT Structures

The previously discussed vertical IGBT structures have the two high-current carrying terminals on opposite sides of the wafer. This is the most appropriate chip architecture for discrete devices with large current-handling capability. For applications that need multiple IGBT devices that must handle relatively low-current levels (~1 A), monolithic integration of all the devices on a single chip is attractive to reduce the packaging cost. Lateral IGBT structures are required to implement this approach. In

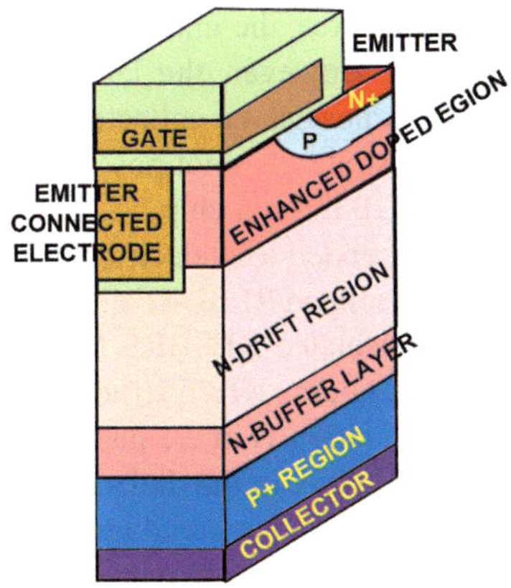

FIGURE 2.11

IGBT with trench electrode-based charge-coupled region.

addition to the challenge of supporting high voltage in the lateral device structure, it is essential that the devices on a single chip are isolated from each other during circuit operation.

The monolithic integration of multiple high-voltage devices can be achieved by using a variety of structural options. The most attractive approach is to utilize two-dimensional charge coupling to improve the electric field distribution in the lateral direction to enhance the breakdown voltage. This approach, called the reduced surface electric field (RESURF), was first proposed for making lateral unipolar devices such as high-voltage MOSFETs [11]. A lateral IGBT structure implemented with this approach is shown in Fig. 2.12(a). This is known as a junction isolated device structure because multiple devices are isolated from each other by P-N junction J_3. As described earlier, the IGBT structure relies upon the injection of a minority carrier (holes in the above structure) from the collector region into the N-drift region. As in

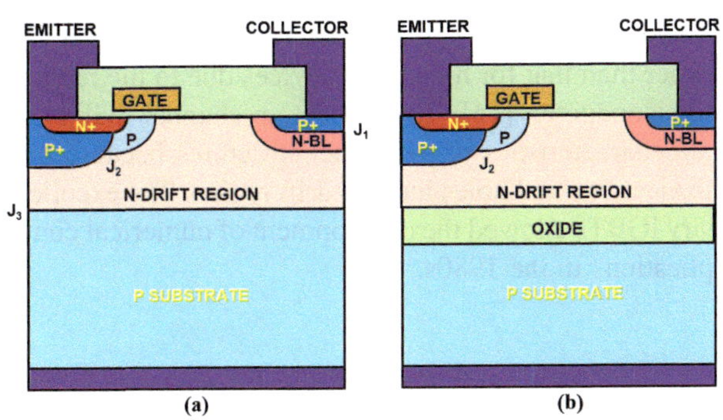

FIGURE 2.12

Lateral insulated gate bipolar transistor structures. (a) Junction isolated IGBT; (b) dielectrically isolated IGBT.

the case of the vertical IGBT structures, the injected holes are collected by junction J_2 in the lateral IGBT structure. However, the lateral IGBT structure contained another junction J_3 at which the holes can be collected. Because the holes collected at junction J_3 flow into the P-substrate, it is not possible to achieve the desired isolation between multiple lateral IGBTs on a chip.

The RESURF concept was extended to dielectrically isolated structures in 1991 [12]. High-voltage dielectrically isolated RESURF diodes [13] were demonstrated in 1992, followed by dielectrically isolated RESURF MOS-bipolar structures [14] in 1993. The dielectrically isolated lateral IGBT structure is shown in Fig. 2.12(b). It can be seen that the substrate junction J_3 in the junction isolated structure is eliminated by the presence of an oxide layer in the dielectrically isolated structure. This prevents the flow of holes into the substrate, greatly improving not only the isolation between devices but also the switching performance of each device. Many publications on power integrated circuits (ICs) have appeared in the literature since the first demonstration of the basic structures. It is worth pointing out that the dielectric isolation technology allows monolithic integration of the fly-back rectifiers with the IGBTs, providing a complete solution for making a motor control power circuit on a single chip.

2.7 Complementary IGBT Structures

Many power delivery and management systems rely upon AC power supplied by utilities. In these applications, it is necessary to control current flow during both the positive and negative half-cycle of the AC waveform. It is possible to do this by connecting two n-channel IGBTs in inverse parallel. However, this requires level shifting the control signal for one of the IGBTs with a reference that varies from zero to the peak voltage of the AC input waveform. Because the shifting of gate drive signals to high voltages is inconvenient and expensive, the p-channel or complementary IGBT shown in Fig. 2.13 was developed [15] in the 1980s. In the case of p-channel silicon power MOSFETs, it has been found that the specific on-resistance is three times larger than that for n-channel devices due to the reduced mobility for holes in the semiconductor [16]. In contrast, p-channel IGBT displays nearly identical on-state characteristics to n-channel structures because of the high-level injection of both electrons and holes into the drift region. The excellent performance of complementary IGBTs allowed the development of numerical controls for factory automation applications in the 1980s.

2.8 Advanced IGBT Structures

There has been a continual effort to try to reduce the on-state voltage drop for the IGBT while maintaining low switching losses. This can be achieved by enhancing the injected carrier concentration near the upper junction. One method to achieve this

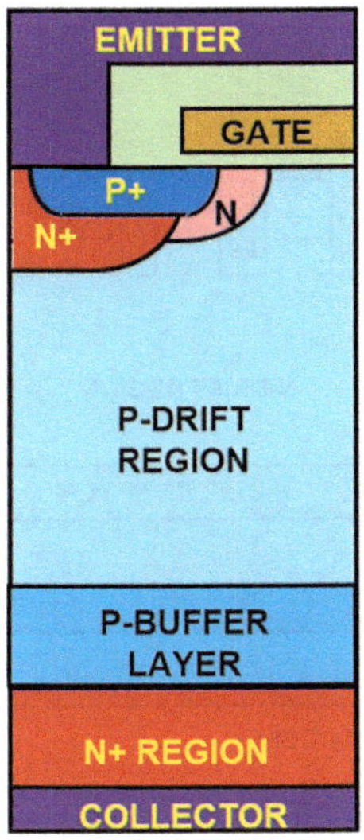

FIGURE 2.13

Complementary insulated gate bipolar transistor structures.

is by using "dummy cells" that do not contain the N^+ emitter region. The first option is to not make a connection of the P-base region in the dummy cell, called a floating dummy cell structure, as shown in Fig. 2.14. The second option is to connect the P-base region in the dummy cell to the emitter electrode, called the shorted dummy cell, shown in Fig. 2.15. The floating dummy cells have been shown to increase the injected carrier density below the P-base regions, leading to a smaller on-state voltage drop than the shorted dummy cell case [17]. The two cases fall on different parts of the trade-off curve between on-state voltage drop and switching losses.

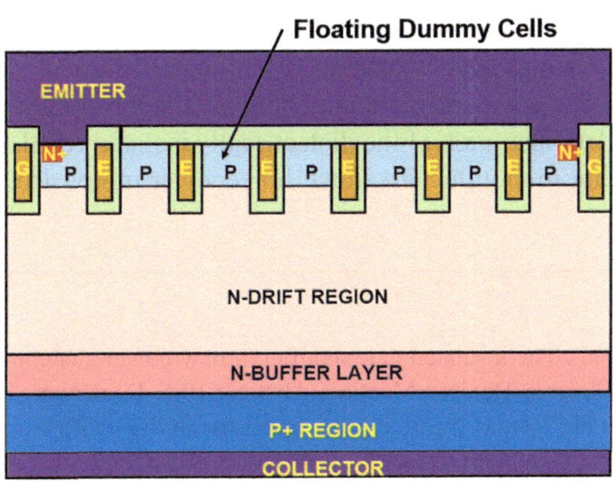

FIGURE 2.14

IGBT structure with floating dummy cells.

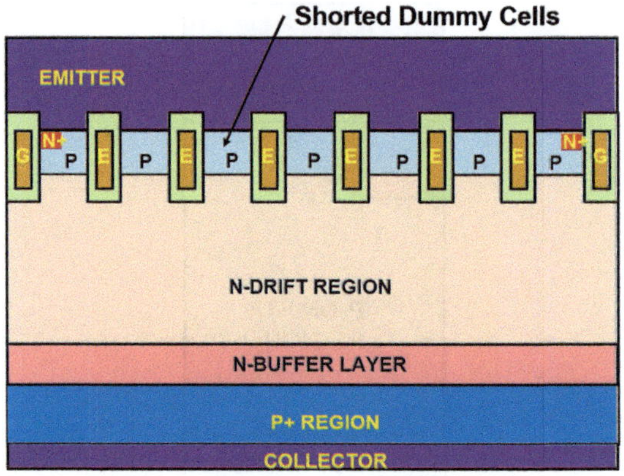

FIGURE 2.15

IGBT structure with shorted dummy cells.

It has been shown that shrinking the IGBT cell size provides many benefits to the performance [18]. The forward biased safe operating area was greatly expanded by using smaller design rules because of suppressing the latch-up of the internal thyristor. This approach was subsequently used to enhance IGBT product performance [19]. The trade-off curve between on-state voltage drop and switching loss improved when the design rule was changed from 3 μm to submicron. The devices were found to exhibit superior ruggedness as well. This approach has been referred to as a fine mesh IGBT structure and a microtrench IGBT structure [20].

2.9 Summary

The various vertical IGBT structures that have evolved after the introduction of the basic IGBT concept have been described in this chapter. All these structures share the common physical principles of a MOSFET controlling current flow in a wide-base bipolar transistor. Each of these structures offers specific benefits in electrical operating characteristics to serve different applications. Lateral IGBT structures became feasible after the demonstration of the dielectrically isolated RESURF concept. This has enabled creating monolithic fractional horsepower motor drive circuits.

References

[1] B.J. Baliga, et al., The insulated gate rectifier: a new power switching device, in: IEEE International Electron Devices Meeting, Abstract 10.6, December, 1982, pp. 264–267.

[2] H.R. Chang, et al., Insulated gate bipolar transistor (IGBT) with a trench gate structure, in: IEEE International Electron Devices Meeting, Abstract 29.5, 1987, pp. 674–677.

[3] M. Kitagawa, et al., A 4500V injection enhanced insulated gate bipolar transistor (IEGT) operating in a mode similar to a thyristor, in: IEEE International Electron Devices Meeting, Abstract 28.3.1, 1993, pp. 679–682.

[4] F. Bauer, et al., A comparison of emitter concepts for high voltage IGBTs, in: IEEE International Symposium on Power Semiconductor Devices and ICs, 1995, pp. 230–235.
[5] M. Antoniou, F. Udrea, F. Bauer, The super-junction insulated gate bipolar transistor optimization and modeling, IEEE Trans. Electron Devices 57 (2010) 594–600.
[6] J. Wei, M. Zhang, K.J. Chen, Super-junction IGBT with conductivity modulation actively controlled by two separate driving signals, IEEE Trans. Electron Devices 67 (2020) 4335–4339.
[7] Y. Wu, et al., 650 V super-junction insulated gate bipolar transistor based on 45 μm ultra-thin wafer technology, IEEE Electron Device Lett. 43 (2022) 592–595.
[8] M. Kitagawa, et al., 4500-V IEGTs having switching characteristics superior to GTO, in: IEEE International Symposium on Power Semiconductor Devices and ICs, 1995, pp. 486–491.
[9] Q. Zhang, et al., Design and characterization of high voltage 4H-SiC p-IGBTs, IEEE Trans. Electron Devices 55 (2008) 1912–1919.
[10] B.J. Baliga, Power Semiconductor Devices Having Improved High Frequency Switching and Breakdown Characteristics. U. S. Patent # 5,998,833, Issued December 7, 1999.
[11] J.A. Appels, H.M.J. Vaes, High-voltage thin layer devices (RESURF devices), in: IEEE International Electron Devices Meeting, Abstract 10.1, 1979, pp. 238–241.
[12] Y.S. Huang, B.J. Baliga, Extension of RESURF principle to dielectrically isolated power devices, in: IEEE International Symposium on Power Semiconductor Devices and ICs, Paper 2.1, 1991, pp. 27–30.
[13] S. Sridhar, Y.S. Huang, B.J. Baliga, Dielectrically isolated high voltage P-i-N rectifiers for power ICs, in: IEEE International Electron Devices Meeting, Abstract 9.6.1, 1992, pp. 245–248.
[14] Y.S. Huang, S. Sridhar, B.J. Baliga, Junction and dielectrically isolated lateral ESTs for power ICs, in: IEEE International Symposium on Power Semiconductor Devices and ICs, Paper 9.3, 1993, pp. 259–263.
[15] M.F. Chang, et al., Comparison of N and P channel IGTs, in: IEEE International Electron Devices Meeting, Abstract 10.6, 1984, pp. 278–281.
[16] B.J. Baliga, Fundamentals of Power Semiconductor Devices, second ed., Springer-Science, New York, 2019.
[17] R. Gejo, et al., Ideal carrier profile control for high-speed switching of 1200 V IGBTs, in: IEEE International Symposium on Power Semiconductor Devices and ICs, 2014, pp. 99–102.
[18] S. Sawant, B.J. Baliga, Impact of VLSI design rules on high voltage (2000 V) IGBTs/ESTs, in: IEEE International Symposium on Power Semiconductor Devices and ICs, 1998, pp. 253–256.
[19] J. Yamashita, et al., The 5th generation highly rugged planar IGBT using sub-micron process technology, in: IEEE International Symposium on Power Semiconductor Devices and ICs, 2001, pp. 421–424.
[20] A. Philippou, et al., A comparative study of oxidized spacer trench and Micro-pattern trench concepts for 1200 V IGBTs, in: IEEE International Symposium on Power Semiconductor Devices and ICs, 2020, pp. 459–461.

CHAPTER

IGBT Structural Design

3

The methodology for designing the internal structure of the insulated gate bipolar transistor (IGBT) is systematically presented in this chapter. It is essential that the internal structure be designed to achieve the desired blocking voltage capability. The blocking voltage requirement for the IGBT is dictated by the maximum voltage impressed upon the device in its intended application. In typical motor control applications with power delivered from a DC bus, the IGBT can be subjected to spikes in collector voltage that are 50% greater than the DC bus voltage. This margin must be taken into account when designing the blocking voltage capability of the IGBT structure.

All high-voltage IGBTs must have a reasonable threshold voltage of 4 V. Smaller threshold voltages can lead to inadvertent turn-on of the IGBT due to spikes in voltage on the gate, which are due to the high rate of change in collector voltage during the IGBT switching transients. Larger threshold voltages are not desirable because they require an increase in the gate drive voltage to avoid a larger on-state voltage drop. The threshold voltage is determined by a combination of the doping concentration of the *P*-base region and the gate oxide thickness.

Various IGBT structures were discussed in the previous chapter. Although two-dimensional (2D) effects differ among these structures, it is possible to create one-dimensional (1D) models for the IGBT because of a homogenous behavior within the drift region over most of its thickness. These 1D models allow analytical solutions to be developed that can be used to estimate the optimum thickness and doping concentration of the drift region before resorting to 2D numerical simulations to refine these parameters. 1D analytical models for the injected carrier distribution can also be generated to provide a deeper understanding of the physics of operation of the IGBT structures. The free carrier concentrations determine not only the on-state voltage drop but also the turn-off transients. In this chapter, 1D models are provided for each of the primary IGBT structures discussed in the previous chapter.

As described in this chapter, the parameters of the internal structure can be adjusted to achieve a trade-off between the on-state voltage drop and the turn-off switching losses. In the case of the symmetric blocking IGBT, this is typically done by varying the lifetime in the drift region. In the case of the asymmetric IGBT structure, this can be accomplished by changing the thickness and doping concentration in the buffer layer as well as by varying the lifetime in the drift region. In the case of the transparent emitter IGBT structure, the trade-off can be performed by varying the doping concentration of the *P*-collector region, the doping concentration and

The IGBT Device. https://doi.org/10.1016/B978-0-323-99912-0.00022-2

thickness of the buffer layer, and the lifetime in the drift region. The choice among these approaches depends upon the voltage rating of the device, the available process technology, and the know-how developed by each IGBT manufacturer.

3.1 Threshold Voltage

The threshold voltage for the IGBT is the minimum gate bias voltage required to observe on-state current flow. Typical IGBT products have a threshold voltage of 4 V with a gate drive voltage of 15 V used for on-state operation. The threshold voltage of these high-voltage devices is typically larger than for low-voltage (e.g., 30 V) MOSFET products because of the larger voltage transients in IGBT applications that can induce larger spurious voltages on the gate terminal. The design methodology used to achieve a desired threshold voltage is the same for all the IGBT structures discussed in this chapter.

Current flow in the IGBT structure begins to occur when an inversion layer forms in the integrated MOSFET portion, which supplies the electrons that serve as the base drive current for the internal PNP bipolar transistor. The formation of an inversion layer at the surface of the *P*-base region is induced by the positive gate bias voltage. In IGBT products, the threshold voltage is defined as the gate voltage at which a certain magnitude of collector current is observed. However, from the standpoint of device physics, the threshold voltage is defined as the gate bias at which the transition occurs from weak inversion to strong inversion. In the absence of charge in the oxide and any work-function difference between the gate electrode and the semiconductor, the threshold voltage is given by [1]:

$$V_{TH} = \frac{\sqrt{4\varepsilon_S kTN_A \ln(N_A/n_i)}}{C_{OX}} + \frac{2kT}{q} \ln\left(\frac{N_A}{n_i}\right) \tag{3.1}$$

where C_{OX} is the specific gate oxide capacitance and N_A is the peak doping concentration of the *P*-base region. The threshold voltage can be estimated using:

$$V_{TH} = \frac{t_{OX}}{\varepsilon_{OX}} \sqrt{4\varepsilon_S kTN_A \ln\left(\frac{N_A}{n_i}\right)} \tag{3.2}$$

where t_{OX} is the gate oxide thickness. Based upon this equation, the threshold voltage increases linearly with increasing gate oxide thickness and approximately as the square root of the doping concentration in the semiconductor. This knowledge can be used to improve the design of IGBT structures, as discussed later in the book.

The doping profile of the *P*-base region in IGBTs produced using the double-diffusion process is illustrated in Fig. 3.1. Both the *P*-base and N^+ emitter regions are formed by using a diffusion process with a surface concentration and junction depth decided by the ion-implant dose and drive-in time. The *n*-type emitter doping compensates the *p*-type base doping near the junction. The net *p*-type doping concentration in the *P*-base region is indicated by the blue dashed line in the figure.

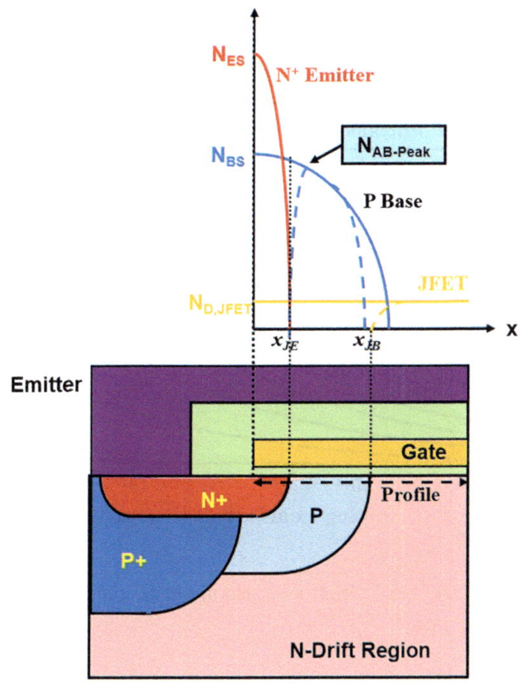

FIGURE 3.1

Doping profile of the channel region in the IGBT structure.

The highest or peak doping concentration ($N_{AB\text{-}Peak}$) in the P-base region has a lower value than the surface concentration (N_{BS}) of the P-base doping profile. The threshold voltage is determined by this doping concentration because of its largest value when compared with the threshold voltage at other portions of the P-base doping profile. The simple 1D analytical solution for the threshold voltage pertains to this portion of the P-base region with the highest net p-type doping level.

For n-channel IGBTs, an n-type polysilicon gate is utilized because this produces a negative shift in the threshold voltage by about 1 V [1]. This negative shift is desirable because it allows for a higher doping concentration of the P-base region to achieve the required threshold voltage. A larger doping concentration for the P-base region suppresses the latch-up of the internal thyristor device within the IGBT structure, making it more rugged in applications. The threshold voltage is also affected by the presence of positive fixed charge in the gate oxide. The fixed charge density is typically 1×10^{11} cm^{-2} for IGBT devices due to the high-temperature diffusion steps required for the formation of the P-base region. This value will be assumed for the computation of the threshold voltages below.

The threshold voltages for silicon n-channel IGBTs obtained by using Eqn (3.1) are provided in Fig. 3.2 as a function of the peak doping concentration of the P-base region. The gate oxide thickness is used as a parametric variable. The design target of

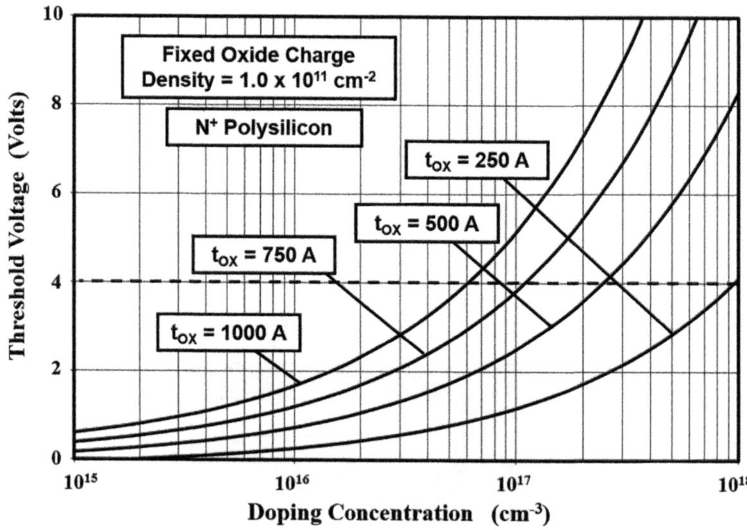

FIGURE 3.2

Threshold voltages for silicon IGBT structures, assuming N^+ polysilicon gate and fixed oxide charge density of $1 \times 10^{12}\,\text{cm}^{-2}$.

a threshold voltage of 4 V is indicated by the dashed line in the figure. During the early days of IGBT development, the gate oxide thickness was maintained at 1000 Å or above to obtain sufficient yield. Modern IGBTs have gate oxide thickness of 500 Å or less. A *P*-base doping concentration to $2.5 \times 10^{17}\,\text{cm}^{-3}$ provides a threshold voltage of 4 V in the case of a gate oxide thickness of 500 Å. The *P*-base doping concentration increases to $1.0 \times 10^{18}\,\text{cm}^{-3}$ to get the same threshold voltage of 4 V in the case of a gate oxide thickness of 250 Å. These doping levels can be obtained by adjusting the diffusion process used to make the *P*-base and N^+ emitter regions, as discussed later in the book.

3.2 Symmetric IGBT Structure

The structure for the symmetric-blocking, double-diffused IGBT was shown on the left side of Fig. 2.1. A symmetric-blocking, trench-gate IGBT structure is similar to that shown in Fig. 2.4 without the *N*-buffer layer. The analyses and design procedures described below are applicable to both these gate structures. The first symmetric blocking IGBT was demonstrated in 1982 [2]. Because the development of IGBT products was mainly focused on motor drive circuits operated from a DC power bus, the symmetric structure lay dormant for nearly 20 years because the asymmetric IGBT structure has much more favorable characteristics for motor drive applications. Interest in the symmetric IGBT structure was revived for application in matrix or cycloconverters [3–6]. The most challenging aspect for development of the symmetric IGBT is a process for edge termination of the reverse blocking junction.

3.2.1 Blocking Voltage

The symmetric IGBT structure can support a high voltage within the N-drift region when the collector is biased either positive or negative by the formation of a depletion region across either junction J_1 or junction J_2. The thickness and doping concentration of the drift region must be optimized to achieve the desired blocking voltage during device structural design. The optimum design depends on the lifetime in the drift region, as shown below.

The maximum voltage that the symmetric IGBT structure can support is determined by open-base transistor breakdown [1] because the current generated by leakage and impact ionization is amplified by the gain of the PNP transistor. The open-base transistor breakdown criterion is given by:

$$\alpha_{PNP} = (\gamma_E \cdot \alpha_T)_{PNP} M = 1 \tag{3.3}$$

where γ_E is the injection efficiency of the PN junction (J_1 or J_2), α_T is the base transport factor, and M is the multiplication factor. The magnitude of α_T and M is a strong function of the collector bias voltage.

The base transport factor (α_T) can be computed using:

$$\alpha_T = \frac{1}{\cosh\left(l/L_p\right)} \tag{3.4}$$

where the width (l) of the undepleted portion of the N-drift region is given by:

$$l = W_N - \frac{\sqrt{2\varepsilon_S V_C}}{qN_D}. \tag{3.5}$$

In this equation, ε_S is the dielectric constant of the semiconductor, V_C is the applied bias to the collector electrode, q is the electron charge, and L_p is the diffusion length for holes in the N-drift region. The width of the undepleted portion of the N-drift region decreases when the collector bias is increased, resulting in an increase in the base transport factor.

The multiplication factor can be determined using:

$$M = \frac{1}{1 - (V_C/BV_{PP})^n} \tag{3.6}$$

where BV_{PP} is the avalanche breakdown voltage of the PN junction. A value of $n = 6$ should be used for the case of an n-channel IGBT structure. The multiplication factor increases with increasing collector voltage.

An example of the design procedure for optimization of the drift region doping concentration and width for the symmetric blocking IGBT structure with a blocking capability of 3000 V is shown in Fig. 3.3. A low-level lifetime of 10 μs was used in the N-drift region during this analysis. In this figure, the blocking voltage capability has been computed for a given drift region doping concentration (e.g., 1.5×10^{13} cm^{-3}

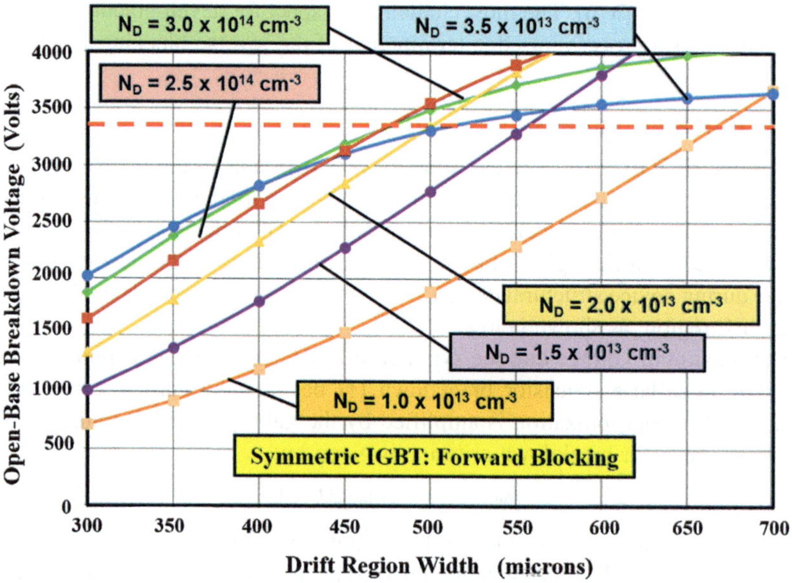

FIGURE 3.3

Open-base breakdown voltage for the symmetric IGBT structure for a lifetime of 10 μs.

purple line) for various drift region thickness values. It can be seen that the blocking voltage increases with increasing drift region thickness. The procedure is then repeated for other drift region doping concentrations. A breakdown voltage of 3300 V, indicated by the red dashed line in the figure, provides a 10% margin during device design. It can be observed that many combinations of the drift region doping concentration and width can be used to obtain this blocking voltage capability. The width of the *N*-drift region decreases from 670 μm for a doping concentration of 1.0×10^{13} cm^{-3} to 475 μm for a doping concentration of 2.5×10^{13} cm^{-3}. The width of the drift region then increases to 510 μm for a doping concentration of 3.5×10^{13} cm^{-3}. It can be concluded that a minimum thickness for the *N*-drift region can be achieved by optimization of the doping concentration. A smaller drift region thickness favors smaller on-state voltage drop and reduced switching losses due to smaller stored charge.

For any desired blocking voltage capability for the symmetric IGBT structure, there is an optimum drift region doping concentration with a corresponding minimum drift region width. These values depend upon the low-level lifetime in the drift region. For the case of a device with blocking voltage capability of 3000 V, the drift region design required to achieve a breakdown voltage of 3300 V is provided in Fig. 3.4 for three values of the low-level lifetime. It can be observed that the optimum drift region width is 565 μm with a doping concentration of 2.0×10^{13} cm^{-3} for a lifetime of 100 μs; the optimum drift region width is 467 μm with a doping concentration of 3.0×10^{13} cm^{-3} for a lifetime of 10 μs; and the optimum drift region width

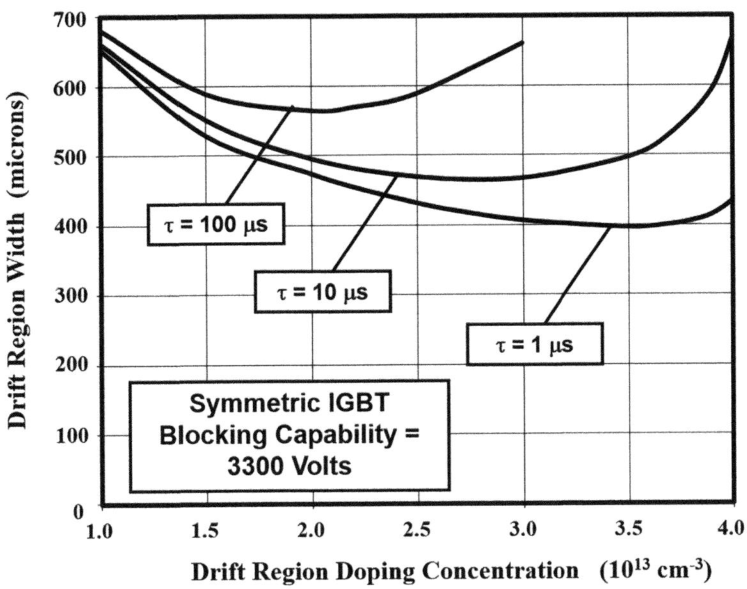

FIGURE 3.4

Optimization of the width and doping concentration of the drift region for the symmetric IGBT structure.

is 396 μm with a doping concentration of 3.5×10^{13} cm^{-3} for a lifetime of 1 μs. The optimum drift region thickness decreases when the lifetime is reduced because of a shorter diffusion length for holes, which reduces the base transport factor.

3.2.2 On-State Characteristics

The on-state voltage drop for the IGBT is lower than that for power MOSFETs with the same voltage rating due to the high-level injection of free carriers into the drift region. The voltage drop contributed by the drift region becomes dominant in devices with fast turn-off characteristics. The free carrier distribution in the drift region of the IGBT is an important factor in determining its on-state characteristics.

A schematic illustration of the free carrier distribution in the IGBT structure in the on-state is shown in Fig. 3.5 at two locations in the basic cell cross-section. The carrier distribution on the left side pertains to the location of the deep P^+ region in the cell. The carrier distribution on the right side pertains to the location of the gate electrode in the cell. It can be observed that the carrier concentration p_0 at the collector junction J_1 and through most of the drift region is the same for both cases. However, the free carrier concentration is reduced to zero at junction J_2 because it is reverse biased during the on-state mode of operation of the IGBT while it becomes enhanced near the gate electrode due to the formation of an accumulation layer produced by the positive gate bias. Because the reduced conductivity modulation of the drift region

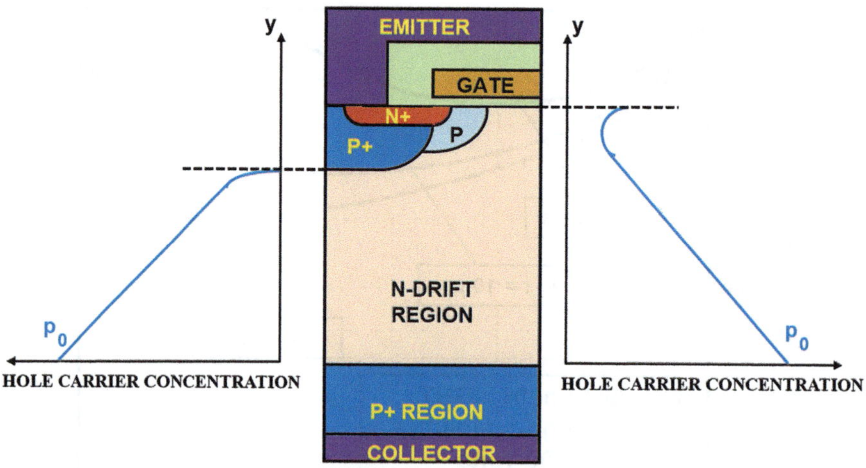

FIGURE 3.5

Free carrier distribution within the IGBT structure.

near the emitter side increases the voltage drop in this portion, it is prudent to utilize the carrier distribution on the left side during on-state analysis to provide a more conservative value for the on-state voltage drop for the IGBT structure.

1D analysis of the free carrier distribution profile under the deep P^+ region is obtained by solving the continuity equation for minority carriers under steady-state high-level injection conditions [1]:

$$\frac{d^2p}{dy^2} - \frac{p}{L_a^2} = 0 \tag{3.7}$$

where L_a is the ambipolar diffusion length in the N-base region given by:

$$L_a = \sqrt{D_a \tau_{HL}} \tag{3.8}$$

The boundary conditions are

$$p(0) = p_0 \tag{3.9}$$

due to the injection of holes from the P^+ collector/N-base junction (J_1), and

$$p(W_N) = 0 \tag{3.10}$$

due to the reverse biased deep P^+/N-base junction (J_2). Using these boundary conditions yields:

$$p(y) = p_0 \frac{\sinh[(W_N - y)/L_a]}{\sinh(W_N/L_a)} \tag{3.11}$$

The hole concentration (p_0) at the P^+ collector/N-base junction (J_1) can be related to collector current (J_C) by using [1]:

$$p_0 = \frac{J_C L_a}{2qD_p}\left(\frac{\mu_p}{\mu_n}\right)\tanh\left(W_N/L_a\right)\left[\left(\frac{\mu_n}{\mu_p}\right)\gamma_{E,ON} - \left(1 + \gamma_{E,ON}\right)\right] \tag{3.12}$$

with the injection efficiency at junction J_1 in the on-state given by:

$$\gamma_{E,ON} = \frac{J_p}{J_C} = 1 - \frac{J_n}{J_C} \tag{3.13}$$

where J_p and J_n are the hole and electron current densities at junction J_1, respectively, and J_C is the total collector current density. The electron current density at junction J_1 can be determined using:

$$J_n = \frac{qD_{nE}p_0^2}{L_{nE}N_{AE}} \tag{3.14}$$

Combining these equations yields a quadratic solution for the hole concentration (p_0) at the P^+ collector/N-base junction:

$$ap_0^2 + bp_0 + c = 0 \tag{3.15}$$

where

$$a = \frac{qD_{nE}}{L_{nE}N_{AE}J_C}\left(1 + \frac{\mu_n}{\mu_p}\right) \tag{3.16}$$

$$b = \frac{2qD_p}{L_a J_C \tanh\left(W_N/L_a\right)}\left(\frac{\mu_n}{\mu_p}\right) \tag{3.17}$$

$$c = -\left(\frac{\mu_n}{\mu_p}\right) \tag{3.18}$$

The hole concentration (p_0) at the P^+ collector/N-base junction can be computed using:

$$p_0 = \frac{-b}{2a}\left[1 - \sqrt{1 - \left(\frac{4ac}{b^2}\right)}\right] \tag{3.19}$$

The injected hole concentration for this structure is shown in Fig. 3.6 for the case of four values of the high-level lifetime (τ_{HL}). The doping concentration and thickness for the drift region for the device were chosen to be 2.5×10^{13} cm^{-3} and 450 μm, respectively. A doping concentration of 1×10^{19} cm^{-3} was used for the P^+ collector region with an electron diffusion length of 0.5 μm inside this region. The highest injected hole concentration (indicated by p_0 in the figure) occurs at the P^+ collector/N-base junction (J_1) with a value of more than 10^{17} cm^{-3}, which is well above the doping concentration of the drift region. The hole concentration reduces to zero at the deep P^+/N-base junction (J_2) at $y = 450$ μm. The injected hole concentration becomes smaller throughout the drift region when the lifetime is reduced.

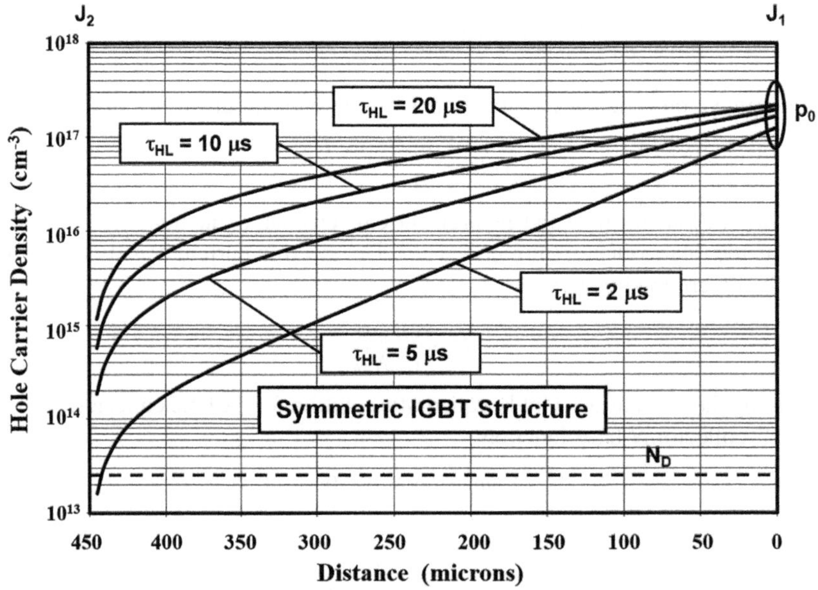

FIGURE 3.6

Distribution of injected holes within the 3000V symmetric IGBT structure.

This implies a larger voltage drop across the drift region, which will also result in a larger on-state voltage drop.

The on-state voltage drop for the IGBT structure can be obtained from the injected carrier distribution shown above after accounting for the voltage drop across the PN junction J_1 and the MOSFET region. These voltage drops can be expressed as:

$$V_{ON} = V_{P+N} + V_{NB} + V_{MOSFET} \tag{3.20}$$

where V_{P+N} is the voltage drop across the P^+ collector/N-base junction (J_1), V_{NB} is the voltage drop across the N-base region after accounting for conductivity modulation due to high-level injection conditions, and V_{MOSFET} is the voltage drop across the MOSFET portion.

The voltage drop across the P^+ collector/N-base junction (J_1) is given by [1]:

$$V_{P+N} = \frac{kT}{q} \ln\left(\frac{p_0 N_D}{n_i^2}\right) \tag{3.21}$$

where n_i is the intrinsic carrier concentration. The voltage drop across the N-base region (V_{NB}) is obtained by integration of the electric field within the N-base region. The voltage drop in the drift region is ($V_{NB1} - V_{NB2}$), where [1]:

$$V_{NB1} = \frac{2L_a J_C \sinh(W_N/L_a)}{q p_0 (\mu_n + \mu_p)} \left\{ \tanh^{-1}\left[e^{-(W_{ON}/L_a)}\right] - \tanh^{-1}\left[e^{-(W_N/L_a)}\right] \right\} \tag{3.22}$$

and

$$V_{NB2} = \frac{kT}{q}\left(\frac{\mu_n - \mu_p}{\mu_n + \mu_p}\right)\ln\left[\frac{\tanh\left(W_{ON}/L_a\right)\cosh\left(W_{ON}/L\right)_a}{\tanh\left(W_N/L_a\right)\cosh\left(W_N/L_a\right)}\right] \tag{3.23}$$

In these equations, W_{ON} is the depletion width across the reverse biased deep P^+/N-base junction (J_2) in the on-state. The voltage drop across the MOSFET portion consists of contributions from the JFET region, the accumulation layer, and the channel. In high-voltage IGBT structures, the doping concentration of the JFET region is increased to reduce its contribution to the on-state voltage drop. The JFET region can be assumed to have an effective (uniform) doping concentration of 5×10^{15} cm^{-3} based upon a Gaussian doping profile with a surface concentration of 1×10^{16} cm^{-3}. The resistivity (ρ_{JFET}) for the JFET region is typically 1.25 Ω-cm.

Based upon the analysis for the power MOSFET structure [1]:

$$V_{JFET} = \frac{J_C \rho_{JFET}(x_{JP} + W_0)W_{CELL}}{(W_G - 2x_{JP} - 2W_0)} \tag{3.24}$$

$$V_{ACC} = \frac{J_C K_A (W_G - 2x_{JP})W_{CELL}}{4\mu_{nA} C_{OX}(V_G - V_{TH})} \tag{3.25}$$

$$V_{CH} = \frac{J_C L_{CH} W_{CELL}}{2\mu_{ni} C_{OX}(V_G - V_{TH})} \tag{3.26}$$

where x_{JP} is the junction depth of the P-base region, W_0 is the depletion width in the JFET region, W_G is the gate width, W_{CELL} is the width of the IGBT cell structure, K_A is the accumulation layer coefficient (typical value 0.6), μ_{nA} is the mobility for electrons in the accumulation layer, C_{OX} is the gate oxide specific capacitance, V_G is the gate bias voltage, V_{TH} is the threshold voltage, and μ_{ni} is the mobility for electrons in the channel.

The on-state voltage drop obtained by using the above equations is shown in Fig. 3.7 for the case of the 3000 V symmetric IGBT structure. The device has the same parameters for the structure that were used to generate Fig. 3.6 for the injected free carrier concentration. The cell pitch (W_{CELL}) of the IGBT structure was 30 μm with a gate electrode width (W_G) of 16 μm. The gate bias and threshold voltages used were 15 and 5 V, respectively. The on-state voltage drop is observed to increase rapidly when the high-level lifetime is reduced below 5 μs. This limits the maximum switching speed for the 3000 V symmetric IGBT structure. The figure also provides all the components contributing to the on-state voltage drop. The voltage drop across the P^+ collector/N-base junction and the MOSFET portion are dominant when the high-level lifetime is large (>20 μs). When the high-level lifetime is reduced, the voltage drop across the N-base region increases and becomes dominant for lifetime value below 10 μs.

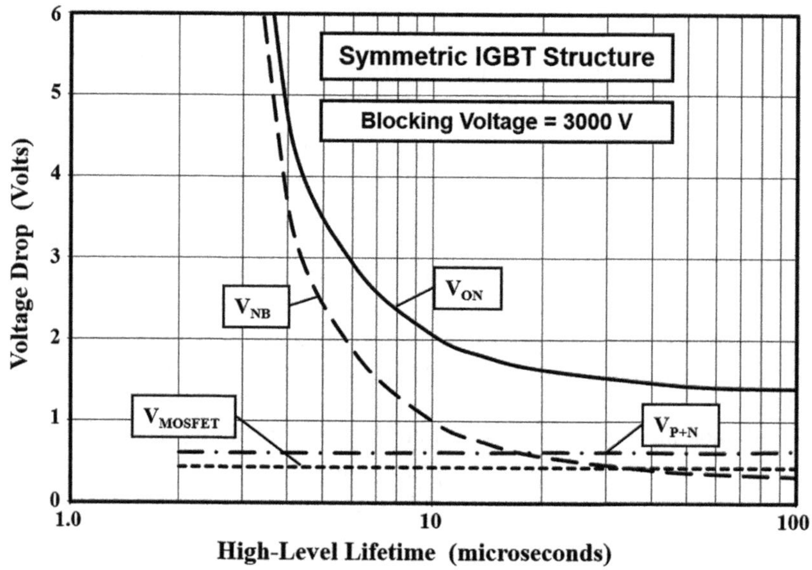

FIGURE 3.7

Components of the on-state voltage drop for the 3000V symmetric IGBT structure.

3.2.3 Stored Charge

The presence of a high concentration of injected carriers within the drift region of the symmetric IGBT structure is regarded as stored charge produced by the on-state current flow. The stored charge must be removed during the turn-off transient to switch the IGBT from the on-state to the forward blocking mode. The stored charge in the drift region can be expected to become smaller when the lifetime is reduced because the hole concentration decreases, as shown in Fig. 3.6.

The stored charge in the drift region can be computed by integration of the free carrier distribution given by Eqn (3.11) [1]:

$$Q_S = \frac{qp_0 L_a}{\tanh\left(W_N/L_a\right)} \tag{3.27}$$

When the lifetime is reduced in the drift region of the symmetric IGBT structure, the stored charge decreases due to a smaller value for the hole concentration p_0 at junction J_1 and due to the smaller ambipolar diffusion length.

The stored charge obtained by using the above equation is shown in Fig. 3.8 for the case of the 3000V symmetric IGBT structure when the high-level lifetime is varied. A rapid growth in the stored charge for lifetime values above 10 μs is beneficial for reducing the on-state voltage drop but is problematic from the switching loss standpoint, as shown below.

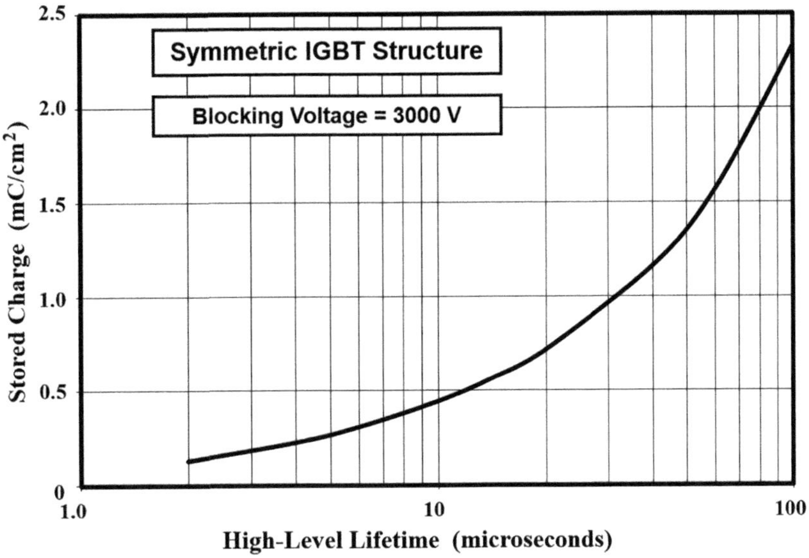

FIGURE 3.8

Stored charge in the 3000V symmetric IGBT structure.

3.2.4 Turn-Off Switching Waveforms

IGBTs are commonly used to control inductive loads such as the winding of motors used in a wide variety of consumer and industrial applications. The power circuit consists of the IGBT and the inductive load connected in series with the DC power source, with a free-wheeling or fly-back diode across the load to transfer the current when the IGBT is turned off. During the turn-off transient, the voltage across the IGBT structure first increases to the collector bias supply voltage before the collector current transfers to the diode.

The collector current initially remains at the on-state value due to the inductive load when the gate voltage is reduced to zero to begin the turn-off transient. The collector current is sustained during this time by the bipolar current in the PNP transistor because the channel current is cut off. During the first phase of the turn-off process, the collector voltage increases until it reaches the collector supply voltage. The voltage is supported within the IGBT structure by the formation of a space-charge region at the *P*-base/*N*-base junction (J_2) with a large concentration of holes in the space-charge layer.

Analysis of the turn-off waveforms for the symmetric IGBT structure can be performed under the assumption that recombination in the *N*-base region can be neglected [1]. In the absence of recombination in the drift region, the free carrier (hole) distribution within the lightly doped portion of the *N*-base region during on-state operation becomes linear, as shown at the bottom of Fig. 3.9:

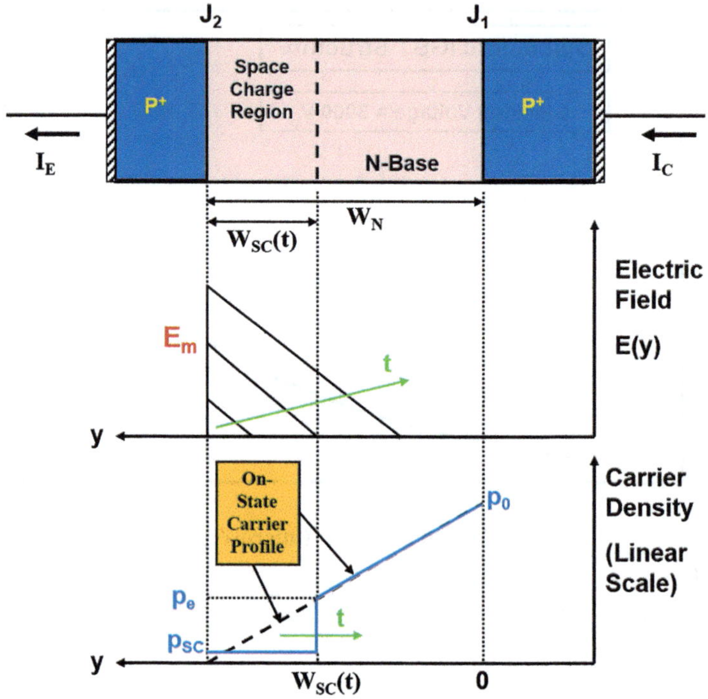

FIGURE 3.9

Stored charge and electric field distributions for inductive-load turn-off conditions in a symmetric IGBT structure.

$$p(y) = p_0\left(1 - \frac{y}{W_N}\right) \tag{3.28}$$

The hole profile does not change during the first phase of the turn-off process. Consequently, the concentration of holes at the edge of the space-charge region (p_e) increases during the turn-off process because the space-charge width increases:

$$p_e(t) = p_0\left[\frac{W_{SC}(t)}{W_N}\right] \tag{3.29}$$

The charge removed by the expansion of the space-charge layer is equal to the charge removed due to collector current flow:

$$J_{C,ON} = qp_e(t)\frac{dW_{SC}(t)}{dt} = qp_0\left[\frac{W_{SC}(t)}{W_N}\right]\frac{dW_{SC}(t)}{dt} \tag{3.30}$$

Integrating this equation with the boundary condition of zero width for the space-charge layer at time zero:

$$W_{SC}(t) = \frac{\sqrt{2W_N J_{C,ON} t}}{qp_0} \tag{3.31}$$

The collector voltage supported by the symmetric IGBT structure can be derived from the space-charge layer width:

$$V_C(t) = \frac{q(N_D + p_{SC})W_{SC}^2(t)}{2\varepsilon_S} \tag{3.32}$$

where the hole concentration in the space-charge layer (p_{SC}) is given by

$$p_{SC} = \frac{J_{C,ON}}{qv_{sat,p}} \tag{3.33}$$

under the assumption that the carriers are moving at the saturated drift velocity ($v_{sat,p}$) in the space-charge layer due to the high electric field in it. Using Eqn (3.31):

$$V_C(t) = \frac{W_N(N_D + p_{SC})J_{C,ON}}{\varepsilon_S p_0}t \tag{3.34}$$

This equation predicts a linear increase in the collector voltage with time during the first phase of the turn-off process for the symmetrical IGBT structure.

The collector voltage reaches the collector supply voltage (V_{CS}) at the end of the first phase of the turn-off process. The time ($t_{V,OFF}$) for the duration of the first phase is:

$$t_{V,OFF} = \frac{\varepsilon_S p_0 V_{CS}}{W_N(N_D + p_{SC})J_{C,ON}} \tag{3.35}$$

The width of the space-charge layer at the end of the first phase is:

$$W_{SC}(t_{V,OFF}) = \frac{\sqrt{2\varepsilon_S V_{CS}}}{q(N_D + p_{SC})} \tag{3.36}$$

This width is less than the width (W_N) of the N-base region because the collector supply voltage is less than the blocking voltage and because the space-charge layer width is reduced by the presence of holes in the space-charge region. Consequently, a large concentration of holes remains in the conductivity-modulated portion of the N-base region at the end of the first phase.

The decay of the collector current after the voltage rise time is determined by the recombination of the excess holes and electrons trapped within the N-base region in the vicinity of the P^+ collector/N-base junction [1]. The continuity equation for holes in the N-base region in the absence of diffusion is given by:

$$\frac{d\delta p_N}{dt} = \frac{-\delta p_N}{\tau_{HL}} \tag{3.37}$$

where δp_N is the excess hole concentration in the N-base region. The solution for this equation is:

$$\delta p_N(t) \approx p_N(t) = p_0 e^{-t/\tau_{HL}} \tag{3.38}$$

The collector current flow that supports the recombination of carriers within the stored charge region can be analyzed by examination of the carrier distribution on both sides of the P^+ collector/N-base junction (J_1). The high concentration of electrons in the N-base region produces the injection of electrons into the P^+ collector region. These injected electrons diffuse away from the junction, producing an exponential decay in concentration [1].

The collector current produced by the diffusion of the injected electrons in the P^+ collector side of the junction is given by:

$$J_C(t) = \frac{qD_{nE}p_0^2}{L_{nE}N_{AE}} e^{-2t/\tau_{HL}} = J_{C,ON}e^{-2t/\tau_{HL}} \tag{3.39}$$

It can be concluded that the collector current decreases exponentially with time with a time constant of one-half of the high-level lifetime in the drift region. The collector current turn-off time ($t_{I,OFF}$), defined as the time taken for the current to decay to one-tenth of the on-state value, is given by:

$$t_{I,OFF} = \frac{\tau_{HL}}{2} \ln(10) = 1.15\tau_{HL} \tag{3.40}$$

The turn-off waveforms for the case of the 3000 V symmetric IGBT structure with N-base region with a width of 450 μm and a doping concentration of 2.5×10^{13} cm^{-3} can be obtained by using the above analytical model. The hole concentration in the space-charge layer is 6.25×10^{13} cm^{-3} for an on-state collector current density of 100 A/cm^2. Three cases of the high-level lifetime are considered here. The hole concentrations (p_0) at the P^+ collector/N-base junction in the on-state are 1.26×10^{17} cm^{-3}, 1.67×10^{17} cm^{-3}, and 1.95×10^{17} cm^{-3} for high-level lifetime values of 2, 5, and 10 μs, respectively. The collector voltage and current transients obtained by using these values are shown in Fig. 3.10. The collector voltage increases

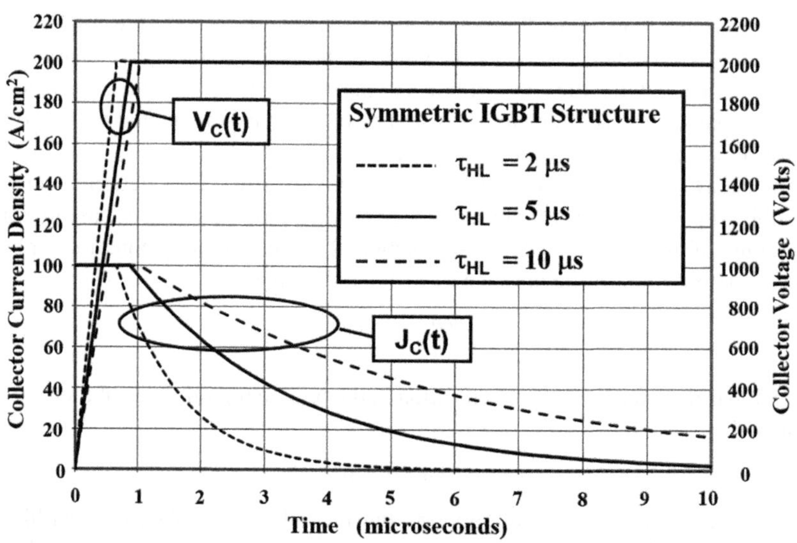

FIGURE 3.10

Inductive load waveforms for the collector current and voltage during the turn-off transient for the symmetric IGBT structure.

linearly with time with a collector voltage rise time ($t_{V,OFF}$) of 0.66, 0.88, and 1.03 μs for high-level lifetime values of 2, 5, and 10 μs, respectively. The collector current then decays exponentially with a collector current turn-off time ($t_{I,OFF}$) of 2.30, 5.75, and 11.5 μs for high-level lifetime values of 2, 5, and 10 μs, respectively. The losses are much greater during the current fall-time for the symmetric IGBT structure. The turn-off switching waveforms and power losses can be controlled by altering the lifetime in the drift region. This is usually accomplished by using high-energy (3 MeV) electron irradiation.

3.2.5 Turn-Off Power Loss

From the switching waveforms shown in Fig. 3.10 for the symmetric IGBT structure, it can be concluded that the transients occur with large values for the collector voltage and current. This produces significant instantaneous power loss. From the device application standpoint, it is useful to compute the energy loss occurring during each turn-off transient. The power loss incurred in the IGBT structure in any application can then be computed by multiplying the energy loss per switching cycle with the operating frequency.

The energy loss per cycle can be derived by integration of the power loss during the voltage and current transients. The collector voltage increases linearly with time during the first phase of the turn-off process while the collector current density remains constant for an inductive load. The energy loss per cycle during the first phase of the turn-off process for the symmetric IGBT structure can be calculated using [1]:

$$E_{V,OFF} = \frac{1}{2} J_{C,ON} V_{CS} t_{V,OFF} \tag{3.41}$$

The collector current decreases exponentially with time while the collector voltage remains at the collector supply voltage during the second phase of the turn-off process. The energy loss during the current transient can be calculated by using [1]:

$$E_{I,OFF} = J_{C,ON} V_{CS} \left(\frac{\tau_{HL,N-Base}}{2} \right) \tag{3.42}$$

The total energy loss per cycle (E_{OFF}) is obtained by the addition of these two terms, $E_{V,OFF}$ and $E_{I,OFF}$.

The energy loss per cycle for the case of the 3000 V symmetric IGBT structure with *N*-base region with a width of 450 μm and a doping concentration of 2.5×10^{13} cm^{-3} can be obtained by using the above analytical model for the case of an on-state collector current density of 100 A/cm^2 and a collector supply voltage of 2000 V. The results calculated for high-level lifetime values ranging from 1 to 100 μs are provided in Fig. 3.11. For the lowest value of the high-level lifetime of 2 μs, the energy loss during the voltage transient is 66 mJ/cm^2 while the energy loss during the current transient is 200 mJ/cm^2. The energy loss during the voltage transient increases gradually to 131 mJ/cm^2 when the high-level lifetime is increased to 100 μs. In contrast, the energy loss during the current transient increases dramatically

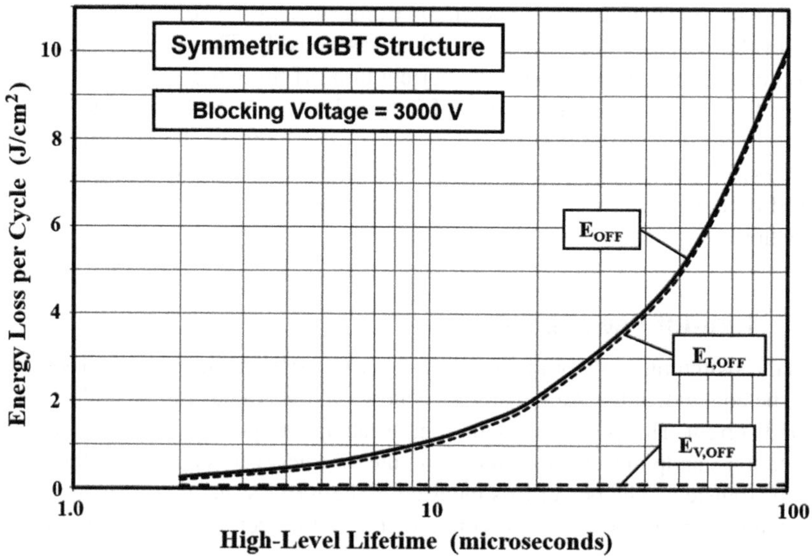

FIGURE 3.11

Energy loss during the turn-off transient for the symmetric 3000V IGBT structure.

to 10,000 mJ/cm^2 when the high-level lifetime is increased to 100 μs. The total energy loss for the turn-off transient is dominated by the energy loss during the current transient. The high-level lifetime must be kept below 10 μs to limit the total energy loss to below 1 J/cm^2.

3.2.6 Power Loss Trade-Off Curve

The optimization of the design of an IGBT requires trading off the on-state power loss with the turn-off switching power loss. For devices designed for working in circuits operating at lower frequencies, it is appropriate to minimize the on-state voltage drop by using a large lifetime in the drift region. In contrast, for devices designed for operating in high-frequency circuits, it is appropriate to minimize the turn-off switching energy per cycle. This trade-off is usually performed by making a trade-off curve between on-state voltage drop and energy loss per cycle.

The trade-off curve for the 3000 V symmetric blocking IGBT structure with 450 μm wide *N*-base region determined by using the equations provided in the previous sections is shown in Fig. 3.12. For the case of an on-state current density of 100 A/cm^2, the energy loss per cycle is found to be quite large—in the range of 1 J/cm^2. If the switching power loss is limited to 100 W/cm^2 by the packaging technology, this would limit the operating frequency of the device to below 100 Hz. The 3000 V symmetric blocking IGBT structure can be operated at a higher frequency if the on-state current density is reduced. This is demonstrated in Fig. 3.12 with trade-off curves for on-state current densities of 25, 50, and 100 A/cm^2. A reduction in the on-state

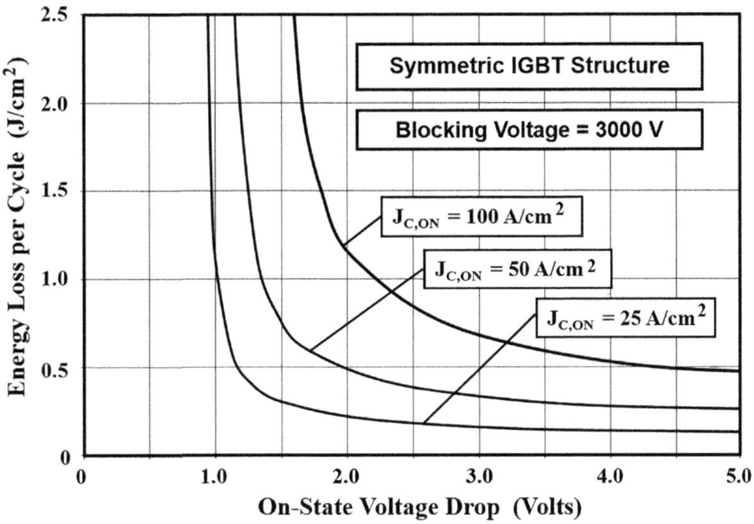

FIGURE 3.12

Trade-off curves for the symmetric 3000V IGBT structure.

current density reduces not only the on-state voltage drop but also the energy loss per cycle because of a reduction of the injected free carrier concentration. The maximum operating frequency can be increased approximately proportional to the reduction in the on-state current density. This will of course require a larger chip active area for a given current rating for the device.

3.3 Asymmetric IGBT Structure

The structure for the asymmetric, double-diffused IGBT was shown on the left side of Fig. 2.3 and an asymmetric, trench-gate IGBT structure was shown in Fig. 2.5. The discussion in this section is applicable for both types of gate structures. The asymmetric IGBT structure was developed soon after the symmetric device structure [7]. The asymmetric structure has a much superior trade-off curve between on-state voltage drop and switching loss when compared with the symmetric IGBT structure. Consequently, most of the worldwide effort in the 1980s and 1990s was focused on the development of the asymmetric device structure [8–11]. The asymmetric IGBT structure has seen widespread applications in motor drives for air conditioning and refrigeration, in lighting, and in transportation. The voltage ratings of silicon devices have been scaled up rapidly from the initial 600V capability reported in 1982 to 6500V capability, as shown in Fig. 1.6.

3.3.1 Blocking Voltage

The asymmetric IGBT structure can support a high voltage within the N-drift region for positive collector voltages. As in the case of the symmetric blocking IGBT structure, the maximum voltage that the asymmetric IGBT structure can support is determined by open-base transistor breakdown using the criterion given by Eqn (3.3). However, the presence of the buffer layer requires alterations of the equations used to determine the thickness of the drift region. In the asymmetric IGBT structure, the injection efficiency of the P^+ collector/N-buffer junction (J_1) is given by [1]:

$$\gamma_E = \frac{D_{pNB}L_{nE}N_{AE}}{D_{pNB}L_{nE}N_{AE} + D_{nE}W_{NB}N_{DNB}} \tag{3.43}$$

where D_{pNB} and D_{nE} are the diffusion coefficients for minority carriers in the N-buffer and P^+ collector regions, respectively; N_{AE} and L_{nE} are the doping concentration and diffusion length for minority carriers in the P^+ collector region, respectively; and N_{DNB} and W_{NB} are the doping concentration and width of the N-buffer layer, respectively.

In the asymmetric IGBT structure, the depletion layer extends through the lightly doped drift region at a relatively low collector voltage well below the breakdown voltage of the structure. Consequently, the minority carriers need to diffuse through only the N-buffer layer. The base transport factor is therefore independent of collector voltage and given by [1]:

$$\alpha_T = \frac{1}{\cosh\left(W_{NB}/L_{p,NB}\right)} \tag{3.44}$$

where $L_{p,NB}$ is the diffusion length for holes in the N-buffer layer and W_{NB} is the width of the buffer layer. The diffusion length for holes ($L_{p,NB}$) in the N-buffer layer depends upon the diffusion coefficient and the minority carrier lifetime in the N-buffer layer, which are both dependent upon the doping concentration in the N-buffer layer [1].

The multiplication factor for the asymmetric IGBT structure is determined by the collector bias relative to the avalanche breakdown voltage of the deep P^+ region/N-base junction (BV_{PP}) without the punch-through phenomenon. The multiplication coefficient can be computed by using [1]:

$$M = \frac{1}{1 - \left(V_{NPT}/BV_{PP}\right)^n} \tag{3.45}$$

where $n = 6$ for the case of a P^+/N diode. In this expression, the non-punch-through voltage (V_{NPT}) can be computed from the applied collector bias by using [1]:

$$V_{NPT} = \frac{\varepsilon_S}{2qN_D}\left(\frac{V_C}{W_N} + \frac{qN_DW_N}{2\varepsilon_S}\right)^2 \tag{3.46}$$

The forward blocking capability of the asymmetric IGBT structure is determined by the collector voltage at which the multiplication factor becomes equal to the reciprocal of the product of the base transport factor and the emitter injection efficiency.

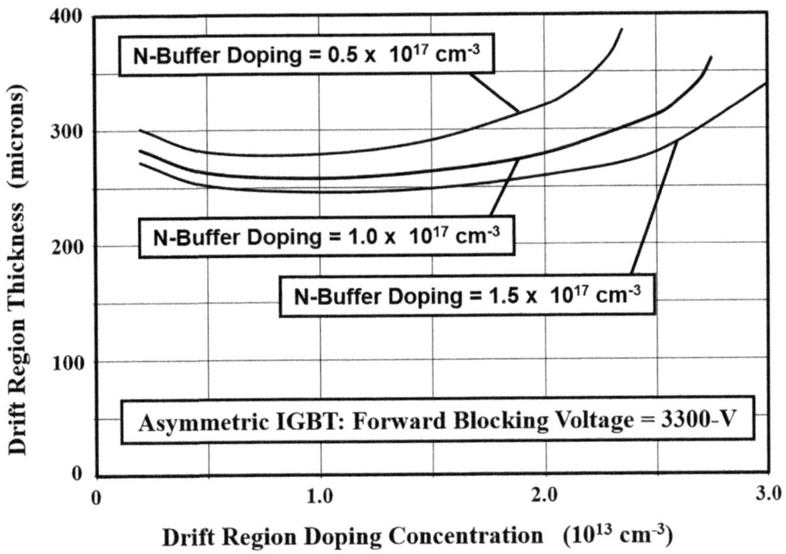

FIGURE 3.13

Optimizing drift region thickness for the asymmetric IGBT structure: *N*-buffer doping as parametric variable.

Any desired forward blocking capability can be achieved in the asymmetric IGBT structure by appropriate combination of the drift region doping, drift region thickness, buffer layer doping, and buffer layer thickness. During the optimization of the design, it is preferable to achieve the smallest thickness for the drift region to reduce the on-state voltage drop and the stored charge. The change in the thickness of the drift region with drift region doping concentration to achieve a forward blocking voltage of 3300 V is shown in Fig. 3.13 for the case of a buffer layer thickness of 10 μm. Here, the buffer layer doping was used as a variable. It can be observed that for each buffer layer doping concentration, the drift region thickness has a minimum value. In all three cases, the minimum drift region thickness occurs at the same optimum drift region doping concentration of 1×10^{13} cm^{-3}. The optimum drift region thicknesses are 279, 258, and 246 μm for the buffer layer doping concentrations of 0.5×10^{17}, 1.0×10^{17} cm^{-3}, and 1.5×10^{17} cm^{-3}, respectively. The smallest drift region thickness is achieved for the highest buffer layer doping concentration because this produces the smallest injection efficiency and base transport factor.

The change in the thickness of the drift region with drift region doping concentration to achieve a forward blocking voltage of 3300 V is shown in Fig. 3.14 for the case of a buffer layer doping concentration of 1.0×10^{17} cm^{-3}. Here, the buffer layer thickness was used as a parametric variable. It can be observed that for each buffer layer thickness, the drift region thickness has a minimum value. In all three cases, the minimum drift region thickness occurs at the same optimum drift region doping concentration of 1×10^{13} cm^{-3}. The optimum drift region thicknesses are

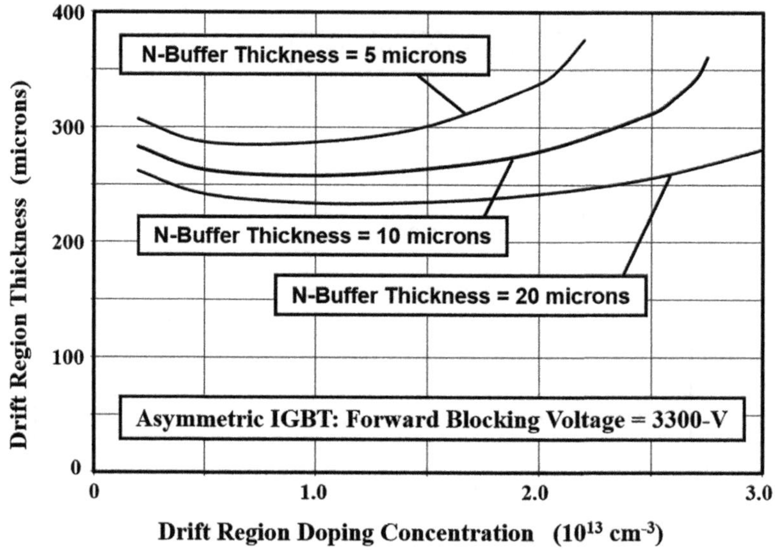

FIGURE 3.14

Optimizing drift region thickness for the asymmetric IGBT structure: *N*-buffer thickness as parametric variable.

287, 258, and 234 μm for the buffer layer thicknesses of 5, 10, and 20 μm, respectively. The smallest drift region thickness is achieved for the largest buffer layer thickness because this produces the smallest injection efficiency and base transport factor.

3.3.2 On-State Characteristics

The free carrier distribution in the asymmetric IGBT structure determines its on-state voltage drop. As in the case of the symmetric IGBT structure, the free carrier distribution in the *N*-drift region can be altered by changing the lifetime. However, in the case of the asymmetric IGBT structure, the free carrier distribution can also be changed by altering the doping concentration and thickness of the *N*-buffer layer. This provides more latitude for optimization of the structure during its design for a particular application. In general, the optimization requires finding the free carrier distribution that provides the combination of on-state voltage drop and turn-off time that produces the lowest power loss in the intended circuit application.

1D analysis of the free carrier distribution profile under the deep P^+ region for this structure can be obtained by solving the continuity equation (see Eqn 3.7) for minority carriers under steady-state high-level injection conditions [1]. The boundary condition on the emitter side is the same as that for the symmetric structure (see Eqn 3.10). However, the boundary condition on the collector side is different for the asymmetric IGBT structure due to the presence of the buffer layer.

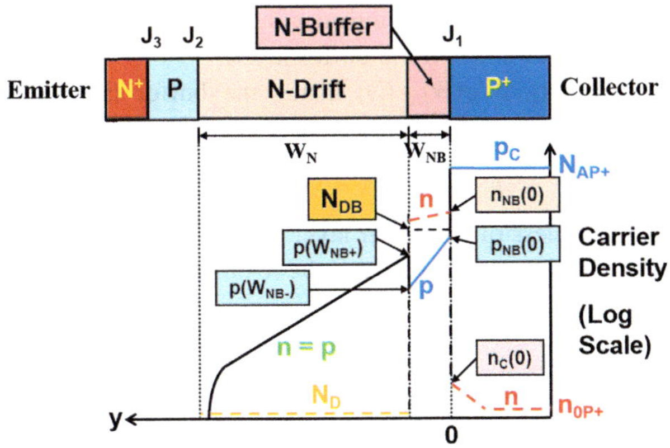

FIGURE 3.15

On-state free carrier distribution within the asymmetric IGBT structure.

An analytical model has been developed for the asymmetric IGBT structure that is valid for any injection level in the buffer layer [12]. In this model, the hole and electron concentrations in the N-base region are assumed to be equal due to charge neutrality and the low doping concentration required for the drift region. The on-state carrier distribution profile for the asymmetric IGBT structure is shown in Fig. 3.15, indicating an abrupt change in the hole concentration at the boundary between the N-drift region and the N-buffer layer. In the N-buffer layer, the hole concentration is comparable to the doping concentration (N_{DB}).

The hole concentration ($p_{NB}(0)$) in the buffer layer at junction (J_1) can be obtained by the solution of the following quadratic equation [12]:

$$p_{NB}^2(0) + \left(\frac{D_{pNB}N_{AP+}L_{nP+} + D_{nP+}N_{DB}L_{pNB}}{D_{nP+}L_{pNB}}\right)p_{NB}(0) - \frac{N_{AP+}L_{nP+}J_C}{qD_{nP+}} = 0 \tag{3.47}$$

The solution of this equation is:

$$p_{NB}(0) = \frac{1}{2}\left(\sqrt{b^2 - 4c} - b\right) \tag{3.48}$$

where

$$b = \frac{D_{pNB}N_{AP+}L_{nP} + D_{nP+}N_{DB}L_{pNB}}{D_{nP+}L_{pNB}} \tag{3.49}$$

and

$$c = -\frac{N_{AP+}L_{nP+}J_C}{qD_{nP+}} \tag{3.50}$$

Here, D_{pNB} is the diffusion coefficient for holes in the buffer layer, N_{AP+} is the acceptor doping concentration of the P^+ collector region, L_{nP+} is the diffusion length for electrons in the collector region, D_{nP+} is the diffusion coefficient for electrons in the collector region, N_{DB} is the donor doping concentration in the buffer layer, and L_{pNB} is the diffusion length for holes in the buffer layer.

The hole concentration ($p(W_{NB}-)$) inside the buffer layer at the boundary between the N-buffer layer and the N-base region is given by [12]:

$$p(W_{NB-}) = p_{NB}(0)e^{-(W_{NB}/L_{pNB})} \tag{3.51}$$

where W_{NB} is the thickness of the buffer layer. The hole concentration ($p(W_{NB+})$) in the N-base region at the boundary between the N-buffer layer and the N-base region can be obtained by using [2]:

$$p(W_{NB+}) = \frac{L_a \tanh[(W_N + W_{NB})/L_a]}{2qD_p} J_p(W_{NB-}) \tag{3.52}$$

with the current density

$$J_p(W_{NB-}) = J_p(0)e^{-(W_{NB}/L_{pNB})} \tag{3.53}$$

The hole concentration profile in the N-drift region can be computed using [2]:

$$p(y) = p(W_{NB+}) \frac{\sinh[(W_N + W_{NB} - y)/L_a]}{\sinh[(W_N + W_{NB})/L_a]} \tag{3.54}$$

The hole carrier distribution obtained by using the above equations is shown in Fig. 3.16 for the case of the 3000 V asymmetric IGBT structure when the lifetime

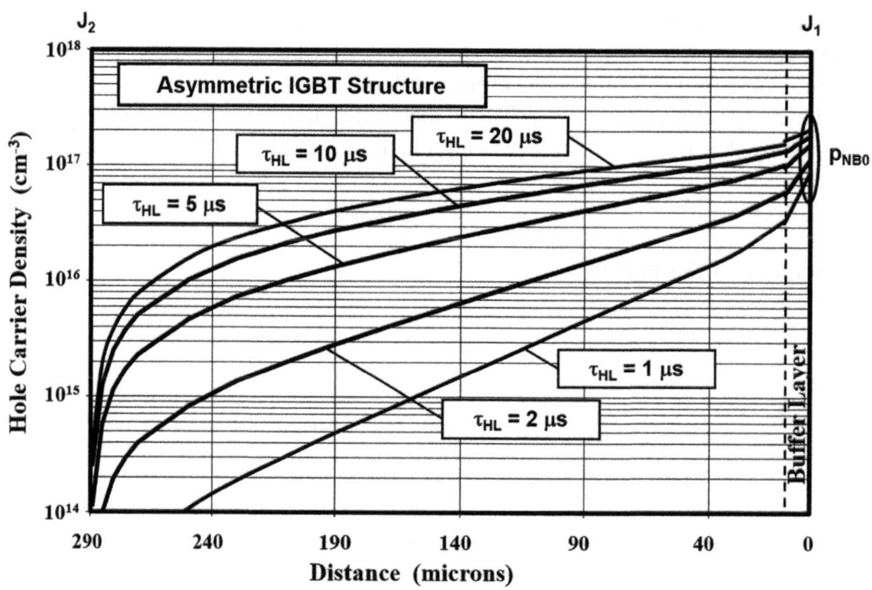

FIGURE 3.16

Hole carrier distribution in the 3000 V asymmetric IGBT structure: N-drift region lifetime dependence.

in the drift region is altered. Based upon the breakdown voltage analysis, an N-base region thickness of 280 μm was chosen, with a buffer layer doping concentration of 1.0×10^{17} cm^{-3} and a buffer layer thickness of 10 μm. The injected hole concentration remains well above the doping concentration of 1×10^{13} cm^{-3} for the N-drift region when the high-level lifetime is above 2 μs. At a high-level lifetime value of 2 μs and below, the hole concentration in the drift region on the emitter side becomes much lower than at the collector side. This produces an increase in the on-state voltage drop.

The on-state voltage drop for the asymmetric IGBT structure can be obtained by using [12]:

$$V_{ON} = V_{P+NBL} + V_B + V_{MOSFET} \tag{3.55}$$

where V_{P+NBL} is the voltage drop across the P^+ collector/N-buffer layer junction (J_1), V_B is the voltage drop across the N-base region after accounting for conductivity modulation due to high-level injection conditions, and V_{MOSFET} is the voltage drop across the MOSFET portion. Because the junction (J_1) between the P^+ collector region and the N-buffer layer operates at neither high-level nor low-level injection conditions, the voltage drop across the junction (J_1) must be obtained using:

$$V_{P+NB} = \frac{kT}{q} \ln\left(\frac{p_{NB}(0)N_{BL}}{n_i^2}\right) \tag{3.56}$$

The voltage drop across the N-base region is given by the sum of two terms, V_{B1} and V_{B2} [12]:

$$V_{B1} = \frac{2L_aJ_C \sinh(W_N/L_a)}{qp(W_{NB+})(\mu_n + \mu_p)}\left\{\tanh^{-1}\left[e^{-(W_{ON}/L_a)}\right] - \tanh^{-1}\left[e^{-(W_N/L_a)}\right]\right\} \tag{3.57}$$

and

$$V_{B2} = \frac{kT}{q}\left(\frac{\mu_n - \mu_p}{\mu_n + \mu_p}\right) \ln\left[\frac{\tanh(W_{ON}/L_a)\cosh(W_{ON}/L_a)}{\tanh(W_N/L_a)\cosh(W_N/L_a)}\right] \tag{3.58}$$

where W_{ON} is the depletion width across the P-base/N-base junction (J_2) in the on-state. The voltage drop across the MOSFET portion is given by the same equations previously provided for the asymmetric blocking IGBT structure.

The on-state voltage drop obtained by using the above equations is shown in Fig. 3.17 for the case of the 3000 V asymmetric IGBT structure when the high-level lifetime is varied from 1 to 100 μs in the N-drift region. The device has the same parameters for the structure that were used to generate Fig. 3.16 for the injected free carrier concentration. The cell pitch (W_{CELL}) of the IGBT structure was 30 μm with a gate electrode width (W_G) of 16 μm. The gate bias and threshold voltages used were 15 and 5 V, respectively. The on-state voltage drop is observed to increase rapidly when the high-level lifetime is reduced below 2 μs. This lifetime is three times smaller than that for the 3000 V symmetric IGBT structure because the drift region width is much larger for the symmetric IGBT structure. The figure also provides all the components contributing to the on-state voltage drop. The voltage drop across the

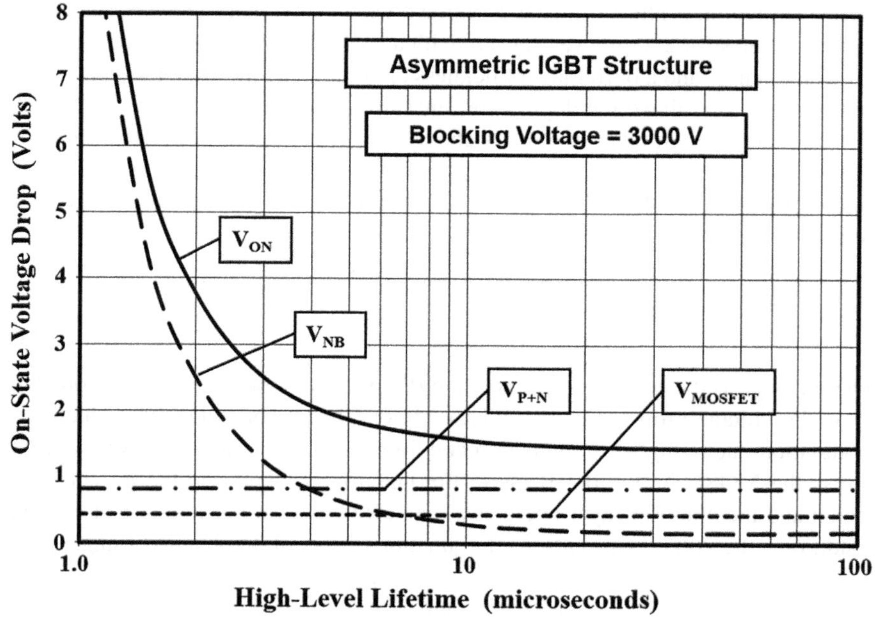

FIGURE 3.17

Components of the on-state voltage drop for the 3000V asymmetric IGBT.

P^+ collector/*N*-base junction and the MOSFET portion are dominant when the high-level lifetime is large (>5 μs). When the high-level lifetime is reduced, the voltage drop across the *N*-base region increases and becomes dominant for lifetime value below 3 μs.

As mentioned earlier, the asymmetric IGBT structure can be optimized by varying the doping concentration in the *N*-buffer layer. This is illustrated in Fig. 3.18 for the case of the 3000 V asymmetric IGBT structure with a high-level lifetime of 5 μs. The buffer layer thickness was kept fixed at 10 μm. It can be observed that the hole concentration [$p_{NB}(0)$] decreases from 9.7×10^{16} to 4.7×10^{16} cm^{-3} when the buffer layer doping is increased from 5×10^{17} to 5×10^{16} cm^{-3}. This produces a reduced injected hole concentration throughout the drift region as well. A smaller hole concentration in the drift region in the on-state will increase the on-state voltage drop and reduce the turn-off time.

The on-state voltage drop (at an on-state current density of 100 A/cm^2) computed for the 3000 V asymmetric silicon IGBT structure by using the above equations is provided in Fig. 3.19 as a function of the doping concentration of the buffer layer. This structure had the same parameters for the drift region as used for the carrier distribution in Fig. 3.18. A high-level lifetime of 5 μs was used for the graph. It can be observed that the on-state voltage drop increases rapidly when the *N*-buffer layer doping is increased beyond 5×10^{17} cm^{-3} due to a rapid increase in the voltage drop across the *N*-drift region.

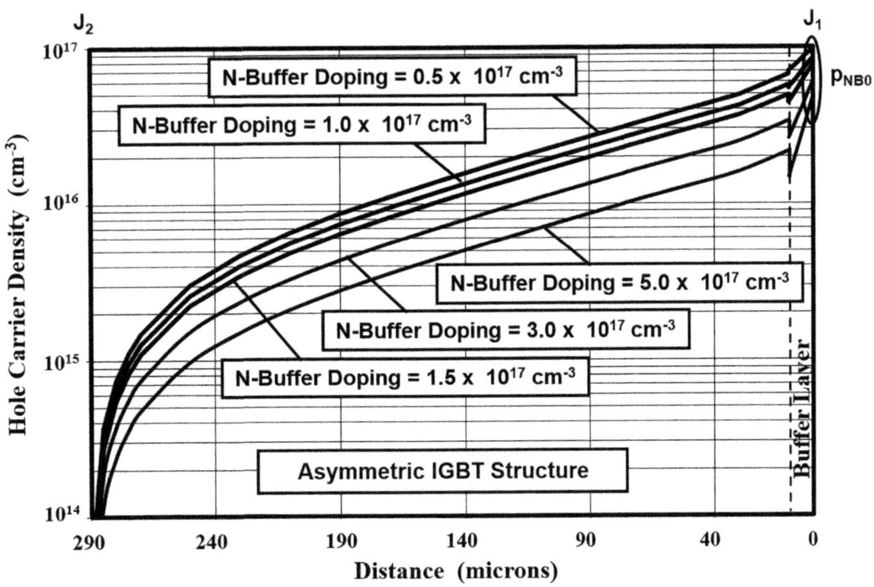

FIGURE 3.18

Hole carrier distribution in the 3000 V asymmetric IGBT structure: buffer layer doping dependence.

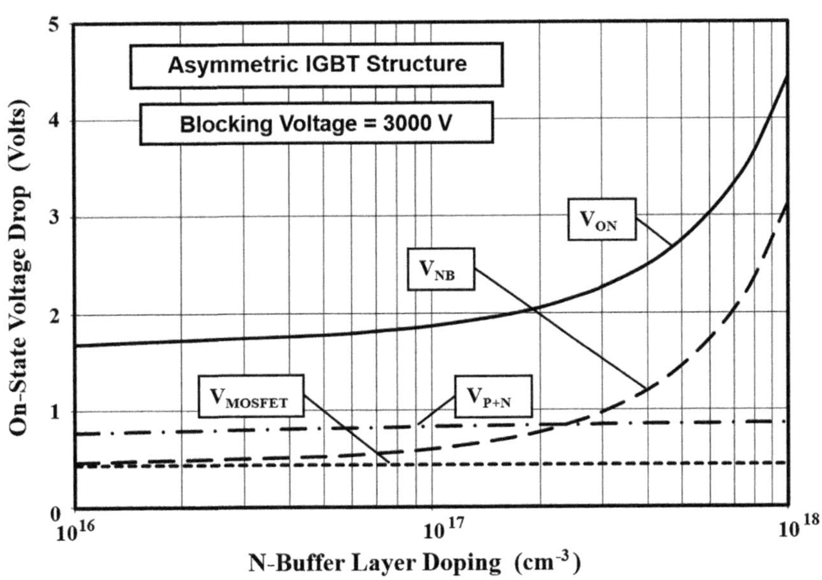

FIGURE 3.19

On-state voltage drop components for the 3000 V asymmetric IGBT structure: *N*-buffer layer doping dependence.

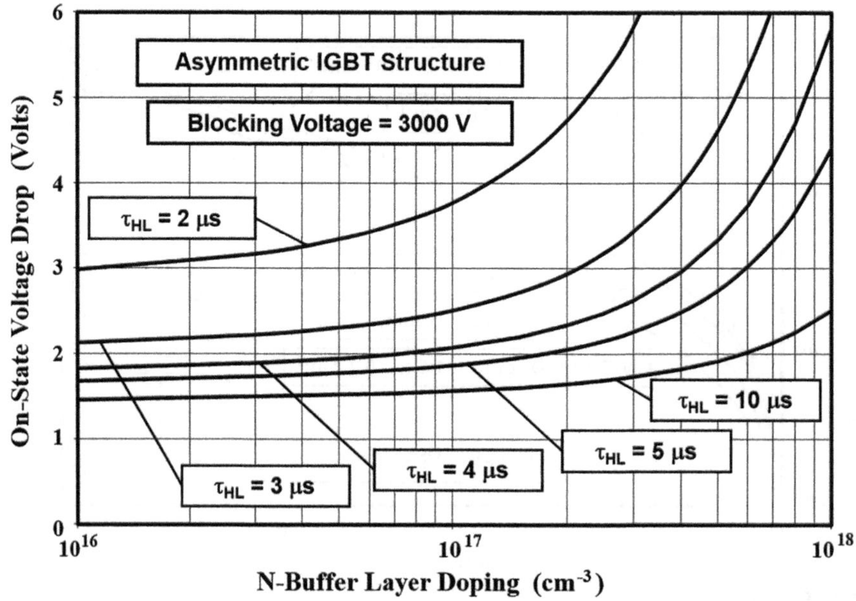

FIGURE 3.20

On-state voltage drop for the 3000V asymmetric IGBT structure: *N*-buffer layer doping dependence.

The on-state voltage drop computed for the 3000 V asymmetric silicon IGBT structure at an on-state current density of 100 A/cm^2 is provided in Fig. 3.20 as a function of the doping concentration of the buffer layer for various values of the high-level lifetime in the *N*-drift region. This structure had a buffer layer thickness of 10 μm. It can be observed that the on-state voltage drop increases rapidly when the *N*-buffer layer doping is increased beyond a certain level for each of the values for the high-level lifetime in the *N*-buffer layer. This level of the *N*-buffer layer doping concentration becomes smaller when the high-level lifetime is reduced.

The asymmetric IGBT structure can also be optimized by varying the thickness of the *N*-buffer layer. This is illustrated in Fig. 3.21 for the case of the 3000 V asymmetric IGBT structure with a high-level lifetime of 5 μs. The buffer layer doping concentration was kept fixed at 1×10^{17} cm^{-3} in this example. It can be observed that the hole concentration [$p_{NB}(0)$] is independent of the buffer layer thickness. However, the hole concentration in the drift region decreases when the buffer layer thickness is increased because of the longer path for the diffusion of holes from junction J_1 to the interface between the buffer layer and the drift region. A smaller hole concentration in the drift region in the on-state will increase the on-state voltage drop and reduce the turn-off time.

The on-state voltage drop (at an on-state current density of 100 A/cm^2) computed for the 3000 V asymmetric silicon IGBT structure by using the above equations is provided in Fig. 3.22 as a function of the thickness of the buffer layer. This structure

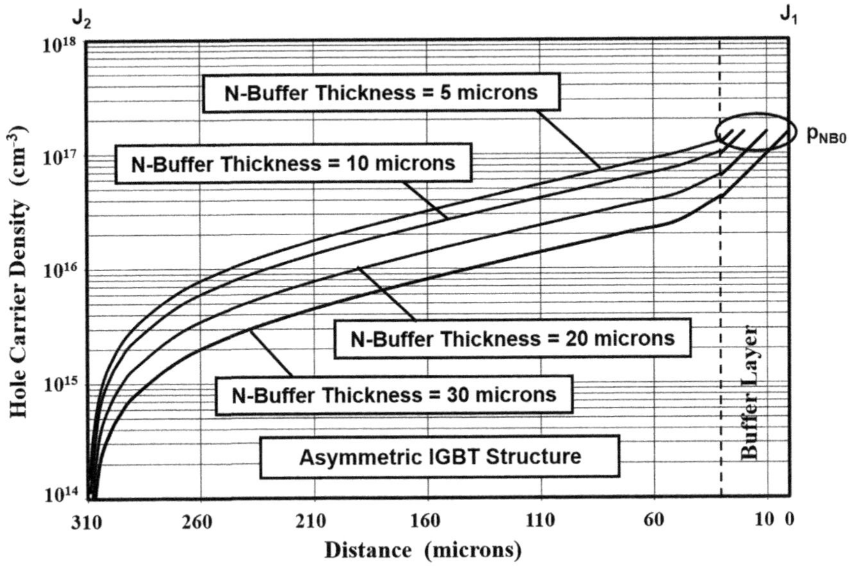

FIGURE 3.21

Hole carrier distribution in the 3000V asymmetric IGBT structure: buffer layer thickness dependence.

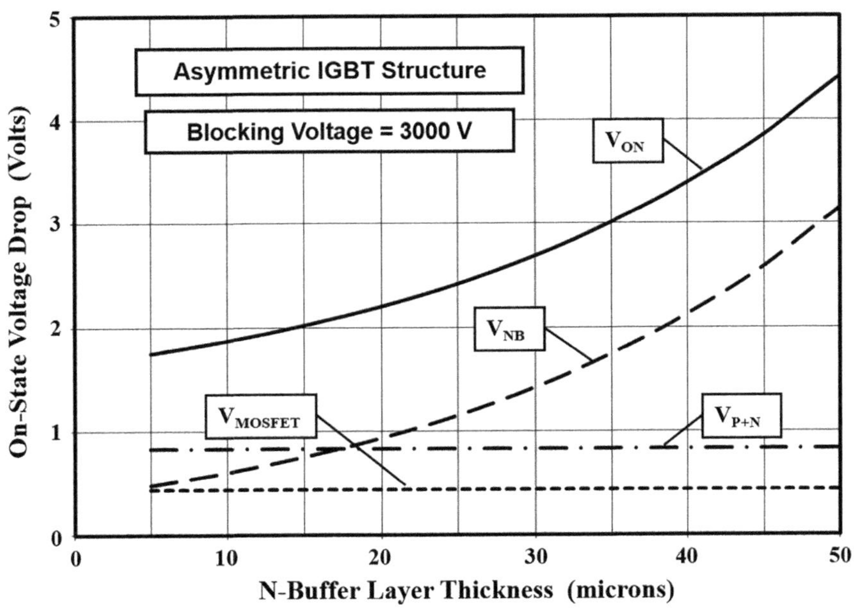

FIGURE 3.22

On-state voltage drop components for the 3000V asymmetric IGBT structure: *N*-buffer layer thickness dependence.

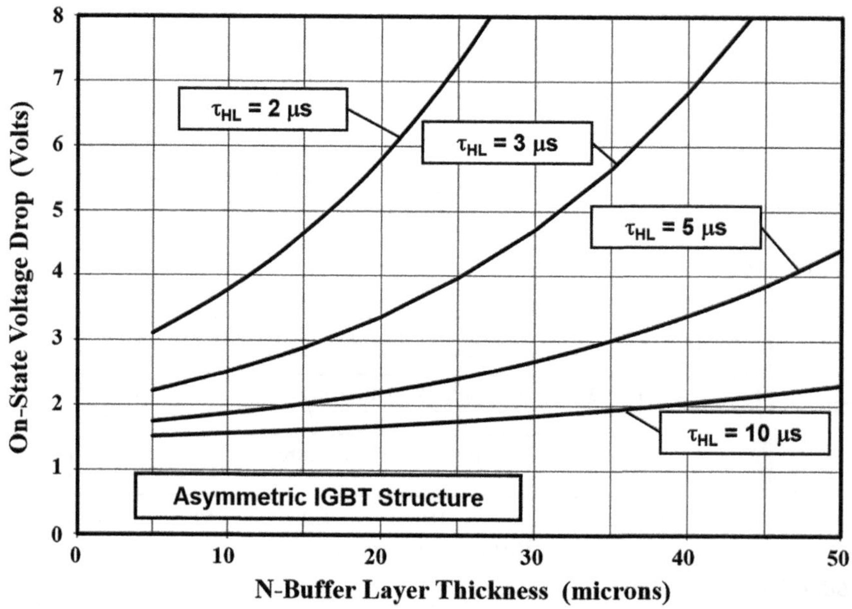

FIGURE 3.23

On-state voltage drop for the 3000V asymmetric IGBT structure: *N*-buffer layer thickness dependence.

had the same parameters for the drift region as used for the carrier distribution in Fig. 3.21. A high-level lifetime of 5 μs was used for the graph. It can be observed that the on-state voltage drop increases gradually when the *N*-buffer layer thickness is increased from 5 to 50 μm.

The on-state voltage drop computed for the 3000 V asymmetric silicon IGBT structure at an on-state current density of 100 A/cm^2 is provided in Fig. 3.23 as a function of the thickness of the buffer layer for various values of the high-level lifetime in the *N*-drift region. This structure had a buffer layer doping concentration of 1×10^{17} cm^{-3}. It can be observed that the on-state voltage drop is weakly dependent on the buffer layer thickness when the high-level lifetime in the drift region is large (10 μs in this example). However, the on-state voltage drop increases rapidly when the *N*-buffer layer thickness is increased when the high-level lifetime in the drift region is reduced to 2 μs. These graphs indicate that all the parameters for the asymmetric IGBT must be adjusted to obtain an optimized structure.

3.3.3 Stored Charge

As in the case of the symmetric IGBT structure, the stored charge in the drift region must be removed during the turn-off transient to switch the asymmetric IGBT from the on-state to the forward blocking mode. The free carrier distribution in the drift

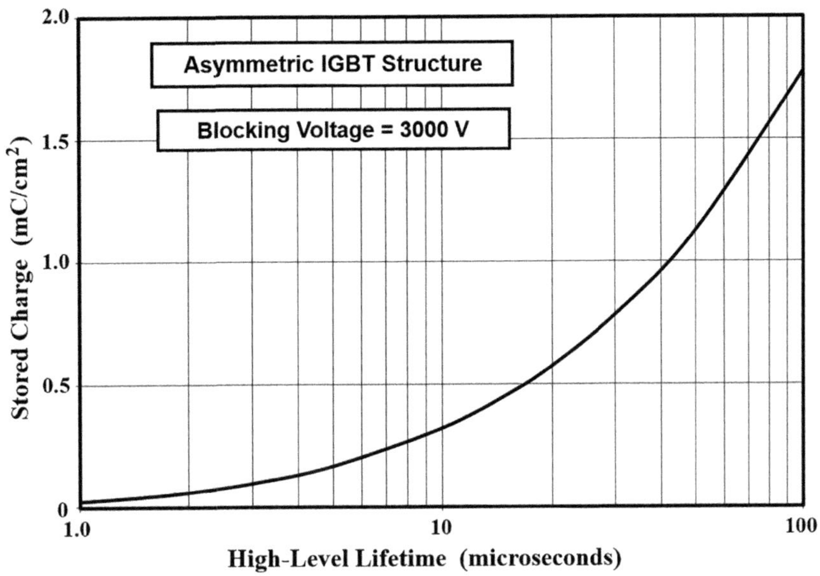

FIGURE 3.24

Stored charge in the 3000V asymmetric IGBT structure: drift region lifetime dependence.

region of the asymmetric IGBT was discussed in the previous section and shown to be dependent on the lifetime in the drift region and the buffer layer doping and thickness. The stored charge in the drift region can be computed by integration of the free carrier distribution given by Eqn (3.54) [1]:

$$Q_S = \frac{qp(W_{NB+})L_a}{\tanh\left(W_N/L_a\right)} \tag{3.59}$$

The stored charge obtained by using the above equation is shown in Fig. 3.24 for the case of the 3000 V asymmetric IGBT structure when the high-level lifetime in the drift region is varied. The parameters used for this device structure are the same as those used to generate the on-state voltage drop in Fig. 3.17. A rapid growth in the stored charge for lifetime values above 10 μs is beneficial for reducing the on-state voltage drop but is problematic from the switching loss standpoint, as shown below.

The stored charge obtained for the case of the 3000 V asymmetric IGBT structure is provided in Fig. 3.25 when the *N*-buffer layer doping concentration is varied. The parameters used for this device structure are the same as those used to generate the on-state voltage drop in Fig. 3.19 with a high-level lifetime of 5 μs in the drift region. It can be observed that the stored charge is a weak function of the buffer layer doping concentration until it exceeds 1×10^{17} cm^{-3}. A rapid reduction in the stored charge occurs when the buffer layer doping concentration exceeds 1×10^{17} cm^{-3} but this is accompanied by a rapid increase in the on-state voltage drop, as observed in Fig. 3.19.

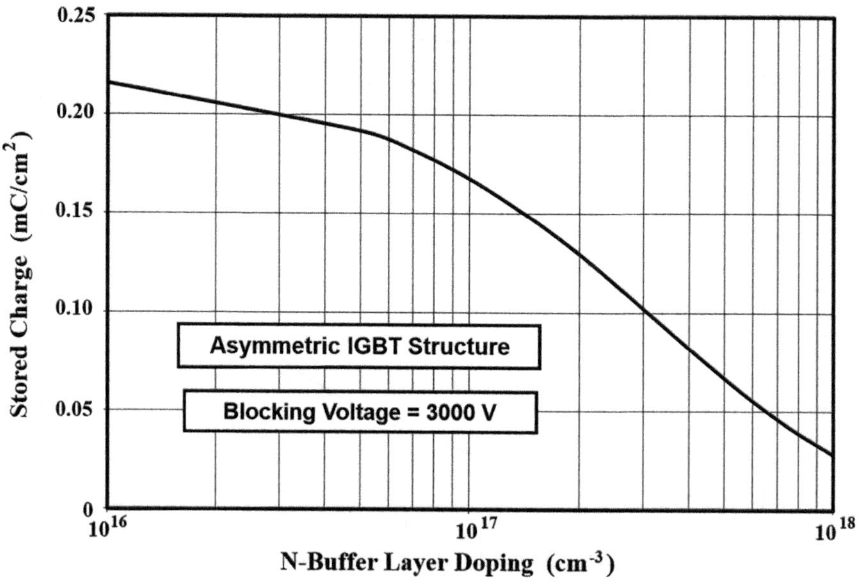

FIGURE 3.25

Stored charge in the 3000V asymmetric IGBT structure: buffer layer doping dependence.

The stored charge for the case of the 3000 V asymmetric IGBT structure is shown in Fig. 3.26 when the N-buffer layer thickness is varied. The parameters used for this device structure are the same as those used to generate the on-state voltage drop in Fig. 3.22 with a high-level lifetime of 5 μs in the drift region. It can be observed that the stored charge can be monotonically reduced by increasing the buffer layer thickness. A gradual change occurs in both the stored charge and on-state voltage drop when the buffer layer thickness is altered, indicating that it is a good design parameter for the optimization of asymmetrical IGBT structures.

3.3.4 Turn-Off Switching Waveforms

The analysis of the turn-off waveforms for the asymmetric IGBT structure can be performed using the same approach as used for the symmetric IGBT structure in Section 3.2.4 [1,12]. For the asymmetric IGBT structure, the hole concentration in the drift region can be approximated by:

$$p(y) = p_{WNB+}\left(1 - \frac{y}{W_N}\right) \tag{3.60}$$

It will be assumed that the above hole distribution does not change during the first phase when the collector voltage increases during the turn-off process. Consequently, the concentration of holes at the edge of the space-charge region (p_e) increases with time as the space-charge width increases:

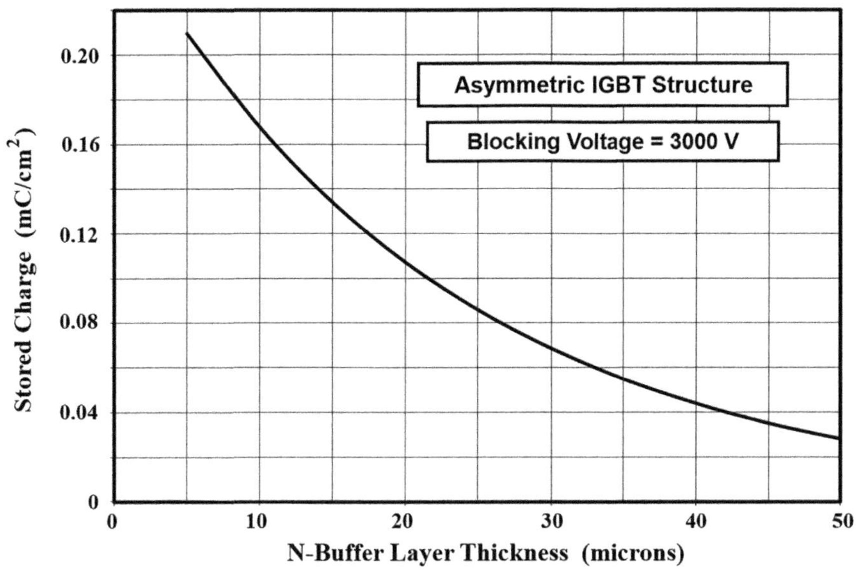

FIGURE 3.26

Stored charge in the 3000V asymmetric IGBT structure: buffer layer thickness dependence.

$$p_e(t) = p_{WNB+}\left[\frac{W_{SC}(t)}{W_N}\right] \tag{3.61}$$

The expansion of the space-charge layer with time, as indicated by the horizontal time arrow in Fig. 3.27, is then obtained using the same approach as for the symmetric structure:

$$W_{SC}(t) = \sqrt{\frac{2W_N J_{C,ON} t}{q p_{WNB+}}} \tag{3.62}$$

The collector voltage supported by the asymmetric IGBT structure can be calculated from the space-charge layer width:

$$V_C(t) = \frac{q(N_D + p_{SC})W_{SC}^2(t)}{2\varepsilon_S} \tag{3.63}$$

where the hole concentration in the space-charge layer is related to the collector current density:

$$p_{SC} = \frac{J_{C,ON}}{q v_{sat,p}} \tag{3.64}$$

Using Eqn (3.62) in Eqn (3.63):

$$V_C(t) = \frac{W_N(N_D + p_{SC})J_{C,ON}}{\varepsilon_S p_{WNB+}} t \tag{3.65}$$

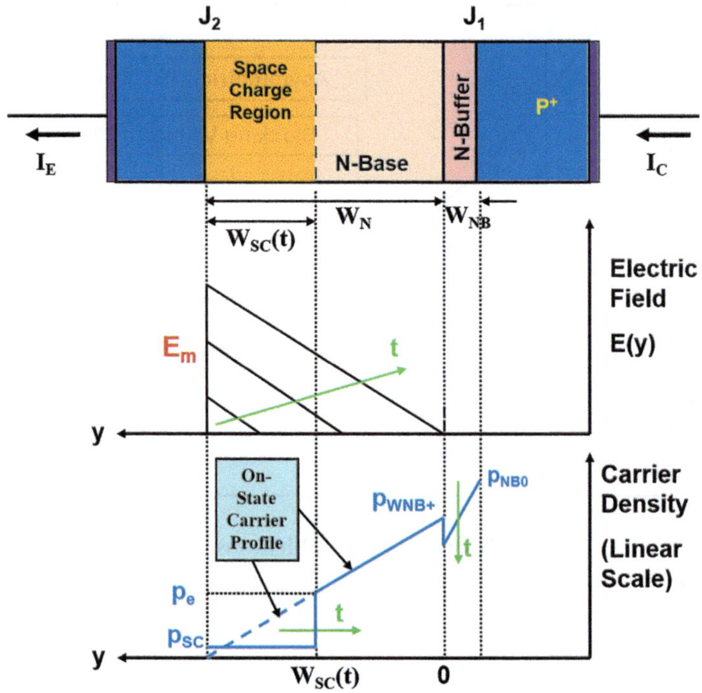

FIGURE 3.27

Stored charge and electric field distributions for inductive-load turn-off conditions in an asymmetric IGBT structure.

which indicates a linear increase in the collector voltage with time. The first phase is completed when the collector voltages reach the collector supply voltage (V_{CS}). This time interval ($t_{V,OFF}$) can be obtained by making the collector voltage equal to the collector supply voltage in Eqn (3.65):

$$t_{V,OFF} = \frac{\varepsilon_S p_{WNB+} V_{CS}}{W_N(N_D + p_{SC})J_{C,ON}} \tag{3.66}$$

The width of the space-charge layer at the end of the first phase depends upon the collector supply voltage. In most asymmetric IGBT designs, the width of the space-charge layer is slightly less than the width of the drift region for typical collector supply voltage values for which the IGBT has been designed.

The collector current decays during the second phase of the turn-off process. The rate of change of the collector current is governed by recombination of the stored charge remaining after the collector voltage transient. There is only stored charge in the buffer layer during the current decay because the stored charge in the drift region has been removed during the voltage rise time. The stored charge in the buffer layer decays by recombination as indicated by the vertical time arrow in Fig. 3.27:

$$\delta p_{NB}(t) \approx p_0 e^{-t/\tau_{p0,NB}} \tag{3.67}$$

because low-level injection conditions prevail in the N-buffer layer during most of the decay. Based upon this, the collector current density is given by [1]:

$$J_C(t) = J_{C,ON} e^{-t/\tau_{p0,NB}} \tag{3.68}$$

The collector current turn-off time ($t_{I,OFF}$) is defined as the time taken for the current to decay to one-tenth of the on-state value. It can be computed using:

$$t_{I,OFF} = \tau_{p0,NB} \ln(10) = 2.3\tau_{p0,NB} \tag{3.69}$$

The turn-off waveforms for the 3000 V asymmetric IGBT structure with N-base region with a lightly doped portion with a width of 280 μm and a doping concentration of 1×10^{13} cm^{-3}, a buffer layer width of 10 μm, and a doping concentration of 1×10^{17} cm^{-3} are shown in Fig. 3.28 for the case of three high-level lifetime values in the drift region. The collector voltage increases linearly with time with collector voltage rise times ($t_{V,OFF}$) of 0.615, 1.06, and 1.40 μs for high-level lifetime values of 2, 5, and 10 μs, respectively. The collector current then decays exponentially with collector current turn-off times ($t_{I,OFF}$) of 0.77, 1.92, and 3.83 μs for high-level drift-region lifetime values of 2, 5, and 10 μs, respectively. This demonstrates that the turn-off switching waveforms and power losses can be controlled by altering the lifetime in the drift region. This is usually accomplished by using high-energy (3 MeV) electron irradiation.

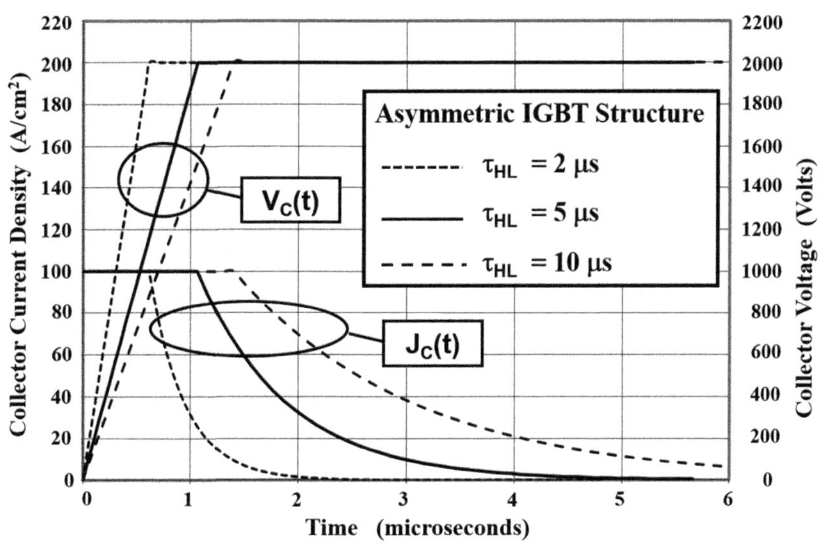

FIGURE 3.28

Collector current and voltage transients during turn-off for the asymmetric IGBT structure with an inductive load: lifetime dependence.

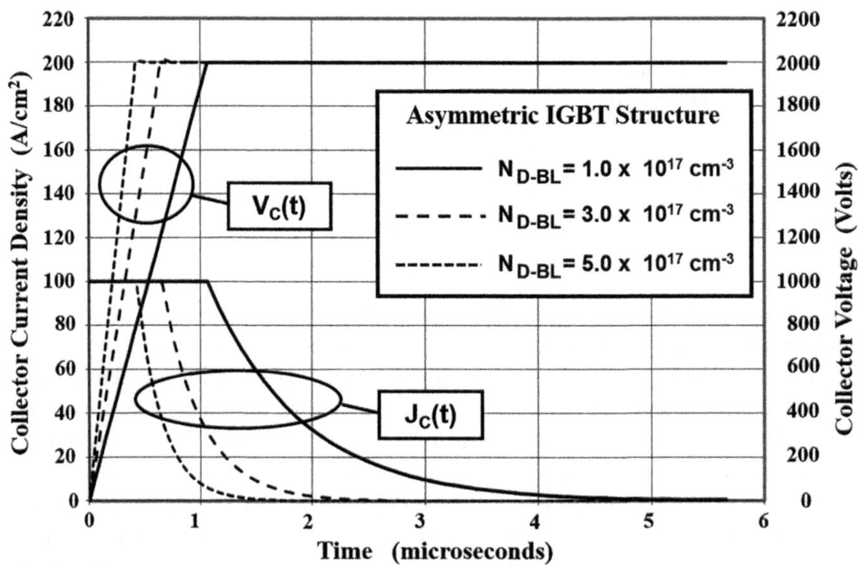

FIGURE 3.29

Collector current and voltage transients during turn-off for the asymmetric IGBT structure with an inductive load: buffer doping dependence.

The switching loss during turn-off for the asymmetric IGBT can also be controlled by varying the doping concentration in the buffer layer. This is demonstrated for the case of the 3000 V asymmetric IGBT structure with *N*-base region with a lightly doped portion with a width of 280 μm and a doping concentration of 1×10^{13} cm^{-3}, a buffer layer width of 10 μm, and a high-level lifetime of 5 μs in the drift region. It can be observed in Fig. 3.29 that the collector voltage rise time ($t_{V,OFF}$) can be reduced from 1.06 to 0.648 to 0.424 μs when the buffer layer doping concentration is increased from 1.0 to 3.0 to 5.0 $\times$ 10^{17} cm^{-3}, respectively. The collector current decays exponentially with collector current turn-off times ($t_{I,OFF}$) of 1.92, 0.821, and 0.523 μs for these values of the buffer layer doping, respectively. This demonstrates that the buffer layer doping concentration can be utilized as a sensitive design parameter for optimization of the turn-off switching waveforms and power losses for the asymmetric IGBT. This is usually accomplished by controlling the doping during epitaxial growth for lower blocking voltage devices or by the addition of a diffused *n*-type layer from the collector side of the wafer.

The turn-off transients for the asymmetric IGBT can also be altered by varying the thickness of the buffer layer. This is demonstrated for the case of the 3000 V asymmetric IGBT structure with *N*-base region with a lightly doped portion with a width of 280 μm and a doping concentration of 1×10^{13} cm^{-3}, a buffer layer doping concentration of 1×10^{17} cm^{-3}, and a high-level lifetime of 5 μs in the drift region. It can be observed in Fig. 3.30 that the collector voltage rise time ($t_{V,OFF}$) can be reduced from 1.06 to 0.679 to 0.434 μs when the buffer layer doping thickness

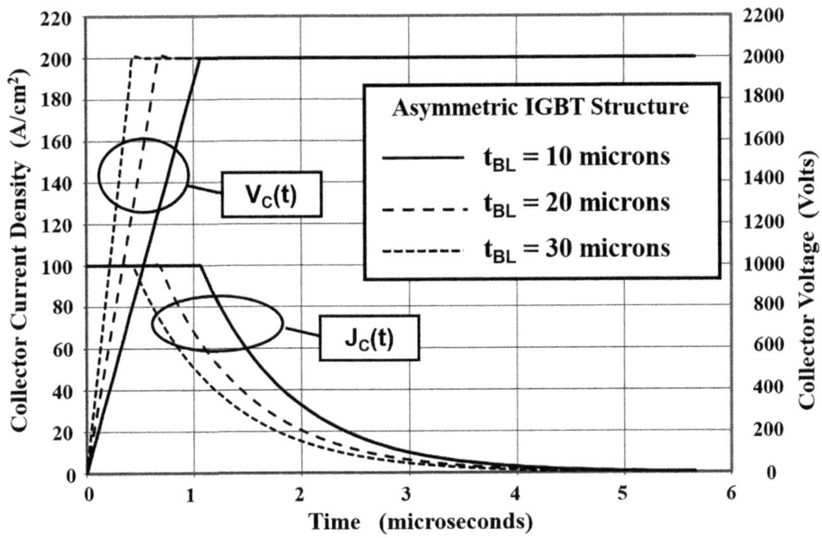

FIGURE 3.30

Collector current and voltage transients during turn-off for the asymmetric IGBT structure with an inductive load: buffer thickness dependence.

is increased from 10 to 20 to 30 μm, respectively. In all these cases, the collector current decays exponentially with a collector current turn-off time ($t_{I,OFF}$) of 1.92 μs because the lifetime in the buffer layer has the same value. This example demonstrates that the buffer layer thickness can be utilized as another design parameter for optimization of the turn-off switching waveforms and power losses for the asymmetric IGBT. The thickness of the buffer layer can be easily controlled during epitaxial growth for lower blocking voltage devices or by adjusting the diffusion time for an *n*-type layer from the collector side of the wafer.

3.3.5 Turn-Off Power Loss

The energy loss per cycle during the first phase of the turn-off process ($E_{V,OFF}$) for the asymmetric IGBT structure can be calculated using Eqn (3.41) because the voltage waveform during turn-off for the asymmetric IGBT is similar to that for the symmetric IGBT structure. The energy loss during the current transient for the asymmetric IGBT can be calculated by using [12]:

$$E_{I,OFF} = J_{C,ON} V_{CS} \tau_{p0,N-Buffer} \tag{3.70}$$

The total energy loss per cycle (E_{OFF}) is obtained by the addition of these two terms ($E_{V,OFF}$ and $E_{I,OFF}$).

The energy loss per cycle for the case of the 3000 V asymmetric IGBT structure with *N*-base region with a width of 280 μm and a doping concentration of 1.0×10^{13} cm^{-3}, a buffer layer thickness of 10 μm, and a doping concentration of

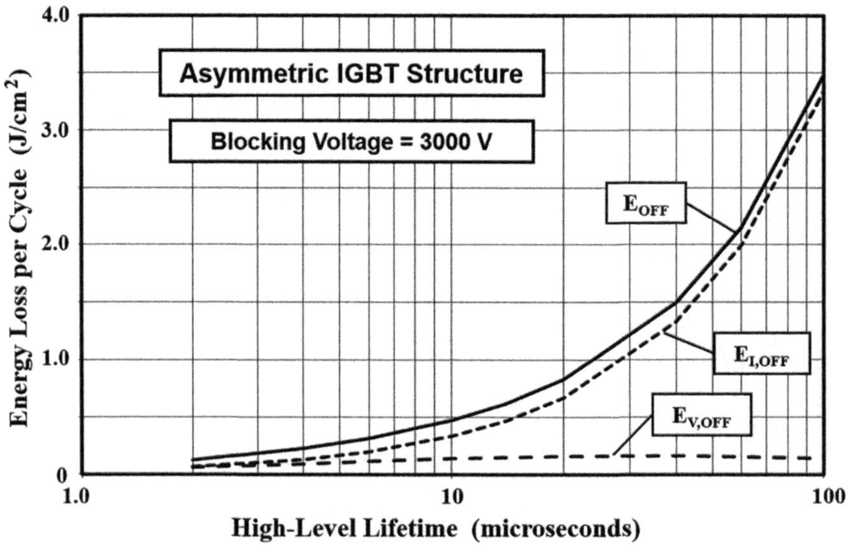

FIGURE 3.31

Energy loss during the turn-off transient for the asymmetric 3000 V IGBT structure: lifetime dependence.

1.0×10^{17} cm^{-3} can be obtained by using the above analytical model for the case of an on-state collector current density of 100 A/cm^2 and a collector supply voltage of 2000 V. The results calculated for high-level lifetime values ranging from 1 to 100 μs are provided in Fig. 3.31. For the lowest value of the high-level lifetime of 2 μs, the energy loss during the voltage transient is 62 mJ/cm^2 while the energy loss during the current transient is 67 mJ/cm^2. The energy loss during the voltage transient increases gradually to 141 mJ/cm^2 when the high-level lifetime is increased to 100 μs. In contrast, the energy loss during the current transient increases dramatically to 3480 mJ/cm^2 when the high-level lifetime is increased to 100 μs. The total energy loss for the turn-off transient is dominated by the energy loss during the current transient. The high-level lifetime must be kept below 10 μs to limit the total energy loss to below 0.5 J/cm^2. The energy loss per cycle for the asymmetric IGBT structure is much less than that for the symmetric IGBT structure (see Fig. 3.11).

The energy loss per cycle for the 3000 V asymmetric IGBT structure can be controlled by changing the buffer layer doping concentration. The energy loss obtained by using the above analytical model for the case of an on-state collector current density of 100 A/cm^2 and a collector supply voltage of 2000 V is shown in Fig. 3.32 for the case of a high-level lifetime of 5 μs in the drift region and a buffer layer thickness of 10 μm. At buffer layer doping levels below 1×10^{17} cm^{-3}, the energy loss during the current transient is found to be dominant. At buffer layer doping levels above 1×10^{17} cm^{-3}, the energy losses during the voltage and current transients are found to be equal. The energy loss monotonically reduces with increasing buffer layer

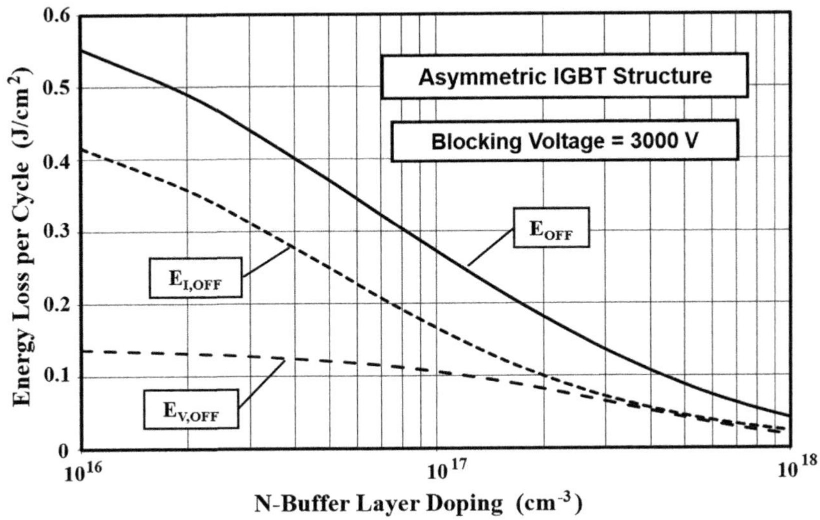

FIGURE 3.32

Energy loss during the turn-off transient for the asymmetric 3000V IGBT structure: buffer layer doping dependence.

doping concentration, providing a good design variable when optimizing the asymmetric IGBT structure.

The energy loss per cycle for the above 3000V asymmetric IGBT structure is shown in Fig. 3.33 as a function of the buffer layer doping concentration for various values for the high-level lifetime in the drift region. The energy loss was calculated

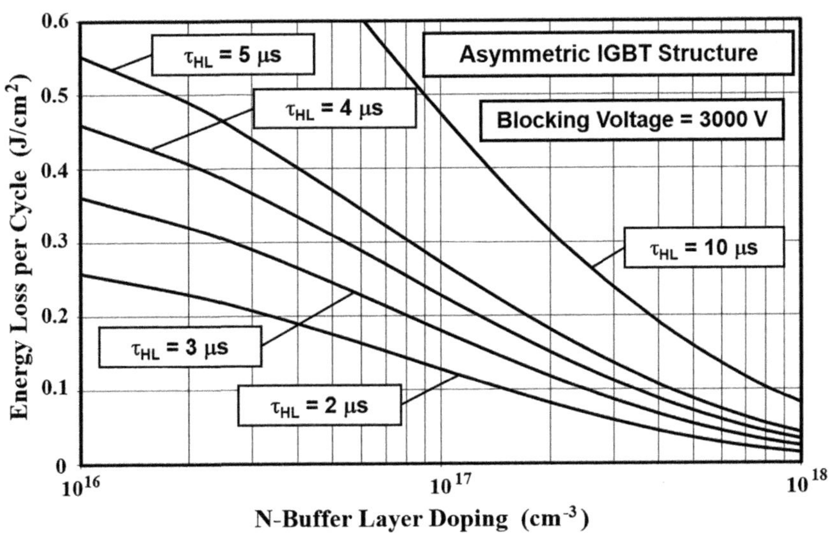

FIGURE 3.33

Energy loss during the turn-off transient for the asymmetric 3000V IGBT structure: buffer layer doping and drift region lifetime dependence.

using the above analytical model for the case of an on-state collector current density of 100 A/cm^2 and a collector supply voltage of 2000 V. In all cases, the energy loss decreases monotonically with increasing buffer layer doping concentration. From these plots, it can be observed that the design of the asymmetric IGBT can be optimized by using an optimum combination of high-level lifetime in the drift region and the buffer layer doping concentration.

The energy loss per cycle for the 3000 V asymmetric IGBT structure can also be controlled by changing the buffer layer thickness. The energy loss obtained by using the above analytical model for the case of an on-state collector current density of 100 A/cm^2 and a collector supply voltage of 2000 V is shown in Fig. 3.34 for the case of various high-level lifetime values in the drift region while using a buffer layer doping concentration of 1×10^{17} cm^{-3}. The energy loss decreases with increasing buffer layer thickness for each value for the high-level lifetime. The gradual monotonic reduction in energy loss with increasing buffer layer thickness allows control of the energy loss using buffer layer thickness as a design parameter.

3.3.6 Power Loss Trade-Off Curve

As mentioned before, the trade-off curve between on-state voltage drop and turn-off energy per cycle allows optimization of the IGBT structure for any particular application. The trade-off curve obtained by varying the high-level lifetime in the drift region for the 3000 V asymmetric blocking IGBT structure is shown in Fig. 3.35 for the case of an on-state current density of 100 A/cm^2 and a collector supply voltage

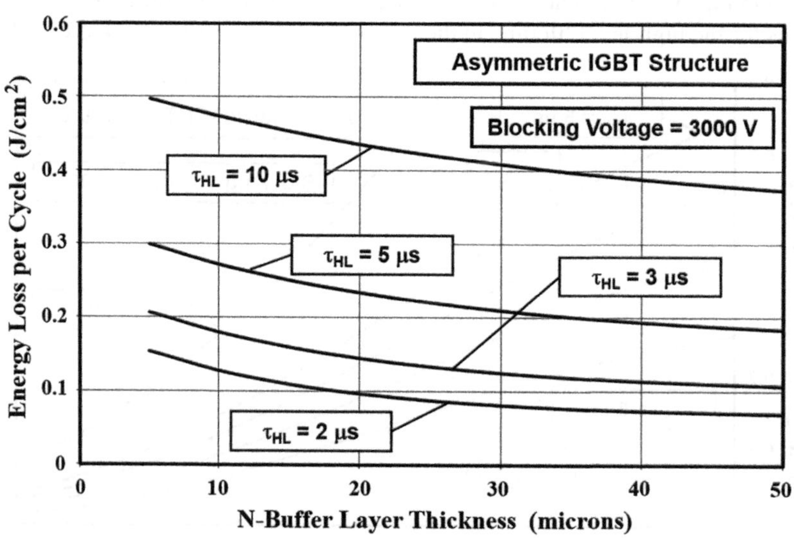

FIGURE 3.34

Energy loss during the turn-off transient for the asymmetric 3000 V IGBT structure: buffer layer thickness dependence.

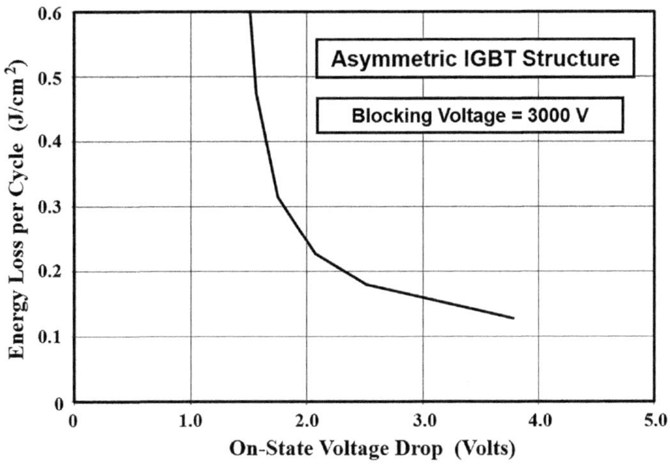

FIGURE 3.35

Trade-off curves for the asymmetric 3000V IGBT structure: drift lifetime variation.

of 2000 V during switching. This device structure has a drift region width of 280 μm and a doping concentration of 1×10^{13} cm^{-3}, a buffer layer width of 10 μm, and a doping concentration of 1×10^{17} cm^{-3}. If the switching power loss is limited to 100 W/cm^2 by the packaging technology, the asymmetric IGBT would be limited to an operating frequency below 500 Hz based upon an energy loss per cycle of 0.2 J/cm^2. The operating frequency for the 3000 V asymmetric IGBT is found to be five times larger than that for the 3000 V symmetric IGBT structure for an on-state current density of 100 A/cm^2. This is the reason asymmetric IGBT structures are preferred for typical motor drive applications.

A trade-off curve between on-state voltage drop and turn-off energy per cycle can be developed for the asymmetric IGBT structure by varying the doping concentration of the buffer layer. The trade-off curve obtained using this approach for the 3000 V asymmetric blocking IGBT structure is shown in Fig. 3.36 for the case of various high-level lifetime values in the drift region. An on-state current density of 100 A/cm^2 and a collector supply voltage of 2000 V during switching was used for generating these trade-off curves. This device structure has a drift region width of 280 μm, a doping concentration of 1×10^{13} cm^{-3}, and a buffer layer width of 10 μm. It can be observed that the trade-off curve shifts toward the left side when the lifetime in the drift region is increased, producing a superior trade-off curve. If the switching power loss is limited to 100 W/cm^2 by the packaging technology, the asymmetric IGBT would be limited to an operating frequency below 1000 Hz based upon an energy loss per cycle of 0.1 J/cm^2. This demonstrates the ability to improve the performance of the asymmetric IGBT structure by the optimum combination of the drift region lifetime and the buffer layer doping concentration. It is worth pointing out that the best combination of on-state voltage drop and energy loss

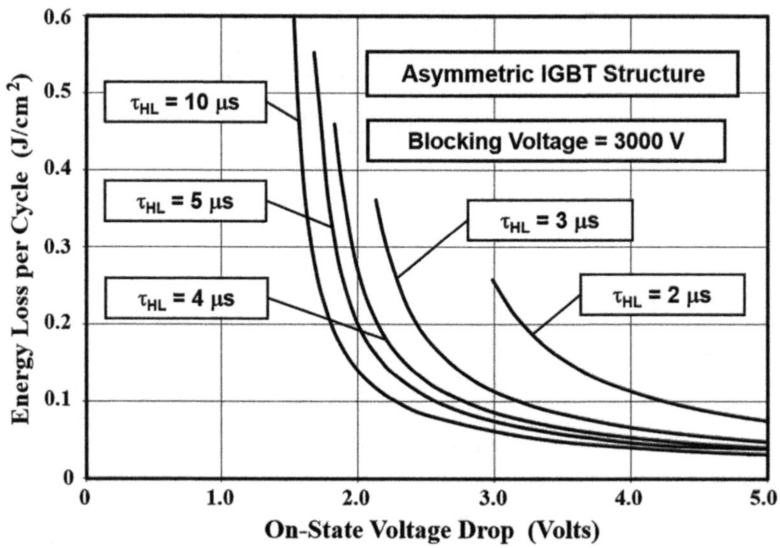

FIGURE 3.36

Trade-off curves for the asymmetric 3000V IGBT structure: buffer doping variation.

per cycle occurs for a high-level lifetime of 10 μs in the drift region with a relatively high buffer layer doping concentration of 5×10^{17} cm^{-3}.

The asymmetric IGBT structure can also be optimized by variation of the thickness of the buffer layer. This is demonstrated in Fig. 3.37 for the case of the 3000 V asymmetric IGBT structure with a drift region width of 280 μm, a doping concentration of

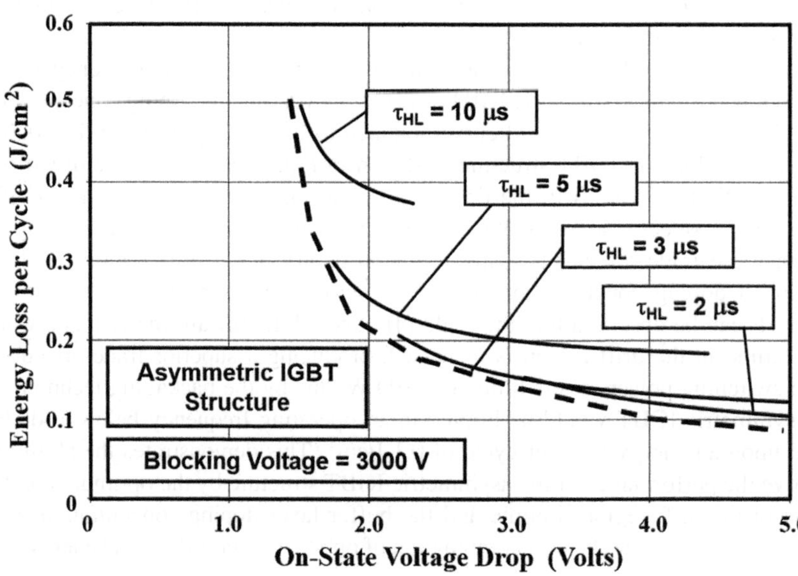

FIGURE 3.37

Trade-off curves for the asymmetric 3000V IGBT structure: buffer thickness variation.

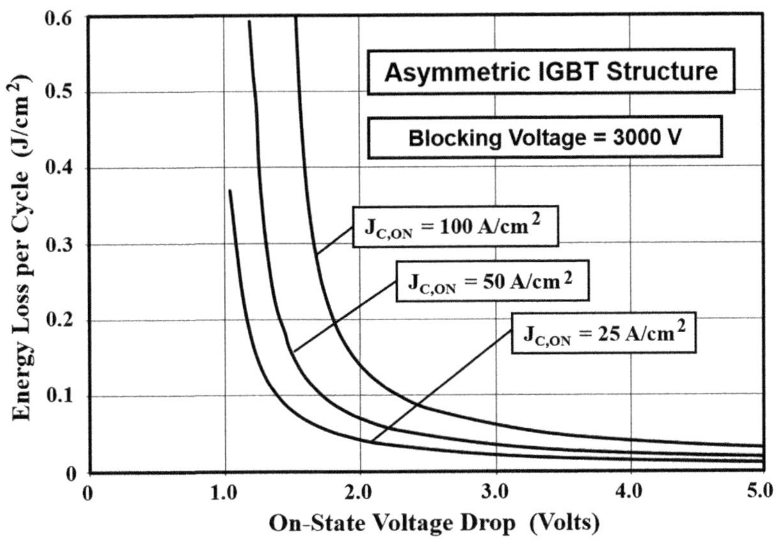

FIGURE 3.38

Trade-off curves for the asymmetric 3000 V IGBT structure: various on-state current densities.

1×10^{13} cm^{-3}, and a buffer layer doping concentration of 1×10^{17} cm^{-3}. Various values for the drift region high-level lifetime were used during this optimization. In this case, a composite trade-off curve can be created, as shown in Fig. 3.37 by the dashed line representing the best trade-off curve obtainable with this approach.

As in the case of the symmetric blocking structure, the asymmetric IGBT structure can operate at a higher frequency if the on-state current density is decreased because both the on-state voltage drop and energy loss per cycle are reduced. This is demonstrated in Fig. 3.38 with trade-off curves achieved by varying the buffer layer doping concentration for on-state current densities of 100, 50, and 25 A/cm^2. In all cases, a high-level lifetime of 10 μs was used for the drift region because this produces the best trade-off curve, as previously shown with the aid of Fig. 3.36. The maximum operating frequency can be increased approximately proportional to the reduction in the on-state current density. This will of course require a larger chip active area for a given current rating for the device.

The maximum operating frequency for the IGBT is limited by the power dissipation, which produces a rise in temperature within the chip. For reliable operation, it is typical to limit the total power dissipation density to 200 W/cm^2 to maintain a junction temperature of 125°C. The total power dissipated by the IGBT is given by the sum of the on-state and switching power loss:

$$P_L = 0.5 J_{C,ON} V_{ON} + E_{OFF} f \tag{3.71}$$

for the case of a 50% duty cycle if the turn-on power loss is neglected. The maximum operating frequency is given by:

$$f_{Max} = \left(\frac{P_{L,Max} - 0.5 J_{C,ON} V_{ON}}{E_{OFF}} \right) \tag{3.72}$$

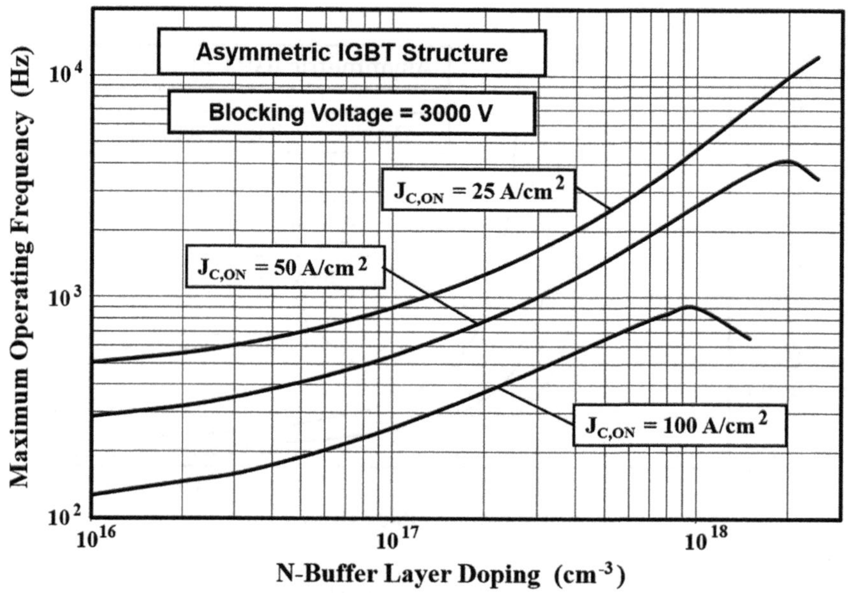

FIGURE 3.39

Maximum operating frequency for the asymmetric 3000V IGBT structure: various on-state current densities.

where $P_{L,Max}$ is the maximum power dissipation and V_{ON} is the on-state voltage drop. The maximum operating frequency for the 3000V asymmetric IGBT structure whose trade-off curves were shown in Fig. 3.38 is provided in Fig. 3.39. For the case of an on-state current density of 100 A/cm^2, a peak of 900 Hz is observed for the maximum operating frequency when the buffer layer doping concentration is increased to 1×10^{18} cm^{-3}. This peak moves to a value of 4000 Hz at a buffer layer doping concentration is increased to 2×10^{18} cm^{-3} when the on-state current density is reduced to 50 A/cm^2. For an on-state current density of 25 A/cm^2, a maximum operating frequency beyond 10 kHz can be achieved for this 3000V asymmetric IGBT structure. The desired increase in operating frequency is achieved at the expense of a larger chip size for the intended application.

3.4 Transparent Emitter IGBT Structure

The structure for the transparent emitter IGBT was shown on the left side of Fig. 2.6 together with the doping profile on the right side. The DMOS-gate structure for the device is similar to the previous structures and can be replaced by the trench-gate structure as well. The unique aspect of the transparent emitter IGBT structure is its very thin and more lightly doped collector region. The term “transparent” refers to the ability of electrons injected into the collector region to diffuse to the collector contact due to its small thickness. The term "emitter" is used because the collector region of the IGBT actually serves as the emitter of the internal PNP transistor. The smaller dopant charge in the collector region reduces the injection efficiency at junction J_1, allowing a reduction of the stored charge in the drift region.

The transparent emitter concept was first applied to the IGBT structure in the 1990s when the effort was directed toward very high (4.5 kV) blocking voltage structures for traction applications [13–17]. When the capability to handle thinner wafers during device processing became possible, the idea was extended to lower blocking voltages as well, including devices with 600 V forward blocking capability fabricated from 60 μm thick silicon wafers.

3.4.1 Blocking Voltage

It is typical to include a buffer layer in the transparent emitter IGBT structure and to use a high lifetime within the drift region. The buffer layer has also been referred to as a "field-stop" layer because the electric field punches through the drift region until its spread is stopped by the higher doping level of the buffer layer. With the inclusion of a buffer layer, the analysis of the blocking voltage for the transparent emitter IGBT becomes similar to that for the asymmetric IGBT structure, as described in detail in Section 3.3.1. The drift region can be optimized using an appropriate combination of the drift region doping concentration and thickness. In the case of a 3000 V IGBT structure, a drift region doping concentration of 1×10^{13} cm^{-3} and a thickness of 280 μm are suitable with a typical buffer layer thickness of 10 μm.

3.4.2 On-State Characteristics

A generally applicable analytical model for the asymmetric IGBT structure that is valid for any injection level in the buffer layer was previously proposed [12]. This analytical model is extended to the transparent emitter IGBT structure with a buffer layer here.

A schematic illustration of the carrier profile in the on-state for the transparent emitter IGBT structure is shown in Fig. 3.40. The thickness of the P-collector region (W_{PC}) is small when compared with the previous structures. Consequently, the diffusion length for the electrons injected into the collector region is much greater than its thickness. The electron profile in the collector region then becomes linear from an injected concentration of $n_C(0)$ at the junction to zero at the metal contact. The hole and electron concentrations in the N-base region are equal due to charge neutrality. The hole concentration in the N-buffer layer is comparable to the doping concentration (N_{DB}) in the buffer layer. Consequently, the electron concentration in the buffer becomes larger than the doping concentration to preserve charge neutrality. The electron concentration ($n_{NB}(0)$) in the buffer layer at the junction (J_1) is given by:

$$n_{NB}(0) = N_{DB} + p_{NB}(0) \tag{3.73}$$

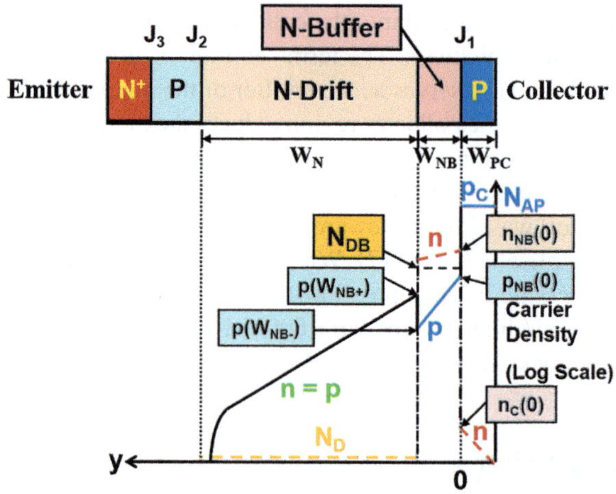

FIGURE 3.40

On-state carrier distribution in the transparent emitter IGBT structure.

Using the Boltzmann quasiequilibrium boundary condition for the carrier densities on both sides of the junction (J_1) together with Eqn (3.73):

$$\frac{p_C}{p_{NB}(0)} = \frac{n_{NB}}{n_C(0)} = \frac{N_{DB} + p_{NB}}{n_C(0)} \tag{3.74}$$

For a uniformly doped collector region, the hole concentration (p_C) is equal to the doping concentration (N_{AP+}) because it is operating under low-level injection conditions. In the case of a diffused collector region, an effective doping concentration is given by:

$$N_{AEP+} = 20\sqrt{N_{AES}N_D} \tag{3.75}$$

should be used where N_{AES} is the surface concentration of the collector diffusion [1].

From Eqn (3.74):

$$n_C(0) = \frac{N_{DB}p_{NB}(0) + p_{NB}^2(0)}{N_{AEP+}} \tag{3.76}$$

For the transparent emitter structure, the linear electron carrier profile results in the electron current density at junction (J_1) given by:

$$J_n(0) = \frac{qD_{nP+}}{W_{P+}} n_C(0) \tag{3.77}$$

where D_{nP+} is the diffusion coefficient for electrons in the P^+ collector region and W_{PC} is its thickness. The hole current density at junction (J_1) is given by:

$$J_p(0) = \frac{qD_{pNB}}{L_{pNB}} p_{NB}(0) \tag{3.78}$$

where D_{pP+} and L_{pP+} are the diffusion coefficient and diffusion length for holes in the buffer layer, respectively. All the parameters in these equations must be calculated after taking into account the reduction of the mobility and lifetime in the highly doped buffer layer and collector. The total collector current density is the sum of these components:

$$J_C = J_n(0) + J_p(0) = \frac{qD_{nP+}}{W_{P+}} n_C(0) + \frac{qD_{pNB}}{L_{pNB}} p_{NB}(0) \tag{3.79}$$

Using Eqn (3.76), an expression for the hole concentration in the buffer layer at junction (J_1) can be derived:

$$p_{NB}^2(0) + \left(\frac{D_{pNB}N_{AP+}W_{P+} + D_{nP+}N_{DB}L_{pNB}}{D_{nP+}L_{pNB}}\right) p_{NB}(0) - \frac{N_{AP+}W_{P+}J_C}{qD_{nP+}} = 0 \tag{3.80}$$

The hole concentration in the buffer layer at junction (J_1) is the solution of this quadratic equation:

$$p_{NB}(0) = \frac{1}{2}\left(\sqrt{b^2 - 4c} - b\right) \tag{3.81}$$

where

$$b = \frac{D_{pNB}N_{AP+}W_{P+} + D_{nP+}N_{DB}L_{pNB}}{D_{nP+}L_{pNB}} \tag{3.82}$$

and

$$c = -\frac{N_{AP+}W_{P+}J_C}{qD_{nP+}} \tag{3.83}$$

The holes concentration ($p(W_{NB}-)$) inside the buffer layer at the boundary between the N-buffer layer and the N-base region is given by:

$$p(W_{NB-}) = p_{NB}(0)e^{-(W_{NB}/L_{pNB})} \tag{3.84}$$

where W_{NB} is the thickness of the buffer layer. Due to high-level injection conditions, the hole concentration profile in the N-base region is given by [1]:

$$p(y) = p(W_{NB+}) \frac{\sinh\left[(W_N + W_{NB} - y)/L_a\right]}{\sinh\left[(W_N + W_{NB})/L_a\right]} \tag{3.85}$$

for $y > W_{NB}$. In this expression, the hole concentration ($p(W_{NB+})$) in the N-base region at the boundary between the N-buffer layer and the N-base region can be obtained by equating the hole current density on the two sides of this boundary [1]:

$$p(W_{NB+}) = \frac{L_a \tanh\left[(W_N + W_{NB})/L_a\right]}{2qD_p} J_p(W_{NB-}) \tag{3.86}$$

with

$$J_p(W_{NB-}) = J_p(0)e^{-(W_{NB}/L_{pNB})} \tag{3.87}$$

The on-state carrier distribution obtained by using the above equations is provided in Fig. 3.41 for the case of the 3000 V transparent emitter IGBT structure with a

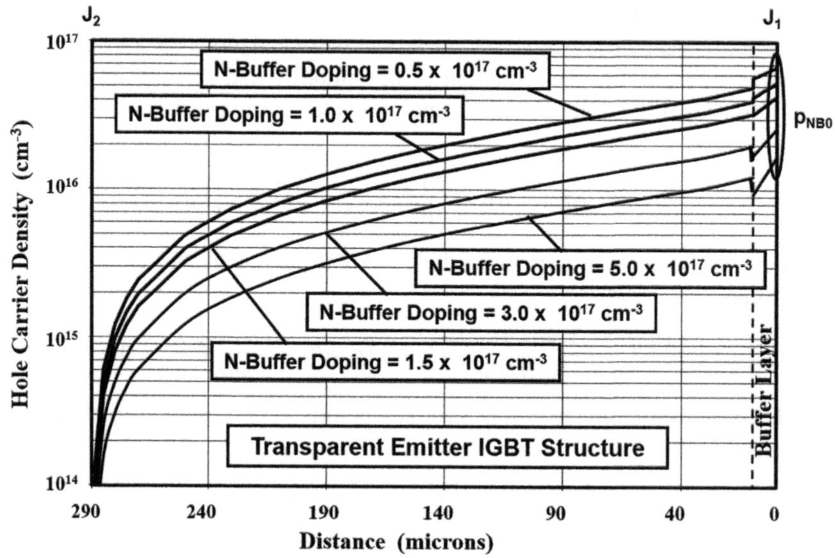

FIGURE 3.41

On-state carrier distribution in the 3000V transparent emitter IGBT structure: buffer layer doping dependence with $N_{AP+} = 1 \times 10^{18}\,\text{cm}^{-3}$.

collector region thickness of 1 μm and a doping concentration of 1×10^{18} cm^{-3}. A drift region thickness of 280 μm, a buffer layer thickness of 10 μm, and a high-level lifetime (τ_{HL}) in the drift region of 20 μs were used in this example. It can be observed that the hole concentrations $p_{NB}(0)$ and p_{WNB+} decrease with increasing buffer layer doping. In comparison with the asymmetric IGBT structure, the hole concentration in the drift region for the transparent emitter structure is reduced for all cases by about a factor of three on the collector side of the drift region in spite of the two times larger lifetime in the drift region. This is due to the reduced injection efficiency of the collector junction created by the transparent emitter. In contrast, the hole concentration in the transparent emitter IGBT structure is about two times larger than for the asymmetric IGBT structure on the emitter side of the drift region due to high value for the lifetime in the drift region.

The on-state carrier distribution can also be adjusted by varying the doping concentration of the collector region in the transparent emitter IGBT structure. This is illustrated in Fig. 3.42 for the case of the 3000 V transparent emitter IGBT structure with a drift region thickness of 280 μm, a buffer layer thickness of 10 μm, a buffer layer doping concentration of 5×10^{16}, and a high-level lifetime (τ_{HL}) of 20 μs in the drift region. It can be observed that the hole concentrations $p_{NB}(0)$ and p_{WNB+} increase with increasing collector doping concentration due to the enhanced injection efficiency of the collector junction.

As in the case of the previous symmetric and asymmetric IGBT structures, the on-state carrier distribution can be adjusted by varying the lifetime in the drift region of the transparent emitter IGBT structure. This is illustrated in Fig. 3.43 for the case of

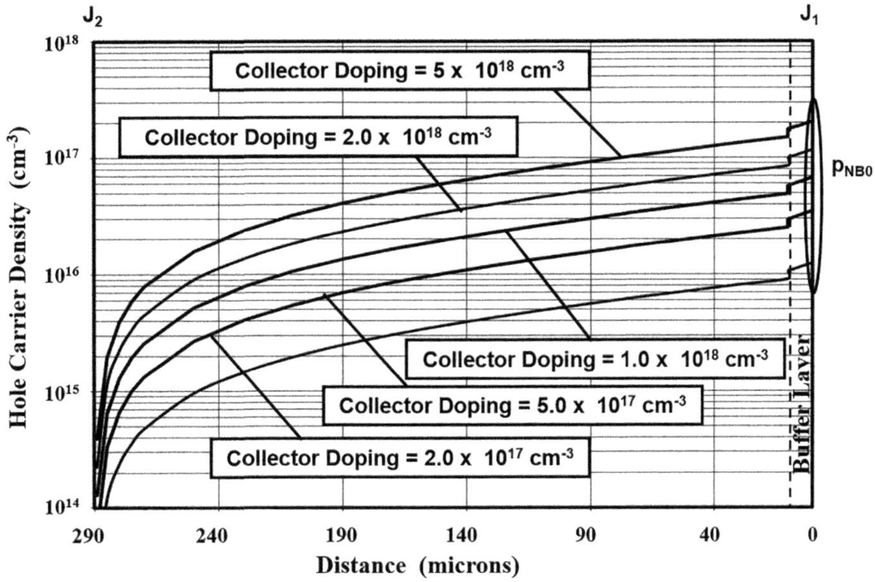

FIGURE 3.42

On-state carrier distribution in the 3000 V transparent emitter IGBT structure: collector doping dependence.

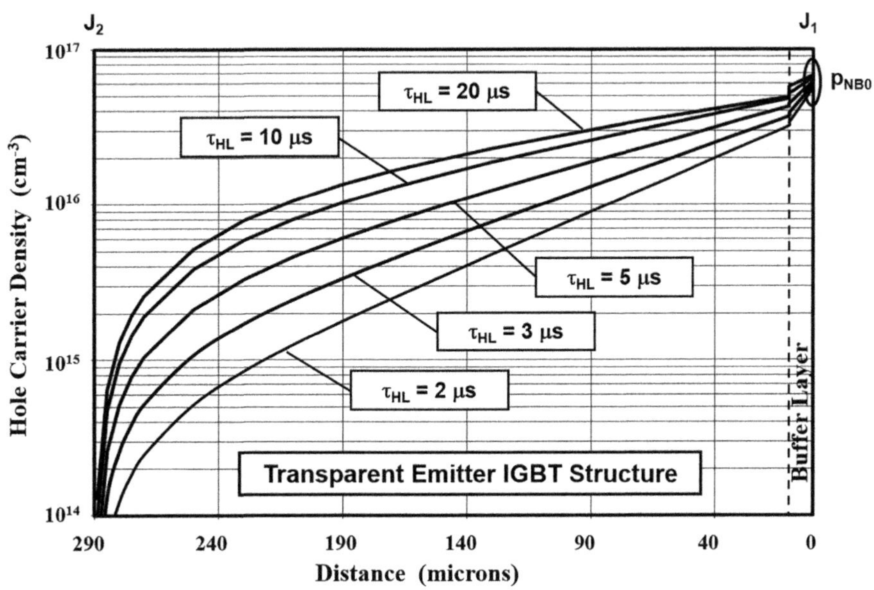

FIGURE 3.43

On-state carrier distribution in the 3000 V transparent emitter IGBT structure: drift region lifetime dependence.

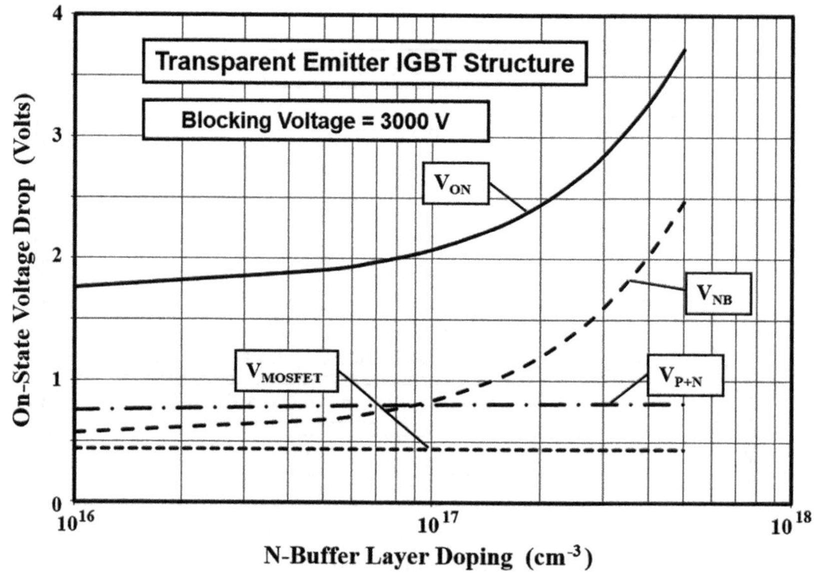

FIGURE 3.44

On-state voltage drop components for the 3000V transparent emitter IGBT structure: buffer layer doping dependence.

the 3000 V transparent emitter IGBT structure with a drift region thickness of 280 μm, a buffer layer thickness of 10 μm, a buffer layer doping concentration of 5×10^{16} cm^{-3}, and a collector doping concentration of 1×10^{18} cm^{-3}. It can be observed that the hole concentrations $p_{NB}(0)$ and p_{WNB+} decrease when the lifetime in the drift region is reduced.

The on-state voltage drop for the transparent emitter IGBT structure can be obtained by using the equations provided in Section 3.2.2 for the asymmetric IGBT structure. The influence of changing the buffer layer doping concentration, the collector doping concentration, and the lifetime in the drift region is discussed here for the case of the 3000 V structure.

The components of the on-state voltage drop for the 3000 V transparent emitter IGBT structure at an on-state current density of 100 A/cm^2 are shown in Fig. 3.44 as a function of the buffer layer doping concentration. This structure has a collector region thickness of 1 μm and a doping concentration of 1×10^{18} cm^{-3}, a drift region thickness of 280 μm, a buffer layer thickness of 10 μm, and a high-level lifetime (τ_{HL}) in the drift region of 20 μs. The on-state voltage drop increases rapidly for this structure when the buffer layer doping concentration exceeds 1×10^{17} cm^{-3} because of the rapid increase in the voltage drop (V_{NB}) across the drift region.

For the case of a high-level lifetime of 20 μs in the drift region, the on-state voltage drop increases relatively slowly with increasing buffer layer doping concentration, as shown in Fig. 3.44. This does not hold true when the lifetime in the drift region is reduced, as shown in Fig. 3.45. It can be observed that the on-state voltage

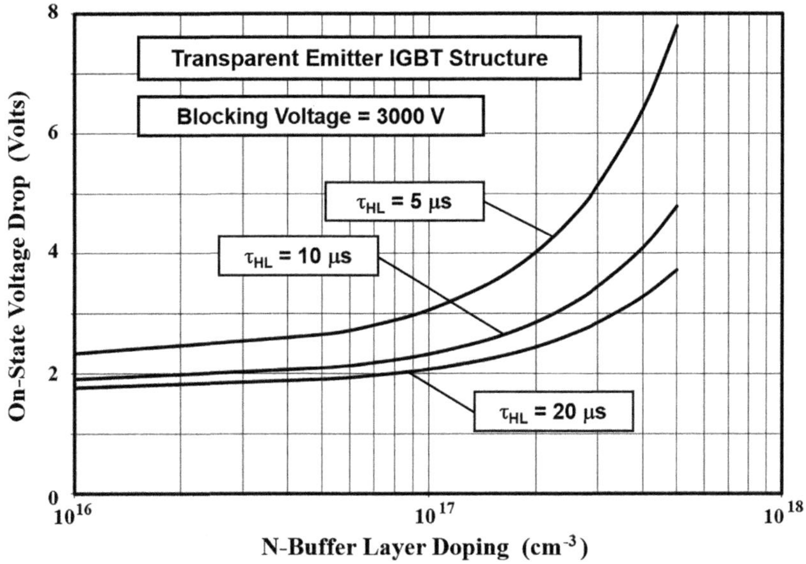

FIGURE 3.45

On-state voltage drop for the 3000V transparent emitter IGBT structure: buffer layer doping dependence.

drop becomes too large at a high buffer layer doping concentration when the high-level lifetime is reduced to 5 μs. This is the reason for using relatively high lifetime values for the drift region in the transparent emitter IGBT structure and utilizing the collector doping concentration to optimize the injected hole concentration in the drift region.

The impact of varying the collector doping concentration on the on-state voltage drop for the transparent emitter IGBT structure is illustrated in Fig. 3.46 for the case of the 3000 V structure. This structure has a collector region thickness of 1 μm, a drift region thickness of 280 μm, a buffer layer thickness of 10 μm, and a high-level lifetime (τ_{HL}) in the drift region of 20 μs. It can be observed that the on-state voltage drop becomes too large at high buffer layer doping concentrations when the collector doping level is reduced below 1×10^{18} cm^{-3}.

3.4.3 Stored Charge

The free carrier distribution in the drift region of the transparent emitter IGBT structure was discussed in the previous section and shown to be dependent on the collector doping concentration, the lifetime in the drift region, and the buffer layer doping concentration. The stored charge in the drift region can be computed by using Eqn (3.59).

The stored charge computed for the case of the 3000 V transparent emitter IGBT structure is provided in Fig. 3.47 when the buffer layer doping concentration is varied for various values of the lifetime in the drift region. The parameters used for this

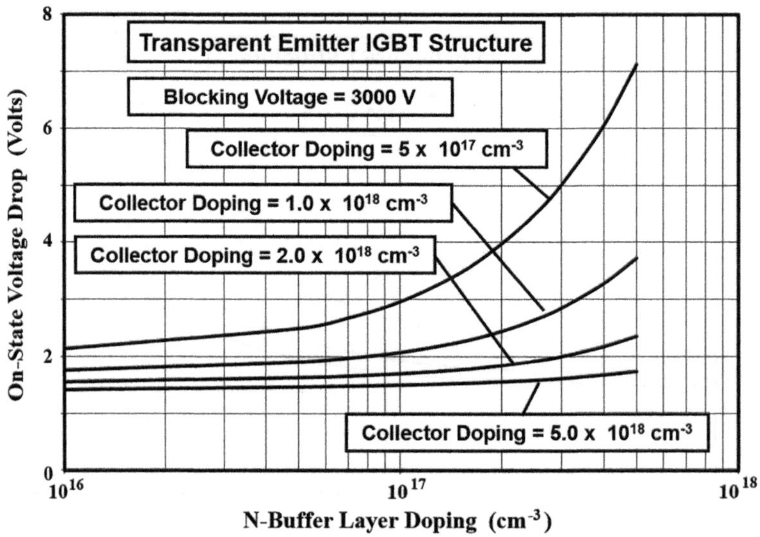

FIGURE 3.46

On-state voltage drop for the 3000V transparent emitter IGBT structure: collector doping dependence.

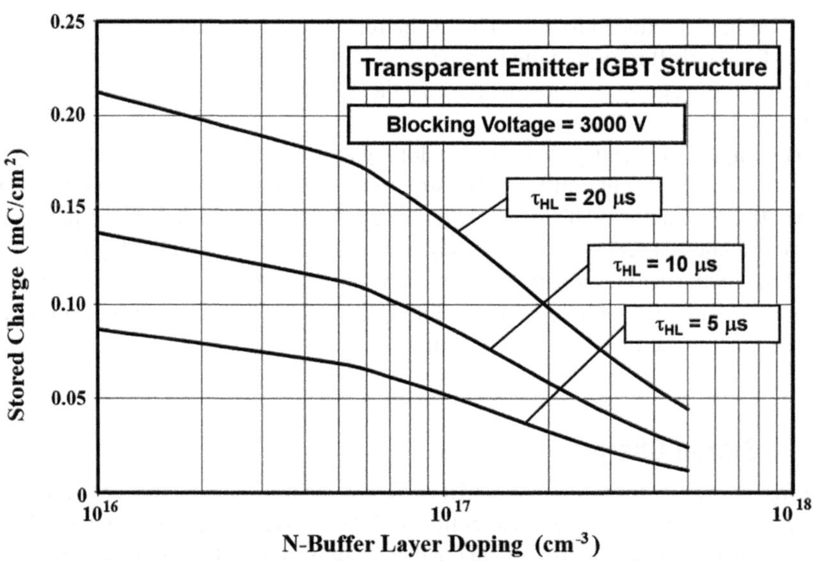

FIGURE 3.47

Stored charge in the 3000V transparent emitter IGBT structure: buffer layer doping dependence with lifetime as parametric variable.

device structure are the same as those used to generate the on-state voltage drop in Fig. 3.45. It can be observed that the stored charge is a weak function of the buffer layer doping concentration until it exceeds 1×10^{17} cm^{-3}. A rapid reduction in the stored charge occurs when the buffer layer doping concentration exceeds 1×10^{17} cm^{-3} but this is accompanied by a rapid increase in the on-state voltage drop, as observed in Fig. 3.45. For the same high-level lifetime of 5 μs in the drift region, the transparent emitter IGBT has a three times smaller stored charge at a buffer layer doping concentration of 1×10^{17} cm^{-3} than the asymmetrical IGBT structure (see Fig. 3.25).

The stored charge in the transparent emitter IGBT can also be controlled by changing the doping concentration in the collector region. This is demonstrated for the 3000 V structure for the case of a drift region lifetime of 20 μs in Fig. 3.48. It can be observed that the stored charge increases when the doping concentration of the transparent emitter is increased. This is due to an improvement in the injection efficiency of junction J_1.

3.4.4 Turn-Off Switching Waveforms

The analysis of the turn-off waveforms for the transparent emitter IGBT structure can be performed by using the same approach used for the asymmetric IGBT structure in Section 3.2.4. The voltage transient is given by Eqn (3.65) and the current transient by Eqn (3.68). The turn-off waveforms for the 3000 V transparent emitter IGBT

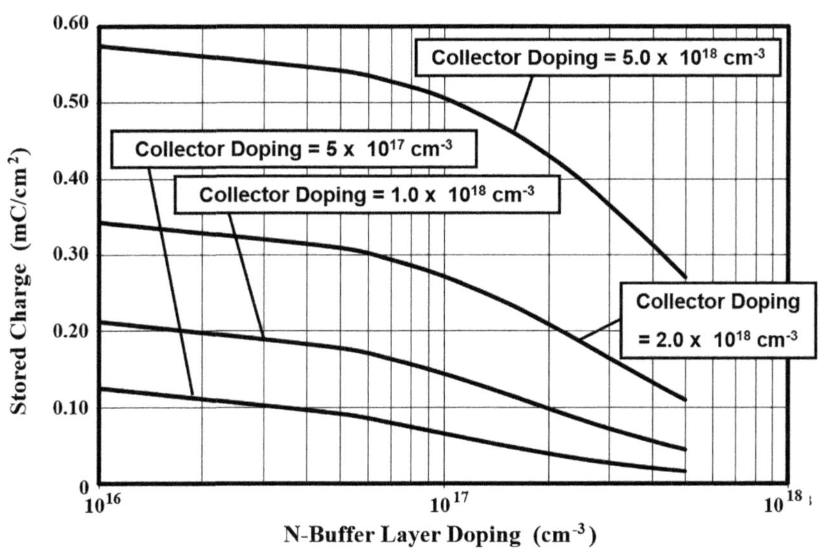

FIGURE 3.48

Stored charge in the 3000 V transparent emitter IGBT structure: buffer layer doping dependence with collector doping as parametric variable.

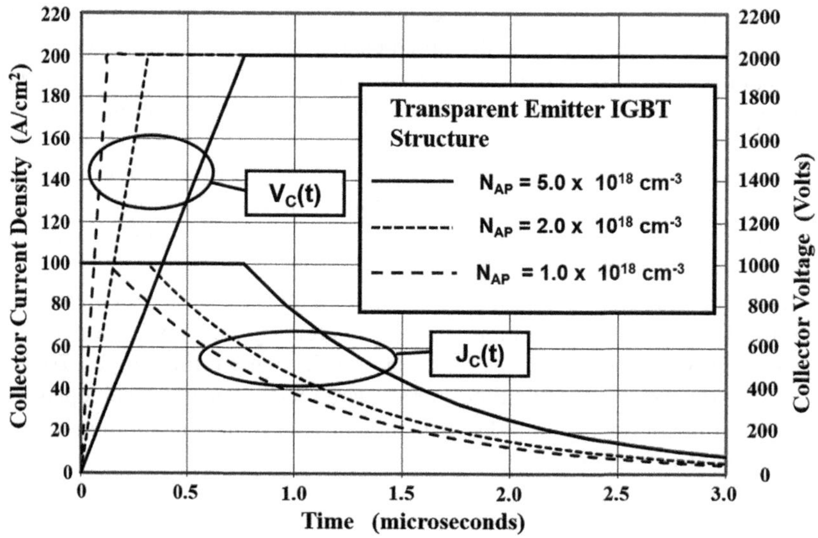

FIGURE 3.49

Collector current and voltage transients during turn-off for the transparent emitter IGBT structure with an inductive load: collector doping dependence.

structure with an *N*-base region width of 280 μm and a doping concentration of 1×10^{13} cm^{-3}, a buffer layer width of 10 μm, and a doping concentration of 5×10^{17} cm^{-3} are shown in Fig. 3.49 for the case of three collector doping concentrations. A high-level lifetime of 20 μs was used in the drift region in this example. The collector voltage increases linearly with time with a collector voltage rise time ($t_{V,OFF}$) that decreases from 0.764 to 0.309 to 0.126 μs when the collector doping concentration is reduced from 5 to 2 to 1×10^{18} cm^{-3}, respectively, because of the reduced injected hole concentration in the drift region. It can be concluded that the collector rise time can be controlled by altering the doping concentration in the collector region for the transparent emitter IGBT structure. The collector current decays exponentially with a collector current turn-off time ($t_{I,OFF}$) of 2.09 μs for all three cases because the lifetime in the buffer layer has the same value.

The switching loss during turn-off for the transparent emitter IGBT structure can also be controlled by varying the lifetime in the drift region. It is worth pointing out that the lifetime in the buffer layer is also reduced due to scaling of the lifetime with doping concentration [1]. This is demonstrated in Fig. 3.50 for the case of the 3000 V structure with *N*-base region width of 280 μm and a doping concentration of 1×10^{13} cm^{-3}, a buffer layer width of 10 μm and a doping concentration of 5×10^{17} cm^{-3}, and a collector region width of 1 μm and a doping concentration of 1×10^{18} cm^{-3}. It can be observed that the collector voltage rise time ($t_{V,OFF}$) can be reduced from 0.126 to 0.105 to 0.074 μs when the drift region high-level lifetime is reduced from 20 to 10 to 5 μs, respectively. The collector current decays

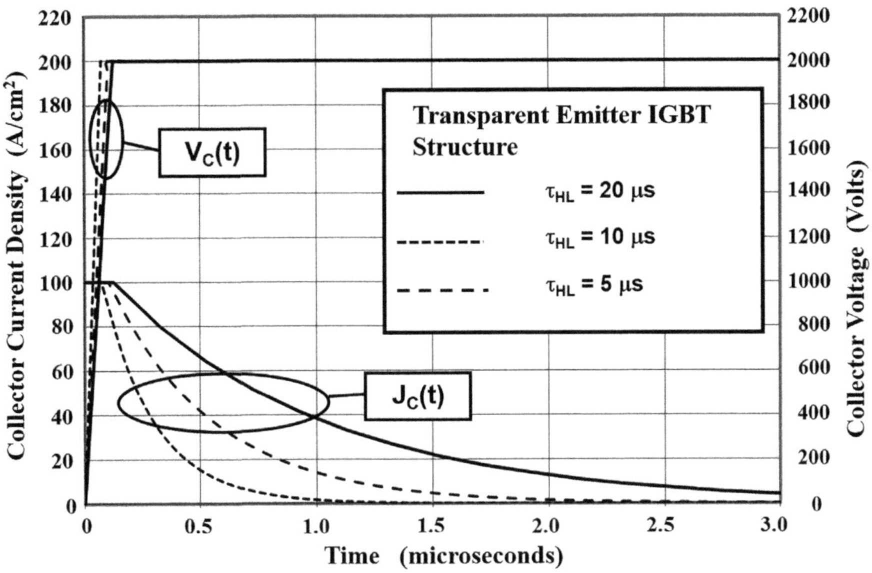

FIGURE 3.50

Collector current and voltage transients during turn-off for the transparent emitter IGBT structure with an inductive load: lifetime dependence.

exponentially with collector current turn-off times ($t_{I,OFF}$) of 2.09, 1.05, and 0.523 μs for these values of the lifetime, respectively. This demonstrates that the drift region lifetime is a sensitive design parameter for optimization of the turn-off switching waveforms and power losses for the transparent emitter IGBT.

The third approach for optimization of the performance of the transparent emitter IGBT is by altering the buffer layer doping concentration. This is demonstrated in Fig. 3.51 for the case of the 3000 V transparent emitter IGBT structure with *N*-base region width of 280 μm and a doping concentration of 1×10^{13} cm^{-3}, a buffer layer thickness of 10 μm, a high-level lifetime of 10 μs in the drift region, and a collector doping concentration of 1×10^{18} cm^{-3}. It can be observed that the collector voltage rise time ($t_{V,OFF}$) increases from 0.105 to 0.254 to 0.387 μs when the buffer layer doping concentration is reduced from 5 to 2 to 1×10^{17} cm^{-3}, respectively, due to an increase in the injected hole concentration in the drift region. The collector current decays exponentially with a collector current turn-off time ($t_{I,OFF}$) that increases from 1.05 to 2.30 to 3.83 μs for these cases because of an increase in the buffer layer lifetime. This example demonstrates that the buffer layer doping concentration can be utilized as another design parameter for optimization of the turn-off switching waveforms and power losses for the transparent emitter IGBT structure. Because the transparent emitter IGBT structure has been employed for very high voltage devices, the buffer layer doping is usually controlled by a diffusion of phosphorus from the back side of the wafer.

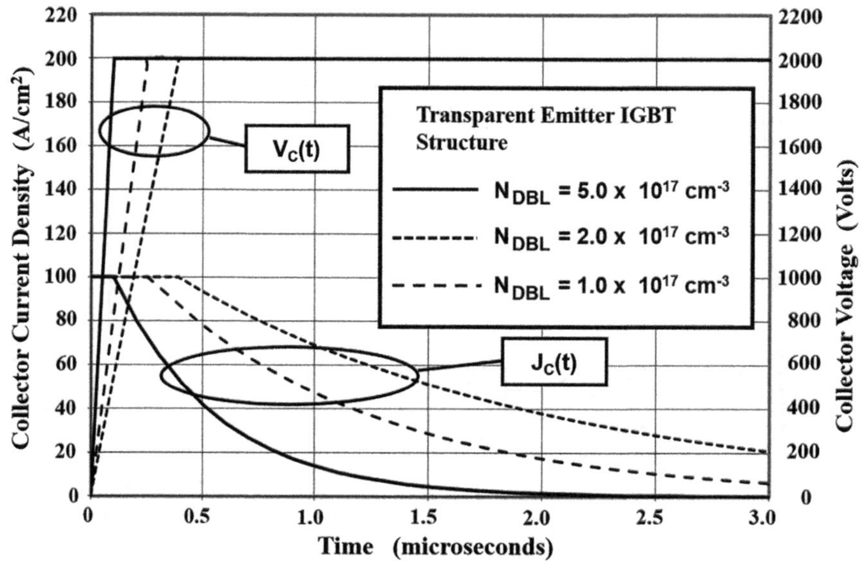

FIGURE 3.51

Collector current and voltage transients during turn-off for the transparent emitter IGBT structure with an inductive load: buffer doping dependence.

3.4.5 Turn-Off Power Loss

The energy loss per cycle for the transparent emitter IGBT structure during the voltage rise time can be computed using Eqn (3.41) developed for the symmetric structure and during the current fall time using Eqn (3.70) for the asymmetric IGBT structure. The energy loss per cycle for the case of the 3000 V transparent emitter IGBT structure with *N*-base region width of 280 μm and a doping concentration of 1.0×10^{13} cm^{-3}, a buffer layer thickness of 10 μm, and a doping concentration of 1.0×10^{17} cm^{-3} obtained by using the above analytical model for the case of an on-state collector current density of 100 A/cm^2 and a collector supply voltage of 2000 V is provided in Fig. 3.52. In this plot, three cases of the high-level lifetime in the drift region have been considered. It can be observed that the collector doping concentration can be used as a design variable to determine the energy loss per cycle. The lowest energy loss per cycle occurs for a low collector doping concentration because of the reduced injected hole concentration in the drift region. The hole concentration is also reduced when the lifetime is decreased, making the energy loss per cycle smaller.

The energy loss per cycle for the transparent emitter IGBT structure can also be controlled by changing the buffer layer doping concentration. The energy loss obtained by using the above analytical model for the case of an on-state collector current density of 100 A/cm^2 and a collector supply voltage of 2000 V is shown in Fig. 3.53 for the case of a collector doping concentration of 1×10^{18} cm^{-3}. Three

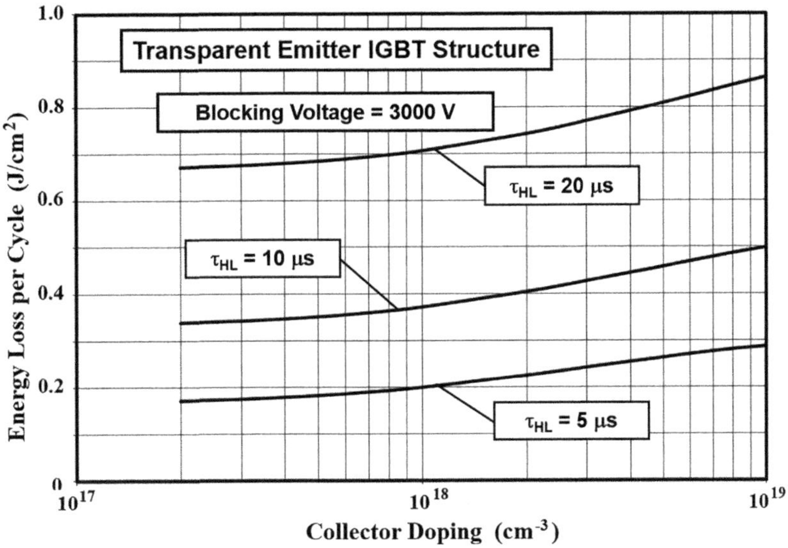

FIGURE 3.52

Energy loss during the turn-off transient for the transparent emitter 3000V IGBT structure: collector doping dependence.

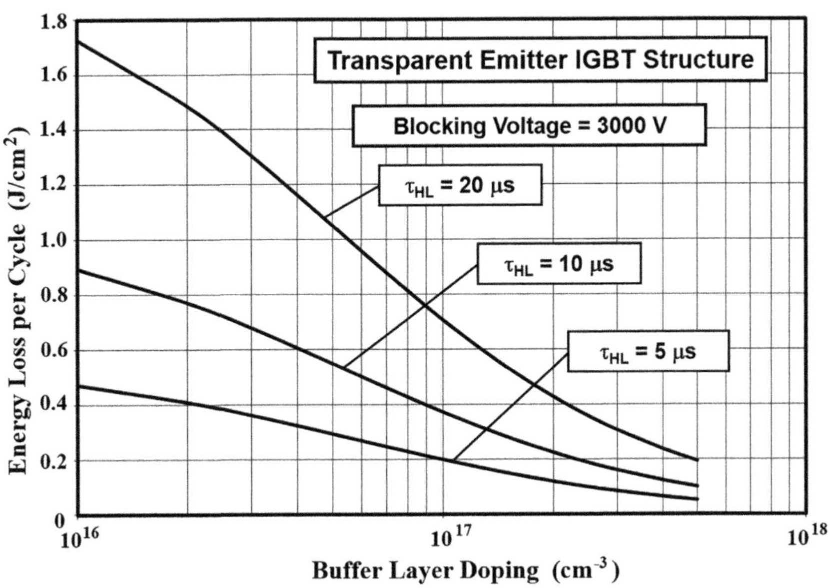

FIGURE 3.53

Energy loss during the turn-off transient for the transparent emitter 3000V IGBT structure: buffer layer doping dependence.

cases of the drift region high-level lifetime are considered in this figure. The energy loss per cycle decreases monotonically with increasing buffer layer doping concentration providing a good design variable when optimizing the transparent emitter IGBT structure. The energy loss per cycle can be greatly reduced by decreasing the lifetime in the drift region.

3.4.6 Power Loss Trade-Off Curve

The trade-off curve for the 3000 V transparent emitter IGBT structure obtained by varying the collector doping concentration is shown in Fig. 3.54 for the case of an on-state current density of 100 A/cm^2 and a collector supply voltage of 2000 V during switching. This device structure has a drift region width of 280 μm and a doping concentration of 1×10^{13} cm^{-3}, and a buffer layer width of 10 μm and a doping concentration of 1×10^{17} cm^{-3}. Three cases of the high-level lifetime in the drift region are considered in the figure. From the perspective of the trade-off curve, the collector doping is not a good variable because the energy loss per cycle is only weakly dependent on the collector doping concentration, as seen in Fig. 3.52. However, it is strongly dependent on the drift region lifetime, as seen by the downward shift of the trade-off curve when the lifetime is reduced.

A trade-off curve between on-state voltage drop and turn-off energy per cycle can be developed for the transparent emitter IGBT structure by varying the doping

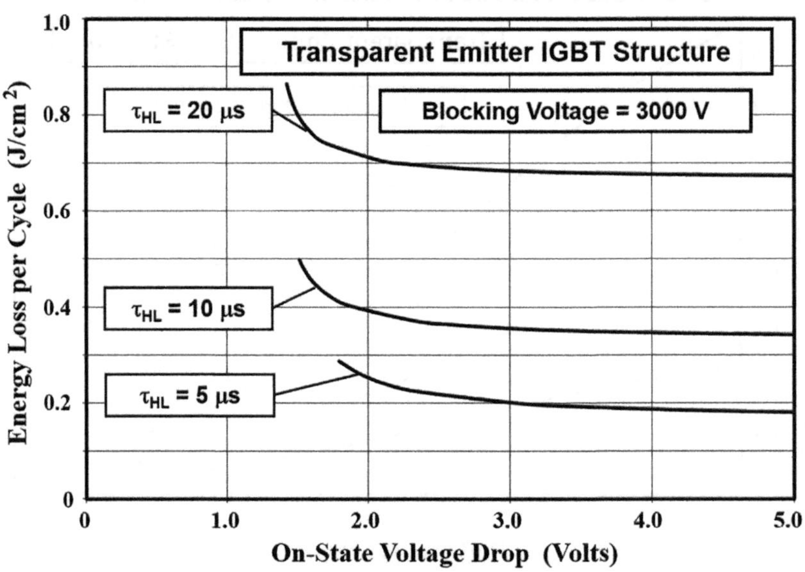

FIGURE 3.54

Trade-off curves for the transparent emitter 3000V IGBT structure: collector doping as parametric variable.

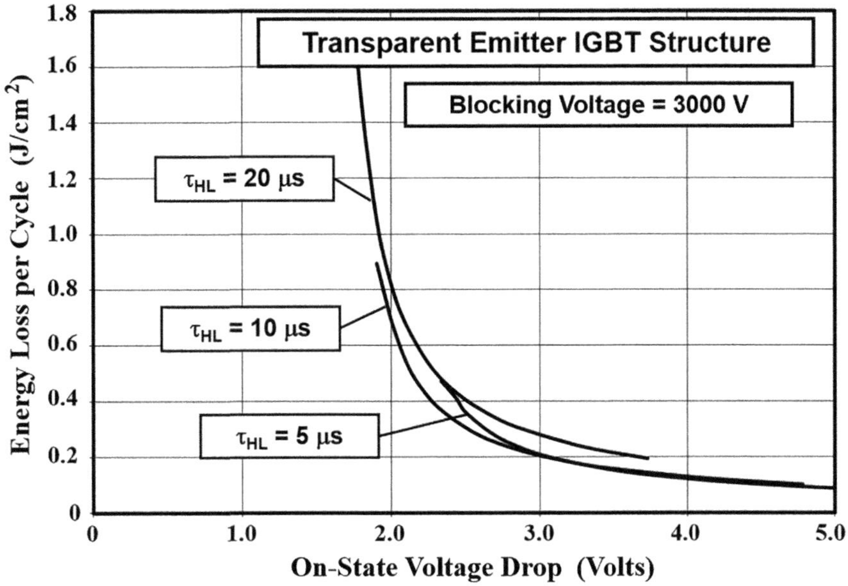

FIGURE 3.55

Trade-off curves for the transparent emitter 3000V IGBT structure: buffer doping as parametric variable.

concentration of the buffer layer. The trade-off curve obtained using this approach for the 3000 V asymmetric blocking IGBT structure is shown in Fig. 3.55 for the case various high-level lifetime values in the drift region. An on-state current density of 100 A/cm^2 and a collector supply voltage of 2000 V during switching was used for generating these trade-off curves. This device structure has a drift region width of 280 μm and a doping concentration of 1×10^{13} cm^{-3}, and a buffer layer width of 10 μm. It can be observed that the trade-off curves for the different lifetime cases overlap. This indicates that the same point on the trade-off curve is produced for multiple combinations of the drift region lifetime and the buffer layer doping concentration.

The maximum operating frequency for the IGBT is limited by the power dissipation, which produces a rise in temperature within the chip. The total power dissipation in the IGBT structure is given by Eqn (3.71) for the case of a 50% duty cycle if the turn-on power loss is neglected. Using this equation, the maximum operating frequency is given by Eqn (3.72). For reliable operation, it is typical to limit the total power dissipation density to 200 W/cm^2 to maintain a junction temperature of 125°C. The maximum operating frequency for the 3000 V transparent emitter IGBT structure, whose trade-off curves were shown in Fig. 3.54, is provided in Fig. 3.56 as a function of the collector doping concentration for three drift region lifetime values. In all three cases, the maximum operating frequency becomes less than zero when the collector doping level becomes too small because the on-state power dissipation

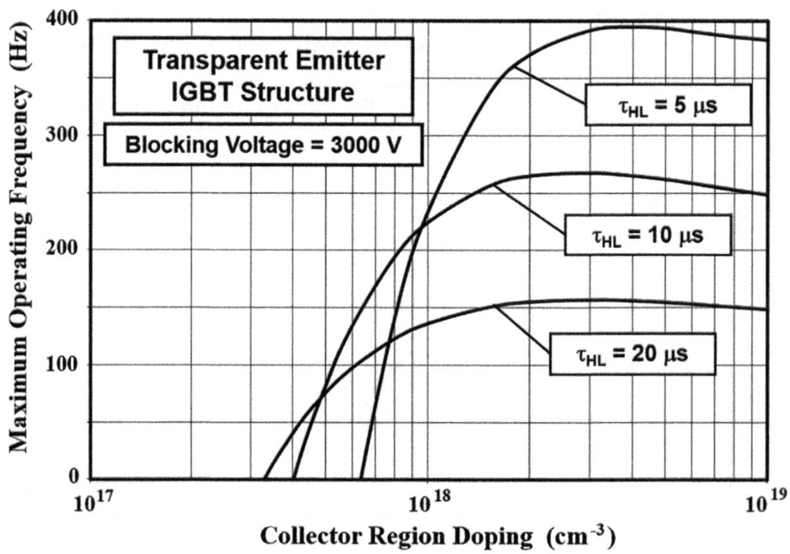

FIGURE 3.56

Maximum operating frequency for the transparent emitter 3000V IGBT structure: as a function of collector doping concentration: high-level lifetime as parametric variable.

exceeds 200 W/cm^2. It can be concluded that a larger maximum operating frequency can be achieved by reducing the drift region lifetime. There is an optimum collector doping level of about 3×10^{18} cm^{-3} at which the maximum operating frequency has its highest value for each lifetime case, but this is a broad maxima giving good design and process latitude when making the transparent emitter IGBT structure. It can also be concluded that a collector doping concentration of 1×10^{18} cm^{-3} is too low because it degrades the maximum operating frequency by half for the case of a high-level lifetime of 5 μs.

The maximum operating frequency for the 3000 V transparent emitter IGBT structure is provided in Fig. 3.57 as a function of the buffer layer doping concentration for three drift region lifetime values. A collector doping concentration of 3×10^{18} cm^{-3} was used for these plots because this was found to improve the maximum operating frequency, as shown in Fig. 3.56. The maximum operating frequency becomes less than zero when the buffer layer doping level becomes too high because the on-state power dissipation exceeds 200 W/cm^2. For each lifetime value, there is a well-defined optimum buffer layer doping where the maximum operating frequency has its highest value. It can be concluded that the largest maximum operating frequency can be achieved by using a drift region lifetime of 10 μs and a buffer layer doping concentration of 6×10^{17} cm^{-3}. This demonstrates the design optimization procedure required to extract the best performance from the transparent emitter IGBT structure.

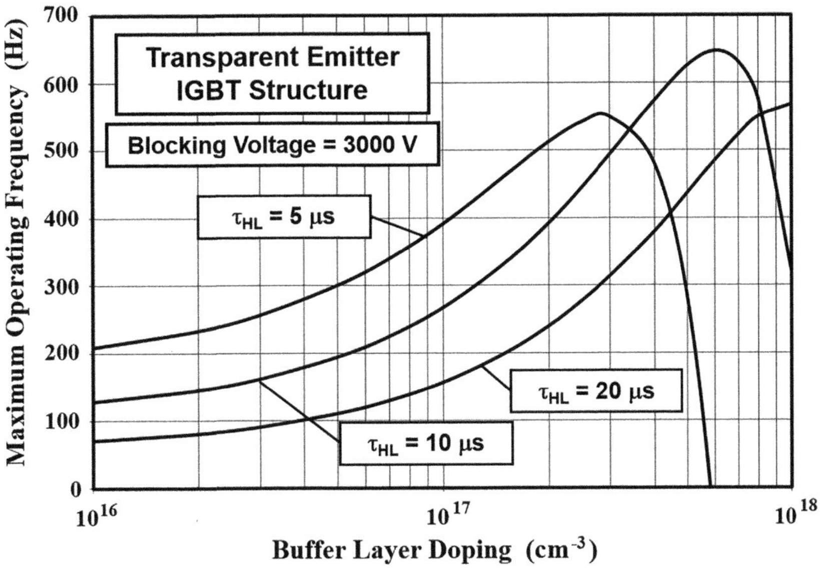

FIGURE 3.57

Maximum operating frequency for the transparent emitter 3000V IGBT structure as a function of buffer layer doping concentration: high-level lifetime as parametric variable.

The maximum operating frequency for the 3000 V transparent emitter IGBT structure can be increased by decreasing the on-state current density because this reduces both the on-state voltage drop and the energy loss per cycle. This requires using a larger chip area to achieve any desired current rating, making the cost of the device greater. The maximum operating frequency for the 3000 V transparent emitter IGBT is provided in Fig. 3.58 as a function of the buffer layer doping concentration for three on-state current density values. But a collector doping concentration of 3×10^{18} cm^{-3} and a drift region high-level lifetime of 10 μs were used for these plots because this was found to improve the maximum operating frequency, as shown in Fig. 3.56. The maximum operating frequency becomes limited to about 1000 Hz for an on-state current density of 100 A/cm^2. It can be increased to 3000 and 6500 Hz by reducing the on-state current density to 50 and 25 A/cm^2, respectively.

3.5 Silicon Carbide IGBT Structures

The development of power devices from silicon carbide (SiC) is motivated by its high breakdown field strength. The specific resistance of the drift region for a vertical power device structure is related to the basic semiconductor properties by [18]:

$$R_{on-specific} = \frac{4BV^2}{\varepsilon_S \mu_n E_C^3} \tag{3.88}$$

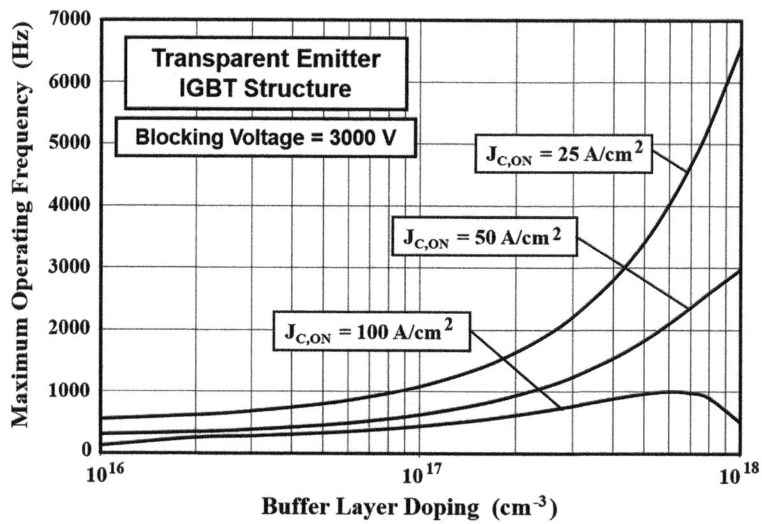

FIGURE 3.58

Maximum operating frequency for the transparent emitter 3000V IGBT structure: on-state current density as parametric variable.

where BV is the breakdown voltage, ε_S is the dielectric constant, μ_n is the electron mobility, and E_C is the critical electric field for breakdown. The denominator of this equation is commonly referred to as Baliga's figure of merit (BFOM) for semiconductors. Based upon available information about the properties of silicon carbide, a BFOM of about 2000 is predicted for the 4H-SiC polytype [19]. This has motivated the development of SiC unipolar devices, such as Schottky rectifiers [20] (including the JBS structure [21–23]) and power MOSFETs [24–26] in the 1990s. These devices are now commercially available from several companies [27–29].

Excellent SiC unipolar power devices can be created with breakdown voltages up to 5000 V [19]. For applications that require devices with even larger blocking voltage capability, it is possible to develop IGBTs from silicon carbide. SiC-based IGBTs have the advantage of a much smaller drift region thickness, which reduces the on-state voltage drop and the stored charge. The development of *n*-channel SiC IGBTs was hindered by the high resistance of the *P*-type substrates that must serve as the collector region. Consequently, *p*-channel, asymmetric SiC IGBTs were first developed with 9–12 kV forward blocking capability using thick *p*-type epitaxial layers grown on highly doped *n*-type substrates [30–32]. A channel mobility of 6.5 cm^2/V-s was measured for holes for these devices. Asymmetric *n*-channel SiC IGBTs were subsequently developed [33] with a blocking voltage of 13 kV. A channel mobility of 18 cm^2/V-s was measured for electrons for these devices. An analysis of 15 kV symmetric blocking IGBTs has demonstrated [34] that these structures can have competitive performance with asymmetric devices. The blocking voltage capability for p-channel asymmetric 4H-SiC IGBTs has been extended to 15 kV [35]. A comparison between the SiC power MOSFET and the n-channel IGBT for typical motor drive applications [36] suggests that the IGBT structure is superior to the power MOSFET structure when the blocking voltage exceeds 9 kV. This analysis predicts the highest MVA rating for the n-channel SiC IGBT structure at a blocking voltage of 20 kV.

3.5.1 *N*-Channel Asymmetric SiC IGBT Structure

The planar, asymmetric, n-channel SiC IGBT structure is shown in Fig. 3.59 together with its doping profile. Unlike the silicon IGBT, the doping concentrations of the various layers are uniform for the SiC devices because they are prepared using epitaxial growth or ion implantation. Because techniques for controlling the lifetime in the drift region are immature, the SiC IGBT structure must be optimized by altering the doping and thickness of the buffer layer [35]. The basic physics of operation for the SiC n-channel asymmetric IGBT structure are identical to those previously discussed for silicon devices in Section 3.2. The equations provided in that section for the blocking voltage design, on-state carrier distribution, on-state voltage drop, and turn-off wave forms can therefore also be applied to the SiC IGBT after taking into account appropriate changes to the basic material properties.

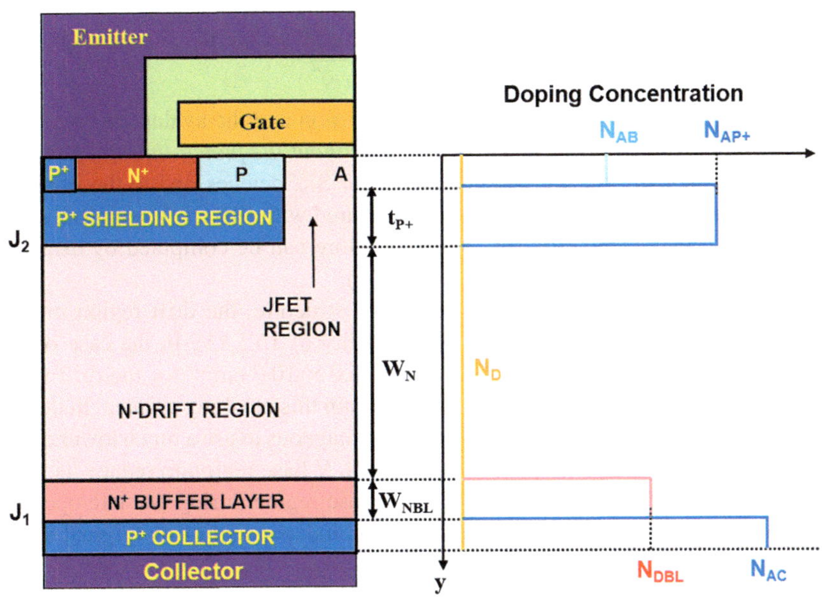

FIGURE 3.59

The SiC n-channel asymmetric IGBT structure with its doping profile.

3.5.2 Blocking Characteristics

As an example, the design of the 15 kV 4H-SiC n-channel asymmetric IGBT structure is considered in this section. The forward blocking capability of the asymmetric SiC IGBT structure is determined by the open-base transistor breakdown phenomenon. The emitter injection efficiency for the P^+ collector/N-buffer junction (J_1) can be obtained by using Eqn (3.43). The diffusion coefficients and diffusion length are dependent on the doping concentrations in the P^+ collector region and the N-buffer layer. Open-base breakdown occurs in the silicon carbide asymmetric IGBT structure when the multiplication coefficient is only slightly above unity. Using the avalanche breakdown criteria when the multiplication coefficient becomes equal to infinity, as assumed in some papers [37], can lead to significant errors during the design of the drift region for the IGBT structure. The minority carrier lifetime has been found to be dependent upon the doping concentration [38] in the case of silicon devices. Although this phenomenon has not been verified for silicon carbide, it is commonly used when performing numerical analysis of SiC devices. The effect can be modeled by using the relationship:

$$\frac{\tau_{LL}}{\tau_{p0}} = \frac{1}{1 + (N_D/N_{REF})} \tag{3.89}$$

where N_{REF} is a reference doping concentration whose value will be assumed to be $5 \times 10^{16}\ \text{cm}^{-3}$ as done for the silicon devices.

As in the case of silicon devices, the multiplication factor for a $P-N$ junction is given by:

$$M = \frac{1}{1 - (V_A/BV_{PP})^n} \tag{3.90}$$

with a value of $n = 6$ for the case of a P^+/N junction and the avalanche breakdown voltage of the P-base/N-base junction (BV_{PP}) *without the punch-through phenomenon*. This breakdown voltage is much larger for SiC devices due to the very high critical electric field for breakdown when compared with silicon. The multiplication coefficient for the asymmetric SiC IGBT structure can be computed by using this non-punch-through voltage given by Eqn (3.46).

For the 15 kV n-channel asymmetric IGBT structure, the drift region must be designed to achieve a blocking voltage capability of 16.5 kV. In the case of avalanche breakdown, there is a unique value of $6.0 \times 10^{14}\ \text{cm}^{-3}$ for the drift region with a width of 170 μm for silicon carbide to obtain this blocking voltage. In the case of the asymmetric SiC IGBT structure, it is advantageous to use a much lower doping concentration for the lightly doped portion of the N-base region to reduce its width. The strong conductivity modulation of the N-base region during on-state operation favors a smaller thickness for the N-base region independent of its original doping concentration.

Any desired forward blocking capability can be achieved in the SiC asymmetric IGBT structure by appropriate combination of the drift region doping, drift region thickness, buffer layer doping, and buffer layer thickness. During the optimization

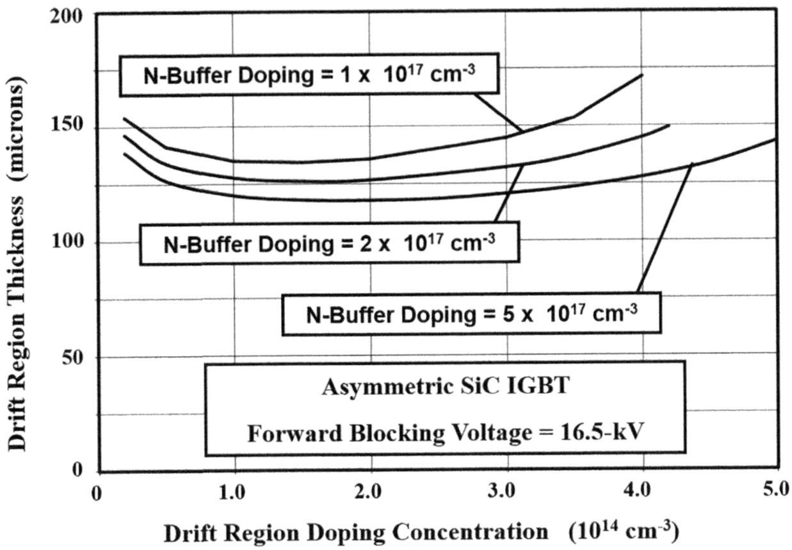

FIGURE 3.60

Optimizing drift region thickness for the SiC asymmetric IGBT structure: *N*-buffer doping as parametric variable.

of the design, it is preferable to achieve the smallest thickness for the drift region to reduce the on-state voltage drop and the stored charge. The change in the thickness of the drift region with drift region doping concentration to achieve a forward blocking voltage of 16,500 V is shown in Fig. 3.60 for the case of a buffer layer thickness of 5 μm. Here, the buffer layer doping was used as a parametric variable. A low-level lifetime of 1 μs was assumed in the drift region. It can be observed that for each buffer layer doping concentration, the drift region thickness has a minimum value. In all three cases, the minimum drift region thickness occurs at the same optimum drift region doping concentration of 2×10^{14} cm^{-3}. The optimum drift region thicknesses are 134, 126, and 118 μm for the buffer layer doping concentrations of 1.0, 2.0, and 5.0×10^{17} cm^{-3}, respectively. The smallest drift region thickness is achieved for the highest buffer layer doping concentration because this produces the smallest injection efficiency and base transport factor.

The change in the thickness of the drift region with drift region doping concentration to achieve a forward blocking voltage of 16,500 V is shown in Fig. 3.61 for the case of a buffer layer doping concentration of 1.0×10^{17} cm^{-3}. Here, the buffer layer thickness was used as a parametric variable. A low-level lifetime of 1 μs was assumed in the drift region. It can be observed that for each buffer layer thickness, the drift region thickness has a minimum value. The minimum drift region thickness occurs at an optimum drift region doping concentration ranging from 1 to 2×10^{14} cm^{-3}. The optimum drift region thicknesses are 156, 134, and 122 μm for buffer layer thicknesses of 2, 5, and 10 μm, respectively. The smallest drift region

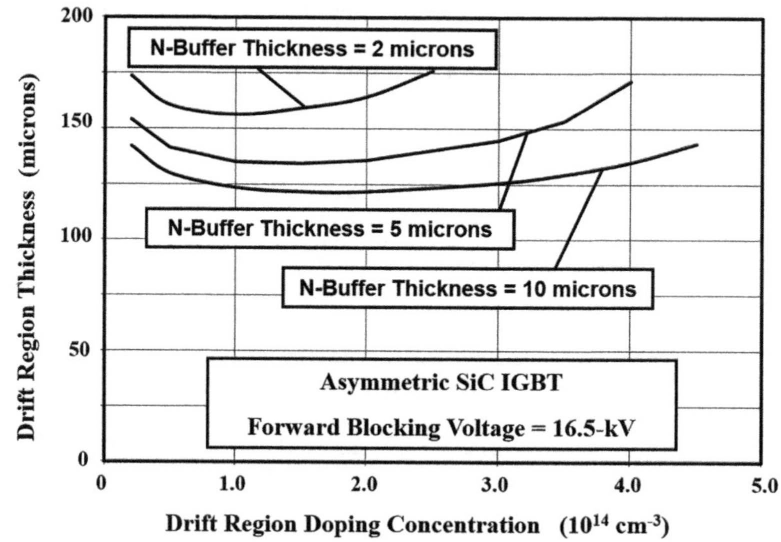

FIGURE 3.61

Optimizing drift region thickness for the SiC asymmetric IGBT structure: *N*-buffer thickness as parametric variable.

thickness is achieved for the largest buffer layer thickness because this produces the smallest injection efficiency and base transport factor.

Based upon the results of the above analysis, it can be concluded that the drift region for an SiC asymmetric IGBT structure should have a doping concentration of 2×10^{14} cm^{-3} and a thickness of 140 μm to achieve the design target of 16,500 V forward blocking capability. This is the value that has been typically used in making actual SiC asymmetric IGBTs [35].

3.5.3 On-State Voltage Drop

The analytical model for the carrier distribution in the drift region and the on-state voltage drop developed for the silicon asymmetric IGBT structure can also be applied to SiC devices. It is of course necessary to apply the appropriate values for the SiC material parameters. One of the major differences between the silicon and SiC structures is that voltage drop across the forward biased *p-n* junction increases from about 0.7 to 3 V. In addition, the lifetime in the drift region of SiC devices has been found to be much lower than in silicon devices, but this is compensated by the smaller thickness of the drift region in SiC devices. As a design example, the 15 kV SiC asymmetric IGBT structure will be analyzed here with a drift region having a doping concentration of 2×10^{14} cm^{-3} and a thickness of 140 μm based upon the analysis described in the previous section.

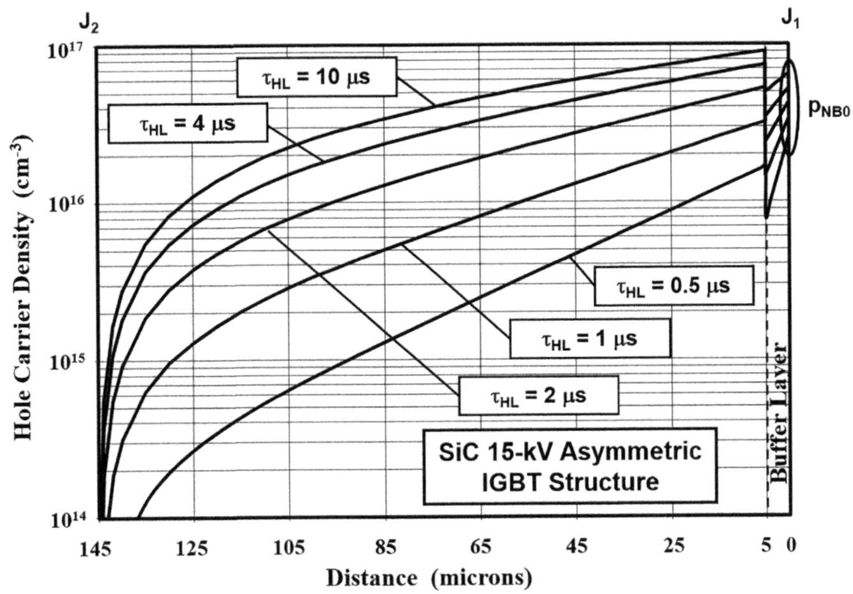

FIGURE 3.62

Carrier distribution in the 15 kV *n*-channel asymmetric SiC IGBT structure: lifetime dependence.

The carrier distribution in the drift region and buffer layer for the 15 kV asymmetric IGBT structure at an on-state current density of 25 A/cm^2 is shown in Fig. 3.62 for the case of various high-level lifetime values. A buffer layer with a doping concentration of 1×10^{17} cm^{-3} and a thickness of 5 μm was assumed here. It can be observed that the carrier concentration falls below the drift region doping level of 2×10^{14} cm^{-3} on the emitter side when the high-level lifetime is reduced below 1 μs. This indicates that an increase in the on-state voltage drop can be expected for this IGBT structure when the high-level lifetime is reduced below 1 μs. The on-state voltage drop for the 15 kV asymmetric IGBT structure obtained by using an analytical model is provided in Fig. 3.63 as a function of the high-level lifetime in the drift region. From the various components of the on-state voltage drop in the figure, it is clear that the voltage drop in the drift region increases rapidly once the high-level lifetime is made less than 1 μs.

As in the case of the asymmetric silicon IGBT, the doping concentration of the buffer layer can be used as a design parameter to control its on-state voltage drop. The carrier distribution in the drift region and buffer layer for the 15 kV asymmetric IGBT structure at an on-state current density of 25 A/cm^2 is shown in Fig. 3.64 for the case of various buffer layer doping concentrations. A drift region high-level lifetime of 0.6 μs and a buffer layer thickness of 5 μm were assumed here. It can be observed that the carrier concentration falls below the drift region doping level of

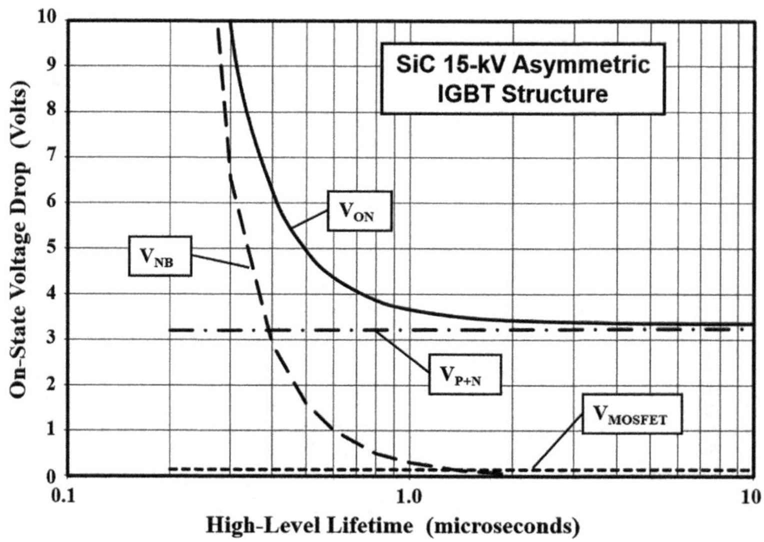

FIGURE 3.63

On-state voltage drop components for the SiC 15 kV asymmetric IGBT structure: lifetime dependence.

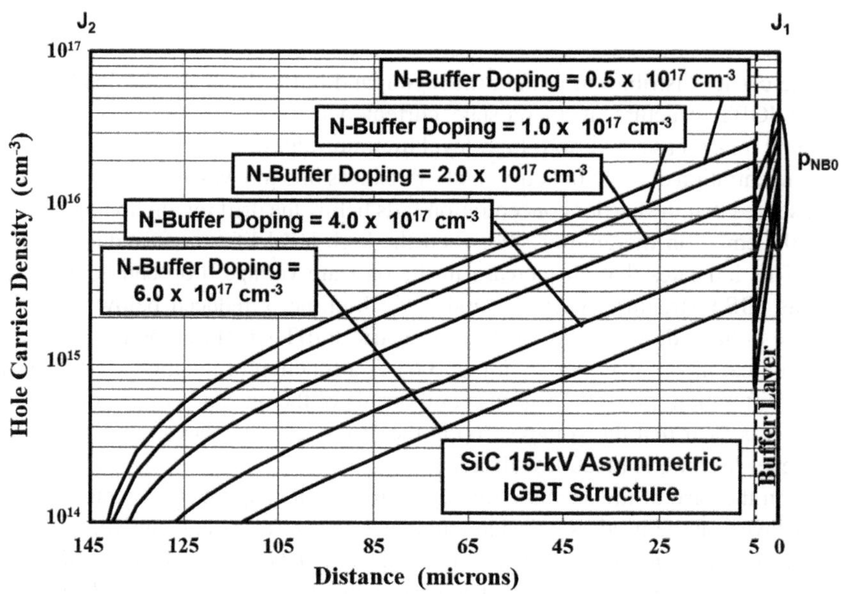

FIGURE 3.64

Carrier distribution in the 15 kV *n*-channel asymmetric SiC IGBT structure: buffer layer doping dependence.

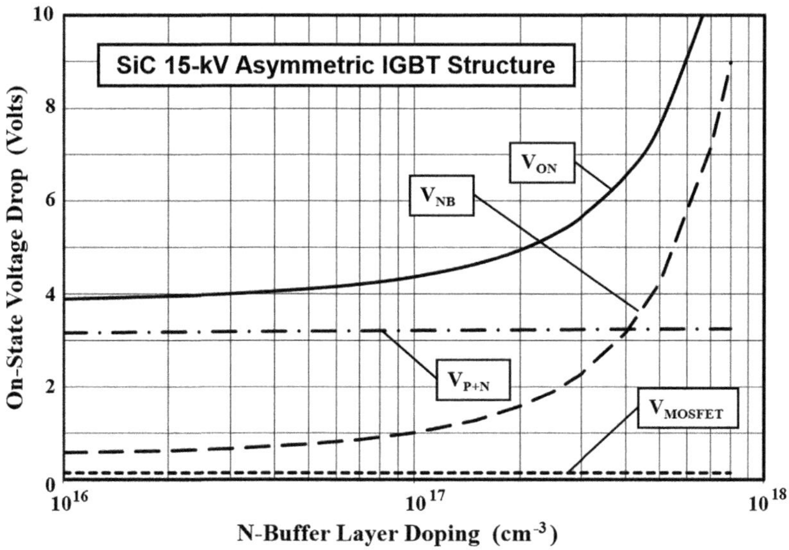

FIGURE 3.65

On-state voltage drop components for the SiC 15 kV asymmetric IGBT structure: buffer layer doping dependence.

2×10^{14} cm^{-3} on the emitter side when the buffer layer doping level exceeds 2×10^{17} cm^{-3}. This indicates that an increase in the on-state voltage drop can be expected for this IGBT structure when the buffer layer doping level exceeds 2×10^{17} cm^{-3}. The on-state voltage drop for the 15 kV asymmetric IGBT structure obtained by using an analytical model is provided in Fig. 3.65 as a function of the buffer layer doping level. From the various components of the on-state voltage drop in the figure, it is clear that the voltage drop in the drift region increases rapidly once the buffer layer doping level exceeds 2×10^{17} cm^{-3}.

The on-state voltage drop for the 15 kV asymmetric IGBT structure at an on-state current density of 25 A/cm^2 obtained by using an analytical model is provided in Fig. 3.66 as a function of the buffer layer doping level for various high-level lifetime values. It can be observed that the buffer layer doping concentration at which the on-state voltage drop increases rapidly becomes smaller when the high-level lifetime is reduced. It is therefore necessary to optimize these parameters together when designing the asymmetric SiC IGBT structure.

The SiC asymmetric IGBT structure can be optimized by altering the thickness of the buffer layer as a design parameter. This is illustrated for the case of the 15 kV structure in Fig. 3.67 where the carrier distribution at an on-state current density of 25 A/cm^2 is shown with a buffer layer thickness ranging from 1 to 15 μm. A buffer layer doping level of 1×10^{17} cm^{-3} and a drift region high-level lifetime of 0.6 μs were used in all these cases. It can be observed that the injected free carrier concentration falls below the doping concentration of the drift region when the buffer layer

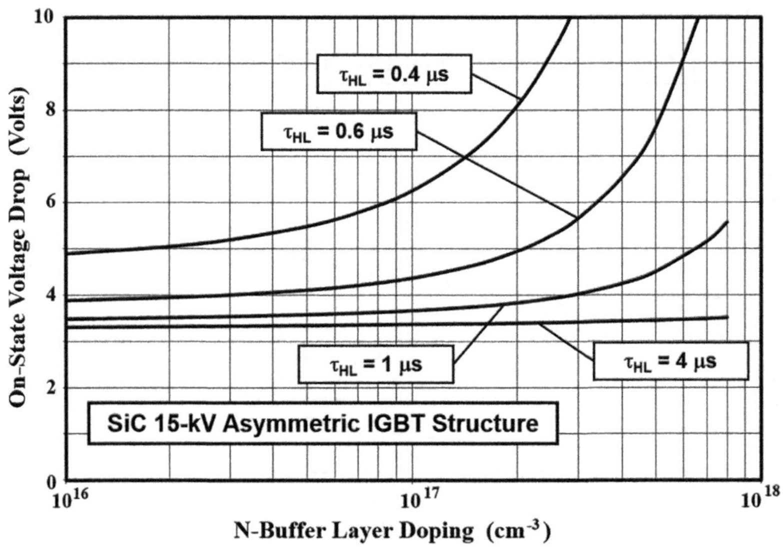

FIGURE 3.66

On-state voltage drop for the SiC 15 kV asymmetric IGBT structure: buffer layer doping dependence.

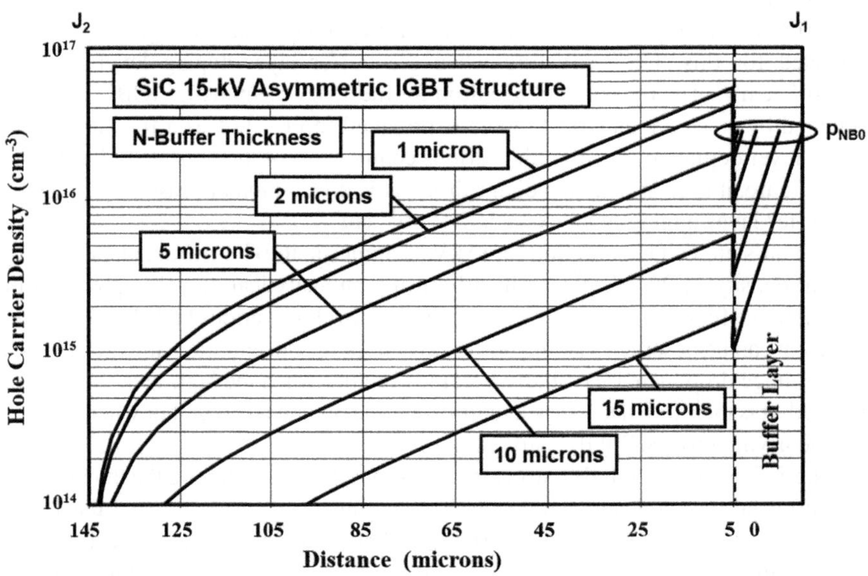

FIGURE 3.67

Carrier distribution in the 15 kV *n*-channel asymmetric SiC IGBT structure: buffer layer thickness dependence.

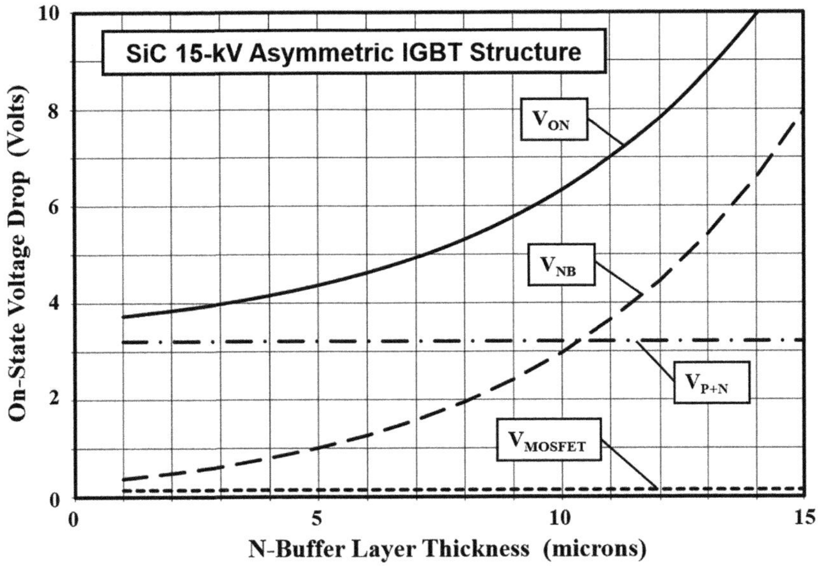

FIGURE 3.68

On-state voltage drop components for the SiC 15 kV asymmetric IGBT structure: buffer layer thickness dependence.

thickness is increased beyond 5 μm. This indicates that a strong increase in the on-state voltage drop will be observed for larger values of the buffer layer thickness. This can be seen in Fig. 3.68 where the on-state voltage drop for the 15 kV asymmetric IGBT structure at an on-state current density of 25 A/cm^2 obtained by using an analytical model is provided as a function of buffer layer thickness for various high-level lifetime values. The voltage drop across the drift region begins to increase rapidly once the buffer layer thickness exceeds 5 μm, resulting in a large on-state voltage drop for the IGBT. This behavior has been experimentally observed in SiC IGBTs fabricated with 2 and 10 μm thick buffer layers [35].

3.5.4 Turn-Off Characteristics

The asymmetric SiC IGBT structure can be expected to switch in a manner similar to that shown earlier for the silicon asymmetric IGBT structure. The hole concentration in the space-charge region during the voltage rise time in the case of the silicon device becomes larger than the doping concentration in the drift region due to the low drift region doping level. Consequently, the space-charge region extends through most of the drift region but does not reach through to the *N*-buffer layer at the end of the voltage rise time. In contrast, the hole concentration in the space-charge region during the voltage rise time for the asymmetric SiC IGBT structure is much smaller than the doping concentration of the drift region due to the high drift region doping level and the smaller on-state current density used for these higher breakdown voltage devices. Consequently, the space-charge region reaches through to the buffer layer when the collector bias is well below the collector supply voltage during the voltage rise time. When the space-charge region reaches through to the buffer layer, the electric field in the *N*-base region takes a trapezoidal shape, allowing the collector voltage to rise at a much more rapid rate until it reaches the supply voltage.

The analysis of the turn-off waveform for the collector voltage transient for the asymmetric SiC IGBT structure can be performed by using the charge control principle described in Section 3.2.4 for the silicon device. However, due to the lower lifetime values for SiC devices, the on-state carrier distribution profile cannot be assumed to be linear [12]. Using the on-state carrier distribution given by Eqn (3.54), the concentration of holes at the edge of the space-charge region (p_e) during the turn-off process is given by:

$$p_e(t) = p(W_{NB+})\frac{\sinh\left[W_{SC}(t)/L_a\right]}{\sinh\left[(W_N + W_{NB})/L_a\right]} \tag{3.91}$$

Because the charge removed by the expansion of the space-charge layer must equal the charge removed due to collector current flow:

$$J_{C,ON} = qp_e(t)\frac{dW_{SC}(t)}{dt} = qp(W_{NB+})\frac{\sinh\left[W_{SC}(t)/L_a\right]}{\sinh\left[(W_N + W_{NB})/L_a\right]}\frac{dW_{SC}(t)}{dt} \tag{3.92}$$

The solution for the evolution of the space-charge region width with time is obtained by integrating this equation:

$$W_{SC}(t) = L_a a\cosh\left\{\frac{J_{C,ON}\sinh\left[(W_N + W_{NBL})/L_a\right]}{qL_a p(W_{NB+})}t + \cosh\left[W_{SC}(0)/L_a\right]\right\} \tag{3.93}$$

The rate of change in the collector voltage supported by the asymmetric SiC IGBT structure can be obtained using:

$$V_C(t) = \frac{q(N_D + p_{SC})W_{SC}^2(t)}{2\varepsilon_S} \tag{3.94}$$

with the hole concentration in the space-charge layer given by Eqn (3.64). In the case of SiC IGBTs designed with drift regions as described in Section 3.4.2, the collector voltage increases until the space-charge region reaches through the *N*-base region. The reach-through voltage is given by:

$$V_{RT}(J_{C,ON}) = \frac{q(N_D + p_{SC})W_N^2}{2\varepsilon_S} \tag{3.95}$$

Based upon a saturated velocity of 8.6×10^6 cm/s for holes in SiC, the hole concentration in the space-charge region (p_{SC}) computed by using Eqn (3.64) is 1.8×10^{13} cm^{-3} at an on-state current density of 25 A/cm^2. In the case of the 15 kV asymmetric SiC IGBT structure with an *N*-base width of 140 μm and a doping concentration of 2.0×10^{14} cm^{-3}, the reach-through voltage is found to be 3980 V after including the above concentration of holes in the space-charge region. The time at which reach through of the space-charge region occurs can be obtained from Eqn (3.93) by setting the space-charge region width equal to the width of the N-base region:

$$t_{RT} = \frac{qL_a p(W_{NB} +)}{J_{C,ON}} \left\{ \frac{\cosh\left[W_N/L_a\right] - \cosh\left[W_{SC}(0)/L_a\right]}{\sinh\left[(W_N + W_{NB})/L_a\right]} \right\} \tag{3.96}$$

All the stored charge in the N-base region is removed by the voltage transient when the space-charge region reaches through the N-base region, leaving a substantial stored charge only in the N-buffer layer. When the space-charge region reaches the edge of the N-buffer layer, the hole concentration at the buffer layer at this point becomes equal to the hole concentration (p_{SC}) inside the space-charge region, which is close to zero. Because the collector current density is held fixed during the voltage rise time, the hole concentration in the N-buffer layer at junction (J_1) changes abruptly when reach through occurs. The distribution of holes in the N-buffer layer is given by [12]:

$$p(y) = \frac{J_{C,ON} L_{pNB}}{qD_{pNB}} \left\{ \frac{\sinh\left[(W_{NBL} - y)/L_{pNB}\right]}{\cosh\left(W_{NBL}/L_{pNB}\right)} \right\} \tag{3.97}$$

During the next phase of the voltage rise time, the electric field in the N-base region increases with a punch-through distribution because of the high doping concentration of the N-buffer layer. A depletion region is formed in the N-buffer layer by a displacement current flow toward the collector contact. The collector voltage waveform after reach through is given by [12]:

$$V_C(t) = V_{RT}(J_{C,ON}) + \frac{J_{C,ON}}{C_{SCR}} \left[1 - \frac{1}{\cosh\left(W_{NBL}/L_{pNB}\right)} \right] t \tag{3.98}$$

The end of the voltage rise time is defined by the collector voltage becoming equal to the collector supply voltage (V_{CS}). The collector voltage rise time is given by [12]:

$$t_V = t_{RT} + \frac{\varepsilon_S}{J_{C,ON} W_N} \left[\frac{\cosh\left(W_{NBL}/L_{pNB}\right)}{\cosh\left(W_{NBL}/L_{pNB}\right) - 1} \right] \left[V_{C,S} - V_{RT}(J_{C,ON})\right] \tag{3.99}$$

After the collector voltage transient, there is stored charge in the N-buffer layer that decays by recombination under low-level injection conditions. The collector current transient is described by:

$$J_C(t) = J_{C,ON} e^{-t/\tau_{BL}} \tag{3.100}$$

where t_{BL} is the low-level lifetime in the N-buffer layer. This lifetime scales with the lifetime in the drift region.

In the case of the 15 kV asymmetric SiC n-channel IGBT structure with an N-base region width of 140 μm and a doping concentration of 2×10^{14} cm^{-3}, an N-buffer layer doping concentration of 1×10^{17} cm^{-3} and a thickness of 5 μm, and a P^+ collector region doping concentration of 1×10^{19} cm^{-3}, the collector voltage and current waveforms predicted by the above analytical model are shown in Fig. 3.69 for

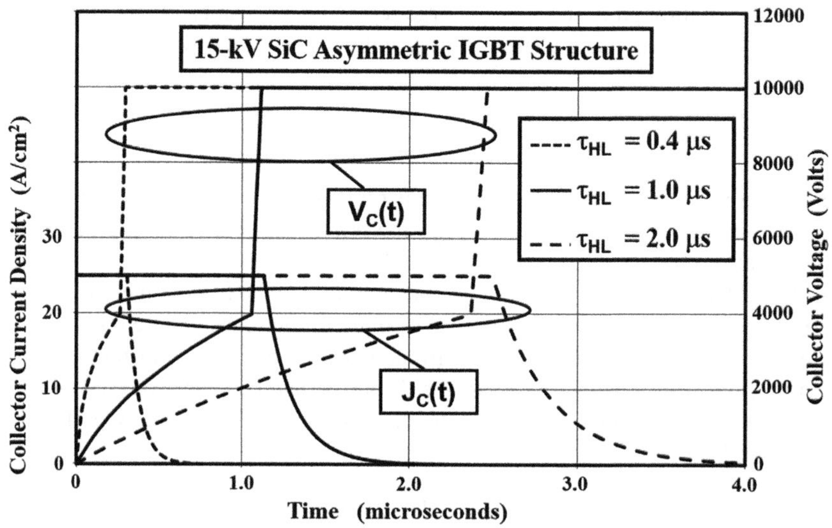

FIGURE 3.69

Collector voltage waveform for the 15 kV SiC asymmetric *n*-channel IGBT structure during inductive load turn-off.

three high-level lifetime values. A collector supply voltage of 10,000 V and an on-state collector current density of 25 A/cm^2 were assumed in this example. It can be observed that the collector voltage increases in a nonlinear manner until a reach-through voltage of 3980 V occurs, independent of the lifetime. The collector voltage then increases linearly with a high [dV/dt]. After the end of the collector voltage transient, the collector current decreases exponentially to zero at a faster rate when the lifetime is reduced.

3.5.5 Switching Energy Loss per Cycle

The turn-off losses in the IGBT during each cycle are associated with the voltage rise time interval and the current fall time interval. Due to the approximately linear increase in collector voltage, the energy loss during the voltage rise time interval until reach through can be computed using:

$$E_{OFF,V1} = \frac{1}{2} J_{C,ON} V_{RT} t_{RT} \tag{3.101}$$

Because the collector voltage increases very rapidly to the supply voltage after reach through, the energy loss during this time can be neglected. The energy loss during the collector current fall time interval can be computed using:

$$E_{OFF,J} = J_{C,ON} V_{C,S} \tau_{BL} \tag{3.102}$$

because of the exponential fall in collector current. The total energy loss per cycle is the sum of these two terms. These energy losses per cycle can be used to create a trade-off curve between on-state voltage drop and switching loss. These trade-off curves are compared with an optimum IGBT design later in the chapter. The energy losses for the optimum IGBT structure are also given by the same equations with the reach-through voltage being equal to the supply voltage.

3.6 Optimum SiC Asymmetric IGBT Structure

An abrupt increase in the collector voltage is observed for the SiC IGBT structure described in the previous section due to punch through of the space-charge region at voltages well below the supply voltage. This behavior is undesirable in applications for IGBTs. It is possible to avoid this problem by reengineering the drift region thickness and doping to obtain the needed forward blocking capability while allowing the space-charge region to punch through to the buffer layer when the collector voltage reaches the supply voltage [12]. In other words, the reach-through voltage is made equal to the collector supply voltage in this design of the asymmetric IGBT structure.

3.6.1 Optimum Structure Design

The reach-through voltage is a function of the width and doping concentration of the drift region as well as the on-state current density through the hole concentration in the space-charge region. Although Eqn (3.95) indicates that the reach-through voltage can be increased by solely increasing the doping concentration of the *N*-base region, this approach results in a reduction of the blocking voltage. In the case of the 15 kV SiC asymmetric IGBT device, the width and doping concentration of the *N*-base region must be optimized together to simultaneously obtain the desired open-base breakdown voltage of 16.5 kV and a reach-through voltage equal to a collector supply voltage of 10 kV.

The possible combinations of the thickness and doping concentrations of the *N*-base region are shown in Fig. 3.70 to achieve a blocking voltage of 16.5 kV based upon using a low-level lifetime of 0.5 μs in the drift region. An *N*-buffer layer doping concentration of 1×10^{17} cm^{-3} and a thickness of 5 μm were assumed here. The reach-through voltage can be computed by using Eqn (3.95) for each combination of the doping concentration and thickness of the drift region in Fig. 3.70. These reach-through voltages are plotted in Fig. 3.71 as a function of the drift region doping concentration. It can be seen that a reach-through voltage of 10 kV is obtained for a drift region doping concentration is 4.8×10^{14} cm $-$ ^{3}. From Fig. 3.70, an open-base breakdown voltage of 16.5 kV is obtained for this drift region doping concentration for a drift region thickness of 147 μm.

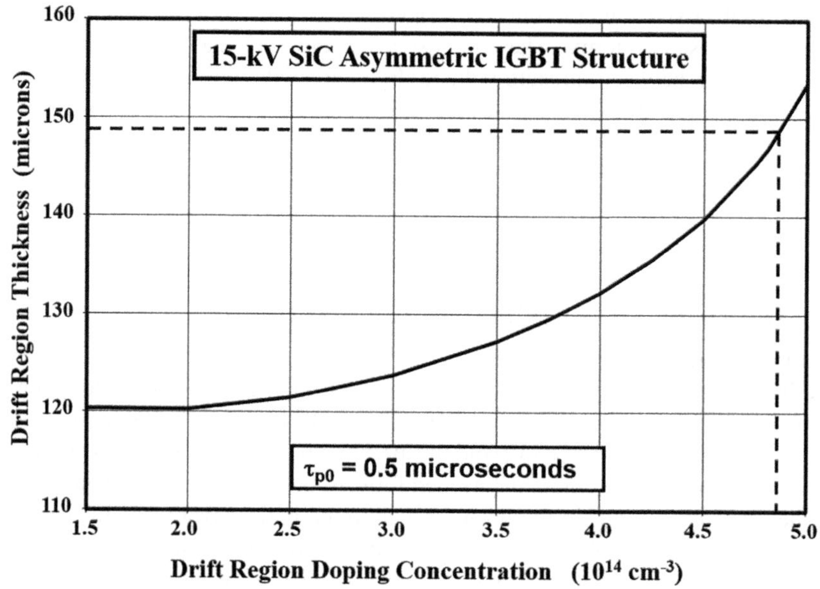

FIGURE 3.70

Optimization of the drift region width for the 15 kV asymmetric 4H-SiC *n*-channel IGBT structure.

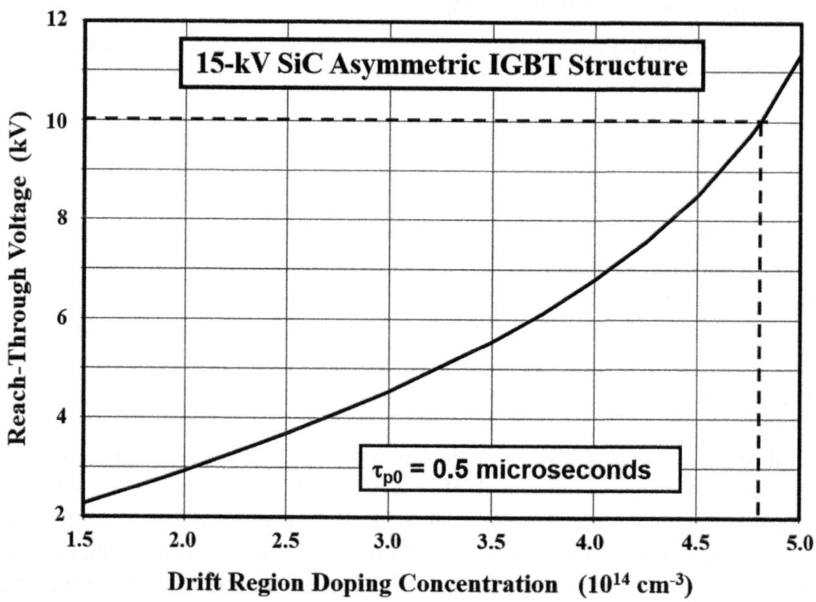

FIGURE 3.71

Reach-through voltages for the 15 kV asymmetric 4H-SiC *n*-channel IGBT structure.

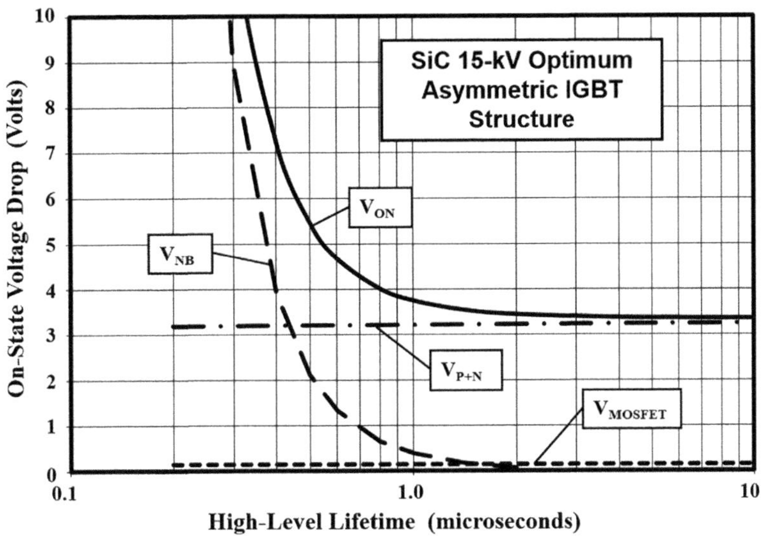

FIGURE 3.72

On-state voltage drop components for the optimum SiC 15kV asymmetric IGBT structure: lifetime dependence.

3.6.2 On-State Voltage Drop

The on-state voltage drop for the optimum SiC asymmetric IGBT structure can computed using the analytical model given in the previous section with the optimum thickness of 147 μm for the drift region with a doping concentration of 4.8×10^{14} cm^{-3}. The variation of the on-state voltage drop (at an on-state current density of 25 A/cm^2) with a high-level lifetime in the drift region is shown in Fig. 3.72 for the case of a buffer layer doping concentration of 1×10^{17} cm^{-3} and a thickness of 5 μm. The various components of the on-state voltage are also provided in the figure. It can be observed that the voltage drop in the drift region increases rapidly when the high-level lifetime is reduced below 0.6 μs. The on-state voltage drop for the optimum structure is slightly larger than that of the structure in the previous section at low values for the lifetime because of its larger drift region width.

The on-state voltage drop for the 15kV optimum asymmetric IGBT structure obtained by using an analytical model is shown in Fig. 3.73 as a function of the buffer layer doping concentration. A high-level lifetime of 1 μs was used for this example with a buffer layer thickness of 5 μm. The various components of the on-state voltage drop are also included in the figure. The voltage drop across the drift region increases rapidly once the buffer layer doping level exceeds 2×10^{17} cm^{-3}. The on-state voltage drop is larger for the optimum structure in comparison with the structure in the

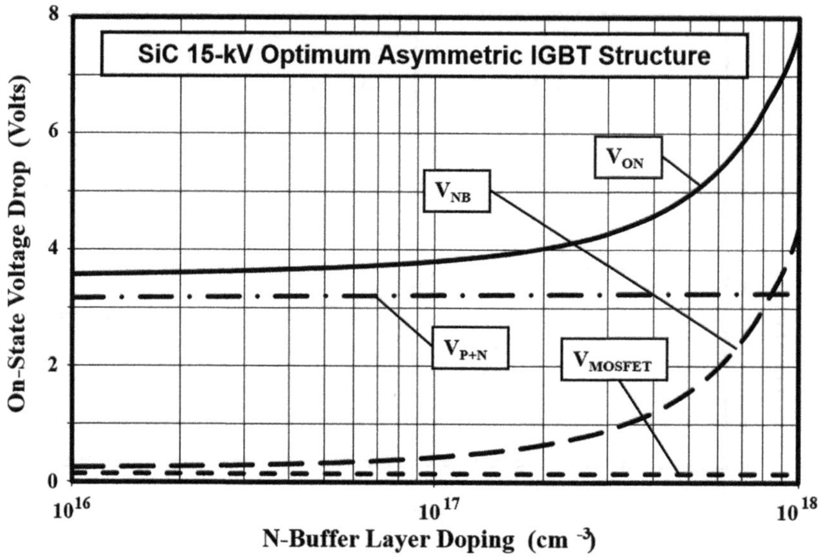

FIGURE 3.73

On-state voltage drop components for the SiC 15 kV optimum asymmetric IGBT structure: buffer layer doping dependence.

previous section. For instance, at a buffer layer doping concentration of 6×10^{17} cm^{-3}, the on-state voltage drop for the optimum structure is 10% larger than for the structure in the previous section.

3.6.3 Turn-Off Characteristics

In the optimum structure, the collector voltage increases monotonically to the collector supply voltage, followed by the collector current decaying exponentially to zero by the recombination of holes in the *N*-buffer layer. The rise in the collector voltage is described by the physics described in the previous section prior to punch through of the space-charge region. Consequently, the time at which the collector voltage transient is completed can be obtained using Eqn (3.96). The current fall occurs with the physics that govern the recombination of holes in the buffer layer, as discussed in the previous section. The on-state carrier distributions within the optimum structure are similar to those within the conventional structure shown in the previous section.

The collector voltage and current waveforms predicted by the above analytical model are provided in Fig. 3.74 for various values of the lifetime in the drift region for the case of the optimum 15 kV SiC asymmetric *n*-channel IGBT structure with an *N*-base region width of 147 μm and a doping concentration of 4.8×10^{14} cm^{-3}. An on-state collector current density of 25 A/cm^2 was used in this example. It can

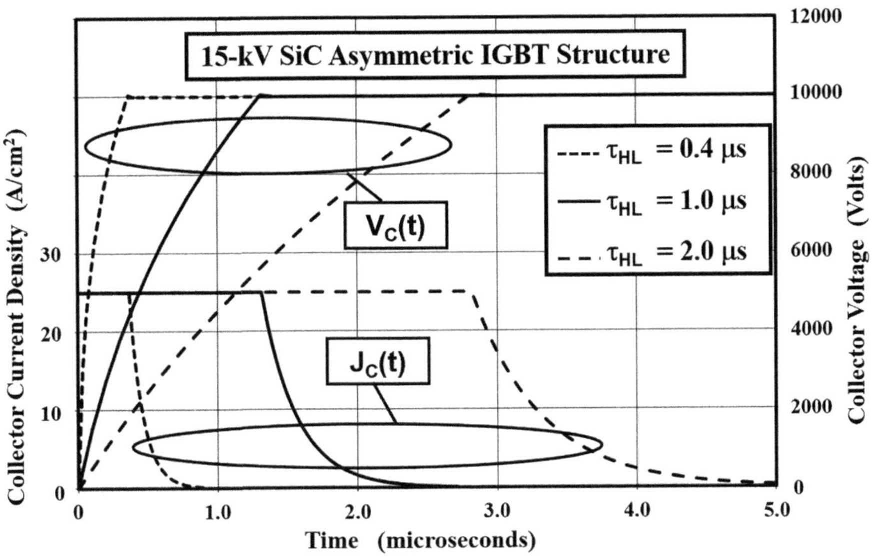

FIGURE 3.74

Waveforms for the collector current and voltage for the optimum 15 kV SiC asymmetric *n*-channel IGBT structure during inductive load turn-off: lifetime dependence.

be observed that the collector voltage increases monotonically until the reach-through time (t_{RT}) where it becomes equal to the collector supply voltage of 10,000 V. Consequently, there is no high [dV/dt] during the collector voltage transient. The collector current then decays exponentially to zero with the time constant determined by the lifetime in the buffer layer. In the case of this optimum IGBT structure, the time taken for the collector voltage to rise to the supply voltage is larger than for the structure discussed in the previous section (see Fig. 3.69). Consequently, the elimination of the high [dV/dt] during turn-off must be traded-off against larger switching losses.

The turn-off behavior for the optimum SiC asymmetric IGBT structure can also be controlled by changing the doping concentration and thickness in the buffer layer. This can be demonstrated using the example of the 15 kV SiC asymmetric optimum *n*-channel IGBT structure with an *N*-base region width of 147 μm and a doping concentration of 4.8×10^{14} cm^{-3}. The impact of changing the buffer layer doping concentration is provided in Fig. 3.75 for the optimum 15 kV SiC asymmetric IGBT structure with a buffer layer thickness of 5 μm and a high-level lifetime of 1 μs. A faster rise of the collector voltage is observed with increasing buffer layer doping concentration. This is due to the reduction of the injected carrier concentration in the drift region. The collector current also turns off faster with increasing buffer doping concentration, but this is due to a reduction of the buffer layer lifetime.

The impact of changing the buffer layer thickness is provided in Fig. 3.76 for the optimum 15 kV SiC asymmetric IGBT structure with a buffer layer doping

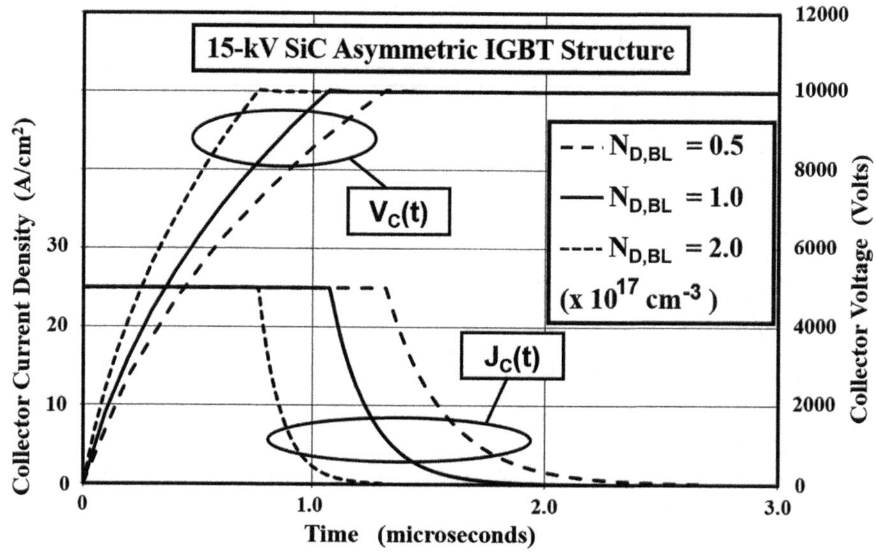

FIGURE 3.75

Waveforms for the collector current and voltage for the optimum 15 kV SiC asymmetric *n*-channel IGBT structure during inductive load turn-off: buffer layer doping dependence.

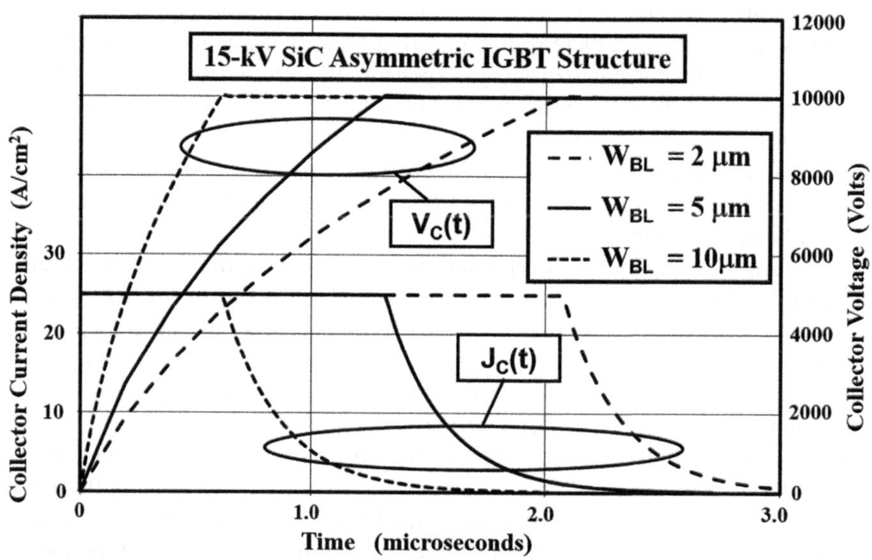

FIGURE 3.76

Waveforms for the collector current and voltage for the optimum 15 kV SiC asymmetric *n*-channel IGBT structure during inductive load turn-off: buffer layer thickness dependence.

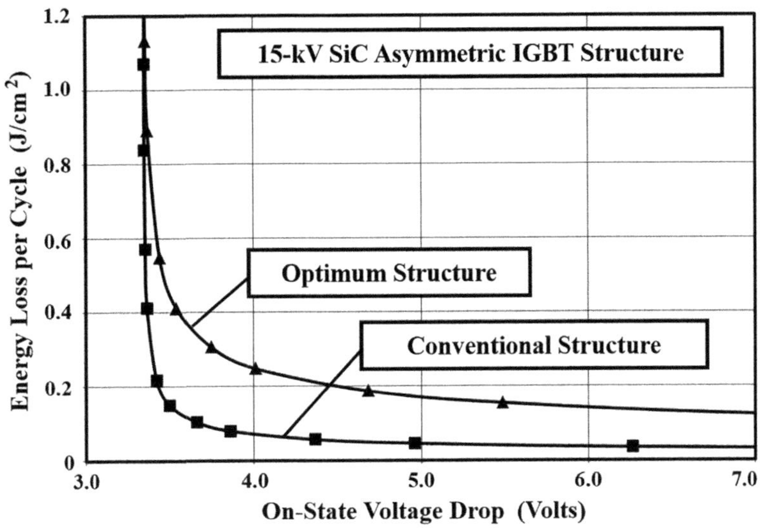

FIGURE 3.77

Trade-off curves for the 15 kV asymmetric *n*-channel 4H-SiC IGBT structures: lifetime dependence.

concentration of $5 \times 10^{16}\ \text{cm}^{-3}$ and a high-level lifetime of 1 μs. A faster turn-off occurs for the collector voltage with increasing buffer layer thickness due to the reduction of the injected carrier concentration in the drift region. However, the collector current fall time does not change because the lifetime in the buffer layer is the same for all three cases.

3.6.4 Power Loss Trade-Off Curves

As mentioned in the previous sections, the IGBT structure must be optimized by trading off the on-state voltage power loss and the switching power loss. This can be performed by making a trade-off curve between on-state voltage drop and the energy loss per switching cycle. The trade-off curves for the conventional and optimum SiC 15 kV asymmetric *n*-channel 4H-SiC IGBT structures obtained by varying the lifetime in the *N*-base region are provided in Fig. 3.77. Devices used in lower-frequency circuits would be chosen from the left side of the trade-off curve while devices used in higher-frequency circuits would be chosen from the right side of the trade-off curve. The trade-off curve for the optimum structure is worse than that for the conventional structure due to its larger on-state voltage drop and switching losses. It can be concluded that the elimination of the high $[dV/dt]$ during turn-off is obtained at the expense of higher power losses.

The trade-off curve for power losses in the 15 kV SiC asymmetric IGBT can also be performed by varying the buffer layer doping concentration. This is

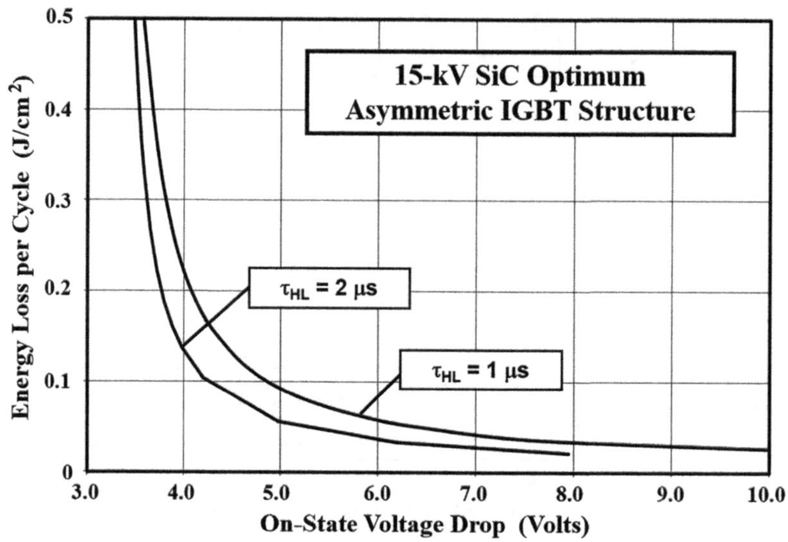

FIGURE 3.78

Trade-off curves for the 15 kV optimum asymmetric *n*-channel 4H-SiC IGBT structures: buffer layer doping dependence.

demonstrated in Fig. 3.78 for the optimum structure with a drift region thickness of 147 μm and a doping concentration of 4×10^{14} cm^{-3}. Two cases for the high-level lifetime in the drift region are considered in this figure. It can be observed that the trade-off curve is better for the case of a larger lifetime in the drift region. These structures require higher doping levels in the buffer layer to achieve the same turn-off losses.

3.6.5 Maximum Operating Frequency

The power dissipation during operation of the 15 kV asymmetric *n*-channel 4H-SiC IGBT structures can be obtained by using:

$$P_{D,TOTAL} = \delta P_{D,ON} + E_{OFF} f \tag{3.103}$$

where δ is the duty cycle and f is the switching frequency. The maximum operating frequency is limited by the power dissipation reaching a value decided by the packaging and heat sink. The maximum operating frequency obtained for the conventional and optimum structures is shown in Fig. 3.79 using a maximum total power dissipation of 200 W/cm^2. It can be concluded that the maximum operating frequency for the optimum structure is much lower than that for the conventional structure. Consequently, the elimination of the high [dV/dt] during the collector voltage transient is obtained at a significant disadvantage in terms of the highest operating frequency.

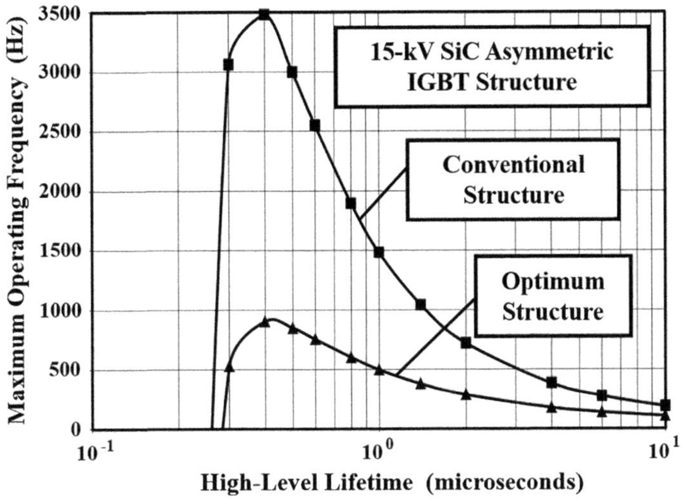

FIGURE 3.79

Maximum operating frequency for the 15 kV asymmetric *n*-channel 4H-SiC IGBT structures.

3.7 Summary

Although many structural variations of the IGBT have been proposed since its introduction in the early 1980s, they can be broadly classified into symmetric blocking and asymmetric blocking devices. In this chapter, it is demonstrated that the design of these structures begins by defining the parameters of the drift region to enable the device to support the desired voltage in the blocking mode. The thickness and doping concentration of the drift region of the IGBT are determined by open-base transistor breakdown physics. For the symmetric blocking device structure, a minimum drift region thickness occurs at an optimum doping concentration for the drift region. This produces the best on-state voltage drop and switching energy loss. For the asymmetric blocking device, the drift region thickness can be minimized by the proper choice of not only the drift region doping concentration but also the doping and thickness of the buffer layer. The trade-off curve between the on-state voltage drop and the switching energy loss per cycle can be optimized by adjusting the buffer layer properties and the lifetime in the drift region. This has been demonstrated by using the example of a 3000 V silicon device structure.

This chapter has also provided a review of the design of the SiC IGBT structure. Although the physics of operation of the SiC devices are similar to those of the silicon devices, the conventional approach to the design that is suitable for silicon structures produces a very high $[dV/dt]$ during the turn-off transient. It has been demonstrated that an optimum SiC device structure can be created without this high $[dV/dt]$ problem by reengineering the drift region to make the reach-through voltage equal to the collector supply voltage. However, the optimum SiC device structure is shown to have a significantly worse trade-off curve between on-state voltage drop and energy loss per switching cycle.

References

[1] B.J. Baliga, Fundamentals of Power Semiconductor Devices, second ed., Springer-Science, New York, 2019.

[2] B.J. Baliga, et al., The insulated gate rectifier: a new power switching device, in: IEEE International Electron Devices Meeting, Abstract 10.6, 1982, pp. 264–267.

[3] M. Takei, Y. Harada, K. Ueno, 600-V IGBT with reverse blocking capability, in: IEEE International Symposium on Power Semiconductor Devices and ICs, Abstract 11.1, 2001, pp. 413–416.

[4] H. Takahashi, M. Kaneda, T. Minato, 1200-V class reverse blocking IGBT for AC matrix converter, in: IEEE International Symposium on Power Semiconductor Devices and ICs, Abstract 3.1, 2004, pp. 121–124.

[5] T. Naito, et al., 1200-V reverse blocking IGBT with low loss for matrix converter, in: IEEE International Symposium on Power Semiconductor Devices and ICs, Abstract 3.2, 2004, pp. 125–128.

[6] N. Tokuda, M. Kaneda, T. Minato, An ultra-small isolation area for 600-V class reverse blocking IGBT with deep trench isolation process, in: IEEE International Symposium on Power Semiconductor Devices and ICs, Abstract 3.3, 2004, pp. 129–132.

[7] B.J. Baliga, et al., The insulated gate transistor: a new three-terminal MOS-controlled bipolar power device, IEEE Trans. Electron Devices ED-31 (1984) 821–828.

[8] A. Nakagawa, et al., Non-latch-up 1200-V, 75-A bipolar-mode MOSFET with large SOA, in: IEEE International Electron Devices Meeting, Abstract 16.8, 1984, pp. 860–861.

[9] H. Shigekane, H. Kirihata, Y. Uchida, Developments in modern high power semiconductor devices, in: IEEE International Symposium on Power Semiconductor Devices and ICs, 1993, pp. 16–21.

[10] G. Majumdar, Advanced IGBT technologies for HF operation, in: IEEE European Power Electronics Conference, 2009.

[11] G. Majumdar, T. Minato, Recent and future IGBT evolution, in: IEEE Power Conversion Conference, 2007, pp. 355–359.

[12] B.J. Baliga, Advanced High Voltage Power Device Concepts, Springer-Science, New York, 2011.

[13] F. Bauer, et al., A comparison of emitter concepts for high voltage IGBTs, in: IEEE International Symposium on Power Semiconductor Devices and ICs, 1995, pp. 230–235.

[14] F. Bauer, et al., Design considerations and characteristics of rugged punch-through (PT) IGBTs with 4.5 kV blocking capability, in: IEEE International Symposium on Power Semiconductor Devices and ICs, 1996, pp. 327–330.

[15] T. Laska, et al., The field-stop IGBT (FS-IGBT): a new power device concept with great improvement potential, in: IEEE International Symposium on Power Semiconductor Devices and ICs, 2000, pp. 355–358.

[16] M. Otsuki, et al., A study of the short-circuit capability of field-stop IGBTs, IEEE Trans. Electron Devices 50 (2003) 1525–1531.

[17] T. Matsudai, et al., New anode design concept of 600-V thin wafer PT-IGBT with very low dose P-buffer and transparent P-emitter, IEEE Proc. Circuits Syst. 151 (2004) 255–258.
[18] B.J. Baliga, Semiconductors for high voltage vertical channel field effect transistors, J. Appl. Phys. 53 (1982) 1759–1764.
[19] B.J. Baliga, Gallium Nitride and Silicon Carbide Power Devices, World Scientific Press, 2017.
[20] M. Bhatnagar, P.M. McLarty, B.J. Baliga, Silicon carbide high voltage (400V) Schottky barrier diodes, IEEE Electron Device Lett. EDL-13 (1992) 501–503.
[21] B.J. Baliga, The pinch rectifier: a low forward drop high speed power diode, IEEE Electron Device Lett. EDL-5 (1984) 194–196.
[22] R. Held, N. Kaminski, E. Niemann, SiC merged p-n/Schottky rectifiers for high voltage applications. Silicon carbide and related materials—1997, Mater. Sci. Forum 264–268 (1998) 1057–1060.
[23] F. Dahlquist, et al., Junction barrier Schottky diodes in 4H-SiC and 6H-SiC. Silicon carbide and related materials—2001, Mater. Sci. Forum 389–393 (2002) 1129–1132.
[24] J.N. Shenoy, J.A. Cooper, M.R. Melloch, High voltage double-implanted power MOSFETs in 6H-SiC, IEEE Electron Device Lett. 18 (1997) 93–95.
[25] A.K. Agarwal, et al., 1.1-kV 4H-SiC power UMOSFETs, IEEE Electron Device Lett. 18 (1997) 586–588.
[26] P.M. Shenoy, B.J. Baliga, The planar 6H-SiC ACCUFET, IEEE Electron Device Lett. 18 (1997) 589–591.
[27] Cree Z-fet™ SiC MOSFET and Z-rec™ SiC Schottky Diodes. www.cree.com.
[28] SiC Schottky Rectifiers. www.genesicsemi.com.
[29] SiC Schottky Barrier Diodes and Power MOSFETs. www.rohm.com.
[30] Q. Zhang, et al., 10-kV trench gate IGBTs on 4H-SiC, in: IEEE International Symposium on Power Semiconductor Devices and ICs, 2005, pp. 159–162.
[31] Q. Zhang, et al., 9-kV 4H-SiC IGBTs with 88 mΩ-cm^2 of $R_{diff,on}$. Silicon carbide and related materials—2006, Mater. Sci. Forum 556–557 (2007) 771–774.
[32] Q. Huang, et al., 12-kV p-channel IGBTs with low on-resistance in 4H-SiC, IEEE Electron Device Lett. 29 (2008) 1027–1029.
[33] M.K. Das, et al., A 13-kV 4H-SiC N-Channel IGBT with Low $R_{diff,on}$ and Fast Switching. Silicon Carbide and Related Materials, 2007.
[34] W. Sung, et al., Design and investigation of frequency capability of 15-kV 4H-SiC IGBT, in: IEEE International Symposium on Power Semiconductor Devices and ICs, 2009, pp. 271–274.
[35] S.-H. Ryu, et al., Ultra-high voltage (>12 kV), high performance 4H-SiC IGBTs, in: IEEE International Symposium on Power Semiconductor Devices and ICs, 2012, pp. 257–260.
[36] R.J. Callanan, et al., Recent progress in SiC DMOSFETs and JBS diodes at CREE, in: IEEE Industrial Electronics Conference, 2008, pp. 2885–2890.
[37] T. Tamaki, et al., Optimization of on-state and switching performances for 15–20 kV 4H-SiC IGBTs, IEEE Trans. Electron Devices 55 (2008) 1920–1927.
[38] B.J. Baliga, M.S. Adler, Measurement of carrier lifetime profiles in diffused layers of semiconductors, IEEE Trans. Electron Devices ED-25 (1978) 472–477.

CHAPTER

Safe Operating Area Design

4

One of the important reasons for the great success of the insulated gate bipolar transistor (IGBT) in a wide variety of applications has been its large safe operating area (SOA). This has provided the device with sufficient ruggedness to be operated without the need for expensive and bulky snubber circuits that were previously required for bipolar power transistors and gate turn-off thyristors.

During the early period of IGBT development, it was believed that the presence of the internal parasitic thyristor within the structure would be a show stopper. Critics of the IGBT expected the parasitic thyristor to latch-up at higher current levels and during rapid switching events, producing destructive failure. This issue was recognized from the very beginning of the development of the IGBT, resulting in an intensive research effort for a period of several years at the General Electric Company to suppress the parasitic thyristor. The success of this research effort, the innovative structural improvements proposed at that time (and since then by other researchers) has been critical to making the IGBT a highly successful product. Today, it is taken for granted that IGBT products are not prone to failure by latch-up of the parasitic thyristor. It is worth pointing out that the very first patent application submitted by General Electric on the IGBT had structures that did not contain a parasitic thyristor [1]. However, the process complexity of achieving the proposed tunneling of carriers from the emitter metal contact to the inversion channel prevented the realization of practical devices in the 1980s. Tunneling currents from metal contacts to the channel in MOSFETs have more recently been demonstrated in Schottky source/drain structures [2].

All IGBT structures being manufactured at this time contain a parasitic thyristor within the structure. Approaches to suppressing the parasitic thyristor in the IGBT structure are described in this chapter. Modern IGBT devices exhibit excellent forward and reverse biased SOA. The significant difference in the SOA of n-channel and p-channel IGBTs is described and design approaches to improving the SOA for p-channel IGBTs are provided. In addition, modern IGBT products have very good short circuit handling capability. These features have made the IGBT the device of choice for hard switching topologies that are widely used for motor control and other applications.

The IGBT Device. https://doi.org/10.1016/B978-0-323-99912-0.00014-3

4.1 Parasitic Thyristor

The n-channel IGBT structure is illustrated in Fig. 4.1 on the left side with the parasitic thyristor indicated by the dashed lines. The parasitic thyristor is a four-layer semiconductor device formed by the P^+ collector region/N-drift region/P-base region/N^+ emitter region. The equivalent circuit for the IGBT is shown on the right side of this figure with the internal NPN and PNP transistors, which taken together constitute the parasitic thyristor. These transistors are coupled together with regenerative feedback, producing latch-up of the parasitic thyristor. Once the thyristor latches up, current flow is sustained by the internal positive feedback between the two bipolar transistors. The current flow is then no longer controlled by the gate bias. This usually leads to destructive failure of the device.

In the IGBT structure, the emitter metal is designed to intentionally overlap the junction J_3 formed between the N^+ emitter diffusion and the P-base region via the deep P^+ region. Consequently, the junction J_3 is shorted at the emitter contact. However, the NPN transistor can still become active if the resistance of the P-base region under the N^+ emitter diffusion is sufficiently large. This resistance is represented by the shunting resistance R_S in the equivalent circuit. If the shunting resistance can be made very small, for instance, by including the P^+ diffusion shown in the device structure, the NPN transistor can be prevented from turning on, allowing suppression of the parasitic thyristor [3].

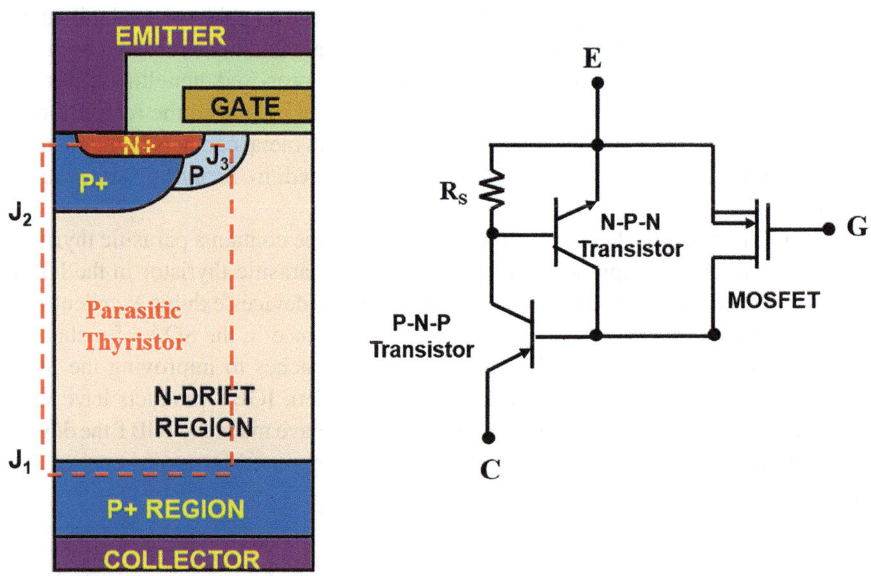

FIGURE 4.1

The insulated gate bipolar transistor structure with parasitic thyristor highlighted.

4.2 Suppressing the Parasitic Thyristor

As described in the previous chapters, the IGBT structure operates as a wide-base bipolar transistor being controlled by the integrated MOSFET structure to provide its base current. When the parasitic thyristor latches up, the current flows vertically through the device, bypassing the MOSFET channel, resulting in a loss of gate control. The voltage drop across the latched-up thyristor is smaller than that observed in the IGBT structure, producing a snapback in the *i*–*v* characteristic, as illustrated in Fig. 4.2. The reduction in the voltage drop across the IGBT structure after the onset of latch-up can produce a large jump in the current flow, resulting in destructive failure of the device. It is therefore necessary to prevent latch-up of the parasitic thyristor under all operating conditions. In general, the latch-up current density ($J_{C,L}$) must be at least 10 times larger than the on-state current density ($J_{C,ON}$) to obtain a broad range of operating current levels in the IGBT mode.

The current gain of the coupled NPN and PNP bipolar transistors within the IGBT structure must exceed unity to produce latch-up of the thyristor [4]. The latch-up of the parasitic thyristor can be suppressed by reducing the current gain of either of these transistors. However, the PNP wide-base transistor conducts a portion of the on-state current. Consequently, the on-state voltage drop increases when the gain of the PNP transistor is reduced. It is preferable to reduce the current gain of the NPN transistor within the n-channel IGBT structure. The approaches that have a strong impact on latch-up suppression are discussed in this section.

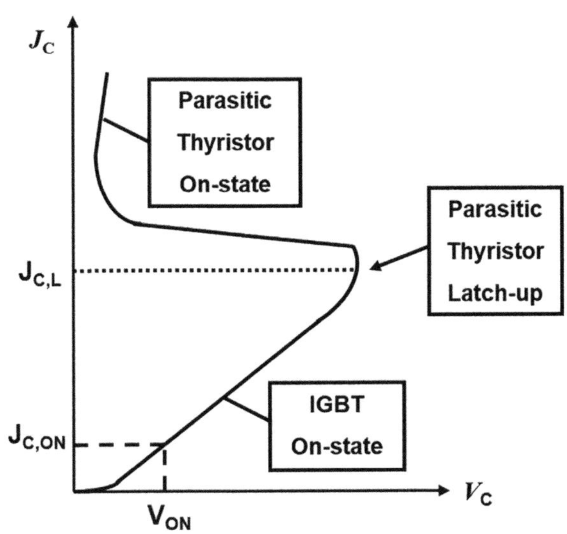

FIGURE 4.2

On-state characteristics observed in the insulated gate bipolar transistor (IGBT) structure with latch-up of the parasitic thyristor.

4.2.1 Deep P^+ Diffusion

The most effective approach to suppressing the latch-up of the parasitic thyristor in the IGBT structure is by reducing the resistance of the P-base region under the N^+ emitter region. This was accomplished by the inclusion of a deep P^+ diffusion to reduce the sheet resistance of the P-base region under the N^+ region without changing the doping in the channel of the MOSFET portion [5]. Advances in device fabrication lithography have allowed making "fine pattern" IGBTs with a much smaller cell size and width of the emitter region (L_p), as shown in Fig. 4.3. This approach can suppress the parasitic thyristor without the use of a deep P^+ region [6,7].

The latch-up current density for the IGBT structure with the deep P^+ region can be analyzed by using the current flow, as shown in Fig. 4.3. The hole current component (I_p) flows through the lightly doped P-base region with resistance R_{B1} and through the deep P^+ region with resistance R_{B2}. These resistances can be computed using:

$$R_{B1} = \frac{\rho_P L_P}{(x_{J,P} - x_{J,N+})Z} = \frac{\rho_{SP} L_P}{Z} \tag{4.1}$$

and

$$R_{B2} = \frac{\rho_{SP+} L_{P+}}{Z} = \frac{\rho_{P+} L_{P+}}{(x_{J,P+} - x_{J,N+})Z} = \frac{L_{P+}}{q\mu_{pP+} N_{AP+}(x_{J,P+} - x_{J,N+})Z} \tag{4.2}$$

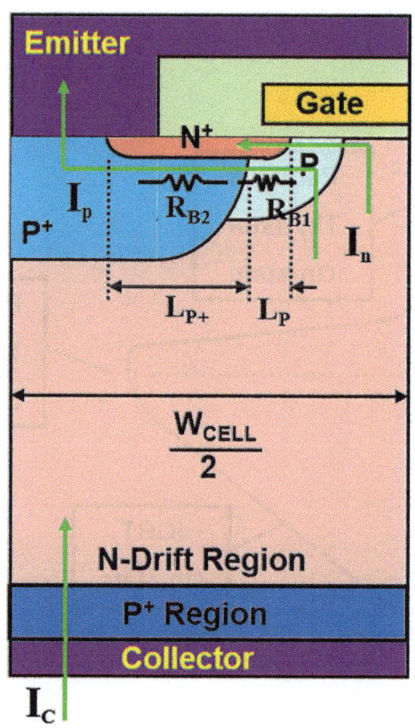

FIGURE 4.3

Insulated gate bipolar transistor structures with the deep P^+ region.

where ρ_P and ρ_{SP} are the resistivity and sheet resistance of the P-base region, respectively; ρ_{P+} and ρ_{SP+} are the resistivity and sheet resistance of the deep P^+ region, respectively; and Z is the length of the IGBT cell orthogonal to the cross-section in the figure.

The NPN transistor becomes active, that is, its emitter begins to inject electrons, when the emitter-base potential exceeds the built-in voltage of the junction (~0.7 V for silicon devices). Consequently, latch-up of the parasitic thyristor occurs when the voltage drop in the P-base region produced by the hole current flow becomes equal to the built-in potential of the P–N junction:

$$I_p(R_{B1} + R_{B2}) = V_{bi} \tag{4.3}$$

The hole current in this equation can be calculated by taking the product of the hole current density and the cell area ($W_{CELL}.Z$). In the on-state, the hole current density is given by:

$$J_p = \alpha_{PNP,ON} J_C \tag{4.4}$$

where $\alpha_{PNP,ON}$ is the common base current gain of the PNP transistor in the on-state with high-level injection conditions. For the asymmetric IGBT structure, the current gain is determined by the injection efficiency at the collector/buffer layer junction J_1 and the base transport factors of the buffer layer and the N-base region [4]:

$$\alpha_{PNP,ON} = \gamma_E \alpha_{T,N-Buffer} \alpha_{T,N-Base,0} \tag{4.5}$$

A typical value for the current gain in the on-state is 0.5. Combining the above relationships yields the latch-up current density:

$$J_{C,L} = \frac{V_{bi}}{\alpha_{PNP,ON}(\rho_{SP} L_P + \rho_{SP+} L_{P+}) W_{CELL}} \tag{4.6}$$

The latch-up current density obtained by using the above equation is plotted in Fig. 4.4 for the asymmetric IGBT structure. For this graph, typical values for the pinch sheet resistance for the P^+ region and for the P-base region of 50 Ω/sq. and 2000 Ω/sq., respectively, were used. The latch-up current density is strongly dependent on the length (L_P) of the N^+ emitter region that extends beyond the lateral diffusion of the P^+ region. This behavior was confirmed in early experiments to suppress latch-up of the parasitic thyristor in IGBT devices [5].

It is worth pointing out that the lateral diffusion of the P^+ region can be extended until its doping becomes comparable to the peak P-base doping concentration, which is further than the junction depth of the P^+ region in the vertical direction. It is possible to make the length (L_P) of the N^+ emitter region that extends beyond the lateral diffusion of the P^+ region as small as 1 μm to achieve a latch-up current density of 8000 A/cm^2, which is far larger than the typical on-state current density of 100 A/cm^2. Unfortunately, the latch-up current density reduces with increasing temperature because of an increase in the current gain and resistance of the P-base and P^+ regions, and a reduction in the built-in potential.

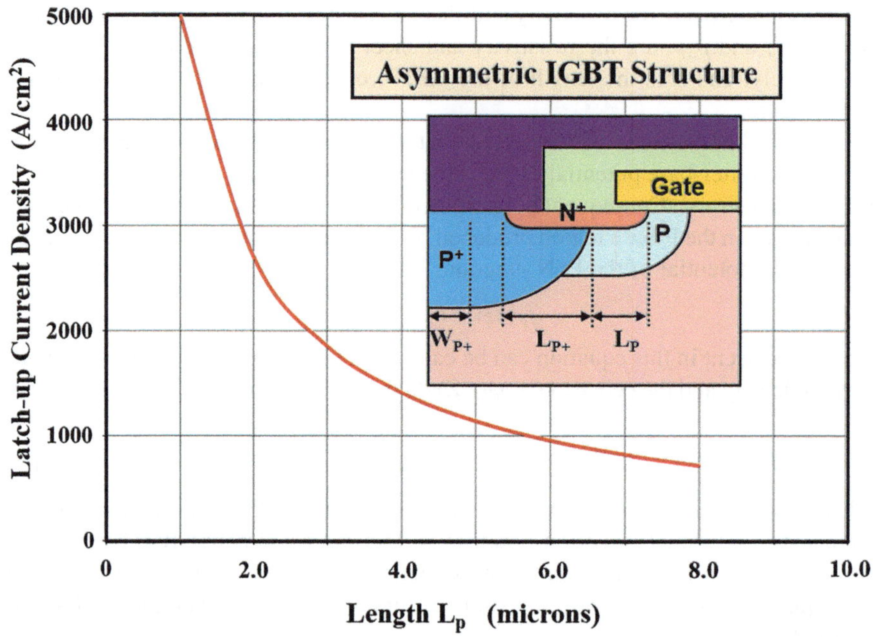

FIGURE 4.4

Latch-up current density for the n-channel insulated gate bipolar transistor (IGBT) structure.

4.2.2 Reducing Gate Oxide Thickness

In the previous section, it was demonstrated that the latch-up current density is strongly dependent on the length of the N^+ emitter over the P-base region in the IGBT structure. It is, therefore, desirable to reduce the sheet resistance of the P-base region by increasing its doping concentration. If the gate oxide thickness is fixed, this would lead to a large threshold voltage. However, the increase in threshold voltage can be compensated for by simultaneously reducing the gate oxide thickness. A reduced gate oxide thickness provides the added benefit of reducing the on-state voltage drop of the IGBT structure.

The threshold voltage for the IGBT structure is given by:

$$V_{TH} = \frac{t_{OX}\sqrt{4\varepsilon_S kTN_A \ln(N_A/n_i)}}{\varepsilon_{OX}} + \frac{2kT}{q}\ln\left(\frac{N_A}{n_i}\right) - \frac{Q_F}{C_{OX}} - WF \tag{4.7}$$

where t_{OX} is the gate oxide thickness, N_A is the peak doping concentration of the P-base region, Q_F is the fixed oxide charge density, and WF is the work function difference between the gate electrode and the P-base region. Using this equation, the peak P-base doping concentration required to achieve a fixed threshold voltage of 2 V is plotted in Fig. 4.5 as a function of the gate oxide thickness. It is clear that a substantial increase in the peak P-base doping concentration becomes possible by reducing the gate oxide thickness. For example, reducing the gate oxide thickness

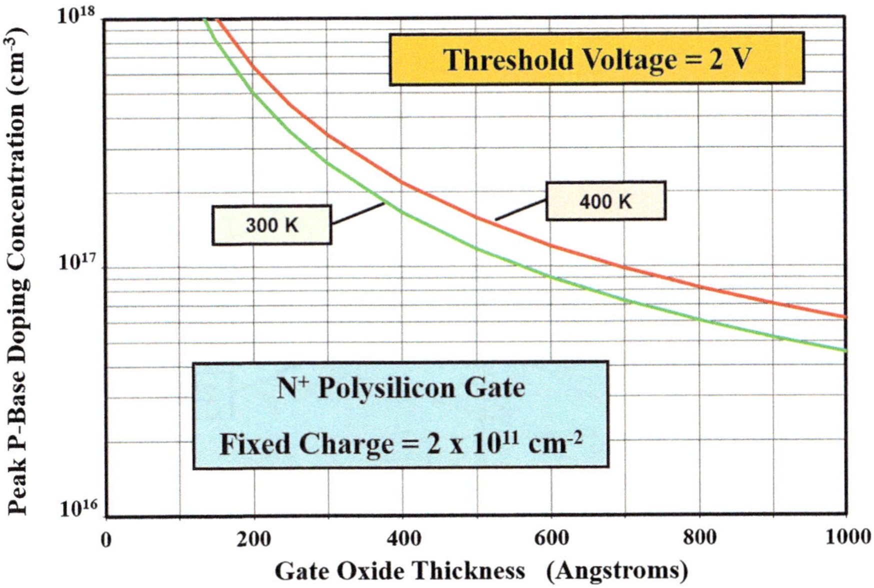

FIGURE 4.5

Dependence of P-base doping on gate oxide thickness to achieve a fixed threshold voltage of 2 V.

from 1000 to 500 Å allows increasing the peak doping concentration from 4.5×10^{16} to 1.2×10^{17} cm^{-3}. It should be noted that the sheet resistance of the P-base region does not reduce in inverse proportion to doping concentration because of a reduction in the mobility with increased doping concentration.

The threshold voltage computed by using Eqn (4.7) is plotted in Fig. 4.6 as a function of the square root of the P-base doping concentration for three values for the gate oxide thickness. It is apparent that the threshold voltage increases linearly with the square root of the P-base doping concentration in all three cases. From these plots, it can be concluded that the first term in the equation is dominant, allowing computation of the threshold voltage by using:

$$V_{TH} = \frac{t_{OX}}{\varepsilon_{OX}} \sqrt{4\varepsilon_S kTN_A \ln\left(\frac{N_A}{n_i}\right)} \tag{4.8}$$

Based upon this equation, the threshold voltage increases linearly with increasing oxide thickness and approximately as the square root of the doping concentration in the semiconductor. Consequently, a reduction of the gate oxide thickness by a factor of two allows increasing the doping concentration of the P-base region by a factor of four while maintaining the same threshold voltage. This should produce an increase in the latch-up current density by a factor of four as well. In practice, the improvement in latch-up current density is smaller due to a reduction of the mobility with increasing doping level.

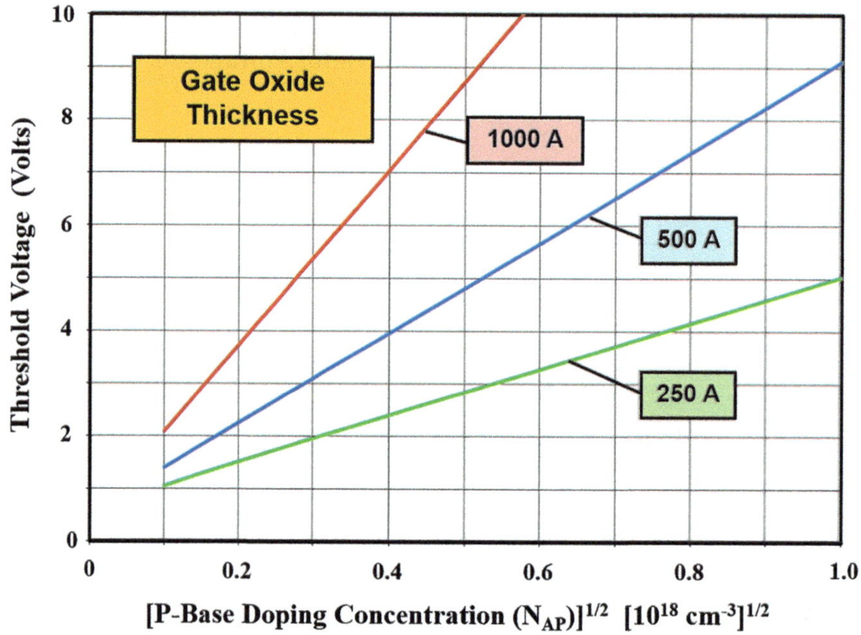

FIGURE 4.6

Dependence of threshold voltage on P-base doping concentration.

The improvement in the latch-up current density that can be achieved by scaling down the gate oxide thickness while maintaining a fixed threshold voltage is illustrated in Fig. 4.7 for the case of the symmetric IGBT structure. This structure has no deep P^+ region, as illustrated in the inset in the figure. This structure has an N-base with a width of 200 μm and a high-level lifetime of 20 μs, resulting in a common base current gain of 0.4. The P-base doping concentration used for the computation is based on using Eqn (4.7) to obtain the maximum P-base doping concentration in the P-base region and dividing this by a factor of two to obtain the effective doping concentration in the P-base region. It can be observed from this figure that the latch-up current density can be increased from about 300 to 1500 A/cm^2 by scaling the gate oxide down from 1000 to 250 Å. The improvements projected by the analytical model were experimentally confirmed during the early stages of IGBT development, providing a strong impetus to scale the gate oxide down in products [8].

The channel resistance in the MOSFET portion of the IGBT structure becomes smaller with a reduction of the gate oxide thickness, producing a reduction in the on-state voltage drop. As an example, the on-state characteristics obtained by using the PiN/MOSFET model for the symmetric IGBT structure are shown in Fig. 4.8 for three cases of gate oxide thickness. This structure has an N-base region with a width of 200 μm, a channel length of 1.5 μm, and a high-level lifetime of 20 μs in the drift region. From this figure, it can be concluded that the on-state voltage drop at an on-state current density of 100 A/cm^2 improves from 1.26 to 1.00 V when the gate oxide

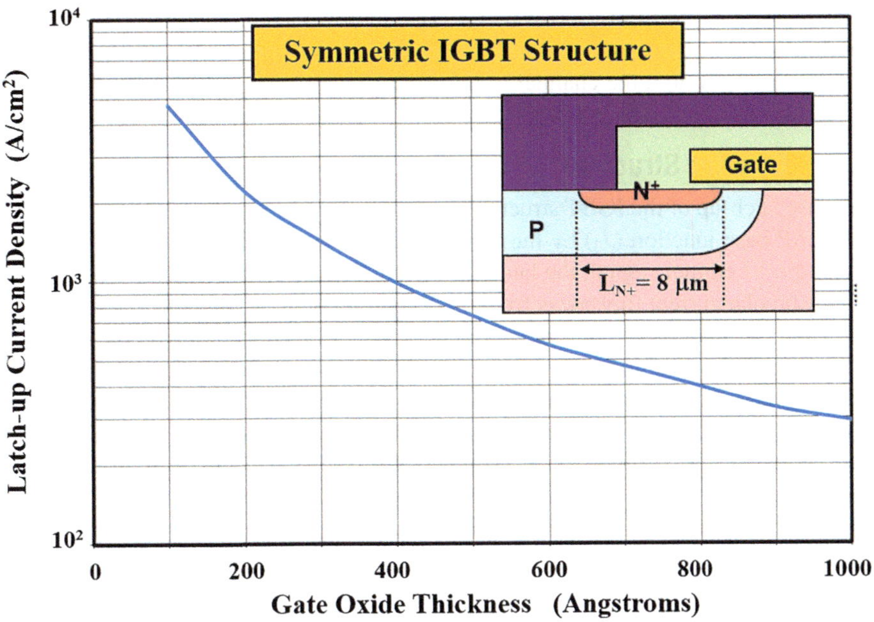

FIGURE 4.7

Latch-up current density for insulated gate bipolar transistor structures with different gate oxide thicknesses.

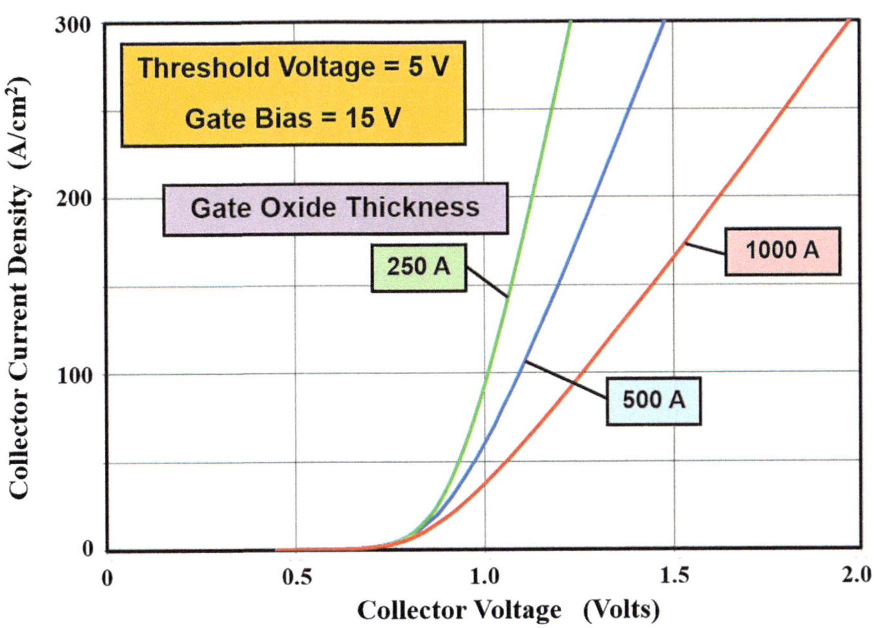

FIGURE 4.8

On-state characteristics of insulated gate bipolar transistor structures with different gate oxide thicknesses.

thickness is reduced from 1000 to 250 Å. The improvement is even greater when the lifetime in the N-base region is reduced because a larger proportion of the collector current flows through the channel in these cases.

4.2.3 Diverter Structure

Because latch-up of the IGBT structure is produced by the forward biasing of the N^+ emitter/P-base junction (J_3) by the flow of the bipolar current in the P-base region under the N^+ emitter region, the latch-up current density can be increased by diverting the bipolar current flow away from the N^+ emitter region. This can be achieved by diverting the bipolar current flowing in the vicinity of the channel with the incorporation of a diverter region [9] near the channel. A cross-section of the structure with the P^+ diverter region is shown in Fig. 4.9. It is worth pointing out that the same metal layer can be used to simultaneously make contact to the N^+ emitter, P-base, and P^+ diverter regions. This allows using a single large area metal electrode at the top of the device (as in the case of the basic IGBT structure), eliminating any fine line patterning of the thick contact metal for the emitter.

A simple approach for the analysis of the latch-up current density of the IGBT structure in the presence of the diverter region is by assuming that the fraction of the bipolar current (I_{P2}) that flows into the diverter is proportional to the area in the cell on the right side of the middle of the gate electrode [4]. Using this current

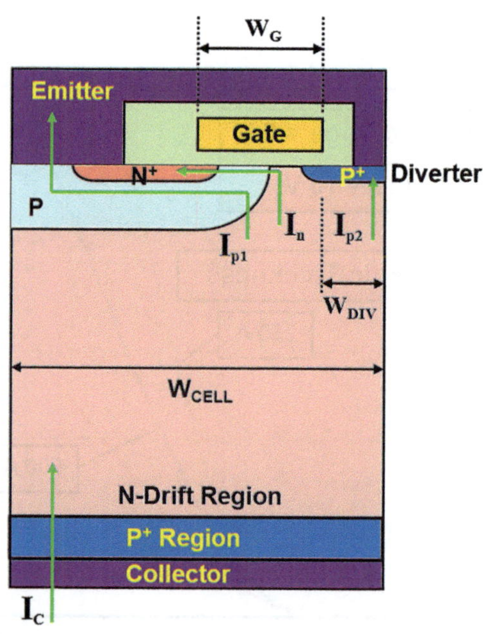

FIGURE 4.9

Insulated gate bipolar transistor structure with the diverter region.

as the basis for creating the forward bias across the N^+ emitter/P-base junction yields the following latch-up current density:

$$J_{C,L}(Div) = \left[\frac{W_{CELL}}{W_{CELL} - W_{DIV} - (W_G/2)}\right] \frac{V_{bi}}{\alpha_{PNP,ON}\rho_{SP}L_{N+W_{CELL}}} \tag{4.9}$$

As an example, using a 1 μm deep P^+ diverter region with a 2 μm wide diverter window (W_{DIV}) in the IGBT structure with a cell width (W_{CELL}) of 16 μm, the latch-up current density can be increased by about 50%. This has been experimentally confirmed [9].

The introduction of the diverter into the IGBT structure produces an increase in the on-state voltage drop because of the constriction of the channel current flow. The presence of the P^+ diverter region reduces the area for spreading the current from the channel into the drift region. For this reason, it is important to maintain a shallow junction depth for the P^+ diverter region.

4.2.4 Cell Topology

The latch-up current density of the IGBT is also impacted by the pattern used for the polysilicon gate electrode. Many topologies for the polysilicon gate layer can be utilized for the IGBT structure, as illustrated in Fig. 4.10. The topology shown in Fig. 4.10(a) with a linear window in a linear array was used for the analyses described in all the previous sections of this chapter. It will therefore be used as a benchmark for comparison of other polysilicon gate patterns.

4.2.4.1 Square window in a square array

From a layout design perspective, the use of a square-shaped polysilicon window in a square array provides a simple design approach for the IGBT structure. For this reason, it was utilized for the very first IGBT devices reported in the literature [10]. This layout significantly degrades the latch-up current density for two reasons. First, the bipolar current that is responsible for promoting latch-up is collected from a greater area at the corners of the polysilicon window, as illustrated in Fig. 4.11. In this figure, the dashed lines represent the edges of the diffusion windows (polysilicon for the N^+ emitter and P-base region, and a photoresist mask for the deep P^+ diffusion). Second, the length (L_P) of the P-base region under the N^+ emitter (see inset cross-section in Fig. 4.4) beyond the edge of the deep P^+ region is larger at the corners because the deep P^+ region is also formed using a smaller square window inside the polysilicon window, as illustrated in Fig. 4.11.

The degradation of the latch-up current density in comparison with the linear cell in a linear array can be analyzed by taking the above two factors into account [4]:

$$\frac{J_{C,L}(SquareCell, SquareArray)}{J_{C,L}(LinearCell, LinearArray)} = f_{LP} f_A \tag{4.10}$$

The length factor obtained using geometrical analysis is given by:

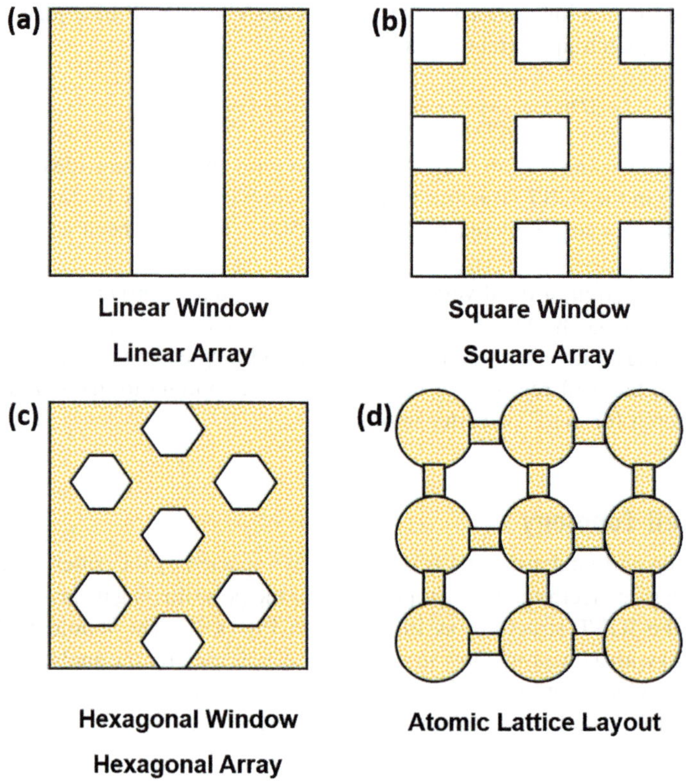

FIGURE 4.10

Cell topologies for the insulated gate bipolar transistor structure.

$$f_{LP} = \frac{L_{PE}}{L_{PC}} = \frac{(W_{POLY} - W_{P+}) - (x_{JP+} - x_{JN+})}{\sqrt{2}(W_{POLY} - W_{P+}) - (x_{JP+} - x_{JN+})} \tag{4.11}$$

The area factor (f_A) can be obtained by assuming that the hole current from the corner of the cell flows into a sector with a radius equal to the average of the junction depths of the N^+ and deep P^+ regions:

$$f_A = \frac{\pi(x_{JP+} + x_{JN+})W_{POLY}}{4W_G^2} \tag{4.12}$$

For typical cell dimensions of 8 μm for the polysilicon window (W_{POLY}), 8 μm for the polysilicon width (W_G), and junction depths of 1 and 5 μm for the N^+ and deep P^+ diffusions, respectively, the length factor is 0.45 while the area factor is 0.59. This leads to a net reduction of the latch-up current density by a factor of 3.8 times. Based upon this analysis, it can be concluded that the topology with a square cell in a square array is highly detrimental to the performance of the IGBT structure. This has been experimentally confirmed in the literature [11].

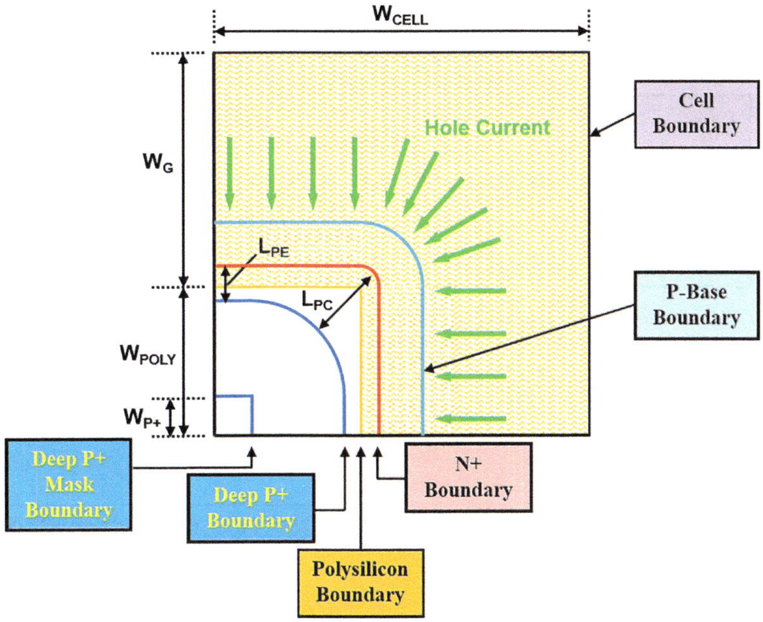

FIGURE 4.11

Distribution of the bipolar current in the insulated gate bipolar transistor structure with square polysilicon window in a square array.

4.2.4.2 Circular window in a hexagonal array

The hexagonal-shaped polysilicon window in a hexagonal array (see Fig. 4.10(c)) has been used for power MOSFETs. These layout topologies can be analyzed by using the hole current distribution shown in Fig. 4.12 using circular symmetry. In this figure, the dashed lines represent the edges of the diffusion windows (polysilicon for the N^+ emitter and P-base region, and a photoresist mask for the deep P^+ diffusion).

This polysilicon topology has the advantage of a uniform current distribution and equal length (L_P) of the P-base region under the N^+ emitter beyond the edge of the deep P^+ region. However, the latch-up current density for this topology is worse than that for the linear cell because the resistance of the bipolar current path is greater due to the circular geometry. Using cylindrical coordinates appropriate for the circular cell geometry, an expression for the latch-up current density can be derived [4]:

$$J_{C,L}(CC, CA) = \frac{2V_{bi}}{\alpha_{PNP,ON}\rho_{SP}W_{CELL}^2 \ln[(W_{POLY} + x_{JN+})/(W_{P+} + x_{JP+})]} \tag{4.13}$$

Using the same device physical parameters as in the linear cell topology, namely junction depths of the deep P^+ region ($x_{J,P+}$) and N^+ region ($x_{J,N+}$) of 5 and 1 μm, respectively; a pinch sheet resistance for the P-base region of 2150 Ω/sq.; a 16 μm cell radius (W_{CELL}) with a polysilicon window radius of 8 μm; and a diffusion

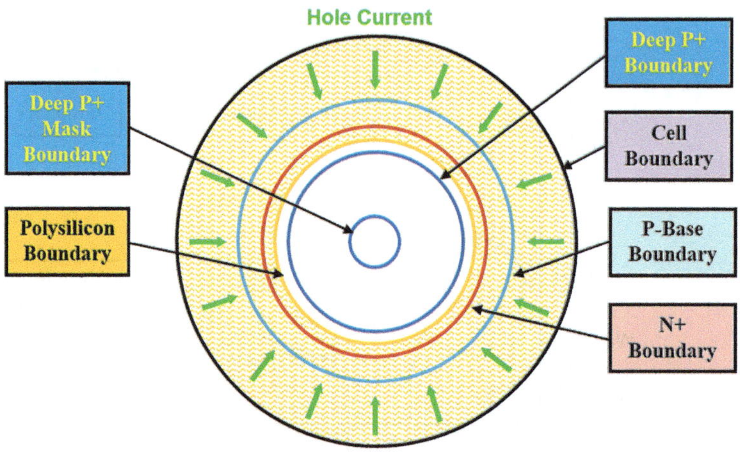

FIGURE 4.12

Distribution of the bipolar current in the insulated gate bipolar transistor structure with circular polysilicon window in a circular array.

of the deep P^+ region from a 2 μm diameter window (W_P^+) in the middle of the polysilicon window, the latch-up current density obtained by using the above analytical model is 2890 A/cm^2. This value is a factor of 1.8 times smaller than that for the linear cell topology. Consequently, the hexagonal layout topology degrades the latch-up performance of the IGBT structure in comparison with the linear cell geometry.

4.2.4.3 Atomic lattice layout

The atomic lattice layout (ALL) topology is illustrated in Fig. 4.10(d). This is conceptually the inverse of the hexagonal topology shown in Fig. 4.10(c). In the ALL topology, circular polysilicon islands are used with the diffusions being performed from outside its boundaries. The polysilicon islands are interconnected using the thin polysilicon bars to allow application of the gate bias. The name "atomic lattice layout" is derived from the resemblance of this topology to atoms in a crystal lattice with bonds between them [12].

As illustrated in Fig. 4.13, the current distribution is uniform around the polysilicon window in the ALL topology. In this figure, the dashed lines represent the edges of the diffusion windows (polysilicon for the N^+ emitter and P-base region, and a photoresist mask for the deep P^+ diffusion). The length (L_P) of the P-base region under the N^+ emitter is also uniform around the polysilicon window. Moreover, the latch-up current density for this topology is superior to that for the linear cell because the resistance of the bipolar current path is reduced due to the current radiating outward. Using cylindrical coordinates appropriate for the ALL cell geometry, an expression for the latch-up current density can be derived [4]:

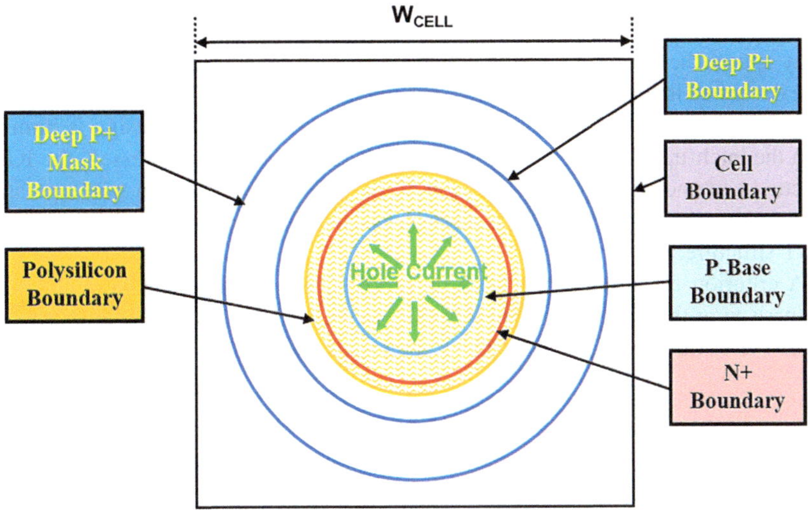

FIGURE 4.13

Distribution of the bipolar current in the insulated gate bipolar transistor structure with atomic lattice layout.

$$J_{C,L}(ALL) = \frac{2V_{bi}}{\alpha_{PNP,ON}\rho_{SP}W_G^2 \ln\left[(W_{P+} - x_{JP+})/(W_{POLY} - x_{JN+})\right]} \tag{4.14}$$

Using the same device physical parameters as in the linear cell topology, namely junction depths of the deep P^+ region ($x_{J,P+}$) and N^+ region ($x_{J,N+}$) of 5 and 1 μm, respectively; a pinch sheet resistance for the P-base region of 2150 Ω/sq.; a polysilicon window radius of 8 μm; and a diffusion of the deep P^+ region from a 14 μm diameter window (W_P^+), the latch-up current density obtained by using the above analytical model is 11,570 A/cm^2. This value is 2.2 times larger than that computed for the linear cell topology. Consequently, the ALL topology provides an enhancement of the latch-up performance of the IGBT structure in comparison with the linear cell geometry. This improvement was first experimentally confirmed for p-channel IGBT structures [12]. Subsequently, the ALL cell topology has been shown to produce n-channel IGBT structures that exhibit current saturation without latch-up, even at 200 °C [13].

The impact of cell topology on the current-induced latch-up of n-channel IGBTs has also been experimentally verified [13]. In this work, 500 V n-channel IGBTs were fabricated with square-cell, linear-cell, hexagonal-cell, and ALL cell geometries. It was found that the square and hexagonal cells exhibited latch-up at 200°C while the ALL cell topology exhibited current saturation without latch-up at this temperature. The experimentally observed difference in the latch-up current levels between the cell topologies is in agreement with the simple latch-up current model described above.

4.2.5 Latch-Up Proof Structure

Many approaches to improving the latch-up current density for the IGBT structure were described in the previous section. During the design of IGBT products, it is prudent to apply a combination of these methods to ensure that the design is immune from the latch-up phenomenon. If the latch-up is sufficiently suppressed, the IGBT structure can undergo current saturation at the gate bias used for on-state operation prior to the onset of latch-up. This device design is referred to as the *latch-up proof* IGBT structure.

A criterion for creating the latch-up proof IGBT structure can be obtained by comparison of the saturation current density to the latch-up current density. The saturated current density for the IGBT structure at a gate bias (V_G) is given by:

$$J_{C,SAT} = \frac{\mu_{ni} C_{OX}}{2 L_{CH} W_{CELL} (1 - \alpha_{PNP})} (V_G - V_{TH})^2 \tag{4.15}$$

and the latch-up current density for the linear structure with a deep P^+ region is given by:

$$J_{C,L} = \frac{V_{bi}}{\alpha_{PNP} \rho_{SP} L_P W_{CELL}} \tag{4.16}$$

if the resistance contributed from the deep P^+ region is neglected in comparison with the much larger resistance of the P-base region. Using the criterion:

$$J_{C,L} > J_{C,SAT} \tag{4.17}$$

leads to the latch-up proof operating condition:

$$(V_G - V_{TH}) < \sqrt{\frac{2 V_{bi} L_{CH}}{\rho_{SP} L_P \mu_{ns} C_{OX}} \left(\frac{1 - \alpha_{PNP}}{\alpha_{PNP}} \right)} \tag{4.18}$$

Because the addition of a diverter region to the IGBT structure leads to a 50% increase in the latch-up current density, the criterion for the latch-up proof IGBT design with the inclusion of a diverter region becomes:

$$(V_G - V_{TH}) < \sqrt{\frac{3 V_{bi} L_{CH}}{\rho_{SP} L_P \mu_{ns} C_{OX}} \left(\frac{1 - \alpha_{PNP}}{\alpha_{PNP}} \right)} \tag{4.19}$$

As an example of a latch-up proof IGBT design, consider the case of a structure with a deep P^+ region that reduces L_P to 1 μm. The maximum gate bias voltage for latch-up proof operation for this structure computed by using Eqn (4.18) is plotted in Fig. 4.14. For a gate oxide thickness of 500 Å, latch-up proof operation is not possible at an on-state gate bias of 15 V, as indicated by the vertical arrow. The gate bias must be reduced to 12 V for latch-up-free operation, as indicated by the green arrow, leading to a significant increase in the on-state voltage drop. The maximum gate bias voltage for latch-up proof operation for the structure when the diverter region is included (as computed by using Eqn (4.19)) is also plotted in Fig. 4.14. For a gate oxide thickness of 250 Å, this structure can be operated at an on-state gate bias of 8 V

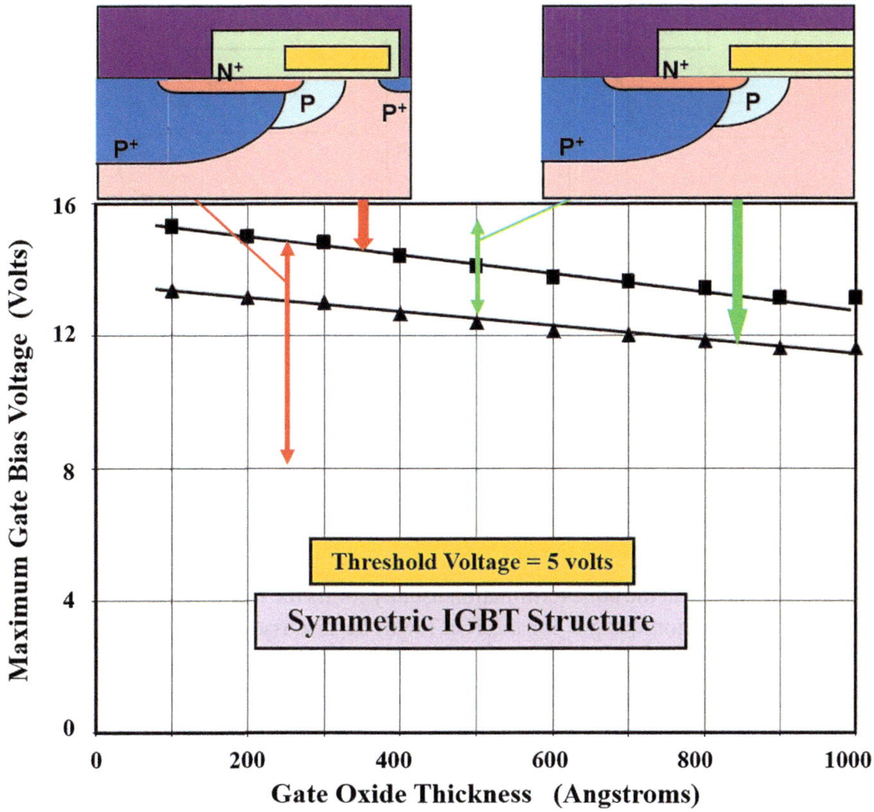

FIGURE 4.14

Criterion for latch-up proof operation of the insulated gate bipolar transistor (IGBT) structure.

due to the larger transconductance. It is therefore possible to obtain latch-up proof operation with a significant margin as indicated by the vertical red arrow.

The pinch sheet resistance for the P-base layer for a gate oxide thickness of 250 Å is 1000 Ω/square for a threshold voltage of 5 V. Using the same structural parameters as in the previous sections, the latch-up current density is found to be 10,000 A/cm^2 for the IGBT structure with the diverter region. The saturated collector current density is 3400 A/cm^2 at an on-state gate bias of 8 V. This indicates that the IGBT design is latch-up proof because the latch-up current density is larger than the saturated collector current density at the reduced on-state gate bias voltage.

4.3 Safe Operating Area

The SOA for a power device is defined as the area of the current–voltage boundary within which the power device can be operated without destructive failure. At low collector bias voltage, the maximum collector current density for the IGBT structure can

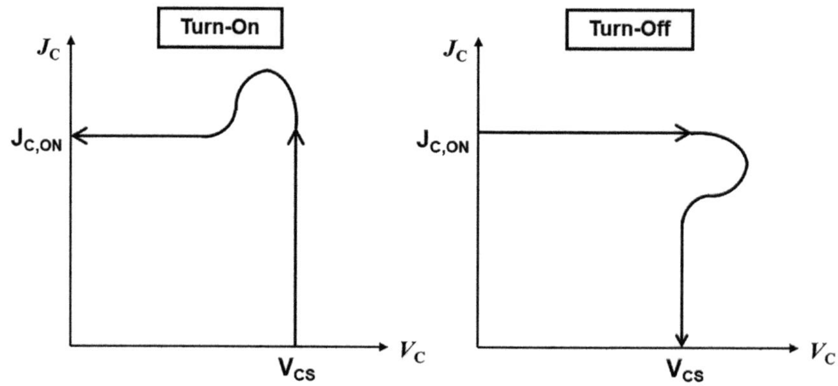

FIGURE 4.15

Hard switching loci for the insulated gate bipolar transistor structure during power switching.

be limited by the onset of the latch-up of the parasitic thyristor. At high collector bias voltages and low current levels, the boundary of operation becomes constrained by breakdown at the edge termination. For the IGBT structure, the open-base transistor breakdown voltage at the edge termination usually limits the maximum operating voltage. Edge termination designs for the IGBT will be discussed in the next chapter.

With reference to Fig. 4.15, it can be observed that the current and voltage become simultaneously large during the turn-on and turn-off switching events with an inductive load such as a motor. The IGBT structure must be capable of operating without destructive failure over the entire area covered by these loci. During turn-on of the IGBT, the current can also overshoot the on-state value due to the large reverse recovery currents associated with PiN fly-back rectifiers. During turn-off of the IGBT, the collector voltage can exhibit an overshoot beyond the DC supply voltage due to stray inductances. Both these phenomena increase the stress on the IGBT, leading to the potential for destructive failure. It is worth pointing out that the destructive failure cannot be avoided by limiting the heat dissipation.

When the IGBT structure is simultaneously subjected to high current and voltage, a phenomenon called *avalanche-induced second breakdown* can occur that leads to destructive failure. This phenomenon can occur during both the turn-on and turn-off transient. During the turn-on transient, it is said to limit the *forward biased SOA (FBSOA)* because the gate bias is turned on. During the turn-off transient, it is said to limit the *reverse biased SOA (RBSOA)* because the gate bias is turned off. These SOA boundaries are different, as discussed below. In addition, a short circuit SOA is also commonly used in IGBT datasheets. This describes the length of time the IGBT can survive a short circuit current while supporting the DC supply voltage.

4.3.1 Forward-Biased SOA

The FBSOA for the IGBT structure limits the current–voltage trajectories when the device is being turned on in a circuit with an inductive load. Because the gate bias is applied during turn-on, the channel of the MOSFET supplies electrons into the

N-base region while holes are supplied to the N-base region by injection from the P^+ collector region. The high collector voltage is supported across a space-charge region formed at the junction (J_2) between the deep P^+ region and the N-base region. Due to the high electric field in the space-charge region, the electrons and holes move at their saturated drift velocities. The concentrations of the holes and electrons in the space-charge region are therefore given by:

$$n_{SC} = \frac{J_n}{qv_{sat,n}} \tag{4.20}$$

$$p_{SC} = \frac{J_p}{qv_{sat,p}} \tag{4.21}$$

where J_n and J_p are the electron and hole current components, respectively, in the N-base region. The net positive charge that determines the electric field profile in the space-charge region is then given by:

$$N^{+}N_D + p - n = N_D + \frac{J_p}{qv_{sat,p}} - \frac{J_n}{qv_{sat,n}} \tag{4.22}$$

The hole current component is larger than the electron current component in the IGBT structure. The hole concentration (p_{SC}) in the space-charge region then exceeds the electron concentration (n_{SC}) because the saturated velocities for holes and electrons are nearly equal in silicon. Consequently, the net positive charge in the current saturation mode becomes larger than the background donor concentration. This enhances the electric field at the junction (J_2) between the deep P^+ region and the N-base region.

Solving for the avalanche breakdown voltage in the N-base region using the net charge in the space-charge region with Fulop's formula for impact ionization yields:

$$BV_{FBSOA} = \frac{5.34 \times 10^{13}}{(N^{+})^{3/4}} \tag{4.23}$$

This breakdown voltage is smaller than the avalanche breakdown voltage determined by the background donor concentration. The breakdown voltage (BV_{FBSOA}) under current saturation decreases with increasing collector current density because the net positive charge increases with increasing collector current density. The avalanche current is amplified by the gain of the PNP transistor in the IGBT structure. Consequently, the FBSOA boundary is limited by the product of the current gain and multiplication factor becoming equal to unity.

The output characteristics of the symmetric blocking IGBT structure are shown in Fig. 4.16 for the case of a drift region width of 200 μm with a lifetime of 10 μs. It can be observed that the output resistance in the current saturation regime is a function of collector bias. When the gate voltage is increased, the collector current increases as expected, but the device is unable to sustain high collector voltages due to the enhanced electric field in the space-charge region associated with the larger hole concentration. The FBSOA boundary is shown by the dashed lines.

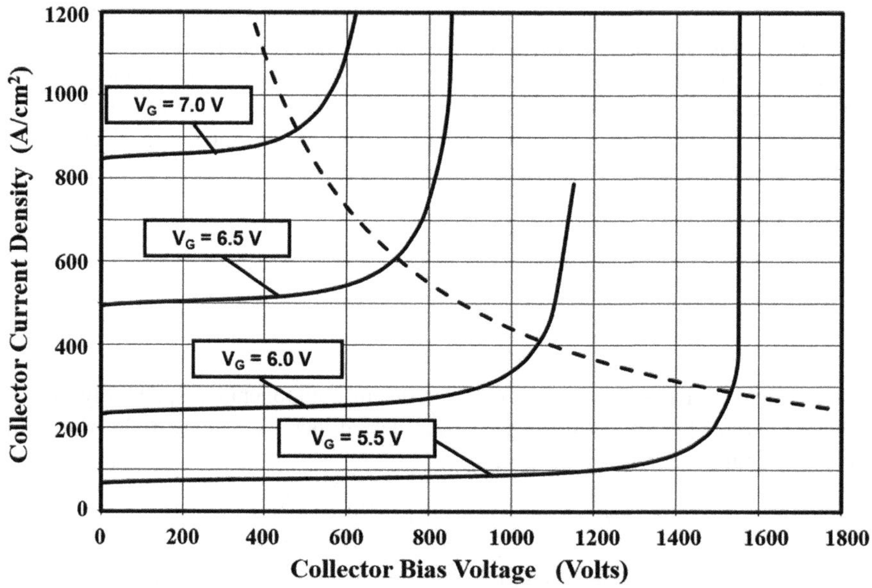

FIGURE 4.16

Forward biased safe operating area for the symmetric blocking n-channel insulated gate bipolar transistor structure.

The output characteristics of the asymmetric IGBT structure are shown in Fig. 4.17 for the case of a drift region width of 100 μm with a lifetime of 1 μs, and a buffer layer with a doping concentration of 1×10^{17} cm^{-3} and a thickness of 10 μm. It can be observed that the output resistance is larger than that observed for the symmetric IGBT structure. When the gate voltage is increased, the collector current increases as expected and the device is capable of supporting a larger collector bias than the symmetric structure. The FBSOA boundary, shown by the dashed lines in the figure, indicates that the asymmetric IGBT has an almost square-shaped FBSOA boundary.

In general, the FBSOA for the IGBT is defined by three limiting factors, as illustrated in Fig. 4.18 for the n-channel (solid lines) and p-channel (dashed lines) IGBT structures [14]. The first limiting factor is current-induced latch-up of the parasitic thyristor at low collector bias voltages. The second limiting factor is breakdown at the edges of the device structure and the third limiting factor is the onset of avalanche breakdown in the presence of current flow under current saturation conditions. It can be seen that the current-induced latch-up limit for the p-channel IGBT structure is twice that for the n-channel IGBT structure. This is because the sheet resistance of the N-base region in the p-channel device is smaller due to the larger mobility for electrons. However, its SOA limit is much smaller because electron current dominates over the hole current within the wide P-base region of the p-channel IGBT structure and the impact ionization coefficient for electrons is an order of magnitude

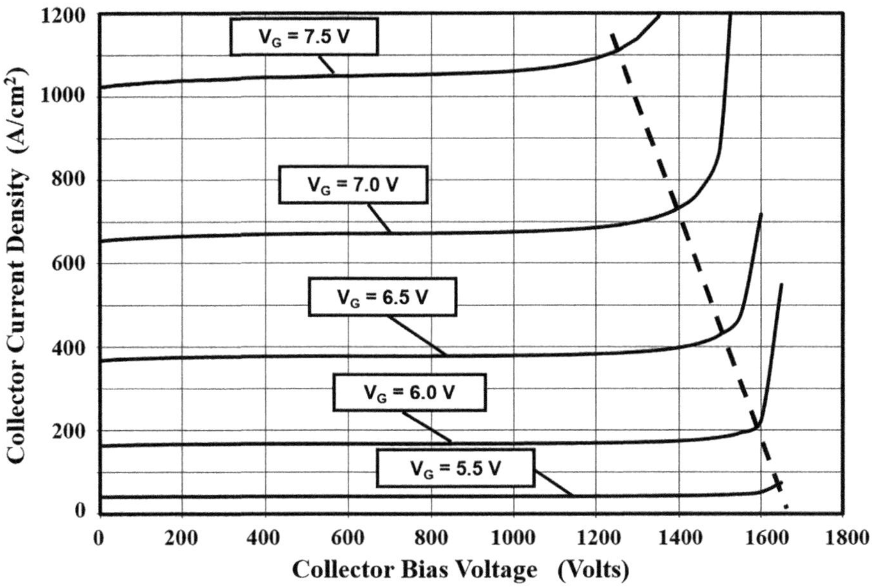

FIGURE 4.17

Forward biased safe operating area for the asymmetric blocking n-channel insulated gate bipolar transistor structure.

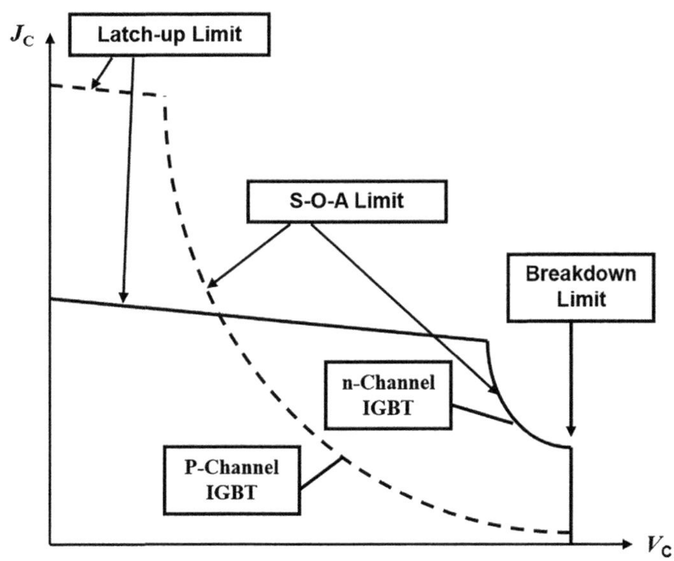

FIGURE 4.18

Comparison of the forward biased safe operating area for the n- and p-channel insulated gate bipolar transistor (IGBT) structures. *SOA*, safe operating area.

greater than that for holes. In p-channel IGBT devices, the breakdown in the SOA mode shifts from edge termination to the cell structure where current transport is occurring. Topologies such as the ALL that suppress impact ionization within the p-channel IGBT cell structure are required to make commercially viable devices.

4.3.2 Reverse Biased SOA

The limit for the loci for the current–voltage trajectories when the IGBT structure is being turned off is defined as RBSOA. Because the gate bias is zero during this mode of operation, the collector current is maintained during the voltage rise time by hole current flow through the N-base region in the n-channel IGBT structure. The collector voltage is supported across a space-charge region formed at the junction (J_2) between the deep P^+ region and the N-base region with the holes moving through the space-charge region at their saturated drift velocities due to the high electric field.

The concentration of the holes in the space-charge region during turn-off with an inductive load is given by:

$$p_{SC} = \frac{J_{C,ON}}{q v_{sat,p}} \tag{4.24}$$

The electric field profile in the space-charge region is then determined by a net positive charge given by:

$$N^+ N_D + p = N_D + \frac{J_{C,ON}}{q v_{sat,p}} \tag{4.25}$$

The avalanche breakdown voltage using the net charge in the space-charge region is:

$$BV_{RBSOA} = \frac{5.34 \times 10^{13}}{(N^+)^{3/4}} \tag{4.26}$$

This breakdown voltage is smaller than the avalanche breakdown voltage determined by the background donor concentration because of the positive charge contributed by the presence of holes in the space-charge region. This breakdown voltage (BV_{RBSOA}) decreases with increasing collector current density because of the increasing net positive charge (N^+). The RBSOA boundary is limited by the product of the current gain of the PNP transistor becoming equal to unity. The breakdown in the RBSOA mode shifts from the edge terminations to the cell structure where current transport is occurring.

Impact ionization within the p-channel IGBT cell structure is affected by the polysilicon layout because of electric field crowding at the nonplanar N-base junction. Consider the cross-section of a p-channel asymmetric IGBT structure illustrated in Fig. 4.19. For a linear cell topology, a cylindrical junction forms at the corner of the N-base region (at location A in the figure) resulting in an enhancement of the electric field when compared to a parallel-plane junction. The gate electrode behaves as a field plate, mitigating some of the electric field enhancement. The RBSOA of the p-channel IGBT structure with a linear cell topology becomes degraded by the

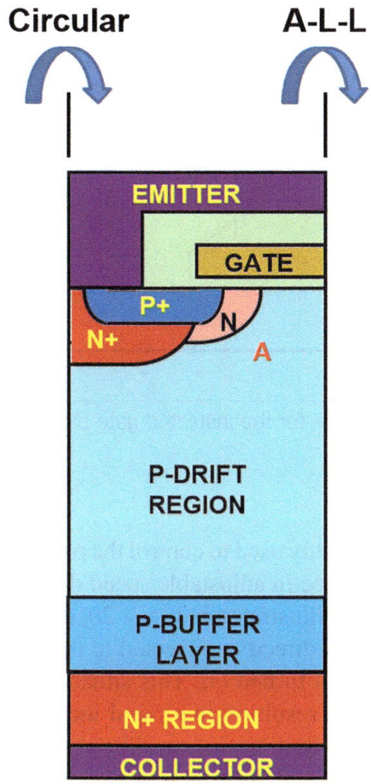

FIGURE 4.19

Comparison of cell topologies for the p-channel insulated gate bipolar transistor structure. *ALL*, atomic lattice layout.

junction curvature. If the ends of the linear fingers are terminated with sharp corners, as is typical in a chip layout where a rectangular-shaped window is opened in the polysilicon window, a spherical junction is formed at the corners of the polysilicon window, degrading the breakdown voltage in the cell even further.

If a hexagonal cell design is used for the p-channel IGBT (as illustrated in Fig. 4.10(c)), it is equivalent to the rotation of the cell around an axis located at the left edge of the cross-section in Fig. 4.19, which creates additional curvature at the junction that enhances the electric field at location A beyond that for the linear cell topology. This will degrade the RBSOA. If the ALL cell design (as illustrated in Fig. 4.10(d)) is used for the p-channel IGBT, it is equivalent to the rotation of the cell around an axis located at the right edge of the cross-section in Fig. 4.19, which creates a saddle junction that reduces the electric field at location A below that for the linear cell topology. This enhances the RBSOA. The ALL cell topology has been demonstrated to provide an improvement by a factor of two for the p-channel IGBT structure when compared with the linear cell topology [12]. This improvement in the RBSOA is critical to making p-channel IGBT products.

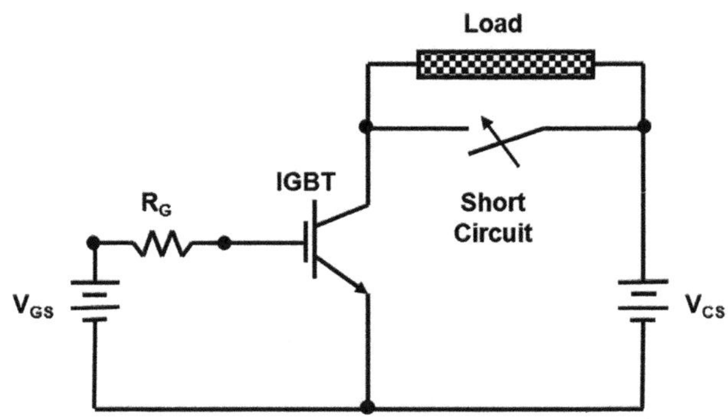

FIGURE 4.20

Short circuit operating conditions for the insulated gate bipolar transistor (IGBT) structure.

4.3.3 Short Circuit SOA

The IGBT structure is commonly used to control the power delivered to windings in a motor from a DC power source in adjustable speed drives. A short circuit can form across the motor winding, as illustrated in Fig. 4.20, due to failure of the insulation. The DC power source is then directly connected to the collector of the IGBT structure while its gate bias is still turned on. This situation results in very high power dissipation within the IGBT, resulting in a rapid increase in the chip temperature. The IGBT can survive the short circuit condition if the short circuit is detected and the IGBT gate bias is switched off before destructive failure [15]. The length of time the IGBT structure can withstand the simultaneous application of high current and voltage is referred to as its *short circuit SOA (SCSOA) or as the short circuit withstand time (SCWT)*. In typical motor drive circuits, the time frame for detection of the fault and subsequent turning off of the gate bias is about 10 μs. The IGBT structure must be capable of tolerating the high current flow under short circuit conditions while supporting the collector power supply voltage for at least 20 μs.

The collector current transient during the short circuit condition is illustrated in Fig. 4.21. The collector current is initially constant ($I_{C,SAT}$) due to the device operating in current saturation mode with a constant gate bias. In practice, the collector current usually decreases with time due to a reduction of the channel mobility due to the temperature rise. The temperature rise is approximately linear, as shown in the figure. It has been found the IGBT undergoes destructive failure when the temperature reaches a critical value (T_{CR}) and the collector current increases abruptly. The time duration that the IGBT structure can withstand the short circuit condition is given by [4]:

$$t_{SCSOA} = \frac{(T_{CR} - T_{HS})W_{Si}C_V}{K_T J_{C,SAT} V_{CS}} \tag{4.27}$$

where T_{HS} is the initial (heat sink) temperature, W_{Si} is the thickness of the drift region, C_v is the volumetric specific heat (1.66 J/cm^3 -°K for silicon), and K_T

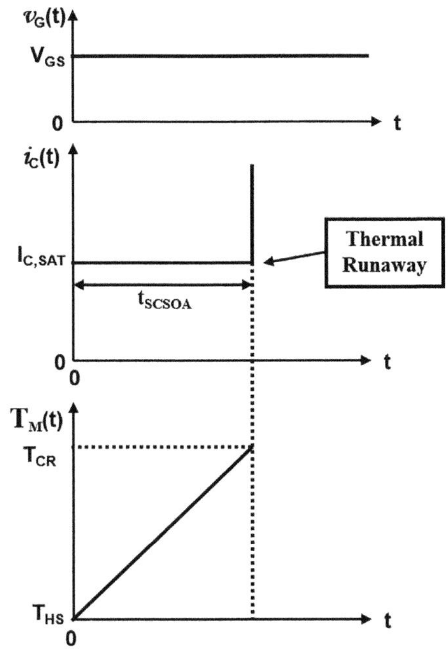

FIGURE 4.21

Gate voltage, collector current, and temperature transients during short circuit conditions.

($\sim$3.5) is a parameter that accounts for nonuniform chip temperature. From this expression, it can be concluded that the SCSOA time becomes smaller as the collector supply voltage and saturated collector current are increased. If an IGBT structure with a drift region thickness of 200 μm is operating at a saturated collector current density of 500 A/cm^2 with a collector supply voltage of 400 V, the SCSOA time is found to be 20 μs using this equation.

4.4 Novel Silicon Device Structures

Several novel IGBT structures have been proposed and demonstrated to reduce the on-state voltage drop while retaining good SOA. The on-state voltage drop in the IGBT structure is increased by the reduction of the injected carrier density near the emitter region due to the reverse bias across the P-base/N-drift junction. This can be compensated by the addition of an N-layer at this junction to improve the conductivity. This enhanced doping is performed in addition to the traditional JFET doping between the P-base regions to reduce the resistance of the MOSFET current path.

A planar IGBT structure with an N-layer at the bottom of the P-base region is illustrated in Fig. 4.22. The enhanced doping at the junction leads to a larger electric

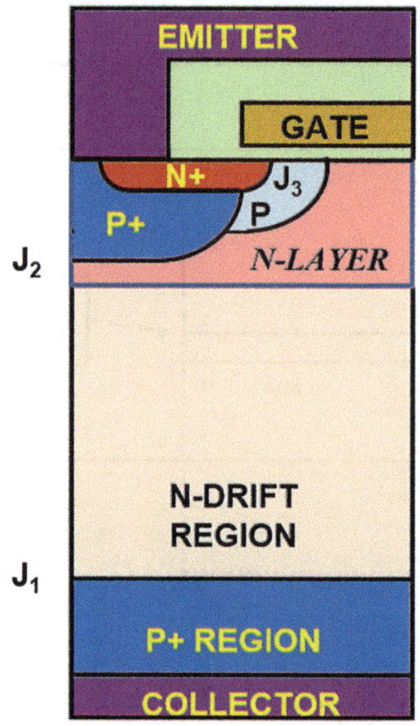

FIGURE 4.22

Novel planar-gate insulated gate bipolar transistor structure to improve trade-off between on-state voltage drop and switching losses while retaining good safe operating area.

field, which could degrade the blocking voltage and the SOA. However, careful optimization of the doping and depth of the N-layer has been demonstrated to produce an improved trade-off curve between on-state voltage drop and energy loss per cycle for 6.5 kV IGBTs while retaining a good short circuit SOA [16]. A similar approach can be applied to the trench-gate IGBT structure, as illustrated in Fig. 4.23. An N-layer formed laterally at the P-base region has also been shown to result in an improved trade-off between on-state voltage drop and switching loss while maintaining a good short circuit SOA in the planar gate IGBT structure [17] capable of blocking 3.3 and 6.5 kV.

4.5 Silicon Carbide Devices

There has been recent growth in developing intelligent microgrids that combine energy generation using renewable sources, energy storage devices to balance power delivery, and fault interruption capability to maintain higher power quality [18]. Solid-state transformers are under development for controlling bidirectional power flow within the microgrid. The semiconductor switches on the primary side of the solid-state transformer must be designed to operate at 15 kV. For this application, even silicon carbide power MOSFETs are inadequate due to the high specific on-

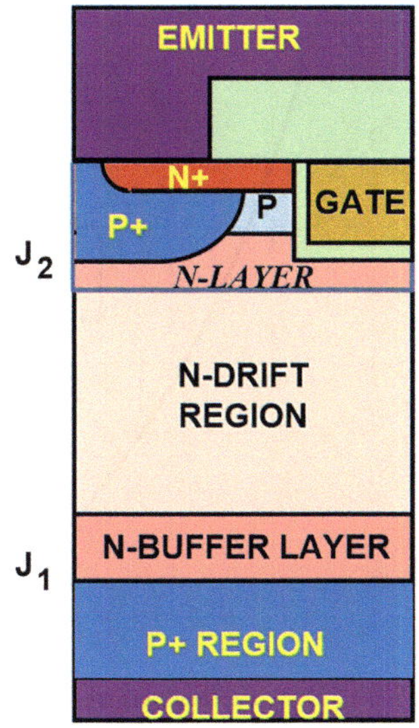

FIGURE 4.23

Novel trench-gate IGBT structure to improve trade-off between on-state voltage drop and switching losses while retaining good SOA.

resistance of the drift region for such high-voltage structures [19]. The best candidate devices for this application are silicon carbide IGBTs.

The operation of the SiC IGBT was discussed in Chapter 3, where it was shown that they have excellent on-state voltage drop and switching performance. The RBSOA for these devices is governed by the same physics described for silicon devices in the previous section. However, the avalanche breakdown parameters for SiC are much superior to those for silicon, indicating better RBSOA as well [20]. This can be demonstrated with a typical example of a 15 kV device. In the case of a silicon asymmetric IGBT, the doping concentration of the drift region would be typically 3×10^{12} cm^{-3}. In contrast to this, the typical doping concentration of the drift region for the SiC IGBT would be typically 3×10^{14} cm^{-3}.

The drift region doping level in SiC devices is therefore comparable to the hole concentration in the space-charge region while that for silicon is far smaller than the mobile charge density. Consequently, the presence of the holes in the space-charge region due to current flow has a much smaller influence on the breakdown voltage for SiC devices, as shown in Fig. 4.24. From this graph, it can be concluded that a silicon IGBT operating with an on-state current density of 100 A/cm^2 would be limited to supply voltages below 2000 V while an SiC IGBT could be operated with a supply voltage of 10–12 kV. This improved SOA for SiC IGBTs has yet to be experimentally confirmed.

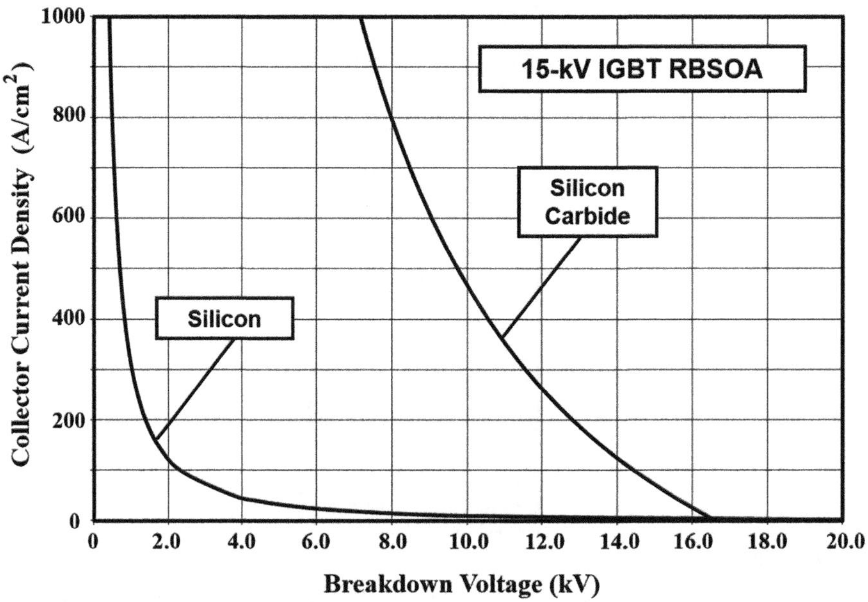

FIGURE 4.24

Comparison of the reverse-biased safe operating area (RBSOA) for SiC and Si insulated gate bipolar transistor (IGBT) devices.

4.6 Summary

One of the main reasons for the popularity of the IGBT device for numerous applications is its excellent SOA. In contrast with previous bipolar power devices, the IGBT can be used in hard switching circuits with no snubbers to protect the device. This has eliminated expensive and bulky components, which also leads to increased power losses and degraded efficiency and reliability. The FBSOA of the asymmetric IGBT structure is much superior to that for the symmetric IGBT structure. In addition, the asymmetric IGBT structure has a much better trade-off curve between on-state voltage and switching losses. For these reasons, most of the commercial development of IGBT products has focused on the asymmetric device structure.

References

[1] B.J. Baliga, Gate Enhanced Rectifier, U.S. Patent 4,969,028, Filed December 2, 1980, Issued November 6, 1990.

[2] L. Knoll, et al., 20 nm gate length schottky MOSFETs with ultra-thin NiSi/epitaxial source/drain, in: IEEE Conference on Ultimate Integration of Silicon, 2011, pp. 1–4.

[3] B.J. Baliga, M.S. Adler, Method of Fabricating a Semiconductor Device With a Base Region Having a Deep Portion, U.S. Patent 4,443,931, Issued April 24, 1984.

[4] B.J. Baliga, Fundamentals of Power Semiconductor Devices, second ed., Springer Science, New York, 2019.
[5] B.J. Baliga, et al., Suppressing latch-up in insulated gate transistors, IEEE Electron Device Lett. 5 (1984) 323–325.
[6] S. Honda, et al., Next generation 600V CSTBTTM with an advanced fine pattern and a thin wafer process technologies, in: IEEE International Symposium on Power Semiconductor Devices and ICs, 2012, pp. 149–152.
[7] H. Xiao, et al., 600A/1200V S3+ IGBT module with fine geometry trench IGBT technology for EV/HEV application, in: PCIM-Asia, 2020, pp. 72–75.
[8] T.P. Chow, B.J. Baliga, M.F. Chang, The effect of channel length and gate oxide thickness on the performance of insulated gate transistors, IEEE Trans. Electron Devices 32 (1985) 2554.
[9] N. Thapar, B.J. Baliga, A new IGBT structure with a wider safe operating area, in: International Symposium on Power Semiconductor Devices and ICs, 1994, pp. 177–182.
[10] B.J. Baliga, et al., The insulated gate rectifier: a new power switching device, in: IEEE International Electron Devices Meeting, Abstract 10.6, 1982, pp. 264–267.
[11] H. Yilmaz, Cell geometry effect on IGT latch-up, IEEE Electron Device Lett. 6 (1985) 419–421.
[12] B.J. Baliga, et al., New cell designs for improved IGBT safe-operating-area, in: IEEE International Electron Devices Meeting, Abstract 34.5, 1988, pp. 809–812.
[13] V. Parthasarathy, et al., Cell optimization for 500-V n-channel IGBTs, in: IEEE International Symposium on Power Semiconductor Devices and ICs, Abstract 3.5, 1994, pp. 69–74.
[14] H. Hagino, et al., An experimental and numerical study of the forward biased SOA of IGBTs, IEEE Trans. Electron Devices 43 (1996) 490–499.
[15] N. Iwamuro, et al., Numerical study of short-circuit safe operating area for p-channel and n-channel IGBTs, IEEE Trans. Electron Devices 38 (1991) 303–309.
[16] M. Rahimo, A. Kopta, S. Linder, Novel enhanced-planar IGBT technology rated up to 6.5 kV for low losses and higher SOA capability, in: IEEE International Symposium on Power Semiconductor Devices and ICs, 2006, pp. 1–4.
[17] K. Nakamura, et al., The next generation of HV-IGBTs with low loss and high SOA capability, in: IEEE International Symposium on Power Semiconductor Devices and ICs, 2008, pp. 145–148.
[18] A.Q. Huang, B.J. Baliga, FREEDM system: role of power electronics and power semiconductors in developing an energy internet, in: IEEE International Symposium on Power Semiconductor Devices and ICs, 2009, pp. 9–12.
[19] B.J. Baliga, Silicon Carbide Power Devices, World Scientific Publishers, 2006.
[20] J. Wang, A.Q. Huang, B.J. Baliga, RBSOA study of high voltage SiC bipolar devices, in: IEEE International Symposium on Power Semiconductor Devices and ICs, 2009, pp. 263–266.

CHAPTER

Chip Design, Protection, and Fabrication

5

The basic physics of operation and the resulting operating characteristics of various types of insulated gate bipolar transistors (IGBTs) were discussed in Chapter 3. In this chapter, these characteristics are used to describe the design of an IGBT chip with a specific current rating suitable for applications. To make devices of finite size, it is necessary to surround the IGBT cells with an "edge termination" that provides the transition from the interior of the chip where current flow occurs in the on-state to the edge of the die where it is separated from other chips on the wafer by using a sawing operation. Cutting out the individual chips by using a saw blade produces considerable damage to the semiconductor crystal. The damaged region must be kept separated from the "active area" of the chip where current flow transpires and the edge termination where high voltages are supported to avoid degradation of the device characteristics. This chapter discusses how to determine the active area of the chip, typical edge-termination structures for IGBTs, and the layout of the chip to create pads for bonding wires for the connection to the emitter and gate regions. This chapter also discusses the design of integrated protection sensors on the IGBT chip for overcurrent, overvoltage, and overtemperature conditions. The typical process steps used to make planar-gate and trench-gate IGBT structures are described to provide an understanding of how IGBTs are manufactured.

5.1 Active Area

The previous chapters discuss the operation and design of the IGBT at the cell level. This performance determines the on-state voltage drop and the energy loss per cycle for the device structure as well as its safe operating area (SOA). The device cells, shown in the cross-sections in Chapter 3, are distributed throughout the active area of the chip. The on-state current flows through this portion of the chip.

The on-state current density for the IGBT is determined by the intersection point between its on-state *i–v* characteristics and the on-state power dissipation limit. This is illustrated in Fig. 5.1 for the symmetric IGBT structure with a blocking voltage rating of 1200 V. The structure has a drift region width of 200 μm and a doping concentration of 5×10^{13} cm^{-3} with four values of the low-level lifetime. As previously discussed in Chapter 3, a reduction of the lifetime in the drift region produces an increase in the on-state voltage drop due to a reduction of conductivity modulation of the drift region.

The IGBT Device. https://doi.org/10.1016/B978-0-323-99912-0.00024-6

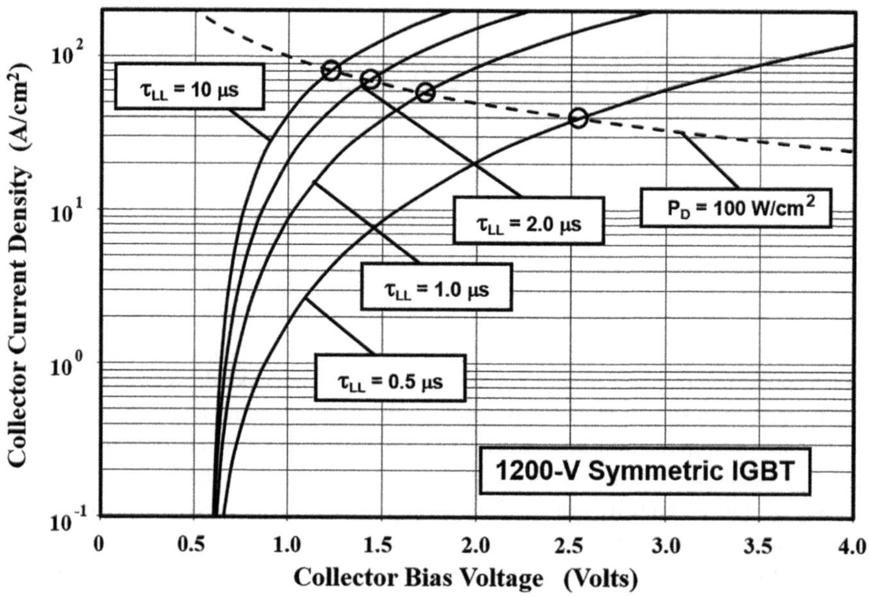

FIGURE 5.1

On-state operating point for the insulated gate bipolar transistor (IGBT) structure.

It is desirable to operate the IGBT structure at a high on-state current density because this reduces the chip size needed to serve a given application. The maximum current density in the on-state is limited by the removal of heat due to the current flow with a finite on-state voltage drop. The maximum junction temperature ($T_{J,Max}$) for the IGBT must be kept below 200°C to achieve long-term reliability. The maximum allowable power dissipation is given by:

$$P_{D,Max} = \frac{(T_{J,Max} - T_A)}{R_\theta} \tag{5.1}$$

where T_A is the ambient temperature and R_θ is the thermal impedance. Using a typical maximum junction temperature and thermal impedance, the maximum power dissipation is about 200 W/cm^2. Assuming that half the power dissipation is associated with the switching losses, the maximum power density for operation of the IGBT in the on-state is 100 W/cm^2. This power dissipation is shown in Fig. 5.1 by the dashed line obtained by keeping the product of the current density and on-state voltage drop constant.

The on-state operating point for the IGBT is obtained by using the intersection point between the on-state *i–v* characteristics and the maximum power dissipation limit. These points are indicated by the circles in the figure for each case of the low-level lifetime. It can be observed that the on-state current density is close to 100 A/cm^2 for the symmetric IGBT with a drift region lifetime of 10 μs. The on-state current density decreases when the lifetime is reduced. For a lifetime of 0.5 μs, the on-state current density drops to 40 A/cm^2. It is worth pointing out that on-state characteristics for the asymmetric IGBT are significantly superior to the symmetric IGBT structure, making the on-state operating current density typically 100 A/cm^2.

The active area for the IGBT structure can be obtained using:

$$A_{Active} = \frac{I_L}{J_{ON}} \tag{5.2}$$

where I_L is the load current and J_{ON} is the on-state current density. As an example, if an IGBT is used in an adjustable speed drive circuit to provide 10 A of current to a motor winding, the active area required would be 0.1 cm^2 if the on-state current density is 100 A/cm^2. For a square-shaped chip design, this corresponds to an active area with sides of 0.316 × 0.316 cm (or 3160 μm × 3160 μm) in size. The IGBT cells are integrated in parallel throughout this active area. If a planar IGBT linear cell design is implemented with a cell width of 15 μm, the number of cells operating in parallel for an active area edge of 0.316 cm is 210. Each of these linear cells has a width of 0.316 cm orthogonal to the cross-section. Consequently, the total width of all the IGBT cells operating in parallel is 66.4 cm (or nearly 1 m). This example demonstrates that the typical IGBT chip has a metal oxide semiconductor field effect transistor (MOSFET) region with a channel length of 1 μm and a width of 6.6 × 10^5 μm or a width-to-length ratio of nearly 1 million. The device fabrication process must ensure that the lithography does not produce any short circuits between the gate and source along this huge width. In addition, the quality of the intermetal gate dielectric between the polysilicon gate and the source metal must be free of pinhole defects over the entire active area.

The typical layout of the polysilicon gate region in an IGBT structure is illustrated in Fig. 5.2. An expanded view of the chip near one edge of the active area is shown on the right side of the figure to show the details of the layout. Note that the polysilicon windows do not extend all the way to the edge of the active area to allow the formation of a horizontal polysilicon stripe that connects all the vertical gate region stripes to each other. This allows conveying the gate signal across all parts of the active area. A cross-section of the IGBT is provided in the lower right side to show its relationship to the layout of the polysilicon layer. It is worth pointing out that the emitter metal extends over the entire active area to interconnect all the emitter regions together. This avoids the need to make thin metal lines and maintains a vertical current path in the chip, which is beneficial for reducing the parasitic resistances and deleterious current crowding effects.

5.2 Gate Pad Design

Because the IGBT is a vertical three-terminal device structure with the collector contact located at the bottom of the chip, it is necessary to include a gate pad on the top surface of the chip, together with the emitter contact, for connection by wire bonding to the pins on the package. Two basic configurations for the gate pad design are

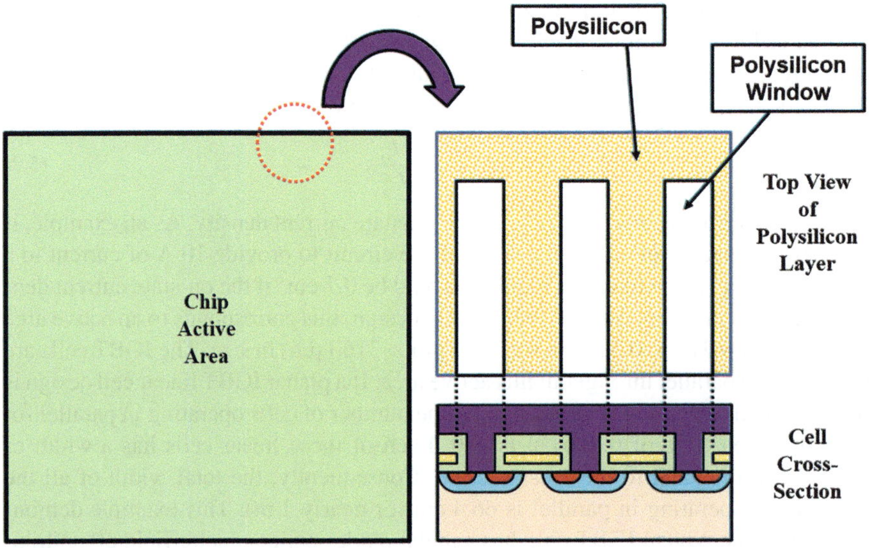

FIGURE 5.2

Design of the active area in an IGBT chip.

commonly used. The first design has a gate pad located at one corner of the chip, as illustrated in Fig. 5.3(a). In the second design, the gate pad is placed in the center of the active area, as illustrated in Fig. 5.3(b). In addition, the gate pad can be placed along one of the edges of the chip, as illustrated in Fig. 5.3(c). The placement of the gate pad is dictated by the wire bonding scheme required within the IGBT package or module. The location of the gate pad can impact the switching performance of the IGBT chip. The portions of the active area that are in the immediate vicinity of the gate pad will receive the gate signal earlier than remote portions of the active area. Consequently, the turn-on and turn-off within the IGBT chip can be nonuniform,

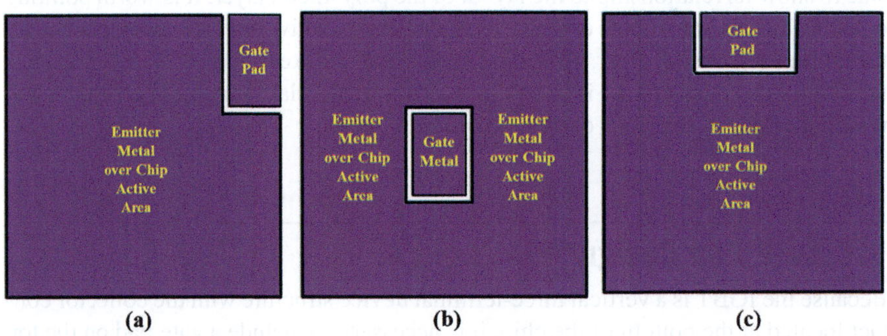

FIGURE 5.3

Designs for the gate pad in an IGBT chip. (a) Corner gate pad design; (b) Center gate pad design.

producing current crowding with enhanced localized heating. For larger chips, gate bus lines extending from the gate bonding pad to further parts of the chip are used to ensure more uniform turn-on and turn-off during switching.

The size of the gate bonding pad is decided by the diameter of the wire and the bonding process. For gold wire bonds, the bonding pad is usually twice the diameter of the wire bond to accommodate the resulting gold ball, as illustrated in Fig. 5.4(a). In the case of aluminum wedge bonds, the design rule for the gate bonding pad is to make its width twice the size of the wire and its length three times the wire diameter, as illustrated in Fig. 5.4(b).

It is typical to connect the emitter to the package terminal by using multiple wire bonds distributed over the entire active area of the chip. This is illustrated in Fig. 5.5 for the case of six IGBTs within a module. Note that the gate pad is located at one

(a) (b)

FIGURE 5.4

Typical gate wire bonds for IGBT chips. (a) Gold ball wire bonds; (b) Aluminum wedge wire bond.

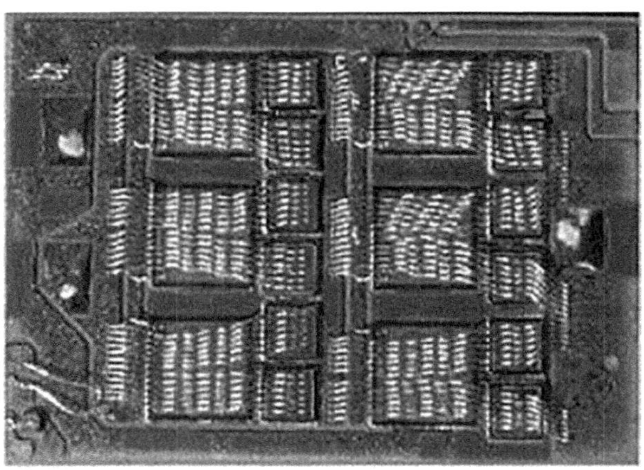

FIGURE 5.5

Typical emitter bonds for IGBTs chips.

corner of the IGBT chips in this design. The use of a large number of emitter wire bonds over the active area reduces the resistance from lateral current flow across the chip through the emitter metallization and maintains a vertical current path for current flow as much as possible.

Some of the pitfalls during the gate pad design that can degrade IGBT performance are worth pointing out with the aid of Fig. 5.6, where several errors can degrade performance. Here, the polysilicon gate electrode is shown as the tan-shaded region. A cross-section of the device at the arrows is provided to describe the problems with this design. First, the gate electrode at the gate pad is placed on top of the gate oxide, as done within the IGBT cells. This adds considerable gate-collector transfer capacitance, which degrades the switching performance. Second, this layout creates a cylindrical junction curvature at location A along the edges of the polysilicon fingers, which enhances the electric field leading to degrading the SOA. This problem is made worse at the sharp corners of the fingers (point B) where spherical junctions are formed. In addition, the current-induced latch-up for the IGBT is degraded because a large amount of hole current enters the cell at location A from under the gate pad.

These problems can be avoided by using the gate pad design shown in Fig. 5.7. Here, the deep P^+ region within the IGBT cell is extended to cover the entire region

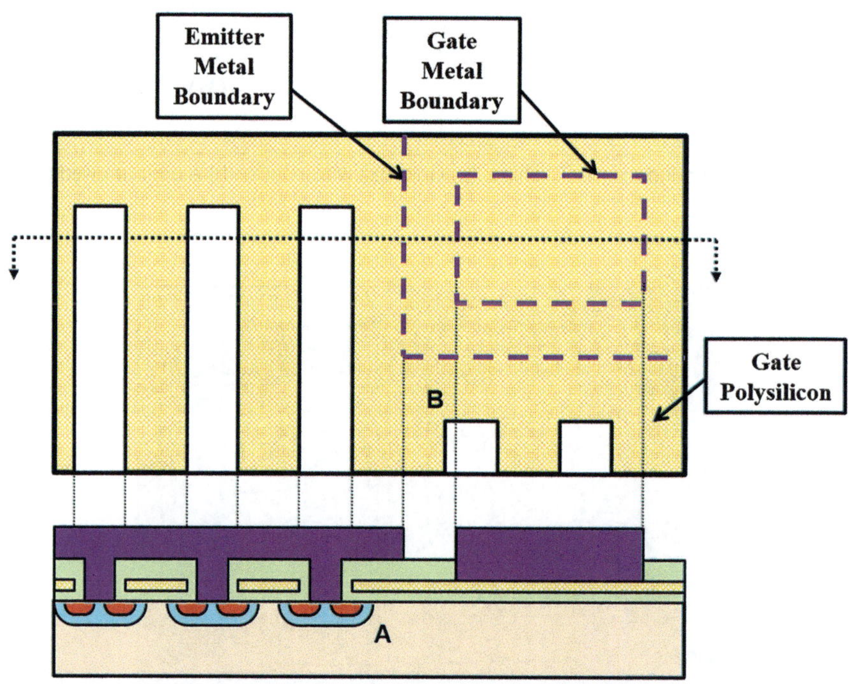

FIGURE 5.6

Flawed gate pad design in an IGBT chip.

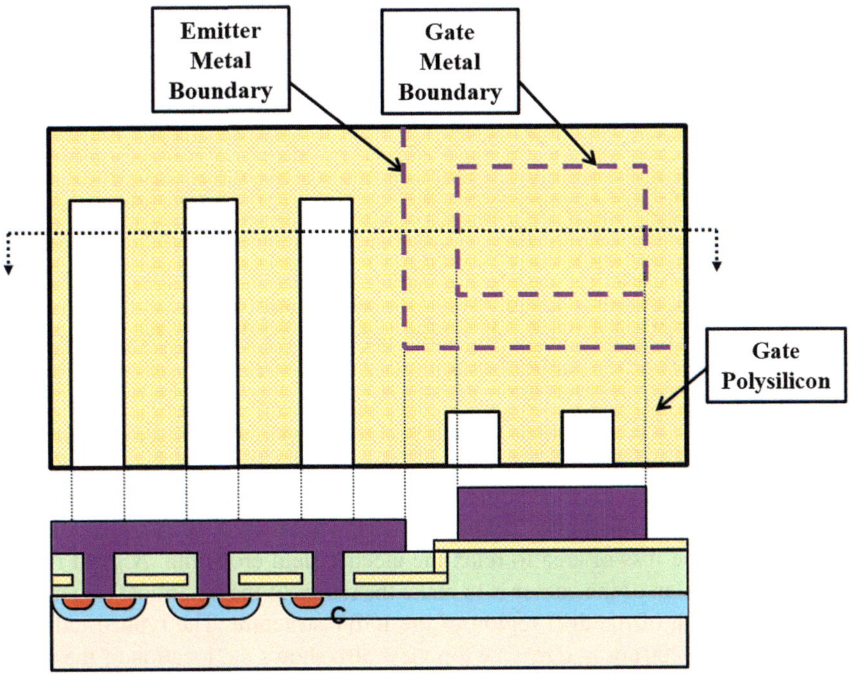

FIGURE 5.7

Corrected gate pad design in an IGBTs chip.

below the gate pad. This eliminates the junction curvature problem and allows collection of the bipolar current under the gate pad and its removal through the emitter contact at point C without degrading the current-induced latch-up.

5.3 Edge Termination Design

Because all IGBT products must be finite in size, they are manufactured by dicing of chips from wafers with diameters of 6–8 in. The silicon crystal is badly damaged when sawing is performed using diamond-coated blades. This damage will degrade the leakage current if it intersects with the depletion layer from the edge termination. Consequently, it is necessary to separate the junctions within the active area of the IGBT and the edge termination from the borders of the chip where the sawing is performed. Special edge termination designs are required between the active area of the IGBT and the saw lanes. The forward blocking capability of the IGBT structure is limited by the design and performance of the edge termination because the electric field is enhanced at the terminations in comparison with the device cells within the active area.

Many edge termination designs for power devices have been proposed in the literature [1]. Because IGBTs are manufactured using relatively shallow junctions (less than 5 μm in junction depth), an edge termination design using similar shallow junctions, compatible with the fabrication of the IGBT cell, is required. Ideally, the edge termination should be created using one of the process steps already needed for making the IGBT cell structure to reduce the fabrication cost. The deep P^+ region in the IGBT cell structure shown in Fig. 3.1 is the most suitable for making planar edge terminations. The most common approach for edge terminations in IGBTs utilizes multiple floating field rings and field plates, as illustrated in Fig. 5.8. The space between the rings must be optimized to make sure that the electric field is as uniform as possible along the x-direction. The analytical solution for the optimum spacing for a single floating field ring is adequate only for devices with low breakdown voltages [2]. Because IGBTs are designed with high breakdown voltages, it is necessary to employ multiple floating field rings with field plates. The spacing of these rings is usually optimized using numerical simulations. More floating field rings are required as the breakdown voltage increases. For example, a 3300 V IGBT has been reported with an edge termination with 22 floating field rings [3].

A top of the IGBT chip at one of its corners is illustrated in Fig. 5.9 where part of the floating field rings is shown in dark blue. It is important to round the corners of the die despite the loss of area to relax the electric field crowding. A good rule of thumb during design of the corner is to make the radius of the innermost ring at least equal to the width of the drift region of the IGBT structure. The typical radius of curvature is 100–200 μm in size. The top view also shows the location of the scribe lane where the chip will be cut for separation of the die after wafer processing. An N^+ diffusion is commonly performed in this portion to serve as a channel stop region. The typical width of the scribe lane between individual chips is about 100 μm.

In the case of a trench-gate IGBT structure, care must be taken to protect the corners of the trenches at the periphery of the chip to avoid the generation of a high electric field in the gate oxide. A simple solution for this problem is to use the configuration shown in Fig. 5.10. Here, the P^+ region connected to the emitter electrode around the periphery of the device overlaps the corner and bottom of the trench on the periphery of the active area shielding the gate [4,5].

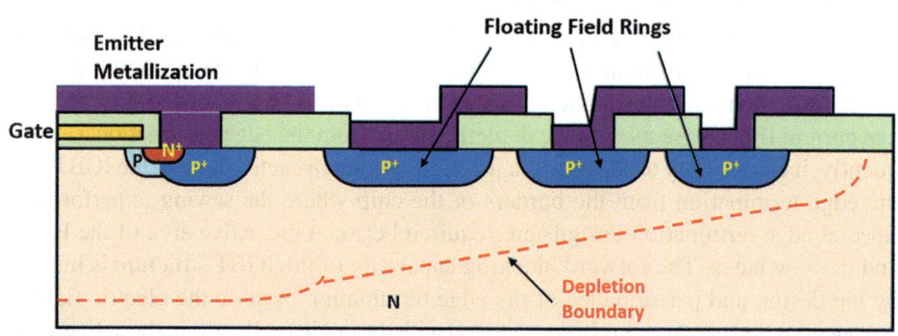

FIGURE 5.8

Cross-section of the typical edge termination design for a planar-gate IGBT.

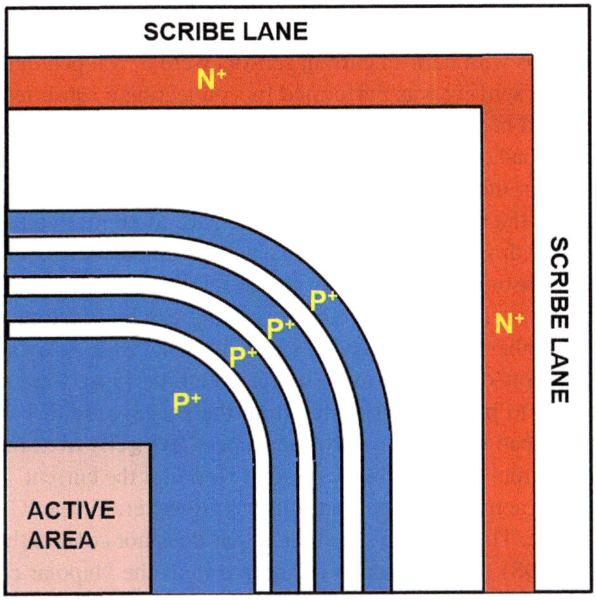

FIGURE 5.9

Edge termination layout for an IGBT chip.

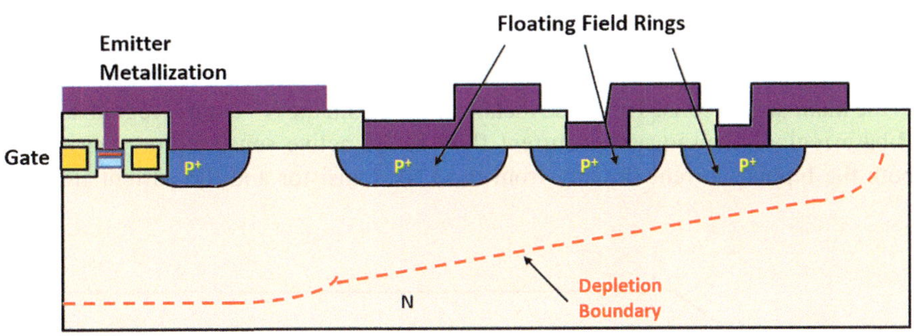

FIGURE 5.10

Cross-section of the typical edge termination design for a trench-gate IGBT.

5.4 Integrated Sensors

When operating the IGBT in applications, it is common to monitor the circuit operation by using sensors embedded with the power electronics. It would be preferable to integrate current, voltage, and temperature sensors within the IGBT chip for protection against these parameters exceeding the specifications. This section discusses sensors that can be integrated into the IGBT chip design.

5.4.1 Overcurrent Protection

The first IGBT chips with built-in current sensors were developed in the late 1980s. Before this, current sensing was performed by connecting a sense resistor in series with the emitter lead of the IGBT. There are two problems with this approach. First, because the entire load current flows through the sense resistor, there is considerable power dissipation in the sense resistor, which reduces the efficiency. Second, the voltage drop across the sense resistor reduces the gate-to-emitter voltage, increasing the on-state voltage drop in the IGBT. These issues can be addressed by incorporation of a current sensor or "pilot" within the IGBT chip.

In the case of power MOSFETs, current sensing can be accomplished by partitioning a small number of the active cells into a sensor with its own source metal contact that is separated from the main source electrode [5,6]. Care must be taken to avoid errors due to parasitic resistance from the package. In the case of IGBTs, the same approach can be taken to partition some IGBT cells from the main active area and use the current from this sense region to monitor the current. This approach is referred to as an "active IGBT current sensor." However, for the IGBT, two other options are possible. The first is a sensor area that does not contain the metal oxide semiconductor (MOS) channel, which is referred to as the "bipolar current sense." The second is a sensor area where contact is made exclusively to the N^+ region, which is referred to as the "MOS current sense." The area of the current sense region is typically chosen so than the sense current is about 10 mA when the IGBT is operating at its nominal on-state current level. This corresponds to an area of 1×10^{-4} cm^2 if the on-state current density is 100 A/cm^2.

A cross-section of the IGBT chip design with an integrated active current sense region is shown in Fig. 5.11. Note that the cell at the current sense region is identical to the main device cells. The sense metal contacts both the N^+-emitter region and the P-base region. Consequently, current flow in the active sense region consists of both the bipolar current flowing from the PNP transistor and the current flowing

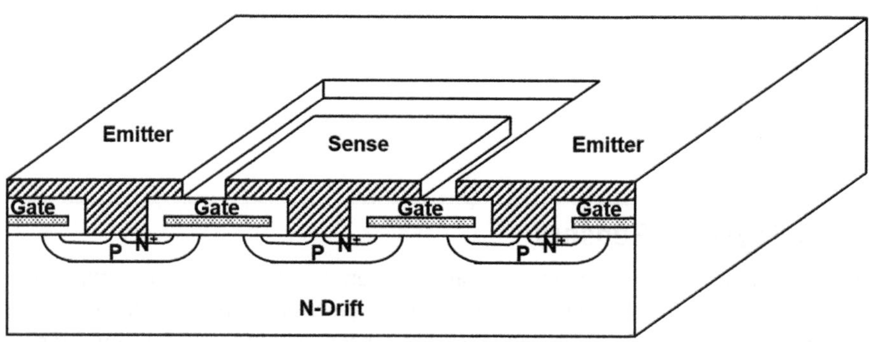

FIGURE 5.11

Typical design for an active current sense region in the IGBT.

through the MOS channel. The gate electrode of the current sense cells is connected to the gate of the main cells so that they are switched on simultaneously.

A cross-section of the bipolar current sense region is shown in Fig. 5.12. Note that the cell at the current sense region is not identical to the main device cells because the N^+ region has been omitted. The current flow in the bipolar sense region consists of only the PNP transistor current. In this case, the sensor collects current flow from the wide base PNP transistor in the IGBT without any MOS channel current. This approach is referred to as a "bipolar IGBT current sensor."

A cross-section of the MOS current sense region is shown in Fig. 5.13. Note that the cell at the current sense region is not identical to the main device cells because the sense metal contacts only the N^+ region without contacting the P-base region. The current flow in the MOS sense region consists of only the MOS channel current. This approach is referred to as an "MOS IGBT current sensor."

Current sensing in IGBT chips is typically used to provide protection from over-current operation during short circuits in the load, such as the motor winding. When

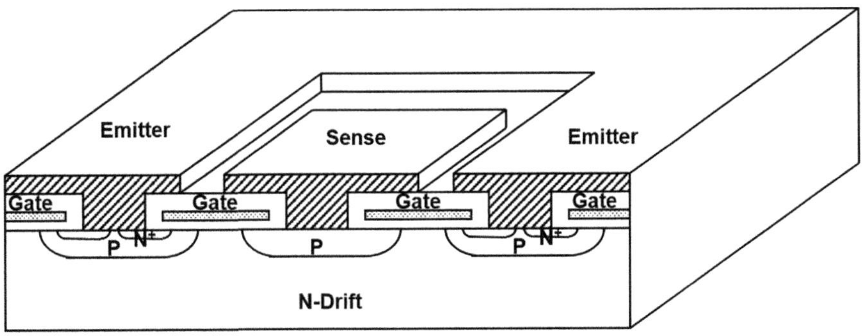

FIGURE 5.12

Typical design for the bipolar current sense region in the IGBT.

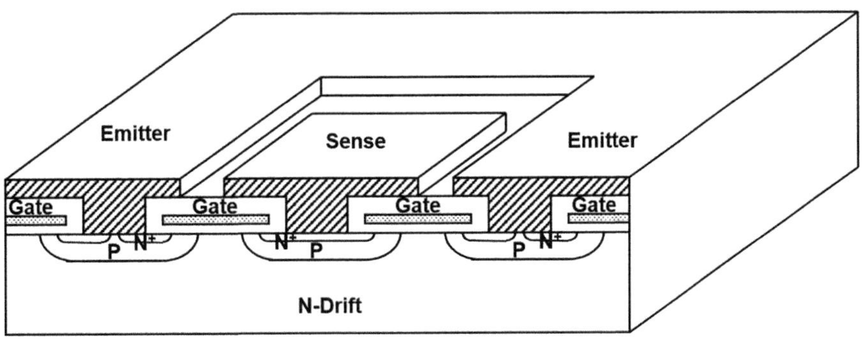

FIGURE 5.13

Typical design for the MOS current sense region in the IGBT.

the IGBT chip current exceeds a level that is several times the on-state rated current, it is beneficial to detect the situation and turn-off the gate drive voltage to avoid exceeding the acceptable maximum junction temperature for the device. A typical totem pole circuit within an H-bridge inverter for adjustable speed motor drives is shown in Fig. 5.14 [7]. The control logic directs the turn-on and turn-off of the IGBTs to produce the desired wave shape for the output voltage (V_{OUT}) that is fed to the motor winding. The control signal to the upper device in the totem pole must be level shifted to the DC supply voltage. The sensing electrode from the IGBTs is connected to the sense resistor (R_S). The voltage across this resistor is monitored within the gate drive circuits. If this voltage exceeds a preset limit, the gate driver can stop the gate drive signal for the IGBTs with a trip signal sent to the microprocessor controlling the motor drive.

For the current sensing to be most effective, it is necessary that the current flowing from the sense region be proportional to the main collector current. The linearity of the current sensing from the three types of sense regions described above has been experimentally studied [8,9]. This work has demonstrated that the active sensor region has the best linearity, as illustrated in Fig. 5.15. In addition, it has been found that the magnitude of the resistor in series with the sensor electrode must be sufficiently small to keep the potential at the sense electrode less than 50 mV. The

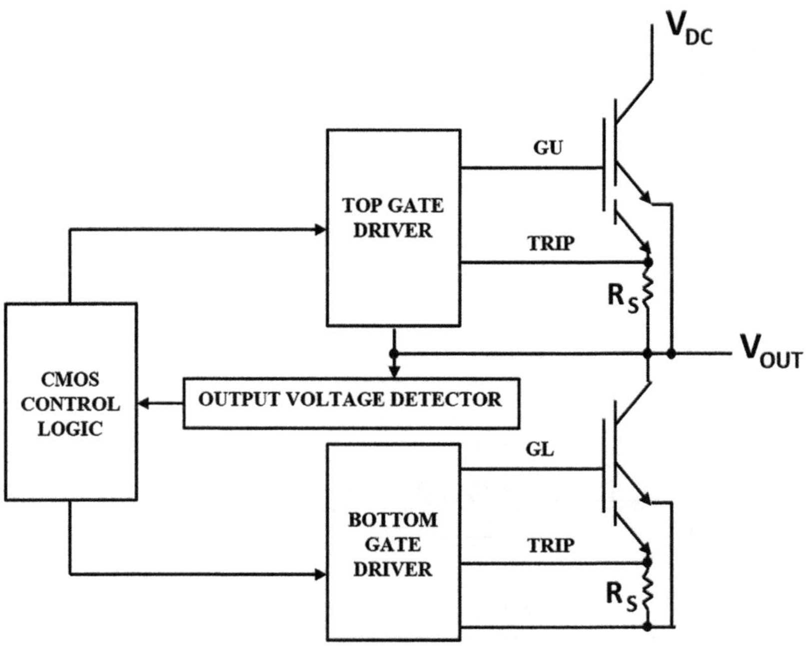

FIGURE 5.14

Current sensing in an IGBT chip.

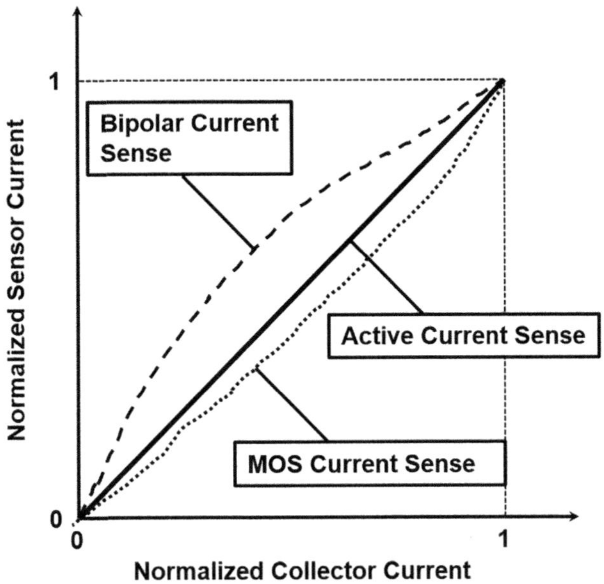

FIGURE 5.15

Current sensing linearity for various current sensors in an IGBT chip.

temperature dependence of the sense current has also been studied [10]. The least variation is observed for the MOS current sensor region.

In the previous discussion, the circuitry used for detecting the overcurrent and turning-off the IGBT was located in the control chip, as shown in Fig. 5.14. A superior design for short circuit protection can be achieved by monolithic integration of the overcurrent protection circuit on the IGBT chip [11]. This approach also allows reduction of the IGBT saturation current during short circuit events, resulting in good short circuit SOA. The monolithically integrated short circuit protection circuit is illustrated in Fig. 5.16 with its equivalent circuit shown in Fig. 5.17. A current sense IGBT is used together with a polysilicon resistor (R_S) to generate a gate voltage for the lateral MOSFET. Under short circuit operation, the sense current increases until the voltage across the sense resistor exceeds the threshold voltage of the lateral MOSFET and turns it on. This clamps the gate voltage of the main (and sense) IGBT to less than the normal gate drive voltage for on-state operation of the IGBT. Typically, the gate voltage is reduced from 15 V for on-state operation to 8 V during the short circuit event. This greatly reduces the saturation current for the main IGBT under short circuit conditions, allowing it to exhibit sufficient short circuit SOA to allow detection of the short circuit and turning off the IGBT without failure. It is convenient to locate the overcurrent sensing and protection circuit around the gate pad. Note that the polysilicon resistor and polysilicon diode are formed on top of a thick field oxide to isolate them from the underlying active area of the IGBT chip.

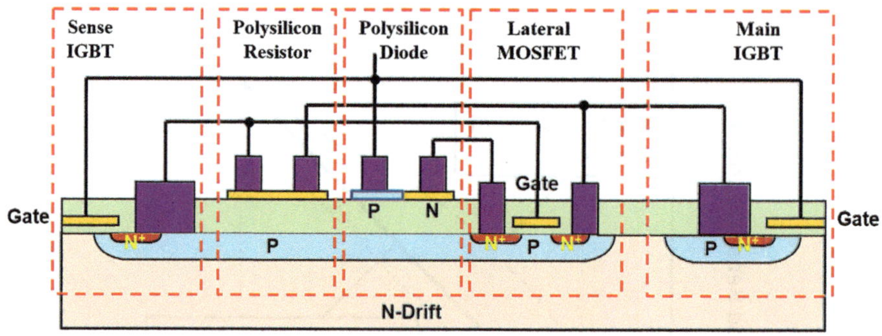

FIGURE 5.16

Monolithically integrated short circuit protection circuit for an IGBT.

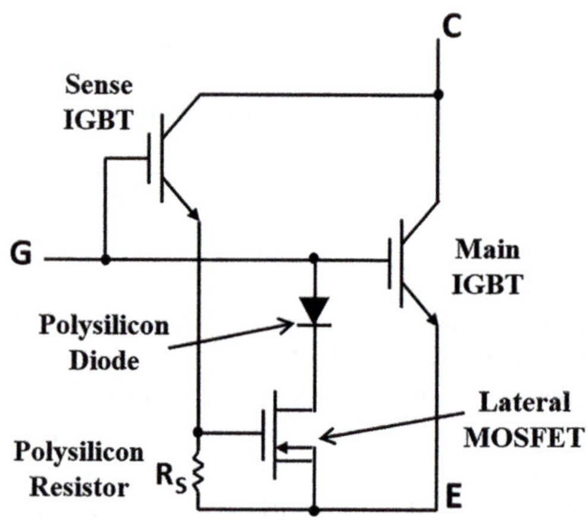

FIGURE 5.17

Circuit for limiting saturation current during short circuit operation.

5.4.2 Overvoltage Protection

The IGBT can be subjected to an overshoot in the collector voltage during circuit operation due to the presence of high voltages produced across stray inductances. This problem becomes worse during higher frequency operation when the device is switching with a large *dI/dt*. The voltage across the stray inductance adds to the DC supply voltage, resulting in the instantaneous collector voltage of the IGBT exceeding its blocking voltage rating. Avalanche breakdown at the edge termination of the IGBT during voltage overshoots is not desirable because it is not designed to absorb significant energy. Such high-voltage overshoots are also encountered when IGBTs are operated in soft-switching circuits. Two approaches for protecting the IGBT during overvoltage conditions are described in this section.

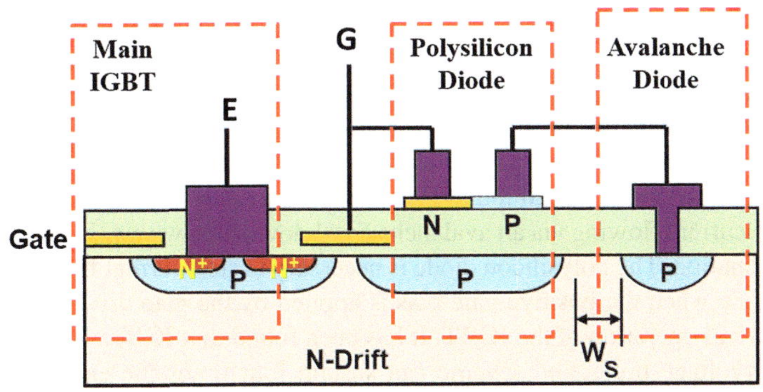

FIGURE 5.18

Monolithically integrated overvoltage protection circuit for an IGBT.

The first method for overvoltage protection is based on incorporation of an avalanche diode [12] within the IGBT structure, as illustrated in Fig. 5.18. The breakdown voltage of the avalanche diode is designed to be slightly lower than the breakdown voltage of the edge termination so that breakdown within the IGBT chip shifts to the avalanche diode. The breakdown voltage of the avalanche diode can be adjusted because it is dependent on the radius of curvature of its junction and the separation between the junctions (W_S) within the avalanche diode [13]. It is also dependent on the radius of the corner of the window through which the diffusion of the P region is performed.

The equivalent circuit for the overvoltage protection scheme is shown in Fig. 5.19. This circuit includes the resistance in the control or drive circuit for the

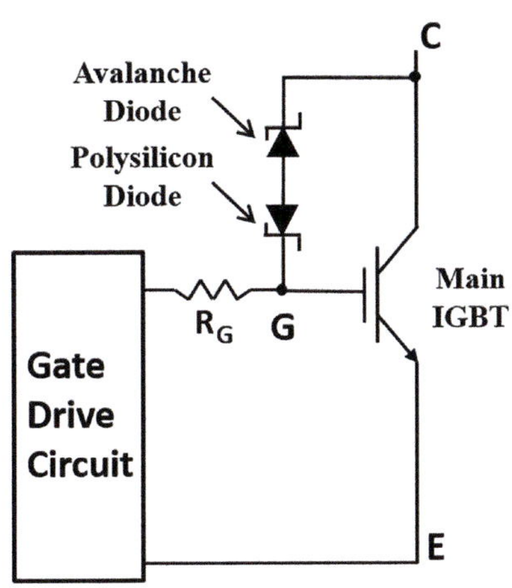

FIGURE 5.19

Circuit for overvoltage protection of the IGBT.

IGBT. When the collector voltage of the IGBT exceeds the breakdown voltage of the avalanche diode, the current flow through the avalanche diode is routed via the forward biased polysilicon diode to the resistance (R_G) of the gate drive circuit. When the voltage across the gate resistance exceeds the threshold voltage of the IGBT, the device turns on, allowing current to flow through the emitter terminal. This dissipates the energy stored in the circuit inductance that produced the voltage overshoot without all the current flowing via an avalanche breakdown path within the IGBT cells or edge termination. The polysilicon diode is necessary to stop current flow via the avalanche diode when the positive gate bias is applied by the gate drive circuit during normal circuit operation of the IGBT. It has been found that IGBTs with this monolithic overvoltage protection scheme can sustain four times the energy compared with unprotected devices [13].

An alternate method for overvoltage protection is illustrated in Fig. 5.20. Here, a junction field effect transistor (JFET) structure with a floating source region is incorporated within the IGBT structure. When the collector bias increases, the depletion regions formed around the *P*-regions of the JFET extend below the N^+ source region and shield it from the collector potential. It has been found that the potential of the floating source region tracks the collector bias voltage [3]. The voltage at the N^+ region can therefore be fed to an off-chip control circuit to monitor an overvoltage condition and send a gate bias to turn on the IGBT to dissipate the energy.

5.4.3 Overtemperature Protection

It is important to maintain the maximum junction temperature within the IGBT chip below 200°C to avoid problems with failures from current-induced latch-up. The device temperature can be monitored by measurement of the forward voltage drop across a P–N junction diode. This can be accomplished using a polysilicon diode formed on top of a thick field oxide to isolate it from the underlying semiconductor region [14]. At a constant on-state current level, the on-state voltage drop of a P–N

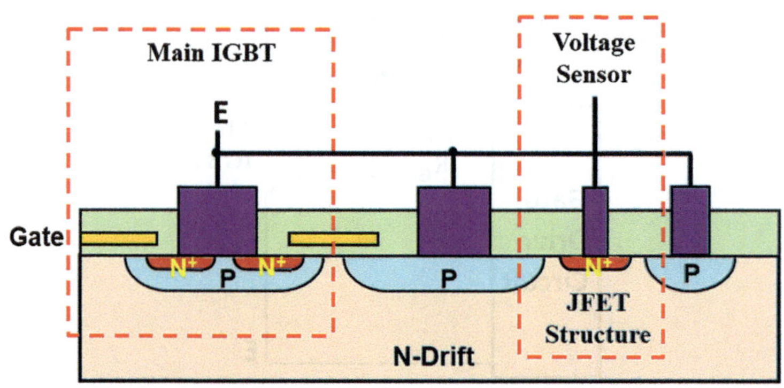

FIGURE 5.20

Alternate monolithically integrated overvoltage protection circuit for an IGBT.

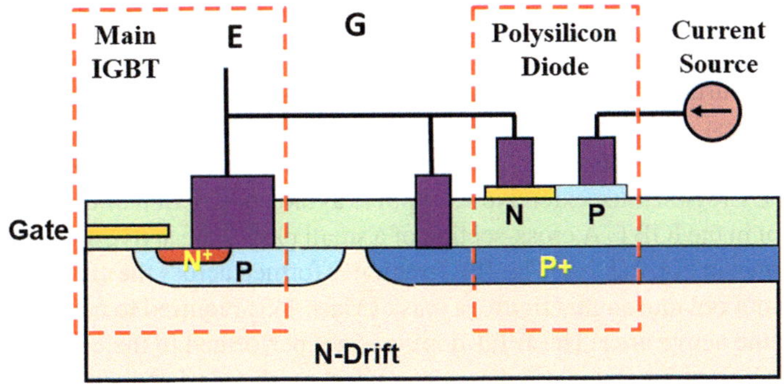

FIGURE 5.21

Integrated overtemperature monitoring circuit for an IGBT.

diode reduces with increasing temperature [1]. A constant current is delivered to the polysilicon diode from an external circuit and the voltage across the diode is monitored in the overtemperature monitoring scheme shown in Fig. 5.21. The gate drive to the IGBT can be shut down when the voltage across the polysilicon diode dips below a preset magnitude corresponding to the desired maximum operating temperature. The overtemperature sensing polysilicon diode is typically located at the periphery of the emitter area to enable easy connection to the current source [14]. It is placed above the P^+ regions at the chip edges that join the edge termination structure.

5.5 Planar-Gate Device Fabrication Process

There are many possible process variations that can be used to manufacture an IGBT chip. Each manufacturer has created a unique process that is proprietary. Only a typical double-diffused metal oxide semiconductor (DMOS) process sequence can be provided in this section. In general, the regions with the largest junction depths must be formed first, followed by regions with successively smaller junction depths. For the first IGBT devices with low breakdown voltages (<1200 V), the starting material was produced by the growth of the *N*-type drift region on a heavily doped *P*-type substrate because the drift region is only 50 μm in thickness. A P^+ substrate with a thickness of 300 μm provides a stable wafer that can be handled by manufacturing equipment without breakage. For asymmetric blocking devices, an *N*-type buffer layer is first grown adjacent to the P^+ substrate, followed by the growth of the *N*-drift region in one step. In the case of fabricating more recent IGBT devices, it has become possible to handle very thin wafers through the manufacturing steps. Consequently, less expensive bulk silicon wafers with a thickness of 50–100 μm are used. The buffer layer is first formed on the backside of the wafer (collector side of the IGBT) by ion implantation of phosphorus, followed by the drive-in step. This approach is also applicable to IGBT chips with breakdown voltages greater than 1200 V because of the large thickness of the drift region. A typical depth for the buffer layer is 5–10 μm with a final surface concentration of 5×10^{17} cm^{-3}.

The deepest diffusions on the front side (emitter side) of the IGBT chip are the JFET and the deep P^+ regions. A typical junction depth for the P^+ region is 4–5 μm and the JFET region must extend to its depth to avoid constriction of the MOS current component in the IGBT. A cross-section of a small part of the active area in the IGBT chip is shown in Fig. 5.22(a). The JFET region is formed across the entire active area. Although not obvious in this figure, a mask (Mask 1) is required to confine the JFET doping to the active area. The JFET doping is not performed in the edge termination to prevent loss of breakdown voltage. A second mask (Mask 2) is used to pattern a photoresist layer to define regions where the P^+ region is formed on the IGBT chip. As discussed in Chapter 4, a deep P^+ region is required in the IGBT cell area to

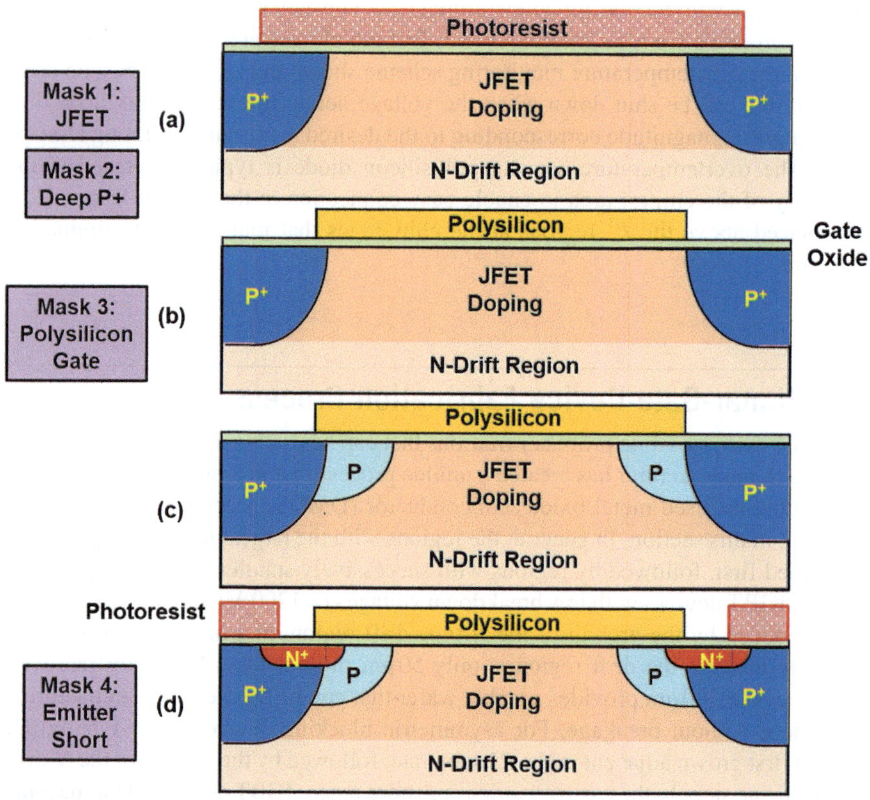

FIGURE 5.22

Fabrication process for the planar-gate IGBT structure: (a) cross-section after deep P^+ diffusion; (b) cross-section after JFET diffusion; (c) cross-section after P-base diffusion; (d) cross-section after N^+ diffusion.

suppress latch-up of the IGBT. The same deep P^+ region can also be utilized to form the floating field rings at the edge termination. In this case, the photoresist defined by Mask 2 can also be employed to mask the etching of the field oxide at the edge termination. The deep P^+ region is typically formed by ion implantation of boron with a high dose in the range of 10^{16} cm^{-2}.

The gate oxide is grown after the diffusion of the P^+ region followed by the deposition of the polysilicon gate layer. The gate oxide thickness has been scaled down over the years from the 1200 Å range to below 500 Å in modern devices due to performance advantages discussed in previous chapters. Usual practices for the growth of high-quality gate oxide are utilized during this step. However, in the IGBT process, high-temperature diffusion of the *P*-base and N^+-emitter regions is required after formation of the gate stack. This can degrade the interface properties of the MOS channel region. The polysilicon layer is then deposited and patterned using Mask 3, as illustrated in Fig. 5.22(b). A good alignment between the polysilicon layer and the previously fabricated P^+ region is required to get good latch-up suppression in the IGBT, as explained in Section 4.2.4 in Chapter 4. The width of the polysilicon gate in the IGBT cell must be optimized to get a good on-state voltage drop by adjusting the JFET width. Various shapes for the polysilicon layer were discussed in Chapter 4 for performing the trade-off between on-state voltage drop and safe operating area.

The *P*-base region is now formed by ion implantation of boron with the polysilicon gate as a mask (i.e., no mask is used during this step), as shown in Fig. 5.22(c). The energy of the boron implantation must be chosen to be sufficiently low to avoid penetration of the polysilicon layer. The dose of the *P*-base boron ion implantation is approximately 10^{14} cm^{-2} to obtain a junction depth of 2.5–3 μm with a surface concentration of 3–5 × 10^{17} cm^{-3} that is required to achieve a threshold voltage of about 4 V. After driving in the *P*-base region, Mask 4 is used to define the photoresist regions, which pattern the N^+-emitter regions, as shown in Fig. 5.22(d). Note that the N^+ emitter must be self-aligned to the polysilicon gate edge. Consequently, the photoresist mask is needed only where the N^+ emitter is shorted to the *P*-base region at the center of the polysilicon window. The N^+-emitter region is typically formed by ion implantation of phosphorus with a high dose in the range of 10^{15} cm^{-2}. Its drive-in cycle is chosen to achieve a junction depth of 1 μm, leading to an MOS channel length of 1.5–2 μm for the IGBT.

An intermetal dielectric is now deposited on the top surface using a low-pressure chemical vapor deposition (LPCVD) process. The dielectric is typically silicon dioxide or phosphosilicate glass. This layer must be free of pinholes to avoid creating gate-source shorts because the emitter metal overlaps the gate electrode in the cells. It must also be conformal and cover the edges of the polysilicon. If the process creates a thinning of the intermetal dielectric at the gate edges, more instances of gate-source shorts occur due to extremely large width (typically many meters) of the polysilicon edge in a typical IGBT chip. This intermetal dielectric is etched using Mask 5, as shown in Fig. 5.23(e) to produce the contact window in the cell. If the contact window is not precisely aligned to the polysilicon edge, the contact window

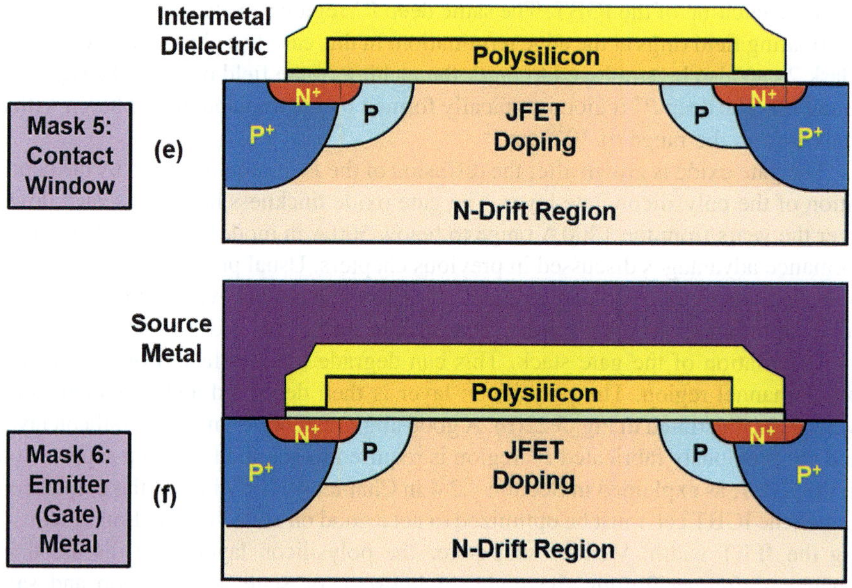

FIGURE 5.23

Fabrication process for the planar-gate IGBT structure (continued): (e) cross-section after intermetal dielectric patterning; (f) cross-section after emitter metal.

can overlap the polysilicon gate, leading to a gate-emitter short circuit. This is usually a critical yield-limiting process step during the manufacturing of IGBTs with large active areas. The space between the edge of the gate and the edge of the window for the contact to the emitter region is determined by the alignment tolerances of the lithography tools. A smaller space is preferable to obtain better latch-up current density but can lead to yield loss due to incidence of gate-emitter shorts arising from the misalignment along the width of the cell. In addition, the contact window must expose both the P^+ region and the N^+-emitter region to allow making a short between them for latch-up suppression.

Aluminum metallization is now deposited to contact the emitter regions in the IGBT cells, as shown in Fig. 5.23(f). The metal is patterned using Mask 6 to define the emitter area and the gate bonding pad. This metallization is typically 4–5 μm in thickness to reduce the resistance to current flow between the bond wires for the emitter of the IGBT. It has to be sufficiently thick to buffer the underlying device structure from being damaged by the force applied by the bonding wire over the active area. The contact is now annealed at about 450°C in forming gas (a mixture of nitrogen and hydrogen) to reduce the contact resistance to the P^+ and N^+ regions. After this, the backside of the wafer is coated with the collector contact. Because the bottom of the IGBT chip is soldered to the package or module, this metal is usually a multilayer stack of titanium (2000 Å), nickel (5000 Å), and gold or silver (2000 Å). The last layer provides good wetting of the solder to the chip during packaging.

A silicon nitride layer is often deposited at the top surface using plasma-enhanced chemical vapor deposition (PECVD) as the final step. This layer is patterned using an additional mask to expose the emitter and gate pad areas for wire bonding.

Modern IGBT chips utilize a submicron scale lithography capability. The small size of the N^+ emitter regions in these structures is very effective in suppressing the latch-up of the parasitic thyristor. It is therefore not necessary to diffuse a deep P^+ region in the cell structure. This allows shrinking the cell size to increase the MOS channel density to achieve a lower on-state voltage drop.

5.6 Trench-Gate Device Fabrication Process

As in the case of planar-gate IGBT products, each manufacturer has created a unique trench-gate process that is proprietary to fabricate their products. Only a typical U metal oxide semiconductor (UMOS) process sequence can be provided in this section. The starting material for these devices is similar to that for the DMOS or planar-gate IGBTs. In general, the regions with the largest junction depths must be formed first, followed by regions with successively smaller junction depths. The deepest diffusion on the front side (emitter side) of the trench-gate IGBT chip is the deep P^+ region. A typical junction depth for the P^+ region is 4–5 μm. The first mask is used to pattern the field oxide to define the edge termination region and remove it from the active area. A mask (Mask 2) is then used to pattern a photoresist layer to define regions where the P^+ region is formed on the IGBT chip. A cross-section of a small part of the active area in the IGBT chip at this stage of the process is shown in Fig. 5.24(a). As discussed in Chapter 4, a deep P^+ region is required in the IGBT cell area to suppress latch-up of the IGBT. The same deep P^+ region can also be utilized to form the floating field rings at the edge termination. The deep P^+ region is typically formed by ion implantation of boron with a high dose in the range of $10^{16}\ \text{cm}^{-2}$.

The *P*-base region is now formed by ion implantation of boron with the already patterned field oxide as a mask (i.e., no mask is used during this step), as shown in Fig. 5.24(b). A screening oxide (not shown) of 250 Å in thickness is usually used to prevent channeling during the ion implantation. The dose of the *P*-base boron ion implantation is approximately $10^{14}\ \text{cm}^{-2}$ to obtain a junction depth of 2.5–3 μm with a surface concentration of $3\text{–}5 \times 10^{17}\ \text{cm}^{-3}$ that is required to achieve a threshold voltage of about 4 V. After driving in the *P*-base region, Mask 3 is used to define the photoresist regions, which pattern the N^+-emitter regions, as shown in Fig. 5.24(b). The N^+-emitter region is typically formed by ion implantation of phosphorus with a high dose in the range of $10^{15}\ \text{cm}^{-2}$. Its drive-in cycle is chosen to achieve a junction depth of 1 μm leading to an MOS channel length of 1.5–2 μm for the IGBT. A thick oxide is grown over the silicon surface during this step for use as a mask during the etching of the trenches.

A fourth mask is used to define the location of the trenches, as shown in Fig. 5.24 (c). This mask must be carefully aligned to the previous N^+ emitter mask. The etching of the trenches is performed using reactive ion etching (RIE). To obtain a vertically

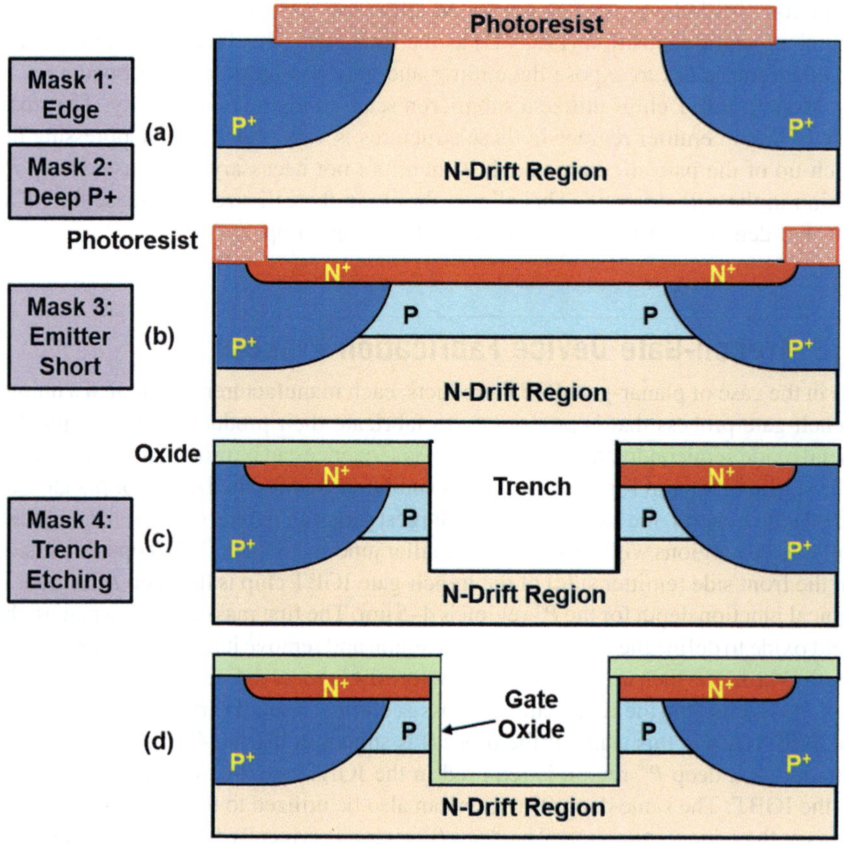

FIGURE 5.24

Fabrication process for the trench-gate IGBT structure: (a) cross-section after deep P^+ diffusion; (b) cross-section after *P*-base and N^+ emitter diffusions; (c) cross-section after trench etching; (d) cross-section after gate oxide growth.

oriented sidewall for the trenches, it is common to use fluorine-based gases such as SF_6 plus oxygen. The process chemistry is optimized for each process tool to avoid producing microtrenching that can enhance electric fields at the corners of the trenches. The RIE process leaves damage on the sidewalls and bottom of the trenches that can severely degrade the MOS channel in the IGBT structure. It has been experimentally demonstrated that the RIE damage can be mitigated by the growth and subsequent removal of a sacrificial oxide [15]. A sacrificial oxide thickness of 2000–4000 Å is typically required to remove the damage depending on the RIE process. This has also been shown to impact the reliability of the MOS devices [16]. The growth of this oxide consumes some silicon on the trench sidewalls. This must be accounted for during the design of masks. Once the trench sidewall has been cleaned to remove the RIE damage, the gate oxide is grown, as shown in Fig. 5.24(d). The gate oxide growth process must be optimized to ensure uniform oxide thickness on the trench sidewall and the bottom of the trench.

The polysilicon gate layer is now deposited using a conformal CVD process, as illustrated in Fig. 5.25(e). The process must ensure the complete refill of the trench without creating voids. The polysilicon is then planarized, that is, its thickness is reduced, as shown in Fig. 5.25(f). A mask (Mask 5) is then used to pattern the polysilicon, as shown in Fig. 5.25(f). Sometimes, the polysilicon is recessed below the top silicon surface so that it goes over the trench corner at the top to avoid reliability problems at this corner. In this case, the polysilicon etch process must be sufficiently controlled so that it does not extend below the depth of the N^{+}-emitter region.

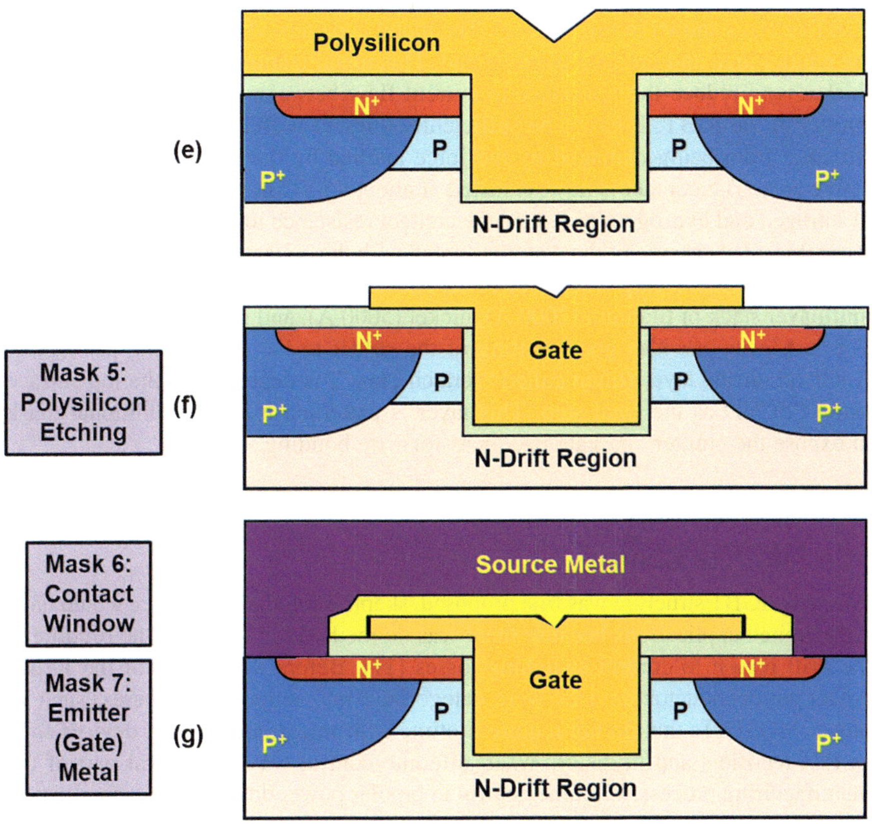

FIGURE 5.25

Fabrication process for the trench-gate IGBT structure (continued): (e) cross-section after polysilicon deposition; (f) cross-section after etching polysilicon; (g) cross-section after emitter metal.

An intermetal dielectric is now deposited on the top surface using an LPCVD process. The dielectric is typically silicon dioxide or phosphosilicate glass. This layer must be free of pinholes to avoid creating gate-source shorts. It must also be conformal and cover the edges of the polysilicon. If the process creates a thinning of the intermetal dielectric at the edges, more instances of gate-source shorts occur due to extremely large width (typically many meters) of the polysilicon edge in an IGBT. This intermetal dielectric is etched using Mask 6, as shown in Fig. 5.25(g) to produce the contact window in the cell. This is usually a critical yield-limiting process step during the manufacturing of IGBTs. If the contact window is not precisely aligned to the polysilicon edge, the contact window can overlap the polysilicon gate, leading to a gate-short short circuit. In addition, the contact window must expose both the P^+ region and the N^+-emitter region to allow making a short between them for latch-up suppression.

Aluminum metallization is now deposited to contact the emitter regions in the IGBT cells, as shown in Fig. 5.25(g). The metal is patterned using Mask 7 to define the emitter area and the gate bonding pad. This metallization is typically 4–5 μm in thickness to reduce the resistance to current flow between the bond wires for the emitter of the IGBT. It has to be sufficiently thick to buffer the underlying device structure from being damaged by the force applied by the bonding wire over the active area. The contact is now annealed at about 450°C in forming gas (a mixture of nitrogen and hydrogen) to reduce the contact resistance to the P^+ and N^+ regions. After this, the backside of the wafer is coated with the collector contact. Because the bottom of the IGBT chip is soldered to the package or module, this metal is usually a multilayer stack of titanium (2000 Å), nickel (5000 Å), and gold or silver (2000 Å). The last layer provides good wetting of the solder to the chip during packaging. A silicon nitride layer (often called "scratch glass") is deposited at the top surface using PECVD as the final step. This layer is patterned using an additional mask to expose the emitter and gate pad areas for wire bonding.

5.7 Lifetime Control

When the IGBT structure was first proposed, skeptics felt that the device would have very limited applications because there was no method for adjusting the lifetime in the drift region to control switching losses [17]. Before the advent of the IGBT, power semiconductor devices were categorized into unipolar structures, such as power MOSFETs that do not require lifetime adjustment, and bipolar devices such as PiN rectifiers and thyristors where lifetime control was an integral part of the manufacturing process. Lifetime control in bipolar power devices was typically performed by the introduction of deep-level impurities such as gold [18] or platinum [19] using diffusion of these elements at 800–900°C. This process had to be performed prior to the aluminum metallization because the diffusion temperature exceeds the melting point of aluminum (660°C). There were many disadvantages in this process such as contamination of production lines and a large variation in the device electrical parameters. In 1975, electron irradiation with an energy of 2 MeV was proposed for controlling the lifetime in silicon power diodes [20]. Electron irradiation had the advantage of being performed at room temperature with excellent precision achieved by monitoring the beam dose during the radiation process. The relative performance of these methods for controlling the characteristics of silicon power didoes was shown to favor electron irradiation [21].

None of these lifetime control processes were considered possible for MOS-gated devices. In the case of the gold and platinum process, these impurities were found to segregate to the interface between the semiconductor and the gate oxide and introduce a negative charge in the gate oxide that alters the threshold voltage [22]. In the case of electron irradiation, it was found that the threshold voltage of n-channel devices becomes negative due to build-up of positive charge at the gate-oxide/semiconductor interface, preventing the device from blocking voltage during circuit operation [23]. Based on this situation, it would have been impossible to control the lifetime in an IGBT structure, relegating the device to only circuits operating at low frequencies.

Fortunately, an electron irradiation process was discovered [24,25] that allowed reducing the lifetime in the silicon and subsequently removing the damage to the gate oxide by annealing the devices at low temperatures (150–200°C). This process was first demonstrated for controlling the reverse recovery of the body diode in power MOSFETs and then applied to controlling the switching speed of IGBTs [26,27]. The electron irradiation process was found to produce on-state characteristics for the IGBT similar to those illustrated in Fig. 5.1. An increase in the on-state voltage drop was observed as the dose for the electron irradiation was increased. Concomitantly, the turn-off time for the collector current in the IGBT was found to reduce with increasing electron irradiation dose [26,27]. This resulted in a reduction of the energy loss per switching cycle for the IGBT, allowing efficient operation at higher frequencies. It is common practice today to perform a trade-off between the on-state voltage drop and the energy loss per switching cycle for IGBTs by using electron irradiation [28].

The ability to control the switching speed of IGBTs was first demonstrated using 3 MeV electron irradiation of symmetric blocking IGBT structures with 600 V capability [27]. The results of this work are illustrated in Fig. 5.26, where the measured turn-off time for the collector current is plotted as a function of the dose for the electron irradiation. The behavior observed in the figure can be described by using:

$$\frac{1}{t_{OFF}} = \frac{1}{t_{OFF,0}} + K_{ER}\phi_{ER} \tag{5.3}$$

where t_{OFF} is the postirradiation turn-off time, $t_{OFF,0}$ is the preirradiation (or zero dose) turn-off time, and ϕ_{ER} is the electron irradiation dose. The coefficient K_{ER} depends on the energy of the electron irradiation and has a value of 0.25 μs^{-1}/Mrad for 3 MeV electron irradiation. The preirradiation lifetime is determined by the residual deep-level impurities in the drift region of the IGBT after fabrication and has a typical value of 10–20 μs.

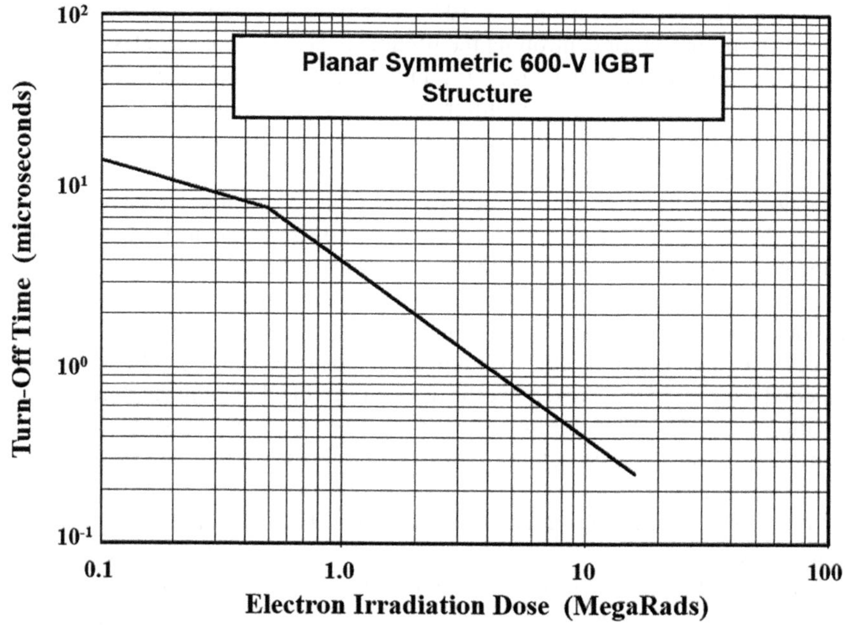

FIGURE 5.26

Electron irradiation process for IGBTs.

In the case of asymmetric IGBTs, the turn-off of the collector current occurs with minority carriers located near or within the buffer layer. In this case, it is advantageous to selectively reduce the lifetime only in this portion of the drift region. This can be achieved by using proton or helium implantation. As an example, a 1200 V asymmetric IGBT has a drift region thickness of about 120 μm with a 10 μm thick buffer layer located beneath it. The lifetime can be selectively reduced in the drift region near the buffer layer by using helium ions with an energy of 18 MeV because they have a range of 128 μm in silicon. The resulting trade-off curve [29] between the on-state voltage drop and the turn-off time is shown in Fig. 5.27. A much superior trade-off curve is produced with this process when compared with electron irradiation, as shown in the figure. However, the helium implantation approach is much more expensive than the electron irradiation process.

5.8 Summary

When novel power semiconductor devices are proposed, their structure is usually quite distinct from any devices already under production by the industry. It is usual practice to create a unique process for the fabrication of the proposed device in a

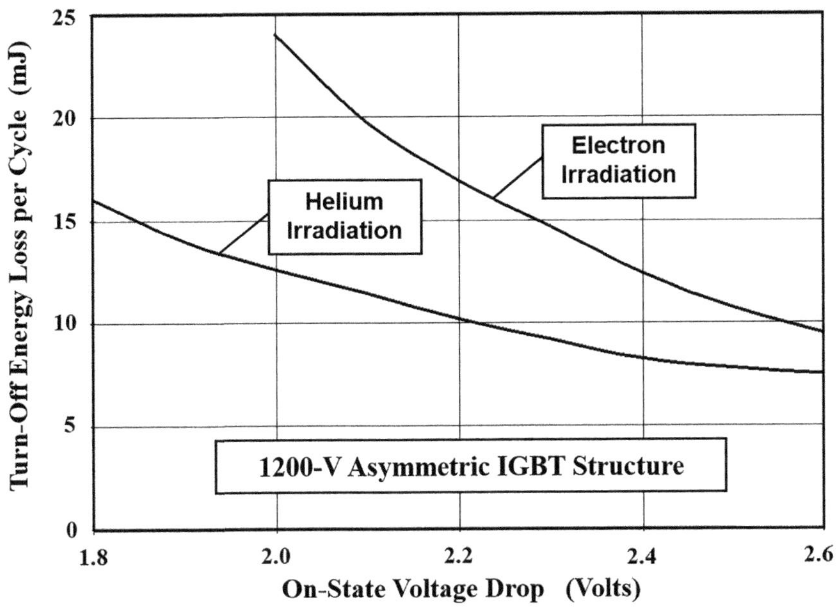

FIGURE 5.27

Helium irradiation process for IGBTs.

research environment to demonstrate the operation of the device and its superior characteristics. This step can often entail several years of effort. After the proposed device is shown to have promising characteristics, a new manufacturing line must then be set up at the product division, which requires substantial investment in equipment cost and several years of effort to bring about the process and obtain high yields. However, the cell structure for the IGBT was carefully and deliberately designed at its conception to be remarkably similar to that for the power MOSFET structure [30]. This was a major factor enabling the manufacturing of the IGBT directly in a production line less than a year after it was conceived at the General Electric Corporate Research and Development Center in 1980 [17,31]. The fabrication of the first IGBT chips in an already existing production process and facility for power MOSFETs allowed sidestepping the expensive and time-consuming steps involved with most power device technology innovations. This was a major factor in its commercial success and rapid widespread adoption for numerous consumer and industrial applications.

The pitfalls during the design of the IGBT cell structure have been discussed in this chapter. An edge termination that achieves close to ideal parallel-plane breakdown voltage is required. A sufficiently wide scribe with an N^+ field stop region is needed to keep the wafer cutting damage from encroaching into the edge termination. The layout of the chip must also be done with care to avoid weak spots at the periphery of the chip and at the bonding pad for the gate.

Modern IGBT products are sometimes designed with self-protection capability. This includes overcurrent, overvoltage, and overtemperature conditions. Unique sensor designs that can be integrated into the IGBT chip have been discussed in this chapter to enable these functions. These sensors can work with off-chip control circuits or by the incorporation of polysilicon diodes and resistors to provide on-chip protection capability.

References

[1] B.J. Baliga, Fundamentals of Power Semiconductor Devices, 2nd, Springer-Science, 2019.

[2] B.J. Baliga, Closed form analytical solutions for the breakdown voltage of planar junctions terminated with a single floating field ring, Solid State Electron. 33 (1990) 485–488.

[3] D. Flores, et al., Investigation on 3.3kV-50A IGBT protection against over-voltage conditions, in: IEEE European Power Electronics Conference, 2009, pp. 1–7.

[4] N. Thapar, B.J. Baliga, Influence of the trench corner design on edge termination of UMOS power devices, Solid State Electron. 41 (1997) 1929–1936.

[5] Y. Xiao, et al., Current sensing trench power MOSFET for automotive applications, in: IEEE Applied Power Electronics Conference, 2005, pp. 766–770.

[6] D.A. Grant, R. Pearce, Dynamic performance of current-sensing power MOSFETs, Electron. Lett. 24 (1988) 1129–1131.

[7] T.M. Jahns, R.C. Becerra, M. Ehsani, Integrated current regulation for a brushless ECM drive, IEEE Trans. Power Electron. 6 (1991) 118–126.

[8] T.P. Chow, et al., Design of current sensors in IGBTs, in: IEEE Device Research Conference, Paper VIA-6, 1992.

[9] T.P. Chow, et al., Modeling and Analysis of Current Sensors for N-Channel, Vertical IGBTs, 1992, pp. 253–256 (Abstract 9.8.1).

[10] S.P. Robb, A.A. Taomoto, S.L. Tu, Current sensing in IGBTs for short circuit protection, in: IEEE International Symposium on Power Semiconductor Devices and ICs, 1994, pp. 81–86 (Paper 3.7).

[11] Y. Seki, et al., A new IGBT with a monolithic over-current protection circuit, in: IEEE International Symposium on Power Semiconductor Devices and ICs, 1994, pp. 31–35 (Paper 2.3).

[12] T. Yamazaki, et al., The IGBT with monolithic over-voltage protection circuit, in: IEEE International Symposium on Power Semiconductor Devices and ICs, 1993, pp. 41–45 (Paper 2.4).

[13] Z.J. Shen, S.P. Robb, C. Cheng, Design and characterization of the 600V IGBT with monolithic over-voltage protection, in: IEEE Power Electronics Specialists Conference, vol. 2, 1996, pp. 1773–1778.

[14] N. Iwamuro, et al., A new vertical IGBT structure with a monolithic over-current, over-voltage, and over-temperature sensing and protecting circuit, IEEE Electron Device Lett. 16 (1995) 399–401.

[15] T. Syau, B.J. Baliga, Mobility study on RIE etched silicon surfaces using SF_6/O_2 gas etchants, IEEE Trans. Electron Devices 40 (1993) 1997–2005.

[16] S.A. Suliman, et al., The impact of trench geometry and processing on the performance and reliability of low voltage power UMOSFETs, in: IEEE Reliability Physics Symposium, 2001, pp. 308–314.
[17] B.J. Baliga, The role of power semiconductor devices in creating a sustainable society, in: IEEE Applied Power Electronics Conference, Invited Plenary Paper, Long Beach, CA, 2013.
[18] J.M. Fairfield, B.V. Gokhale, Gold as a recombination center in silicon, Solid State Electron. 8 (1965) 685–691.
[19] K.P. Lisiak, A.G. Milnes, Platinum as a lifetime control deep impurity in silicon, J. Appl. Phys. 46 (1975) 5229–5235.
[20] K.S. Tarneja, J.E. Johnson, Tailoring the recovered charge in power diodes using 2 MeV electron irradiation, in: Electrochemical Society Meeting, 1975 (Paper 261RNP).
[21] B.J. Baliga, E. Sun, Comparison of gold, platinum, and electron irradiation for controlling lifetime in power rectifiers, IEEE Trans. Electron Devices 24 (1977) 1103–1108.
[22] D.R. Collins, D.K. Schroder, C.T. Sah, MOS capacitance evaluation of gold diffusion in silicon and thermally grown silicon dioxide, IEEE Trans. Electron Devices 13 (1966) 673.
[23] E.H. Snow, A.S. Grove, D.J. Fitzgerald, Effects of ionizing radiation on oxidized silicon surfaces and planar devices, Proc. IEEE 55 (1967) 1168–1185.
[24] B.J. Baliga, J.P. Walden, Power MOSFET integral diode reverse recovery tailoring using electron irradiation, in: IEEE Device Research Conference, 1982 (Paper IV-A-7).
[25] B.J. Baliga, J.P. Walden, Improving the reverse recovery of power MOSFET integral diodes by electron irradiation, Solid State Electron. 26 (1983) 1133–1141.
[26] B.J. Baliga, Fast switching insulated gate transistors, IEEE Electron Device Lett. 4 (1983) 452–454.
[27] B.J. Baliga, Switching speed enhancement of insulated gate transistors by electron irradiation, IEEE Trans. Electron Devices 31 (1984) 1790–1795.
[28] C. Tadokoro, et al., Carrier lifetime control optimization for high speed IGBT based on electrical and physical analysis, in: IEEE International Symposium on Power Semiconductor Devices and ICs, 2010, pp. 145–148 (Paper HV-P6).
[29] Y. Konishi, et al., Optimized local lifetime control for the superior IGBTs, in: IEEE International Symposium on Power Semiconductor Devices and ICs, 1996, pp. 335–338 (Paper 14.3).
[30] B.J. Baliga, Gate Enhanced Rectifier, U.S. Patent 4,969,028, Filed December 2, 1980, Issued November 6, 1990.
[31] B.J. Baliga, IGBT: the GE story, IEEE Power Electron. Mag. 2 (2015) 16–23.

CHAPTER

Package and Module Design

6

It is necessary to place the insulated gate bipolar transistor (IGBT) chip within a package for many reasons beyond physical support of the thin breakable silicon structure. First, it is necessary to access the current flow within the semiconductor device and relay it to the power electronic circuit. Second, it is necessary to remove the heat generated within the IGBT chip due to on-state and switching losses to maintain a reasonable junction temperature. Third, it is necessary to protect the chip from the ambient to avoid long-term degradation due to moisture, corrosion, and electrical charges.

Historically, the first IGBT chips were made available as a discrete power switch in a three-terminal package. Subsequently, the IGBT chip was married with a free-wheeling or flyback diode colocated within the package to provide a three-terminal package suitable for voltage source inverter applications. Because the H-bridge topology for hard-switched power circuits used in adjustable speed motor drives became increasing popular, manufacturers began to supply multiple IGBT chips and fly-back diodes within a power module. The power module has grown from containing a single totem pole to an entire H-bridge to a module containing six IGBTs and six diodes for three-phase motor drive applications. In addition, manufacturers have begun to integrate the drive and control circuits inside the module, creating the intelligent power module (IPM) concept.

Many books and articles have been published on packaging technologies and novel innovations for enhancing heat removal. This chapter provides a review of only basic approaches used for packaging IGBTs due to limited space.

6.1 Discrete Device Package

A popular package for the IGBT as a discrete transistor is the TO-220 design shown in Fig. 6.1. Another similar package used for discrete IGBT products is the TO-3PN. In these packages, a copper tab serves as the base for mounting the IGBT chip using solder. The silicon die must have a "solder-compatible" back-side metallization for the P^+ collector region of the IGBT chip. The back-side metallization is typically composed of titanium–nickel–silver, chromium–nickel–silver, or aluminum–titanium–nickel–silver. The titanium, chromium, or aluminum layers are chosen to make the ohmic contact with the heavily doped P-type collector region of the IGBT because of their low contact resistance with a value of typically less than

The IGBT Device. https://doi.org/10.1016/B978-0-323-99912-0.00013-1

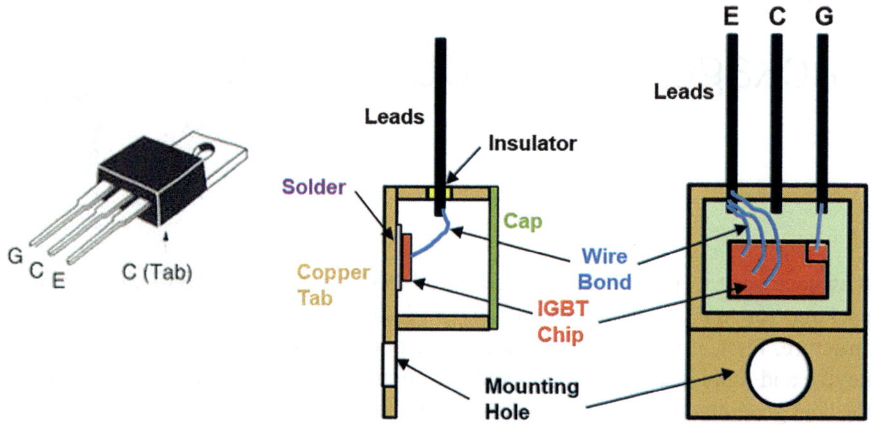

FIGURE 6.1

TO-220 package for the insulated gate bipolar transistor (IGBT) chip.

$10^{-5}\ \Omega\ cm^2$. The silver layer is chosen to promote wetting of the solder to obtain contact over the entire chip area. The other metal layers are used as barriers to mitigate penetration of the solder during mounting of the chip on the substrate.

The solder must be chosen with a melting temperature as low as possible to minimize the thermal stress caused by any mismatch in the expansion coefficients for the silicon and the metals. However, because the junction temperature for IGBT chips is typically rated at up to 150°C, the solder melting point must be at least 160°C to prevent remelting and detachment of the IGBT die during normal circuit operation. Eutectic solder alloys have a single melting temperature but noneutectic alloys have a "plastic" phase between the solid and liquid temperatures. This plastic temperature range must be as narrow as possible. The solder alloy must have a low elastic modulus to absorb the thermal stress due to the mismatch in the expansion coefficients for the silicon and metals. It must have a high thermal conductivity to allow the heat generated within the IGBT chip to flow down into the copper baseplate in the package. The resistivity of the solder must be low to reduce the ohmic voltage drop through the solder layer. Some typical solder compositions that have been widely used by the industry are listed in Table 6.1 with their pertinent properties. The melting point of the lead-based solders can be increased by reducing the tin content. Because the European Union banned lead-based solders in 2006, the packaging industry has moved away from lead-based solders due to environmental concerns. The typical lead-free solders listed in the table provide a more environmentally favorable choice for the industry.

Typically, the composition of the solder is chosen to achieve a melting temperature of about 200°C [1]. The die attach assembly is performed with a chip mounting thermal cycle consisting of an initial ramp-up to a preheat temperature (~150°C) for 100 s, followed by a second ramp-up to a temperature exceeding the melting point of the solder. The time above the liquidus is typically 20–90 s in duration. Too short a

Table 6.1 Typical Solders Used for Mounting the Insulated Gate Bipolar Transistor Chip

Alloy Composition	Liquidus (C)	Solidus (C)	Plastic Range (C)
Lead-based solders			
60% Sn/40% Pb	190	183	7
25% Sn/75% Pb	266	183	83
15% Sn/82.5% Pb/2.5% Ag	280	275	5
Nonlead-based solders			
71% Sn/25% Bi/4% Ag	180	180	0
96.5% Sn/3.5% Ag	221	221	0

duration leads to poor wetting and incomplete joints for the chip while intervals that are too long produce intermetallic compounds that degrade the resistance. The temperature is then ramped down at a constant rate. The temperature profile for this die mounting process is illustrated in Fig. 6.2.

After the die attach step, wire bonds are applied to make connection to the emitter and gate terminals of the IGBT device. The wires can be either gold or aluminum. Aluminum wires are generally preferred for IGBT chips due to their lower cost. Aluminum wire with 1% magnesium has superior resistance to fatigue. Large emitter currents can be routed to the device terminals by using multiple aluminum wire bonds distributed uniformly across the active area of the chip. The diameter of

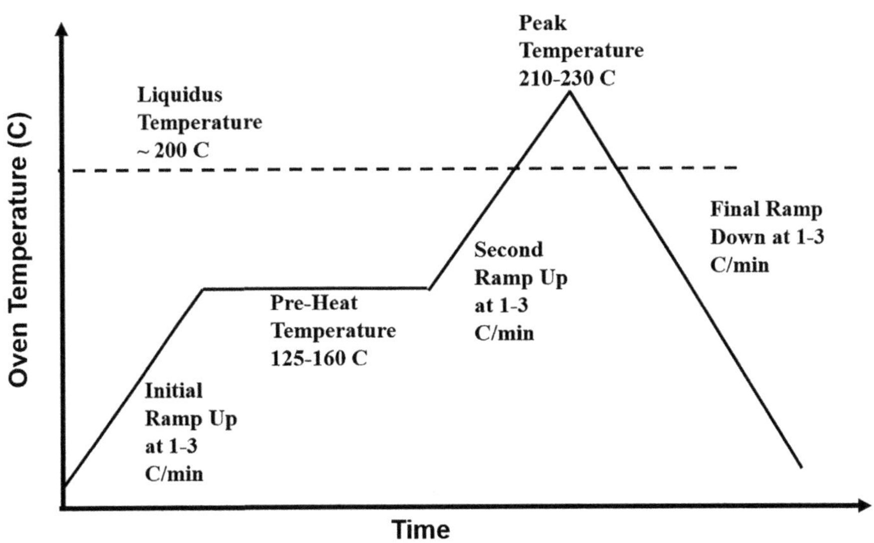

FIGURE 6.2

Typical solder mounting cycle for the IGBT chip.

Table 6.2 Current-Handling Capability of Aluminum Wires

Aluminum Wire Diameter (mils)	Maximum Current (A)
1	0.5
2	1.5
5	6
12	25
22	60

the aluminum wires can vary from 1 mil (0.001 in) to 22 mils (0.022 in). The current carrying capability of these wires is provided in Table 6.2. A smaller aluminum wire diameter is appropriate for smaller IGBT die sizes.

As previously discussed, the power-handling capability of the IGBT chip is decided by limiting the junction temperature to a maximum value based on reliability considerations. The junction temperature is determined by the power dissipation and the thermal resistance. The package design has a strong influence on the thermal resistance. The typical thermal resistance of the TO-220 or TO-3PN package under steady-state operation is 1 °C/W. The thermal resistance is dependent on the duty factor under pulsed mode operation, as is typical for pulse-width-modulated motor control circuits. A typical example of the thermal impedance is illustrated in Fig. 6.3 for pulse mode operation. It is worth noting that the thermal impedance becomes equal to the steady-state thermal resistance when the pulse width exceeds 0.1 s. The thermal impedance becomes smaller as the duty factor becomes smaller, that is, for short pulses with long periods.

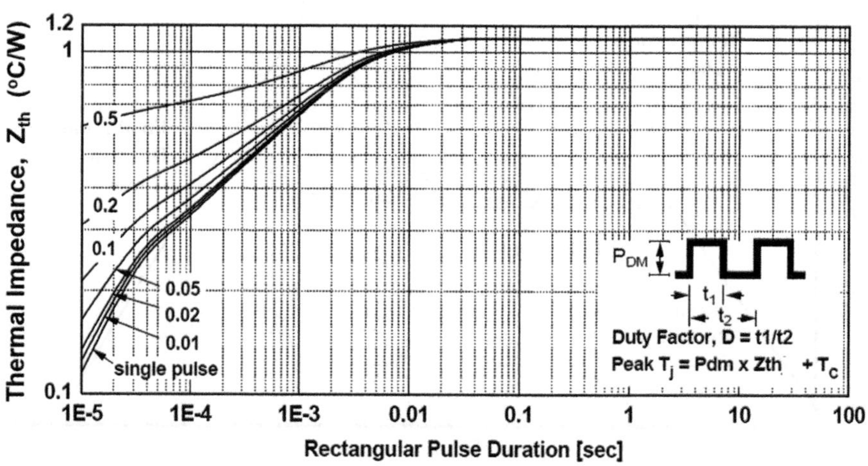

FIGURE 6.3

Thermal resistance of the IGBT package.

6.2 Improved Discrete Device Package

The discrete device package discussed in the previous section has several drawbacks. First, wire bonds are used to access the current flow within the IGBT chip. The wire bonds contribute to the series resistance and series inductance. Second, the wire bonds have been proven to limit the reliability, which will be discussed in the section on power modules. Third, the use of wire bonds restricts the removal of heat from the chip via only the bottom or collector side of the device. This results in a relatively large thermal resistance.

These problems can be addressed by replacing the aluminum wire bonds for the emitter and gate contacts with a ball grid array (BGA) [2]. This allows creating a chip scale package that has a much smaller form factor and better electrical and thermal characteristics. A typical layout for the BGA on an IGBT chip is shown in Fig. 6.4, together with a cross-section below it. The solder bumps are placed on pads of about 1 mm in size. The emitter pads are distributed over the active area of the emitter on the IGBT device. Six emitter pads are shown in the illustration. The gate pad also has a solder bump attached to it. The construction of the BGA requires an under bump

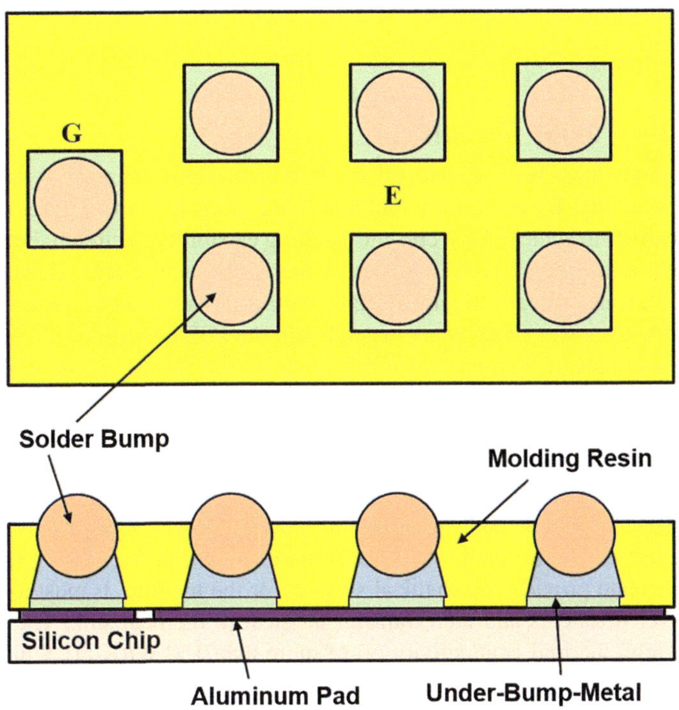

FIGURE 6.4

IGBT package based on using a ball grid array.

metal (UBM) pad consisting of titanium/nickel/silver on top of the aluminum metallization.

The solder bump formation process begins by evaporation of the UBM on the aluminum metallization. The UBM is patterned by photolithographic liftoff or chemical etching processes. A patternable solder mask is applied to the wafer using spin coating and the solder bump pads are then defined using photolithography. Stencil printing is used to form the solder bumps. This consists of pushing solder paste (96.5% tin/3.5% silver) onto the pads through a stencil with a squeegee. After prebaking the solder paste to form a solder cap, solder balls are dropped through windows in a stencil on top of the solder paste. The next reflow process step is performed at 250°C. This temperature exceeds the melting point of 221°C for the solder cap but is below the melting point of 268°C of the (10% tin/90% lead) solder ball. The reflow operation is done in a reducing atmosphere (96% nitrogen/4% hydrogen) to avoid oxidation of the surfaces.

When compared with the conventional IGBT package using wire bonds, the BGA package has been found to reduce the on-state voltage drop of the IGBT by 20% and the series resistance by 30%. In addition, the solder bump provides a large contact area at the emitter side to extract heat. An improvement in thermal resistance by 30% is expected from the two-sided cooling of the IGBT chip [2]. Thermal cycling tests have demonstrated stable operation of devices with BGA connections.

6.3 Basic Power Module

Power modules have been developed for IGBTs to allow increasing the current-handling capability. By combining multiple IGBT chips in parallel within the module, it is possible to achieve high current-handling capability to circumvent the problem of yield reduction of IGBT chips with increasing active area. Because motor drive applications typically require IGBTs with a flyback diode, the modules are engineered with power rectifiers colocated with the IGBT chips.

Some examples of high-current IGBT modules are shown in Fig. 6.5, together with the basic cross-section below. The module consists of a baseplate to which a ceramic substrate is attached. The silicon chips (IGBTs and power rectifiers) are mounted on the ceramic substrate using solder. Wire bonds are used to interconnect the chips and make connections to the terminals of the package. A significant (~50%) reduction in the package parasitic inductance was achieved by replacing S-shaped electrodes with the L-shaped electrodes shown in the figure [3].

The baseplate provides mechanical support for the module. It must absorb heat during power transients and then transmit the heat to the heat sink. The baseplate must have high thermal conductivity (λ) of more than 150 W/(K m), a high degree of surface smoothness of better than 2 μm, and reproducible warpage to allow intimate contact to the heat sink. The coefficient of thermal expansion of the baseplate should be compatible with the insulating substrate or ceramic used for providing electrical isolation for the silicon chips. The baseplate should have high tensile

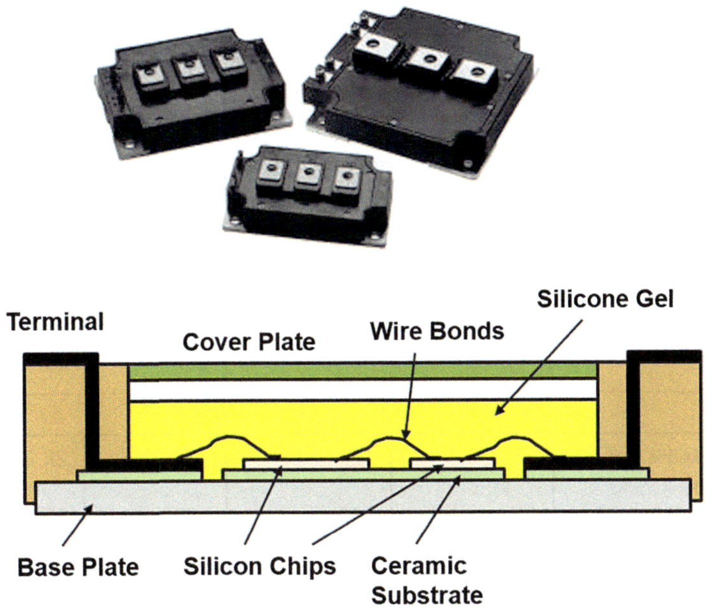

FIGURE 6.5

High current IGBT module.

strength and high flexural strength to allow it to be shaped and bent [1]. Its density (ρ) should be low to reduce the weight of the module. The baseplate is typically made from metal matrix composites [4] such as 37% aluminum/76% silicon carbide or 60% copper/40% carbon. Some of the relevant properties of these materials are provided in Table 6.3.

The ceramic layer is used to provide electrical isolation of the silicon chips from the baseplate and heat sink. The most suitable ceramics for power modules are alumina (Al_2O_3), aluminum nitride (AlN), and silicon nitride (Si_3N_4), whose properties are provided in Table 6.3. Alumina is the most commonly used ceramic for substrate material and is a mature, low-cost technology. It can be easily metallized and machined, and is nontoxic. However, its thermal conductivity is inadequate for high-power applications and it has a significant mismatch in thermal expansion coefficient compared with silicon. Aluminum nitride offers six times better thermal conductivity, which makes it attractive for high-power applications. Its thermal expansion coefficient is a better match with silicon. It is also easily machinable and nontoxic. However, it is four times more expensive than alumina. Silicon nitride has a very good thermal expansion match to silicon and a high thermal conductivity. It can be easily metallized and machined, and is nontoxic. It is twice as expensive as alumina. It is more suitable as a stand-alone substrate without the baseplate.

The choice of the solder used for the module assembly is based on considerations similar to those already discussed for the packaging of discrete devices. The melting

Table 6.3 Properties of Materials Used in Insulated Gate Bipolar Transistor Modules

Material	λ (W/(K m))	C (J/(kg K))	ρ (kg/m^3)	CTE (10^{-6}/K)
Baseplate material				
37% Al/63% SiC	175	740	4000	7.9
60% Cu/40% C	300	420	6100	8.5
Silicon and metallization				
Si	146	750	2330	2.5
Al	220	880	2700	24
Cu	400	380	8850	17
Solder				
29.5% Pb/5% Sn/ 2.5% Ag	35	129	11,300	29
96.5% Sn/3.5% Ag	33	200	7360	30
Ceramics				
Al_2O_3	24	765	4000	6.0
AlN	170	750	3300	4.5
Si_3N_4	60	800	3290	3.3

CTE, *coefficient of thermal expansion.*

point of the solder should be as low as possible to minimize the thermal stress due to mismatch in the coefficients of thermal expansion of the various materials in the module. The highest temperature used during assembly is kept below 350°C. The lower limit to the melting point for the solder is 160°C based on a maximum junction temperature of 150°C for the IGBT chips. Unlike the case of discrete devices, two types of solder are required for the IGBT module. The silicon chip is first mounted onto the ceramic substrate using a higher melting point solder. The ceramic substrate is subsequently mounted onto the baseplate using a lower melting point solder. A difference in melting point of at least 40°C is required for the solders to ensure that the chip is not detached during mounting of the ceramic substrate [1]. These requirements can be met by using the 96.5% Sn/2.5% Ag solder with a melting point of 220°C to attach the silicon chip to the ceramic substrate and the 29.5% Pb/5% Sn/ 2.5% Ag solder with a melting point of 180°C to attach the ceramic substrate to the baseplate.

After mounting the ceramic substrate with the IGBT and power rectifier chips on the baseplate, wire bonding is performed to interconnect the chips and connect them to the electrodes of the module. Aluminum wires are preferred in IGBT modules because a wedge bond can be formed on the aluminum metallization on the chips using large-diameter wires to carry large currents. The low electrical resistivity

($2.65 \times 10^{-10}\ \Omega$ cm) for aluminum allows the wires to handle high currents, as shown in Table 6.2.

The silicon chips must be protected against moisture, which can produce significant corrosion and contamination from mobile ions. This protection is achieved by using an encapsulant inside the module. The encapsulant also provides insulation between the wires if its dielectric strength is more than 250 V/mil. Typical encapsulants used by the industry include silicone gel (as illustrated in Fig. 6.5), epoxy, polyurethane, and acrylic. The encapsulation should not shirk during curing to avoid damaging the wires and should not crack during thermal cycling of the module. Silicone gel is the most popular encapsulant due to its high purity, chemical resistance, and low toxicity. Its softness reduces stress on the chips and wires.

The plastic case and cover provide the final protection of the semiconductor devices against mechanical and environmental abuse. The design of the terminals on the case must take into account high-voltage creepage and arcing for high-voltage applications.

6.4 Flat-Pack Power Module

The basic power module described in the previous section utilizes wire bonding to interconnect the chips and connect them to the electrode. Although this technology is in widespread use in the industry, it has been demonstrated that the wire bonds create significant parasitic inductance that impacts the switching performance of the IGBT [5]. A single 24 mm long wire with a diameter of 20 mils has a self-inductance of 10–16 nH. This parasitic inductance is reduced by using multiple wire bonds operating in parallel and distributed over the emitter area of the IGBT. The wire bonds have also been shown to create nonuniformity in current distribution between paralleled IGBT chips. Wire bond failure has also been identified as a major reliability problem in modules.

The flat-pack module concept was developed to overcome the shortcomings of wire bonds in IGBT power modules [6]. In this approach, the package configuration looks like that previously used for packaging gate turnoff thyristors. The IGBT and rectifier chips are mounted using a solder joint to a collector side molybdenum baseplate. The contact to the emitter of the IGBT chip is made using pressure contact with a molybdenum contact plate. Molybdenum is chosen because its coefficient for thermal expansion is close to that for silicon. This eliminates the use of wire bonds and allows extraction of heat from both sides of the chip. A wire bond is used only for the gate electrode, which is brought out at the side of the package.

An example of the flat-pack IGBT module is shown in Fig. 6.6 with its cross-section below. Eleven IGBT chips and five rectifier chips are used to achieve a current-handling capability of 1800 A with a blocking voltage capability of 2.5 kV. The termination at the edges of the IGBT chip is covered with polyimide for protection.

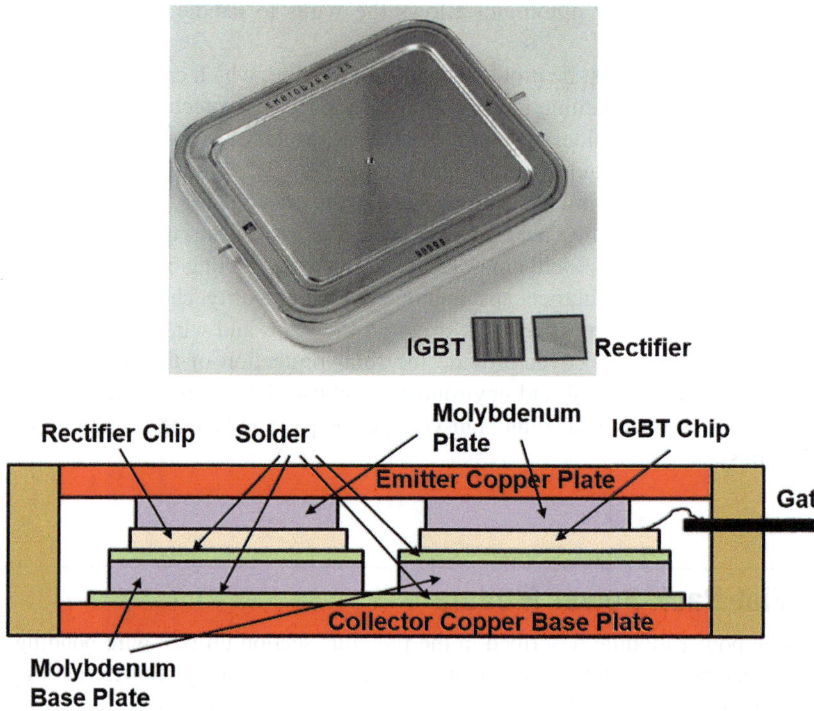

FIGURE 6.6

High current flat pack IGBT module.

One of the problems with achieving high-current handling capability in IGBT modules is loss of yield due to short circuits between the gate and source regions on the chip. This problem can be overcome by using wafer repair methods. One approach is illustrated in Fig. 6.7 where the flat-pack module is shown at the top without the emitter contact plate. This module contains nine IGBT chips and three rectifier chips to achieve a current-handling capability of 1000 A with a voltage blocking capability of 2.5 kV. Each IGBT chip is 20 mm × 20 mm in size. To obtain a good yield, the IGBT chip is partitioned into 36 units, each with its own gate and emitter contact area. Testing is performed to find gate source short circuits to locate the bad units. A polyimide layer is applied and patterned to allow the emitter contact to contact the emitters of all the units and short the gate and source of the bad units, as shown at the bottom of Fig. 6.7. The polyimide pattern is also designed to allow the gate bus to contact the gate pads of only the good units, as shown in the figure. In this wafer repair approach, each IGBT unit is designed to operate at 20% above the current rating of the chip so that up to four defective units out of the 36 total units can be eliminated. The thermal impedance of this module is 40% smaller than the conventional module with wire bonds [6]. The concept has been extended to modules capable of supporting 4.5 kV for use in industrial, traction,

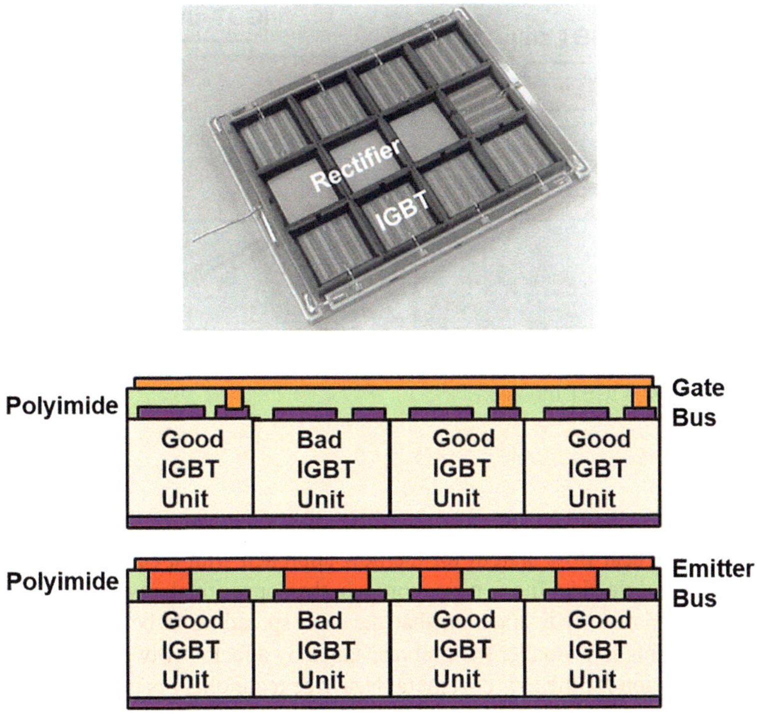

FIGURE 6.7

Flat pack IGBT module with chip repair.

and power transmission applications [7]. An alternate approach for wafer repair by the use of fusible links to automatically isolate IGBT units with gate to source shorts has also been demonstrated [8].

6.5 Metal Baseplate Free Power Module

It is desirable to reduce the weight and cost of IGBT modules for applications. One approach toward this goal is the elimination of the metal baseplate. Direct copper bonded (DCB) ceramic substrates are commonly used for modules. The ceramic substrate containing the IGBT and rectifier chips can be directly mounted to the heat sink. However, it is necessary to increase the thickness of the DCB substrate for obtaining adequate mechanical strength. This results in a substantial increase in the thermal resistance.

An approach to reducing the thermal resistance while achieving good mechanical strength is shown in Fig. 6.8 [9]. This approach is based on better heat spreading within copper than in the ceramic (alumina) substrate. The arrows in the figure indicate the heat

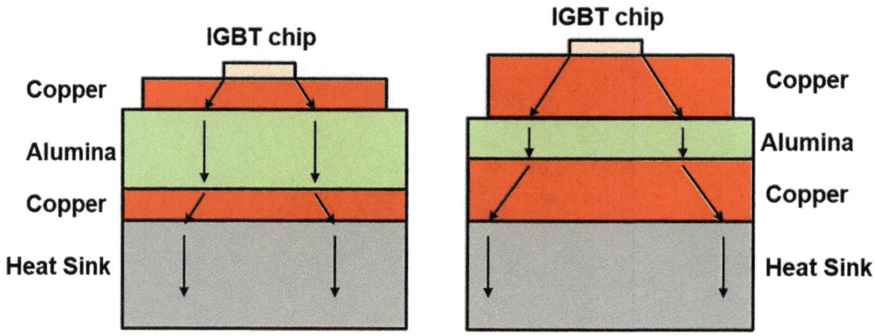

FIGURE 6.8

Metal baseplate free IGBT module.

flow pattern for the conventional approach with a ceramic thickness of 0.635 mm on the left side and the heat flow pattern for the new approach with a ceramic thickness of 0.32 mm on the right side. In the new approach, the copper thickness is also increased from 0.25 to 0.5 mm. It is apparent that the heat spreading is better with the new approach, resulting in a smaller thermal resistance by a factor of two times.

One application for the low-cost metal base free technology is in electric vehicles (EVs) and hybrid-electric vehicles (HEVs). These modules require a voltage blocking capability of 650 V and a current-handling capability of 300 A. The relatively low blocking voltage for these IGBTs is achieved using the asymmetric device structure with a wafer thickness of only 70 μm. Such thin wafers are prone to warpage when the solder bump is applied to the emitter and collector sides [10]. Despite these issues, metal baseplate free modules with 70 μm thick IGBT and rectifier chips have been successfully developed with 200 kVA capability.

The direct liquid cooled (DLC) module has become an industry standard for HEV and EV applications. It allows double-sided cooling of the IGBT and free-wheeling diode chips, which reduces the thermal resistance in half compared with the conventional modules shown in Fig. 6.5. This allows for a reduction of module size and weight, which is important for mobile applications. The DLC module structure is illustrated in Fig. 6.9. It provides the desired electrical, thermal, mechanical, and reliability criteria for the stringent automotive market [11]. Planar connections to the high current emitter and collector electrodes of the IGBT are made by soldering the IGBT collector side of the chip to a direct bonded copper (DBC) layer at the bottom. The emitter is also soldered to a molybdenum/copper post connected to the top copper (DBC) layer in the package, eliminating the use of wire bonds. The same approach is applied to the free-wheeling-diode (FWD) inside the package. The top and bottom copper (DBC) layers are connected to the bus bar connections extending out of the package. External heat sinks can be attached to both sides of this package due to the planar architecture.

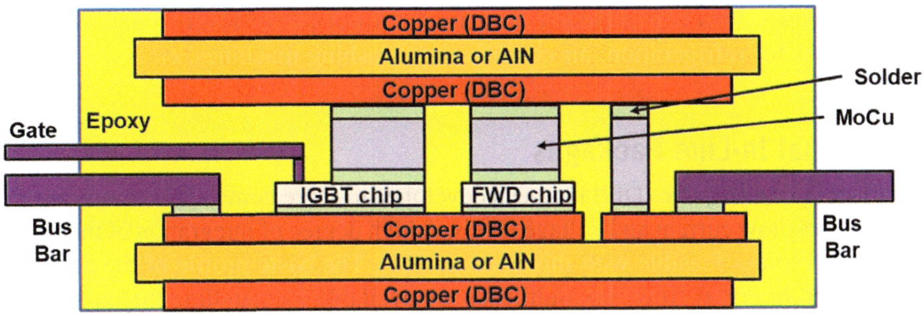

FIGURE 6.9

Direct bonded copper technology for IGBT modules.

Modern IGBT modules with high power density and reliability have been constructed using the DBC technology shown in Fig. 6.9. The bottom copper layer of the DBC is soldered to the heat sink. The top copper layer of the DBC is patterned to accommodate the IGBT chip and the fly-back diode chip. The ceramic layer is comprised of either alumina (Al_2O_3) or aluminum nitride (AlN). The advantages of the DBC substrate are: (a) high thermal and electrical conductivity with high current-carrying capability of the copper layers; (b) excellent electrical insulation, thermal conductivity, and mechanical strength of the ceramic layer; and (c) ability to etch patterns in the copper layer to create versatile module designs. AlN is the preferred ceramic material due to its thermal expansion coefficient matching that of Si. The chip is sintered to the bottom DBC layer and the Mo/Cu connectors using nanosilver. The Mo/Cu connectors not only provide the electrical connection to the emitter of the IGBT, but also provide an additional heat removal path. This two-sided cooling has been shown to reduce junction temperature by 47% compared to single-sided cooling [12].

6.6 Smart Power Modules

The advent of the IGBT heralded the second electronic revolution by enabling smart power technology. Smart power technology combines high power management capability using IGBTs as power switches with intelligence to control the switching and protection against adverse overcurrent, overvoltage, and overtemperature circuit operating conditions brought about by malfunctions in the load [13]. Smart power technology was initially based on multiple discrete IGBTs and diodes packaged separately. The system was then assembled on a printed circuit board (PCB). The next generation of smart power technology was the placement of all the power chips (IGBTs and diodes) within a single module with the rest of the protection and control circuits located on the PCB. In order to reduce cost and size, the latest trend has been to package the power chips together with the control and protection chips within a single package. This is particularly attractive at lower power levels found in home applications for refrigeration, air conditioning, washing machines, etc.

6.6.1 Dual In-Line Packages

The dual in-line package (DIP) has been developed to serve home applications due to the relatively low current and voltage levels. A high degree of integration of the functionality becomes feasible with this package [14]. The basic profile of the package can be seen in Fig. 6.10 where multileads are available for handling the power and communications needs of the applications. To allow removal of heat from the power

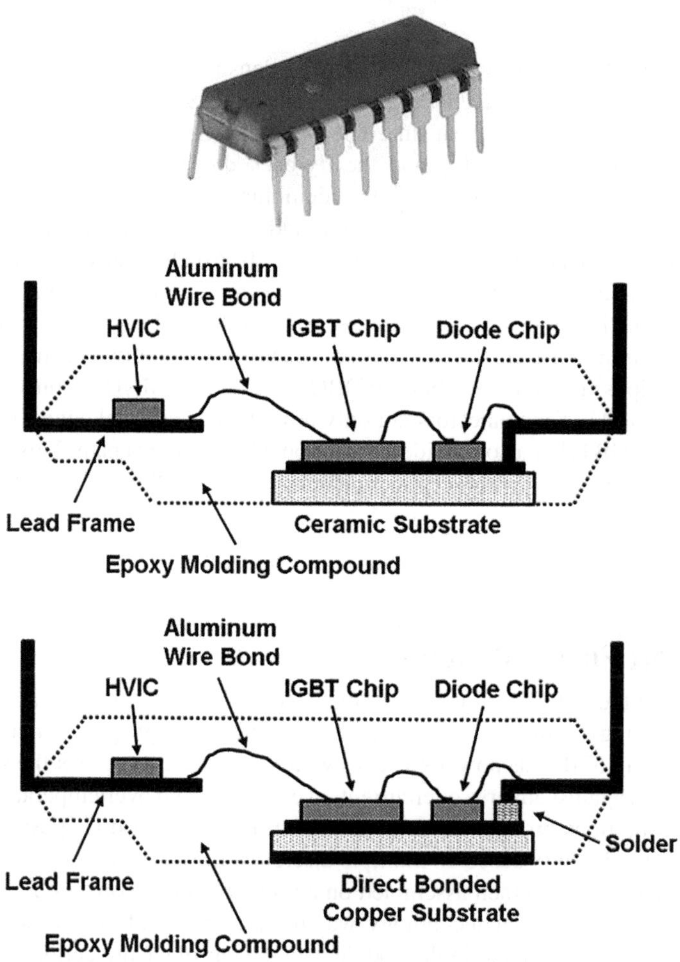

FIGURE 6.10

Dual in-line package for intelligent IGBT power module. *HVIC*, high-voltage IC.

devices (typically IGBTs and fly-back diodes) inside the package, they are mounted on ceramic substrates, as illustrated in the figure. The direct bonded ceramic substrate is a popular approach. Passive components and control chips, including the high-voltage IC (HVIC) for level shifting the gate signal, can be assembled inside the DIP package to minimize the number of parts. This approach provides a compact power management solution convenient for home appliances. Typical power ratings are 3–30 A of current capability with 600 V operating capability. The thermal resistance of the DIP package varies from 4.5 °C/W for a current rating of 3 A to 1 °C/W for a current rating of 30 A, as shown in Fig. 6.11.

A complete three-phase inverter bridge with the gate drive and protection circuits can be achieved that can directly interface with microcontrollers and digital signal processing (DSP) chips. The block diagram for this inverter in a DIP package is shown in Fig. 6.12. The six IGBTs and their companion diodes are located on the right side. The control chips are located on the left side.

An improved version of the above smart power module in the compact DIP package for home appliances has been developed with a power-handling capability of 7 kW [15]. These modules are intended for home applications such as air conditioners and refrigerators. The IGBTs in the module include current sense capability. The control circuits not only provide overvoltage, overcurrent, and overtemperature capability, but include undervoltage lockout protection when the drive supply voltage is inadequate. An HVIC is used for level shifting the signal to the IGBTs on the upper side of the bridge circuit to eliminate the need for optocouplers.

A similar compact IGBT based package, called SLIMDIP, has been reported for applications in home air conditioners, washing machines, and refrigerators [16]. The RC-IGBT structure is employed here because the integration of the fly-back diode into the IGBT device provides significant space savings. The conventional package with a size of 24 mm × 38 mm contained six IGBT chips and six fly-back diode chips,

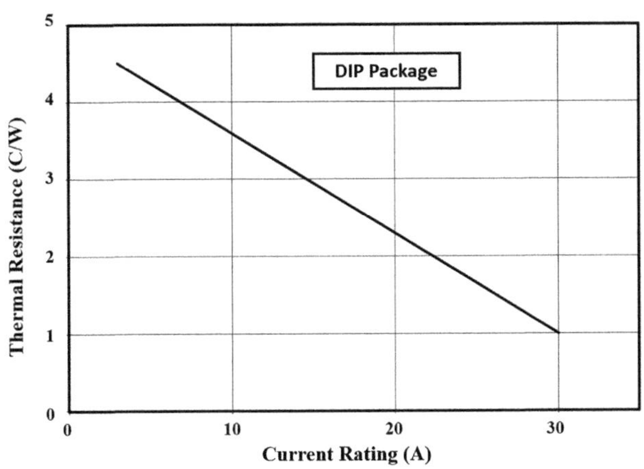

FIGURE 6.11

Thermal resistance for the IGBT dual in-line package.

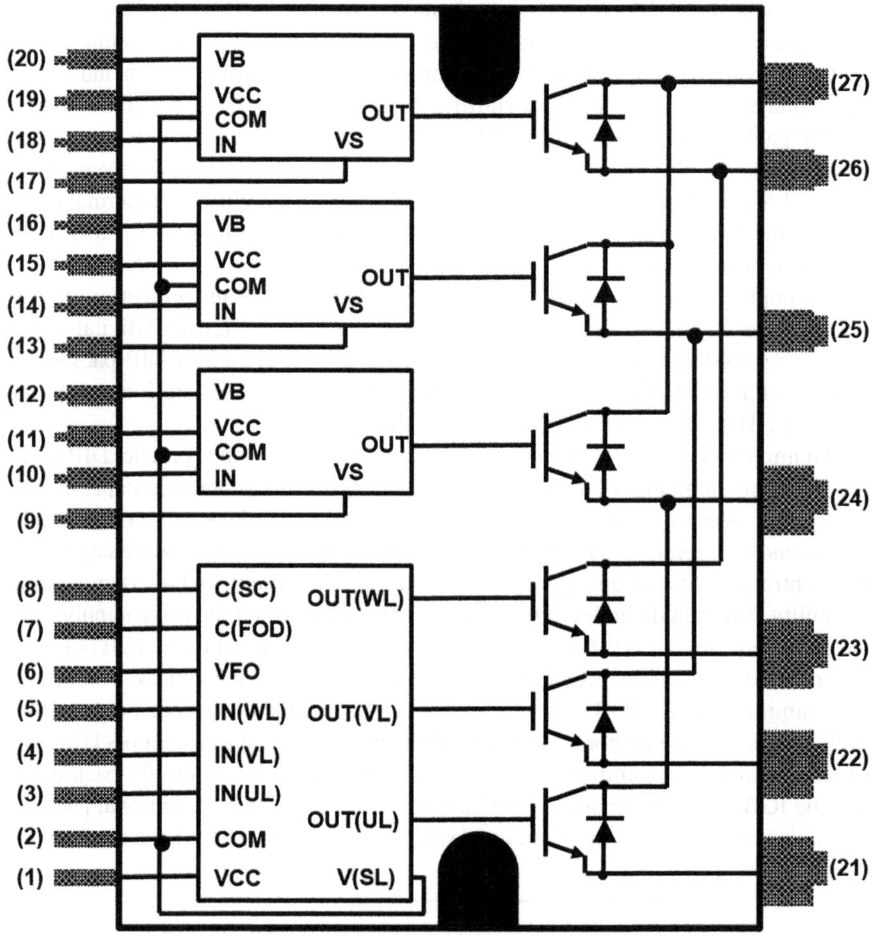

FIGURE 6.12

Block diagram for IGBT dual in-line package for motion control applications.

plus the HVIC as a control chip with gate drive and protection features. Reducing the number of power chips in half with the "superthin" RC-IGBT chips allowed reducing the package size to 18.8 mm × 32.8 mm. The superthin RC-IGBT has a 50% smaller on-state voltage drop, allowing a reduction of the package size from a thermal standpoint as well. The package has a rating of 15 A/600 V, which is suitable for motor drives of up to 1.5 kW.

6.6.2 Intelligent Power Modules

The term intelligent power module (IPM) has been coined to describe modules that can handle large power levels with significant self-protection and load-monitoring capability. One of the important applications for this technology is in EVs and HEVs for driving the motors [17]. These modules contain not only the power devices, namely IGBTs and free-wheeling diodes, but a circuit board with control and protection circuits. An example [17] of an IPM designed for EV/HEV applications, called an intelligent power drive unit, is provided in Fig. 6.13. The block diagram shown at the top of the figure shows the basic components within the module. The module contains the DC–DC converter for supplying power to the control chips, a predriver for generating the gate drive voltages for the IGBTs, and the DC link capacitor. An important innovation is the replacement of the previously used electrolytic capacitors with a solid capacitor whose volume is one-10th in size. Placement of the capacitor within the module in close proximity to the IGBTs reduces the parasitic inductance, leading to smaller voltage overshoots during the switching of the power devices.

IPMs with even larger power-handling capability for electric street cars and trains have also been developed by many companies. One example is the SEMiX series that utilizes spring contacts for the IGBTs to eliminate the reliability problems associated with wire bonds [18].

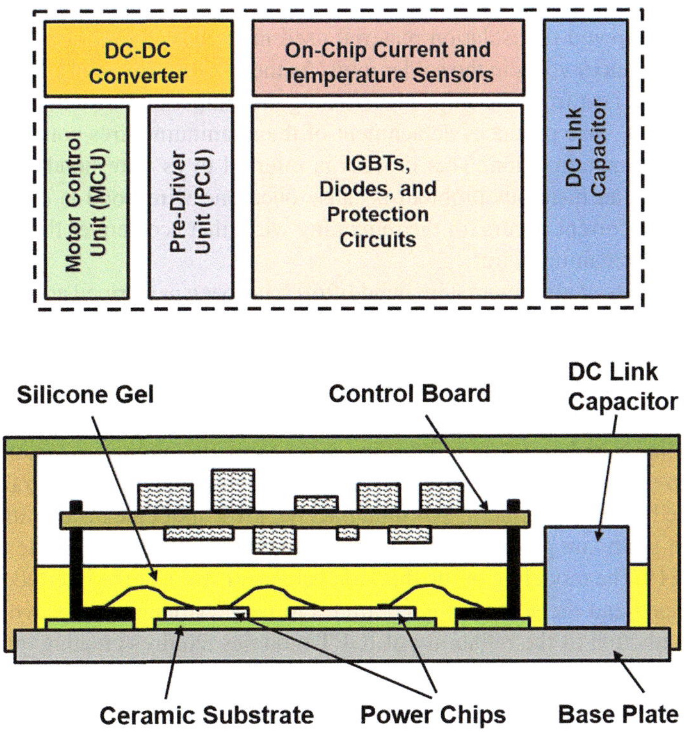

FIGURE 6.13

Block diagram and package for IGBT intelligent power unit (IPU) for electric and hybrid-electric vehicle applications.

6.7 Reliability

Modules containing IGBTs and rectifiers are attractive for inverters used for motor drives in many applications. Among them, modules used for traction (subways, urban and suburban trains, and high speed rail) have severe reliability constraints with a specified failure rate of 100 FITs over a 30-year operational time frame. The use of power modules for traction applications requires prediction of the time required for a population of devices to reach defectivity and the mean time to failure (MTTF) of a module after taking into account the application stress cycle. Prediction of the MTTF for a particular module design when operated under specific application conditions is also needed.

Since the beginning of the construction of IGBT modules, failures due to the lift-off of the aluminum wire bonds and voids in the solder joints for the chips were identified as leading reliability concerns [19]. These failure mechanisms are generated because the modules are constructed using materials with different thermal expansion coefficients. During operation in traction applications, the IGBT modules are subjected to thermal cycling. As an example, in a tramway car operated for 52 weeks per year with 130 h of service per week, the IGBT module must endure more than 300,000 cycles [19]. The difference in the thermal expansion coefficient of the silicon IGBT chip and the isolation material used in the DBC substrate creates shear stress that produces voids in the solder joint. As the void increases in size, it impedes the removal of heat from the chip, exacerbating the temperature rise in the chip. The thermal cycling also produces detachment of the aluminum wires connected to the IGBT emitter metallization. This process is referred to as wire bond liftoff. Wire bond liftoff is an insidious problem because once one wire bond is disconnected, an increase in current occurs for the remaining wires that accelerates the subsequent liftoff of the remaining wires.

Many studies of aluminum wire bond liftoff have been performed and approaches to reducing this problem have been explored. One method used to suppress the wire bond liftoff is to fill the module with a polymer or silicone gel, as illustrated in previous sections (see Fig. 6.13). Another method that can improve reliability is to add a molybdenum strain buffer plate under the IGBT chip [20]. It has also been found that the aluminum wire bonds can fail due to cracks formed in the wire. A metallurgical solution to this problem is to suppress crack propagation along small grain boundaries at the bonding interfaces. To counteract the growth of cracks, aluminum wires with larger grains are proposed with the wire bonding performed at high temperature (423 K) [21]. The modeling of the thermomechanical stress at the interface between the wire bond and the silicon chip continues to be a challenging endeavor [22].

The prediction of the reliability of IGBT modules requires creating quantitative models that take into account the above failure mechanisms. A simple bimetallic model for the thermomechanical stress at the interface of the joint between the aluminum wire and the silicon chip can be used with the Coffin–Manson Law for plastic strain to yield an expression for the mean number of cycles to failure [23]:

$$N_f = a(\Delta T)^{-n} \tag{6.1}$$

where a is an empirically determined constant and ΔT is the temperature excursion incurred during thermal cycling. From the power cycling of single aluminum wire bonds, an average value for the parameter a has been found to be 1.04×10^{15} with an average value for n of 5.36. This model can be utilized to analyze the MTTF for a module with multiple IGBTs, each containing many wire bonds. This model predicts a decrease in the MTTF from 100 to 18 to 5 years when the temperature excursion increases from 20 to 30 to 40 K, an exponential reduction with increasing cycling temperature.

The reliability problems associated with aluminum wire bonds in IGBT modules prompted the development of the flat-pack or press-pack modules discussed in the previous sections. In these packages, the high current contacts to the IGBT chips are made by soldering to a molybdenum plate on the collector side and pressing a molybdenum plate to the emitter side. In this approach, the thermal stress is reduced because the thermal expansion coefficients for silicon and molybdenum are relatively close (3 and 5×10^{-6} per K, respectively). The thermal stress is also reduced by making each IGBT chip 2.15 cm × 2.15 cm in size [6]. No change in the device characteristics were observed for this IGBT module after 10,000 power cycles with a temperature change of 70 K.

6.8 Summary

The packaging approaches for IGBTs have been reviewed in this chapter. Discrete devices are available in a wide variety of three terminal packages (TSON-8, SOP-8, TO-220, TO-3P, etc.). These devices have typical current ratings of 1–10 A and breakdown voltages of 600–1200 V. For higher current- and voltage-handling capability, IGBTs are available in a variety of modules. It is common practice to package the free-wheeling diode with the IGBT so that both devices can be optimized together to reduce power losses. The free-wheeling diode has been integrated into the IGBT structure to form the RC-IGBT so that space can be saved in compact low-power (< 2 kW) modules for home goods. The latest trend is to eliminate the wire bonds in the module by using a molybdenum plate pressed on to the chip emitter contact to enhance reliability.

When a compact solution is desired for home appliance applications, the IGBTs are copackaged with diodes, the gate drive circuit, and the protection circuits within a DIP package. For higher power applications, such as EVs or HEVs, an IPM is created by placing the control board together with the IGBTs and diodes. These strides in packaging have allowed scaling the current-handling capability of IGBT modules to above 1000 A with blocking voltages as high as 6.5 kV, allowing their use in an increasingly broader array of applications.

References

[1] W.W. Sheng, R.P. Colino, Power Electronic Modules: Design and Manufacture, CRC Press, New York, 2004.

[2] X. Liu, G.-Q. Lu, D^2BGA chip-scale IGBT package, in: IEEE Appl. Power Electron. Conf, vol. 2, 2001, pp. 1033–1039.
[3] D. Medaule, et al., Latest technology improvements of Mitsubishi IGBT modules, in: IEEE Colloq. New Dev. Power Semicond. Devices, 1996, pp. 5/1–5/5.
[4] A. Zeanh, et al., Proposition of IGBT modules assembling technologies for aeronautical applications, in: IEEE Integr. Power Syst. Conf, 2008, pp. 1–8.
[5] K. Xing, F.C. Lee, D. Boroyevich, Extraction of parasitics within wire-bond IGBT modules, in: IEEE Appl. Power Electron. Conf, vol. 1, 1998, pp. 497–503.
[6] H. Kirihata, et al., Investigation of flat-pack IGBT reliability, in: IEEE Ind. Appl. Conf, vol. 2, 1998, pp. 1016–1021.
[7] T. Fujii, et al., 4.5 kV-2000 A power pack IGBT, in: IEEE Int. Symp. Power Semicond. Devices ICs, 2000, pp. 33–36.
[8] P. Venkatraman, B.J. Baliga, Large area MOS-gated power devices using fusible link technology, IEEE Trans. Electron Devices 43 (1996) 347–351.
[9] Y. Nishimura, et al., New generation metal base free IGBT module structure, in: IEEE Int. Symp. Power Semicond. Devices ICs, 2004, pp. 347–350 (Paper 4.3).
[10] H.-R. Chang, et al., 200 kVA compact IGBT modules with double-sided cooling for HEV and EV, in: IEEE Int. Symp. Power Semicond. Devices ICs, 2012, pp. 299–302.
[11] Y. Wang, et al., High power compact automotive IGBT module with planar technology, in: European Conference on Power Electronics, 2017. P.1–P.6.
[12] H. Wang, et al., Structure design and thermal simulation analysis of DBC substrate for high-power IGBT module, in: IEEE Int. Conf. on Electronic Packaging Technology, 2020, pp. 1–4.
[13] B.J. Baliga, Smart power technology: an elephantine opportunity, in: IEEE Int. Electron Devices Meet, 1990, pp. 3–6 (Plenary Invited Paper 1.1.1).
[14] Y. Liu, Power Electronic Packaging, Springer, New York, 2012.
[15] J.-B. Lee, et al., Improved smart power modules for up to 7 kW motor drive applications, in: IEEE Power Electron. Spec. Conf, 2006, pp. 1–5.
[16] N. Clark, E. Motto, S. Shibata, New SLIM package intelligent power modules (SLIM-DIP) with thin RC-IGBT for consumer goods applications, in: IEEE Energy Conversion Conference and Exposition, 2017, pp. 4510–4512.
[17] Y. Kuramoto, et al., Inverter miniaturizing technologies for EV/HEV applications, in: IEEE Int. Veh. Electron. Conf, 2001, pp. 261–264.
[18] T. Stockmeier, Y. Manz, J. Steger, Novel high power semiconductor module for trench IGBTs, in: IEEE Int. Symp. Power Semicond. Devices ICs, 2004, pp. 343–346 (Paper 4.2).
[19] P. Jacob, M. Held, W. Wu, Reliability testing and analysis of IGBT power semiconductor modules, in: IEE Colloq. IGBT Propul. Drives, 1995, pp. 4/1–4/5.
[20] A. Hamidi, S. Kaufmann, E. Herr, Increased lifetime of wire bonding connection for IGBT power modules, in: IEEE Appl. Power Electron. Conf, vol. 2, 2001, pp. 1040–1044.
[21] J. Onumki, M. Koizumi, M. Suwa, Reliability of thick Al wire bonds in IGBT modules for traction motor drives, IEEE Trans. Adv. Packag. 23 (2000) 108–112.
[22] K.B. Pedersen, K. Pedersen, Bond wire lift-off in IGBT modules due to thermomechanical induced stress, in: IEEE Int. Symp. Power Electron. Distrib. Gener. Syst, 2012, pp. 519–526.
[23] M. Ciappa, W. Fichtner, Lifetime prediction of IGBT modules for traction applications, in: IEEE Int. Reliab. Phys. Symp, 2000, pp. 210–216.

CHAPTER

Gate Drive Circuit Design

7

The success of the insulated gate bipolar transistor (IGBT) as a widely used power switch for a large variety of applications is not only due to the high power-handling capability per unit area of the device itself, but also to the simplification of the gate drive circuit. Prior to the development of the IGBT, the power bipolar transistor was the only device available for control of low- and medium-power motors. The power bipolar transistor exhibits a low current gain due to the wide base region required for high-voltage operation [1]. Consequently, a relatively large base drive current (e.g., 10 A) is required for operating the transistor at a collector current of 100 A. Such high base currents cannot be delivered using integrated circuits, resulting in bulky drive circuits that are expensive to build and prone to reliability problems. The IGBT was the first power transistor suitable for high voltage operation with an MOS gate structure. The device could be controlled using low gate drive voltages, typically 10–15 V, which can be delivered using integrated circuits. Unlike the power bipolar transistor, the MOS gate structure of the IGBT does not require any steady-state current to keep the device operating in its blocking or on-state. The gate drive circuit must only provide the transient currents required to charge and discharge the MOS structure. The combination of the small IGBT chip size with high power-handling capability and the small integrated control circuit allowed a huge reduction in the volume, weight, and cost for the power electronics used in consumer, industrial, transportation, and other applications.

This chapter begins with the simple gate drive circuits used for the IGBT at its initial commercialization and reviews the evolution of more sophisticated gate drive circuits in modern IGBT modules. The gate drive circuit can control the dI/dt and dV/dt during switching of the IGBT. The ability to perform fault protection using IGBTs is also dependent on the control circuit used with the device. It is necessary to trade-off electromagnetic interference (EMI) generation and switching power loss by optimization of the gate drive signal.

7.1 Basic Gate Drive

The first paper to discuss gate drive circuits [2] for the IGBT was published soon after the release of the product by the General Electric Company (GE) [3]. Resistive load and inductive load switching behavior of the IGBT was described in this paper. The IGBT devices were classified into three types: Type A with 15 μs turn-off time,

The IGBT Device. https://doi.org/10.1016/B978-0-323-99912-0.00008-8

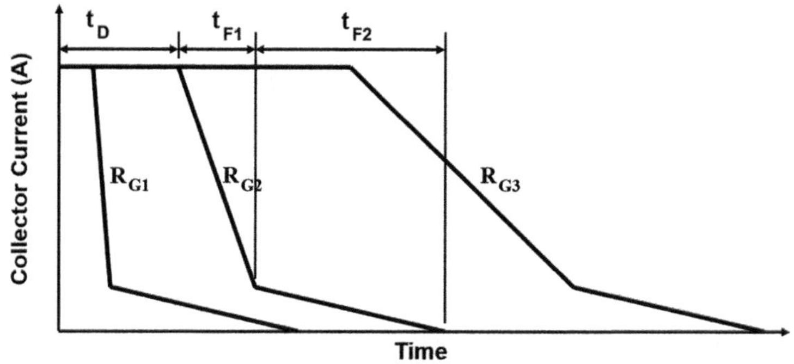

FIGURE 7.1

Insulated gate bipolar transistor turn-off waveforms versus gate resistance.

Type B with 1 μs turn-off time, and Type C with 0.25 μs turn-off time. The turn-off waveform for the IGBT with gate resistor R_{G2} was broken into two segments, as illustrated in Fig. 7.1, with the first segment (t_{F1}) controllable by adjusting the resistance (R_G) in the gate drive circuit. The second segment (t_{F2}), also referred to as the current tail, was found to be independent of the gate drive resistance but dependent on the type A, B, or C classification.

The current turn-off time for the IGBT can be described as consisting of three components, as illustrated in Fig. 7.1. The first is the delay time (t_D), the second is the fall time t_{F1}, and the third is the fall time t_{F2}. The delay time and the turn-off time t_{F1} are dependent on the gate resistance, as shown for a typical case in Fig. 7.2. The variation shown in the figure can be modeled as an R–C time constant

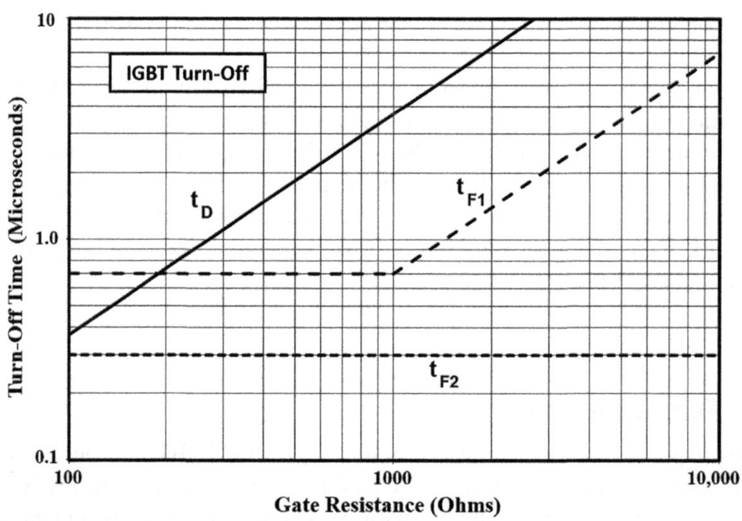

FIGURE 7.2

Insulated gate bipolar transistor turn-off times versus gate resistance.

with a constant capacitance. The tail time t_{F2} is independent of the gate resistance, as shown in Fig. 7.2. A shorter value for this time occurs when changing from IGBT device type A to type B to type C. These shorter times were obtained by increasing the electron irradiation dose used to reduce the lifetime in the drift region of the IGBT structure.

7.2 Asymmetric Gate Drive

The simple gate drive circuit [2] originally used for the IGBT is shown in Fig. 7.3. This is referred to as an asymmetric gate drive circuit because of the diode in series with the resistor R_S. In this circuit, the rate of turn-on of the IGBT, which determines the *dI/dt* during turn-on, can be controlled by the resistance R_S. The turn-off of the IGBT is controlled by resistance R_{GE}, which is used to discharge the input capacitance of the IGBT.

7.3 Two-Stage Gate Drive

In the hard-switched power converters, typically used for adjustable speed motor control, one of the problems encountered is the voltage overshoot created by high rates of change of the IGBT collector current flowing through stray inductances during its turn-off. The stray inductances consist of the inductance in the DC link capacitor and the bus bars within IGBT modules. The overshoot voltage appears across the IGBT, leading to avalanche breakdown within the device. It is possible to reduce the *dI/dt* by increasing the gate resistance, as discussed in previous sections, but this leads to a substantial increase in the power dissipation. A gate drive circuit that allows independent control of the rate of rise of the collector voltage and the rate of fall of the collector current [4] is shown in Fig. 7.4. Initially, both transistor T_1 and T_2 are turned on to discharge the input capacitance of the IGBT rapidly. When the IGBT collector voltage reaches the breakdown voltage of the Zener diode D_1, current flows via capacitor C_1 to trigger the monoflop (monostable multivibrator). The monoflop turns off transistor T_1 for a short time. During this period, the collector

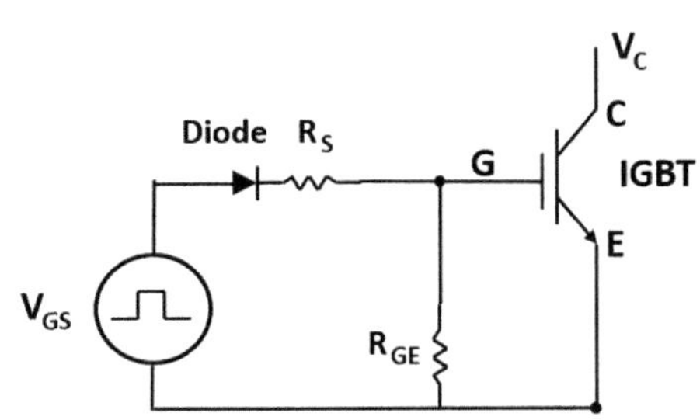

FIGURE 7.3

Asymmetrical gate drive circuit for the insulated gate bipolar transistor chip.

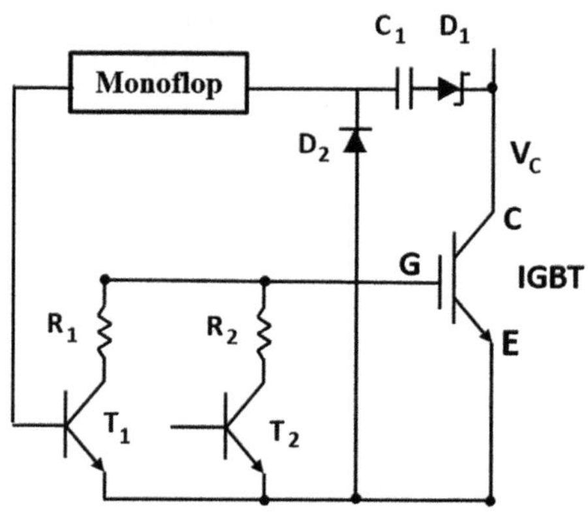

FIGURE 7.4

Two-stage gate drive circuit for the insulated gate bipolar transistor.

current falls with a smaller *dI/dt* due to the larger gate resistance R_2. After a short time, transistor T_1 is turned back on, producing faster turn-off of the IGBT. This method has been found to reduce the overvoltage by 60% [4] while maintaining low switching losses.

7.4 Active Gate Voltage Control

The rate of rise of the collector voltage and the rate of fall of the collector current can be independently controlled using an active gate voltage control (AGVC) circuit [5], as illustrated in Fig. 7.5. The basic principle of the AGVC method is to apply a two-level gate drive voltage during the turn-on and turn-off of the IGBT.

The two-level gate voltage waveform used for controlling the *dI/dt* during the turn-on of the IGBT is illustrated in Fig. 7.6. Note that reducing the *dI/dt* during turn-on of the IGBT reduces the current overshoot incurred from the reverse recovery of the fly-back diodes [1]. The two-level gate drive consists of increasing the gate

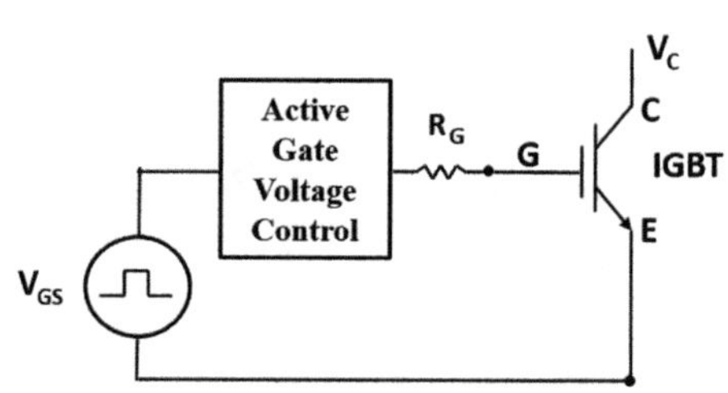

FIGURE 7.5

Active gate voltage control for the insulated gate bipolar transistor.

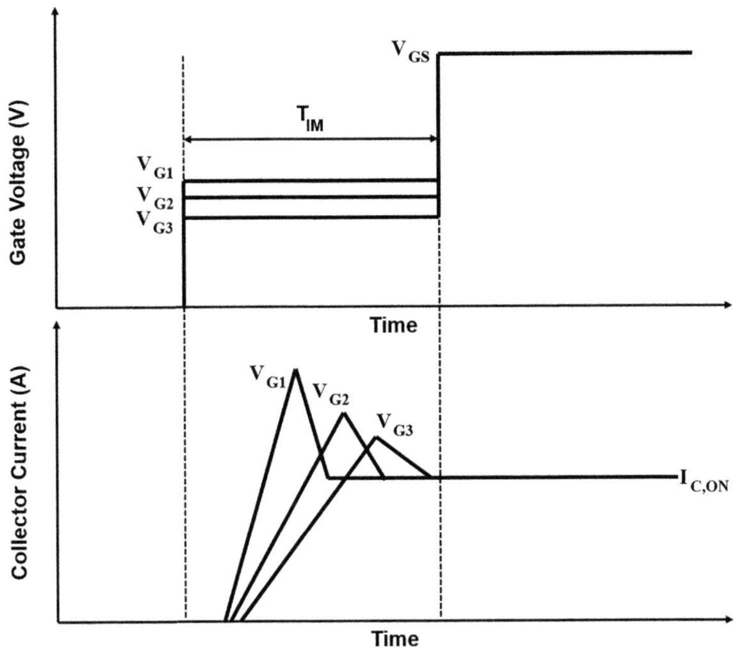

FIGURE 7.6

Controlling *dI/dt* during turn-on of the insulated gate bipolar transistor with active gate voltage control (AGVC).

voltage to an intermediate level (V_{G1}, V_{G2}, or V_{G3}) followed by increasing it to the on-state gate drive voltage (V_{GS}) after a time interval T_{IM}. It can be observed that the turn-on dI/dt can be reduced by decreasing the gate voltage used during the first step. The intermediate gate voltage must be larger than the threshold voltage (V_{TH}) of the IGBT for this approach to be effective, and the time duration for this intermediate level (T_{IM}) must be sufficient to complete the reverse recovery process of the fly-back diode.

The two-stage gate voltage waveform can also be used for controlling the dV/dt during the turn-off of the IGBT, as illustrated in Fig. 7.7. The two-level gate drive consists of first turning off the gate voltage at time T_1 to begin the process, whereby the collector voltage increases due to the formation of a space-charge region [1]. The collector voltage increases more gradually during the initial stages of the turn-off process. Just before the collector voltage begins to increase rapidly, the gate voltage is turned back on to an intermediate level (V_{G1}, V_{G2}, or V_{G3}), followed by reducing it back to zero after a time interval T_{IM}. It can be observed that the turn-off dV/dt can be reduced by decreasing the gate voltage used during the intermediate step. The intermediate gate voltage must be less than the threshold voltage (V_{TH}) of the IGBT for this approach to be effective, and the time duration for this intermediate level (T_{IM}) must be sufficient to allow the collector voltage to reach the collector supply voltage.

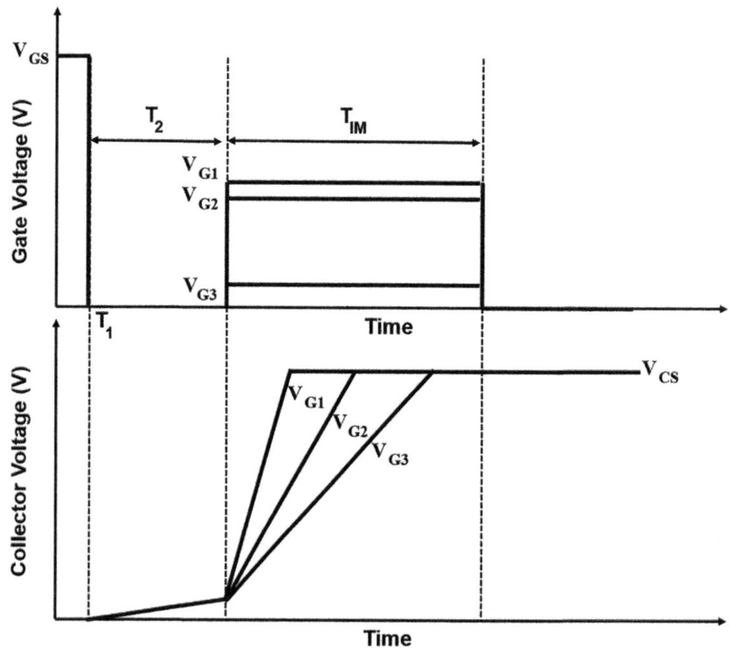

FIGURE 7.7

Controlling *dV/dt* during turn-off of the insulated gate bipolar transistor with AGVC.

The AGVC method allows controlling the rate of rise of the IGBT collector current during turn-on and the collector voltage during turn-off. This increases the switching time duration for the turn-on and the turn-off switching events, producing greater switching losses in the IGBT. However, these losses are smaller than those incurred if the reduction of the *dI/dt* and *dV/dt* was achieved by using larger gate resistances. This type of gate drive has been implemented in an integrated circuit [6].

7.5 Variable Gate Resistance Drive

As discussed previously, it is desirable to reduce the overshoot in the collector voltage of the IGBT during device turn-off. The voltage overshoot is produced by a high *dI/dt* for the collector current flowing through the stray inductance in the loop between the IGBT and the DC link capacitor. Although the *dI/dt* can be reduced by increasing the gate resistance, as discussed in Section 7.1, this produces an increase in power dissipation. A better combination of a smaller spike in collector voltage and switching power loss can be achieved by varying the gate resistance [7] during the turn-off event. This is illustrated in Fig. 7.8, which shows the gate voltage waveform together with the collector current and voltage waveforms. During the time interval from t_0 to t_1, it is desirable to have a low gate resistance to allow the gate

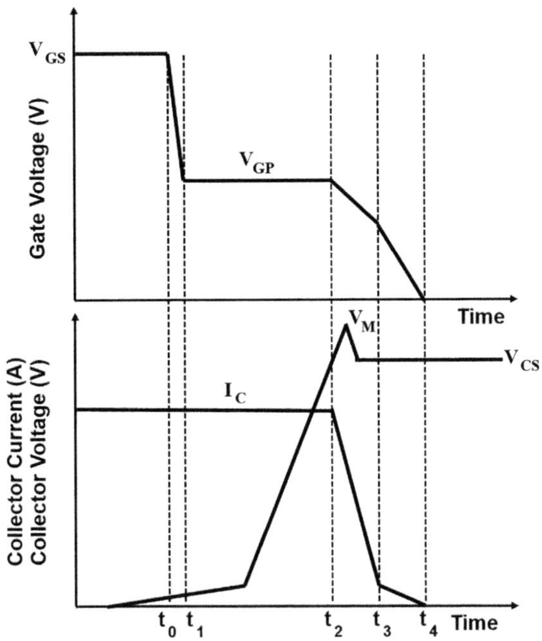

FIGURE 7.8

Turn-off waveforms for the variable resistance insulated gate bipolar transistor gate drive.

voltage to rapidly fall to the plateau voltage (V_{GP}). During the time interval from t_1 to t_3, it is desirable to have a high gate resistance to reduce the rate of rise (dV/dt) of the collector voltage and the rate of fall (dI/dt) of the collector current. During the time interval from t_3 to t_4, it is desirable to have a low gate resistance to allow discharging the gate input capacitance.

The variable gate resistance can be implemented using the circuit shown in Fig. 7.9. During the time interval from t_0 to t_1, a low gate resistance (R_3) is achieved by turning on transistor T_3. During the time interval from t_1 to t_3, a high gate resistance (R_2) is achieved by turning off transistor T_3. During the time interval from t_3 to t_4 after the collector voltage is high, a low gate resistance (R_1) is achieved by turning on transistor T_2. This approach has been found to reduce the maximum (V_M) in the collector voltage. A two-stage circuit [8] for adjusting the gate resistance between a low and a high voltage can also be used to derive the same benefits of reducing the overvoltage transient during turn-off of the IGBT.

It is also important to control the rate of rise of the IGBT collector current during the turn-on transient to avoid producing very large reverse recovery current in the fly-back diodes. The diode reverse recovery current produces an overshoot (I_M) in the IGBT current, as illustrated in Fig. 7.10. The turn-on dI/dt can be reduced by using a large gate drive resistance but this increases the turn-on energy loss. A better combination of turn-on energy loss and maximum IGBT current during turn-on can

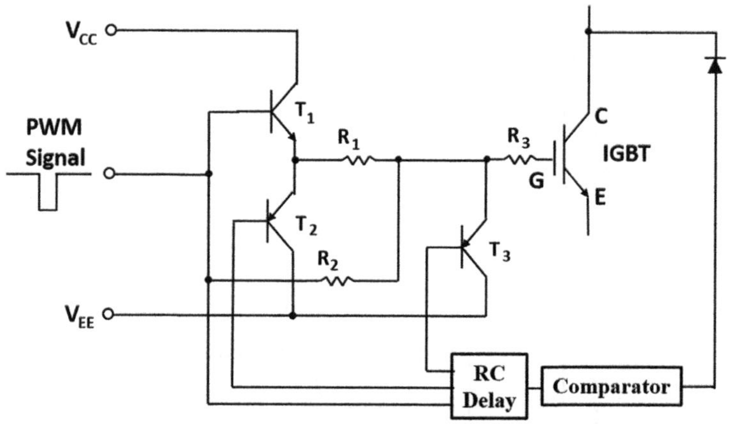

FIGURE 7.9

Variable resistance turn-off gate drive circuit for the insulated gate bipolar transistor. Pulse width modulation (PWM).

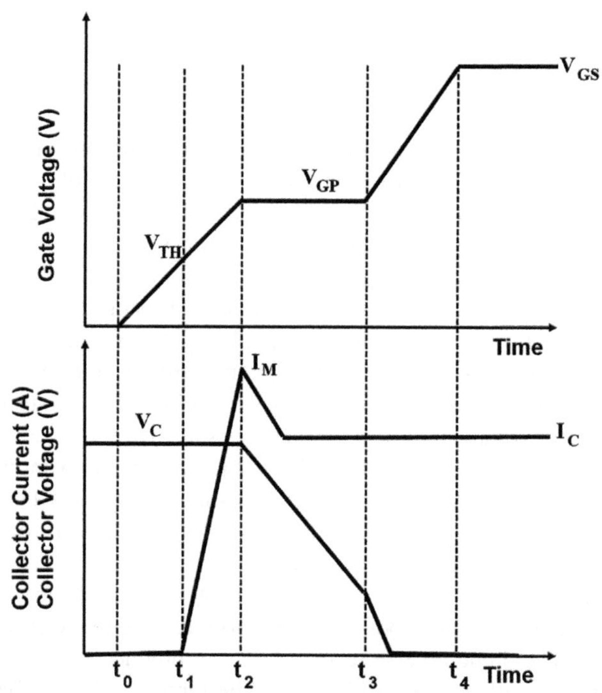

FIGURE 7.10

Turn-on waveforms for the variable resistance insulated gate bipolar transistor gate drive.

be produced by varying the gate resistance during the turn-on switching event [9]. An integrated circuit has been developed that allows varying the gate resistance with minimal input to output propagation delay. To reduce the overshoot in collector current without significantly increasing the turn-on energy loss, the rate of rise of the gate voltage is kept high at all times during the turn-on except from time t_2 to t_3, that is, during the gate plateau stage.

7.6 Digital Gate Drive

More detailed tailoring of the rate of fall of the collector voltage and rate of rise of the collector current in IGBTs during turn-on and turn-off can be achieved by using a digital gate drive unit. In this approach, the key IGBT device parameters (V_{CE}, V_{GE}, and I_C) are monitored and the gate drive voltage is adaptively controlled during the entire switching event to tailor the shape of the collector current and voltage to any desired shape.

A simplified illustration [10] of the implementation of the digital gate drive unit is shown in Fig. 7.11. The collector and gate voltage are continuously monitored using the resistor dividers, and the collector current is monitored using the current probe. This information is continuously fed into the microcontroller field programmable gate array (FPGA) to generate the desired gate signal. The gate current for driving the IGBT module can supply up to 5 A of current with a high slew rate [11]. Any desired combination of dI/dt and dV/dt can be maintained during the switching of the IGBT to minimize the power dissipation and voltage overshoots. The large delay

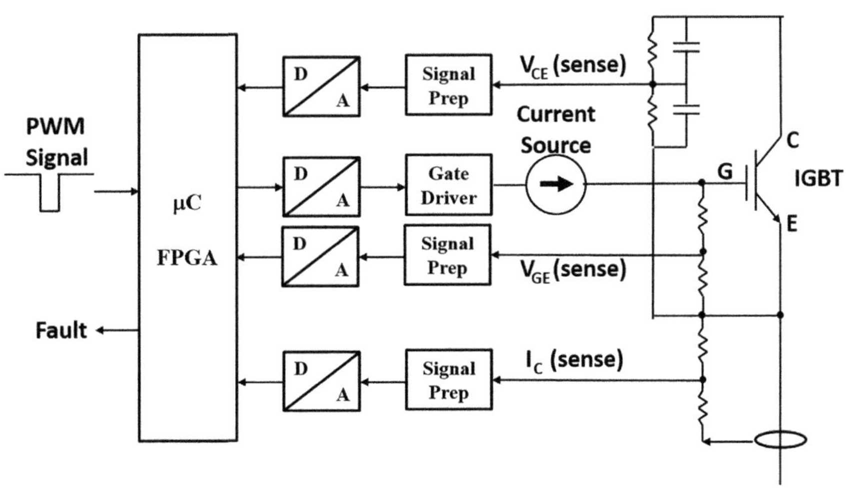

FIGURE 7.11

Digital gate drive schematic for the insulated gate bipolar transistor.

times of the D/A and A/D converters limit this approach to IGBTs with switching transients faster than a few microseconds [12].

Advanced gate drive circuits [13,14] that allow reduction of the *dI/dt* during turn-off under normal operation and short circuit conditions have been developed using the principle of varying the gate resistance after detection of the fault. Due to the widespread use of IGBTs for motor control, many manufacturers have developed gate driver products. Some examples are Texas Instruments UCC27531 [15], Mitsubishi Electric M81738FP [16], and Toshiba Electronics TLP700H [17] gate drivers.

7.7 Short Circuit Protection

One of the major IGBT features that has been important for adoption in motor drives is its excellent short circuit-handling capability with appropriate design. Without protection, a short circuit developed in a motor winding due to failure of the winding insulation would result in destruction of the expensive IGBT-based power stage. The DC bus voltage appears directly across the IGBT device during a short circuit in the motor winding. If the gate drive voltage was turned-on at the onset of the short circuit event, the current becomes limited by the saturation current of the IGBT. Very high power dissipation occurs in the IGBT due to the combination of the high saturation current and the large supply voltage. Adiabatic heating during the short circuit can raise the junction temperature above the melting point of the aluminum metallization, resulting in catastrophic failure of the IGBT, as discussed in Chapter 4. The saturation current of the IGBT is determined by the channel density. The low channel density of planar gate IGBT structures reduces the saturation current level, allowing them to survive a short circuit for at least 10 μs. This became an industry standard for IGBTs used for motor control. Trench gate IGBTs have much larger channel density than planar gate devices, which allows reducing their on-state voltage drop. However, this can degrade the short circuit time for trench gate devices below 10 μs. The design of modern IGBTs is done with a trade-off among on-state voltage drop, switching losses, and short circuit capability.

It is estimated that about 40% of faults in variable speed drives are attributed to power devices. These faults can be classified as open circuit faults, short circuit faults, or intermittent faults [18]. Among these, the short circuit fault is the most problematic. IGBT protection during a short circuit event can be achieved by detection of the short circuit, followed by shutting off the gate drive signal before destructive failure. The most commonly used method for detection of the short circuit is the DESAT approach shown in Fig. 7.12. Commercial IGBT gate drivers are available specifically designed for this function [19]. In this approach, the IGBT collector emitter voltage ($V_{CE,SAT}$) is continually monitored by using diode D_{SS}. This diode isolates the DESAT control circuit from high collector emitter voltage across the IGBT during the normal blocking mode of circuit operation. This diode must have fast reverse recovery behavior. Although 1000 V fast recovery diodes are readily available, this can become a problem with increasing blocking voltage for the

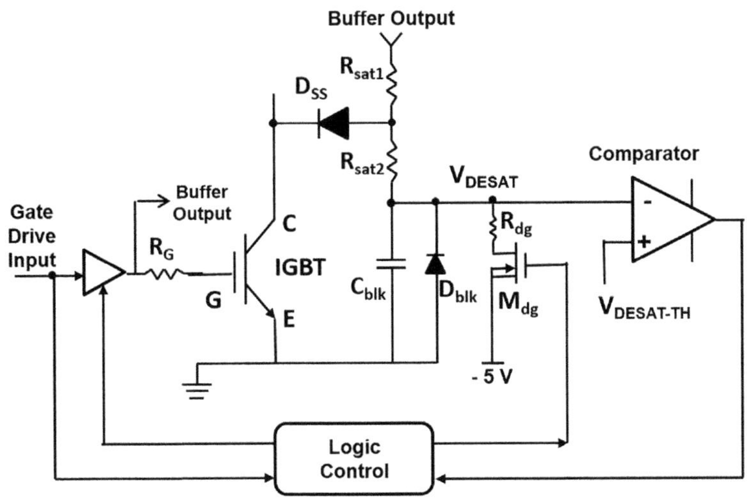

FIGURE 7.12

DESAT protection circuit for the insulated gate bipolar transistor.

IGBT. A blanking time must be implemented to avoid nuisance triggering of the DESAT protection during normal pulse width modulation switching circuit operation because the IGBT collector emitter voltage swings from the on-state value to the DC supply value. The value of the banking capacitor C_{blk} is chosen so that the blanking time is about 2.5 μs. The gate signal is shut-off after the detection of the short circuit in about 5 μs. The IGBTs are designed to tolerate the short circuit conditions for typically 10 μs.

7.8 Magnetically Coupled Gate Drive

It is important to isolate the control circuit for the IGBT from the large voltage transients occurring across its terminals during normal power circuit operation. This can be accomplished using optical isolators [20,21]. A magnetically coupled isolator offers an alternate approach. The topology for the magnetically coupled digital isolator is illustrated in Fig. 7.13. This has been implemented using face-to-face stacked coils that transmit and receive signals [22]. The alignment of the coils introduces a 20% variation in performance. An isolation of 7 kVrms was demonstrated with a working voltage of 1.5 kV.

The use of a planar integrated magnetic digital isolator has been reported by the development of an IGBT structure with very low input capacitance to reduce gate drive current [23]. The control signal can be transferred across the magnetic digital isolator but driving the IGBT usually requires significant power due to its large input capacitance. This issue was addressed by creating a modified IGBT structure to reduce its input capacitance. The modified IGBT structure contains a P^+ region

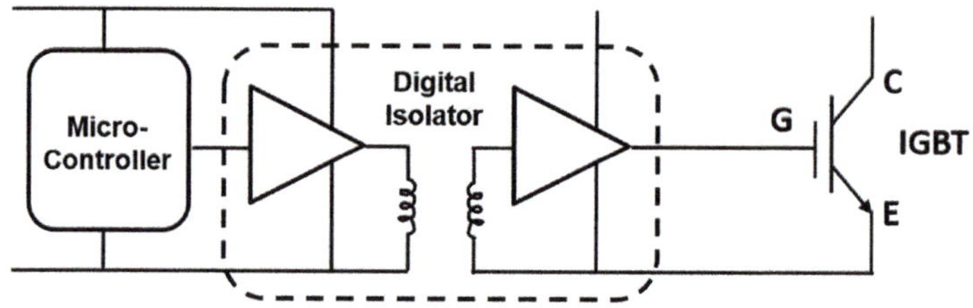

FIGURE 7.13

Magnetic digital isolation gate drive for the insulated gate bipolar transistor.

located at the bottom of the trench gate region of the conventional IGBT, as shown in Fig. 7.14(a). This introduces a junction field effect transistor (JFET) in series with the MOS channel of the IGBT structure. A contact to the P^+ region is made using another trench, which has a highly doped *p*-type polysilicon refill, as illustrated in Fig. 7.14 (b), to makes an ohmic contact to it. The JFET is designed to pinch off at about 30 V to shield the upper N-P-N transistor in the IGBT. The P^+ region of the JFET also diverts the holes arriving at the emitter side to suppress latch-up of the IGBT parasitic thyristor. The authors report a reduction of the input capacitance by a factor of $\sim 200\times$, which is beneficial for reducing the gate drive power requirements.

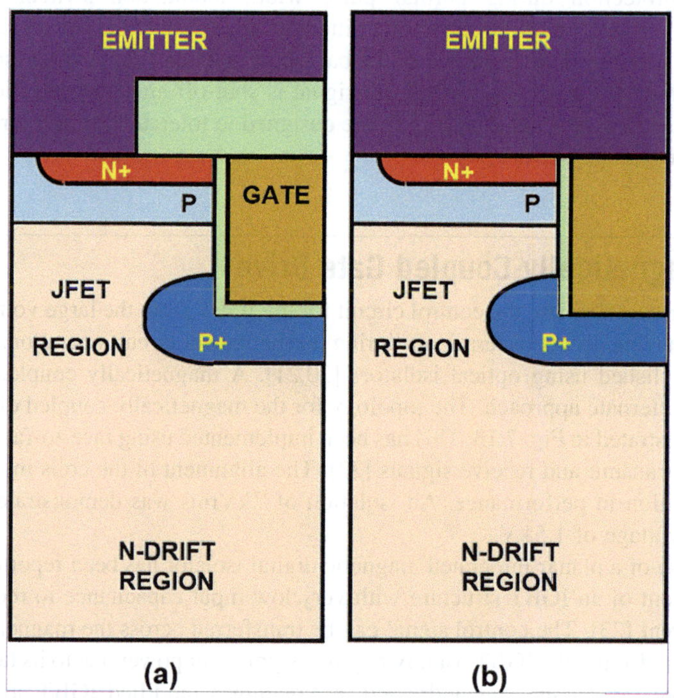

FIGURE 7.14

Very low input capacitance structure for the insulated gate bipolar transistor.

7.9 Posicast Gate Drive

A novel method of optimizing the gate drive signal for the IGBT is with the posicast feed-forward approach. The posicast method was previously used to dampen oscillations in mechanical systems. It was adopted to prevent current overshoots and ringing during device switching transients [24]. The basic concept is illustrated in Fig. 7.15. The posicast controller is interposed between the conventional gate driver and the IGBT gate terminal. The gate driver bias switches from − 15 to +15 V during turn-on and vice versa during turn-off. The posicast controller injects a signal, as shown in red in Fig. 7.16, during the turn-on and turn-off of the IGBT device. The net applied gate bias to the IGBT is shown in blue. The reduction in positive gate bias for a short duration as shown at the top reduces overshoots in current from the reverse recovery of the fly-back diode. The reduction in the magnitude of the negative gate signal present oscillations of the collector current. The authors have found that this approach does not produce an increase in switching losses, despite reducing the *dI/dt* during the switching transients.

7.10 EMI Reduction Gate Drive

One of the problems encountered in building high-performance converters is to satisfy EMI regulations. It is desirable to increase the *dI/dt* and *dV/dt* during the IGBT switching transients to reduce switching power losses. However, this leads to the generation of greater EMI noise. A trade-off between switching losses and EMI noise becomes consequently mandatory during the converter design. A digital programmable gate driver can control the turn-on and turn-off gate signal to bring the EMI within desired limits [25]. The concept of the digital gate drive is illustrated in Fig. 7.17. The digital integrated circuit (IC) contains numerous P-MOS and N-MOS transistors that can be turned on to generate any desired stepped gate drive signal to the IGBT. Eight 200 ns time steps were used to generate the gate drive signal for the IGBT. The V_{CE} spectrum was then generated to evaluate the EMI while also

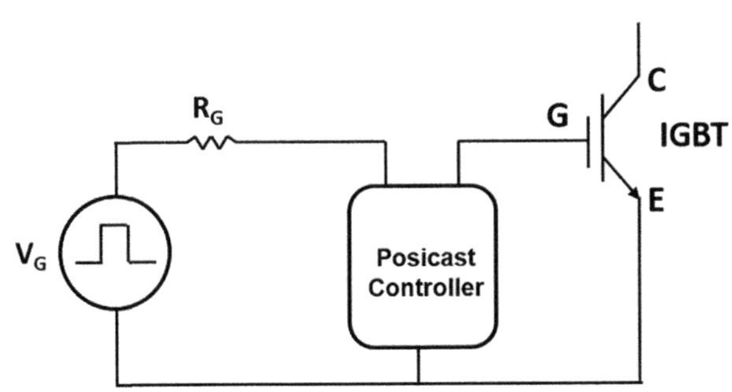

FIGURE 7.15

Posicast-based gate drive for the insulated gate bipolar transistor.

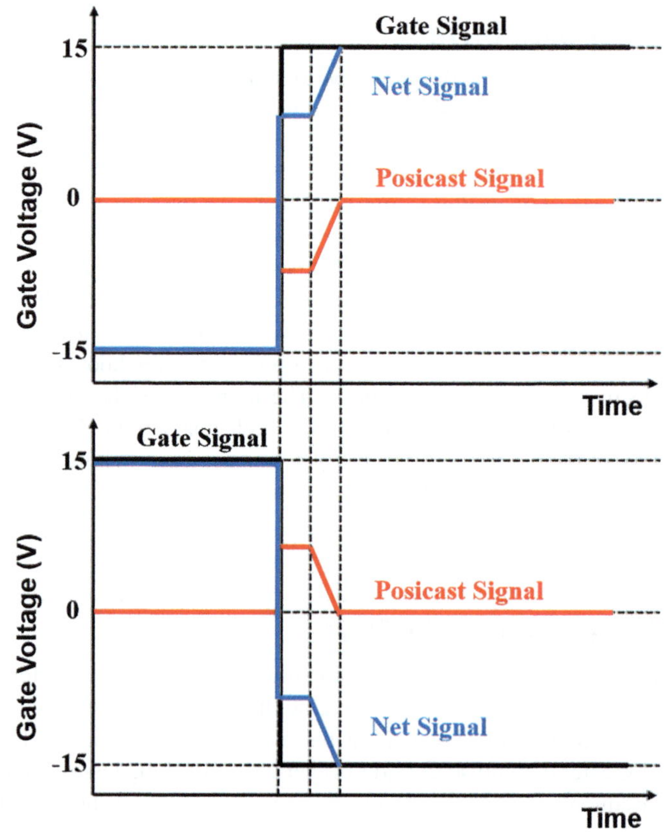

FIGURE 7.16

Posicast-based gate drive signal for the insulated gate bipolar transistor.

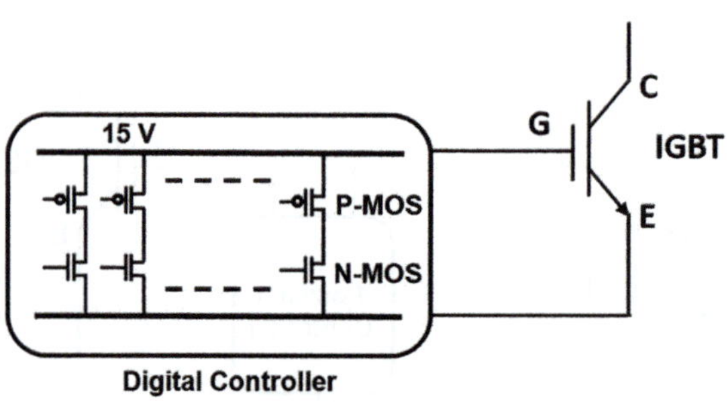

FIGURE 7.17

Digital controlled gate drive for the insulated gate bipolar transistor.

obtaining the switching energy loss. For the case of the 200 ns time steps optimized to reduce EMI in the 7.5–10 MHz range, the EMI amplitude ($A_{SPECTRUM}$) was reduced by 46% and the maximum EMI was reduced by −105 dBμV with the same switching energy loss as the conventional gate drive. The time step needs to be reduced to achieve EMI mitigation at higher frequencies.

Digital control over the gate drive signal can also be derived by using a "self-regulating voltage-source gate drive (SRVSD)" generated with FPGAs [26]. Reductions of turn-on loss by 58% and turn-off loss by 31% were observed with identical current and voltage overshoots as the conventional gate drive case.

7.11 The BaSIC Topology

The **Ba**liga **S**hort circuit **I**mprovement **C**oncept (BaSIC) topology was originally proposed for achieving improved short circuit capability for SiC power metal oxide semiconductor field effect transistor (MOSFET) [27]. It was subsequently applied to extending the short circuit capability of 1.2 kV silicon IGBTs [28] and 1.7 kV silicon IGBTs [29]. The proposed idea [30] consists of a gate-source-shorted (GSS) silicon depletion-mode power MOSFET (DMM) connected to the emitter terminal of the IGBT, as shown in Fig. 7.18, to create a composite three-terminal transistor. Because the Si GSS-DMM operates within the gate drive loop of the IGBT, the maximum voltage applied to its drain terminal is the gate drive voltage of the IGBT, typically 15 V. Consequently, the Si DMM can have a blocking voltage capability of 20–30 V. These devices can be fabricated with a low on-resistance when operated at zero gate bias, that is, with the gate shorted to the source. The on-state operating point for the Si

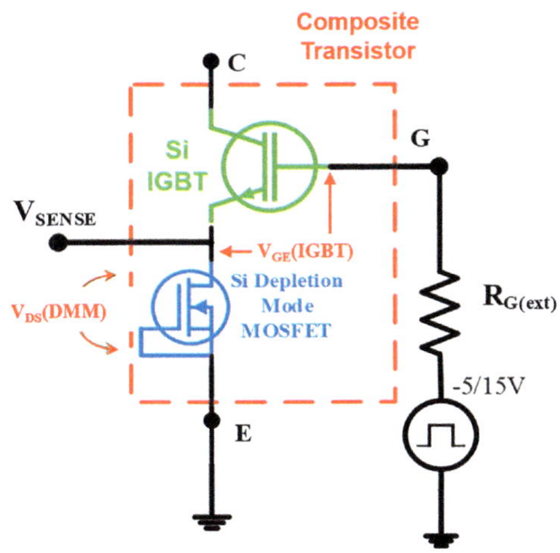

FIGURE 7.18

BaSIC topology-based gate drive for the insulated gate bipolar transistor.

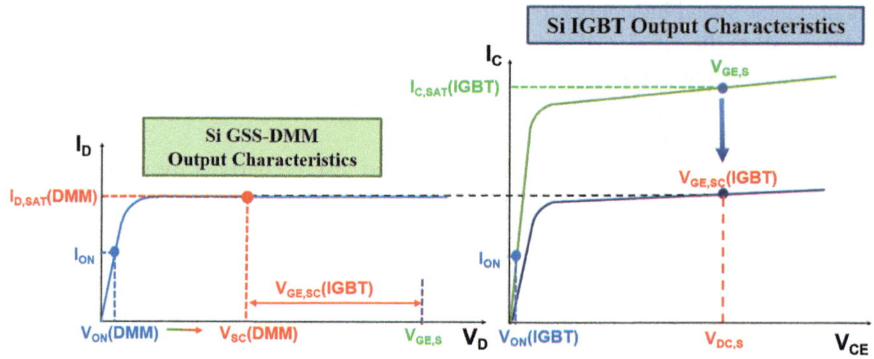

FIGURE 7.19

BaSIC topology operating principle for the insulated gate bipolar transistor.

GSS-DMM is shown in Fig. 7.19 on its output characteristics. It is worth pointing out that these output characteristics extend to only just above the gate supply voltage ($V_{GE,S}$), that is, to about 20V. The voltage drop across the Si GSS-DMM (V_{ON}(DMM)) can be much smaller than the on-state voltage drop of the IGBT (V_{ON}(IGBT)) if its resistance is low. Note that the x-axis scale for the IGBT output characteristic extends to high voltages (e.g. 1000V for a 1200V rated IGBT operating with a DC supply of 800V). This minimizes the impact of the Si GSS-DMM on on-state power loss. If the Si GSS-DMM resistance is small, its presence does not substantially reduce the transconductance of the composite transistor. This is important for minimizing its impact on switching losses.

When a short circuit event occurs for the IGBT, its on-state gate bias ($V_{GE,S}$) produces a very large (typically 10× on-state current level) saturation current ($I_{C,SAT}$(IGBT)), as shown by the green line in its output characteristics in Fig. 7.19. The short circuit operating point for the IGBT is shown in Fig. 7.19 with the device subjected to the power dissipation produced by the combination of the large saturation current ($I_{C,SAT}$(IGBT)) and the DC bus voltage ($V_{DC,S}$). The power dissipation produces adiabatic heating with a rapid temperature rise to above the melting point (660°C) of the aluminum metallization within a few μs [1]. It is necessary to reduce the saturation current to decrease the power dissipation to achieve a larger short circuit withstand time. This can be done by reducing the on-state gate bias for the IGBT to obtain the output characteristics shown by the dark blue line in Fig. 7.19. However, this will substantially increase the on-state voltage drop and switching losses for the IGBT.

In the case of the composite transistor created by the BaSIC topology, the current under short circuit conditions becomes limited to $I_{D,SAT}$(DMM) by the output characteristics of the Si GSS-DMM device, as indicated in its output characteristics in Fig. 7.19. The Si GSS-DMM is chosen so that its saturation current $I_{D,SAT}$(DMM) is much smaller than the IGBT saturation current $I_{C,SAT}$(IGBT) at the on-state gate bias. The drain-source voltage across the Si GSS-DMM increases under short circuit conditions to V_{SC}(DMM), as shown in Fig. 7.19. This voltage opposes the gate drive voltage, reducing the gate emitter voltage across the IGBT to $V_{GE,SC}$(IGBT), as indicated in the figure. This reduction of gate bias for the IGBT brings down its saturation current until it matches the saturation current of the GSS-DMM. It is worth emphasizing that the voltage across the GSS-DMM (V_{SC}(DMM)) *automatically adjusts* until the IGBT saturation current matches the saturation current of the GSS-DMM. This is an important principle of operation of the BaSIC topology. The smaller saturation current $I_{D,SAT}$(DMM) for the composite transistor results in a smaller power dissipation during the short circuit, extending the short circuit withstand time. In the BaSIC topology, the GSS-DMM is not subjected to any high voltages during either normal power switching operation or a short circuit event. Its drain-source voltage remains below the gate drive voltage of the IGBT (typically 15 V). The GSS-DMM temperature does not increase substantially during the short circuit event because of the small voltage V_{SC}(DMM) across it.

As mentioned previously, IGBT products are designed while making a trade-off among on-state voltage drop, switching losses, and short circuit time. The manufacturers now build two classes of IGBT: (a) IGBTs with good (> 10 μs) short circuit capability for motor drive applications that have a relatively large on-state voltage drop; and (b) IGBTs with low on-state voltage drop for resonant circuits such as induction cooktops that have poor short circuit capability. The BaSIC topology can be implemented using the commercially available IGBT with low on-state voltage drop and poor short circuit capability to achieve a composite transistor that has a superior combination of good short circuit withstand capability, low on-state losses, and low switching losses. This has been experimentally demonstrated for the case of 1.2 kV rated IGBTs [28] and 1.7 kV rated IGBTs [29]. The short circuit time was increased from 4.8 μs for the stand-alone 1.2 kV IGBT to more than 10 μs for the composite transistor with a 15% increase in on-state voltage drop [28]. In the case of the 1.7 kV IGBT, the short circuit time was increased from 5.7 μs for the stand-alone device to 27 μs for the composite transistor with a 15% increase in on-state voltage drop [29]. The impact on on-state voltage can be reduced by optimization of the GSS-DMM device structure to simultaneously achieve a low saturation current level and low on-resistance [31].

One of the features of the BaSIC topology is that a fourth terminal can be added, as shown in Fig. 7.18, called the sense terminal. The sense terminal can be used to monitor the voltage drop across the GSS-DMM device. Because this voltage is proportional to the on-state current level, as determined by the resistance of the GSS-DMM device, it can be used as a current sensor for the IGBT during normal circuit operation. This eliminates the need to add Rogowski coils, sense resistors, or integrated current sense elements in the IGBT. The sense voltage can be acquired for each of the IGBTs in the H-bridge circuits or within power modules to monitor current distribution between various devices in the conduction mode or during current switching.

The sense terminal voltage increases from a low value (<0.5 V) to a larger value (typically >5 V) under short circuit conditions, as illustrated in Fig. 7.19.

This change in voltage has been found to occur within 0.5 μs after the advent of the short circuit [28,29]. Consequently, the sense voltage can be used to turn-off the gate drive signal in less than 1 μs to protect the IGBT. Unlike the DESAT approach discussed in a previous section, the voltage at the sense node always remains below 10 V. This makes interfacing the sense node to the control circuits much less complex than the DEAT method.

7.12 Summary

Gate drive circuits for IGBTs have evolved from a simple choice of the resistance in the gate drive circuit to a more sophisticated dynamic variation of the gate drive resistance during the switching event. These improved methods allow a reduction of collector current and voltage overshoots during the IGBT switching events while minimizing the switching power loss. This improves the efficiency of the IGBT-based inverter circuits. In addition, digital control of the gate drive voltage during the turn-on and turn-off switching events can also improve the trade-off between switching energy loss and EMI.

Modern IGBT modules are also designed to withstand short circuit conditions in the load. During a load short circuit, the IGBT collector current can rise to many (typically 10 ×) times its steady-state value. This produces very high power dissipation, resulting in failure of the IGBT within a few μs. Manufacturers must perform a trade-off between the short circuit capability and on-state/switching power losses. At present, the most popular method to detect the short circuit is with a DESAT technique. The recently proposed and demonstrated BaSIC topology provides an elegant approach to extend the short circuit withstand time while allowing monitoring on-state current and short circuits events without high-voltage isolation.

References

[1] B.J. Baliga, Fundamentals of Power Semiconductor Devices, second ed., Springer-Science, New York, 2019.

[2] B.J. Baliga, et al., The insulated gate transistor (IGBT)—a new power switching device, in: IEEE Industry Applications Society Meeting, 1983, pp. 794–803.

[3] GE Datasheet for Product D94FQ4, R4 Power-MOS IGT—18-A, 500-V Insulated Gate Transistor, June 1983.

[4] B. Weis, M. Bruckmann, A new gate driver circuit for improved turn-off characteristics of high current IGBT modules, in: IEEE Industrial Applications Conference, vol. 2, 1998, pp. 1073–1077.

[5] N. Idir, R. Bausiere, J.J. Franchaud, Active gate voltage control of turn-on di/dt and turn-off dv/dt in insulated gate transistors, IEEE Trans. Power Electron. 21 (2006) 849–855.

[6] L. Dulau, et al., A new gate drive integrated circuit for IGBT devices with advanced protections, IEEE Trans. Power Electron. 21 (2006) 38–44.

[7] J.H. Kim, et al., An active gate drive circuit for high power inverter system to reduce turn-off spike voltage of IGBT, in: IEEE International Conference on Power Electronics, 2007, pp. 127–131.
[8] D. Boris, et al., Double-stage gate drive circuit for parallel connected IGBT modules, IEEE Trans. Dielectr. Electr. Insul. 16 (2009) 1020–1027.
[9] A. Shorten, et al., A segmented gate driver IC for the reduction of IGBT collector current over-shoot at turn-on, in: IEEE International Symposium on Power Semiconductor Devices and ICs, 2013, pp. 73–76. Paper 3–3.
[10] L. Dang, H. Kuhn, A. Mertens, Digital adaptive driving strategies for high-voltage IGBTs, IEEE Trans. Ind. Appl. 49 (2013) 1628–1636.
[11] L. Dang, H. Kuhn, A. Mertens, Digital adaptive driving strategies for high-voltage IGBTs, in: IEEE Energy Conversion Congress and Exposition, 2011, pp. 2993–2999.
[12] Y. Lobsinger, J.W. Kolar, Closed-loop IGBT gate drive featuring highly dynamic di/dt and dv/dt control, in: IEEE Energy Conversion Congress and Exposition, 2012, pp. 4745–4761.
[13] P.J. Grbovic, Gate driver with feed forward control of turn-off performances of an IGBT in short circuit conditions, in: IEEE European Conference on Power Electronics and Applications, 2007, pp. 1–10.
[14] L. Chen, F.Z. Peng, Active fault protection for high power IGBTs, in: IEEE Applied Power Electronics Conference and Exposition, 2009, pp. 2050–2054.
[15] Texas Instruments Inc. www.ti.com.
[16] Mitsubishi Electric Corp. www.mitsubishielectric.com.
[17] Toshiba Semiconductors Corp. www.toshiba-components.com.
[18] M. Sathik, et al., Short circuit detection and fault current limiting method for IGBTs, IEEE Trans. Device Mater. Reliab. 20 (2020) 686–693.
[19] Avago Technologies Application Note 5324 for gate drive product ACPL-333J, Desaturation Fault Detection, 2012, pp. 1–7.
[20] R. Herzer, S. Pawel, J. Lehmann, IGBT driver chipset for high power applications, in: IEEE International Symposium on Power Semiconductor Devices and ICs, 2002, pp. 161–164.
[21] J. Werthen, M. Cohen, Optical Power Isolation Enables Novel IGBT and Sensor Applications, PCIM Europe, 2017, pp. 446–450.
[22] S. Uchida, et al., A Face-to-Face Chip Stacking 7kV RMS Digital Isolator for Automotive and Industrial Motor Drive Applications, IEEE International Symposium on Power Semiconductor Devices and ICs, 2014, pp. 442–445.
[23] Y. Coyne, et al., Low gate drive IGBT enabling direct control through digital isolator power, in: IEEE Conf. Bipolar/BiCMOS Circuits and Technology, 2016, pp. 134–137.
[24] H. Ghorbhani, et al., Performance of a New Gate Drive Controller for Improving IGBT Switching Trajectory, IEEE Industrial Electronics Society Conference, 2016, pp. 3547–3551.
[25] R. Morikawa, et al., New gate driving technique using digital gate driver IC to reduce both EMI in specific frequency band and switching loss in IGBTs, in: IEEE Int. Power Electronics and Motion Control Conf, 2020, pp. 644–651.
[26] Y. Ling, Z. Zhao, Y. Zhu, A self-regulating gate driver for high-power IGBTs, IEEE Trans. Power Electron. (2021) 3450–3461.
[27] A. Kanale, B.J. Baliga, Enhancing Short Circuit Capability of 1.2 kV SiC Power MOSFETs Using a Gate-Source Shorted Si Depletion-Mode MOSFET in Series With the Source, IEEE Conference on Power Electronics and Drive Systems, 2019, pp. 1–6.

[28] A. Kanale, B.J. Baliga, Enhancing Short Circuit Capability of 1.2-kV Si IGBT Using a Gate-Source Shorted Si Depletion Mode MOSFET in Series With the Emitter, Vol. 35, IEEE Transactions on Power Electronics, 2020, pp. 6350–6361.
[29] A. Kanale, B.J. Baliga, Enhancing short circuit ruggedness of 1.7 kV IGBTs using a gate-source-shorted depletion-mode MOSFET in series with the Emitter, in: IEEE Int. Conf. on Industrial Electronics and Applications, 2020, pp. 651–656.
[30] B.J. Baliga, U.S Patent Filed and Pending.
[31] A. Kanale, B.J. Baliga, Theoretical optimization of GSS Si DMM device for improving the BaSIC(DMM) topology used to enhance SiC power MOSFET short circuit capability, IEEE Access (2021) 70039–70047, https://doi.org/10.1109/ACCESS.2021.3078134.

CHAPTER

IGBT Circuit Models

8

Good models are required for simulation of the insulated gate bipolar transistor (IGBT) in power circuits to understand the principles of its operation in the blocking mode, the on-state, and during switching between these states. Analytical models for the device provide a level of insight into the IGBT structure that is often neglected with the availability of modern computer simulation software. Extensive analytical models for all the operating states for the IGBT have been developed and verified using numerical simulations in the textbook [1]. The development of analytical models for the physics of operation of the IGBT has also been recently reviewed [2]. The physics-based analytical models for the IGBT were discussed in Chapter 3 for the design of blocking characteristics, on-state characteristics, and switching characteristics.

For the analysis of the operation of power electronic circuits containing IGBTs, it is necessary to construct models that can be implemented in circuit simulators. There are two basic approaches to create such models: by using a physics-based approach and by using a parameter extraction or fitting approach. This chapter reviews the development of these types of models for the IGBT. Reviews of the methodology for creating IGBT device models for circuit implementation have been periodically published in the literature [3,4].

8.1 Physics-Based Circuit Model

As discussed in Chapter 2, the *n*-channel IGBT structure can be represented as an *n*-channel metal oxide semiconductor field effect transistor (MOSFET) providing the base drive current to a P-N-P transistor (see Fig. 2.2). Although this equivalent circuit can be easily implemented in circuit simulators by using existing models for the MOSFET and the bipolar transistor, the outcome is not accurate because the P-N-P transistor within the IGBT has a large width with a low doping concentration, resulting in strong conductivity modulation effects that are not accounted for in traditional models for bipolar transistors in circuit simulators. A physics-based model for the low gain P-N-P transistor operating with high-level injection in the base region is required [5].

The IGBT Device. https://doi.org/10.1016/B978-0-323-99912-0.00006-4

8.1.1 SABER NPT-IGBT Circuit Model

A phenomenological representation of the symmetric or nonpunch-through (NPT) IGBT is shown in Fig. 8.1 with the detailed equivalent circuit superposed on the structure. The primary elements in the equivalent circuit are the MOSFET that provides the base drive current and the wide base P-N-P transistor that carries the bipolar current within the IGBT structure. Conductivity modulation of the base resistance (R_b) of the P-N-P transistor is indicated and taken into account in the model. The capacitances C_{oxs} and C_{oxd} have a constant value but the other capacitances C_{gd}, C_{cer}, C_{dej}, C_{ebj}, and C_{ebd} are nonlinear and a strong function of the collector voltage.

The various device parameters are modeled as follows for implementation in SABER modeling software [3] are shown below:

$$V_{dg} = V(drain) - V(gate) \tag{8.1}$$

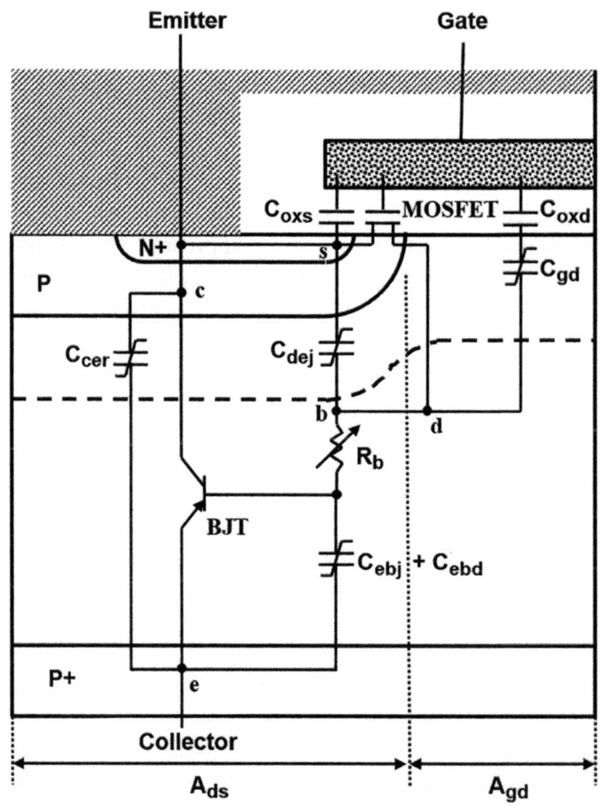

FIGURE 8.1

Phenomenological insulated gate bipolar transistor (IGBT) circuit model. Bipolar junction transistor (BJT).

$$V_{ge} = V(gate) - V(emitter) \tag{8.2}$$

$$V_{ds} = V(drain) - V(source) \tag{8.3}$$

$$V_{eb} = V(emitter) - V(base) \tag{8.4}$$

$$N_{scl} = N_B + N_{sat} \tag{8.5}$$

$$W_{gdj} = \sqrt{2\varepsilon_S\left(V_{dg} + V_{Td}\right)/qN_{scl}} \tag{8.6}$$

$$W_{dsj} = \sqrt{2\varepsilon_S(V_{ds} + 0.6)/qN_{scl}} \tag{8.7}$$

$$W_{bcj} = \sqrt{2\varepsilon_S(V_{bc} + V_{Td})/qN_{scl}} \tag{8.8}$$

$$W = W_B - W_{bcj} \tag{8.9}$$

$$Q_{gs} = C_{gs}V_{gs} \tag{8.10}$$

$$Q_{ds} = A_{ds}\sqrt{2\varepsilon_S(V_{ds} + 0.6)qN_{scl}} \tag{8.11}$$

$$Q_B = qAWN_{scl} \tag{8.12}$$

$$Q_{bi} = A\sqrt{2\varepsilon_S qN_B 0.6} \tag{8.13}$$

$$C_{bcj} = A\varepsilon_S/W_{bcj} \tag{8.14}$$

$$C_{cer} = QC_{bcj}/3Q_B \tag{8.15}$$

$$C_{dsj} = \left(A - A_{gd}\right)\varepsilon_S/W_{dsj} \tag{8.16}$$

$$C_{gdj} = A_{gd}\varepsilon_S/W_{gdj} \tag{8.17}$$

$$C_{gd} = C_{oxd} \quad \text{for} \quad V_{ds} \le \left(V_{gs} - V_{Td}\right) \tag{8.18}$$

$$C_{gd} = C_{oxd}C_{gdj}/\left(C_{oxd} + C_{gdj}\right) \quad \text{for} \quad V_{ds} \ge \left(V_{gs} - V_{Td}\right) \tag{8.19}$$

$$\mu_{nc} = 1/(1/\mu_n + 1/\mu_c) \tag{8.20}$$

$$\mu_{pc} = 1/\left(1/\mu_p + 1/\mu_c\right) \tag{8.21}$$

$$\mu_{eff} = \mu_{nc} + \mu_{pc}Q/(Q + Q_B) \tag{8.22}$$

$$D_c = 2(kT/q)\mu_{nc}\mu_{pc}/\left(\mu_{nc} + \mu_{pc}\right) \tag{8.23}$$

$$L = \sqrt{D_c \tau_{HL}} \tag{8.24}$$

$$P_0 = Q/\left(qAL \tanh \frac{W}{2L}\right) \tag{8.25}$$

$$\delta p = P_0 \sinh(W/2L)/\sinh(W/L) \tag{8.26}$$

$$n_{eff} = \frac{\frac{W}{2L}\sqrt{N_B^2 + P_0^2 \mathrm{csch}^2(W/L)}}{\mathrm{arctanh}\left[\frac{\sqrt{N_B^2 + P_0^2 \mathrm{csch}^2(W/L)}\tanh(W/2L)}{N_B + P_0 \mathrm{csch}(W/L)\tanh(W/2L)}\right]} \tag{8.27}$$

$$R_b = W/(q\mu_{nc}AN_B) \quad \text{for} \quad Q < 0 \tag{8.28}$$

$$R_b = W/\left(q\mu_{eff}AN_{eff}\right) \quad \text{for} \quad Q \geq 0 \tag{8.29}$$

$$V_{ebj} = 0.6 - (Q - Q_{bi})^2/\left(2qN_B\varepsilon_S A^2\right) \tag{8.30}$$

$$V_{ebd} = \frac{kT}{q}\ln\left[\left(\frac{P_0}{n_i^2} + \frac{1}{N_B}\right)(N_B + P_0)\right] - \frac{D_c}{\mu_{nc}}\ln\frac{P_0 + N_B}{N_B} \tag{8.31}$$

$$V_{ebq} = V_{cbj} \quad \text{for} \quad Q < 0 \tag{8.32}$$

$$V_{ebq} = \min\left(V_{ebj}, V_{cbd}\right) \quad \text{for} \quad Q_{bi} > Q \geq 0 \tag{8.33}$$

$$V_{ebq} = V_{cbd} \quad \text{for} \quad Q \geq Q_{bi} \tag{8.34}$$

$$BV_{cb0} = BV_f 5.34 \times 10^{13} N_{scl}^{-0.75} \tag{8.35}$$

$$M = 1/\left[1 - (V_{cb}/BV_{cb0})^n\right] \tag{8.36}$$

$$I_T = V_{ce}/R_b \tag{8.37}$$

$$I_{css} = \left(\frac{1}{1+b}\right)I_T + \left(\frac{b}{1+b}\right)\frac{4D_p}{W^2}Q \tag{8.38}$$

$$I_c = I_{css} + C_{cer}\left(\frac{dV_{cc}}{dt}\right) \tag{8.39}$$

$$I_{bss} = \frac{Q}{\tau_{HL}} + \frac{Q^2}{Q_b^2}\frac{4N_{scl}^2}{n_i^2}I_{snc} \tag{8.40}$$

$$I_{mos} = 0 \quad \text{for} \quad V_{gs} < V_T \tag{8.41}$$

$$I_{mos} = \frac{K_{Plin}\left[(V_{gs} - V_T)V_{ds} - \frac{K_{Plin}V_{ds}^2}{2K_{Pscl}}\right]}{\left[1 + \theta(V_{gs} - V_T)\right]} \quad \text{for} \quad V_{ds} \le (V_{gs} - V_T)\frac{K_{Pscl}}{K_{Plin}} \tag{8.42}$$

$$I_{mos} = \frac{K_{Pscl}(V_{gs} - V_T)^2}{2\left[1 + \theta(V_{gs} - V_T)\right]} \quad \text{for} \quad V_{ds} > (V_{gs} - V_T)\frac{K_{Pscl}}{K_{Plin}} \tag{8.43}$$

$$I_{gen} = \frac{qn_iA}{\tau_{HL}}\frac{\sqrt{2\varepsilon_S V_{bc}}}{qN_{scl}} \tag{8.44}$$

$$I_{mult} = (M-1)(I_{mos} + I_c) + MI_{gen} \tag{8.45}$$

The expression I_{mos} for current flow through the MOSFET channel includes the diffusion effect due to nonuniform doping. The reduction of channel mobility with increasing transverse electric field is accounted for using the parameter θ. The capacitances are modeled as combinations of the fixed oxide capacitance and the variable semiconductor capacitance. Due to conductivity modulation of the wide base region of the P-N-P transistor, the excess carrier base charge Q is much larger than the background base charge Q_B. The expression I_{css} for the collector current in the bipolar transistor consists of the nonquasistatic component and the charge-control component. The expression I_{bss} for the base current in the bipolar transistor consists of the component due to recombination in the base and the component due to injection of electrons into the emitter. The emitter–base voltage includes the voltage across the modulated base resistance R_b and the emitter–base diffusion capacitance C_{ebd} or the depletion capacitance C_{ebj}. The model must include the reduction in mobility at high injection levels due to carrier–carrier scattering [1], velocity saturation in the depletion region, and carrier multiplication due to impact ionization in the drift region.

The IGBT model is included in the SABER library. The IGBT model is formulated to provide the currents between the terminal nodes for the device based on the above Eqs. A schematic of the current components between the IGBT terminals is shown in Fig. 8.2. This analog circuit representation of the IGBT model shows the bipolar transistor replaced by its base and collector current sources. Because the voltage drops due to drift and diffusion are distributed throughout the wide base region of the P-N-P transistor and because it has a low current gain, both the base and collector current contribute to the resistive voltage drop V_{ce}. This is represented by the modulated base resistance R_b. I_{css} is the component of the collector current that does not depend on time rate of change of the base–collector voltage of the P-N-P transistor while I_{ccer} is the component of the collector current that does depend on time rate of change of the base–collector voltage. Similarly, I_{bss} is the component of the base current that does not depend on time rate of change of the base–collector voltage of the P-N-P transistor while I_{ceb} is the component of the base current that does depend on time rate of change of the base–collector voltage. I_{mult} is current associated with avalanche multiplication at high bias voltages. This model has been implemented in the SABER software platform and shown to accurately describe the static and dynamic characteristics of the IGBT [3].

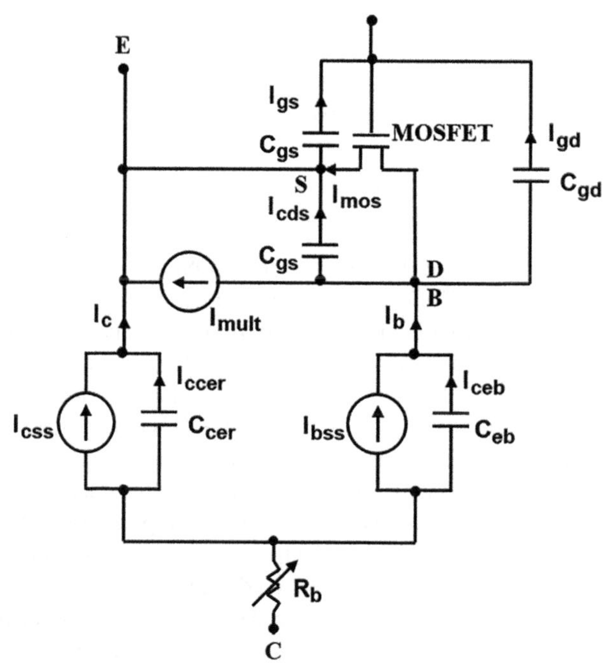

FIGURE 8.2

Analog circuit representation of the IGBT model.

8.1.2 SABER PT-IGBT Circuit Model

The asymmetric or punch-through (PT) IGBT structure contains a buffer layer to allow shortening of the thickness of the drift region, as discussed in Chapter 3. The dynamics of the switching process during turn-off for the asymmetric IGBT structure are different from the symmetric structure because the space-charge layer can expand through the entire lightly doped portion of the drift region during the voltage transient. The current fall then occurs by removal of the stored charge in the buffer layer.

The components of current flow for the asymmetric IGBT structure are the same as those described in the previous section for the symmetric IGBT structure. However, the equations given in the previous section are modified due to the segmentation of the drift region into a low-doped portion with width W_D and a buffer layer with width W_{BL}. The appropriate equations for the asymmetric IGBT structure are given below [6]:

$$N_{scl} = N_L + N_{sat} \tag{8.46}$$

$$N_{sat} = \frac{I_C}{qAv_{psat}} - \frac{I_{mos}}{qAv_{nsat}} \tag{8.47}$$

$$V_{rt} = \frac{qW_D^2 N_{scl}}{2\varepsilon_S} - 0.6 \tag{8.48}$$

$$W_{gdj} = \sqrt{2\varepsilon_S (V_{dg} + V_{Td})/qN_{scl}} \quad \text{for} \quad W_{gdj} \le W_D \tag{8.49}$$

$$W_{dsj} = \sqrt{2\varepsilon_S (V_{ds} + 0.6)/qN_{scl}} \quad \text{for} \quad W_{dsj} \le W_D \tag{8.50}$$

$$W_{bcj} = \sqrt{2\varepsilon_S (V_{bc} + 0.6)/qN_{scl}} \quad \text{for} \quad W_{bcj} \le W_D \tag{8.51}$$

$$W_{nrt} = \sqrt{2\varepsilon_S (V_{bc} + 0.6)/qN_{scl}} \tag{8.52}$$

$$W = W_D - W_{bcj} \tag{8.53}$$

$$W_{eff} = \sqrt{W^2 + W_{BL}^2 \frac{D_C}{D_{pBL}} + \frac{W_{BL}^2}{2D_{pBL}} \frac{WC_{bcj}}{3qN_D A} \frac{dV_{bc}}{dt}} \tag{8.54}$$

$$\gamma = \frac{Q_D^n}{Q_L^n + Q_B} \tag{8.55}$$

$$Q_{gs} = C_{gs} V_{gs} \tag{8.56}$$

$$Q_B = qAWN_{scl} \tag{8.57}$$

$$Q_{bi} = A\sqrt{2\varepsilon_S qN_D 0.6} \tag{8.58}$$

$$Q_{ds} = A_{ds}\sqrt{2\varepsilon_S(V_{ds} + 0.6)qN_{scl}} \quad \text{for} \quad V_{ds} \leq V_{rt} \tag{8.59}$$

$$Q_{ds} = qA_{ds}W_D N_{scl} + A_{ds}\varepsilon_S(V_{ds} - V_{rt})/W_D \quad \text{for} \quad V_{ds} > V_{rt} \tag{8.60}$$

$$Q_D = \frac{W^2}{W_{eff}^2}\left(Q_T - \frac{\gamma I_T}{(1+b)}\frac{W_{BL}^2}{2D_{pBL}}\right) \tag{8.61}$$

$$Q_{BL} = Q_T - Q_D \tag{8.62}$$

$$C_{bcj} = A\varepsilon_S/W_{bcj} \tag{8.63}$$

$$C_{dsj} = \left(A - A_{gd}\right)\varepsilon_S/W_{dsj} \tag{8.64}$$

$$C_{gdj} = A_{gd}\varepsilon_S/W_{gdj} \tag{8.65}$$

$$C_{gd} = C_{oxd} \quad \text{for} \quad V_{ds} \leq \left(V_{gs} - V_{Td}\right) \tag{8.66}$$

$$C_{gd} = C_{oxd}C_{gdj}/\left(C_{oxd} + C_{gdj}\right) \quad \text{for} \quad V_{ds} \geq \left(V_{gs} - V_{Td}\right) \tag{8.67}$$

$$C_{cer} = \frac{W^2}{W_{eff}^2}\frac{Q_T}{Q_B}C_{bcj} \tag{8.68}$$

$$\mu_{nc} = 1/(1/\mu_n + 1/\mu_c) \tag{8.69}$$

$$\mu_{pc} = 1/\left(1/\mu_p + 1/\mu_c\right) \tag{8.70}$$

$$\mu_{eff} = \mu_{nc} + \gamma\mu_{pc} \tag{8.71}$$

$$D_c = 2(kT/q)\mu_{nc}\mu_{pc}/\left(\mu_{nc} + \mu_{pc}\right) \tag{8.72}$$

$$L = \sqrt{D_c\tau_{HL}} \tag{8.73}$$

$$P_{D0} = Q_D^n / \left(qAL \tanh \frac{W}{2L} \right) \tag{8.74}$$

$$P_{BL0} = \frac{2Q_{BL}}{qAW_{BL}} + \frac{P_{D0}(P_{D0} + N_D)}{N_{BL}} \tag{8.75}$$

$$n_{eff} = \frac{\frac{W}{2L}\sqrt{N_B^2 + P_0^2 \mathrm{csch}^2(W/L)}}{\mathrm{arctanh}\left[\frac{\sqrt{N_B^2 + P_0^2 \mathrm{csch}^2(W/L)}\tanh(W/2L)}{N_B + P_0 \mathrm{csch}(W/L)\tanh(W/2L)}\right]} \tag{8.76}$$

$$R_b = W/(q\mu_{nc}AN_B) \quad \text{for} \quad Q_T < 0 \tag{8.77}$$

$$R_b = W/\left(q\mu_{eff}AN_{eff}\right) \quad \text{for} \quad Q_T \geq 0 \tag{8.78}$$

$$V_{ebj} = 0.6 - (Q_T - Q_{bi})^2 / \left(2qN_{BL}\varepsilon_S A^2\right) \tag{8.79}$$

$$V_{ebd} = \frac{kT}{q} \ln\left[\left(\frac{P_{D0}N_{BL}}{n_i^2}\right) + 1\right] - \frac{D_c}{\mu_{nc}} \ln \frac{P_{D0} + N_D}{N_D} \tag{8.80}$$

$$V_{ebq} = V_{cbj} \quad \text{for} \quad Q_T < 0 \tag{8.81}$$

$$V_{ebq} = \min\left(V_{ebj}, V_{cbd}\right) \quad \text{for} \quad Q_{bi} > Q_T \geq 0 \tag{8.82}$$

$$V_{ebq} = V_{cbd} \quad \text{for} \quad Q_T \geq Q_{bi} \tag{8.83}$$

$$V_{nrt} = \frac{V_{bc}}{\frac{W_{bcj}}{W_{nrt}}\left(2 - \frac{W_{bcj}}{W_{nrt}}\right)} \tag{8.84}$$

$$BV_{cb0} = BV_f 5.34 \times 10^{13} N_{scl}^{-0.75} \tag{8.85}$$

$$M = 1/[1 - (V_{cb}/BV_{cb0})^n] \tag{8.86}$$

$$I_T = V_{ce}/R_b \tag{8.87}$$

$$I_{css} = \frac{W^2}{W_{eff}^2} \frac{\gamma I_T}{(1+b)} + \frac{Q_T D_c}{W_{eff}^2} \tag{8.88}$$

$$I_c = I_{css} + C_{cer}\left(\frac{dV_{cc}}{dt}\right) \tag{8.89}$$

$$I_{bss} = \frac{Q_D}{\tau_{HL}} + \frac{Q_{BL}}{\tau_{BL}} + \frac{Q_D^2}{Q_b^2}\frac{4N_{scl}^2}{n_i^2}I_{snc} \tag{8.90}$$

$$I_{mos} = 0 \quad \text{for} \quad V_{gs} < V_T \tag{8.91}$$

$$I_{mos} = \frac{K_pK_f\left[(V_{gs} - V_T)V_{ds} - \frac{K_fV_{ds}^2}{2}\right]}{\left[1 + \theta(V_{gs} - V_T)\right]} \quad \text{for} \quad V_{ds} \le \frac{(V_{gs} - V_T)}{K_f} \tag{8.92}$$

$$I_{mos} = \frac{K_p(V_{gs} - V_T)^2}{2\left[1 + \theta(V_{gs} - V_T)\right]} \quad \text{for} \quad V_{ds} > \frac{(V_{gs} - V_T)}{K_f} \tag{8.93}$$

$$I_{gen} = \frac{qn_iA}{\tau_{HL}}\frac{\sqrt{2\varepsilon_S V_{bc}}}{qN_{scl}} \tag{8.94}$$

$$I_{mult} = (M - 1)(I_{mos} + I_c) + MI_{gen} \tag{8.95}$$

8.1.3 SABER IGBT Electrothermal Circuit Model

The junction temperature of the IGBT devices is well above the ambient temperature due to self-heating during operation in power circuits. The self-heating is generated by power dissipation within the chip due to the finite on-state voltage drop of the device in the conduction duty cycle and switching losses during turn-on and turn-off. The modeling of the device in circuit simulators must include the self-heating effect and the temperature dependence of the device characteristics. These electrothermal models must couple the device characteristics with heat removal from the chip through the heat sink.

A schematic diagram [7] for the electrothermal model for the IGBT is shown in Fig. 8.3. The basic concept is to couple the temperature-dependent characteristics of the IGBT chip to the heat flow through the silicon chip and then via the case (package) and heat sink. The modeling can be performed by assuming either a uniform temperature within the silicon chip for simplicity or by using a distributed temperature within the silicon chip, which is more realistic but computation-intensive. The latter requires implementing the heat diffusion equation inside the silicon chip. This allows more accurate representation of the temperature of the MOSFET channel in the IGBT where the temperature is the largest and the conductivity modulation of the drift region where the temperature is lower.

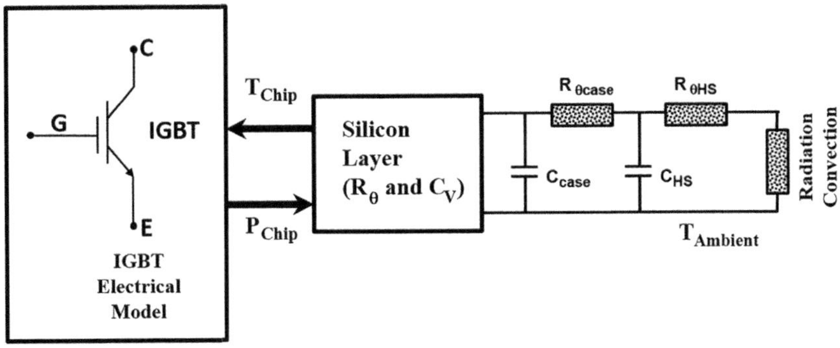

FIGURE 8.3

Electrothermal model for the IGBT.

Typical temperature-dependent parameters are [1,8] shown below:

$$\mu_{n,p}(T) = \mu_{n0,p0}\left(\frac{T_0}{T}\right)^{2.5} \tag{8.96}$$

$$n_i(T) = 3.87 \times 10^{16} T^{3/2} e^{-7020/T} \tag{8.97}$$

$$\tau_{HL}(T) = \tau_{HL0}\left(\frac{T}{T_0}\right)^{\beta} \tag{8.98}$$

$$V_T(T) = V_{T0} - C_{VT0}(T - T_0) \tag{8.99}$$

The temperature dependence of the IGBT characteristics can also be extracted from the datasheets supplied by manufacturers. Examples are shown in Figs. 8.4 and 8.5 of the temperature dependence of the on-state characteristics and the energy loss per cycle for a typical IGBT. There is a gradual increase in the on-state voltage drop in the IGBT with increasing temperature at the on-state current level. The switching losses increase with temperature as well due to an increase in the high-level lifetime in the drift region and buffer layer.

8.1.4 SABER IGBT1 Model

The IGBT1 model implemented in SABER has been validated to predict the on-state, current saturation, and switching behavior of the device [9]. It is described here because of widespread use for power circuit simulation. The equivalent circuit for this model is shown in Fig. 8.6. The diode represents the knee in the on-state characteristics of the IGBT devices. The parameters with a subscript "tail" account for the slow decay of the collector current in the IGBT during the end of the turn-off period.

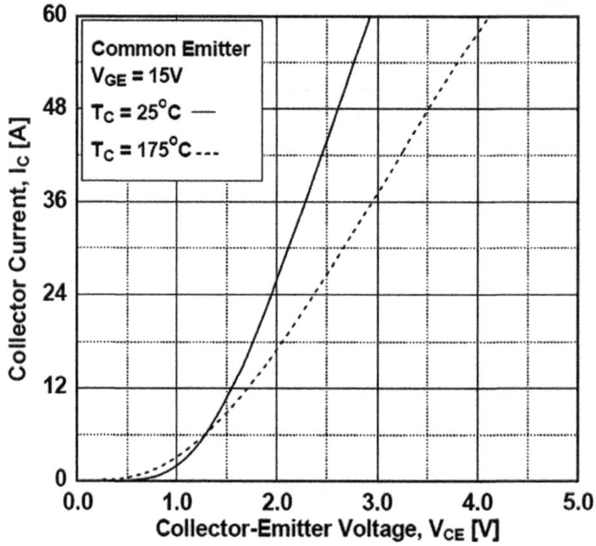

FIGURE 8.4

Temperature dependence of the on-state characteristics of an IGBT product.

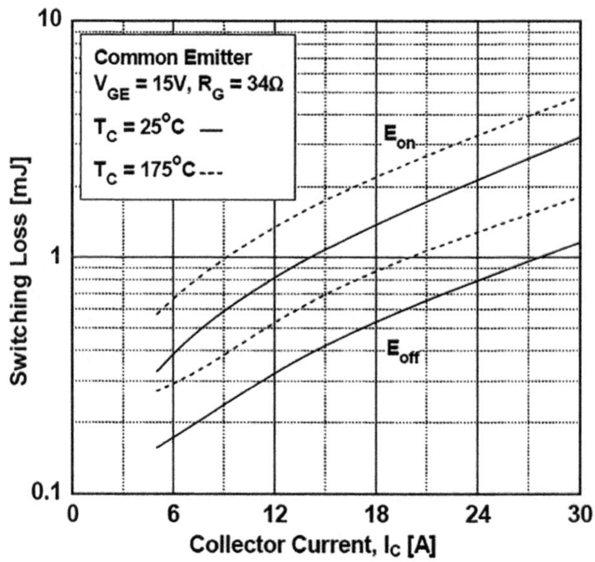

FIGURE 8.5

Temperature dependence of the switching characteristics of an IGBT product.

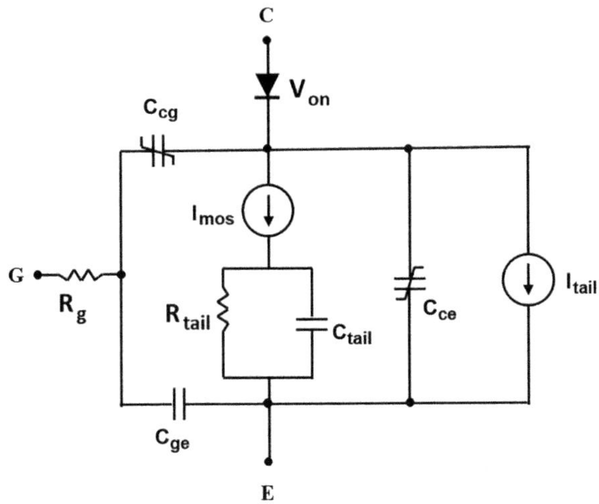

FIGURE 8.6

Basic IGBT 1 SABER model.

The terminal capacitances (C_{cg}, C_{ce}, and C_{ge}) can be related to the capacitances provided in the datasheets (C_{rss}, C_{oss}, and C_{iss}) by IGBT manufacturers:

$$C_{res} = C_{cg} \tag{8.100}$$

$$C_{oes} = C_{cg} + C_{ce} \tag{8.101}$$

$$C_{ies} = C_{cg} + C_{ge} \tag{8.102}$$

An example of the capacitances provided in an IGBT datasheet is shown in Fig. 8.7. It is worth noting that the capacitance C_{ies} is relatively constant but the other capacitances are a strong function of the collector bias. The sudden reduction in the capacitances C_{oes} and C_{res} is due to depletion of the JFET region within the IGBT cells when the collector voltage exceeds 3 V for this device.

The IGBT1 model also requires the gate charge Q_1, Q_2, and Q_3, which can be extracted from the manufacturer's datasheets. A gate charge plot is shown in Fig. 8.8 for a typical IGBT product. The various components of the gate charge (Q_g, Q_{ge}, and Q_{gc}) are defined in Chapter 6 of Ref. [1] and in manufacturer's datasheets. The capacitances and gate charges are important parameters that define the transient response of the IGBT and consequently the turn-on and turn-off switching losses.

8.2 IGBT Analog Behavioral Model

An analog behavioral model (ABM) for a semiconductor device describes its behavior from an external circuit point of view rather than from the physics of operation of the structure as done in the previous section. The device can be regarded a black

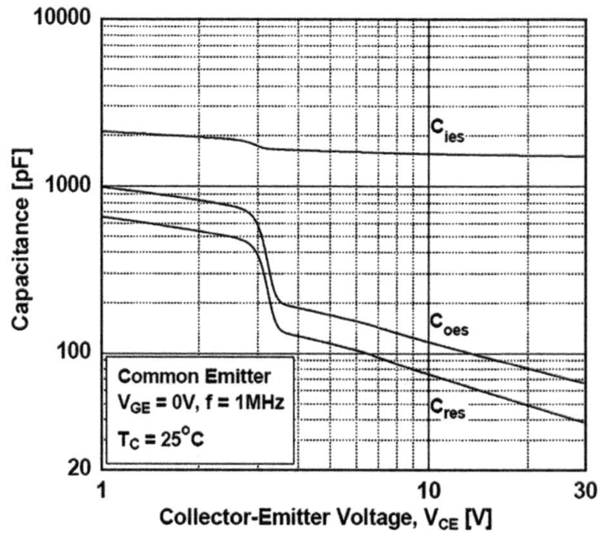

FIGURE 8.7

Typical capacitances for an IGBT product.

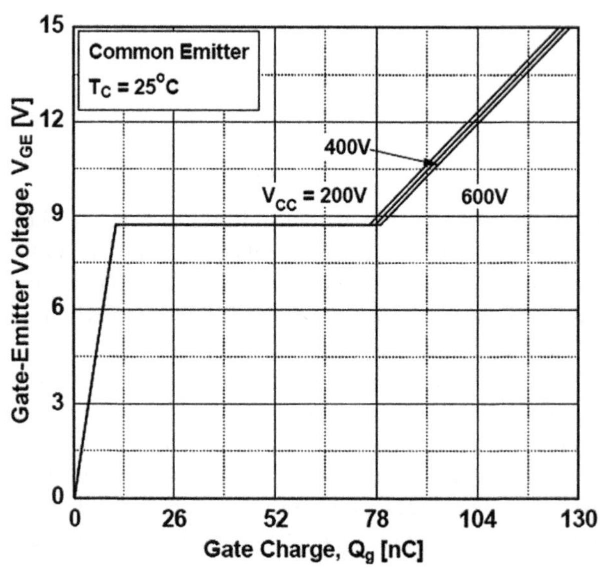

FIGURE 8.8

Typical gate charge plot for an IGBT product.

box within which polynomial-controlled sources are utilized to approximate the static and dynamic characteristics of the device. A popular ABM-based circuit simulation program is PSPICE. The n-channel IGBT consists of a wide base P-N-P transistor driven by a MOSFET integrated within the structure. The equivalent circuit for the IGBT is shown in Fig. 2.2. The device can be modeled by using SPICE models for the bipolar transistor and the MOSFET. The basic SPICE model for the bipolar transistor [10] is created using the Ebers-Moll large signal model, as shown in Fig. 8.9. The charges in the device are modeled using nonlinear capacitances. The diodes account for the forward-biased emitter–base junction and the reverse biased base–collector junctions. The current source in the model is given by:

$$I_{CT} = I_{CC} - I_{EC} = I_S\left(e^{qV_{BE}/kT} - e^{qV_{BC}/kT}\right) \tag{8.103}$$

The SPICE model for the MOSFET [8] is shown in Fig. 8.10. The voltage-controlled current sources account for the gate bias-induced drain current and the drain current arising from the body-bias effect. The nonlinear capacitances are a function of the gate and drain bias voltages [11].

The IGBT model for SPICE simulations can be created by combining the above models for the bipolar transistor and the MOSFET structures, as shown in Fig. 8.11. The bipolar transistor model must be enhanced to account for the wide lightly doped base region for the bipolar transistor in the IGBT structure. A voltage- and current-dependent variable resistance is needed to take conductivity modulation of the base region into account. The model for the wide base P-N-P transistor in the IGBT consists of the collector–base diode, which injects electron current ($I_{n(x=0)}$) into the base region, and two current sources ($I_{p(x=0)}$ and $I_{p(x=W)}$) to account for the hole currents at the edges of the wide base region. Stray inductances in the gate and emitter path are

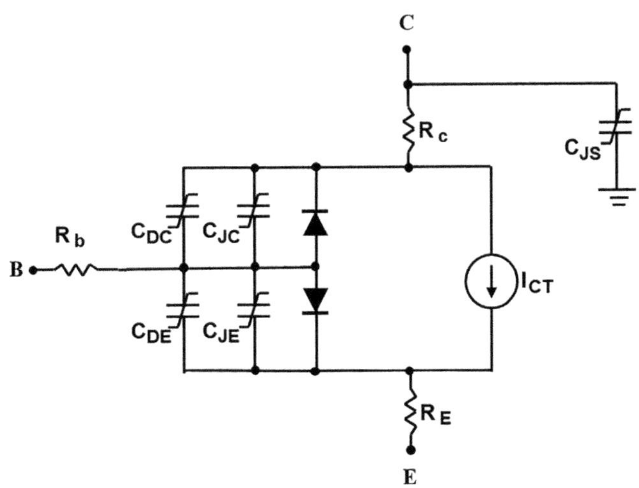

FIGURE 8.9

Ebers-Moll large signal model for the NPN bipolar transistor.

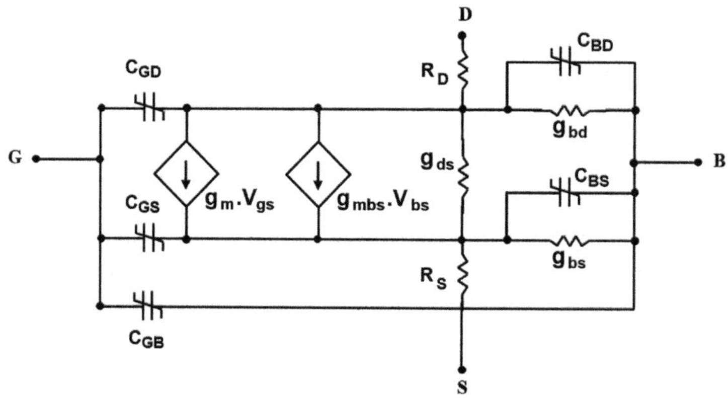

FIGURE 8.10

Linearized equivalent circuit for the MOSFET.

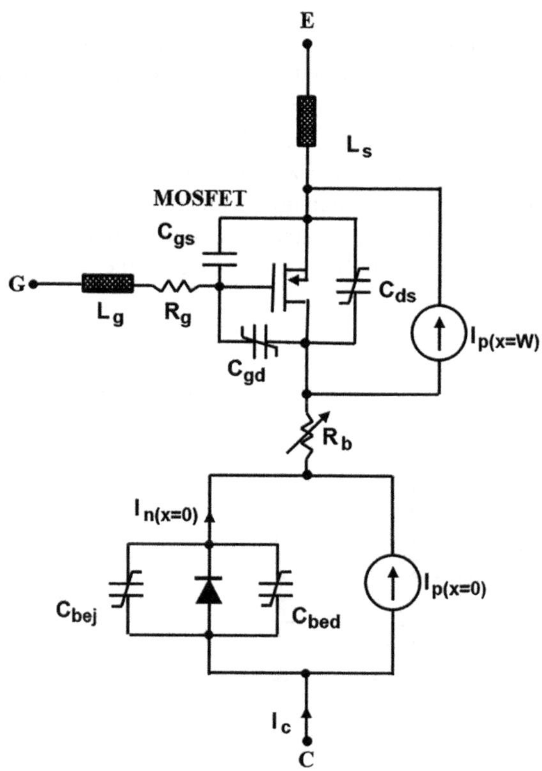

FIGURE 8.11

SPICE model representation of the IGBT.

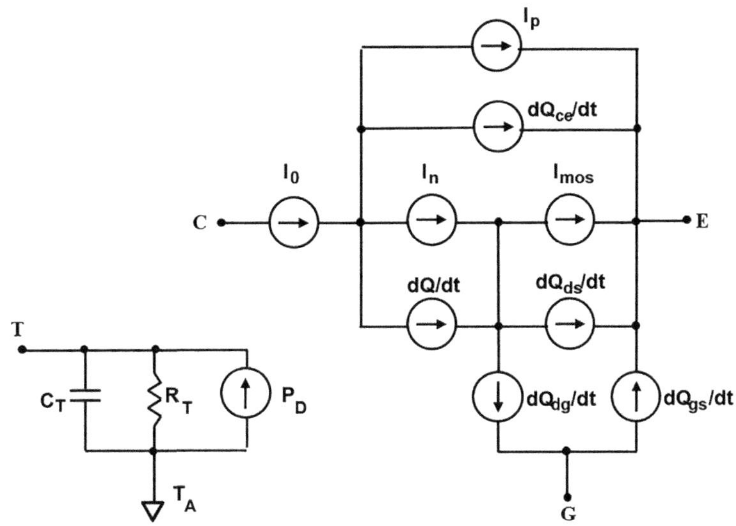

FIGURE 8.12

SPICE model for the IGBT with temperature variation.

also included for improved transient simulations. The model is generated by creating a set of algebraic equations and inhomogeneous differential equations with variable coefficients. The equations are then converted into electrical circuits for solution by SPICE by using controlled sources [12]. To model different operating phases during switching or different operating regimes in the output characteristics (like the linear and saturation regions), it becomes necessary to switch between different sets of equations. This is performed by using IF and THEN logic operators [13].

In IGBT applications, the power dissipated within the device produces an increase in the internal temperature. The rise in temperature leads to alterations in the device characteristics due to changes of the electron and hole mobilities and the recombination lifetime [14]. The SPICE model can be enhanced by adding a sub-circuit [15] that allows taking into account the rise in junction temperature due to the power dissipation, as illustrated in Fig. 8.12. The thermal circuit includes the thermal resistance and the specific heat for the device.

8.3 Model Parameter Extraction

The modeling of the IGBT in the circuit simulator requires extraction of the parameters that are pertinent to the particular device being used in the application. The model for the IGBT requires extraction of 11 parameters for the symmetric (NPT) structure and 13 parameters for the asymmetric (PT) structure [16]. Five of these parameters that pertain to the metal oxide Semiconductor (MOS) gate structure can be extracted using the manufacturer's datasheet. These parameters are: (1) the threshold voltage; (2) the transconductance; (3) the short channel parameter; (4) the gate-emitter capacitance; and (5) the gate oxide capacitance. The last parameter is equal to the gate-collector capacitance at small collector bias voltages (see Fig. 8.7).

The die active area can be estimated from the rated on-state current in the datasheet, assuming an on-state current density of 100–150 A/cm^2. The lifetime in the drift region must be obtained by performing switching measurements [16]. The IGBT collector current exhibits a tail whose time constant is a measure of the lifetime, as discussed in Chapter 3. For the symmetric blocking structure, the decay of the tail current is governed by recombination in the wide base region. The lifetime is then independent of the collector voltage and current levels [16]. For the asymmetrical IGBT structure, the decay of the tail current is governed by the lifetime in the lightly doped portion of the base region at small collector voltages. At large collector voltages, the recombination of carriers occurs primarily in the buffer layer, allowing extraction of the lifetime in this region [16].

8.4 Summary

A brief review of the models developed for the simulation of IGBT-based power circuits has been provided in this chapter. The approaches include physics-based models and those derived using an equivalent circuit approach. In both cases, it is necessary to perform parameter extraction for the particular IGBT device being used in the application. The parameters must be validated using a match to the static characteristics provided in the manufacturer's datasheets and basic switching measurements for a single device. These parameters then allow the simulation of complex power circuits such as inverters containing typically six IGBTs and six antiparallel diodes for adjustable speed motor drives. This approach has been utilized over a very broad range of power levels, including 4.5 kV, 2 kA IGBT modules used for traction applications [17].

References

[1] B.J. Baliga, Fundamentals of Power Semiconductor Devices, second ed., Springer-Science, 2019 (Chapter 9).

[2] B.J. Baliga, Analytical modeling of IGBTs: challenges and solutions, IEEE Trans. Electron Devices 60 (2013) 535–543.

[3] R. Kraus, H. Mattausch, Status and trends of power semiconductor device models for circuit simulation, IEEE Trans. Power Electron. 13 (1998) 452–465.

[4] K. Sheng, B.W. Williams, S.J. Finney, A review of IGBT models, IEEE Trans. Power Electron. 15 (2000) 1250–1266.

[5] A.R. Hefner, D.M. Diebolt, An experimentally verified IGBT model implemented in the saber circuit simulator, IEEE Trans. Power Electron. 9 (1994) 532–542.

[6] A.R. Hefner, Modeling buffer layer IGBT's for circuit simulation, IEEE Trans. Power Electron. 10 (1995) 111–121.

[7] A. Amimi, et al., Modeling and characterization of the IGBT electrothermal transistor for circuit applications, in: IEEE 39th Midwest Symposium on Circuits and Systems, vol. 1, 1996, pp. 281–284.
[8] A.R. Hefner, D.L. Blackburn, A dynamic electro-thermal model for the IGBT, in: IEEE Industrial Applications Society Meeting, 1992, p. 1094.
[9] M. Jun, et al., Application of Saber's simulation model IGBT1 in solid-state switch design, in: International Conference on Electrical Machines and Systems, vol. 3, 2005, pp. 2013–2017.
[10] P. Antognetti, G. Massobrio, Semiconductor Device Modeling With SPICE, McGraw-Hill Book Company, 1988.
[11] B.J. Baliga, Fundamentals of Power Semiconductor Devices, second ed., Springer-Science, 2019 (Chapter 6).
[12] C.S. Mitter, et al., Insulated gate bipolar transistor (IGBT) modeling using IG-SPICE, IEEE Trans. Ind. Appl. 30 (1993) 24–33.
[13] M. Cotorogea, Implementation of mathematical models of power devices for circuit simulation in PSPICE, in: IEEE Workshop on Computers in Power Electronics, 1998, pp. 17–22.
[14] B.J. Baliga, Fundamentals of Power Semiconductor Devices, second ed., Springer-Science, 2019 (Chapter 2).
[15] O. Apeldoorn, S. Schmitt, R.W. De Doncker, An electrical model for a NPT-IGBT including transient temperature effects realized with PSPICE device equations modeling, in: IEEE International Symposium on Industrial Electronics, vol. 2, 1997, pp. 223–228.
[16] X. Kang, et al., Parameter extraction for a physics-based circuit simulator IGBT model, in: IEEE Applied Power Electronics Conference and Exposition, vol. 2, 2003, pp. 946–952.
[17] M. Nawaz, et al., Simple SPICE based modeling platform for 4.5 kV power IGBT modules, in: IEEE Energy Conversion Congress & Expo, 2013, pp. 279–286 (Paper 10.4).

CHAPTER

IGBT Applications: Transportation

9

As pointed out in the introductory chapter of this book, one of the major applications for the insulated gate bipolar transistor (IGBT) is in the transportation sector of the economy. The transportation sector comprises all methods for the movement of people and goods in our society. The transportation sector contributes 11% to the gross domestic product of the United States—amounting to about $1 trillion. The transportation system in the United States is used by about 270 million consumers for nearly 5 trillion passenger miles and nearly 4 trillion ton miles for freight movement every year [1]. In addition, rail and marine transportation each contribute about 11% to the movement of freight in the United States. Transportation is performed on our roads by using cars and trucks, on our seaways by using boats and ships, and in the air by using airplanes and spacecraft. This chapter discusses the impact of the IGBT on automobiles, electric trams or streetcars, high-speed locomotives, ship propulsion, and all-electric drives for modern aircraft.

9.1 Gasoline-Powered Vehicles

At present, gasoline-powered cars and trucks still comprise the majority of vehicles used in our society for commuting, recreation, shopping, and transportation of merchandize. Approximately 150 billion gallons of gasoline are consumed annually in the United States, with about four times as much consumed around the world. Because the burning of each gallon of gasoline produces 19.64 pounds of carbon dioxide, gasoline-powered vehicles have a large impact on our environment. Many approaches have been undertaken by the automotive industry to improve the fuel efficiency of cars and trucks. Among these, the electronic ignition system (or engine management system) has been an important contributor to raising the corporate average fuel economy (CAFE) standards for automobiles. The introduction of the electronic ignition system required a highly reliable semiconductor power switch that could tolerate the adverse environment under the hood of automobiles [2]. The article states: "A primary constraint to the adoption of electronics innovations was the ability to make components durable enough to withstand the heat and vibration factors of an automobile in daily use." The most cost-effective choice for the switch in the electronic ignition system was found to be the IGBT, creating a very large market for these devices since the 1990s.

The IGBT Device. https://doi.org/10.1016/B978-0-323-99912-0.00007-6

9.1.1 Kettering Mechanical Ignition System

Until the late 1980s, the ignition system for the internal combustion engine (ICE) used in gasoline-powered vehicles was based on using the mechanical distributor developed by Charles Kettering for controlling the timing for the spark plugs. The Kettering ignition system is illustrated in Fig. 9.1. The spark plug must be ignited by using the energy available from the 12 V battery in automobiles. Because the formation of the arc at the spark plug requires between 12,000–40,000 V, an ignition coil is required to amplify the voltage by a factor of 1000–3300 times. The ignition coil is a transformer with the car battery connected to the primary side and the spark plug to the secondary side. Prior to igniting the spark plug, the contact breaker is closed, allowing the establishment of a current in the primary of the ignition coil (transformer). A cam connected to the engine shaft opens the contact breaker when one of the cylinders of the engine reaches the top, compressing the fuel mixture. The collapse of the magnetic field in the ignition coil induces a very large voltage on the secondary side. This voltage is directed to the spark plug for that engine cylinder by the distributor to ignite the fuel mixture.

9.1.2 Electronic Ignition System

The mechanical ignition system, used until the 1980s, was prone to poor control of the duration of the spark and poor timing of the spark relative to the cylinder position due to wear out of the contacts in the breaker and the distributor. The availability of

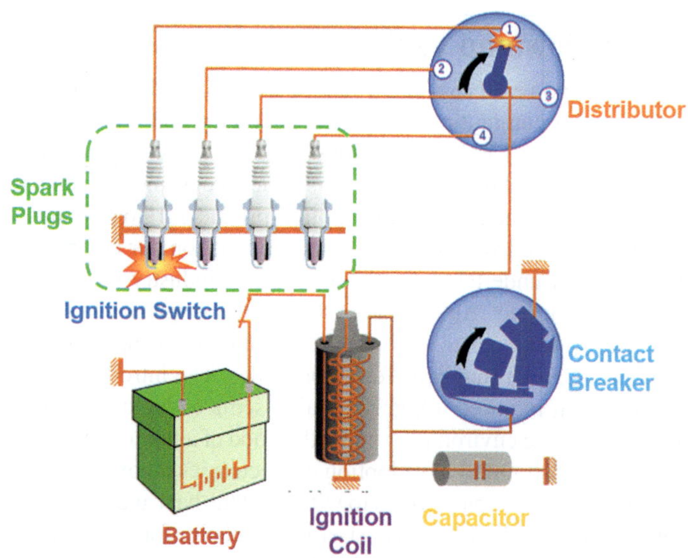

FIGURE 9.1

Mechanical distributor-based ignition system.

the IGBT in the 1980s enabled the introduction of reliable, distributor-less electronic ignition systems with no moving parts. The IGBT was the first power switch with sufficient high-temperature operation and ruggedness attributes to allow cost-effective and reliable implementation of the distributor-less electronic ignition system.

The basic circuit for the electronic ignition system is illustrated in Fig. 9.2. The IGBT is used as the switch on the primary side of the transformer (ignition coil) with the high spark plug voltage generated on the secondary side. When the IGBT is turned on with a positive gate bias, current begins to flow through the primary side of the ignition coil. The rate of rise of the collector current is determined by the inductance of the coil (L_C):

$$\frac{dI_C}{dt} = \frac{V_{DC}}{L_C} \tag{9.1}$$

where V_{DC} is the battery voltage. Typical waveforms for the IGBT collector current and voltage are illustrated in Fig. 9.3. After turn-on of the IGBT at time t_1, the collector current increases linearly with time at a rate determined by Eqn (9.1). The current in the primary side of the coil increases during the dwell time from t_1 to time t_2. This time interval is chosen to obtain enough energy storage in the coil to initiate the ignition of the spark plug:

$$E_{Coil} = 0.5 L_C I_{C\max}^2 \tag{9.2}$$

where $I_{C\max}$ is the peak collector current at time t_2. A typical peak current is 8 A. For a typical coil inductance of 7 mH, the energy stored in the coil at time t_2 is then 224 mJ.

The IGBT is turned off at time t_2 resulting in a rapid increase in the collector voltage to about 300 V (V_{C1}). This produces the high voltage (about 20 kV) across

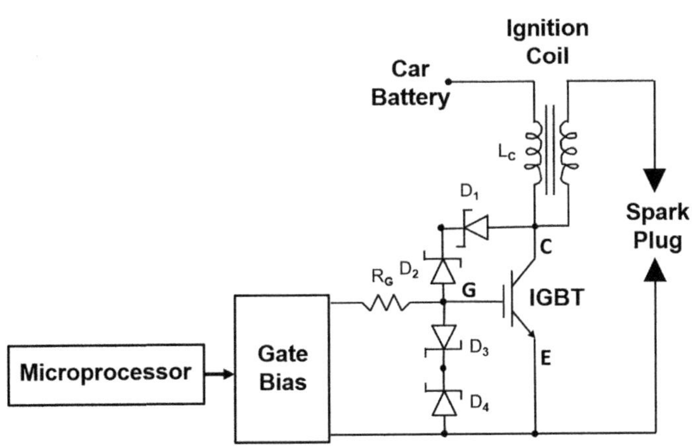

FIGURE 9.2

Insulated gate bipolar transistor (IGBT)-based electronic ignition system.

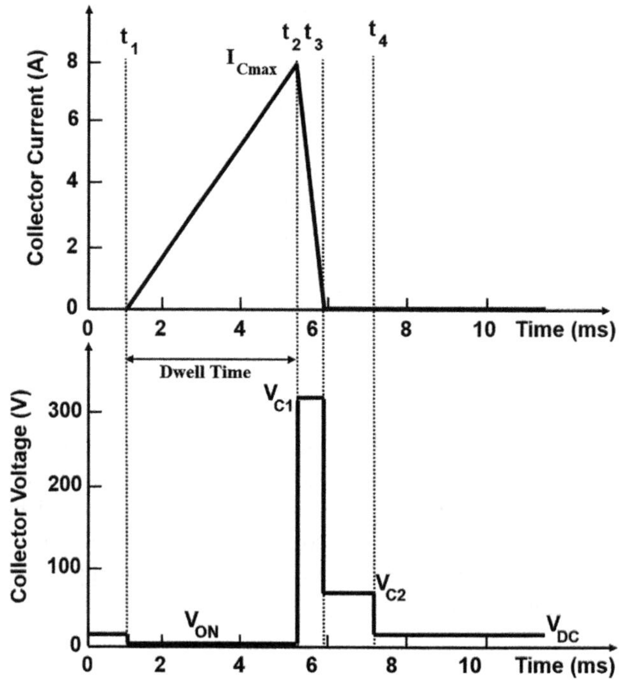

FIGURE 9.3

Ignition system current and voltage waveforms.

the secondary side of the ignition coil to initiate discharge of the spark plug during the time interval from t_2 to t_3. During this time, the collector current reduces rapidly at a rate given by:

$$\frac{dI_C}{dt} = \frac{-V_{C1}}{L_C} \tag{9.3}$$

as shown in the figure. After a short time interval from t_2 to t_3, the arc across the spark plug is maintained with a much smaller voltage (V_{C2}). Consequently, the voltage across the primary side and the IGBT reduces to about 50 V (V_{C2}). The IGBT collector voltage returns to the 12 V DC battery voltage (V_{DC}) once the fuel is consumed, resulting in extinguishing of the spark. During the time t_2 to t_3, the IGBT has to sustain a high voltage and current simultaneously. This requires rugged IGBT designs with good reverse biased safe operating area (RBSOA) as discussed in Chapter 4.

9.1.3 Ignition IGBT Design

The IGBT designs for the electronic ignition system have different requirements than those for motor control applications. The automotive battery voltage can fall below 10 V under cold-cranking conditions. To charge the ignition coil, it is advantageous to have a low on-state voltage drop for the IGBT. This is achieved by using a high lifetime in the drift region and relatively low buffer layer doping levels for the asymmetric IGBT structure. The increased switching losses in the IGBT due to the high lifetime and low buffer layer doping are not a limitation due to the low switching frequency in this application. The reduced buffer layer doping also enables the IGBT to support a limited reverse (negative) voltage applied to the collector. This is beneficial for preventing current flow through the IGBT if the automotive battery connection is inadvertently reversed. The reverse blocking capability required within the IGBT is only in excess of the DC automotive battery voltage, that is, about 20 V.

One of the harshest conditions that the IGBT can be subjected to in this application occurs when the spark plug lead is disconnected. In this case, the energy stored in the ignition coil cannot be dissipated in the spark. The voltage across the primary side will then rise until it reaches the avalanche breakdown voltage of the IGBT. To avoid destructive failure of the IGBT under these conditions, it is preferable to use dynamic clamping [3] of the collector voltage by using Zener diodes connected between the collector and gate of the IGBT, as shown in Fig. 9.2 (diodes D_1 and D_2). With this circuit, when the IGBT voltage exceeds a value, V_{C1}, required to initiate the spark plug under normal operating conditions, gate drive current is provided to the IGBT via the Zener diodes. This current flows through the gate resistance (R_G in Fig. 9.2), producing a positive bias at the gate of the IGBT that turns it on. Current flow through the IGBT dissipates the energy stored in the coil with low power dissipation in the IGBT.

The protection diodes can be monolithically integrated [4] with the IGBT structure, as illustrated in Fig. 9.4. The diodes are fabricated from polysilicon deposited on a thick oxide layer to isolate them from the silicon substrate. One end of the string is connected to the gate electrode of the IGBT while the other end is tied to the collector terminal of the IGBT. A string of 50 back-to-back (BTB) diodes provides a clamping voltage of 350 V, which is greater than the spark plug ignition voltage and below the breakdown voltage of the IGBT (typically 450–500 V). The same technology can be utilized to also provide electrodstatic discharge (ESD) protection

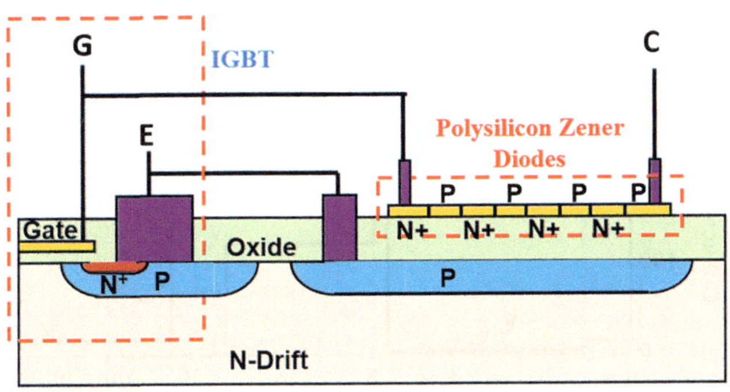

FIGURE 9.4

Ignition IGBT with integrated protection diodes.

for the gate of the IGBT structure, as illustrated in Fig. 9.2 using diodes D_3 and D_4. Two BTB diodes are sufficient to clamp the gate voltage to 14 V.

9.1.4 Dual-Voltage Clamped Ignition IGBT Design

As discussed in the previous section, a high voltage (corresponding to V_{C1} on the primary side of the coil in Fig. 9.3) is needed to initiate the arc at the spark plug and the spark plug voltage subsequently decreases to a lower value (corresponding to V_{C2} on the primary side of the coil in Fig. 9.3). In the IGBT configuration shown in Fig. 9.2, the collector voltage is clamped at a high value ($V_{C1} = 350$ V) for a long duration (typically 140 μs) before decreasing to a V_{C2} of about 50 V. This results in a high peak power during the time the collector voltage has a large value, resulting is a significant rise in the IGBT junction temperature. The self-clamped inductive switching energy (SCIS) can be reduced by making the collector voltage smaller during the first phase (t_2 to t_3) of the spark ignition [5].

The IGBT collector current and voltage waveforms for the dual-voltage self-clamped ignition circuit are illustrated in Fig. 9.5. In this design, the initial phase from time t_2 to t_3 for igniting the spark plug at a high-voltage V_{C1} (350 V on the primary side of the ignition coil) is made shorter (30 μs). This is followed by a second

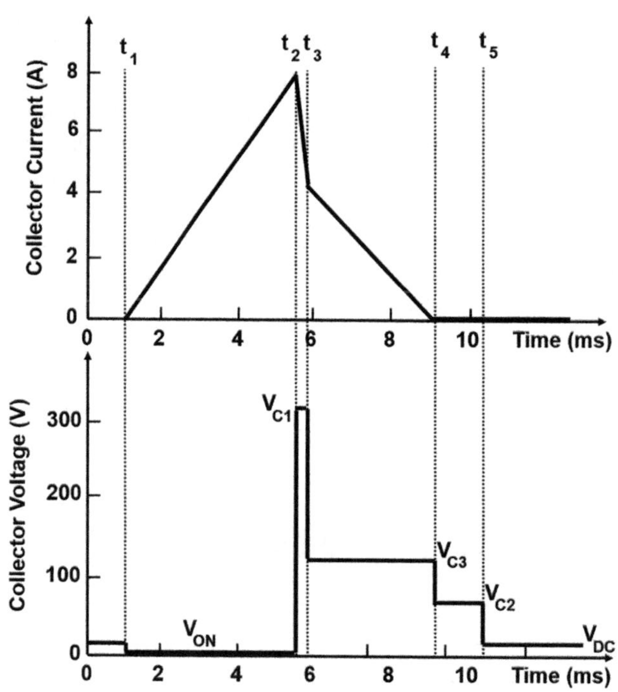

FIGURE 9.5

Dual voltage self-clamped IGBT waveforms.

phase of longer duration from t_3 to t_4 (670 μs) with a voltage V_{C3} (80 V on the primary side of the ignition coil). With these changes in the voltages, the collector current falls to 75% of the peak value during the initial phase and then down to zero during the second phase. This results in an improvement of 35% in the SCIS energy [5].

The dual-voltage clamping capability for the ignition IGBT has been implemented using an auxiliary IGBT integrated into the basic IGBT structure [5], as shown in Fig. 9.6. In addition, two strings of polysilicon diodes are employed to create a low breakdown voltage diode (LBVD) and a high breakdown voltage diode (HBVD). The breakdown voltage of the LBVD is chosen at V_{C3} (~80 V) by using 10–12 polysilicon diodes in series. The breakdown voltage of the HBVD is chosen at V_{C1} (~350 V) by using 50 polysilicon diodes in series. When the IGBT is turned on by the positive gate signal at time t_1, only the main IGBT is turned on because diode D_1 blocks the gate signal to the auxiliary IGBT. The buildup in the ignition coil current occurs via the main IGBT. When the gate bias is turned off at time t_2, the collector voltage begins to rise rapidly. The collector voltage first exceeds the breakdown voltage of the LBVD, allowing current to flow through resistors R_1 and R_2. The voltage across R_1 provides a gate drive voltage to the auxiliary IGBT via resistor R_3. The resistor R_3 is chosen such that the R–C time constant delays turn-on of the auxiliary IGBT for 30 μs. Meanwhile, the collector voltage exceeds the breakdown voltage of the HBVD, leading to a gate bias applied to the main IGBT via the voltage developed across R_G. The collector current then falls during the time interval from t_2 to t_3, as shown in Fig. 9.5. At time t_3, the auxiliary IGBT turns on, producing a voltage drop across R_5. This provides additional gate drive current to the

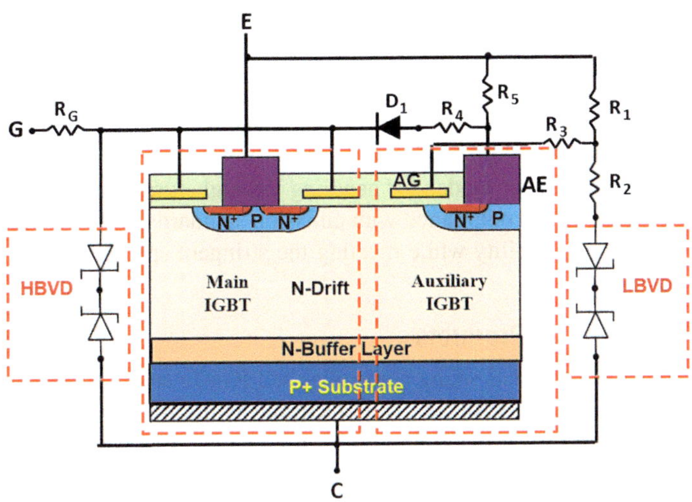

FIGURE 9.6

Ignition IGBT with auxiliary IGBT.

main IGBT via resistor R_4 and diode D_1. The reduced voltage during the time after t_3 is sufficient to maintain the spark once it is ignited by the high voltage.

In the case of an open circuit on the secondary side of the ignition coil due to a disconnected spark plug, the clamping of the collector voltage at a high value for a shorter time period using the dual-voltage circuit produces a smaller power density and lower junction temperature for the IGBT.

9.1.5 Smart Ignition IGBT Design

Automotive ignition systems are being cost optimized with higher levels of complexity and integration. Concomitantly, the ignition coil has also evolved into an improved design called the pencil coil. Locating the electronic ignition circuitry in close proximity to the coil eliminates high current cables that must be designed to operate at high voltages. It also reduces electromagnetic interference noise. In this approach, the ignition IGBT and its protection and diagnostic circuits are placed close to the pencil coil. The engine control module provides the digital signals to control the ignition module based on timing information for each engine cylinder, emissions, etc.

The smart IGBT ignition module must be able to handle a maximum junction temperature of 175°C because of its proximity to the engine. The limited mounting space is a major consideration for this application. These requirements can be met by using chip-on-chip technology [6]. In this approach, the diagnostic and closed loop local control for protection is performed using a bipolar complementary metal oxide semiconductor (CMOS) double-diffused MOS (DMOS) (BCD) logic technology. The BCD chip is mounted on the IGBT chip to create a compact form factor. The block diagram in Fig. 9.7 illustrates the features of the technology. The protection and diagnostic features include: (1) current limiting during faults; (2) soft shutdown when spark is not detected; (3) current feedback to monitor output current; (4) temperature sensing to detect overtemperature conditions; (5) bidirectional interface with multiplexed input and diagnostic capability; and (6) I_{on} current for monitoring combustion events. The IGBT itself has all the features discussed in previous sections with integrated Zener diodes for handling the high current and voltage stress. These capabilities allow compliance with emission standards to get better fuel efficiency with greater reliability while meeting the stringent space constraints [6].

9.1.6 Ignition IGBT Products

Due to the large market size, many companies sell IGBTs specifically designed for use in electronic ignition systems. The electronic ignition systems market represented one of the fastest growth opportunities for IGBTs [7]. By September 1999, ON Semiconductor reported the sale of more than 100 million IGBTs [8]. These devices were used in the Visteon power control modules by the Ford Motor Company. Similarly, by June 2000, INTERSIL reported the shipment of more than 60 million IGBTs for this market [9]. The ignition IGBTs have also been optimized for small engines used in lawn mowers and snow blowers [10].

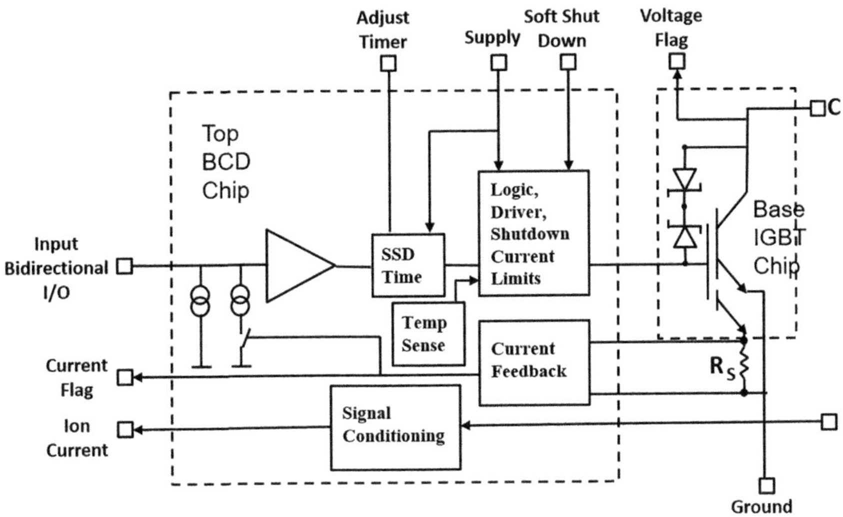

FIGURE 9.7

Smart ignition IGBT concept.

Some examples of IGBT products specifically designed for automotive electronic ignition systems are the Fairchild ISL9V5045S3ST-F085 EcoSPARK n-Channel Ignition IGBT [11], the On Semiconductor MGP15N40CL Ignition IGBT [12], and the ST Microelectronics STGB10NB37LZ internally clamped IGBT [13]. These devices are designed with a breakdown voltage (V_{CES} clamped) of 400 V and a collector current rating of 10 A. The IGBT structure is specifically designed with a relatively low threshold voltage (~1.0 V) for logic-level gate drive capability. The devices have a very low on-state voltage drop of 1.0–1.3 V with a relatively long turn-off time of several microseconds. The devices are rated for operation at relatively high temperatures of 150–175°C.

These IGBTs include the gate emitter ESD protection and temperature-compensated gate collector voltage clamps that are shown in Fig. 9.2. In addition, the datasheets include a specification for the reverse blocking capability of about 20 V to handle the reverse battery connection problem. This specification is usually not found in other IGBT datasheets because most IGBTs are designed and rated for motor control applications where a fly-back diode is connected across the IGBT.

9.2 Auxiliary Automotive Drives

The automotive industry uses millions of auxiliary drives. More than 100 such auxiliary drives are present in mid-size and large executive cars [14]. They are utilized for e-compressors, oil/water pumps, and super/turbo chargers. An automotive smart power module (ASPM) has been created to serve this market [15]. It contains six IGBTs with fly-back diodes within a direct bonded copper (DBC) package, as previously shown in Fig. 6.12. The ASPM also contains one low-voltage IC drive and three high-voltage IC drive chips. The technology is capable of performing under-voltage lock-out, overcurrent protection, and temperature sensing as well as provide noise immunity with optimized filtering.

9.3 Electric and Hybrid Electric Vehicles

The development of electric vehicles predates the development of gasoline-powered cars [16]. The first commercial electric vehicles were a fleet of taxis in New York City built by the Electric Carriage and Wagon Company of Philadelphia. At the end of the 19th century, electric cars had the advantages of less vibration and noise, no smell from gasoline fumes, and no manual effort to start the car, unlike the manual hand crank required for gasoline power cars. The development of mass production technology using assembly lines for gasoline-powered cars by the Ford Motor Company reduced their cost to less than half the electric car, leading to their decline and disappearance from the marketplace.

Concerns with environmental pollution from gasoline-powered cars encouraged the US Congress to pass the Electric Vehicle Development Act in 1966. This provided funds to universities and automakers to develop electric vehicle technology. The state of the art in battery and power electronics technology at that time proved inadequate to create a commercially viable vehicle. The California Air Resources Board (CARB) began to push for more fuel-efficient automobiles to combat air pollution in the 1990s. This encouraged automakers to develop electric cars, such as the Chrysler TEVan, the Ford Ranger electric vehicle (EV) pickup truck, and the GM EV1 and S10 EV pickup truck. An anecdotal story of the development of the GM EV1 car points out that the replacement of 48 power metal oxide semiconductor field effect transistors (MOSFETs) in the inverter with a few IGBTs was critical to the success of the first vehicles in the early 1990s [17]. The book states, "The IGBT was a beautifully compressed piece of solid state circuitry." Ironically, it also states, "(Harry) King did not quite figure out why a new transistor happened along just when it did to replace the MOSFETs…it was sort of a solution looking for a problem." Nothing could have been further from the truth because GE had already applied the IGBT to develop improved products in the consumer, industrial, lighting, and medical industries by the 1990s, as described in Chapter 1. By 1992, the IGBT became the device of choice for electric vehicle inverters [18] and has remained so until now. Recently, silicon carbide power MOSFETs have become commercially available. They allow a reduction of the size and weight of the power conversion unit, which is an important factor in electric vehicles. Although a few EV manufacturers have employed these devices, most manufacturers rely on the IGBT due to its lower cost and proven reliability.

The number of electric vehicles on the road in the United States grew from 300,000 in 2015 to 900,000 in 2018. All electric car manufacturers have now committed to replacing their gasoline power fleets with electric cars by 2030 to reduce carbon emissions. This provides an enormous growth opportunity for IGBTs in the future. The IGBT power module market reached $6 billion in 2020 and is anticipated to grow to $11 billion by 2026.

This section of the chapter reviews the basic inverter topology used in electric and hybrid electric cars. Specific IGBT designs that are optimized for this application are described. The circuitry used for engineering the regenerative braking in EVs is discussed because it is commonly used to improve the range of cars. Battery charger designs that are required in the case of electric cars are also included because they utilize IGBTs.

9.3.1 EV Inverter Design

The inverter is responsible for controlling the flow of power from the battery to the motors in electric vehicles to achieve propulsion. The design of the inverter depends on the power rating of the vehicle. As the power rating is increased, it becomes necessary to scale up the DC bus voltage to maintain reasonable current levels. The rated power levels for various classes of electric vehicles [19] are shown in Fig. 9.8. The power level has been scaled up over the years to provide consumers with an exhilarating driving experience due to the high acceleration of electric motor drives. The 2021 Porsche Taycon is designed to deliver 390kW (522hp) with an overboost capability of 420kW.

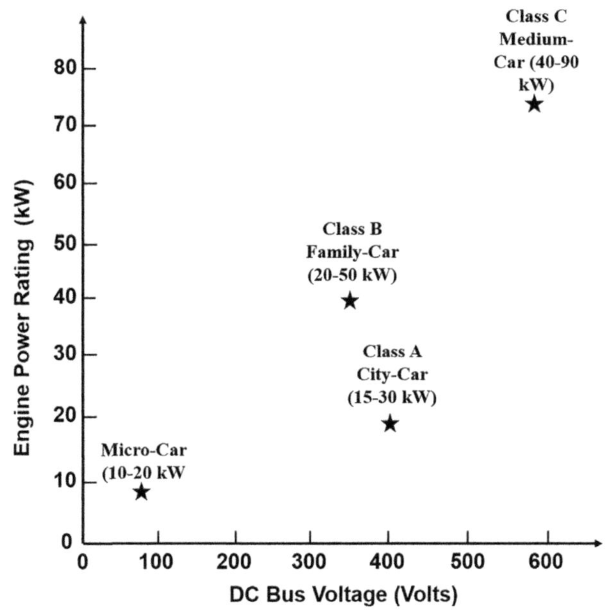

FIGURE 9.8

Electric vehicle classes.

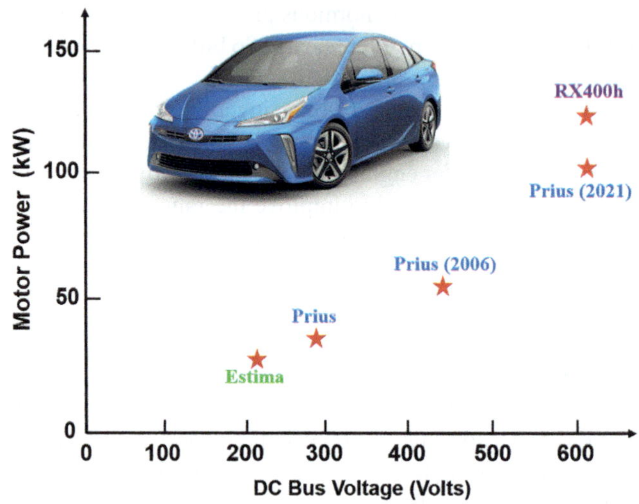

FIGURE 9.9

Toyota's electric vehicles.

The road map for the development of the most commercially successful hybrid electric vehicles by Toyota [20] is shown in Fig. 9.9. The first generation Prius cars from 1997 had a motor power rating of 30 kW, which allowed a relatively low DC bus voltage of 274 V. The new Prius from 2003 had a larger motor power rating of 50 kW, enabled by a larger DC bus voltage of 500 V. Increasing the motor power rating to 120 kW requires raising the DC bus voltage to 650 V, as shown in the figure. For the development and commercialization of all Prius cars on the road today, the authors [21] state, "Si insulated gate bipolar transistors (Si-IGBTs) are widely used in the inverters." This will continue to be the case until transistors from wide bandgap semiconductors, such as SiC and GaN, can be commercialized with sufficiently low cost and high reliability. The 2021 Toyota Prius has a motor power rating of 96 hp or 72 kW.

The most commercially successful electric vehicle (Toyota Prius) in the market utilizes the series/parallel hybrid drive called the THS system [22]. The basic configuration for the THS system is illustrated in Fig. 9.10. This system enables combining energy delivery to the wheels of the vehicle from both an ICE and an electric motor. The THS system introduced in 2003 includes a battery voltage boosting capability to allow increasing the overall available power with the same motor size. The booster allows the inverter to work from voltages ranging between 202 V of the battery to a maximum of 500 V. The basic modes of operation are:

1. Starting and low-speed motion: Only the electric motor is used to drive the wheels using energy delivered from the battery.
2. Normal driving speed: The generator is used as a motor to increase the engine speed. When sufficient speed is reached, IGBTs are used to ignite the spark

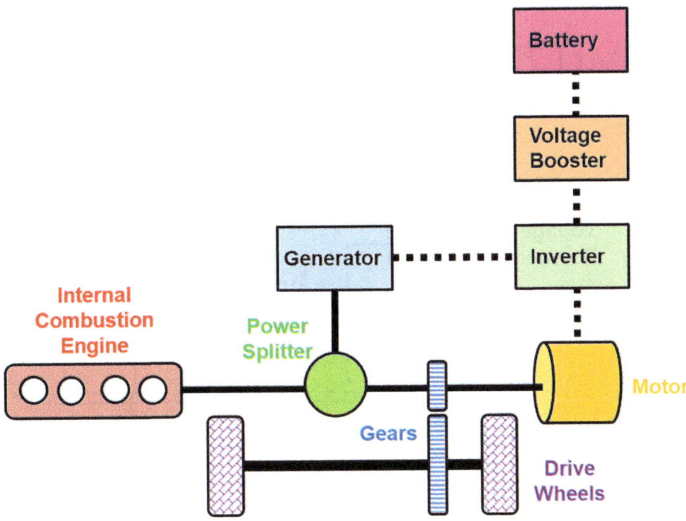

FIGURE 9.10

Toyota series/parallel hybrid electric vehicle system.

plugs in the ICE to gain power from gasoline. The power delivered from the ICE is fed through the power splitter to drive the wheels and to charge the battery via the generator.

3. Full throttle acceleration: The THS system allows combining energy delivered from the ICE and the electric motor to boost the available power to the wheels during full-throttle acceleration and under heavy load.
4. Battery charging: The energy taken from the battery to run the electric motor must be replenished continuously. This is performed with computer-controlled charging that is designed to maintain a constant charge in the battery by delivering power from the ICE to the battery via the generator.

The inverter topology [19] is shown in Fig. 9.11, including the battery boost capability to increase the DC bus voltage as discussed earlier. The use of IGBT-based inverters for providing variable frequency motor drives will be discussed in detail in a subsequent chapter. The totem-pole configuration used in this design requires a fly-back diode across each IGBT. Because the diode conducts current in the reverse direction across the IGBT, no reverse blocking capability is required, allowing the use of the asymmetric structure discussed in Chapter 2.

As already mentioned, the IGBT became the device of choice for the inverter from the earliest days of modern electric vehicle development. A survey of experts [23] on electric vehicles published in 1994 found that a vast majority already favored the IGBT for the inverters over all other technology options. Another article from 1994 on power modules for electric vehicles [24] states, "IGBTs are the device of choice for electric vehicles." By 1995, an intelligent IGBT power module was

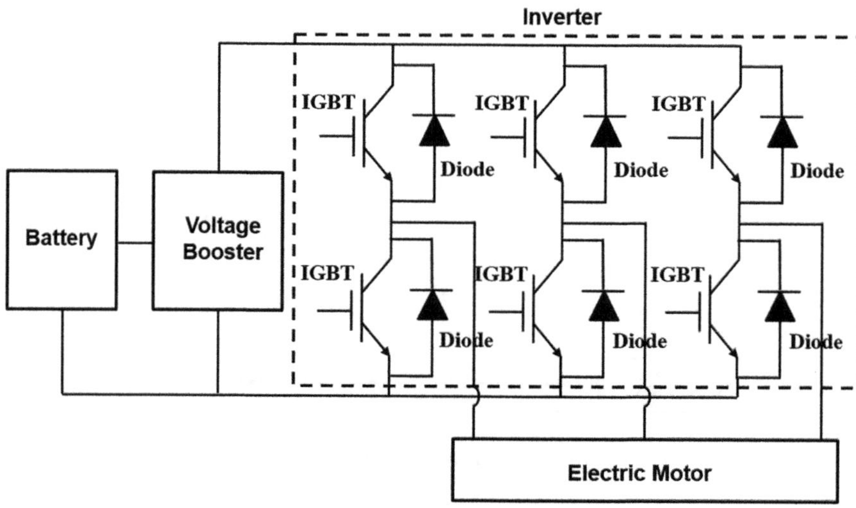

FIGURE 9.11

Inverter topology for the electric vehicle.

developed for Toyota EVs with short circuit protection, overtemperature protection, gate supply undervoltage, and fault detection signals [25]. The IGBTs used in the module had integrated current sensors, as described in Chapter 5. In the 1980s, the MOS-controlled thyristor (MCT) was proposed [26] as a superior alternative to the IGBT because of a lower on-state voltage drop. This device was compared with the IGBT in 1997 for EV applications [27]. The authors concluded: "It is very hard to protect the overcurrent in MCT. Thus, MCT is not suitable for this application."

The THS system also makes use of IGBTs in a boost converter [21] to enhance the DC bus voltage, leading to more power available for the electric motor. The topology for the boost converter is illustrated in Fig. 9.12. This boost converter can raise the DC link voltage from a battery voltage of 202 to 500 V. The increased DC bus voltage is supplied to the IGBT-based inverters that drive the generator and the motor. A total power of 50 kW can be supplied to the motor by deriving 20 kW instantaneous power from the battery and 30 kW from the generator. This method allows increasing both the torque and power output of the motor by 2.5 times without increasing the motor current. These IGBTs are made using the planar (DMOS) cell structure with a chip area of 0.225 cm^2 (1.5 × 1.5 cm chips). Two IGBTs are used in parallel in each arm of the boost converter.

9.3.2 EV IGBT Chip Design

The IGBT used in the EV inverter must be designed to obtain a high power transfer efficiency from the DC bus to the electric motor. This requires designs with low on-state voltage drop and fast switching speed that also exhibit good avalanche

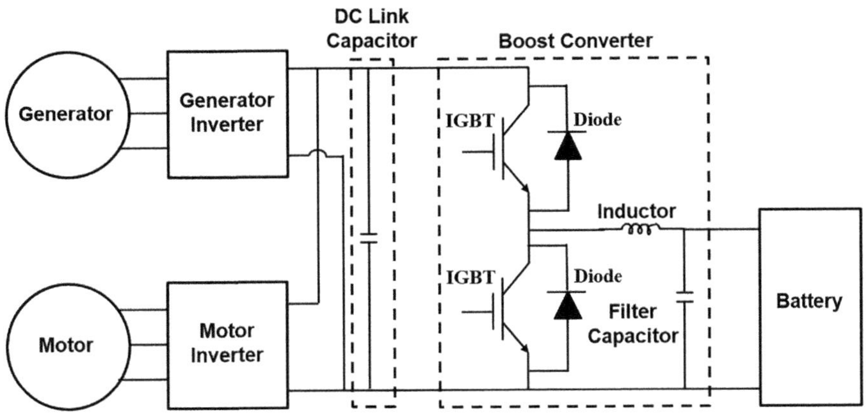

FIGURE 9.12

Variable DC bus voltage for the electric vehicle using boost converter.

capability. The IGBTs used in the first generation of Toyota hybrid EVs were 600 V, 200 A planar cell designs with a chip size of 1.3 × 1.3 cm. They had an on-state voltage drop of 2.1 V at the on-state current density of 145 A/cm^2, a current fall time of 0.8 μs, and an avalanche capability of 1 J at room temperature [28]. For subsequent generations of Toyota hybrid EVs, the company developed trench gate IGBTs because of the superior trade-off curve between on-state voltage drop and switching losses, as discussed in Chapter 2. The IGBT cell had a pitch of 4 μm with a drift region thickness of 70 μm and a buffer layer thickness of 15 μm [27]. Both the on-state voltage drop and the current-induced latch up current density were found to improve with smaller cell pitch, consistent with the models in the textbook [29]. Trench gate IGBTs with a chip size of 1.23 × 0.93 cm had an on-state voltage drop of 1.55 V at an on-state current density of 250 A/cm^2 with similar switching times and avalanche capability as the planar gate devices. A local lifetime control process with He irradiation of the buffer layer was employed [30].

One of the problems encountered when switching off the IGBTs in the inverter is a surge voltage produced when the space-charge region punches through to the buffer layer. This problem is exacerbated with the THS systems containing the boost converter. The problems can be resolved by increasing the thickness of the drift region [31]. Another consideration when designing IGBTs for EVs is the prospect of cosmic ray-induced failures at high-altitude mountainous areas. Cosmic rays are generated by supernova explosions and from solar flares. The cosmic rays interact with atmospheric oxygen and nitrogen to create protons, mesons, photons, and neutrons. Although the charged particles are deflected by the geomagnetic field, the neutrons arrive at the Earth's surface and produce single event burnout (SEB) in power devices. A greater tolerance to SEB in IGBTs has been demonstrated for punch-through structures with wider drift regions [32].

There has been a trend in the industry to reduce the wafer thickness when manufacturing IGBTs. Most recently, the wafer thickness has been reduced from

70 to 40 μm [33]. Although this reduced the breakdown voltage of the device from 650 to 400 V, optimization of the injection efficiency of the collector–base junction has allowed reduction of the turn-off *dI*/*dt* from 5.9 to 5.3 kA/μs. With the smaller *dI*/*dt*, the voltage overshoot due to stray inductance in the package was also reduced, allowing operation of the thin wafer devices at a DC bus voltage of 200 V. These IGBTs are suitable for the "mild hybrid" vehicles that operate at lower power levels. The cell structure used for these IGBTs is shown in Fig. 9.13 [34]. It defers from the trench gate structure shown in Chapter 2 by the extension of the gate over the top surface. The packaging of the IGBTs together with the flywheel diodes is done in power modules where the chips are attached to direct-bonded copper substrates (see Chapter 6) [35]. Due to the high currents (typically 450 A), heavy aluminum wires with a diameter of 350 μm are used to connect the chips to the DBC substrate. A wire bondless package [36] based on solder bump technology has also been developed to reduce the on-state voltage drop of the IGBTs for EV applications. This approach allows double-sided cooling of the chips and eliminates the reliability problems associated with wire bonds.

The reverse-conducting (RC) IGBT structure (see Fig. 2.7) integrates the fly-back diode within the IGBT cells. This reduces component count, providing space savings in electric vehicles. The power loss associated with the reverse recovery of the integrated fly-back diode can be reduced by using a three-level gate bias control strategy [37].

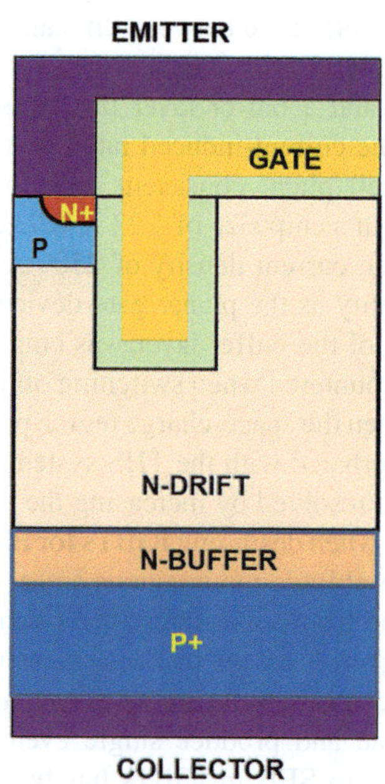

FIGURE 9.13

Optimized trench gate IGBT cell structure with thin wafer technology.

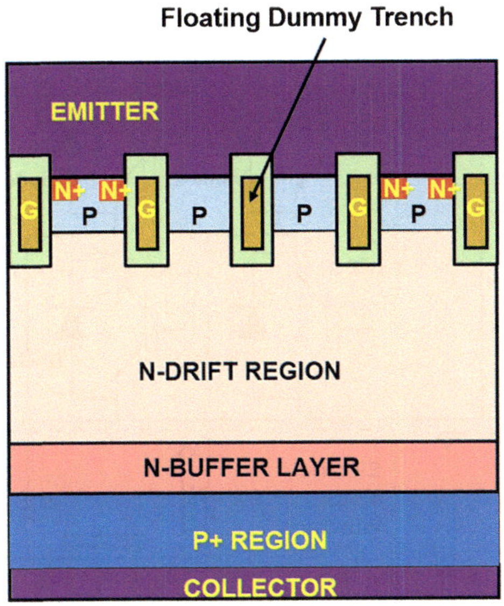

FIGURE 9.14

IGBT structure with floating dummy trench to improve *dI/dt* control.

Optimization of the IGBT for electric vehicle applications requires minimizing switching loss while also controlling turn-on *dI/dt* to avoid voltage overshoots through stray inductances. The structure shown in Fig. 9.14 has trench regions where the electrode is left floating [38] as compared with the structures shown in Chapter 2. It was found that this allows greater control over the *dI/dt* when the gate resistance is changed. A superior trade-off curve between turn-on *dI/dt* and turn-on energy loss was obtained with the floating trench design.

Today, two separate cooling systems are employed in hybrid electric vehicles (HEVs)—one as an engine coolant at 105°C and one for the power converter system at 65°C. The second cooling system could be eliminated if the power devices could be operated at a junction temperature of 200°C. Characterization of trench gate, field stop (or punch-through (PT)) IGBTs has demonstrated 200°C operation of the devices, albeit with larger switching losses [39]. The devices were found to have adequate latch-up immunity and short circuit capability. However, external overvoltage snubbers were needed because the avalanche sustaining capability was inadequate at 200°C.

9.3.3 EV Regenerative Breaking

One of the shortcomings of electric vehicles is their short range because of the limited amount of energy stored in the batteries. Regenerative braking is commonly employed in all EVs to improve the range by transferring the vehicle's kinetic energy

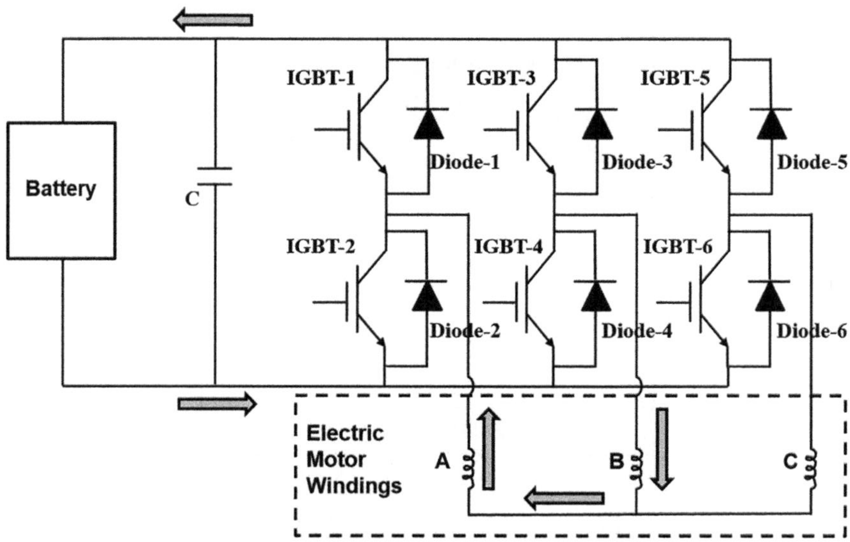

FIGURE 9.15

Regenerative braking circuit for electric vehicles.

to the battery when slowing down the vehicle instead of wasting the energy as heat in the brake pads. The typical circuit topology for regenerative braking [40] is shown in Fig. 9.15, where six IGBTs and fly-back diodes are used to transfer power to the battery. The diagram indicates the current flow path (arrows) from winding A and B of the electric motor to the battery. This is achieved by pulse width modulation (PWM) control of IGBT-2 and utilizing the fly-back diode-1 across IGBT-1 and fly-back diode-4 to form a boost converter.

The importance of regenerative braking for improving the range of EVs was recognized from the early days of its development [18]. An improvement of the range of the vehicle by more than 10% is projected by using regenerative braking [40] with fuzzy logic-based power management [41]. Regenerative braking has been employed even in compact three-wheel EVs that were used in Rome in 1997 [42]. It is worth pointing out that the topology for driving the motor as shown in Fig. 9.11 is similar to the regenerative braking circuit shown in Fig. 9.15. Consequently, a bidirectional power convertor can be designed to serve both functions, allowing a reduction of the number of components and the cost of the power electronics.

9.4 EV Charging Stations

The true promise of electric vehicles was reducing urban pollution. Intolerable levels of smog in large metropolises, such as Los Angeles, Paris, and Tokyo, associated with the increasing number of gasoline-powered automobiles prompted action in the early 1990s. The pure electric vehicle was envisioned as the most promising solution to the health of the community. It was recognized that EVs stood little chance of being successfully embraced by the populace unless a refueling infrastructure was available [43]. Electricite de France, the Japan Electric Vehicle Association (JEVA), and the CARB began to push for the EV mandate. The standard electric charging station was modeled along the lines of the gas pump familiar to consumers. Consumer safety was a major concern because the general public would be asked to make high-power electrical connections on a daily basis in all kinds of weather conditions. The benefits to the electric utilities of charging EVs during low demand, off-peak hours were obvious. This prompted the development of chargers that could be installed in the garages of residences. Today, the replacement of gasoline-powered cars with EVs is being encouraged in Europe [44] and China [45]. In 2010, there were about 80 million gasoline-powered cars in China. It is estimated that the 1 million EVs on the road in China in 2015 will grow to 60 million vehicles by 2030, making them a significant fraction of vehicles on the road [46].

9.4.1 EV Charging Requirements

A desired consumer driving range between charges of 100–150 miles (160–240 km) requires a battery capacity of 20–50 kWh. The battery can be charged overnight with a nominal 6 kW source. This is feasible in North America using the 240 V single phase electrical supply in the home delivering 30 A of current (similar to that needed for clothes dryers) [43]. Two examples of modern EVs in the market are the Mitsubishi i-MiEV and the Nissan LEAF five-door hatchbacks. The Mitsubhsi i-MiEV utilizes lithium ion batteries that can store 16 kWh of energy while the Nissan LEAF battery capacity is 24 kWh [47]. The chargers located in homes are called Level-1 chargers suitable for operating from the 120 V/15 A duplex outlet.

The charging topology for EVs can be configured in three ways, as illustrated in Fig. 9.16. They all perform the essential function of converting the alternating current input source to a direct current source to charge the battery. The physical connection of the charger to the source also has three basic configurations: (1) off-board cord and connector with connector attached to the supply; (2) adaptor cord set with independent cord containing the plug to the source and the connector to the charger; and (3) on-board cord and plug stored in the vehicle. The first approach was favored by EV manufacturers because it is similar to the familiar gasoline pump configuration.

An inductive coupler with a paddle-like connector has also been developed [43]. In this approach, the 60 Hz input power source has to be converted into a high-frequency (typically 100 kHz) power source, as shown in Fig. 9.17, to reduce the size of the transformer (paddle).

It is also desirable to create an infrastructure for rapid charging of EVs similar to gasoline fuel stations widely available for automobiles today. These chargers are referred to as Level-3 chargers. They are designed to replenish at least 50% of the energy in the battery within 15 min [43]. In addition, Level-2 chargers with power capability up to 19.2 kW are being considered with charging times of 1–2 h [48]. These charges are suitable for shopping centers where consumers may spend

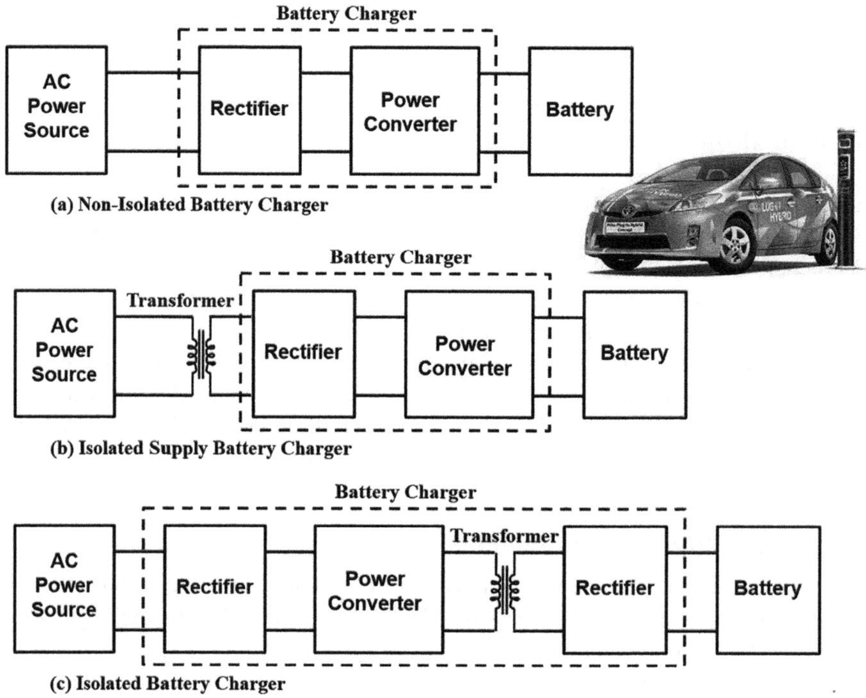

FIGURE 9.16

Battery charging topologies for electric vehicles.

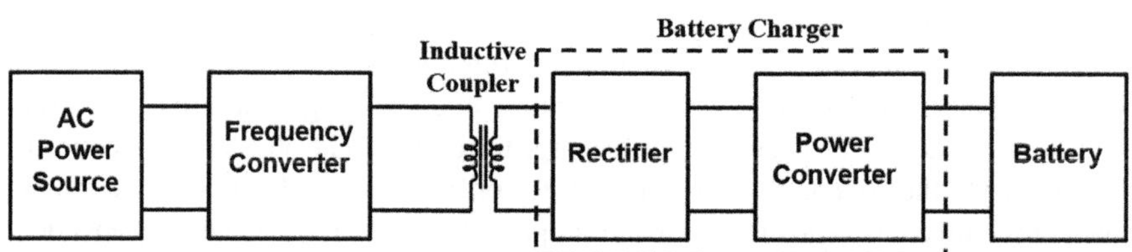

FIGURE 9.17

Inductive coupler battery charger for electric vehicles.

a few hours at most. A pictorial representation of various charging scenarios is the charging pyramid [49] shown in Fig. 9.18. The type of charger that is appropriate for each scenario is shown in the figure. It can be argued that the ubiquitous 120 V outlet can be utilized at home and work with existing utility load control devices to satisfy the immediate needs of the public [45]. This Level-1 charger operates at a power level of 1.4–1.9 kW [48]. In addition, DC fast chargers are being designed with 32–100 kW output power from a 400–500 V line to charge electric vehicles in 15–30 min [48].

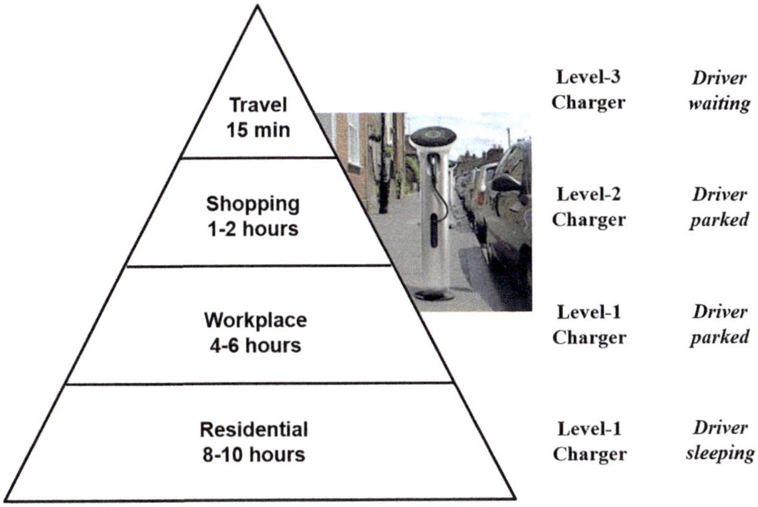

FIGURE 9.18

Charging pyramid for electric vehicles.

9.4.2 EV Charging Circuit

The typical nonisolated buck/boost circuit topology [50] used for charging the battery in EVs is shown in Fig. 9.19. The boost stage is used to raise the DC voltage derived by rectifying the input AC source to above the nominal battery operating voltage and to provide power factor correction (PFC). The operating PWM frequency for the boost stage is about 20 kHz. Under normal operating conditions (most

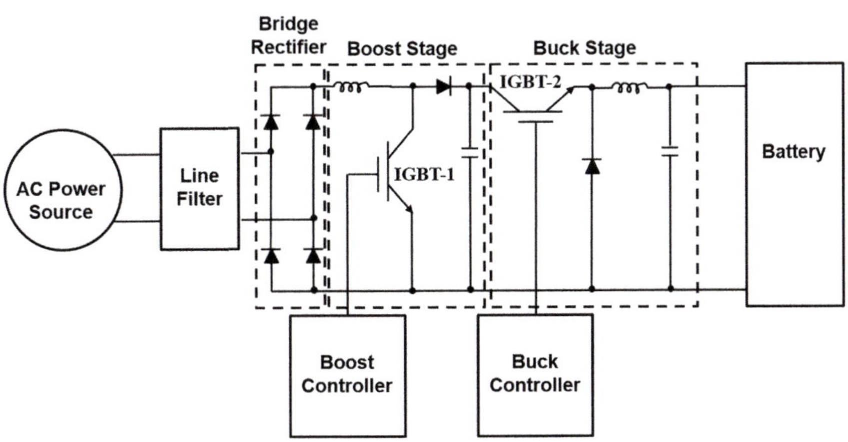

FIGURE 9.19

Charging circuit for electric vehicles.

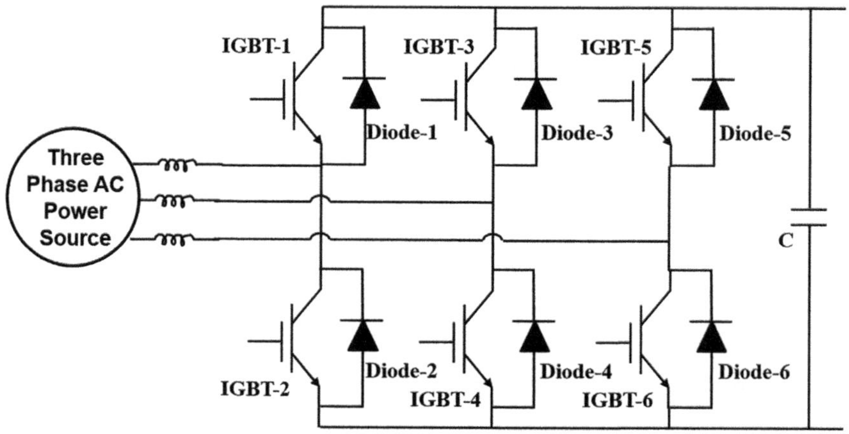

FIGURE 9.20

Power factor correction circuit for electric vehicle chargers.

of the time), the battery voltage exceeds the peak of the AC input voltage. The charger is then operated in the boost mode and the buck stage is turned on fully. Under low battery voltage conditions, the peak of the AC input voltage becomes larger than the battery voltage. The charger is then operated in the buck mode with the boost stage turned off [51].

With the expectation that a large number of EVs will be plugged into the grid in the future, distortion of the line voltage and current is a serious concern from the utility standpoint. PFC must be performed in EV chargers similar to that already mandated for many electrical and electronic products. A three-phase full-bridge PFC circuit is shown in Fig. 9.20 suitable for the high power and voltage levels of the EV chargers. An overall efficiency of greater than 97% can be obtained for the PFC circuit by using high-speed IGBTs and fast recovery diodes [52].

9.4.3 Modern EV Charging Station

Modern electric vehicle charging stations are expected to derive power from various sources and have the capability to store energy to be delivered to cars. The architecture consists of a DC link to which the sources and charging stations are all tied together [53], as illustrated in Fig. 9.21. The renewable energy sources, such as solar panels and wind power, are preferable to avoid power transmission from fossil fuel plants. Energy storage capability is derived from auxiliary batteries, ultracapacitors, or fuel cells, which can be charged by the renewable energy sources when power is not required for the charging stations. A charging station topology with IGBTs can be used to achieve close to unity power factor and low harmonic distortion [54].

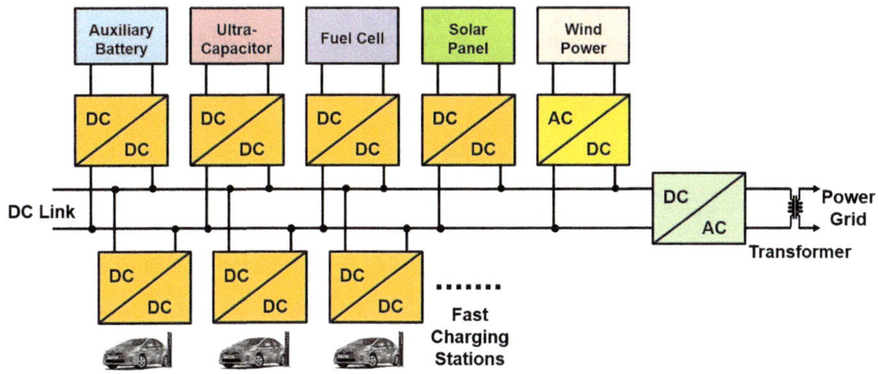

FIGURE 9.21

Architecture for a modern electric vehicle charging station with various energy sources and energy storage approaches.

9.5 Electric Transit Bus

According to the US Bureau of Transportation Statistics [55], the number of passenger miles traveled in buses is 163 billion miles, which is far smaller than the 4520 billion miles traveled by cars and motorcycles. However, it is still desirable to replace gasoline-powered vehicles with electric vehicles to reduce urban pollution. There are more than 25 manufacturers of electric buses. Some examples shown in Fig. 9.22 are the Tindo (Aboriginal for "sun") solar power bus [56] deployed in Adelaide, Australia; and the BE35 electric bus [57] deployed in Los Angeles, Sacramento, San Francisco, and San Hose, California. A supercapacitor energy storage system-based urban bus [58] has been deployed in the Heilongjiang province of China.

In contrast to passenger vehicles, the city bus has 10-fold longer operating time and the frequent stops and starts impose large temperature cycles on the power

FIGURE 9.22

Electric buses: (a) California transit; (b) Australian Tindo.

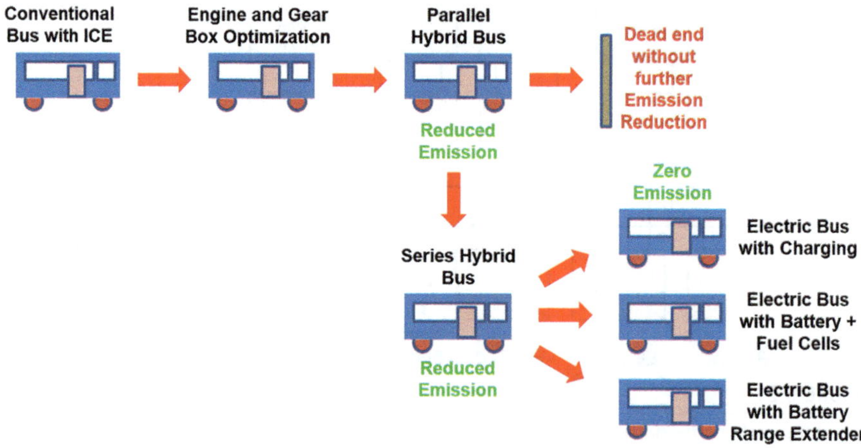

FIGURE 9.23

Electric bus evolution strategy from conventional fossil-fuel based vehicles to zero-emission vehicles.

devices. They operate over a wide speed range, require large start-up torque for acceleration, and need high efficiency to cover the route range. The IGBT modules must fulfill the following requirements [59]: (1) DC link voltage ranging from 500–800 V; (2) switching frequencies up to 10 kHz; (3) motor output power up to 200 kW; (4) overvoltage and overcurrent shut down capability; (5) ambient temperatures ranging from −40°C to 90°C; (6) immunity to vibrations; and (7) coolant temperature ranging from −40°C to 70°C.

The evolution of city buses from conventional gasoline-powered vehicles that use ICEs to zero emission vehicles can be done in stages based on available cost-effective technology, as shown in Fig. 9.23. The approaches of optimization of the conventional bus and then adding an electric drive to create a hybrid bus leads to a dead end in terms of achieving the ultimate goal of zero emissions [60]. The series-hybrid approach offers great benefits for city traffic where the average speed is low, there are long stop durations, and frequent starts and stops. Fuel and carbon emission reduction by 40% is projected. A 600 V DC bus has been used with 1.2 kV IGBTs.

9.5.1 Electric Bus Control Circuits

The control circuit for the Chinese electric bus is shown in Fig. 9.24 [61]. When the electric bus starts or accelerates, the ACCEL signal received by the controller locks out the brake signal and shuts off the IGBT-2. A PWM signal is transmitted to IGBT-1 at a high frequency. This creates a path from the battery through IGBT-1 to the motor to provide the drive power. During braking operation, a BRAKE signal is received by the controller, which locks out IGBT-1 and supplies a PWM signal to

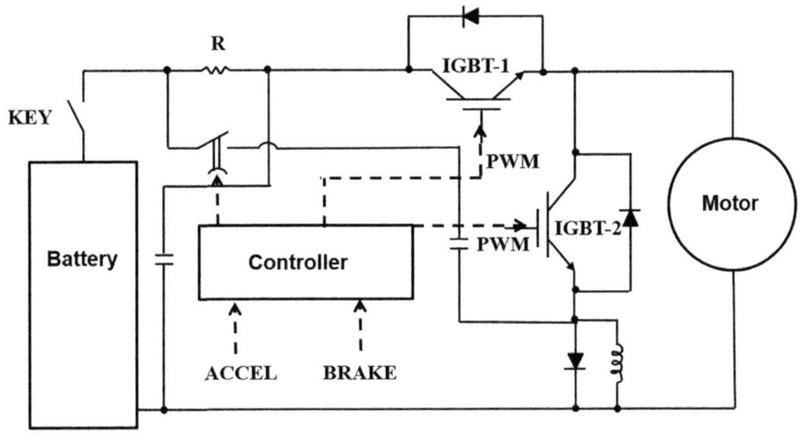

FIGURE 9.24

Control circuit for the Chinese electric bus.

IGBT-2. The main power loss occurs due to switching the IGBTs at high frequencies. The BJD6110-EV electric bus has a maximum speed of 75 kmph and acceleration from 0 to 60 kmph in less than 1 min.

The first generation of ELFA converters with IGBTs was installed in European buses in 1998 [59]. These vehicles have traveled more than 400,000 km in 10 years. The second generation converter design for the European electric buses is shown in Fig. 9.25. The 1200 V trench gate field stop (FS) IGBTs used for these converters have the cross-section shown in Fig. 2.4. Switching losses in the converter were found to be dominated by the large reverse recovery current of the silicon fast

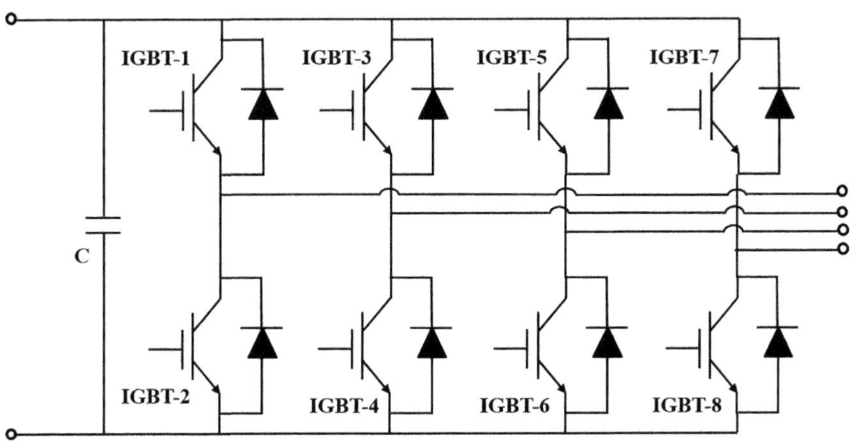

FIGURE 9.25

ELFA converter for the European electric bus.

recovery fly-back diodes. Replacing them with high-voltage SiC Schottky diodes [62] was found to significantly reduce the IGBT turn-on power loss [63].

9.5.2 Electric Bus Charging

Efficient and rapid charging of electric buses is required to maximize their range and continual operability. The infrastructure for charging buses must be reasonable from the point of view of cost and size. The charging capability must also be acceptable from the utility side. Three-phase charging is superior to single-phase circuits because the current from the grid is symmetrical and balanced regardless of the number of vehicles. The loss-pass filter components in the three-phase circuit are also smaller in size and lower in cost.

The block diagram for a three-phase electric bus charger [64] is shown in Fig. 9.26. Power is derived from a three-phase, 60 Hz, 208 V line-to-line voltage source. The IGBT-based voltage source converter (VSC) has a bank of chokes (R-L network) on the AC source side and a bank of electrolytic capacitors (C) on the DC side. The VSC is a boost converter that raises the DC bus voltage to 500 V, which is above the peak (294 V) of the AC line-to-line voltage. The desired unity power factor is obtained for the AC source by controlling the switching of the IGBTs, allowing the total harmonic distortion (THD) to be maintained at below 5% per IEEE standard 519. The choke has an inductance of 10 mH per phase and the electrolytic DC link capacitor is 900 μF. 1200 V, 100 A IGBT modules are required with a KP3-C series snubber and ceramic disc capacitors across each module to absorb high-frequency spikes. The current-controlled bidirectional DC–DC converter is used to manage the charging rate of the battery. A single-phase, full-bridge

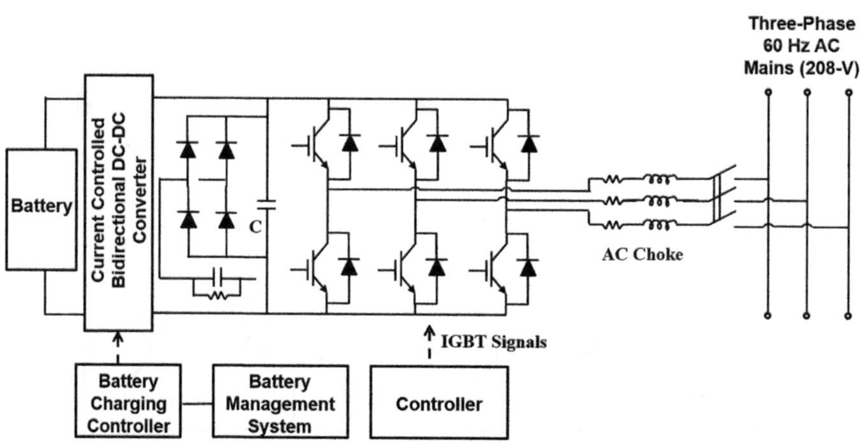

FIGURE 9.26

Charging system for the electric bus.

diode rectifier is also used to precharge the DC link to the peak line-to-line voltage (294 V) through a resistor to limit the diode inrush current.

9.5.3 Inductive Electric Bus Charging

The need to periodically charge electric buses can create significant downtime that impacts the financial viability of public transportation. One elegant solution is to perform inductive (or wireless) power charging while the electric bus is stationary at its many stops. Inductive power charging utilizes a coil on board the bus and a second coil embedded in the ground at the bus stop. The basic topology for the wireless charging station is illustrated in Fig. 9.27. The AC input power is first rectified and fed into the IGBT H-bridge [65]. Because the charger must be rated at several kW, it is necessary to use IGBTs due to their high power ratings. A resonant converter is used to deliver power from the input side to the battery inside the vehicle. A high resonant frequency is desirable to reduce the size of the passive components (transformer and capacitors). However, the IGBT switching frequency must be kept at about 20 kHz due to its high switching losses at higher frequencies. The H-bridge produces a quasisquare waveform but the band-pass nature of the resonant circuit passes only one harmonic frequency to the secondary side. A third harmonic system with a resonant power transfer frequency of 60 kHz was found to be optimum in terms of size and efficiency. A 1 kW charger has been demonstrated with an efficiency of 83% at a 20 cm airgap between the primary coil in the road and the secondary coil in the electric bus.

9.6 Electric Trams and Trolleys

Electric trams and trolleys have been widely used in cities in Europe for more than a century. The availability of the IGBT has enabled the design of these vehicles with a lower floor, as shown in Fig. 9.28, which is more convenient for passengers. By 1991, it was already apparent that IGBTs were preferable power devices when compared with the gate turn-off thyristors (GTOs) that were in use before then. Weschta from Siemens states [66]: "for this application, the IGBT offers the following technical advantages: lower number of components for the drive circuit, lower power

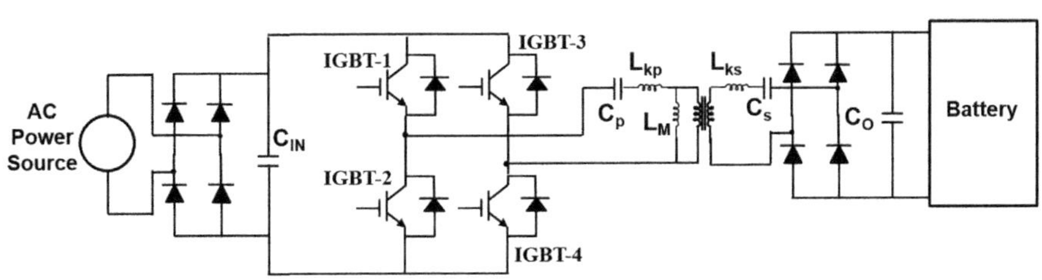

FIGURE 9.27

Wireless inductive charging topology for electric buses.

FIGURE 9.28

Electric trams: (a) Hannover, Germany; (b) Lyon, France.

rating of the drive circuit, better performance under short circuit conditions, better performance of paralleled semiconductor devices, feasibility of higher blocking voltages (necessary for the 750 V overhead line)." The city of Frankfurt ordered 20 trams using IGBTs for driving the traction motor. Low-floor trams require installation of the power electronics on rooftop containers.

The most cost-effective power circuit topology for driving the traction motors in the electric trams is a PWM inverter connected to the overhead power line through an L–C filter [66], as shown in Fig. 9.29. For a typical line voltage of 750 V, IGBTs with

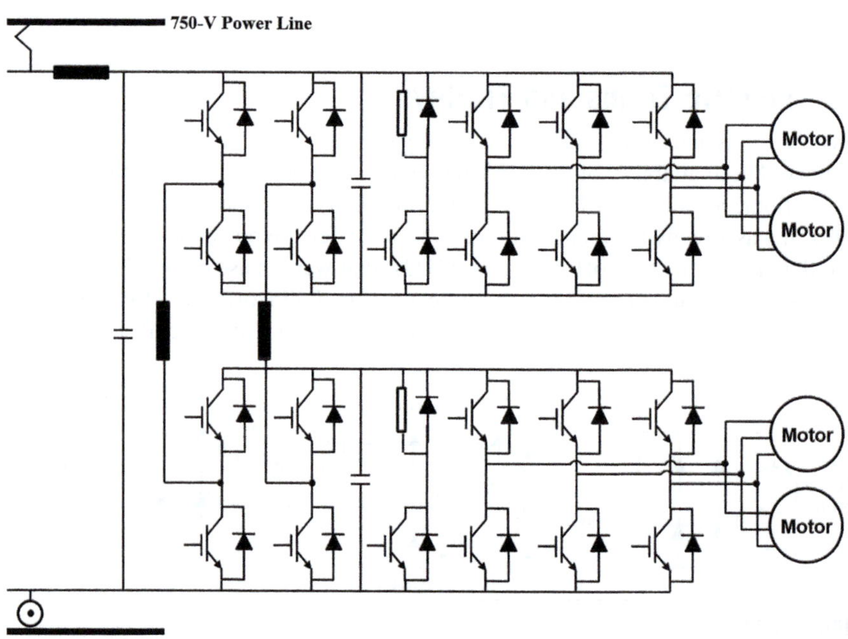

FIGURE 9.29

Power circuit for Frankfurt electric trams.

a blocking capability of 1600 V are required to take into account voltage overshoots. This topology developed has an input chopper designed to create a steady 750 V DC-link voltage for using IGBTs with lower breakdown voltages. With the availability of IGBTs with 1600 V capability in 1993, the input chopper was no longer required, reducing the cost of the inverter. A unique aspect of the design is higher currents encountered during the braking mode (power of 730 kVA) when compared to the driving mode (power of 430 kVA). The IGBT modules have to be selected to tolerate the higher currents, which flow through the fly-back diodes during the braking mode. In addition, the modules have to survive shocks to 3G and vibrations to 1G at 10–100 Hz. Because the traction motors require water cooling, this could also be employed for the power modules. The author concludes: "IGBTs have meanwhile attained such a high level of performance that IGBT converters can be used for rolling stock."

A similar conclusion was made by AEG Westinghouse Transport Systems in 1993 when designing a single-axle low-floor tram [67]: "Today's IGBT modules can match the power demands of single-axle drives in AC technology." The three-level IGBT inverter topology used in trams for Berlin, Halle, Jena, and Mainz is shown in Fig. 9.30. In the three -level inverter, a symmetric voltage distribution is achieved using the line capacitors through control of the midpoint voltage. Advantages derived using IGBT modules in place of GTOs are cited as: "higher switching frequencies due to shorter switching times and lower switching losses, low drive power due to MOS gates, short circuit protection without snubbers, and module isolation." A protection unit supervises over- and undervoltages and overtemperature conditions. In case of a motor short circuit, the IGBTs are switched off within 10 μs to avoid damage to the inverter.

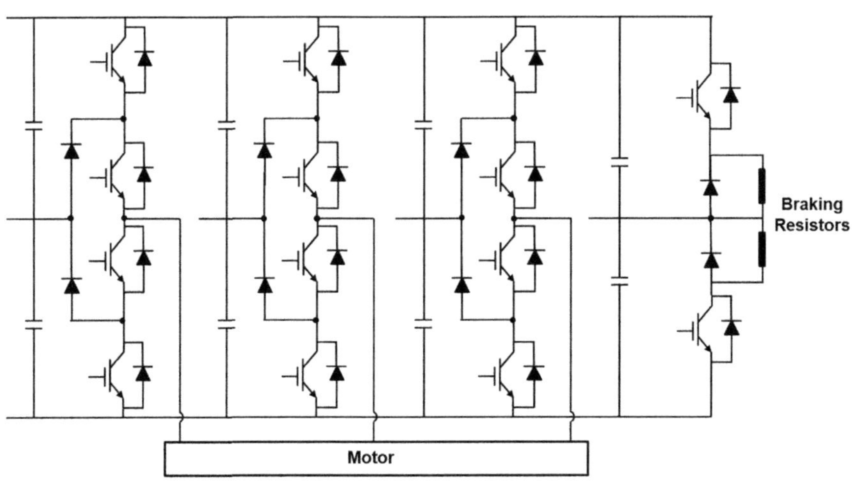

FIGURE 9.30

Three-level IGBT inverter used for German electric trams.

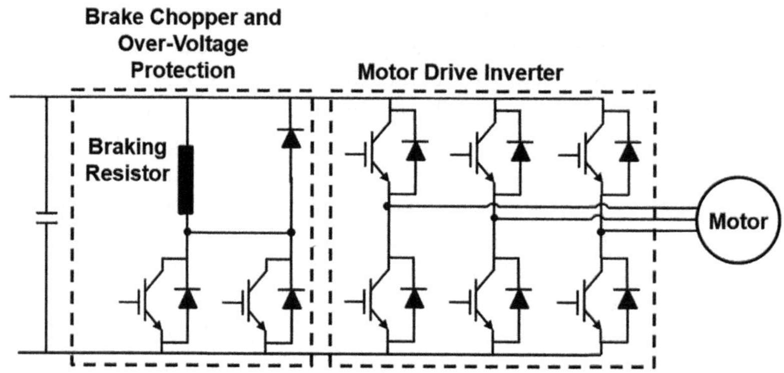

FIGURE 9.31

Two-level IGBT inverter developed by ABB for electric trams.

In 1995, Asea Brown Boveri (ABB) reported the development of a roof-mounted, water-cooled, two-level inverter using IGBTs for use in the low floor Variotram of ABB-Henschel [68]. The two-level inverter topology is shown in Fig. 9.31, which includes not only the drive for the motor but also a braking chopper and overvoltage protection. This inverter with eight 1800 V/800 A IGBT modules was able to replace the previous three-level design with 28 bipolar power transistors. The main switching losses for the inverter were found to be related to the reverse recovery current from the diodes.

By 1996, the light rail industry had migrated from GTO-based inverters to IGBT-based inverters. For example, GEC-Alsthom developed their ONIX traction drive for trams [69] using IGBTs in place of GTOs. They state: "The major technical advantages that have been incorporated into the ONIX drive are the use of Insulated gate Bipolar Transistors (IGBTs) for power conversion switches." The designers favor this power semiconductor switch over the GTO device for many reasons: (1) Simpler gate drive circuit via low voltage circuits of the control electronics, (2) elimination of snubber circuits because the IGBT can tolerate overcurrents far larger than the GTO, (3) higher frequency operation by 2.5 times, leading to less vibration of the motor and reduction of its weight, (4) superior durability due to the ability of IGBTs to withstand short circuits, (5) cheaper to manufacture than GTOs, and (6) smaller and lighter packaging. The authors state, "GEC Alsthom has experienced an overall weight reduction of 30% and (a) corresponding 50% reduction in volume for the 800 Series ONIX IGBT inverter versus an MP89 GTO inverter for Paris Metro cars of comparable power rating."

IGBT-based technology has also been adopted in trolleys commonly used in eastern European countries (Romania, Czech Republic, Moldova, etc.) [70,71] and Russia. The braking chopper using IGBTs that is shown in Fig. 9.31 was utilized to recover braking energy in these vehicles. The trolley serves large cities such as Bucharest, Prague, and Kiev. The authors state "Development of power electronics, particularly IGBT transistors, allows developing traction converters with outstanding technical performance, dimensions, and weights 10 times smaller than those of the traction motor."

9.7 Subway and Airport Trains

Rapid transit electric rail transportation is commonplace in large metropolitan cities and at major airports with multiple terminals. The largest and second-largest subway systems are located in New York City in the United States (Fig. 9.32(a)) and Seoul in South Korea (Fig. 9.32(b)). Examples of electric transit systems at large airports are the Airtran system at JFK International Airport (Fig. 9.32(c)) and Skytrain in Vancouver, Canada (Fig. 9.32(d)). Toshiba has published a history for the delivery of IGBT-based variable-voltage, variable-frequency (VVVF) inverters for these applications since 1996 [72]. They supplied these drives for the Cairo Metro Line in 1996 and 1999, the Irish Rail and Caracas Urban Railway in 2000, JFK (Airtran) and Vancouver (Skytrain) in 2001, the LRTA Manila Light Rail Transit and the Taiwan Railway Authority in 2002, the Chinese Tianjing Railroad and Wuhan City Tracked Traffic in 2003, and the Minneapolis Metropolitan System and Logistica Brazilian system in 2004.

In 1993, the Hitachi company provided its perspective on technology developments for electric railcars, which is reproduced in Fig. 9.33 [73]. In the 1970s, DC motors were placed with induction (AC) motors. This required the use of VVVF inverters that were first implemented using GTOs. The GTOs began to be replaced

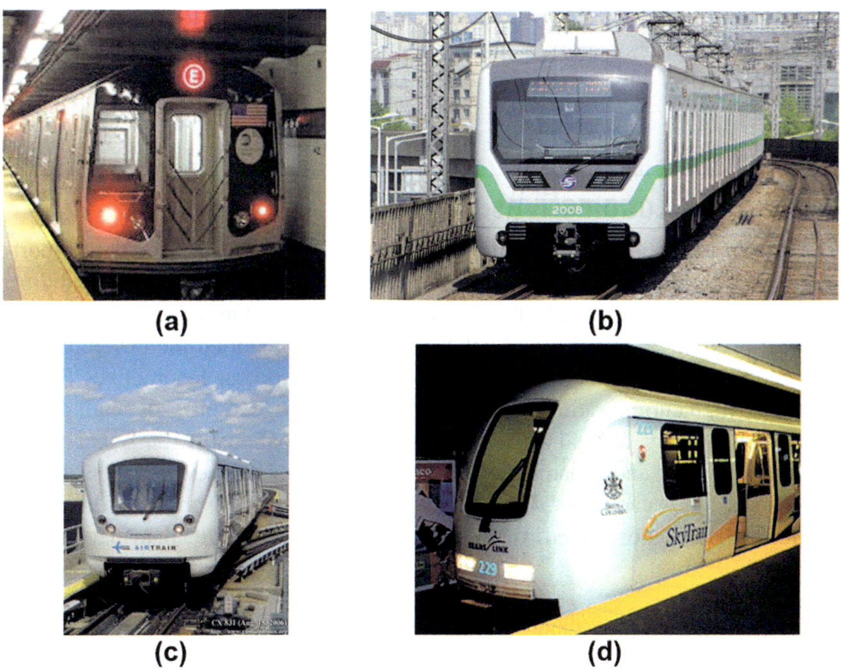

FIGURE 9.32

(a) New York City Subway; (b) Seoul Metro Subway; (c) JFK Airport Airtran; (d) Vancouver Skytrain.

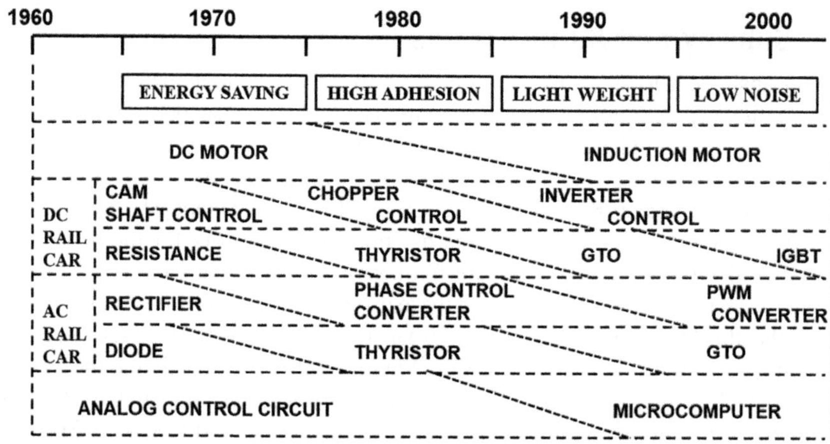

FIGURE 9.33

Hitachi perspective of technology trends for subway drives.

by IGBTs in 1992. They state: "Large-capacity IGBTs have been applied to a prototype model of subway railcars to make lighter, smaller, and lower level of noise inverters." The three-level inverter using IGBTs for subway railcars is shown in Fig. 9.34. Due to the 1550 V DC power line, it was necessary to use 2000 V IGBT modules. The authors state "In the near future, almost (all) subway railcars will be driven by IGBT inverters."

9.8 Electric Locomotives

Electric locomotives are commonly used around the world for long-distance transport of passengers (Fig. 9.35(a)) and freight (Fig. 9.35(b)). As the power-handling capability of IGBT modules was expanded, they became increasingly popular for use in higher-powered locomotives. By 1995, ABB Semiconductors reported the development of 1200 A, 2500 V IGBT modules that can power higher-power locomotives [74]. To achieve the high current ratings, they had to parallel a large number of small chips. A subassembly was first formed with a current-handling capability of 150 A that could be fully tested for static and dynamic parameters before combining seven of them to obtain the desired 1200 A capability. Using aluminum nitride (AlN) substrates during construction of the module resulted in a thermal impedance even smaller than that for GTOs.

9.8.1 DC Power Bus

By 2005, Siemens stated [75]: "The GTO has been replaced by the IGBT in mass transit applications and on auxiliary inverters"..."As soon as 1700 V IGBTs were available, these complicated inverter topologies were no longer necessary and the

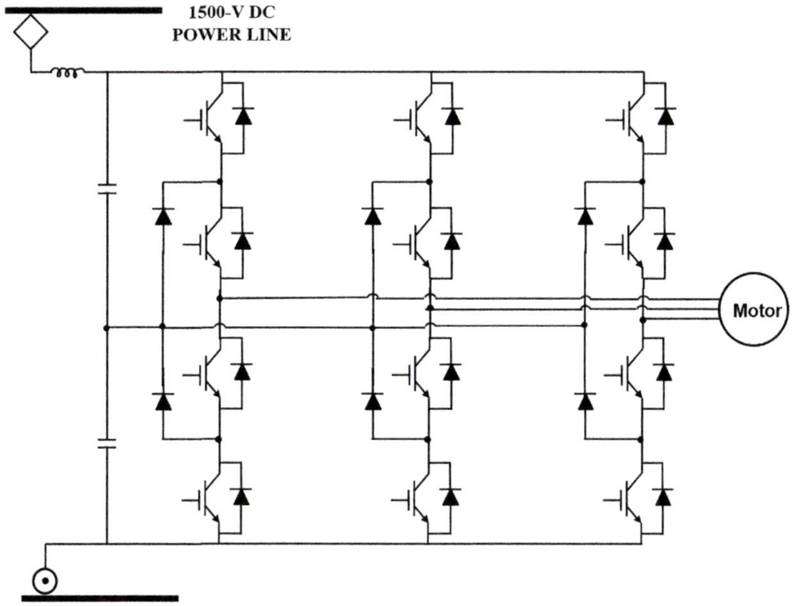

FIGURE 9.34

Subway IGBT inverter drive.

(a) (b)

FIGURE 9.35

(a) Passenger electric train; (b) Freight electric train.

simple two-level inverter, directly connected to the line, was used." The two-level inverter topology is shown in Fig. 9.36. Until the development of IGBTs with high-blocking voltage capability, three-level inverters were required. With the availability of IGBTs with higher blocking voltages, two-level inverters can be implemented by using 1700 V IGBTs for the 750 V DC power line, 3300 V IGBTs for the 1500 V DC power line, and 6500 V IGBTs for the 3000 V DC power line. The two-level inverters reduce the cost of the inverters significantly because of the

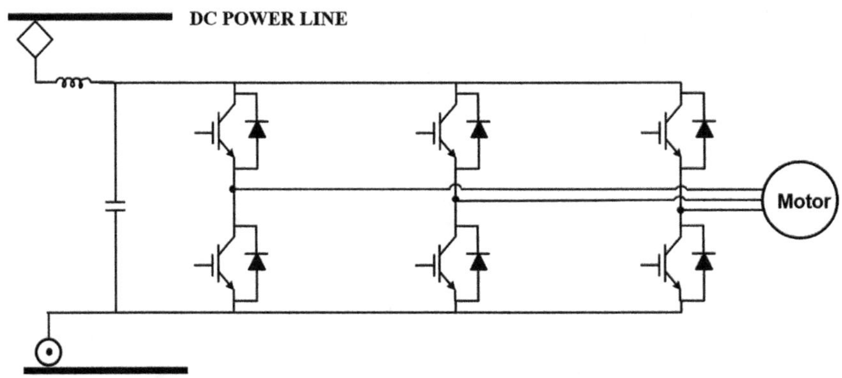

FIGURE 9.36

Two-level IGBT inverter drive for DC power bus.

reduction in the number of semiconductors, DC link capacitors, overvoltage protection networks, and chokes (compare Fig. 9.36 with Fig. 9.34). The reduced power losses in the devices improve the drive efficiency and the smaller number of components leads to improved reliability.

9.8.2 AC Power Bus

The two-level IGBT-based inverter can also be utilized for electric rail systems with AC power lines. In this case, a transformer is used to reduce the magnitude of the AC bus voltage and an AC–DC converter is then used to create a DC bus voltage across the DC link capacitor, as shown in Fig. 9.37. Reducing the DC link voltage allows utilizing lower blocking voltage IGBTs that have smaller on-state voltage drops and switching losses. However, this increases the current in locomotives with more than 1 MW output power, leading to large ohmic losses and thermal problems in the motors. Raising the DC link voltage eliminates this problem, but the losses in the IGBTs become prohibitive. A good design compromise [75] is to use a DC link voltage of 1800 V with IGBTs capable of blocking 3300 V.

9.8.3 Multisystem Electric Trains

Railway lines that utilize both AC and DC power sources need multisystem converters [75]. One example is the Deutsche Bahn BR 189 train shown in Fig. 9.38. A chopperless, IGBT-based converter for these trains is shown in Fig. 9.39. The four-quadrant AC–DC converter is used to create the DC link when connected to an AC power line. For AC and 750 V DC links, 1700 V IGBTs are optimum devices; for AC and 1500 V DC links, 3300 V IGBTs are optimum devices; for AC and 3000 V DC links, 6500 V IGBTs are required. Siemens has developed the SIBC series of modular converters for this application with the capability to serve the 750, 1500, and 3000 V DC links.

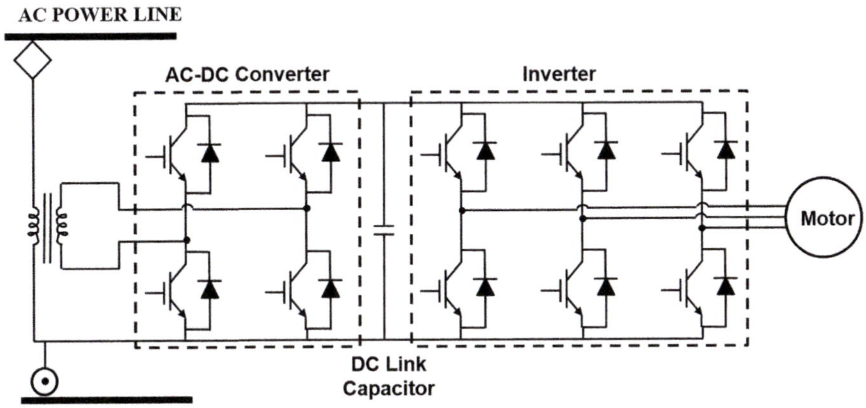

FIGURE 9.37

Two-level IGBT inverter drive for AC power bus.

FIGURE 9.38

Multisystem Deutsche Bahn BR 189 electric train.

9.9 Diesel–Electric Locomotives

Diesel–electric locomotives (two examples are shown in Fig. 9.40) are in widespread use around the world. They have also been impacted by the availability of IGBTs with high power ratings. In these trains, diesel fuel is used to generate electricity. This AC power source is then rectified as shown in Fig. 9.41 to create the DC bus. The considerations when designing the diesel electric drive are similar to those discussed in the previous section. The number of turns in the three-phase synchronous diesel generator can be adjusted to obtain the optimum DC link voltage of 1800 V. This allows utilizing 3300 V IGBTs with good efficiency.

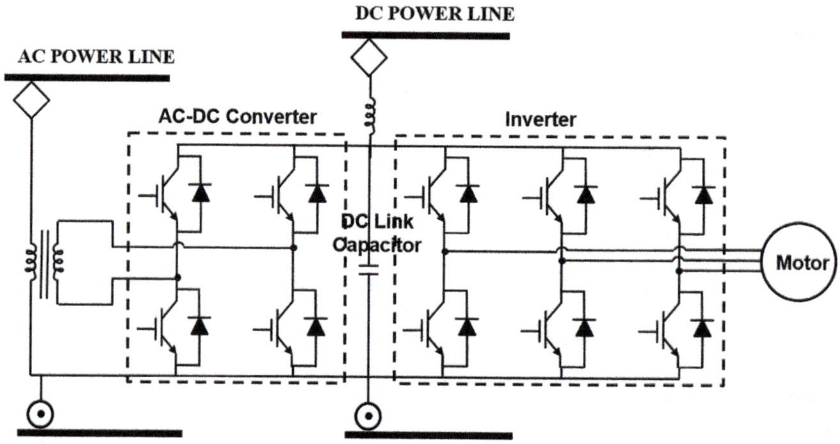

FIGURE 9.39

Multisystem chopperless IGBT converter.

FIGURE 9.40

Diesel–electric locomotives in (a) the United States; (b) India.

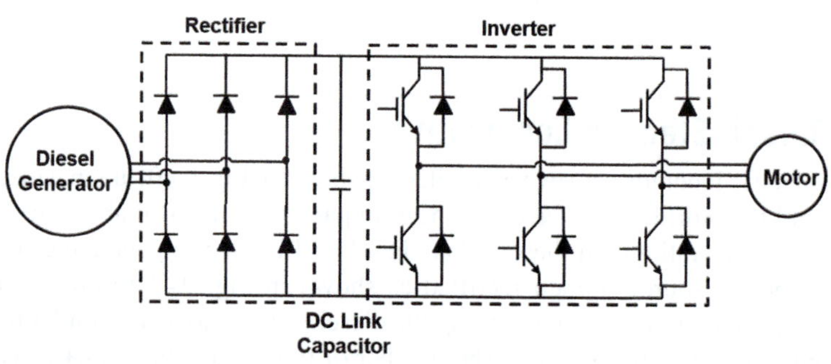

FIGURE 9.41

Two-level IGBT inverter drive for diesel-electric locomotives.

Although electric trains are widely deployed in Europe due to extensive electrification, the diesel electric locomotive is more prevalent in other parts of the world [76]. The voltage source inverter (VSI) was first applied to diesel–electric trains in 1971 using the GTO. By 2000, the VSI transitioned to IGBT-based drives. For electric locomotives, a similar transition from GTOs to IGBTs has occurred because of an increase in power density, smaller weight, and greater efficiency of the converter. The availability of 6.5 kV IGBTs allowed simplification to a two-level inverter directly connected to the 3 kV DC power line.

9.10 High-Speed Electric Trains

There are 34 megacities around the world, each having a population of more than 10 million residents. It is estimated that 8% of the world's population (~600 million people) lives in mega cities. High-speed electric trains are desirable for long-distance travel between major cities with large populations. They have been deployed extensively in Japan, Europe, the United States, and China, as shown by the examples in Fig. 9.42. The European Union Directive 96/48/EC defines the criteria for

FIGURE 9.42

High-speed trains (a) Japanese Shinkansen, (b) European TGV, (c) United States Acela, (d) CRH2C China.

high-speed trains [77] as having ungraded track for high-speed travel with a minimum speed of 250 km/h (155 mph). Although high-speed rail development can be dated to 1899 in Germany, the successful implementation of the Shinkansen (meaning *new trunk line*) bullet trains in Japan in 1964 for the Tokyo Olympic Games [78] paved the way for more widespread deployment of high-speed rail. Since 1964, more than 10 billion passengers have traveled on the Shinkansen trains. The increase in oil prices provided the impetus to deploy the TGV high-speed trains [79] in France in 1981. The high-speed network in France has grown to more than 2000 km while that in Spain exceeds 2500 km. The high-speed rail network now extends throughout the European Union. In the United States, the Amtrak Authorization and Development Act of 1992 resulted in setting up the *Acela Express* high-speed rail [76] to link Boston, New York City, Philadelphia, and Washington, DC. Other countries that have followed are South Korea [80] and Taiwan [76]. High-speed rail introduced in China in 2007 has become the world's most heavily used network [81] with more than 1.3 million passengers per day. The high-speed rail network in China was expanded to 39,700 km (24,000 miles) by the end of 2020. The world's longest high-speed rail route was opened in China in 2012, covering the distance between Beijing and Guangzhou at the speed of 300 km/h.

9.10.1 Motor Drive Topology

The high-speed rail power topology developed by Alstom [82] is shown in Fig. 9.43 for use with the 15 kV, 16.67 Hz AC power line. Eight cascaded 6.5 kV IGBT modules are used on the line side to convert the 16.67 Hz input frequency to 5 kHz. This

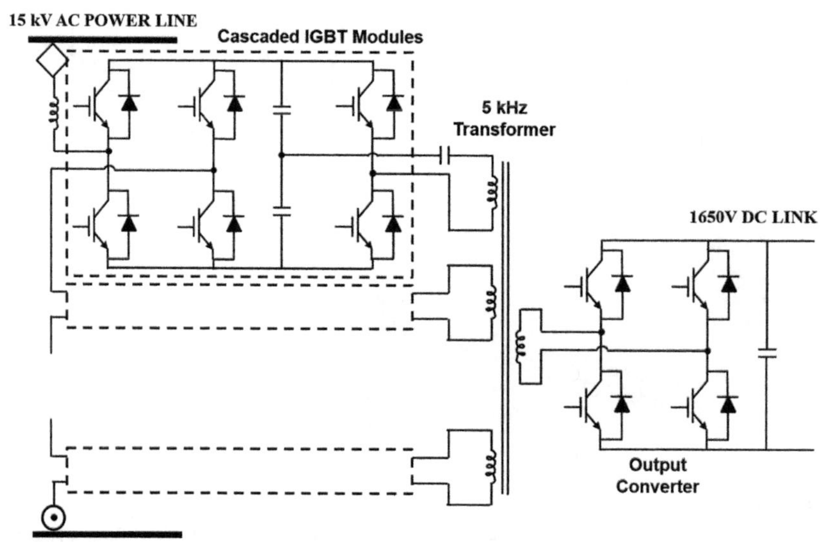

FIGURE 9.43

Alstom high-speed train circuit topology.

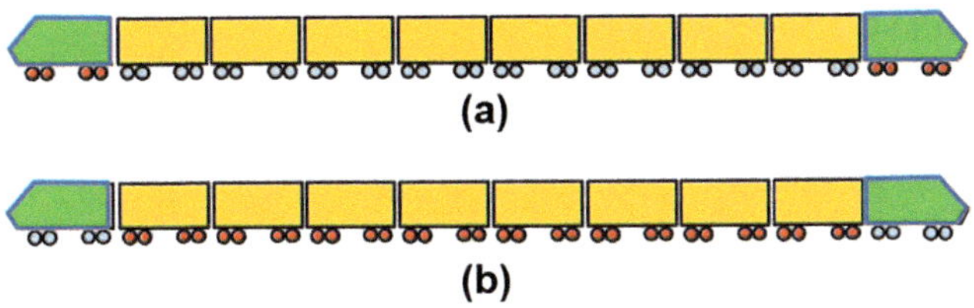

FIGURE 9.44

Comparison of: (a) concentrated and (b) distributed high speed rail traction system.

reduces the size and weight of the transformer while keeping the switching power losses at a reasonable level. The secondary side of the transformer is connected to a converter to form the 1650 V DC link. The converter requires IGBTs with 3.3 kV blocking capability. An IGBT-based propulsion drive is then fed by the DC link. These IGBT-based inverters are similar to those previously shown. Due to the poor efficiency and large weight of the transformer operated at 16.67 Hz, a transformer-less approach has also been investigated [83].

The first-generation high-speed trains were designed with the concentrated traction system, as illustrated in Fig. 9.44(a). It consists of two green electric locomotives with the motor drive supplied to the red-shaded wheels located at the ends of the train and the ochre passenger compartments located between them. In contrast, the Japanese Shinkansen bullet train has migrated to a distributed traction system, as illustrated in Fig. 9.44(b) [84]. In this design, the motor drive is supplied to the red-shaded wheels located under all the ochre passenger compartments. The distributed traction system has the advantages of low maximum axle loads and good adhesion, but was considered to degrade passenger comfort and reliability.

The availability of the IGBT in 1997 with sufficient power-handling capability allowed the distributed traction system to become practical for the Series 700 Shinkansen [84]. The higher operating frequencies of the IGBTs and their smaller losses when compared with GTOs have reduced the noise due to harmonics from the underfloor traction equipment. The weight/power ratio was reduced by a factor of two between the Series 0 and Series N700 Shinkansen trains using IGBT modules in 2007. This approach has also been adopted in the German InterCity Express (ICE3) high-speed trains since 2000 and the French Automotrice Grande Vitesse (AGV) high-speed trains more recently. Some advantages attributed to the new system are: (1) more effective regenerative braking capability with mechanical brakes used for only 3% of the total braking energy, (2) better adhesion between the wheels and the rails allowing higher operating speeds due to the larger number of powered axles, and (3) more space in the passenger compartments by locating the power electronics under the floor. The authors state, "thanks to the advent of the new IGBT with lower switching loss and analysis of airflow under the floor at high speeds, the use of power converters with train draft cooling systems on Shinkansen trains entered the realm of possibility."

The traction system for the Shinkansen Series N700 consists of three and four motor car units [84]. The configuration for the four motor car units is illustrated

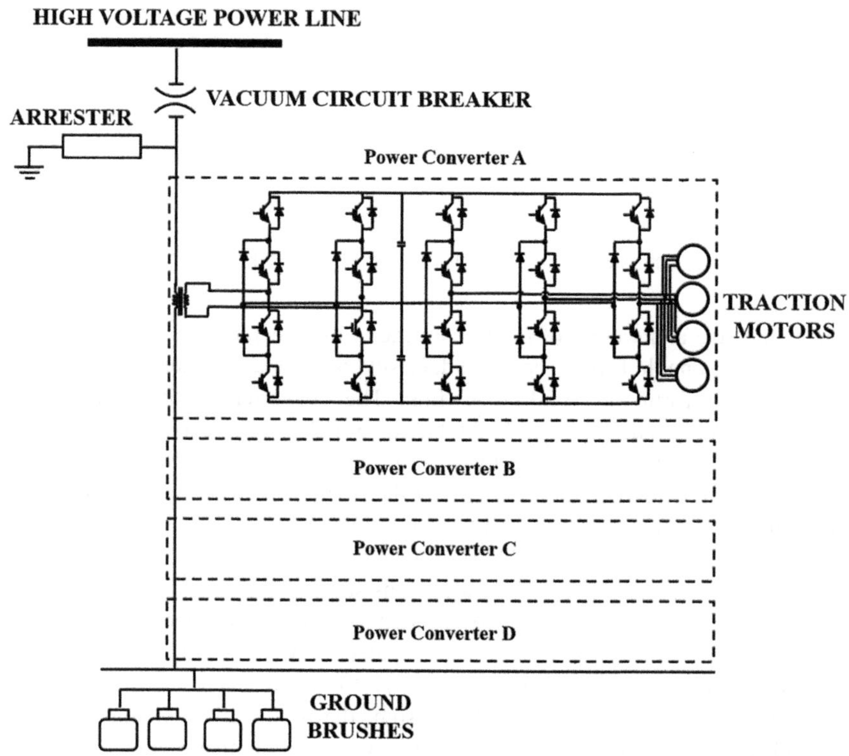

FIGURE 9.45

Shinkansen Series N700 distributed traction system for four-motor car units.

in Fig. 9.45. When running at 220 km/h between Tokyo and Shin-Osaka, the Series N700 trains consume about half the energy used previously by the Series 0 trains. This is due to improved efficiency of the IGBT-based power converters, their smaller weight, and the utilization of regenerative braking.

9.10.2 IGBT Module Design

The high power motor drive capability required for the Shinkansen trains was first achieved using IGBTs in 1997. The press-pack or flat-pack IGBT module described in Section 6.4 enabled paralleling of multiple 2.5 kV IGBT chips to attain the needed current-handling capability. Each IGBT chip was 2.15 × 2.15 cm in size—the largest die size in the world at that time [85]. To obtain sufficient yield during module production, redundant dies were required with isolation of chips with gate-source short circuits using a dielectric polymer layer, as discussed in Section 6.4. Each chip was fabricated using bulk FZ silicon wafers to achieve the 2.5 kV forward blocking capability. The optimization of the IGBT chip for high-speed traction applications is different than that used for industrial applications because the inverters operate at relatively low frequencies where on-state power loss is more important. For the press-pack IGBT module, the on-state voltage drop was 4.8 V at an on-state current of 1800 A with a turn-off time of 2.2 μs at a collector supply voltage of 1300 V.

The IGBT structure has been continuously optimized to reduce the on-state voltage drop for devices suitable in traction applications. The evolution of the 3.3 kV IGBT chip cell structure [86] is illustrated in Fig. 9.46. The first-generation devices had the symmetric or NPT structure where the N-drift region was sufficiently wide to prevent punch-through of the electric field to the collector junction. To reduce the on-state voltage drop, the second-generation IGBT structure was developed with the FS layer (N-buffer layer). The asymmetric or PT IGBT structure has a buffer layer, called the field stop layer, with a relatively low doping and a thin transparent P^+ emitter layer to reduce injection efficiency (see Chapter 2) [87]. The third-generation IGBT structure included a carrier-storage (CS) layer with higher doping concentration around the *P*-base region to reduce the resistance in the upper portion of the IGBT structure. In the fourth-generation IGBT structure, a trench gate structure was used to eliminate the JFET region and enhance the carrier density at the top portion of the drift region. A concomitant improvement in the power rectifier technology was also performed to reduce the reverse recovery current. In recent years, novel IGBT structures have also been proposed for traction applications. Two examples are the reverse blocking IGBT structure [88] and the RC IGBT structure [89]. The RC-IGBT has been found to have low thermal resistance, high current density, and elimination of a fly-back diode. Reverse recovery losses are reduced by using desaturation control [90]. China's high-speed rail has developed from the

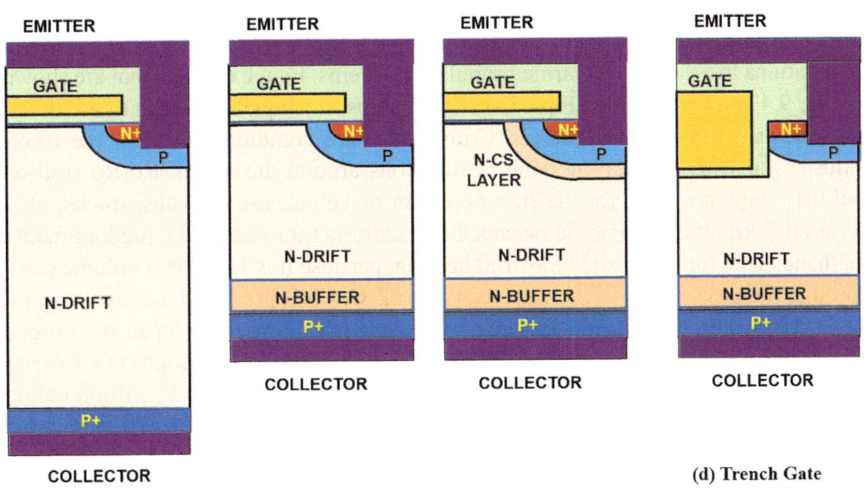

(a) NPT-IGBT (b) FS-IGBT (c) CS-FS-IGBT (d) Trench Gate FS-IGBT

FIGURE 9.46

Evolution of IGBT structures for traction applications.

first-generation CRH1, 2, 3, and 5 series to the second-generation CRH380A, B, and CL series to the third-generation CR400AF and BF series. The RC-IGBTs are being applied to the latest version of the rail system.

9.11 Freight Trains

Freight trains haul large quantities of material over long distances. It is advantageous to be able to transport the merchandize rapidly for commercial and industrial use. Many countries around the world are investing in high-speed rail lines with electric locomotives to serve this market. A recent example from 2018 to 2020 is the $4.9 billion joint venture between India and Alsthom (France) to build the world's most powerful freight locomotive (WAG12B) in Madhepura in Bihar, India. The locomotive utilizes a 12,000 hp (9000 kW) engine powered using an IGBT-based three-phase motor drive. It will haul freight along the eastern corridor of the country. The train has a Bo-Bo design with four axels in two bogies to achieve 22.5 tonnes capacity by pulling 118 wagons at 120 kmph. The shift to an electric locomotive is expected to reduce CO_2 emissions by up to 44 billion pounds per year [91].

9.12 Marine Propulsion

Ship owners and shipyard operators want to migrate to electric propulsion systems to ease maintenance, enhance efficiency, and improve reliability. The electrical propulsion system has to be compact, robust, easily installable, and cost effective. These features have been enabled due to the availability of press-pack IGBTs with blocking voltages of 3.3–6.5 kV and a current-handling capability of 1000 A, extending their applications to a variety of marine propulsion systems. Four examples that are shown in Fig. 9.47 are: (a) cruise ships, (b) Ro-Ro ships, (c) liquefied natural gas (LNG) carriers, and (d) crude oil tankers. Cruise ships are commonly used by the travel industry for trips to many popular destinations around the world. Ro-Ro (roll-on roll-off) ships are used for the transportation of vehicles (cars, vans, trucks, etc.) across the Atlantic and Pacific oceans. LNG carriers transport LNG, predominantly methane, used for home and industrial heating, because it is 1/600th in volume compared to the gaseous state when cooled to −162°C. LNG produced in Angola, Indonesia, Malaysia, and Australia is transported using LNG carriers to Japan and Europe [92]. Crude oil tankers deliver unrefined crude oil from extraction points to refineries for production of petroleum products. In 2005, oil tankers moved 12 trillion metric ton miles of crude oil from loading ports in Western Asia, North Africa, and the Caribbean to refineries in North America, Europe, and Japan [93].

In the late 1990s, press-pack IGBT modules became available with sufficiently high voltage and current-handling capability to be viable alternatives to the GTO and the insulated gate commutated thyristor (IGCT). The performance of these power devices has been compared for a variety of applications, including ship propulsion

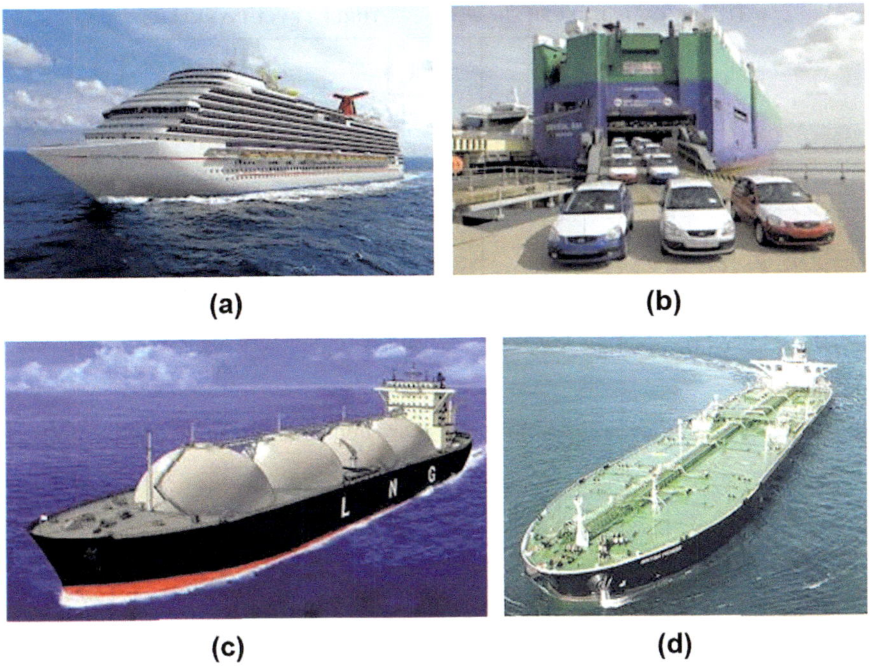

FIGURE 9.47

Marine propulsion: (a) cruise ship, (b) Ro-Ro ship, (c) liquefied natural gas carrier, (d) oil tanker.

[94]. The authors concluded: "Now with the availability of high-voltage IGBTs, the three-level inverter seems to be the most promising practical solution. It provides a very simple circuit arrangement consisting of only 12 or 24 (for 3.3 and 4.16 kV) transistors." One of the big advantages of the IGBTs was a low gate drive current of 5 A compared with 1000 A required for the GTO and the 20-plus paralleled power MOSFETs required with very low inductance connection in series with the GTO to create the IGCT. The three-level inverter circuit using IGBTs is shown in Fig. 9.48 with an active front end rectifier. The three-level inverter is able to feed a nearly perfect sinusoidal current to the motor and the active front end rectifier achieves nearly perfect sinusoidal current from the three-phase AC power source.

9.12.1 Ro-Ro Ships

By 2004, the IGBT-based shaft alternator was available for commercial Ro-Ro ships [95]. The shaft alternator is used to supply main electrical power to the ship. The diesel generator produces variable frequency output power. This source is converted to a regulated 60 Hz power supply for the ship using the circuit topology shown in Fig. 9.49. The PWM inverter is operated at 2 kHz and delivers 1440 kW of output

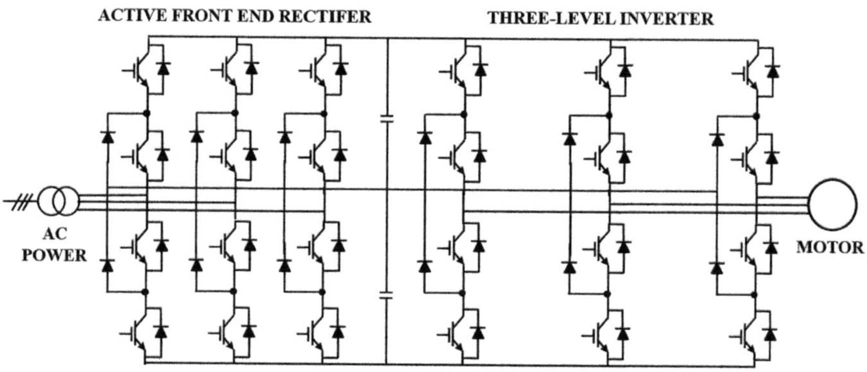

FIGURE 9.48

Three-level IGBT inverter for motor drive with active front end rectifier.

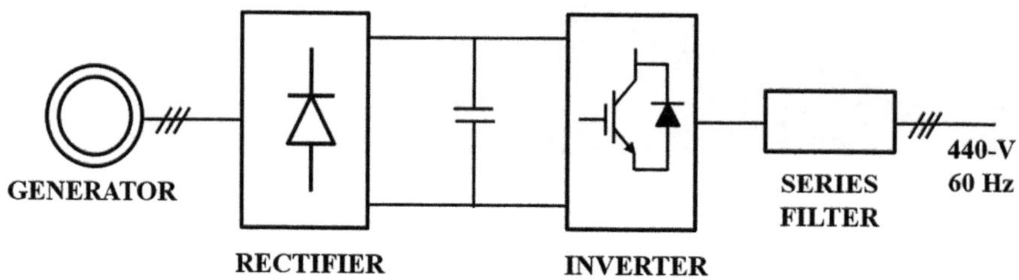

FIGURE 9.49

Shaft alternator for the Ro-Ro ships.

power for the ship using a 440 V AC bus. Five (214 m, 56,799 GRT) Ro-Ro car carrier vessels were in continuous operation on the North and South Atlantic routes by 2004 using this electrical system.

9.12.2 Cruise Ships

It is desirable to reduce fuel consumption in cruise liners to reduce weight and space [96]. The authors state: "To achieve higher power density, higher switching frequencies, and improved fault current limitation capabilities, high-performance press-pack IGBT devices were used in the PWM converter realization." The MV7000 frequency converter, shown in Fig. 9.50, delivers medium to high power (4–33 MW) at voltages of either 3.3 or 6.6 kV. The shift to an IGBT-based PWM inverter allows using induction motors in place of synchronous motors previously required with thyristor drives. Induction motors are far superior in terms of reliability, reduced size, and acoustic noise.

Cruise ships require a fully redundant propulsion architecture for each shaft line. The solution is to drive induction motors with two completely independent 6.6 kV electrical chains created with two 12-pulse transformers feeding two MV7612 converters, as shown in Fig. 9.51 [96]. By using phase-shifted windings, a 24-pulse

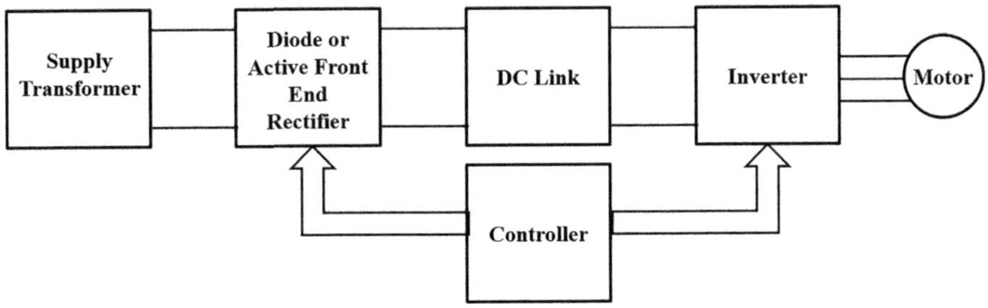

FIGURE 9.50

MV7000 frequency converter for cruise ships.

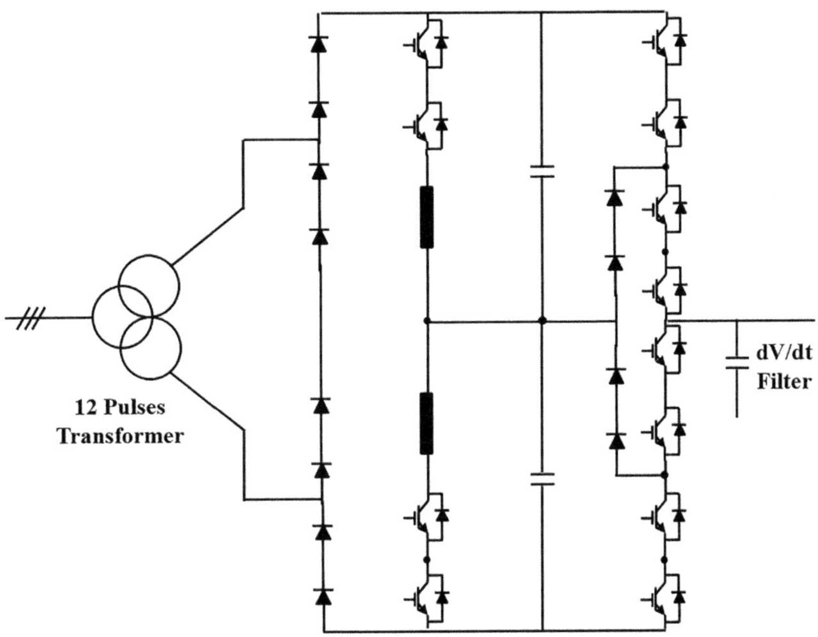

FIGURE 9.51

MV7612 converter for cruise ships.

network is formed that removes harmonics without the need for filters. Two 6.6 kV inverters are connected in parallel to feed a single winding induction motor.

9.12.3 LNG Carriers

LNG carriers are usually designed with a single skeg hull shape with a redundant electrical propulsion architecture using a single-stage gearbox feeding two medium-speed motors. Two independent medium-speed induction motors fed from

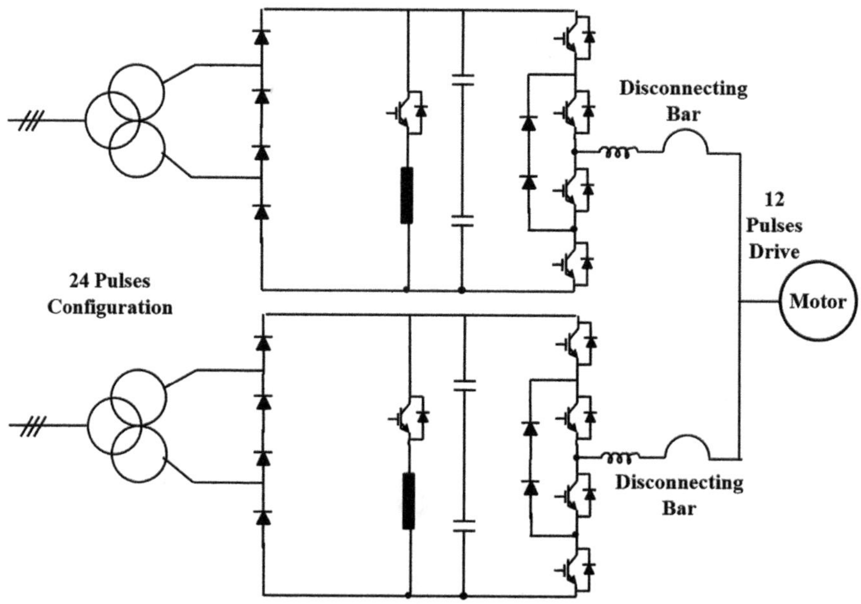

FIGURE 9.52

MV7316 converter for liquefied natural gas carriers.

3.3 kV drive power are used for redundancy [96]. Each motor is powered using the configuration shown in Fig. 9.52 with the two separate MV7316 power converters operated in parallel for redundancy. The power converters consist of a diode front end, DC link capacitors, and an IGBT-based PWM inverter. Two 12-pulse transformers are used at the input to create a 24-pulse arrangement with 12 MW of power. This eliminates the use of harmonic filters. The output power from the converters is fed to the motor at 3.3 kV. This scheme provides a high level of redundancy to deal with failures anywhere in the power chain. The disconnecting bars can be used to isolate the faulty converter, enabling operation with a propeller torque of at least 75%.

9.12.4 Circuit Breakers for Ships

Shipboard power distribution is preferably performed using a DC bus because it eliminates front-end AC/DC conversion [97]. This architecture is similar to that of DC micro grids with multiple power generation sources and loads on a common bus. Fault protection becomes an important design element for these systems with predictable system response and fast fault interruption as desirable attributes. Solid-state protective devices (SSPDs) are required to provide the ability to respond to faults within microseconds. The fault can then be isolated into a zone, leaving the rest of the ship fully functional.

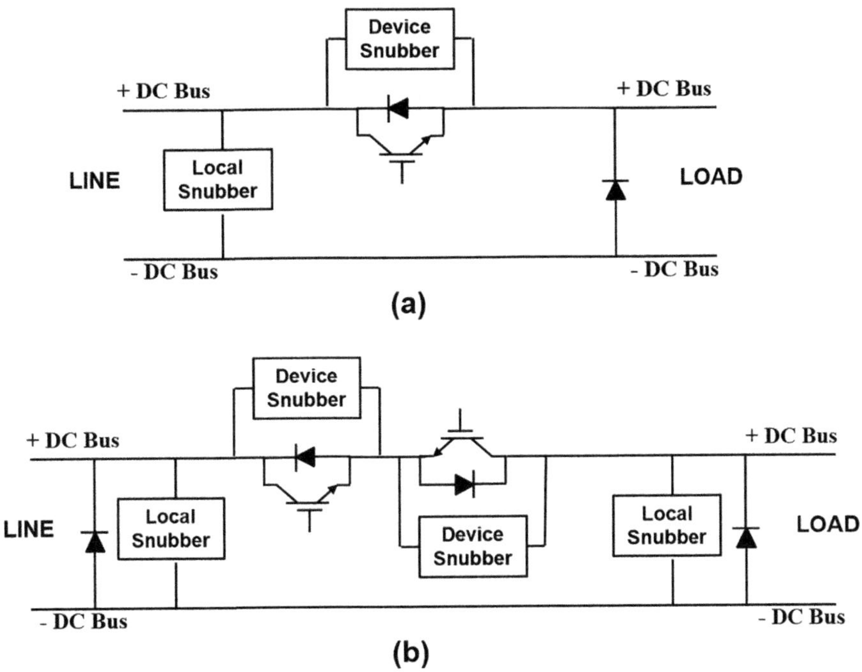

FIGURE 9.53

Solid-state protective device with: (a) unidirectional and (b) bidirectional capability.

The SSPDs can be designed for unidirectional or bidirectional fault current interruption, as shown in Fig. 9.53. These designs utilize the commercially available asymmetric blocking IGBT modules containing fly-back rectifiers that have 1000 V, 1800 A capability. During the fault, the current can rise at the rate of 100 A/μs, leading to a peak current of twice the steady-state current in the SSPD. In the future, the DC bus voltage is expected to increase from 3.3 to 6 kV.

9.13 More Electric Aircraft

A shift from the hydraulic system to the all-electric system in recent years for the development of commercial aircraft, as shown in Fig. 9.54, has been motivated by reductions in the size and weight of the flight control apparatus. Improvements in pneumatic systems stalled around 1995, encouraging the airline industry to embrace power electronic systems, which were experiencing a steep performance and cost renaissance due to the availability of the IGBT [98]. Electrical pressurization of the cabin has reduced the empty weight of the aircraft by 1000–2000 pounds, reducing fuel consumption by several percent. Deicing of the wings is being

(a) (b)

FIGURE 9.54

All electric aircraft: (a) Boeing 787 Dreamliner; (b) Airbus A380.

performed with electrical heating instead of bleed air. Electrical starters have replaced pneumatic starters for the aircraft engines.

Hydraulic power has been historically used on aircraft for actuators of large loads such as the landing gear, nose-wheel steering, braking, and cabin doors [99]. However, hydraulic systems in aircraft are expensive, have high maintenance costs, and are fire hazards due to leakage. Even in 1990, the industry began to explore transitioning to electrically driven actuators despite an extremely conservative approach taken by the industry. The pneumatic power used for engine starting and control of flight surfaces needed to be replaced by electrical systems. Progress in power electronics and motor design has produced a very high power-to-weight ratio, making this approach attractive for fuel savings as well. By 1990, 92% of the functions in an aircraft were being served by electrical systems with hydraulics used mainly for flight control surfaces.

The power system serves two basic loads—namely for vehicle operation and for the comfort of passengers in the cabin [100]. The tasks performed include flight control surfaces, deicing operations, cabin environment management, electrically actuated brakes, fuel pumping, landing gear, and engine starting. Power electronic converters that are required include AC/DC rectifiers, DC/AC inverters, and DC/DC choppers. Electrical technology promises superior cabin air quality while reducing fuel consumption [101]. Electrical engine starters can replace the traditional oil-based system, removing weight, reducing cost, and improving reliability.

By 2006, power electronics had advanced sufficiently for Boeing to launch the 787 Dreamliner using an all-electric methodology [102]. In the Boeing 787, the article states: "Virtually everything that has traditionally been powered by bleed air from the engines has been transitioned to an electric architecture." This includes the engine start, wing ice protection, cabin pressurization, and hydraulic pumps. The no-bleed design significantly reduces maintenance cost and improves reliability. Replacing hydraulic brakes with electrical brakes provides four independent brake actuators per wheel, which can be monitored electronically for fault detection and brake wear.

9.13.1 DC–DC Converter

The DC–DC converters in the all-electric aircraft must operate off a 270 V DC power bus. A series resonant input inverter topology is used to increase the operating frequency to 120 kHz with zero current switching of the IGBTs to reduce switching losses [103]. The circuit diagram for the DC–DC converter is shown in Fig. 9.55. Schottky rectifiers are used on the output side of the high-frequency transformer. The high operating frequency allows shrinking the size of the transformer and reducing its weight. This converter has a power rating of 5.6 kW and provides an output voltage of 28 V DC with a weight of nine pounds and a volume of 160 cubic inches. All the power components are mounted on ceramic substrates with direct bonded copper technology (see Chapter 6).

9.13.2 DC–AC Inverter

The DC–AC inverter for the all-electric aircraft is also based on a high-frequency (120 kHz) resonant circuit to reduce the size and weight of the transformer [103]. The IGBTs are operated using zero current switching to minimize switching losses. Four isolated secondary windings are connected to half-bridge rectifiers to make a 40 V DC power source that is connected in series to produce the 115 V, 400 Hz AC output with an field effect transistor (FET)-based inverter stage. This multistep inverter shown in Fig. 9.56 produces a sinusoidal output power with less than 2% THD. This converter has a power rating of 8 kVA and provides an output voltage of 115 V, 400 Hz, three-phase AC with a weight of 16 pounds and a volume of 240 cubic inches.

9.13.3 Electromechanical Aircraft Rudder Actuator

Electromechanical actuators (EMAs) are replacing hydraulic actuators due to improved weight and reliability. One example is the aircraft rudder control [104]. The matrix converter topology shown in Fig. 9.57 is attractive because it eliminates the bulky and unreliable DC link electrolytic capacitors used in other topologies

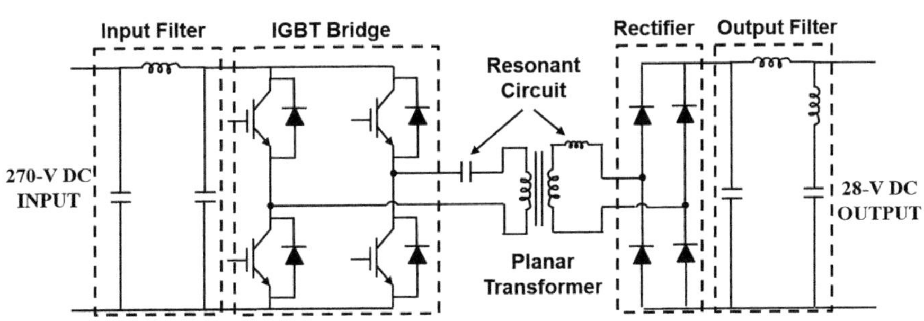

FIGURE 9.55

DC–DC converter for the all-electric aircraft.

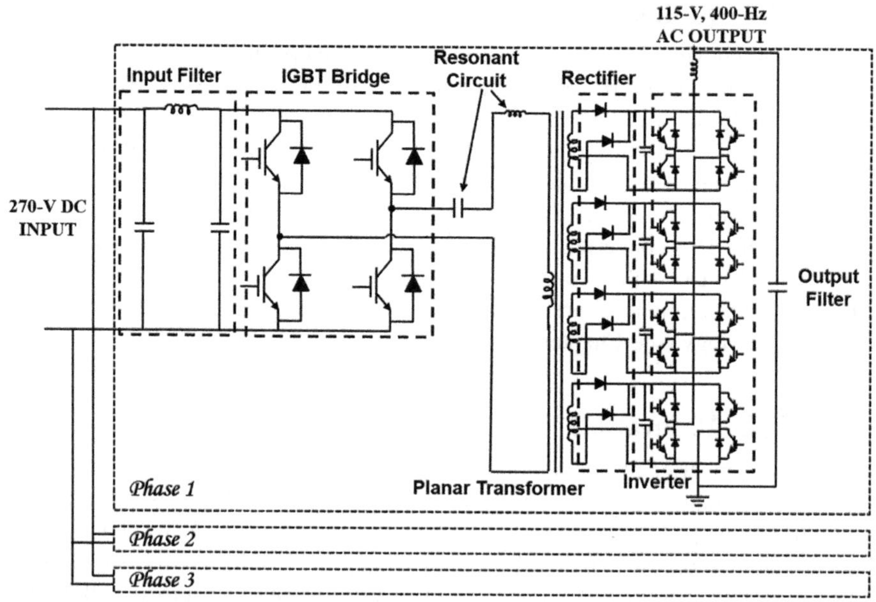

FIGURE 9.56

DC–AC inverter for the all-electric aircraft.

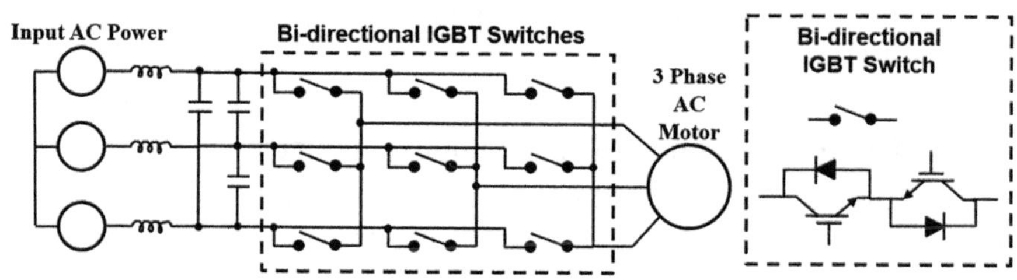

FIGURE 9.57

Electromechanical actuator for the rudder in the all-electric aircraft.

(such as those shown in the above sections). The EMA drive is used to control a three-phase AC motor that controls the ball-screw of the rudder actuator. The bidirectional switches can be created by connecting the IGBTs with fly-back diodes in a back-to-back configuration, as illustrated on the right side of Fig. 9.57. Each output phase of the matrix converter is constructed using 600 V, 300 A IGBT modules.

Another approach for the design of an EMA is shown in Fig. 9.58. Here, an H-bridge configuration is utilized with redundancy achieved by using an H-bridge for each winding of the motor [105]. This configuration doubles the number of IGBTs in the drive but allows each phase of the motor to operate independently. Because the devices must support only the phase voltage rather than the line voltage, IGBTs with

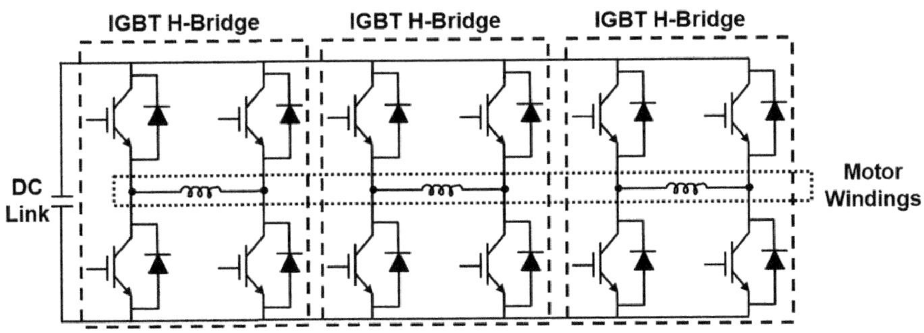

FIGURE 9.58

Redundant H-bridge electromechanical actuator for the all-electric aircraft.

smaller blocking voltage ratings can be used. This reduces the on-state and switching losses of the drive. The reduced losses allow shrinking the size and weight of the heat sinks.

9.13.4 Brushless DC Motor Drives

As discussed already, flight control actuators, fans, engine fuel pumps, and brakes in the all-electric aircraft require electronic control of motors. The high-speed permanent magnet brushless DC motor is attractive for this application due to its smaller size and weight. The control for the motor can be provided using an IGBT-based three-phase two-level inverter [106], as illustrated in Fig. 9.59. The brushless direct current (BLDC) motors are the best choice for electric fuel pumps, actuators, and taxi and braking systems. They are powered off a 270 or 540 V DC electric bus in the aircraft.

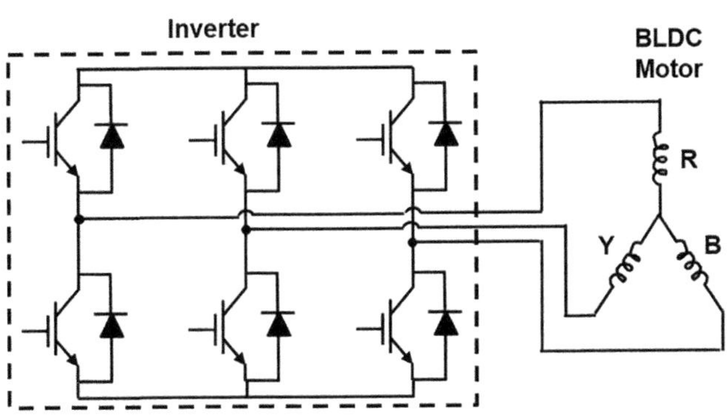

FIGURE 9.59

Three-phase two-level inverter for BLDC motor control in the all-electric aircraft.

9.14 All-Electric Aircraft

Growth in worldwide travel has led to a great expansion of civil aviation. Carbon-dioxide emissions from aircraft now comprise 3% of all global emissions and 12% of emissions from the transportation sector [107]. The European Council on Aviation Research and Innovation has set near-term goals (Vision 2020) of reducing CO_2 emissions by 50%, NO_x emissions by 80%, external noise by 50%, and fuel consumption by 50%. For the medium- to long-term goals, Vision 2050 has set goals of reducing CO_2 emissions by 75%, NO_x emissions by 90%, and external noise by 65%. One solution to achieve these goals is conversion to all-electric aircraft that no longer use fossil fuels. United Airlines has announced the purchase of all-electric commuter planes from Heart Aerospace [108]. These 19-seater ES-19 electric planes are targeted for short commuter flights of up to 250 miles. United Airlines plans to reduce its greenhouse gas emissions by 100% by 2050.

9.14.1 Civil Tilt Rotorcraft

The next generation civil tilt rotorcraft (NextGen CTR) is based on a ±270 V DC distribution bus [109]. The power architecture includes three 90 kW generators and two 50 kW generators with rectifiers. Power converters are fed from the DC bus to deliver energy to the loads. These converters utilize IGBTs. The architecture also needs circuit breakers, which are built using IGBTs because the breaker is normally on with high conduction losses. The authors have compared various power device options, including SiC power MOSFETs, and conclude: "Out of three main devices, the IGBT would be a suitable selection for this application due to its reasonable good reliability, tripping time, cost, weight, and volume."

9.14.2 ANPC Inverter Drive

The creation of all electric aircraft requires minimizing the weight associated with the electrical power delivery and the electronics for driving the propulsion motors. Medium-voltage DC distribution is the preferred architecture to reduce power cable weight [110]. The smaller current obtained with larger DC bus voltage reduces resistive copper wire losses and eliminates skin and proximity effect AC power distribution losses. The five-level active neutral point clamped (ANPC) topology is shown in Fig. 9.60 to deliver three-phase AC power to loads. It contains three inverters—one for each phase of the AC output. Each inverter requires eight IGBT devices. They can be partitioned into three cells as delineated by the purple dashed lines. Low switching frequency is used for the devices in cell 1 and 2, with higher switching frequency for devices in cell 3. This produces the highest efficiency. The authors suggest using SiC power MOSFETs in place of IGBT 7 and 8 due to the high switching frequency used for cell 3.

Larger aircraft for commuters and commercial cargo between small cities are being vigorously pursued by many start-up companies around the world [111]. Their

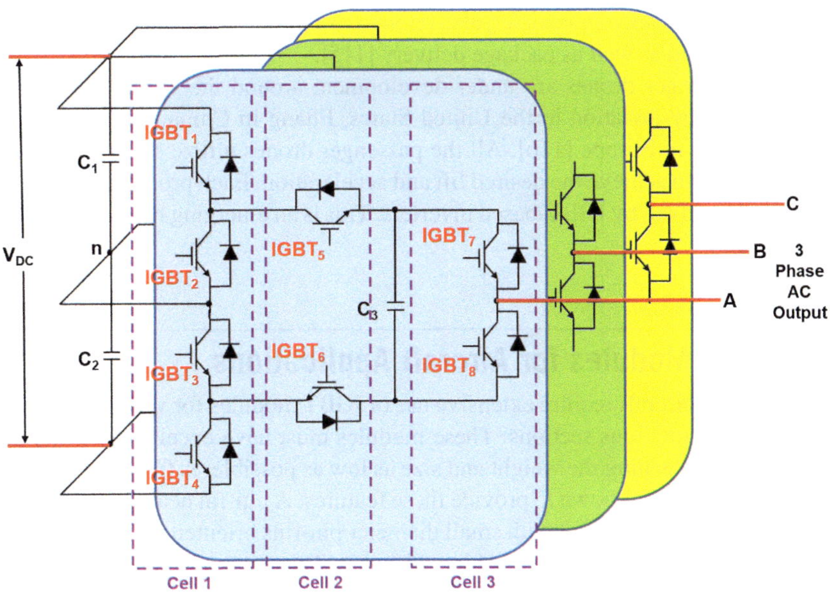

FIGURE 9.60

Active neutral point converter for medium-voltage DC to three-phase AC power distribution.

range is limited to about 100 miles due to the capacity of batteries that can be used without compromising the weight. The power level for operating these aircraft is in the megawatt scale. A MW-scale ANPC converter has been reported for this purpose [112]. It makes use of SiC power MOSFETs for cell 1 with devices switching at 20 kHz and silicon IGBTs for cell 2 switching at 1.4 kHz. A rated power of 1.2 MVA was achieved using a 1.2 kV DC bus.

9.14.3 Passenger Drones

Traffic congestion on the streets of large cities has become an endemic problem detrimental to commuting time and productivity. One of the solutions proposed to overcome this problem is utilizing passenger drones with vertical take-off and landing to minimize the footprint of the hub. The drones will be deployed between specific locations within the city (hubs) with autonomous electric vehicles used by consumers to access these terminals [113]. A case study has been performed for the Bavarian region, which shows that 70% of the population of 12.6 million people living over a 70,000 square-km area can be served using an optimized hub distribution with 30-minute autonomous feeder service. Another study has been performed for the city-state of Singapore with an area of 725 square km and a population of 5.7 million. The proposed seven-hub distribution plan can serve 400,000 orders per day (150 million orders per year), generating a revenue of \$3 billion per year [114]. Unmanned autonomous aerial delivery vehicles, also referred to as drone taxis, can be used for consumer commuting as well as package delivery [115].

Many passenger drones are under development around the world: the Flystro, Archer, and Joby Aviation in the United States, Ehang in China, and City Airbus and Volocopter in Europe [116]. All the passenger drones utilize several (typically 4–8) propellers to achieve the desired lift and acceleration. Each propeller is operated using a motor driven by IGBT-based inverters. This is an emerging market for IGBTs in the future.

9.15 IGBT Modules for Aircraft Applications

The all electric aircraft require extensive use of IGBT modules for various functions, as described in previous sections. These modules must have excellent thermal performance while keeping the weight and size as low as possible. A Col-Max heat sink design [117] has been shown to provide these features. A pin-fin heat sink structure is used in the Col-Max concept with small diameter pin-fins oriented in the direction of heat flow to obtain a high heat transfer coefficient. This approach has been shown to reduce the thermal resistance by a factor of two.

9.16 IGBT Cosmic Ray Failures

Cosmic ray-induced failures [118] in GTOs used for traction applications were discovered and reported as far back as 1995. Cosmic rays interact with the oxygen and nitrogen atoms in the atmosphere, creating secondary particles not only in the upper atmosphere but even at sea level. The highly energetic neutrons generated by this process can penetrate the IGBT power device packages. The neutrons transiting through the silicon chip can strike the lattice, producing electron–hole pairs. The impact on the IGBTs has been discovered to be different than that for GTOs because the junction depths in the IGBT structures are nearly two orders of magnitude smaller in size. The influence of neutron irradiation on IGBTs has been studied using an energetic neutron source at the Los Alamos National Labs. The data indicate the mean time between failures (MTBF) for IGBTs of 19 years [118].

9.17 Summary

The development and commercialization of the IGBT has had a transformative impact on the power electronics used for the transportation sector of the economy [119]. The authors state: "IGBTs have emerged as the preferred power switches in new road vehicle electric traction drives, just as they have in rail traction drives." The development of the first modern electric vehicles in the 1980s and 1990s relied heavily on the availability of IGBTs with blocking voltages in the range of 400–

600 V. With scaling up of the voltage ratings [120,121] of the IGBTs and innovations in packaging multiple chips in modules to achieve high current-handling capability, the IGBTs first became essential components for trams and electric streetcars and by 1997 even for high-speed bullet trains. They are expected to play a dominant role in the evolution of power electronics for traction applications into the foreseeable future.

With growth in the power-handling capability of IGBT modules with larger voltage and current-handling capability, the application of the IGBT has been extended to marine propulsion in cruise ships, Ro-Ro vehicle transporters, LNG carriers, and crude oil tankers. In addition, the desire to migrate from hydraulic systems to more reliable and lightweight electrical systems has been enabled by IGBTs to create the modern all-electric aircraft. They will be essential components required for passenger drones deployed in the future for consumer commuting and the delivery of packages. The growth and challenges facing e-mobility over a broad range from electric vehicles to hybrid-electric aircraft has been recently reviewed [122]. Topics covered by the article relevant to this chapter include electric vehicles, fast chargers, and wireless charging.

References

[1] Transportation in the United States, nationalatlas.gov website http://nationalatlas.gov/transportation.html.

[2] R.G. Amey, Automotive component innovation: development and diffusion of engine management technologies, Technovation 15 (1995) 211–222.

[3] J.M. Donnelly, K. Gauen, New IGBTs and simulation models simplify automotive ignition system design, in: IEEE Power Electronics Specialists Conference, 1993, pp. 473–481.

[4] L. Mamileti, et al., IGBT designed for automotive ignition systems, in: IEEE Power Electronics Specialists Conference, 1996, pp. 1907–1912.

[5] Z.J. Shen, S.P. Robb, A study of a dual-voltage self-clamped IGBT for automotive ignition applications, in: IEEE International Symposium on Power Semiconductor Devices and ICs, Paper 9.31, 2001, pp. 295–298.

[6] H. Estl, C. Preuschoff, J. Darrow, Smart IGBTs for advanced distribution ignition systems, in: IEEE Conference on Power Electronics in Transportation, 2004, pp. 49–52.

[7] M. Towse, Senior Vice President, Fairchild Semiconductor Corp., Speaking at JP Morgan Technology and Telecom Conference, 2003.

[8] On Semiconductor Ships 100 Millionth Ignition IGBT, September 13, 1999. www.onsemi.com.

[9] Intersil Press Release Intersil Ignition IGBT Shipments Drive Past 50 Million Mark, December 6, 2000.

[10] Fairchild Semiconductor Offers the Industry First Standard Current Sense Ignition IGBT for Automotive Applications, May 30, 2012. www2.electronicproducts.com.

[11] ISL9V5045S3ST_F085 EcoSPARK n-Channel Ignition IGBT Fairchild Semiconductor Datasheet, February 2012. www.fairchildsemi.com.

[12] MGP15N40CL Ignition IGBT on Semiconductor Datasheet, December 2005. www.onsemi.com.

[13] SPGB10NB37LZ Internally Clamped IGBT ST Microelectronics Datasheet, November 2009. www.st.com.

[14] A. Muetze, Auxiliary automotive drives revolutionized by power electronics, in: IEEE Applied Power Electronics Conf, 2021.

[15] K. Lee, et al., Introduction of automotive smart power module series for up to 7.5kW automotive high voltage auxiliary motor drive applications, Power Conversion and Intelligent Motion Asia, 2019, pp. 104–107.

[16] Electric Car. en.wikipedia.org/wiki/Electric_car.

[17] M. Shnayerson, The Car That Could, Random House, New York, 1996, pp. 40–42.

[18] M.J. Riezenman, Pursuing efficiency, IEEE Spectr. 29 (11) (1992) 22–24.

[19] R. John, et al., Semiconductor technologies for smart mobility management, in: IEEE European Design, Automation, and Test Conference and Exhibition, 2013, pp. 1749–1752.

[20] H. Ueda, et al., Wide-bandgap semiconductor devices for automotive applications, in: CS MATECH Conference, 2006, pp. 37–40.

[21] M. Sugimoto, et al., Wide-bandgap semiconductor devices for automotive applications, Int. J. High Speed Electron. Syst. 17 (2007) 3–9.

[22] A. Kawahashi, A new-generation hybrid electric vehicle and its supporting power semiconductor devices, in: IEEE International Symposium on Power Semiconductor Devices and ICs, Paper 3, 2004, pp. 23–29.

[23] L. Chang, Comparison of AC drives for electric vehicles—a report on expert's opinion survey, IEEE Aerosp. Electron. Syst. Mag. 9 (1994) 7–11.

[24] K. Berringer, High current power modules for electric vehicles, in: IEEE Conference on Power Electronics in Transportation, 1994, pp. 59–65.

[25] T. Nakajima, et al., New intelligent power module for electric vehicles, in: IEEE Industrial Applications Society Conference, vol. 2, 1995, pp. 954–958.

[26] V.A.K. Temple, MOS controlled thyristors (MCTs), in: IEEE International Electron Devices Meeting, Abstract 10.7, 1984, pp. 282–295.

[27] X. Jing, I. Celanovic, D. Borojevic, Device evaluation and filter design for 20 kW inverter for hybrid electric vehicle applications, in: IEEE Conference on Power Electronics and Drive Systems, vol. 2, 1997, pp. 804–809.

[28] K. Hamada, et al., A 600 V, 200 A low loss high current density trench IGBT for hybrid vehicles, in: IEEE International Symposium on Power Semiconductor Devices and ICs, Paper 13.3, 2001, pp. 449–452.

[29] B.J. Baliga, Fundamentals of Power Semiconductor Devices, second ed., Springer Science, New York, 2019, pp. 984–995 (Chapter 9).

[30] B.J. Baliga, Fundamentals of Power Semiconductor Devices, 2nd, Springer-Science, New York, 2019, pp. 1012–1013 (Chapter 9).

[31] K. Hotta, et al., An abnormal turn-off surge suppression concept for PT-IGBTs at high voltage operation, in: IEEE Power Electronics Specialists Conference, vol. 1, 2004, pp. 591–596.

[32] S. Nishida, et al., Cosmic ray ruggedness of IGBTs for hybrid vehicles, in: IEEE International Symposium on Power Semiconductor Devices and ICs, Paper HV-P2, 2010, pp. 129–132.

[33] H. Boving, et al., Ultrathin 400 V FS IGBT for HEV applications, in: IEEE International Symposium on Power Semiconductor Devices and ICs, 2011, pp. 64–67.

[34] D. Graovac, A. Christmann, M. Munzer, Power semiconductors for hybrid and electric vehicles, in: IEEE International Conference on Power Electronics, Paper ThD2-1, 2011, pp. 1666–1673.
[35] C.-K. Liu, et al., IGBT power module packaging for EV applications, in: IEEE International Conference on Electronic Materials and Packaging, 2012, pp. 129–132.
[36] H.-R. Chang, et al., 200 kVA compact IGBT modules with double-sided cooling for HEV and EV, in: International Symposium on Power Semiconductor Devices and ICs, 2012, pp. 299–302.
[37] X. Huang, et al., Research of the loss and gate desaturation control for RC-IGBT used in vehicle power converters, in: IEEE Int. Conf. on Industrial Technology, 2017, pp. 201–206.
[38] C. Zhu, et al., A floating dummy trench gate IGBT (FDT-IGBT) for hybrid and electric vehicle (HEV/EV) applications, in: IEEE European Conf. on Power Electronics, 2017. pp. P.1–P.7.
[39] Z. Xu, et al., Investigation of Si IGBT operation at 200 °C for traction applications, IEEE Trans. Power Electron. 28 (2013) 2604–2615.
[40] J. Cao, et al., Regenerative braking sliding mode control of electric vehicle based on neural network identification, in: IEEE International Conference on Advanced Intelligent Mechatronics, 2008, pp. 1219–1224.
[41] X. Li, et al., Power management and economic estimation of fuel cell hybrid vehicle using fuzzy logic, in: IEEE Vehicle Power and Propulsion Conference, 2009, pp. 1749–1754.
[42] O. Honorati, et al., Lightweight, compact, three-wheel electric vehicle for urban mobility, in: IEEE International Conference on Power Electronic Devices and Energy Systems for Industrial Growth, vol. 2, 1998, pp. 797–802.
[43] C.B. Toepfer, Charge! EVs power up for the long haul, IEEE Spectr. 35 (11) (1998) 41–47.
[44] C.M. Portela, et al., A flexible and privacy friendly ICT architecture for smart charging of EVs, in: IEEE International Conference on Electricity Distribution, Paper 199, 2013, pp. 1–4.
[45] Y. Wang, et al., An investigation into the impacts of the crucial factors on EVs charging load, in: IEEE Innovative Smart Grid Technologies-Asia, 2012, pp. 1–4.
[46] Z. Li, et al., GPF-based method for evaluating EVs free charging impacts in distribution system, in: IEEE Power and Energy Society General Meeting, 2012, pp. 1–7.
[47] T. Tanaka, et al., Smart chargers for electric vehicles with power quality compensator on single-phase three-wire distribution feeders, in: IEEE Energy Conversion Congress and Exposition, 2012, pp. 3075–3081.
[48] M. Prokin, et al., Optimal control of chargers for electric vehicles, in: IEEE Mediterranean Conf. on Embedded Computing, 2019, pp. P1–P4.
[49] R. Bruninga, J.A.T. Sorensen, Charging EVs efficiently now while waiting for the smart grid, in: IEEE Green Technologies Conference, 2013, pp. 1–7.
[50] B.J. Masserant, T.A. Stuart, On-line computation of TJ for EV battery chargers, in: IEEE Workshop on Computers in Power Electronics, 1996, pp. 152–156.
[51] D. Ouwerkerk, T. Han, J. Preston, Efficiency improvement using hybrid power module in 6.6 kW non-isolated on-vehicle charger, in: IEEE Vehicle Power and Propulsion Conference, 2012, pp. 284–288.

[52] A. Taylor, et al., Design of a 97% efficiency 10 kW power factor correction for fast electric chargers of plug-in hybrid electric vehicles, in: IEEE Conference on Transportation Electrification, 2012, pp. 1–6.
[53] M. Bayati, M. Abedi, G.B. Gharehpetian, A new electric vehicle charging station based on pulse current charging method, in: Iranian Conference on Electrical Engineering, 2017, pp. 967–972.
[54] G. Zhemerov, O. Plakhtii, A. Mashura, Efficiency analysis of charging station for electric vehicles using the active rectifier in microgrid system, in: IEEE 4th International Conference on Intelligent Energy and Power Systems, 2020, pp. 37–42.
[55] Transportation in the United States. en.wikipedia.org/wiki/Transportation_in_the_United_States.
[56] When the Sun Shines Down Under. www.treehugger.com/files/2007/12/when_the_sun_sh.php.
[57] Electric Bus Recharges in Just 10 Minutes. www.matternetwork.com/2009/2/energy-efficient-transit.cfm.
[58] C. Zhu, et al., The development of an electric bus with super-capacitors as unique energy storage, in: IEEE Vehicle Power and Propulsion Conference, 2006, pp. 1–5.
[59] M. Helsper, B. Brendel, Challenges for IGBT modules in hybrid buses, in: European Conference on Power Electronics and Applications, 2009, pp. 1–8.
[60] M. Helsper, N. Ruger, Requirements of hybrid and electric buses—a huge challenge for power electronics, in: European Conf. on Power Electronics, 2014, pp. 1–12.
[61] G. Shao, C. Zhang, Research on a new motor drive control system for electric transit bus, in: IEEE Power Electronics and Motion Control Conference, vol. 2, 2006, pp. 1–5.
[62] B.J. Baliga, Silicon Carbide Power Devices, World Scientific Press, 2005 (Chapter 5 and 6).
[63] B.J. Baliga, Fundamentals of Power Semiconductor Devices, Springer-Science, New York, 2008 (Chapter 10).
[64] X. Lu, et al., Development of a bi-directional off-board level-3 quick charging station for electric bus, in: IEEE Transportation Electrification Conference, 2012, pp. 1–6.
[65] H. Zeng, S. Yang, F.Z. Peng, Design consideration and comparison of wireless power transfer via harmonic current for PHEV and EV wireless charging, IEEE Trans. Power Electron. (2017) 5943–5952.
[66] A. Weschta, Power converters with IGBTs for the new light rail vehicle for the city of Frankfurt, in: European Conference on Power Electronics and Applications, vol. 5, 1993, pp. 229–234.
[67] T. Schutze, V. Stronisch, Low floor trams with IGBT 3-level inverter, in: European Power Electronics Conference, vol. 6, 1993, pp. 92–96.
[68] C. Kehl, W. Lienau, Design of a propulsion converter with high voltage, high current IGBT for light rail vehicles, in: IEE Colloquium on IGBT Propulsion Drives, 1995, pp. 6/1–6/10.
[69] R.W. Schreyer, D.A. Dreisbach, A. Ducourret, IGBT based propulsion improving vehicle reliability and availability, in: IEEE Joint Railroad Conference, 1996, pp. 41–58.
[70] P.M. Nicolae, D.G. Stanescu, Modern urban transportation system based on a PWM inverter, in: International Symposium on Power Electronics, Electrical Drives, Automation, and Motion, 2008, pp. 1008–1012.
[71] I. Nuca, P. Todos, V. Esanu, Urban electric vehicles traction: achievements and trends, in: International Conference and Exposition on Electrical and Power Engineering, 2012, pp. 76–81.

[72] Toshiba Transportation System History List. www.toshiba.co.jp/sis/railwaysystem/en/about/history.htm.

[73] T. Ohmae, K. Nakamura, Hitachi's role in the area of power electronics for transportation, in: IEEE Conference on Industrial Electronics, Control, and Instrumentation, 1993, pp. 714–718.

[74] T. Stockmeier, et al., Reliable 1200 Amp 2500 V IGBT modules for traction applications, in: IEE Colloquium on IGBT Propulsion Drives, 1995, pp. 3/1–3/13.

[75] H.-G. Eckel, et al., A new family of modular IGBT converters for traction applications, in: IEEE European Conference on Power Electronics and Applications, 2005. pp. P.1–P.10.

[76] M.M. Bakran, H.G. Eckel, Power electronics technologies for locomotives, in: IEEE Power Conversion Conference, 2007, pp. 1362–1368.

[77] High Speed Rail. en.wikipedia.org/wiki/High-speed_rail.

[78] Shinkansen. en.wikipedia.org/wiki/Shinkansen.

[79] High Speed Rail in Europe. en.wikipedia.org/wiki/High-speed_rail_in_Europe.

[80] Korean Train Express. en.wikipedia.org/wiki/Korea_Train_Express.

[81] High Speed Rain in China. en.wikipedia.org/wiki/High-speed_rail_in_China.

[82] J. Taufiq, Power electronics technologies for railway vehicles, in: IEEE Power Conversion Conference, 2007, pp. 1388–1393.

[83] S. Dieckerhoff, S. Bernet, D. Krug, Evaluation of IGBT multilevel converters for transformerless traction applications, in: IEEE Power Electronics Specialists Conference, 2003, pp. 1757–1763.

[84] K. Sato, et al., Traction systems using power electronics for shinkansen high-speed electric multiple units, in: IEEE International Power Electronics Conference, 2010, pp. 2859–2866.

[85] Y. Uchida, et al., Development of high power press-pack IGBT and its applications, in: IEEE International Conference on Microelectronics, vol. 1, 2000, pp. 125–129.

[86] M.M. Bakran, et al., Next generation IGBT-modules applied to high power traction, in: IEEE European Conference on Power Electronics and Applications, 2007, pp. 1–9.

[87] J.G. Bauer, T. Duetemeyer, L. Lorenz, New IGBT development for traction drive and wind power, in: IEEE International Power Electronics Conference, 2010, pp. 768–772.

[88] G.J. Su, P. Ning, Loss modeling and comparison of VSI and RB-IGBT based CSI in traction drive applications, in: IEEE Transportation Electrification Conference, 2013, pp. 1–7.

[89] R. Hermann, E.U. Krafft, A. Marz, Reverse-conducting IGBTs—a new IGBT technology setting new benchmarks in traction converters, in: IEEE Power Electronics and Applications Conference, 2013, pp. 1–8.

[90] X. Huang, D. Chang, T.Q. Zheng, Investigation of the RC-IGBT application in high speed railway converters, in: IEEE Energy Conversion Congress and Exposition, 2017, pp. 2062–2067.

[91] https://www.cnbctv18.com/infrastructure/railways-gets-worlds-most-powerful-alstom-12000-bhp-locomotives-5960011.htm.

[92] Liquefied Natural Gas. en.wikipedia.org/wiki/Liquefied_natural_gas.

[93] Oil Tanker. en.wikipedia.org/wiki/Oil_tanker.

[94] P. Bhooplapur, B.P. Schmitt, G. Neeser, HV-IGBT drives and their applications, in: IEEE International Conference on Power Electronics and Drive Systems, vol. 2, 1999, pp. 585–590.

[95] G. Marina, E. Gatti, Large power PWM IGBT converter for shaft alternator systems, in: IEEE Power Electronics Specialists Conference, vol. 5, 2004, pp. 3431–3436.

[96] P. Manuelle, B. Singam, S. Siala, Induction motors fed by PWM MV7000 converters enhance electric propulsion performance, in: IEEE European Conference on Power Electronics and Applications, 2009, pp. 1–9.
[97] R. Schmerda, et al., Shipboard solid-state protection, IEEE Electrific. Mag. 1 (1) (2013) 32–39.
[98] M.A. Dornheim, Boeing's 787 Dreamliner an All-Electric Airplane. www.tourismandaviation.com/news-4624–Boeing_s_787_Dreamliner_an_all_electric_airplane.
[99] M.J.J. Cronin, The all-electric aircraft, IEE Rev. (1990) 309–311. September.
[100] A. Emadi, M. Ehsani, Aircraft power systems technology, state of the art, and future trends, IEEE AES Syst. Mag. (2000) 28–32. January.
[101] M. Howse, All electric aircraft, IEE Power Eng. (2003) 35–37. August/September.
[102] H.J. Boeing, 787 From the Gound Up, 2006, pp. 17–23. 4th QuarterBoeing.com/commercial/aeromagazine.
[103] W.G. Homeyer, et al., Advanced power converters for more electric aircraft applications, in: IEEE Energy Conversion Conference, vol. 1, 1997, pp. 591–596.
[104] L. de Lillo, et al., A 20 kW matrix converter drive system for an electro-mechanical aircraft (EMA) actuator, in: IEEE European Power Electronics and Applications Conference, 2005, pp. P1–P6.
[105] A. Garcia, et al., Reliable electro-mechanical actuators for aircraft, IEEE A&E Syst. Mag. (2008) 19–25. August.
[106] S. De, et al., Low inductance axial flux BLDC motor drive for more electric aircraft, in: IEEE Aerospace Conference, 2011, pp. 1–11.
[107] J. Chen, C. Wang, J. Chen, Investigation on the selection of electric power system architecture for future more electric aircraft, IEEE Trans. Transp. Electrif. (2018) 563–576.
[108] United Airlines to Buy 100, 19-Seat Electric Planes From Heart Aerospace. https://www.reuters.com/business/sustainable-business/united-airlines-buy-100-19-seat-electric-planes-heart-aerospace-2021-07-13/.
[109] J. Adhikari, et al., Thermal analysis of high power high voltage DC solid state power controller (SSPC) for next generation civil tilt rotor-craft, in: IEEE Int. Conf. on Electrical Systems for Aircraft, Railways, Ship Propulsion, and Road Vehicles, 2018, pp. 1–6.
[110] M.T. Fard, et al., Hybrid five-level active neutral point clamped inverter for electric aircraft propulsion drives, in: IEEE Int. Conf. Industrial Applications Society, 2020, pp. 3727–3732.
[111] Top Electric Aircraft Start-ups. Tracxn.com/d/trending-themes/Startups-in-Electrc-Aircraft.
[112] J. He, D. Zhang, D. Pan, PWM strategy for MW-Scale "SiC+Si" ANPC converter in aircraft propulsion applications, IEEE Trans. Ind. Appl. (2021) 3077–3086.
[113] C. Maget, S. Gutmann, K. Bogenberger, Model-based evaluations combining autonomous cars and a large-scale passenger drone service: the Bavarian case study, in: IEEE Int. Conf. on Intelligent Transportation System, 2020, pp. P1–P6.
[114] Y. Zeng, et al., Future demand and optimum distribution of droneports, in: IEEE Int. Conf. on Intelligent Transportation Systems, 2020, pp. P1–P5.
[115] J.W. Yun, et al., Autonomous aerial mobility learning for drone-taxi flight control, in: IEEE Int. Conf. on Information and Communication Technology Convergence, 2021, pp. 329–332.
[116] Passenger Drone. https://en.wikipedia.org/wiki/Passenger_drone.

[117] D.R. Newcombe, et al., Reliability and thermal performance of IGBT plastic modules for the more electric aircraft, in: IEEE International Symposium on Power Semiconductor Devices and ICs, 2003, pp. 118–121.
[118] H.R. Zeller, Cosmic ray induced failures in high power semiconductor devices, Solid State Electron. 38 (1995) 20141–22046.
[119] T.M. Jahns, V. Blasko, Recent advances in power electronics technology for industrial and traction machine drives, Proc. IEEE 89 (2001) 963–975.
[120] T.P. Chow, B.J. Baliga, Comparison of 300, 600, and 1200 V n-channel insulated gate transistors, IEEE Electron Device Lett. EDL-6 (1985) 161–163.
[121] B.J. Baliga, Fundamentals of Power Semiconductor Devices, second ed., Springer Science, New York, 2019, pp. 737–741 (Chapter 9).
[122] A. Adib, et al., E-mobility—advancements and challenges, IEEE Access 7 (2019) 165,226–165,240.

CHAPTER

IGBT Applications: Industrial

10

The creation of the IGBT has had an enormous impact on the industrial sector of the economy. The industrial sector is classified as the secondary sector of the economy [1]. It is responsible for the manufacturing of products using resources supplied by the primary sector. The economic output of the worldwide industrial sector is more than $30 trillion [2]. The industrial revolution began with the use of steam-powered factories but its output was greatly expanded by conversion to electrically powered factories. In modern societies, the industrial sector consumes nearly 40% of the generated electrical power [3], as shown in Fig. 10.1.

Two-thirds of the electricity in the United States is used to operate motor-driven equipment [4]. Industrial applications for motors [5] include: (1) the light and medium industries such as auto parts assembly, food production, semiconductor manufacturing, and light machining; (2) the process industries such as pulp and paper mills, chemical plants, oil refineries, and steel mills; and (3) the heavy industries such as mining, oil and gas production, and power plants.

10.1 Industrial Motor Drives

IGBTs have found widespread applications in the industry [6]. A typical large industrial motor is shown in Fig. 10.2. Squirrel-cage induction motors are dominant for AC industrial drives around the world due to their ruggedness and low cost. Three-phase voltage source inverters are the overwhelming favorite topology for industrial drive application rated for less than 2 MW. Jahns and Blasko state [6]: "During the past decade, IGBTs packaged in compact plastic power modules have rapidly expanded their voltage and current ratings so that they now dominate nearly all of the new industrial drive inverter designs.…In fact, many new IGBT inverters are designed without any snubbers at all. This approach has the advantage of saving cost, space, and losses in the inverter—all positive effects." For industrial systems with power ratings of up to 25 MW, "IGBTs have become completely dominant for low-voltage converters and are increasingly used for medium-voltage converters" according to Siemens [7].

The benefits of the IGBT for industrial applications have been recognized now for several decades [8]. They include: (1) low power input MOS drive; (2) low switching loss, high-speed commutation; (3) low steady-state on-state losses; (4) ease of paralleling devices; and (5) rectangular reverse-biased safe operating area

The IGBT Device. https://doi.org/10.1016/B978-0-323-99912-0.00021-0

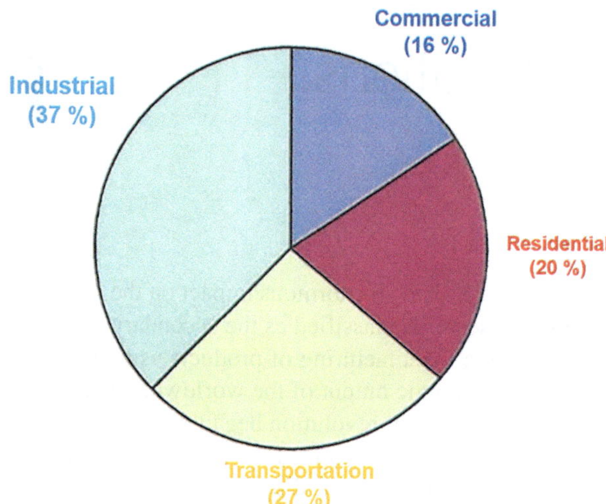

FIGURE 10.1

Electrical energy consumption by economic sector.

FIGURE 10.2

Typical large industrial motor.

(RBSOA) with no second breakdown problem. This makes them ideal for voltage source induction motor control.

10.2 Adjustable Speed Drives for Motor Control

Most applications utilize induction motors due to their simple, low-cost construction with highly reliable operation for long periods of time. An induction motor operates at a fixed rotational frequency (rpm) that is a multiple of the AC frequency of the input power supplied to its windings [9]. The control of any process operated using

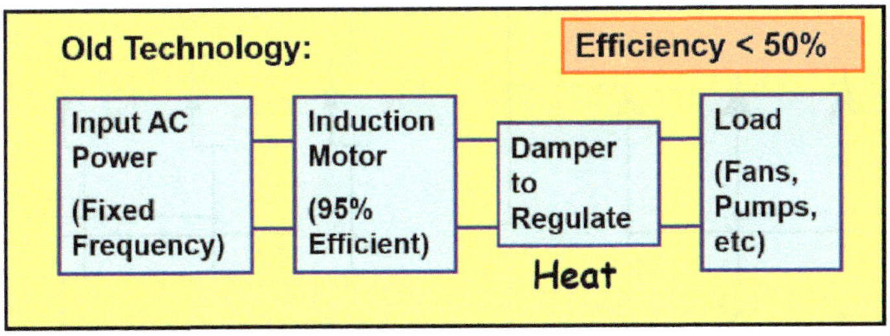

FIGURE 10.3

Induction motor drive with control using dampers.

the motor must be accomplished by using dampers, as illustrated in Fig. 10.3. The damper regulates the output by converting the rest of the power delivered by the motor into wasted heat. Although the efficiency for conversion of electric power to mechanical rotation by the induction motor is very high (~95%), the overall efficiency of the system falls to below 50% because of conversion of energy into heat by the damper when it regulates the output of pumps or fans.

The adjustable speed drives (ASDs), also referred to as the variable speed drives (VSDs) and variable frequency drives (VFDs), are based on controlling the output speed of motors by supplying them with a variable frequency input power. This requires conversion of the 60 Hz (or 50 Hz in some countries) AC power supplied by the electric utilities into variable-frequency AC power, as illustrated in Fig. 10.4. The operating efficiency of the ASD depends upon the efficiency of the inverter stage that is used to convert the input 60 Hz power to the variable-frequency output power. ASDs were transformed into an affordable and dependable technology after the availability of the IGBTs [10]. In fact, the first application for IGBTs developed at the General Electric Research Center in 1982 was ASDs for use in heat pumps for air conditioning. This work was described at the Motor Drives Conference in October 1983 [11]. Since that time, ASDs have been widely utilized to improve efficiency in the residential and commercial sectors [3].

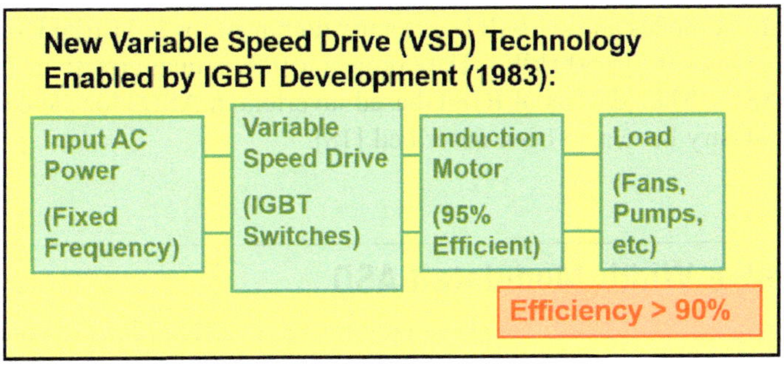

FIGURE 10.4

Induction motor drive using adjustable speed control.

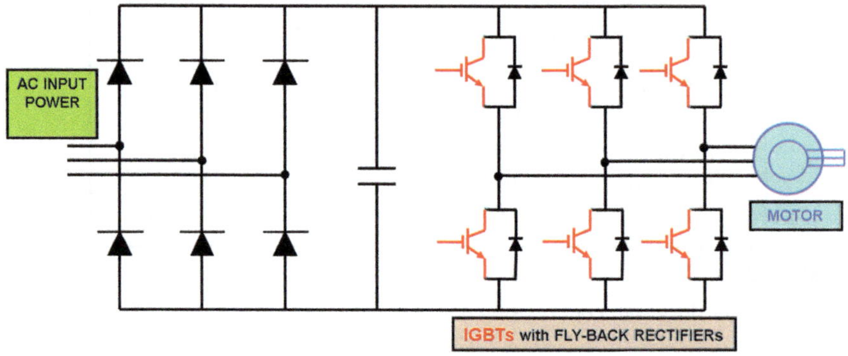

FIGURE 10.5

Typical adjustable speed drive using insulated gate bipolar transistors (IGBTs).

VFD technology has been extensively discussed in many books [12–14]. The most commonly used circuit topology [15] is shown in Fig. 10.5. It consists of an input rectifier stage to convert the fixed-frequency input power delivered by the utility to a DC source across the capacitor. The variable-frequency output power fed to the three-phase motor windings is then synthesized using an inverter stage containing IGBTs and fly-back rectifiers. The high power-handling capability of the IGBT with snubberless operation enabled the development of compact, low-cost, and reliable adjustable speed motor drives. A large number of publications have described the use of ASDs based upon IGBTs in the literature. In tracing the history of motor drive technology, Sawa and Kume state [16]: "Variable speed motor drives require power electronics and it can be said that this work has contributed to the prosperity of today's world-wide economy and allows people to enjoy a more complete social life.... The voltage source type PWM control IGBT inverters with 15 kHz carrier frequency was produced in 1990...After this, IGBTs were widely used and became the standard power devices for general purpose inverters. This continues to this day."

The significant reduction (more than 10 times) in the complexity, size, and weight of ASDs derived after the availability of the IGBT has been documented [17]. Blaabjerg and Thoegersen state: "Presently, the Voltage Source Inverter architecture with six diodes and six IGBTs is totally dominant for Industrial Drives in the power range from 0.5 to 500 kW." They project a 10% annual growth in the market for these ASDs. The benefits of IGBT-based inverters in ASDs for the cement and minerals industry have also been described [18].

10.3 Pulse Width Modulated ASD

The basic adjustable-speed motor drive requires rectification of the available 60 Hz (or 50 Hz in some countries) utility power supply to a DC power bus and subsequent synthesis of sinusoidal power of variable (or adjustable) frequency that can be fed to the winding of the induction motor. The speed of rotation (n_S) of the induction motor (synchronous speed) is given by:

$$n_S = \frac{120f}{P} \quad (10.1)$$

where f is the frequency of the sinusoidal power source and P is the number of poles in the motor. Consequently, the rotational speed for the motor can be adjusted if the frequency of the sinusoidal power source can be adjusted.

A sinusoidal power source can be generated from the DC power bus by using the basic H-bridge circuit shown in Fig. 10.6 for one of the motor windings. If a current (I_M) is flowing through the motor winding in the direction illustrated in the figure during the positive half cycle of the sinusoidal waveform, its magnitude can be increased by turning on IGBT-1 and IGBT-4 for a short time. In contrast, the motor current can be reduced by turning on IGBT-3 and IGBT-2 for a short time.

A sinusoidal waveform for the motor current can be produced with pulse width modulated (PWM) gating of the IGBTs using a triangular carrier wave superimposed on a reference sinusoidal waveform of the desired frequency [19]. This is illustrated in Fig. 10.7 for the positive half cycle of the sinusoid. The triangular carrier wave is chosen to have a much higher frequency, which is a multiple of the frequency of the sinusoid. The PWM signals are generated at the intersection points between the triangular carrier waveform and the sinusoidal reference waveform.

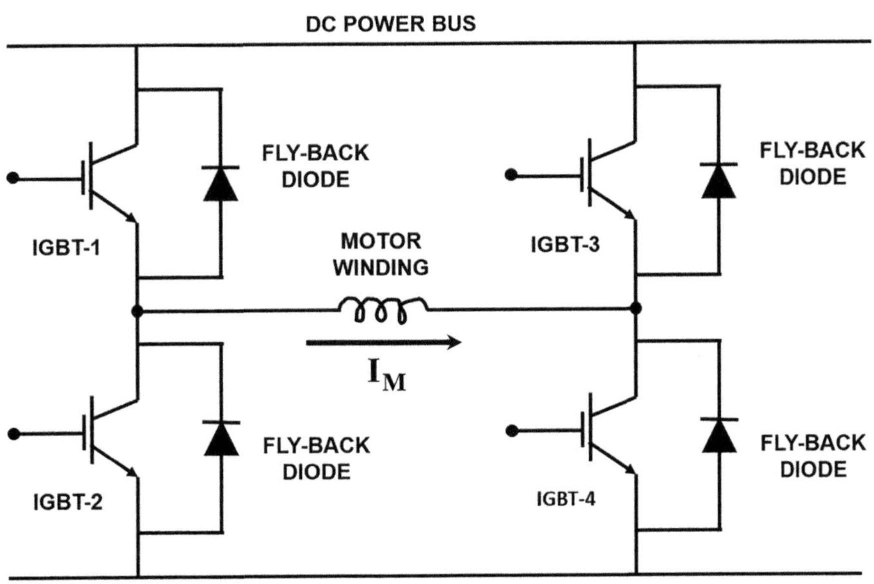

FIGURE 10.6

Basic H-bridge circuit topology.

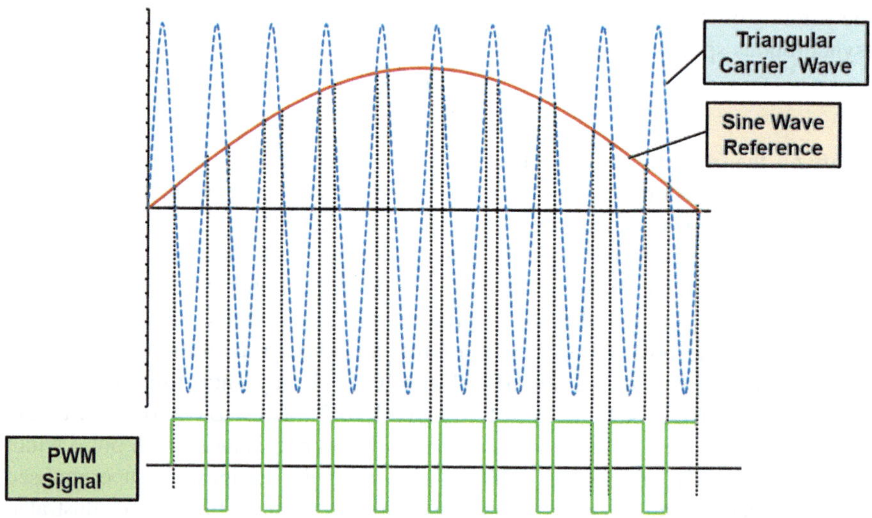

FIGURE 10.7

Sinusoidal pulse width modulation.

10.3.1 PWM Waveforms

The current and voltage waveforms for the IGBT and the fly-back diode (P-i-N rectifier) are illustrated in Fig. 10.8 for one cycle of the PWM carrier wave. These waveforms have been linearized for simplification of the analysis [20]. Prior to time t_1, the IGBT is supporting the DC supply voltage and the motor current is flowing through a fly-back diode. The cycle begins at time t_1 when the IGBT is turned on by its gate drive voltage. When the IGBT gate bias is applied, the motor current is transferred from the fly-back diode to the IGBT during the time interval from t_1 to t_2. P-i-N rectifiers are not able to support voltage until the stored charge in the drift region is removed [21]. The P-i-N rectifier must first undergo its reverse recovery process. A substantial reverse current flow occurs through the rectifier with a peak value I_{PR} reached at time t_2 during reverse recovery. The current in the IGBT at time t_2 is the sum of the motor winding current I_M and the peak reverse recovery current I_{PR}. This produces substantial power dissipation in the IGBT during the turn-on transient. The power dissipation in both the transistor and the diode are therefore governed by the reverse recovery characteristics of the power rectifier.

The IGBT is turned off at time t_4, allowing the motor current to transfer from the transistor to the diode. In the case of an inductive load, such as motor windings, the voltage across the transistor increases before the current is reduced, as illustrated in Fig. 10.8, during the time interval from t_4 to t_5. Subsequently, the current in the transistor reduces to zero during the time interval from t_5 to t_6. The turn-off durations are governed by the physics of the transistor structure, as discussed in Chapter 3.

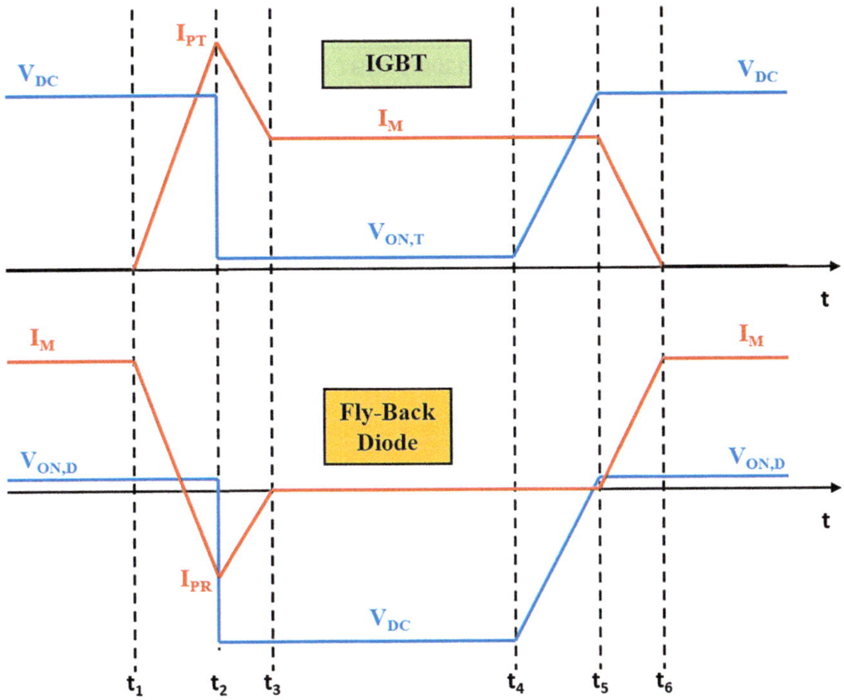

FIGURE 10.8

Single pulse width modulated (PWM) carrier cycle switching transients for the IGBTs.

Consequently, the power dissipation in both the transistor and the diode during the turn-off event are determined by the transistor switching characteristics.

10.3.2 Power Loss Trade-Off Curves

In addition to the power losses associated with the two basic switching events within each cycle, power losses occur within the diode and the IGBT during their respective on-state operation due to a finite on-state voltage drop. It is common practice to trade off a larger on-state voltage drop to obtain a smaller switching loss in the bipolar power devices. Consequently, the on-state power loss cannot be neglected, especially if the operating frequency is low. The leakage current for the devices is usually sufficiently small so that the power loss in the blocking mode can be neglected.

The trade-off curves for energy loss per cycle and the on-state voltage drop, for the three types of IGBT structures discussed in Chapter 3 with 1200 V forward blocking capability [22], are shown in Fig. 10.9 using an on-state current density of 100 A/cm^2. From the figure, it can be observed that the lowest combination of on-state voltage drop and energy loss per cycle is obtained for the asymmetric IGBT

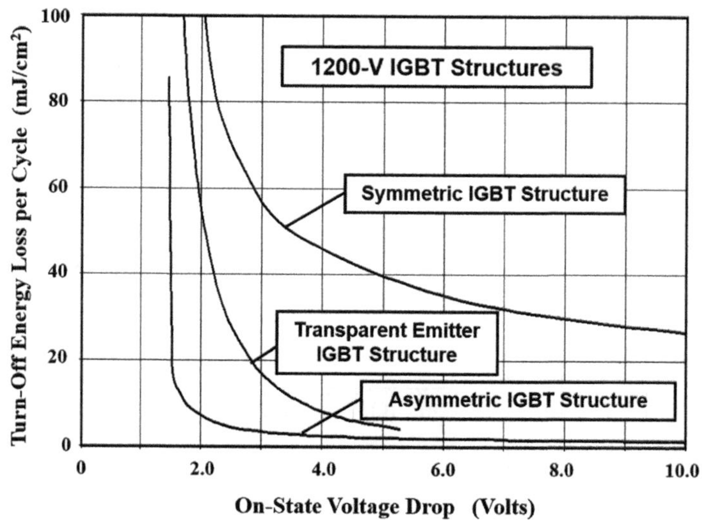

FIGURE 10.9

Trade-off between on-state and switching loss for 1200V IGBTs.

structure. This design is therefore most commonly employed for industrial motor drives.

A similar trade-off curve between energy loss per cycle and on-state voltage can be generated using the information provided in Chapter 3 for the 3000 V structures. The trade-off curves for the symmetric, asymmetric, and transparent-emitter IGBT structures are provided in Fig. 10.10 using an on-state current density of 100 A/cm^2 and a collector supply voltage of 2000 V. The on-state voltage drop of these structures is larger than that for the corresponding 1200 V devices because of their thicker drift regions. The energy loss per cycle for these devices is larger by an order of magnitude because of the increased stored charge and the larger collector supply voltage. These devices are of course also controlling a much larger power level at the load. It is clear that the best trade-off curve is obtained using the asymmetric IGBT structure. This has been the main focus for device development by the industry.

Historically, many papers have been published by various companies to highlight improvements in the trade-off curve between on-state voltage drop and energy loss per cycle resulting from structural innovations for the IGBT. Some examples are 600 V IGBT structures by the Fuji Electric Company [23] and Toshiba Corporation [24]; 1200 V IGBT structures by Fairchild Semiconductor [25]; and 6500 V IGBTs by ABB Semiconductors Limited [26].

A good example of relating the various IGBT structures to the trade-off curves has been published by Fuji for devices with 600–800 V blocking capability [27]. They have compared the conventional IGBT structure with a structure made with

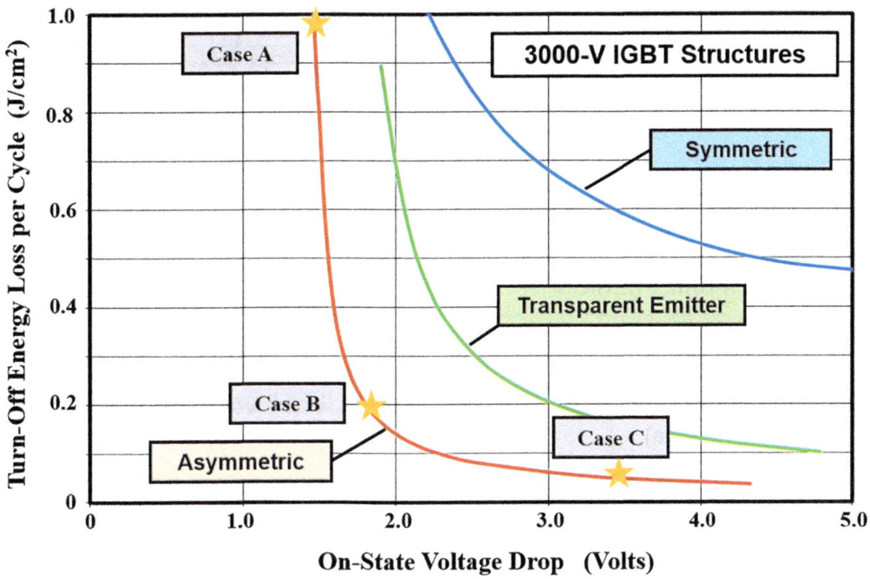

FIGURE 10.10

Trade-off between on-state and switching loss for 3000V IGBTs.

a fine structural pattern (shallower junctions and lithography dimensions). In addition, they have compared these structures with the trench gate device.

The conventional asymmetrical (punch-through (PT)) structure for the Fuji Electric IGBT, given in Fig. 10.11(a), is similar to that shown in Chapter 3 with a cell pitch of about 30 μm. The fine pattern Fuji IGBT structure has shallower junctions and a smaller cell size, as shown in Fig. 10.11(b). The shallower junctions allow reduction of the voltage drop across the JFET and channel regions of the IGBT, making the overall on-state voltage drop smaller than for the conventional IGBT structure. It was found that the on-state voltage drop can also be reduced by increasing the JFET doping concentration consistent with the analytical model given in Chapter 3. The optimum JFET surface concentration was found to be $1 \times 10^{16}/cm^3$ beyond which the breakdown voltage begins to degrade [25]. The same fine pattern cell structure has also been used by Fuji to make a nonpunch-through (NPT) structure, as shown in Fig. 10.11(c), with wafer thickness reduced to 100 μm. The Fuji IGBT with the trench gate structure is shown in Fig. 10.11(d). This device was fabricated with a cell pitch of 4.5 μm, a trench depth of 8 μm, and a trench width of 1 μm. This creates a much larger channel density than the conventional or even fine pattern planar IGBT structures, and eliminates the JFET component leading to a reduction in its on-state voltage drop.

The trade-off curve between the on-state voltage drop and energy loss per cycle was performed for the above Fuji IGBTs using a collector supply voltage of 300 V and a collector current of 75 A [23]. This current corresponds to an on-state current

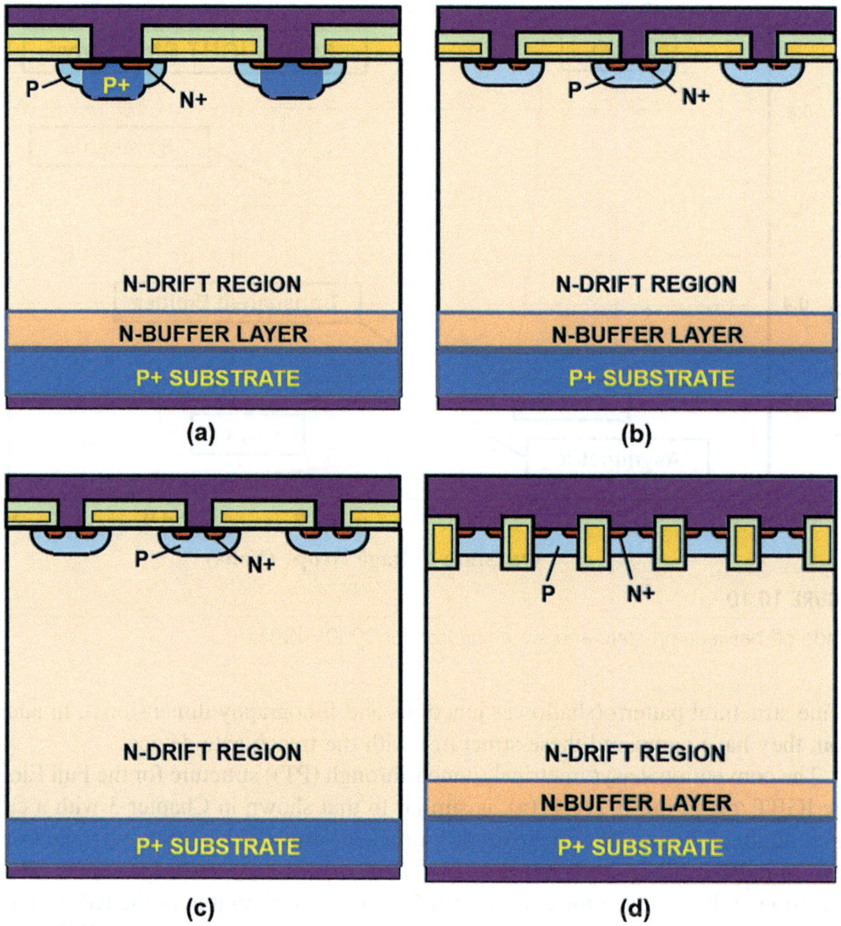

FIGURE 10.11

Fuji electric IGBT structures.

density of 130 A/cm^2. A comparison of the performance of these IGBT structures is shown in Fig. 10.12. For devices with the same blocking voltage rating (780–820 V), it can be observed that a significant reduction (~0.5 V) in on-state voltage drop is obtained for the same switching energy loss by replacing the conventional PT IGBT structure ((a) 820 V) with the fine pattern PT IGBT structure ((b) 820 V). A further reduction in on-state voltage drop by 0.25 V is obtained by replacing the fine pattern PT IGBT with the trench gate PT IGBT structure. The NPT IGBT structure ((c) 730 V dashed line in the figure) exhibits the same trade-off curve as the trench-gate PT IGBT structure, but its blocking voltage is slightly lower. An improvement in the on-state voltage drop of 0.4 V at the same switching loss can also be obtained for the fine pattern PT IGBT structure by reducing its blocking voltage to 630 V. It is

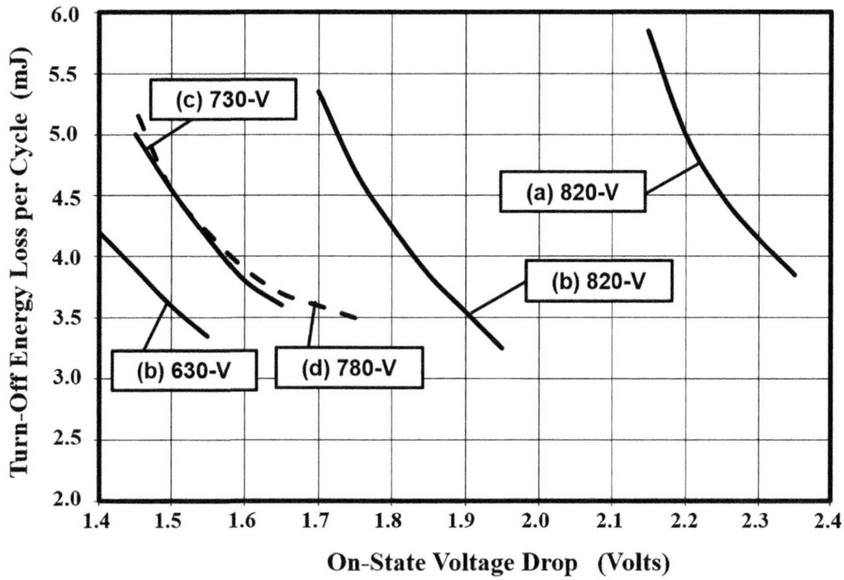

FIGURE 10.12

Trade-off between on-state and switching loss for Fuji electric IGBTs.

therefore important not to overdesign the blocking voltage capability for these applications.

10.3.3 Power Loss Analysis

The total power loss incurred in the power transistor can be obtained by summing four components: (1) turn-on power loss during time t_1 to t_3 in Fig. 10.8; (2) on-state power loss during time t_3 to t_4 in Fig. 10.8; (3) turn-off power loss during time t_4 to t_6 in Fig. 10.8; and (4) off-state power loss during the rest of the period of the PWM carrier wave. The turn-on power loss depends on the reverse recovery characteristics of the fly-back rectifier. For simplicity, it will be assumed here that the turn-on loss is equal to the turn-off loss. In addition, the off-state loss will be neglected under the assumption that the leakage current of the IGBT is small. The total power loss is then given by:

$$P_L(total) = P_L(on) + P_L(switching) \quad (10.2)$$

If the PWM waveform is operating at a duty cycle of 50%, the power loss incurred in the IGBT is given by:

$$P_L = [0.5{*}V_{ON}{*}J_{ON}] + [E_{OFF}{*}f] \quad (10.3)$$

where V_{ON} is the on-state voltage drop, J_{ON} is the on-state current density, E_{OFF} is the switching energy loss per cycle, and f is the frequency of the PWM carrier wave.

One of the features of the IGBT that was recognized at the time of conception was that it can be optimized for each application depending on the operating frequency [11]. This can be illustrated here by considering three asymmetric IGBT options shown in Fig. 10.10 as cases A, B, and C by the star symbols. The asymmetric IGBT structure for case A has a low on-state voltage drop and a high switching loss per cycle. The asymmetric IGBT structure for case C has a low switching loss per cycle and a high on-state voltage drop. The asymmetric IGBT structure for case B falls between these structures and may appear to be the best design by only examining the trade-off curve. The total power loss when operating with a collector supply voltage of 2000 V and an on-state current density of 100 A/cm^2 is shown in Fig. 10.13 for the three structures. The power loss for the IGBT case A is the lowest only at very low operating frequencies (below 10 Hz). At an operating frequency of 30–400 Hz, the IGBT structure B produces the lowest total power loss. At frequencies above 1000 Hz, the IGBT structure C has the lowest total power loss. It can be concluded that minimizing switching energy loss becomes more important than retaining a low on-state voltage drop for high operating frequencies.

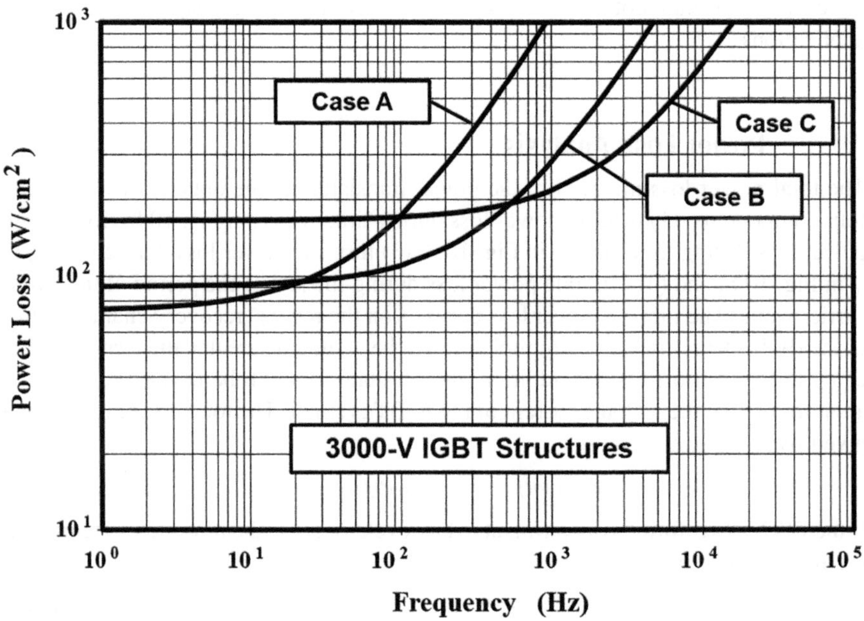

FIGURE 10.13

Power loss in IGBTs with increasing frequency.

10.4 Factory Automation

Factory automation is defined as the use of control systems for the operation of machinery and processes [28]. It is widely employed in the manufacturing of chemicals, plastics, fertilizers, paper products, automobile assembly, aircraft production, and food processing, as some examples. The benefits of automation include increased productivity, improved quality and consistency, and reduced human labor costs. The use of power electronic controllers with microprocessors allows for numerical control of industrial machinery. Numerical controls are defined as the automation of machine tools with precise computer-programmed commands [29]. Computer numerical control systems are commonly used today for performing a series of precise movements and operations such as laser cutting, welding, bending, gluing, sewing, placement, and routing.

At General Electric (GE), the development of numerical controls was initiated in 1982 using the IGBT to create a compact, low-cost, rugged, and reliable platform. The deployment of a numerical controls product, called the Genius I/O, for factory automation by GE was commemorated in 1986, as shown in Fig. 10.14(a). This product is still available from GE, as shown in Fig. 10.14(b). The GE website [30] states: "By providing distributed control on the factory floor, Genius I/O systems offer fewer terminations to document, dramatically shorter wiring runs, and simpler, more effective troubleshooting. Genius I/O blocks automatically provide diagnostic information on field wiring, power conditions and loads, as well as the state of the communication network, blocks and circuits. Genius diagnostics sharply reduce the time needed for initial control and debugging." These capabilities required integration of current sense capability into the IGBT chip, as discussed in Chapter 5.

(a) (b)

FIGURE 10.14

GE genius I/O product (b) using complementary IGBTs (a).

10.4.1 Complementary IGBTs

In numerical controls, it is necessary to control and monitor the current flow during both half-cycles of the AC power cycle. This can be performed by using two *n*-channel IGBTs, as illustrated in Fig. 10.15(a). One of the IGBTs (IGBT T_1 in the figure) can be driven using a medium-voltage integrated circuit because its emitter (E_1) is at the ground reference point. However, the potential at emitter (E_2) of IGBT-2 fluctuates as the AC voltage increases. The gate bias required to turn-on IGBT-2 must exceed the AC potential. This requires a high-voltage integrated circuit to supply the gate drive voltage for IGBT-2. It is preferable to create a complementary pair of transistors to enable control of power during both half-cycles of the AC power cycle without the complexity of creating a gate drive circuit whose potential was tied to the source voltage.

Fortunately, analysis of the IGBT in 1983 indicated that, unlike power metal oxide semiconductor field effect transistors (MOSFETs), *p*-channel devices would have nearly equivalent on-state characteristics to *n*-channel devices, making the IGBT an excellent candidate for creating a complementary pair. The first *p*-channel IGBTs were successfully built directly in the GE semiconductor manufacturing facility in 1984 [31]. This work provided the first experimental demonstration of excellent characteristics for complementary IGBTs.

10.4.2 *p*-Channel IGBT Design

The *p*-channel IGBT structure is shown in Fig. 10.16(a) with its equivalent circuit in Fig. 10.16(b). This IGBT structure is identical to that for the *n*-channel device with all the semiconductor layers changed to their complements. The equivalent circuit for the *p*-channel IGBT is a *p*-channel MOSFET providing base drive current for the wide-base N-P-N transistor. If the IGBT is modeled as a P-i-N rectifier with a MOSFET in series [22], the on-state voltage drop of the *p*-channel device can be

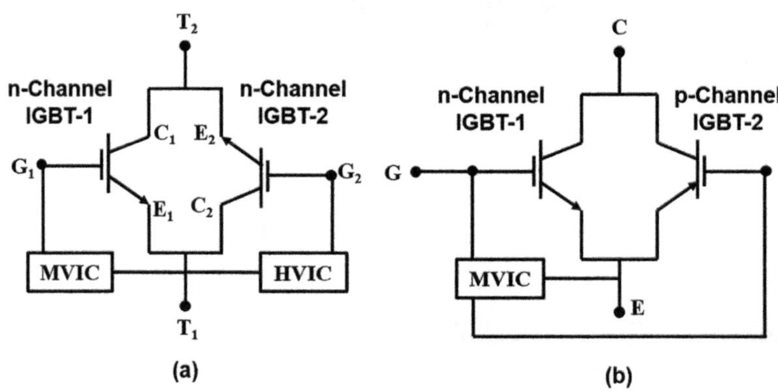

FIGURE 10.15

Gate drive for: (a) *n*-channel IGBTs and (b) complementary IGBTs.

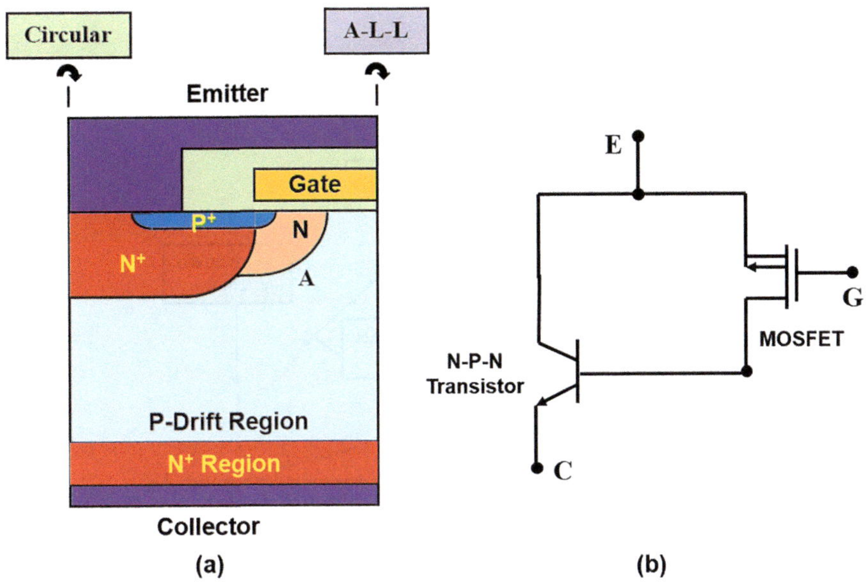

FIGURE 10.16

p-Channel IGBT structure (a) and its equivalent circuit (b).

expected to be only slightly larger than that for the *n*-channel IGBT structure. This is in contrast to the threefold increase in on-resistance of *p*-channel MOSFETs when compared with *n*-channel MOSFETs.

The first 600 V *p*-channel IGBTs were manufactured [31] at GE in 1984. It was found that the on-state voltage drop for the *p*-channel IGBTs was actually slightly lower than that for the *n*-channel counterparts. This is because of the larger gain of the N-P-N transistor within the *p*-channel IGBT, which overcompensates for the larger channel resistance. These devices were developed for use in the GE Genius I/O numerical controls.

One of the differences between the *p*-channel and *n*-channel IGBT structures is the safe operating area (SOA). The forward biased safe operating areas (FBSOAs) for the structures are compared in Fig. 10.17. The latch-up current level at low collector bias voltages is much higher for the *p*-channel IGBT because the sheet resistance of its *N*-base region is smaller than that of the *n*-channel structure due to the larger mobility for electrons. However, its FBSOA limit at high collector bias voltages is much inferior to that of the *n*-channel IGBT structure because of the larger impact ionization coefficient for electrons that transit its *P*-base region.

An improvement in the FBSOA and RBSOA for the *p*-channel IGBT structure can be achieved by reduction of the electric field within the device cell structure to suppress impact ionization. In the typical linear cell (corresponding to no rotation of the cell cross-section in Fig. 10.16(a)) and the circular cell layout (corresponding to rotation of the cell cross-section in Fig. 10.16(a) at the left edge), an enhanced

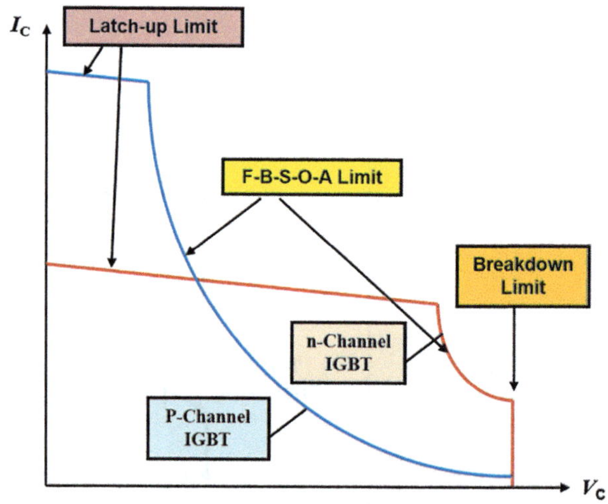

FIGURE 10.17

Forward-biased safe operating area (FBSOA) of complementary IGBT structures.

electric field is produced at location A, which reduces the SOA boundary. The layout geometry for the polysilicon gate layer in the circular cell topology design is shown in Fig. 10.18(a). This problem can be ameliorated by rotation of the cell cross-section in Fig. 10.16(a) at the right edge. The polysilicon gate layer for this cell topology is shown in Fig. 10.18(b) and referred to as the atomic lattice layout [32,33]. This topology creates a saddle junction in the IGBT cell, which reduces the electric field at location A (see Fig. 10.16(a)). The measured RBSOA of *p*-channel IGBTs has been found to be expanded by a factor of 2 with this design [34].

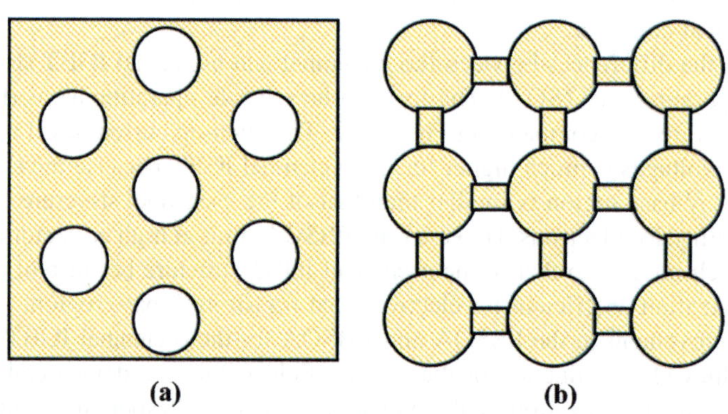

FIGURE 10.18

Chip design topology for *p*-channel IGBT: (a) circular and (b) atomic lattice layout.

10.5 Robotics

Robots are in widespread use in assembly lines for building products ranging from automobiles to washing machines. Industrial automation is now heavily reliant on robots. Typical applications for factory robots are welding, painting, drilling, assembly, packaging, palletizing, and material handling. They improve the quality and speed of production while reducing cost in warehouses, shops, and facilities worldwide. Industrial robots were first installed in 1961 to perform a die-casting application in an automotive plant. Modern factory robots are fully integrated into the factory operations. Robots performing the tasks of palletizing and assembly are shown in Fig. 10.19(a) and Fig. 10.19(b), respectively.

10.5.1 Cableless Power Supply

Robots require a wide range of motions to perform a greater variety of tasks. The motion can be impeded by the movement of power supply and communication cables. An ingenious solution to this problem is a rotatable transformer [34] shown in Fig. 10.20. The transformer size and weight can be reduced by using an AC-to-AC converter to generate a rectangular voltage 25 kHz power source from the 60 Hz AC power main supply as shown by the block diagram in the figure. A variable-frequency AC voltage is then generated at the secondary side of the transformer by using PWM signals to drive the motors for axial control of the robot. The secondary of the rotatable transformer is also connected to the primary of the next rotatable transformer to control the axial position of the next joint of the robot. This is repeated to generate full axial position control capability for the robot.

The IGBT-based bridge converter used with the rotatable transformer is shown in Fig. 10.21. Each bridge converter contains four IGBTs with fly-back diodes. A series resonant circuit is created in the converter by addition of the resonant capacitors C_R.

(a) (b)

FIGURE 10.19

Industrial robots: (a) KUKA robots palletizing; (b) ABB robots welding cars.

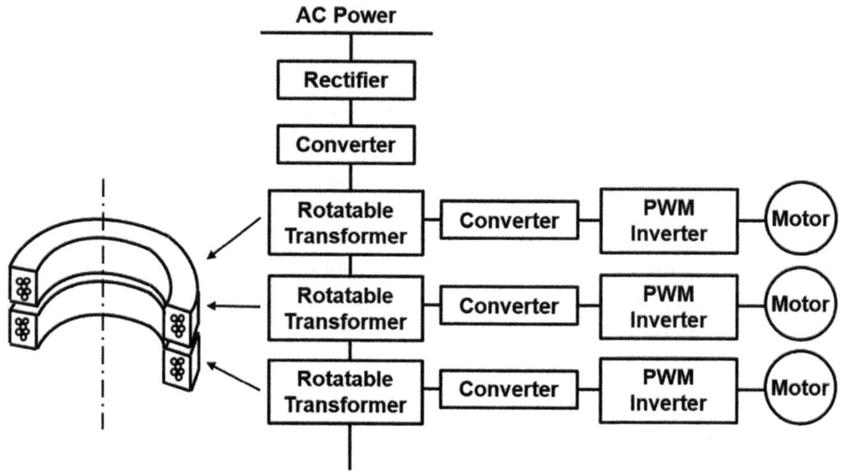

FIGURE 10.20

Rotatable transformer-based robotic drive.

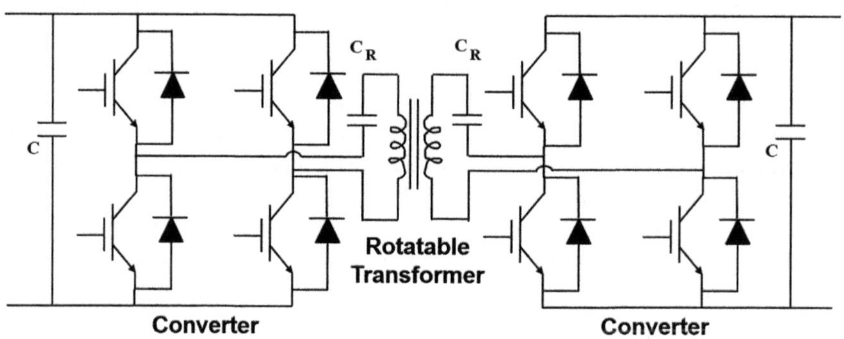

FIGURE 10.21

Bridge converter topology used with the rotatable transformer.

The switching losses in the IGBTs are reduced by performing zero-current switching to allow operation of the circuit at a high frequency (25 kHz). The IGBT modules have ratings of 1000 V and 50 A.

10.5.2 Industrial Robot Controller

Industrial robots are commonly used to perform machining operations. They are controlled by using a system-level controller and a servo-loop controller. The servo controller has been found to be the main bottleneck determining the precision of the robot operations [35]. The improvement in performance of the servo controller for the Cincinnati Milacron Inc. T3-776 heavy-duty industrial robot achieved by

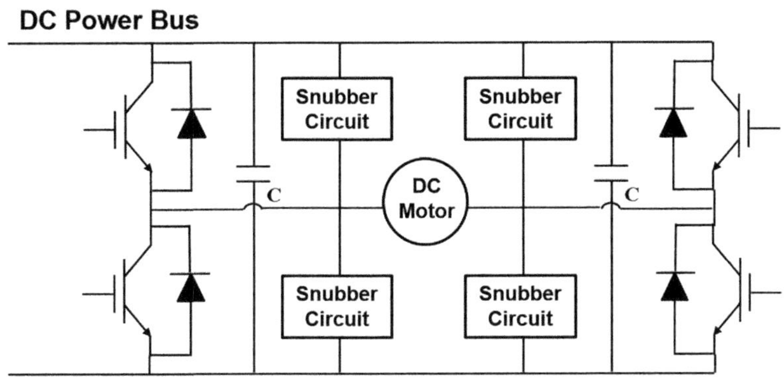

FIGURE 10.22

IGBT-based PWM amplifier for robot servo amplifier.

replacing the silicon-controlled rectifier (thyristor)-based drive with an IGBT-based drive has been documented. This large industrial robot has six degrees of freedom and is capable of lifting and positioning a 150-pound load with 0.01-in. repeatability.

The thyristor-based servo amplifiers create 180 Hz harmonics in the motor outputs that are superimposed on the desired motion of the robot. This problem was overcome using the IGBT-based amplifiers [36] shown in Fig. 10.22. The PWM driver contains a comparator, a triangular wave generator, and gate drive circuits for the IGBTs. A 20 V floating power supply is required to drive the upper IGBTs. Protective circuits are included in the driver for undervoltage and short circuit faults. The PWM circuits can be switched at up to 16 kHz. The IGBT-based controller eliminated the 180 Hz pulsations observed with the thyristor controller and improved the robot response time by a factor of three.

10.5.3 Linear Actuators

Electromagnetic linear actuators are used to create and regulate motion in a straight line [36]. They are employed in machine tools and other manufacturing machinery. The construction of a linear actuator is illustrated in Fig. 10.23(b). The linear actuator can be driven using the circuit [37] shown in Fig. 10.23(a). An energy storage capability is included using the capacitor and a Zener diode. The capacitor charging is regulated using an IGBT. The H-bridge containing IGBTs and the fly-back diode is used with PWM signals to produce fast movement of the actuator.

10.5.4 Mobile Gantry Crane Robots

Rubber-tiered gantry (RTG) cranes are used for movement, loading, and stacking of large, heavy containers at shipping terminals and storage yards. The shipping containers can weigh as much as 50 tons. The RTG cranes are powered using a diesel generator to provide the electricity via inverters for variable-speed AC motors. A

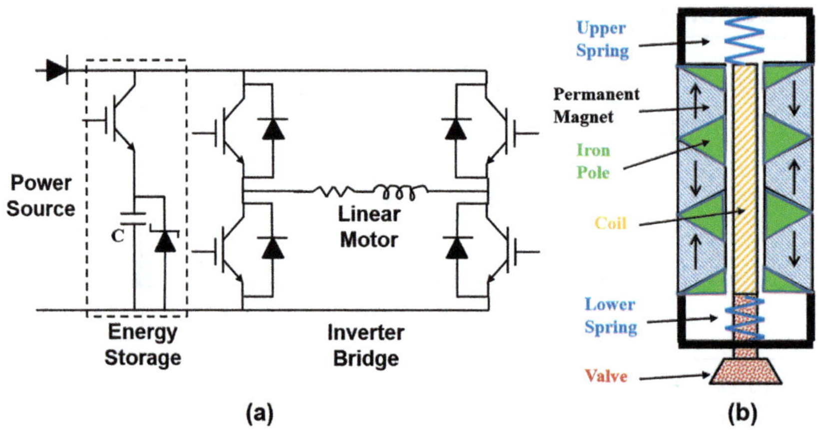

FIGURE 10.23

IGBT-based linear motor actuator.

hoist motor is used to move the containers up and down, a trolley motor is used to move the container from one side of the crane to the other, and a gantry motor is used to move the crane across the shipyard. Significant energy is consumed when lifting the heavy containers. When lowering the containers, the potential energy is sent to a resistor bank and converted into waste heat.

The wasted energy can be saved using a flywheel energy storage system [38] or a kinetic energy recovery system (KERS). A block diagram for the drive is shown in Fig. 10.24. During the lift operation, the diesel generator set provides energy to the hoist motor via the IGBT-based inverter drive. During lowering of the containers, the energy can be fed either into the brake resistor via the brake chopper or stored in the

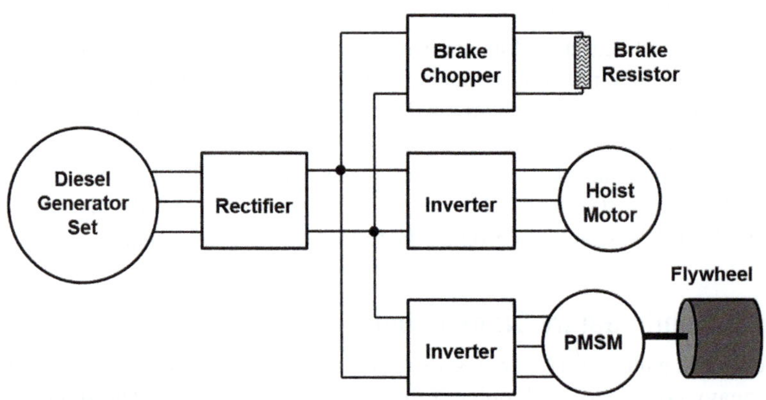

FIGURE 10.24

Kinetic energy recovery system for mobile gantry cranes.

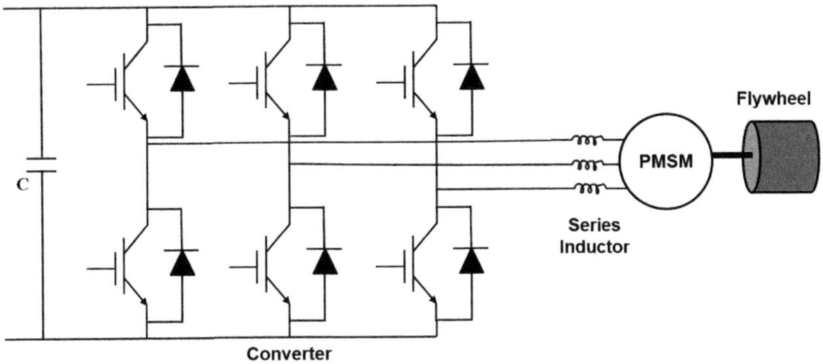

FIGURE 10.25

Kinetic energy recovery system using IGBT converter.

flywheel using the KERS. The KERS circuit diagram is shown in Fig. 10.25 where six IGBTs are used to control the transfer of energy to the flywheel by determining its speed of rotation. The external inductors are used to reduce harmonics and permanent magnet synchronous motor (PMSM) rotor losses.

10.6 Welding

Welding is the process of joining materials by melting the workpieces [39]. Robot welding is commonplace in industrial plants such as those used to manufacture automobiles. Welding can be performed using either consumable or nonconsumable electrodes (tungsten inert gas—TIG welding). The welding can be performed with an inert ambient (referred to as MIG for metal inert gas using argon or helium) or in the presence of some oxygen (referred to as MAG for metal active gas using argon/oxygen or argon/carbon dioxide mixtures). The electrical arc that generates the high temperatures needed to melt the metals can be sustained using either constant current or constant voltage power supplies. The length of the arc is determined by the voltage while the heat generated by the arc depends on the current. Constant current power supplies are favored for manual welding because it is difficult to manually hold the electrode at a steady distance from the workpiece. Constant voltage power supplies are favored for automated welding because the separation between the electrode and the workpiece can be precisely controlled. A typical power supply for arc welding is shown in Fig. 10.26(a). A power supply for tube or pipe orbital welding is shown in Fig. 10.26(b). Many manufacturers specifically tout the use of IGBT-based power supplies for welding. Some examples are the IGBT-MIG, IGBT arc welding machine, and MIG/MAG welding machine with IGBT power supplies produced in China [40–42]; and the IGBT-based TIG welding machine and inverter IGBT controlled MIG produced in India [43,44].

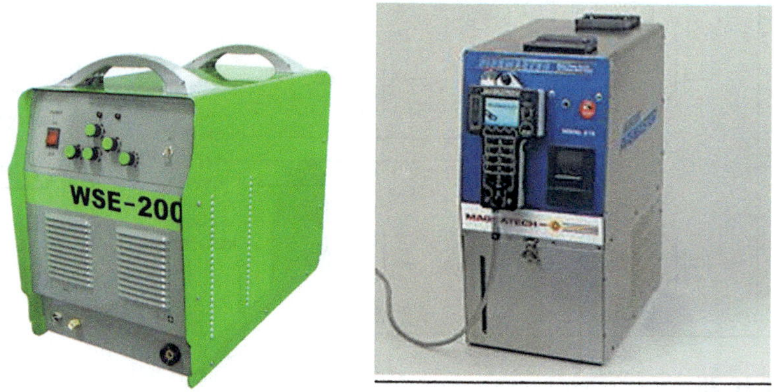

(a) (b)

FIGURE 10.26

Welding power supplies: (a) arc; (b) pipe.

10.6.1 Step-Down Buck Converter

As mentioned earlier, the quality of the weld is dependent on the arc current. Typically, the arc welder power supply needs to output a maximum current of 160 A at 50 V. A simple buck converter, shown in Fig. 10.27, implemented using an IGBT has been used to develop control of the arc current [45] during manual welding. The power circuit operating at 8 kW (40 V, 200 A) is suitable for use with welding machines. The peak IGBT voltage during operation is close to 300 V.

10.6.2 Transformer-Coupled Power Supply

A transformer-coupled arc welding power supply [46] circuit is shown in Fig. 10.28. The 300 V DC power source is produced from the rectified AC mains power. The desired voltage (50 V) at the arc is obtained using a transformer. A high-frequency (80 kHz) transformer design allows reducing the size and weight of the magnetics. High-frequency operation is achieved by switching the IGBT with a low duty cycle of about 10% to prevent excessive heating.

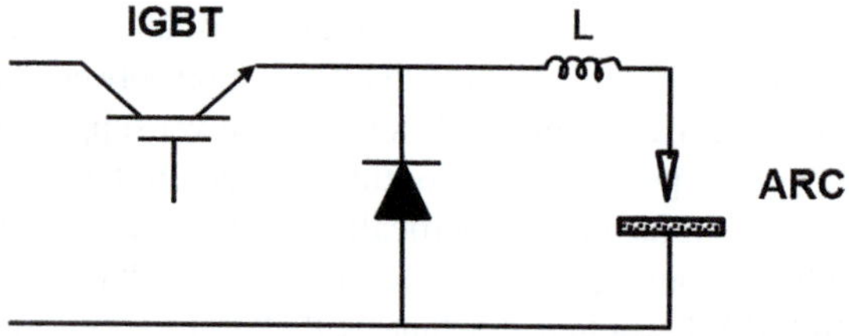

FIGURE 10.27

IGBT-based buck converter for arc welding.

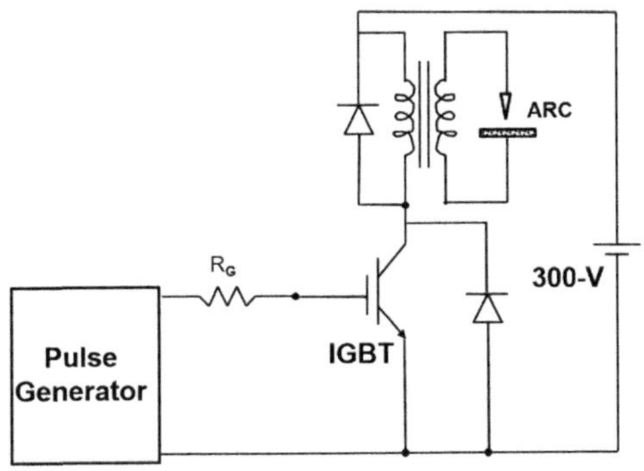

FIGURE 10.28

Transformer-coupled IGBT-based arc welding power supply.

10.6.3 Dual Utility Power Supply

Manufacturers of welding power supplies must serve a global market to increase product sales. The industrial utility power distribution in Japan, Korea, and Taiwan is done using 200–220 V rms power while that in Europe, China, and the United States is performed using 380–400 V rms power. A welding power supply product that can be easily reconfigured using the same internal components has been developed using a resonant circuit technology [47]. The power supply configuration for the 200 V rms utility market is shown in Fig. 10.29. It utilizes a Fuji IGBT module containing two reverse conducting IGBTs (RC-IGBT: an IGBT with an integrated fly-back diode) in series with the DC power bus with voltage E operated as active PWM switches. The IGBT full-bridge inverter is used to generate high-frequency (40 kHz) power on the primary side of the transformer. The AC power is rectified on the secondary side of the transformer using the diodes to supply the arc voltage. The transformer size and weight are greatly reduced by operation at the high

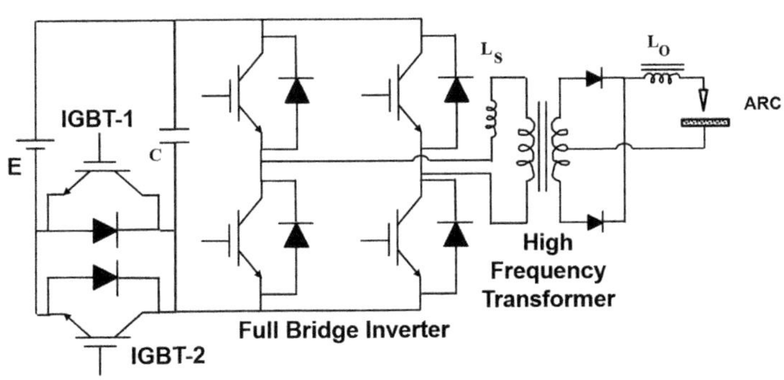

FIGURE 10.29

Dual utility 200 V rms IGBT-based arc welding power supply.

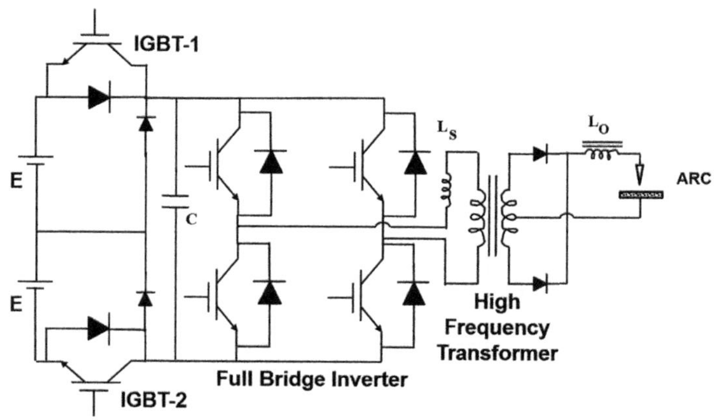

FIGURE 10.30

Dual utility 400V rms IGBT-based arc welding power supply.

frequency. In order to maintain low switching losses in the IGBTs at the high frequency of operation, zero-voltage and zero-current switching are performed.

To reconfigure the welding power supply for the 400 V rms utility market, the same components are used as shown in Fig. 10.30. The larger DC bus voltage is divided into two equal parts of the same size as the previous circuit. This allows using the same IGBT modules for both markets. Power losses in the IGBTs are reduced by using soft-switching, resulting in a volume reduction of 59% and a weight reduction of 47% compared with the hard-switching approach. Dynamic welding performance is improved due to operation at 40 kHz when compared with 13 kHz with hard switching.

10.6.4 Robot Arc Welding

Automation of the arc welding process using robots improves the uniformity and quality of the joint. This requires precise control of the power applied to the arc, which can be achieved with high-frequency inverters built with modern IGBTs. Multiple parallel IGBT units have been used under microprocessor control to achieve low-cost batch production [48].

10.6.5 Consumable Electrode Welding

Electric solders are utilized for the assembly of metal structures in industrial manufacturing. This process results in consumption of the electrode (solder wire) during the welding process conducted within an inert gas ambient (MIG). In order to perform the welding operation with a consumable electrode, it is necessary to feed the solder wire continuously to maintain the appropriate distance from the workpiece.

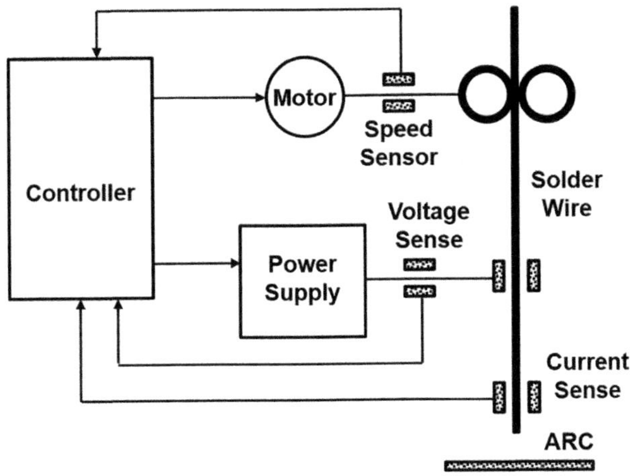

FIGURE 10.31

Consumable electrode arc welding power supply.

A block diagram of the controllable solder-based welding process [49] is illustrated in Fig. 10.31. A stable process is established by matching the speed of consumption of the solder electrode and the speed of feeding the solder wire using the motor. The IGBT-based power supply maintains the appropriate voltage and current to produce a uniform arc.

10.6.6 IGBT Optimization for Welding

As described in the previous sections, the power circuits for welding applications operate at high frequencies to reduce the size of the transformer. It is important to optimize the IGBT structure to reduce switching power losses even if this results in a larger on-state voltage drop. This has been done by Mitsubishi Electric Company in their fifth- and sixth-generation NFH-series IGBT [50] to obtain high performance in resonant circuits with 10–60 kHz operating frequency.

The cross-section of the fifth-generation Mitsubishi IGBT structure is shown in Fig. 10.32(a). It is an asymmetric or PT design with a thin P^+ collector region. A carrier storage (CS) layer is incorporated under the *P*-base region to reduce the resistance to current flow in the upper portion of the device. A trench gate structure is employed to increase channel density and reduce the on-state voltage drop. In addition, dummy trenches are used with the polysilicon electrode in these trenches connected to the emitter electrode. A cross-section of the sixth-generation Mitsubishi IGBT structure is shown in Fig. 10.32(b). It is similar to the fifth-generation IGBT structure with a smaller spacing between the trenches. More dummy trenches are placed between the trenches with the polysilicon electrodes in these trenches connected to the emitter terminal. In addition, an optimized retrograde doping profile is used for the CS layer. These innovations have been shown to improve the

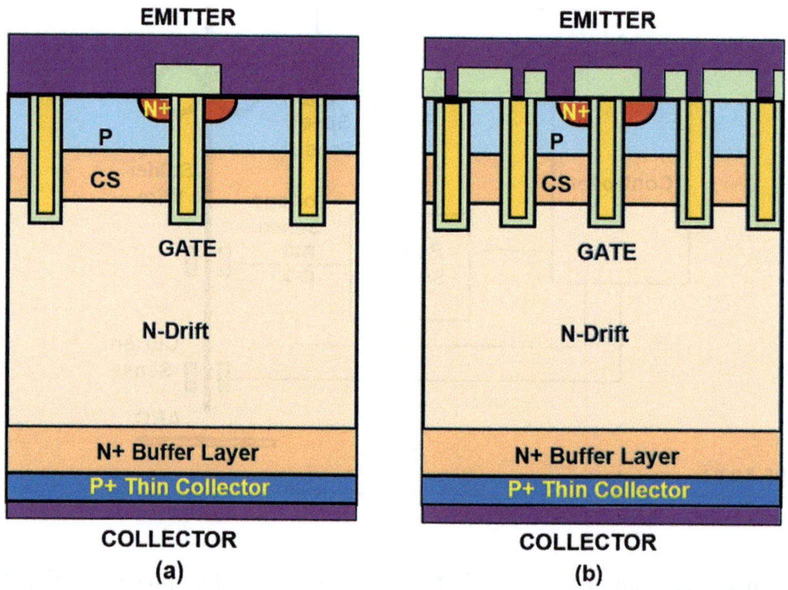

FIGURE 10.32

IGBT structures optimized for arc welding power supply.

trade-off curve between the on-state voltage drop and switching energy loss per cycle. For the same switching energy loss, the on-state voltage drop for the sixth-generation device is 0.6 V smaller than that of the fifth-generation device.

10.7 Induction Heating

Induction heating is based upon transfer of energy from a primary coil into a conducting object (usually metal), producing heating by the generation of eddy currents [51]. Induction heating is used in the industrial manufacturing sector for performing a variety of tasks such as surface hardening, melting, and forging. Induction hardening [52] allows heat treatment of selective areas of metal parts, especially those made from steel. The shaping of metal parts by the application of localized compressive force is referred to as forging [53]. Induction heating allows a localized increase in the temperature of metal parts during forging.

The processing of materials in the industry often requires heating of fluids (liquids and gases) with high uniformity and throughput. The electromagnetic induction-based heating of fluids is applicable to boilers, dryers, hot-air producers, oil heaters, etc., in industrial and chemical plants. An induction heating-based compound dryer for industrial applications made by Nordson is shown in Fig. 10.33(a). It occupies only a fraction of the floor space of conventional dryers and is specifically designed for production lines with speeds of 500–2200 ends per minute [54]. An induction

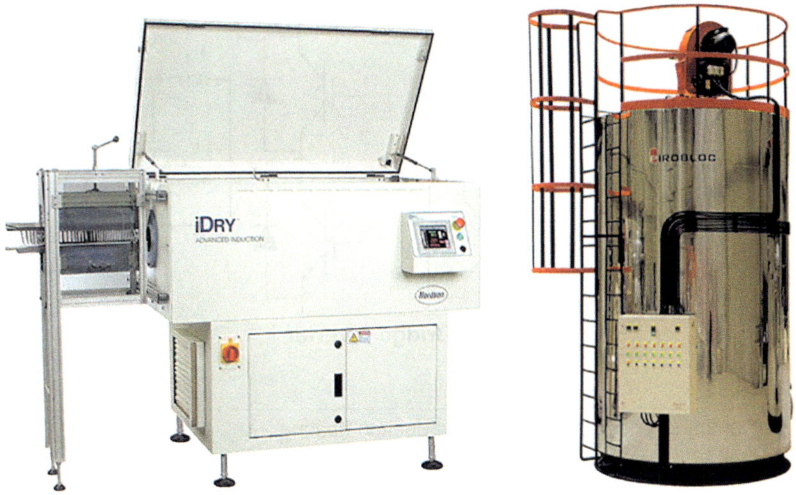

(a) (b)

FIGURE 10.33

(a) Induction dryer; (b) induction oil heater.

heating-based oil heater [55] made by Pirobloc is shown in Fig. 10.33(b). An IGBT superaudio heating equipment product [56] is available for melting precious metals. An IGBT-based dryer for insulation materials in power distribution transformers has been reported [57].

10.7.1 Forging, Annealing, and Tube/Pipe Welding

A series resonant inverter is commonly used for induction heating in forging and pipe welding operations in the industry. Typical power ratings range from 50 kW to several megawatts. The high frequency (typically 50–100 kHz) required to deliver the power to the load is generated using the H-bridge (or full-bridge) configuration [58,59] illustrated in Fig. 10.34. To enable IGBTs to operate at these high frequencies, it is necessary to reduce switching losses by using a resonant topology with zero-voltage switching. The resonant frequency (f_R) is decided by the selection of resonant capacitor (C_R) and the inductor L_1 after taking into account the load inductance L_2:

$$f_R = \frac{1}{2\pi\sqrt{C_R L_1 L_2/(L_1 + L_2)}} \tag{10.4}$$

The load inductance is dependent on the shape of the heating coil and the workpiece as well as the spacing between them.

A 50 kW piece of forging equipment has been built using the above approach with IGBTs operating at a resonant frequency of 50 kHz. An induction heater for

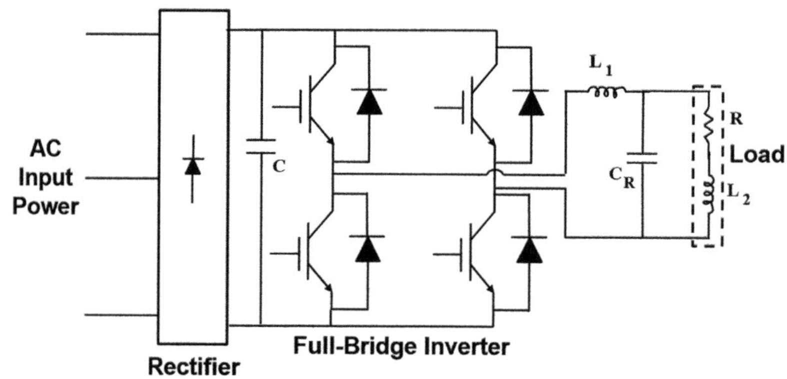

FIGURE 10.34

Series resonant inverter using IGBTs for induction heating.

industrial steel strip surface annealing has been implemented using the above approach [59] at a resonant frequency of 100 kHz operating with 1.2 kV IGBT modules. A 50 kW series resonant power supply for melting, forging, and surface-hardening applications has also been reported with a resonant frequency of 150 kHz [60].

10.7.2 Fluid Heating

The heating of fluids (liquids or gases) in industrial plants is required during their transport and processing. Induction heating of the vessel used for the fluid provides a contactless method for generating the heat. Induction heating provides the advantages of fast response, precise temperature control, cleanliness, compactness, and absence of combustion byproducts.

A typical induction heating system based on using IGBTs for heating fluids [61] is shown in Fig. 10.35. The fluid flows through a nonmetallic vessel with a coil around it. A flow-through metal assembly is placed inside the vessel with multiple fins to create a large surface area for heating the fluid. The eddy current established in the metal fins produces heat that is transferred to the fluid. The fins are designed to produce turbulence to maximize the heat transfer. A constant frequency, phase-shifted, PWM series resonant inverter with IGBTs is used for producing the high-frequency power fed to the working coil.

The advantages of the above technology include: (1) contactless heating, (2) fast heating response, (3) precise temperature control, (4) wide range of temperature regulation, (5) unity power factor operation, and (6) high efficiency due to reduced switching losses in the IGBTs. The inductive heating method can be used for boiling water and in tankless water heaters in homes as well.

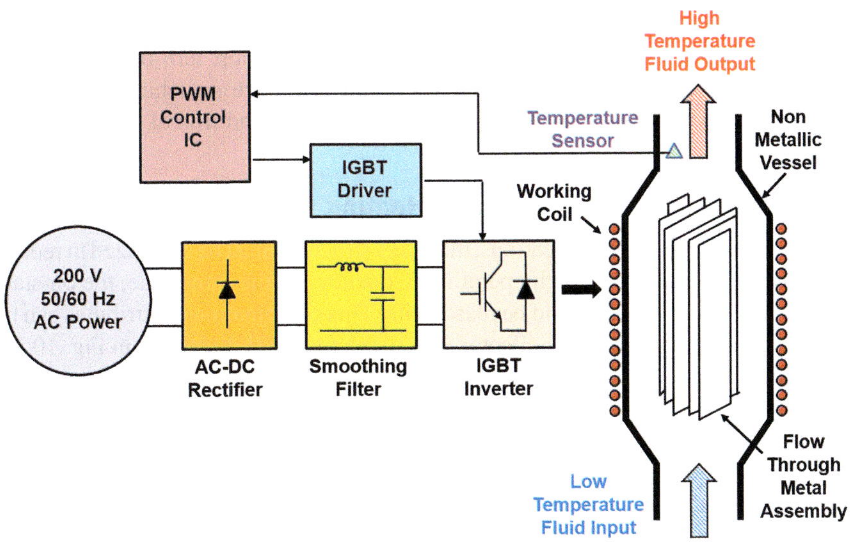

FIGURE 10.35

Induction heater for fluids.

10.7.3 Metal Melting Furnace

Melting metals is required for casting objects of various shapes and sizes in the industry. With sufficient energy supplied using the high-frequency resonant IGBT inverters previously shown in Fig. 10.34, it is possible to melt metals such as steel. To achieve the high power level (250 kW) in a steel foundry, the circuit topology shown in Fig. 10.36 has been developed [62]. Due to high power levels for this application, the rectified DC input voltage is increased using the boost converter. The high-frequency power is generated using the IGBT full-bridge circuit, which is operated in the series resonant mode. A protection circuit with a thyristor is used against

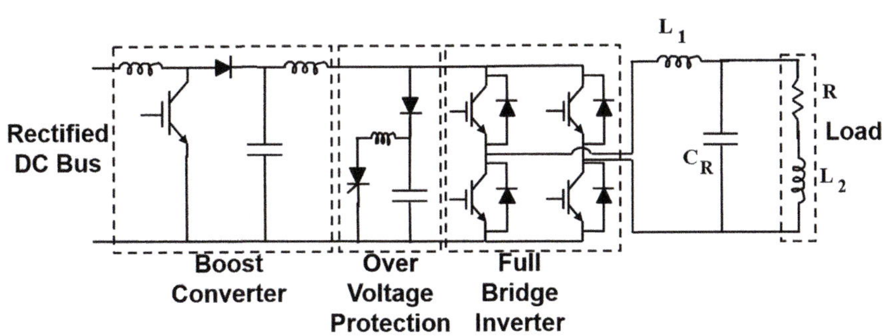

FIGURE 10.36

Induction furnace heater for melting metals.

overvoltages across the IGBTs. Advantages cited for this IGBT-based technology are: (1) operation at exactly the resonant frequency, (2) short turn-on and turn-off times for the IGBTs, and (3) resonant operation to reduce switching losses. The induction furnace has been used to melt steel at 1650°C and iron at 1400°C.

10.7.4 IGBT Design for Induction Heating

For the induction heating applications, the IGBT structure must be optimized to reduce switching losses due to the high operating frequency. At the same time, the on-state voltage drop of the IGBT should be reasonable. This combination of attributes can be achieved by using the shorted collector IGBT structure [63] illustrated in Fig. 10.37. The field stop layer is the same as the buffer layer previously discussed in Chapter 3 for the asymmetric blocking IGBT structure. The IGBT structure with collector shorting regions has also been referred to as the RC-IGBT structure [64,65]. IGBT products with a trench gate, field stop design that are optimized for induction heating applications are available on the market from many companies [66].

10.8 Milling and Drilling Machines

Milling is commonly performed in industrial settings to prepare parts of precise shape and size. The milling operation is performed using a rotating milling cutter to remove material from the workpiece. Unlike drilling, the cutter moves

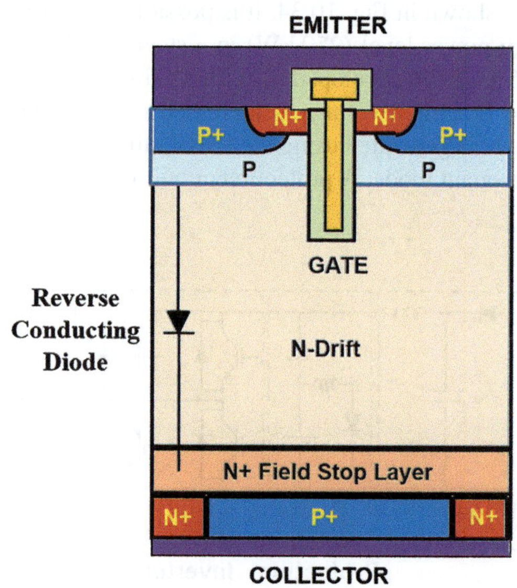

FIGURE 10.37

IGBT structure optimized for induction heating.

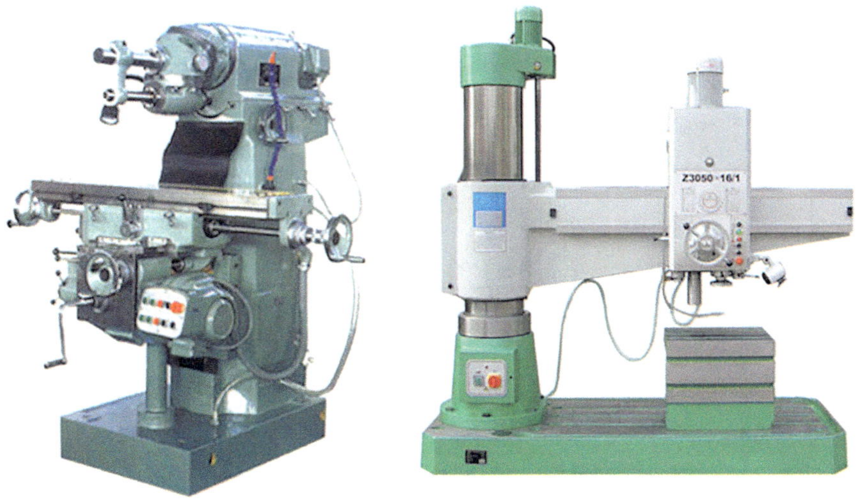

(a) (b)

FIGURE 10.38

(a) Milling machine; (b) high-speed drilling machine.

perpendicular to its axis of rotation, removing material at its circumference [67]. Drilling is performed to remove material along the axis of rotation of the cutter or drill bit. This allows creating holes in workpieces. Automated, high-speed drilling machines are needed for the expeditious processing of products in the industry. Examples of a milling and high-speed drilling machine are shown in Fig. 10.38(a) and Fig. 10.38(b), respectively.

10.8.1 High-Speed Milling Machine

High-speed milling requires an efficient cutting process. The cutting rate improves with an increase in the speed of the cutter. Milling spindles must be capable of extremely high rotational speed with high stiffness. High-speed milling spindles utilize active magnetic bearing (AMB) to achieve this capability [68]. Two counteracting electromagnets are used as shown in Fig. 10.39(a) to create the AMB.

To achieve a high rotational speed, AMBs utilize large magnetic forces to lift the motor/spindle shaft to suspend it and create a frictionless bearing. The dynamic performance and stiffness of the AMB are determined by the maximum slew rate during transients. The power amplifier driving the motor must provide sufficient output voltage to produce the specified slew rate. A high stiffness in the milling spindle is produced at a drive voltage much greater than the voltage drop of the magnetic coil of the electromagnets [69]. A reliable and cost-effective power amplifier has been developed by using PWM control of IGBTs, as shown in Fig. 10.39(b). When

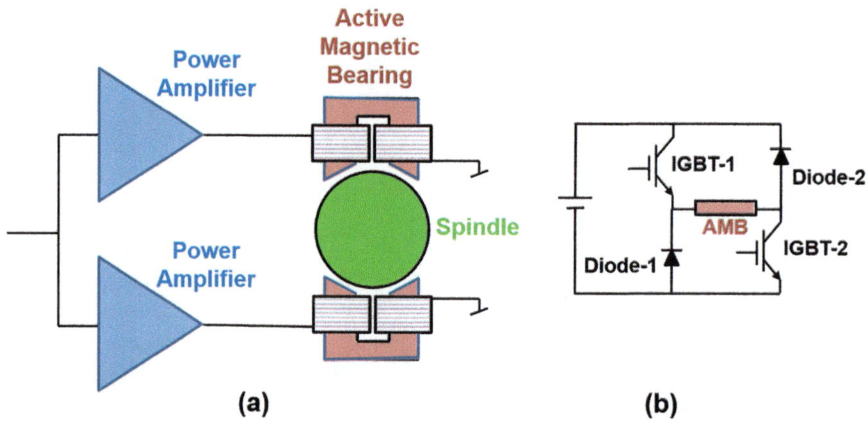

FIGURE 10.39

(a) Active magnetic bearing (AMB) and (b) IGBT drive circuit for AMB.

IGBT-1 and IGBT-2 are both turned on, the current in the electromagnet is increased by energy supplied by the DC source. If only IGBT-1 is turned on, the electromagnet current flows through it and diode-2 to allow maintaining a constant current. If only IGBT-2 is turned on, the electromagnet current flows through it and diode-1 to allow maintaining a constant current. If both IGBTs are turned off, energy flows from the electromagnet back to the DC power source. The current required in the electromagnet coil can be controlled by the IGBT power amplifier at the request of the position controller.

The IGBT power amplifier has been designed to run a high-speed milling machine spindle at 40,000 rpm using 35 kW of power to control five axes [69]. The switching frequency for the PWM control was 20 kHz.

10.8.2 High-Speed Drilling Machine

High-speed automated drilling operations require rapid replacement of tool bits on the machines. Although mechanical tool holders were used for a long time in the industry, faster machining requires larger forces for holding the tool in place. Tool collets based on thermally controlled diameter changes offer this capability. In this approach, the tool holder (collet) has a solid hollow body, as illustrated in Fig. 10.40 (a), that can be heated to expand its inner diameter, enabling the insertion of the tool, followed by rapid cooling to contract the collet to firmly hold the tool. Induction heating of the collet allows a rapid rise in its temperature, expediting the tool change process.

In the inductive shrinking collet, the tool holder is located inside a water-cooled copper tube that serves as the primary of a transformer with the collet as the secondary [69]. Due to the low load impedance of the copper tube, a high-frequency transformer is used with a series capacitor to form a resonant circuit, as shown in

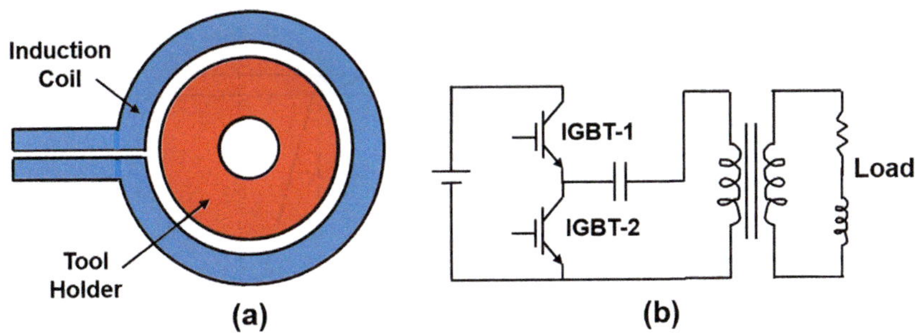

FIGURE 10.40

(a) Inductive shrinking tool holder; (b) IGBT inductive heater for tool holder.

Fig. 10.40(b). This arrangement is used to supply 10 kW of power to the collet at a resonant frequency of 100 kHz.

10.8.3 High-Speed Electrical Discharge Machining

High hardness materials are difficult to machine using conventional abrasive tools. Electrical discharge machining (EDM) or spark erosion machining allows removal of high-hardness conductive material by using repeated electrical discharges [70]. The basic process is illustrated in Fig. 10.41(a). A hollow copper electrode is placed in the vicinity of the workpiece with a dielectric fluid (typically water) injected into the space between them. The application of a high-voltage pulse produces breakdown of the dielectric fluid, producing a spark between the tool and the workpiece. The current density in the spark can reach 10,000 A/cm^2, raising the temperature to tens of thousands of degrees Celsius. The spark creates a shock wave in the dielectric that produces a tensile force sufficient to tear off particles of the molten and softened

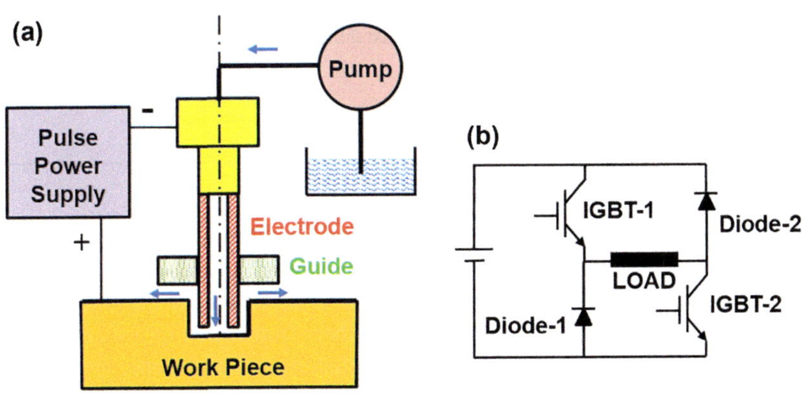

FIGURE 10.41

(a) Electrical discharge machining (EDM) process; (b) EDM power supply.

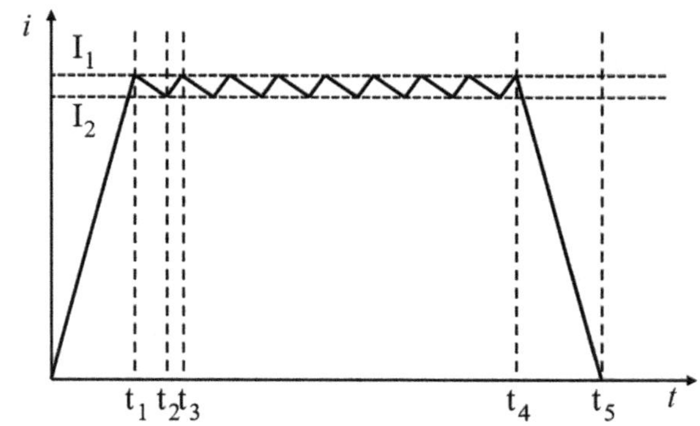

FIGURE 10.42

Electric discharge machining pulse waveform.

metal from the workpiece [71]. The fluid dielectric not only flushes out the debris produced during the drilling but acts as a coolant. An automated servo control system is used to adjust a uniform discharge gap during the drilling sequence, allowing high-speed manufacturing rates of up to 6 cm/min. The metal removal rate, surface finish, and electrode wear depend on the magnitude and duration of the current pulse.

An IGBT-based power supply for the EDM process is shown in Fig. 10.41(b). It is similar to that already described in Section 10.8.1. However, the details of operation of this circuit for the EDM process are different. When both the IGBT-1 and IGBT-2 are turned on, the current in the gap increases rapidly to a level I_1, as shown in Fig. 10.42, during the time from 0 to t_1. IGBT-1 is turned off at time t_1, resulting in the spark current flowing via IGBT-2 and diode-1. This discharges the load, producing a drop in the current level to I_2 at time t_2. This sequence is repeated as illustrated in Fig. 10.42 to maintain an average current of $(I_1 + I_2)/2$ at the load. After the desired drilling interval, both IGBTs are turned off at time t_4, leading to the energy in the load being fed back to the DC bus capacitor and the current reducing rapidly to zero at time t_5. This approach does not require a current-limiting resistor, making the control more precise and efficient.

10.9 Metal and Paper Mills

Steel mills are industrial plants used to manufacture steel from iron ore by smelting it with coke or limestone [72]. The molten iron containing carbon and other impurities is cast into ingots and slabs, as shown in Fig. 10.43(a), and made into sheets by hot/cold rolling. A paper mill converts vegetable fibers from wood into sheets of paper [73]. Large rolls of paper are required in the publishing and newspaper businesses, as shown in Fig. 10.43(b).

The use of IGBTs is motivated by the energy savings potential offered by variable-speed drives in rolling mills. IGBT-based variable-speed drives in textile

FIGURE 10.43

(a) Steel mill; (b) paper mill.

mills have been found to reduce noise while providing the benefits of reduced drive size, reduced cost of operation, and increased drive performance [74]. The transition from DC motors to AC motors by the metal processing industry has also been documented. Applications, such as finishing mills, roughing mills, roller tables, loopers, shears, etc., can now all be served using AC motors. Bhooplapur [75] states: "The early generation of AC drives were lacking in dynamic and static accuracy, power ratings were low, expensive when compared with DC motor drives. These problems were overcome with the availability of vector-control and IGBT-based AC drives." Both hot mills and cold mills now utilize PWM IGBT inverters for the operation of AC motors.

Companies have developed drives tailored for the metal industry and pulp/paper mills. An example discussed in the literature by the Toshiba Mitsubishi-Electric Industrial Systems Company is a two-level IGBT converter/inverter for drives with up to 2.7 MW power levels [76]. The design consists of one large front-end converter to transform the available AC power (at 460/575/690 V to serve the global market) to a DC bus across the capacitor, as shown in Fig. 10.44. The DC bus feeds multiple

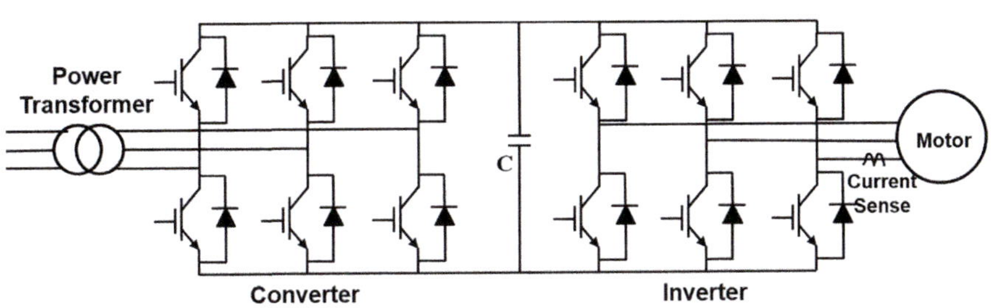

FIGURE 10.44

Two-level motor drives used in metal and paper mills.

motor drives using inverters (only one drive is shown on the right side of the figure). The PWM operating frequency of the active front end converter is optimized to reduce the harmonics introduced into the AC power source. The authors state: "These low-voltage drive systems contribute(s) to industrial application in the world, to save energy and to protect user safety, with reliable quality and durable design for both personnel and machinery."

10.9.1 Metals Industries

The drives for mills were originally based on line-commutated thyristors in cyclo-converter circuits [77]. They had the disadvantages of open-loop operation with poor dynamic operation and accuracy. These problems were overcome with the availability of IGBTs with vector control of speed, torque, and tension. The new drives provided the benefits of high efficiency, high reliability, broad power ratings, stringent control, stiffness, and low maintenance cost. For lower power levels, the two-level inverter previously shown in Fig. 10.5 is adequate. However, for high power loads, the three-level inverter for a VSD, shown in Fig. 10.45, is used in hot and cold mill applications such as the approach table, run-out table, pinch rollers, wrapper rollers, and mandrels.

The three-level inverter allows utilization of available 3.3 and 4.5 kV IGBT modules with only 12 transistors. When compared with the previous thyristor implementation, the IGBT-based inverter has simple gate drive circuitry and no snubbers. By using an active front end AC-to-DC rectifier, the three-level inverter pulls nearly perfect sinusoidal current from the power source, leading to a high power factor. The IGBT-based technology has been implemented for finishing block mills, cold rolling

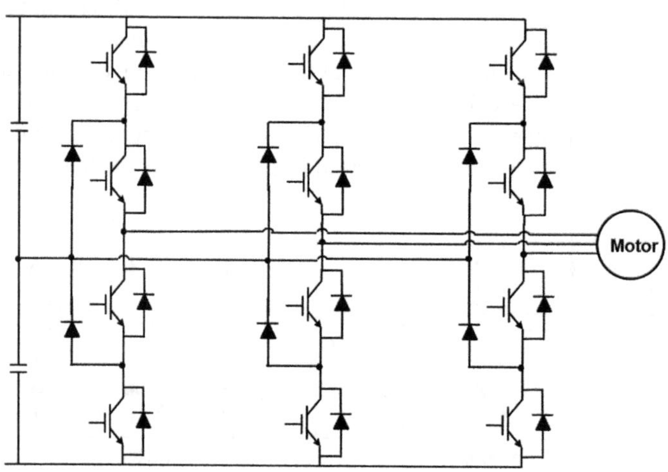

FIGURE 10.45

Three-level motor drives used in hot and cold mills.

in mills, and descaling pumps in steel plants. Other applications include mixers in rubber processing plants, and stirrers and mixers in chemical plants.

10.9.2 Pulp and Paper Industries

The paper and newsprint industry makes use of a large number of IGBT-based inverters to control paper flow during manufacturing of newsprint. A detailed description of the installation of IGBT-based AC winder drives to replace thyristor-based DC drives has been published [78,79]. The installation was performed between December 1994 and March 1995 at the Rainy River Forest Products Plant in Kenora, Ontario, Canada. The winder is designed to cut paper delivered from large shipping rolls into strips of specific width for the production of newsprint. The authors state: "At the start of these Winder upgrade projects, everyone involved was concerned with the motor voltage and IGBT current and peak voltage ratings…AC Winder drives today are a proven technology and, as industry experience grows, are becoming more and more acceptable over DC…The Kenora Winders AC digital drives have dramatically improved the runnability of the Winders and have had a significant impact on the runnability of the finished rolls in the mill's customer press rooms."

The AC drive configuration used at the Kenora plant for the production of newsprint is shown in Fig. 10.46. An 800 V common DC power bus is created using a synchronous rectifier (or converter as previously illustrated in Fig. 10.44) from a

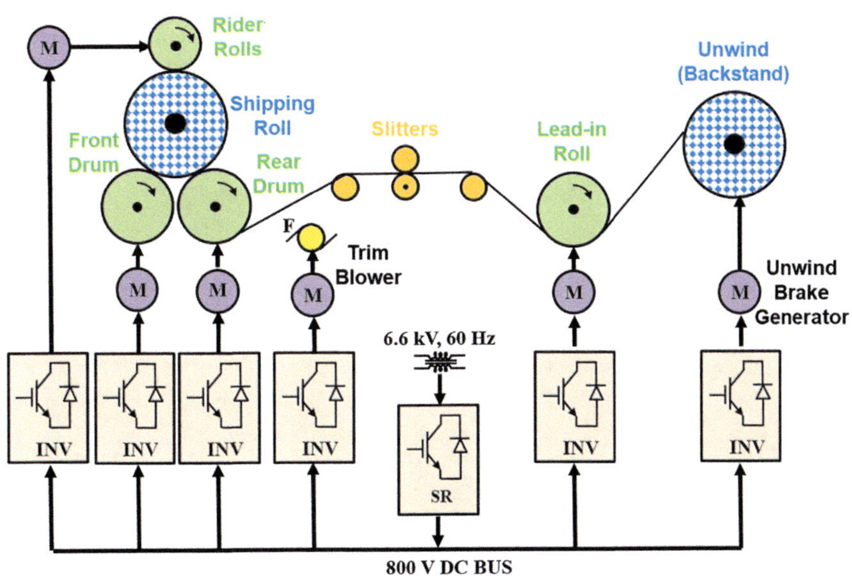

FIGURE 10.46

Newsprint winder AC drive configuration.

6.6 kV, 60 Hz power source that is first stepped down to 480 V using a 500 kVA transformer. This allows maintaining a near unity power factor to the utility power source and regenerates power back into the AC power line. Multiple inverters (similar to those shown in Fig. 10.44) are then used to drive a variety of motors, as shown in the figure for the drums, the slitters, and the trim blower. Vector-controlled PWM modulation with a carrier frequency of 4 kHz is used in the inverters to drive the 1200 V IGBTs and generate the variable voltage and variable frequency power. The inverter has output short circuit protection capability by switching off the IGBTs within 10 μs after detection of the fault.

The paper machine winder operates with acceleration for 1 min, winding for 8 min, deceleration for 1 min, and set change for 2 min. The drums accelerate the core, wind up the paper, and maintain tension. Regenerative power from the unwind drive flows into the DC power bus and is used to drive the drums and lead-in/rider rolls. The winder operates at 6500 RPM, providing a high throughput for the newsprint. The paper mill is immune to voltage disturbances due to lightning strike due to system inertia.

10.10 Electrostatic Precipitators

Electricity-generating power plants that utilize fossil fuels, such as coal and natural gas, produce environmental pollution through the emission of particles into the atmosphere. Chemical processing plants and oil refineries also release particles into the atmosphere as effluent gasses. Particles with diameter below 10 μm (classified as PM_{10}) are able to settle inside human lungs, leading to diseases and poor health outcomes. Electrostatic precipitators, such as those shown in Fig. 10.47, are used to remove particles from exhaust to mitigate the impact on the environment.

(a)

(b)

FIGURE 10.47

Electrostatic precipitators at (a) chemical plant; (b) power plant.

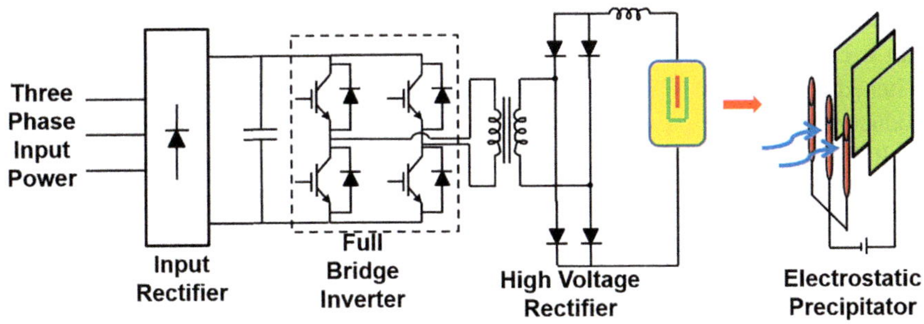

FIGURE 10.48

IGBT inverter for electrostatic precipitators.

The electrostatic precipitator removes particles from a gas stream by first charging the contaminants and then removing them using electrostatic forces [80]. The basic construction of the electrostatic precipitator consists of a row of thin vertical rods and a stack of flat parallel metal plates with a spacing of 1–18 cm apart, oriented as illustrated in Fig. 10.48. The contaminated gas (illustrated by arrows) flows through the space between the rods and then between the plates. A large negative (40–100 kV) voltage is applied between the rods and the metal plates to ionize the gas as it passes through. The ionized particles (contaminants) are attracted to the grounded metal plates, removing them from the gas stream.

Electrostatic precipitators were originally developed using thyristor-based high-voltage power supplies. Replacement of these power supplies with IGBT-based inverters has greatly improved the performance [81]. The ionization of the gas is done with pulses, causing flashovers in the gas. The shorter switching time for IGBTs (0.2 μs) when compared with thyristors (10 ms) prevents local ionization during the flashovers. The IGBT-based inverter shown in Fig. 10.48 operates at a switching frequency of 10 kHz to generate a 500 Hz current in the transformer to optimize its size. Higher operating frequencies for the transformer are not possible because of the large thickness of insulation in the oil-filled high-voltage transformer whose secondary side is developing 40–100 kV across the electrostatic precipitator after rectification. The short switching time for IGBTs allows for a faster repetition rate for flashovers, leading to lower dust emissions. Measurements of the quality of the gas stream show a 25–50% improvement in opacity [82]. A three-stage electrostatic precipitator has been operating in a sinter plant [82]. The authors state: "It could be proved that the most effective gas cleaning could be achieved with an IGBT inverter power supply in the first field."

Large power and chemical plants require electrostatic precipitators with power consumption in the megawatt range because more particles are removed with greater power [82]. The electrostatic precipitator for large industrial applications contains 20–40 zones, each with its own power supply. The power for each zone must be optimized to achieve the most efficient cleaning of the effluent gas. A comparison between the hard-switched topology shown in Fig. 10.48 and the resonant topology with a capacitor added (as shown in Fig. 10.21) has been performed. The hard-switched approach was found to be better in this application with the capability to produce 300 kVA of power at 150 kV. The waveform for the pulses produced by the IGBT inverter is similar in appearance to that shown in Fig. 10.42. The soft PT IGBT structure is suitable for achieving a high-inverter operating frequency of 10 kHz.

10.11 Textile Mills

With the large worldwide demand for clothing, the manufacturing of textiles is a major industry. The manufacturing of fabrics requires the production of yarn using a large number of spindles spinning at a high speed [83], as shown in Fig. 10.49. The energy consumption at a textile mill is one of the significant cost factors during cloth manufacturing. Textile production is responsible for 2.4 trillion pounds of carbon dioxide emission annually. Motor drives consume about 30% of the energy used at a textile plant. ASDs can reduce energy consumption of motors by as much as 60%. Because each meter of cloth requires an expenditure of about 0.5 kWh of electricity, the use of ASDs provides significant cost savings to the textile manufacturer. The use of ASDs for spindle motors with specially designed induction motors has been found to greatly improve the power efficiency and reduce cost [84].

The typical configuration of the ASD used for the spindle drive is shown in Fig. 10.50. The input power from the utility is rectified to create a DC bus using the smoothing filter. The IGBT-based inverter is then used to generate the variable-frequency AC power for the induction motor. The speed of the spindle can be controlled by adjusting the frequency of the AC power delivered to the motor.

FIGURE 10.49

Textile mill.

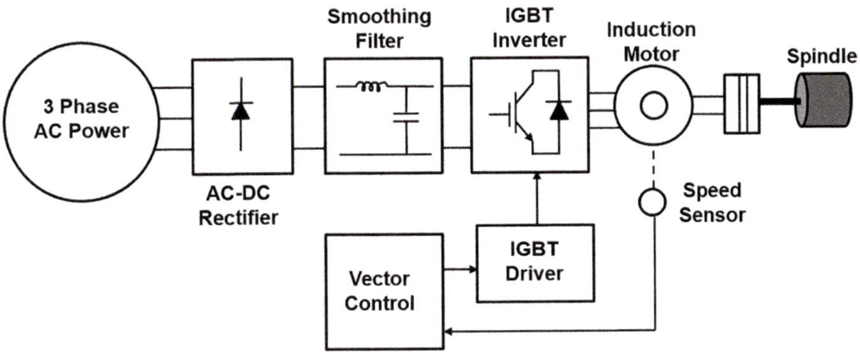

FIGURE 10.50

Adjustable speed drive for textile spindles.

10.12 Mining and Excavation

Bucket-wheel excavators, shown in Fig. 10.51, are used for large-scale surface mining operations [85]. The excavator uses electric shovels to gather the mining material and deliver it to conveyors or trucks. The electric shovel requires the control of the hoist, swing, crowd, and propel motions. All the motions are performed using synchronous motors fed by vector-controlled IGBT inverter drives [86]. The propel motion is performed by two independently controlled motors on either side of the shovel, allowing steers by varying the speeds of the two motors. The propel motors are powered by the hoist and crowd inverters. A transfer switch is used to allow sharing the inverter's hoist and propel 1 functions; or the crowd and propel 2 functions. A schematic drawing of the transfer switch arrangement that allows using only two inverters to perform the control of four motors in the excavator is shown in Fig. 10.52.

FIGURE 10.51

Mining excavator.

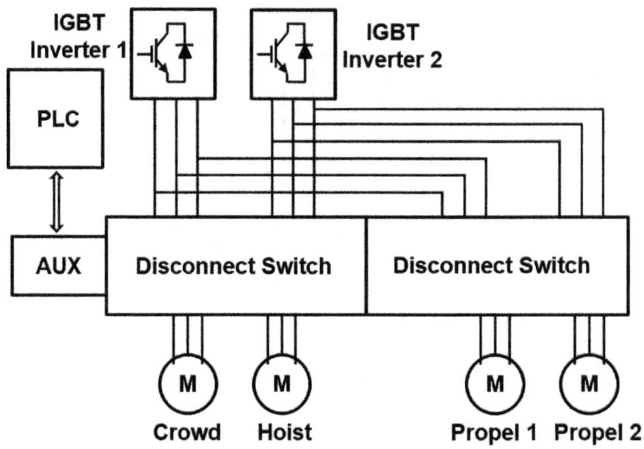

FIGURE 10.52

Transfer switch for electric shovel.

10.13 IGBT Optimization for Industrial Applications

The development of the IGBT was originally motivated by industrial applications at GE. Although the very first devices were fabricated with the symmetric blocking design [87], the development of asymmetric structures was almost immediately undertaken [88] because the significant improvement in the trade-off between on-state voltage drop and switching energy loss per cycle was clear from the beginning. Since that time, a large effort has been undertaken by the power semiconductor community to improve the trade-off curve between on-state voltage drop and energy loss per cycle. The biggest structural improvement was achieved by replacing the planar gate architecture with a trench gate structure [89]. The evolution of IGBT technology at one company, Mitsubishi Electric, is provided in Table 10.1 as an example.

Powerex/Mitsubishi Electric has published the evolution of IGBT chip designs for industrial applications from 1988 to 2000 through five generations of technology enhancements [90], as shown in Table 10.1. The first generation of devices made in 1988 utilized the planar gate architecture (see Fig. 10.11(b)) with 5 μm design rules using the asymmetric or PT vertical structure. Improvements to the PT planar gate structure (second and third generations) were derived by using smaller geometries and shallower diffusions. This resulted in reducing the on-state voltage drop from 3.5 to 2.5 V while simultaneously reducing the chip area from 1.44 to 1 cm^2. The next step in the evolution of the IGBT technology was derived by replacing the planar gate architecture with the trench gate architecture (Fig. 10.53(a)) in the fourth generation using 1 μm design rules. However, the high channel density of the trench gate architecture resulted in a high saturation current, which degraded the short circuit SOA (SCSOA) from 15 to 5 μs. To recover a good SCSOA capability, the fifth-generation IGBT structure was developed with gate electrodes in alternate trenches,

Table 10.1 Evolution of Powerex Inc. Insulated Gate Bipolar Transistor Design

IGBT Generations for 100 A, 1200 V Rating					
	First Gen. 1988	**Second Gen. 1990**	**Third Gen. 1992**	**Fourth Gen. 1998**	**Fifth Gen. 2000**
Family name	–	E-series	H-series	F-series	NF-series
Wafer	Epitaxial	Epitaxial	Epitaxial	Epitaxial	Float zone
Pattern gate	5 μm planar	5 μm planar	3 μm planar	1 μm trench	1 μm trench
Vertical	PT	PT	PT	PT	LPT
$V_{CE(sat)}$	3.5 V	2.8 V	2.5 V	1.9 V	1.9 V
Chip size	12 × 12	12 × 12	10 × 10	9.1 × 11.5	9.1 × 11.5
SCSOA	15 μs	15 μs	20 μs	5 μs	20 μs

Abbreviations: IGBT, *insulated gate bipolar transistor;* PT, *punch through;* SCSOA, *short circuit safe operating area;* LPT, *light punch through;* $V_{CE(sat)}$, *on-state voltage drop.*

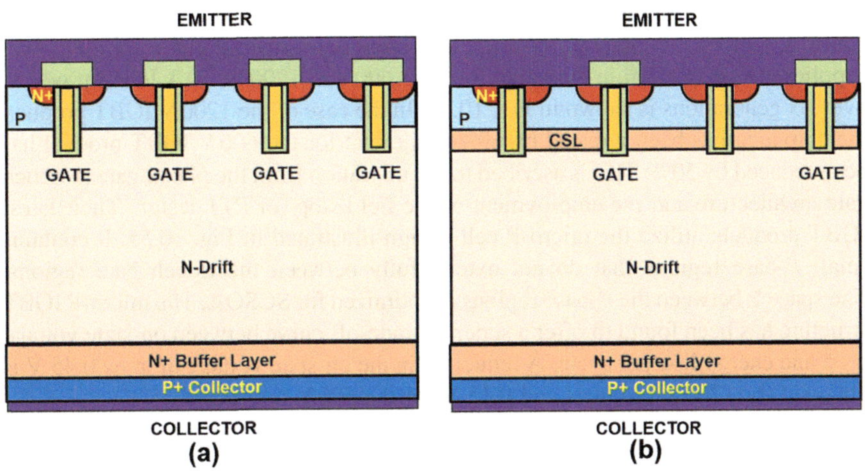

FIGURE 10.53

Powerex Inc. IGBT structures: (a) fourth generation; (b) fifth generation.

as illustrated in Fig. 10.53(b). This structure includes a CS layer below the *P*-base region to enhance the carrier concentration during current conduction in the on-state. The reduction in on-state voltage drop from 2.5 to 1.9 V by replacing the third-generation planar gate IGBT structure with the fourth-generation trench gate structure is noteworthy. Migrating from epitaxial wafers to float-zone bulk material with the fifth-generation IGBT provided significant cost savings. This required equipment tailored to handle thin wafers in the production line without yield loss from breakage.

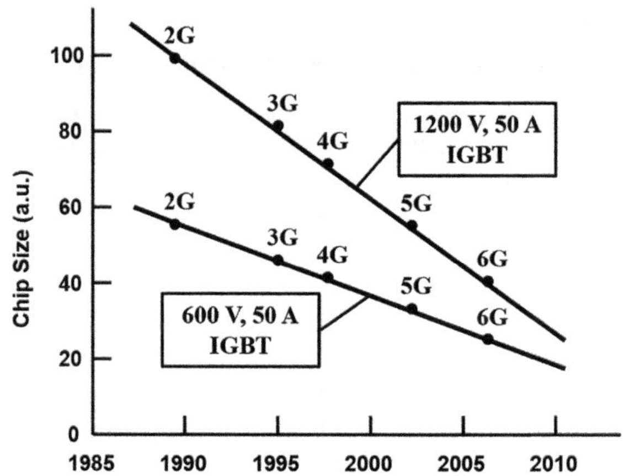

FIGURE 10.54

Reduction of Fuji electric IGBT chip size.

Another example of trends in IGBT product development for industrial applications is that published by Fuji [91]. These products are tailored for use in numerical controls, factory automation, robotics, air conditioning, and uninterruptible power supplies. The reduction in chip size for their 600 and 1200 V, 50 A IGBT products over six generations is shown in Fig. 10.54. In the case of the 1200 V IGBT product, the chip area has been reduced by 60% while that for the 600 V IGBT product has been reduced by 50%. This is ascribed to the migration from the planar gate to trench gate architecture and the employment of the field stop (or PT) design. Their latest IGBT products utilize the micro-P cell design illustrated in Fig. 10.55. It contains small *P*-base regions that do not extend fully between the trench gate regions. The space L between the *P*-base regions is optimized for SCSOA. The micro-P IGBT structure has been found to offer a superior trade-off curve between on-state voltage drop and energy loss per cycle. A reduction in the on-state voltage drop by 0.45 V is obtained for 1200 V products and 0.15 V for the 600 V products.

An IGBT module for industrial applications has been reported using 1.7 kV RC-IGBTs [92]. With the X series of products, Fuji has reduced the on-state voltage from 2.6 to 2.3 V for the same switching energy loss by using smaller design rules and applying thin-wafer production capability.

10.14 Low Power IPM

There are many industrial applications that operate at low power levels that can benefit from a compact form factor for the inverter. This has been achieved with an intelligent power module (IPM) containing six IGBTs with fly-back diodes inside a single package [93]. The IGBT structure shown in Fig. 10.53(b) was used. The carrier storage layer (CSL) was found to be important to reduce on-state voltage drop by improving the injected carrier distribution on the emitter side. The low-voltage IC

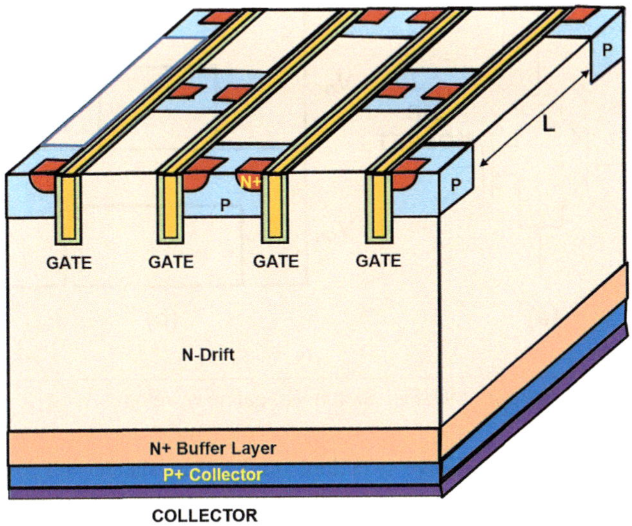

FIGURE 10.55

Micro-P sixth-generation Fuji electric IGBT structure.

for driving the IGBTs connected to the reference bus and the high-voltage IC with level shifted signals for driving the IGBTs connected to the DC power bus were incorporated into the package. The technology named DIPIPM+ was capable of handling 25 A at 1200 V and 50 A at 600 V. The IGBT-based regenerative braking circuit for the motors was also added to the module.

10.15 Dead-Time Compensation

The ASD architecture shown in Fig. 10.5 and others shown in subsequent figures for motor drives are prone to a shoot-through problem where high current can flow across the DC bus if both IGBTs in any totem-pole are turned-on simultaneously during the PWM operation. The solution to this problem is the incorporation of dead time in the gating signals for the two IGBTs. It has been found that the inclusion of dead time can lead to oscillations in the motor winding current [94]. Adding a trapezoidal compensating waveform to the PWM signals has been shown to suppress the oscillations. This approach can be generally used for all VSI inverters with IGBTs for industrial motor drives.

10.16 Hybrid Si IGBT/SiC MOSFET Switches

Si IGBTs have been developed with excellent on-state voltage drop and switching performance. However, their switching losses are worse than unipolar devices. Silicon carbide power MOSFETs have become commercially viable during the last

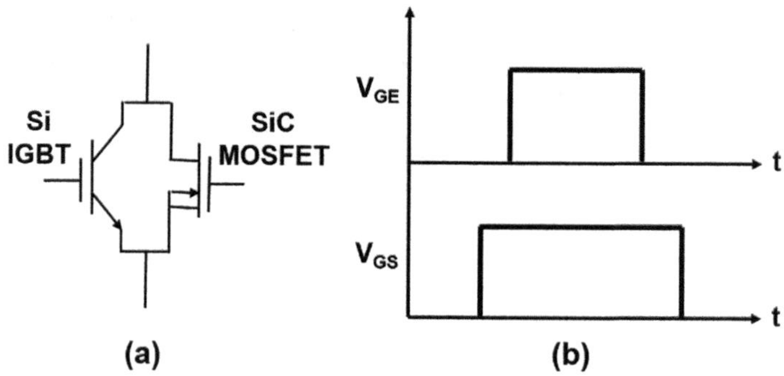

FIGURE 10.56

(a) Hybrid Si IGBT/SiC Power MOSFET Switch; (b) gating waveforms.

decade [95]. These devices have excellent on-state resistance and low switching losses. However, the SiC devices are much more expensive (~ 3–5 times) compared with Si IGBTs. A more cost-effective approach to creating a power switching with low on-state voltage drop and low switching losses is by using the SiC power MOSFET in parallel with the Si IGBT, as shown in Fig. 10.56(a) [96]. The gate bias application scheme for the composite switch is illustrated in Fig. 10.56(b). During the turn-on switching event, the SiC power MOSFET is first switched on, followed by the Si IGBT. During the turn-off switching event, the Si IGBT is first switched off, followed by the SiC power MOSFET. This reduces the switching losses to those observed for the SiC power MOSFET. The Si IGBT carries most of the current during the on-state due to its low on-state voltage drop. This keeps the on-state power loss close to that for the Si IGBT. Although the paper is focused on an EV motor drive, the composite switch can be used in industrial motor drives as well.

10.17 Summary

The most widespread applications for the IGBT are found in the industrial sector due to the utilization of a large number of motors. The availability of the IGBT in the early 1980s enabled the development of cost-effective ASDs for motors. These drives have proliferated during the last three decades. Adjustable speed motor drives reduce energy consumption by more than 40%. Because two-thirds of the electricity in the world is used to run motors, the IGBT technology has had a huge impact on society, as will be documented in a later chapter. The penetration of VSDs in home appliances doubled from 2010 to 2015 [97].

This chapter provides many examples of the use of IGBTs in the industrial sector such as numerical controls, robotics, etc. The tremendous diversity of industrial applications described in this chapter reinforces the benefits of the IGBT technology to modern society.

References

[1] Industry. http://en.wikipedia.org/wiki/industry.
[2] Secondary Sector of the Economy. http://en.wikipedia.org/wiki/Secondary_sector_of_the_economy.
[3] A.D. Little, Opportunities for Energy Savings in Residential and Commercial Sectors With High Efficiency Electric Motors, U.S. Department of Energy Contract No. DE-AC01-90CE23821, December 1, 1999.
[4] U.S. Department of Energy, Industrial Technologies Program. http://www1.eere.energy.gov/industry/bestpractices/motors/html.
[5] Variable Frequency Drives, Report #00-054 Eaton Consultants, Inc, Northwest Energy Efficiency Alliance, June 2000. www.nwalliance.org.
[6] T.M. Jahns, V. Blasko, Recent advances in power electronics technology for industrial and traction machine drives, Proc. IEEE 89 (2001) 963–975.
[7] M. Hiller, R. Sommer, M. Beuermann, Converter topologies and power semiconductors for industrial medium voltage converters, in: IEEE Industry Applications Society Annual Meeting, 2008, pp. 1–8.
[8] E. Gatti, et al., Large power voltage source IGBT inverters for industrial applications, in: IEEE Power Electronics Specialists Conference, vol. 2, 1996, pp. 1059–1064.
[9] A.R. Hambley, Electrical engineering, in: AC Machines, fifth ed., Prentice Hall, 2011, pp. 821–864 (Chapter 17).
[10] Variable Frequency Drives. http://www.rowan.edu/colleges/engineering/clinics/cleanenergy/rowan_university_clean_energy_program/Energy_Efficiency_Audits/Energy_Technology_Case Studies/files/Variable Frequency_Drives.pdf.
[11] B.J. Baliga, The new generation of MOS power devices (Invited Paper), in: Drives/Motors/Controls Conference, October 1983, pp. 139–141.
[12] B.K. Bose, Power Electronics and Variable Frequency Drives, IEEE Press Book, 1997.
[13] J. Vithayathil, Power Electronics, McGraw-Hill, 1995.
[14] J.G. Kassakian, M.F. Schlecht, G.C. Verghese, Principles of Power Electronics, Addison-Wesley Publishing Company, 1991.
[15] A.M. Jungreis, A.W. Kelley, Adjustable speed drive for residential applications, IEEE Trans. Ind. Appl. 31 (1995) 1315–1322.
[16] T. Sawa, T. Kume, Motor drive technology—history and visions for the future, in: IEEE Power Electronics Specialists Conference, 2004, pp. 2–9.
[17] F. Blaabjerg, P. Thoegersen, Adjustable speed drives—future challenges and applications, in: Power Electronics and Motion Control Conference, 2004, pp. 36–45.
[18] R. Hoppler, et al., A Team of Drives, May 2009, pp. 71–76. www.worldcement.com.
[19] B.K. Bose, Adjustable speed AC drives—a technology status review, Proc. IEEE 70 (1982) 116–135.
[20] B.J. Baliga, Power semiconductor devices for variable-frequency drives, Pro. IEEE 82 (1994) 1112–1122.
[21] B.J. Baliga, Fundamentals of Power Semiconductor Devices, second ed., Springer-Science, New York, 2019 (Chapter 5).
[22] B.J. Baliga, Fundamentals of Power Semiconductor Devices, second ed., Springer-Science, New York, 2019 (Chapter 9).
[23] M. Otsuki, et al., The 3rd generation IGBT toward a limitation of IGBT performance, in: IEEE International Symposium on Power Semiconductor Devices and ICs, 1993, pp. 24–29.

[24] M. Matsudai, et al., Advanced 60 micron thin 600 V punch-through IGBT concept for extremely low forward voltage drop and low turn-off loss, in: IEEE International Symposium on Power Semiconductor Devices and ICs, 2001, pp. 441–444.
[25] P.M. Shenoy, S. Shekhawat, B. Brockway, Application specific 1200 V planar and trench IGBTs, in: IEEE Applied Power Electronics Conference, 2006, pp. 160–164.
[26] M. Rahimo, A. Kopta, S. Linder, Novel enhanced-planar IGBT technology rated up to 6.5 kV for lower losses and higher SOA capability, in: IEEE International Symposium on Power Semiconductor Devices and ICs, 2006, pp. 1–4.
[27] Y. Onishi, et al., Analysis on device structures for next generation IGBT, in: IEEE International Symposium on Power Semiconductor Devices and ICs, Paper 4.3, 1998, pp. 85–88.
[28] Automation. http://en.wikipedia.org/wiki/Factory_automation.
[29] Numerical Controls. http://en.wikipedia.org/wiki/Numerical_control.
[30] www.ge-ip.com/products/family/genius-io.
[31] B.J. Baliga, et al., Comparison of N and P channel IGBTs, in: IEEE International Electron Devices Meeting, Abstract 10.6, 1984, pp. 278–281.
[32] B.J. Baliga, et al., Vertical Double Diffused Metal Oxide Semiconductor (VDMOS) Device Including High Voltage Junction Exhibiting Increased Safe Operating Area, U.S. Patent 4, 823,176 (Filed April 3, 1987, Issued April 18), 1989.
[33] B.J. Baliga, et al., New cell designs for improved IGBT safe-operating-area, in: IEEE International Electron Devices Meeting, Abstract 34.5, 1988, pp. 809–812.
[34] A. Esser, H.-C. Skudelny, A new approach to power supply for robots, in: IEEE Industrial Applications Society Meeting, 1990, pp. 1251–1255.
[35] J.W. Geisinger, M.P. Aalund, D. Tesar, The design and development of a control system for a T3-776 robot, in: IEEE Applied Power Electronics Conference, 1993, pp. 333–339.
[36] Linear Actuators. http://en.wikipedia.org/wiki/Linear_actuator.
[37] P. Mercorelli, N. Kubasiak, S. Liu, Multilevel bridge governor by using model predictive control in wavelet packets for tracking trajectories, in: IEEE International Conference on Robotics and Automation, vol. 4, 2004, pp. 4079–4084.
[38] J. Xu, J. Yang, J. Gao, An integrated kinetic energy recovery system with peak power transfer in 3-DOF mobile crane robot, in: IEEE International Symposium on System Integration, 2011, pp. 330–335.
[39] Welding. http://en.wikipedia.org/wiki/Welding.
[40] http://www.tradett.com/products/u40994p330952/igbt-mig.html.
[41] http://www.tradett.com/products/u61715p524833/igbt-arc-welding-machine.html.
[42] http://www.tradett.com/products/u38096p302559/mig-mag-welding-machine-with-igbt.html.
[43] http://www.indiamart.com/thakurindustries/inverters/inverter-welding-machine.html.
[44] http://www.vjmarcons.in/welding-machine.html.
[45] S. Marques, et al., Step down converter with hysteretic current control for welding applications, in: International Conference on Industrial Electronics, Controls, and Instrumentation, vol. 2, 1997, pp. 676–681.
[46] A. Maouad, et al., New design method for controlling power stages based on IGBT switching ferrite transformers: applied to an 8 kW small size light weight electric welding machine, in: IEEE Electronics, Circuits, and Systems Conference, vol. 2, 2000, pp. 802–804.

[47] K. Morimoto, et al., Dual utility AC voltage line operated soft switching PWM DC-DC power converter with high frequency transformer link for arc welding equipment, in: IEEE Conference on Electrical Machines and Systems, 2005, pp. 1084–1089.

[48] D. Dong, et al., Structure and control of an inverter type power source for robot arc welding, Tsinghua Sci. Technol. 3 (2) (June 1998).

[49] P. Vieira, et al., Mathematical modelling and digital control for power supplies of current pulsed for welding machine, in: IEEE European Conference on Power Electronics and Applications, 2005, pp. P1–P8.

[50] G. Majumdar, Advanced IGBT technologies for HF operation, in: IEEE European Conference on Power Electronics and Applications, 2009, pp. 1–26.

[51] Induction Heating. http://en.wikipedia.org/wiki/Induction_heating.

[52] Induction Hardening. http://en.wikipedia.org/wiki/Induction_hardening.

[53] Forging. http://en.wikipedia.org/wiki/Forging.

[54] www.nordson.com.

[55] http://industrials-heatr.com/industrial-heaters/thermal-oil-heater.html.

[56] http://www.tradett.com/products/u64394p570306/igbt-superaudio-heating-equipment.html.

[57] A.R. Quintas, et al., IGBT transistor switching in an unique industrial application, in: IEEE International Conference on Power Electronics and Variable Speed Drives, 1994, pp. 550–553.

[58] E.J. Dede, et al., On the design of a high power IGBT series resonant inverter for induction forging applications, IEEE Africon 1 (1996) 206–208.

[59] U. Schwarzer, R.W. DeDoncker, Power losses of IGBTs in an inverter prototype for high frequency inductive heating applications, in: IEEE Industrial Electronics Society, 2, 2001, pp. 793–798.

[60] V. Esteve, et al., Using pulse density modulation to improve efficiency of IGBT inverters in induction heating applications, in: IEEE Power Electronics Specialists Conference, 2007, pp. 1370–1373.

[61] Y. Uchihori, et al., The state-of-the-art electromagnetic induction flow-through pipeline package type fluid heating appliance using series resonant PWM inverter with self-tuning PID controller-based feedback implementation, in: IEEE Industrial Automation and Controls Conference, 1995, pp. 14–21.

[62] R. Fuentes, P. Lagos, J. Estrada, Self-resonant induction furnace with IGBT technology, in: IEEE Industrial Electronics and Applications Conference, 2009, pp. 1371–1374.

[63] J.-E. Yeon, et al., Field stop shorted anode trench IGBT for induction heating applications, in: IEEE Industrial Electronics Society Conference, 2012, pp. 422–426.

[64] B.J. Baliga, Semiconductor Device Having Rapid Removal of Majority Carriers from an Active Base Region Thereof at Device Turn-Off and Method of Fabricating This Device, U.S. Patent 4,782,379 (Issued November 1), 1988.

[65] H. Takahashi, et al., 1200-V reverse conducting IGBT, in: IEEE International Symposium on Power Semiconductor Devices and ICs, 1995, pp. 174–179.

[66] 650 V IGBT HB Series. www.st.com.

[67] Milling (machining). http://en.wikipedia.org/wiki/Milling_(machining).

[68] J. Zhang, N. Karrer, IGBT power amplifiers for active magnetic bearings of high speed milling spindles, in: IEEE Conference on Industrial Electronics, Control and Instrumentation, vol. 1, 1995, pp. 596–601.

[69] J. Walter, et al., High frequency generator for an inductive shrinking unit, in: IEEE European Conference on Power Electronics and Applications, 2005, pp. P1–P9.
[70] Electrical Discharge Machining. http://en.wikipedia.org/wiki/Electrical_discharge_machining.
[71] H. Huang, et al., A zero current switching half-bridge power supply for high speed drilling electrical discharge machining, in: IEEE Power and Energy Engineering Conference, 2009, pp. 1–5.
[72] Steel Mill. http://en.wikipedia.org/wiki/Steel_mill.
[73] Paper Mill. http://en.wikipedia.org/wiki/Paper_mill.
[74] G. Skibinski, W. Maslowski, J. Pankau, Installation considerations for IGBT AC drives, in: IEEE Textile, Fiber and Film Industry Technical Conference, 1997, pp. 1–12.
[75] P. Bhooplapur, AC drives in metal industries, in: IEEE International Conference on Power Electronics and Drives, 1997, pp. 823–828.
[76] T. Okamoto, et al., Development of low voltage 2 level IGBT inverter and converter for industrial applications, in: IEEE International Conference on Power Electronics, Paper WeP1-009, 2011, pp. 2466–2473.
[77] P. Bhooplapur, B.P. Schmitt, G. Neeser, HV-IGBT drive and their applications, in: IEEE International Conference on Power Electronics and Drive Systems, vol. 2, 1999, pp. 585–590.
[78] P. Fransen, Upgrade of paper machine winders with AC drives and the synchronous rectifier, in: IEEE Pulp and Paper Industry Technical Conference, 1996, pp. 94–116.
[79] P. Fransen, Upgrade of winders with AC drives and the synchronous rectifier, IEEE Ind. Appl. Mag. (1997) 20–33. September/October.
[80] Electrostatic Precipitator. http://en.wikipedia.org/wiki/Electrostatic_precipitator.
[81] N. Grass, W. Hartmann, M. Klockner, Application of different types of high voltage supplies on industrial electrostatic precipitators, in: IEEE Industry Applications Society Conference, vol. 1, 2002, p. 270276.
[82] N. Grass, 150 kV/300 kW high voltage supply with IGBT inverter for large industrial electrostatic precipitators, in: IEEE Industry Applications Society Conference, 2007, pp. 808–811.
[83] Textile Manufacturing. http://en.wikipedia.org/wiki/Textile_manufacturing.
[84] P. Paliwal, N. Kumar, T.R. Cheliah, Energy conservation studies on a spinning drive motor with scalar and vector controllers, in: IEEE Third International Conference on Advanced Computing and Communication Technologies, 2013, pp. 115–120.
[85] B-Wheel Excavator. http://en.wikipedia.org/wiki/Bucket-wheel_excavator.
[86] J. Mazumdar, Application of transfer switch in mining converters, in: IEEE Industry Applications Society Meeting, 2010, pp. 1–5.
[87] B.J. Baliga, et al., The insulated gate rectifier: a new three terminal MOS-controlled bipolar power device, in: IEEE International Electron Devices Meeting, Abstract 10.6, 1982, pp. 264–267.
[88] B.J. Baliga, et al., The insulated gate transistor: a new three terminal MOS-controlled bipolar power device, in: IEEE Trans. Electron Devices ED-31, 1984, pp. 821–828.
[89] H.-R. Chang, B.J. Baliga, 500-V n-channel insulated gate bipolar transistor with trench gate structure, in: IEEE Trans. Electron Devices ED-36, 1989, pp. 1824–1829.
[90] E. Motto, J.F. Donlon, The latest advances in industrial IGBT module technology, in: IEEE Applied Power Electronics Conference, vol. 1, 2004, pp. 235–240.

[91] T. Yamazaki, et al., Advanced IGBT chip technology for industrial motor drive applications, in: IEEE International Power Electronics Conference, 2010, pp. 783–789.
[92] A. Yamano, et al., 1,700 V 7th-Generation "X Series" RC-IGBT Modules for Industrial Applications, PCIM Europe, 2019, pp. 168–174.
[93] Y. Wang, et al., A new multi-functional compact IPM for low power industrial application, in: IEEE Applied Power Electronics Conf, 2017, pp. 3072–3075.
[94] P. Mishra, et al., Dead-time compensation and soft-starting of VSI fed high-power induction motor drive, in: IEEE Int. Conf. on Power Electronics and Energy, 2021, pp. P1–P6.
[95] B.J. Baliga, Gallium Nitride and Silicon Carbide Power Devices, World Scientific Press, Singapore, 2017.
[96] C. Tan, et al., Performance evaluation of an electric vehicle traction drive using Si/SiC hybrid switches, in: IEEE Int. Power Electronics and Motion Control Conf, 2021, pp. 278–283.
[97] Variable speed drives in appliances to double, Power Electronics Technology Magazine (January 2010) 9.

CHAPTER

IGBT Applications: Lighting

11

The availability of electric lighting has had a profound impact on society, improving our productivity and quality of life. The International Energy Agency estimates that nearly one-fifth (20%) of global electricity production is used for lighting [1]. According to the US Energy Information Administration, 219 billion kWh of electricity were consumed in the United States for lighting in 2020 [2]. The Department of Energy (DOE) estimates that an average residential building contains more than 50 lamps while the average industrial or commercial building has 300 lamps [3]. About 72% of the energy used for lighting is consumed in the commercial sector while 28% was used by the residential sector. Within the typical home, 13% of the electricity is used for lighting [4], the second largest amount, as shown in Table 11.1.

The incandescent lamp powered with electricity has provided an excellent light source in homes and factories for more than a century. The incandescent lamp, shown in Fig. 11.1(a), consists of a wire filament encased within a glass globe [5]. The passage of an electrical current through the filament produces heating because of the resistance of the wire. In a 60 W, 120 V incandescent lamp, the filament (shown in Fig. 11.1(b)) is composed of a tightly wound coil consisting of a tungsten wire with a diameter of 0.0046 cm and a length of nearly 6 m [5]. The filament is protected from oxidation by using an inert ambient inside the glass enclosure.

Attempts to produce light using electrical current can be traced back to Sir Humphry Davy in 1802. Many attempts were performed to prevent the filament from burning out in a short time. It was Thomas Edison who set out to not only create a practical light bulb with a long-lasting filament but provide the infrastructure for delivering electrical power to the lamps. By 1885, 300,000 light bulbs were in service. The creation of ductile tungsten for the filament by the General Electric Company in 1906 was another milestone in making the practical light bulb feasible. This accelerated the sales of incandescent light bulbs from 88 million in 1914 to 800 million by 1945.

Unfortunately, more than 95% of the electrical energy is converted into heat in an incandescent lamp to bring the temperature of the filament to about 3000°C. The luminous efficacy of energy radiated from a lamp is measured in lumens per watt. A typical 60 W incandescent bulb produces 14.5 lm/W. The low efficiency of incandescent lamps produces not only a large amount of wasted power but heats the local environment, leading to larger costs for air conditioning. In addition, the life of a typical incandescent lamp is only 2000 h.

The IGBT Device. https://doi.org/10.1016/B978-0-323-99912-0.00010-6

Table 11.1 Electrical Energy Consumption Within the Home

End Use	Quadrillion BTU	Billon kWh	Percent Share
Space cooling	0.93	273	19
Lighting	0.63	186	13
Water heating	0.45	131	9
Refrigeration	0.38	110	8
Color TV	0.32	93	7
Space heating	0.27	79	6
Total	4.86	1424	100

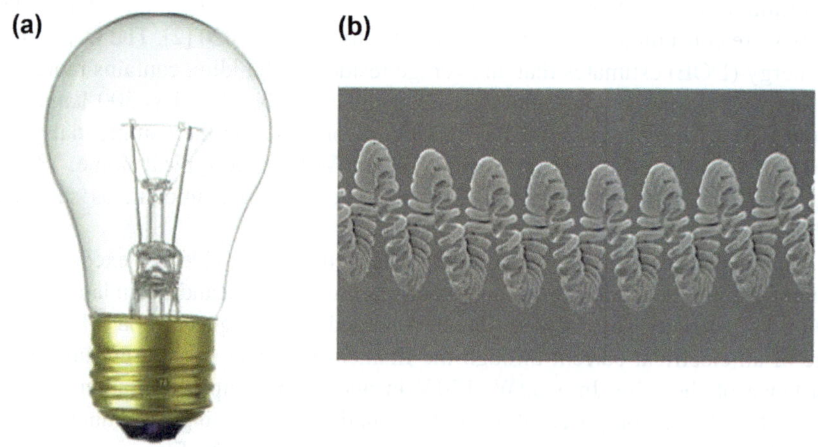

FIGURE 11.1

(a) Incandescent light bulb; (b) filament.

The poor efficiency of incandescent lamps and growing concerns about electricity generation on the environment prompted worldwide action to replace them with more efficient light bulbs. Governments across the globe began to phase out the availability of incandescent bulbs in 2005 [6]. The first countries to begin phasing out incandescent bulbs were Brazil and Argentina in 2005. In the United States, the Energy Independence and Security Act of 2007 was enacted to implement replacement of incandescent bulbs. Australia has disallowed incandescent bulbs since 2010. The European Union effort (Ecodesign) to phase out incandescent lamps has been implemented since 2011. China banned the import and sales of incandescent bulbs starting in October 2012 and India implemented the "Bharat Yojana" program to replace 400 million incandescent bulbs with compact fluorescent lamps (CFLs) by 2012. All CFLs utilize IGBTs in the electronic ballast.

The creation of the blue light-emitting diode (LED) allowed the development of LED lamps for residential and commercial environments. The stability of LEDs was poor at the initial stages, impeding commercialization. Once this "droop" issue was resolved, LED lamps became more widely commercialized after 2015. The LED is based on a P-N junction formed in a wide bandgap semiconductor material (GaN). It requires maintaining a constant current through the junction with a voltage of about 3 V. A string of LEDs in series is usually required to generate sufficient light for residential and commercial settings. The electrical energy to drive the LEDs is provided from the AC power supplied by utilities with power electronics to convert the high AC voltage to a lower DC voltage. This function can be performed efficiently using IGBTs.

11.1 TRIAD Incandescent Lamps

An early example of the effort to develop more energy efficient lamps with IGBTs for residential use is the TRIAD project at the General Electric Company [7]. The authors stated goal for the project initiated in 1981 was "*to develop a low-cost incandescent lamp with substantially improved efficacy, life, and color that could be introduced into the residential market as a premium product.*" It utilized a low-cost glass halogen filament tube with reflective IR films and a low-voltage filament. Unlike the incandescent lamp, the halogen lamp contains a smaller quartz envelop with halogen gas that combines with evaporating tungsten atoms and redeposits them on the filament. This allows operation of the filament at higher temperatures to obtain more light per input energy while producing longer life for the filament [8,9].

The lamp configuration GE chose for product development in 1986 was a parabolic aluminized reflector (PAR) lamp [10], widely used for commercial and residential recess lights, locomotive headlamps, and aircraft landing lights. The PAR lamp assembly is shown in Fig. 11.2(a) with the circuit board containing IGBTs shown in Fig. 11.2(b) [7]. The electronics to control the power fed to the lamp filament must be housed in the piece between the reflector and the screw-in socket.

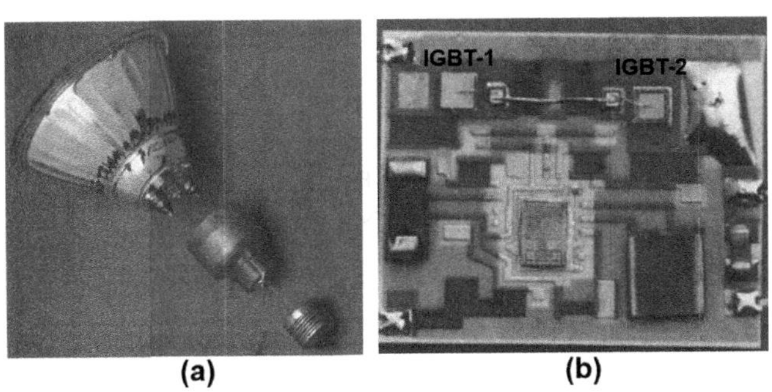

FIGURE 11.2

TRIAD lamp: (a) assembly; (b) electronic ballast. *IGBT*, insulated gate bipolar transistor.

The author's state: "*The IGBT was invented in this same time frame and was selected as the most appropriate power device from an efficiency point of view.*"

The availability of the IGBT allowed creating a new power control strategy called "reverse phase control." Conventional phase control is performed using thyristors or triacs, which turn off when the voltage reverses across the device in an AC power circuit. The power delivered to the load is controlled by triggering the thyristor on during a fraction of the sinusoid when the anode voltage is positive. This generates a high in-rush current in the device and produces high electromagnetic interference (EMI) because of the resulting large dI/dt. With the IGBT, reverse phase control becomes possible by turning on the device when the AC voltage swings from negative to positive and then turning off the IGBT during the positive half-cycle to provide the required load voltage. For a 36 V, 60 W lamp, a 25% conduction time during each cycle is needed to obtain 36 V from a 120 V line power source.

The reverse phase control circuit is shown in Fig. 11.3. It contains two IGBTs with fly-back diodes to achieve forward and reverse blocking capability. The load current is monitored by using the on-state voltage drop of the IGBT. This takes advantage of the fact that the forward drop of the IGBT is relatively insensitive to temperature rise in the chip [11]. The rate of rise of the IGBT collector voltage during its turn-off is limited to 5 V/μs by using gate resistance (R_G) to achieve the desired limit of conducted EMI established by the Federal Communications Commission (FCC).

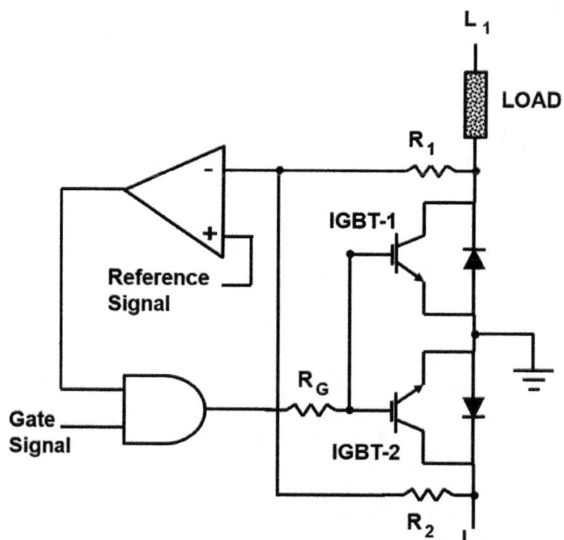

FIGURE 11.3

TRIAD lamp circuit for reverse phase control. *IGBT*, insulated gate bipolar transistor.

11.2 Compact Fluorescent Lamps

There has been a worldwide movement toward replacing incandescent bulbs due to their very poor efficiency (2%) in the conversion of electric power into usable light. The most cost-effective option for replacement of the incandescent lamp from 1990 to 2015 was the compact florescent lamp (CFL) [12]. They remain cost competitive with LEDs even today. The form factor of the CFL is designed to be compatible with the space and fixture used previously for incandescent lamps to make consumer adoption easier. The CFL contains a curved or folded tube, as shown in Fig. 11.4 (a), with a screw-in base compatible with lamp fixtures in homes and offices. The electronic ballast to operate the CFL, shown in Fig. 11.4(b), must be designed to fit in the small space between the tube and the screw-in base. The electronics are located close to the gas discharge tube, creating an elevated temperature at the base. The electronic ballasts were created using IGBTs because of their excellent performance at elevated temperatures [11].

A typical CFL consumes only 15 W of power for producing the same amount of light (lumens) when replacing a 60 W incandescent bulb [13] and lasts eight times longer [12] (typically 6000–15,000 h). The luminous efficacy (number of lumens produced for each watt of electrical power) is 50–70 lm/W for the CFL compared with only 1.5–2.5 lm/W for the incandescent lamp. In addition, the CFLs have a much longer life of 10 years, making them attractive for residential and consumer installations.

Since 1999, CFL sales have grown rapidly in the United States as reported by various sources [14]. Walmart has promoted the sale of CFLs by dropping prices and in November 2006 announced a goal of selling 100 million CFLs by the end of 2007, a target that was met by October 2007. Unfortunately, CFL sales in the United States dropped by 25% in 2008 from their 2007 levels and by 49% in

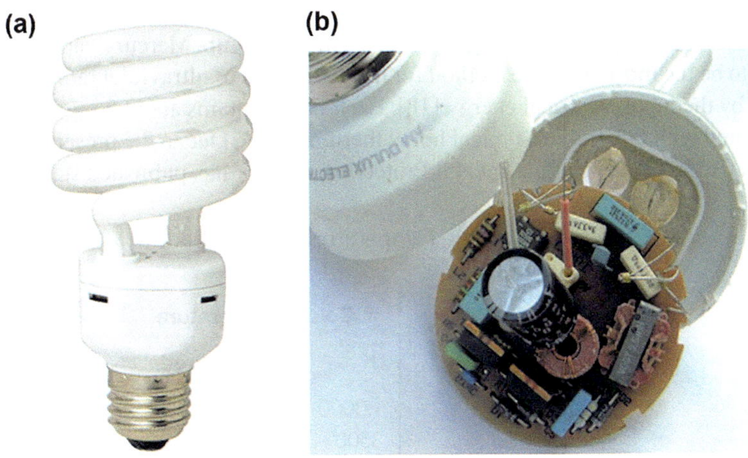

FIGURE 11.4

(a) Compact fluorescent lamp; (b) electronic ballast.

2009 from their 2007 levels [15]. In contrast, the adoption of CFLs in the rest of the world has been progressing monotonically [16]. The Worldwatch Institute states: "*In the United States, CFLs accounted for more than 20% of sales in 2007. But other wealthy nations have shown much higher CFL use rates for quite some time, including 80% of households in Japan and 50% in Germany by 1996. Many developing countries have shown strong CFL market share in recent years as well: 14% of sales in China in 2003 for instance and 17% in Brazil in 2002.*" Australia became the first country to ban the sale of incandescent bulbs in 2007. The European Union, Ireland, and Canada have passed similar legislation. Worldwatch states: "*In total, more than 40 countries have announced plans to follow suit*".

Even India has launched a program [17], called Bachat Lamp Yojana, to replace the 400 million incandescent bulbs in use in the country with CFLs. In 2008, 199 million CFLs were sold in India. More than 20 billion CFLs were sold and in use worldwide by 2020.

11.2.1 CFL Light Emission

The light emitted from the CFL tube is determined by the phosphor coating on the bulb. The mix of phosphors is optimized based on the desired color of the emitted light, the efficiency target, and the cost of production. Typically, a mixture of 3–4 phosphors is used by manufacturers to obtain while light. The quality of the emitted light is described by its color temperature—the temperature of a black body source with the same chromaticity as the light source. The color temperature is quantified in Kelvins (the temperature of the equivalent black body source) or mired (which is calculated by dividing 1 million by the color temperature in Kelvins). The names given to CFLs by manufacturers correspond to various color temperatures, as shown in Table 11.2.

The emission of white light in the CFL from the phosphor coating is stimulated by ultraviolet radiation generated inside the gas within the bulb. Mercury vapor has been found to be the most effective method for the generation of ultraviolet radiation stimulated by the input electrical energy [18]. Each CFL contains about 5 mg of mercury (compared with 500 mg in a typical home thermometer). Due to the presence of mercury in CFLs, the US Environmental Protection Agency has published standards for

Table 11.2 Color Temperature for Compact Fluorescent Lamps

CFL Name	Color Temperature	
	°K	Mired
Warm/soft white	<3000	>333
Bright white	3500	286
Cool white	4000	250
Daylight	>5000	<200

disposal of these types of lamps. Many home supply stores (such as Home Depot and Lowes) provide free recycling of CFLs as a convenience for consumers.

Filaments coated with boron are located at each end of the tube in the CFL. An electrical discharge is produced when a voltage is applied across a tube containing a gas at a reduced pressure. The gas pressure must be between 5 and 15 Torr to allow significant current flow through the gas [19]. The flow of current through the tube occurs by the acceleration of electrons from the cathode electrode to the anode electrode. The electrons collide with the mercury atoms, transferring some of their energy to create a mercury atom in an excited state. The excited mercury atom then returns to its ground state by releasing energy at a wavelength of 254 nm. Consequently, the electrical discharge of mercury vapor in a CFL produces ultraviolet radiation. The ultraviolet radiation is absorbed by the phosphor coating on the CFL lamp surface and reemitted as white light.

11.2.2 Half-Bridge Ballast Topology

The electronic ballast for the CFL must perform several operations in a carefully controlled sequence. In the first step, the filaments must be preheated to improve the electrical discharge of the gas in the tube. This step also increases the life of the filament and the lamp. During the second step, gas discharge is initiated by producing a high voltage across the electrodes. After this, the CFL can operate continuously with current flow in the lamp regulated by the control circuit.

Three options for the ballast topology for the CFL application have been considered [20] and compared in Table 11.3. Among these, the half-bridge configuration with a voltage-fed series resonant topology is the best option from a cost and design standpoint. It requires power devices with the lowest voltage rating of 700 V for the European 220 V AC power source, which reduces the die size and cost of the devices.

The commonly used half-bridge circuit topology for providing power to the CFL tube is shown in Fig. 11.5. A series resonant circuit is formed using the inductor L and the capacitor C_{R1}, which is connected across the terminals of the tube. The resonant frequency of the circuit is:

Table 11.3 Topologies for Compact Fluorescent Lamp Ballast

	BV (CES)	Peak Current	Complexity	Cost	Concept
Fly-back	1200 V	0.4 A	High	Medium	Parallel resonant
Push-Pull	1600 V	0.4 A	Medium	Medium	Parallel resonant current fed
Half-bridge	700 V	0.35 A	Low	Low	Series resonant voltage fed

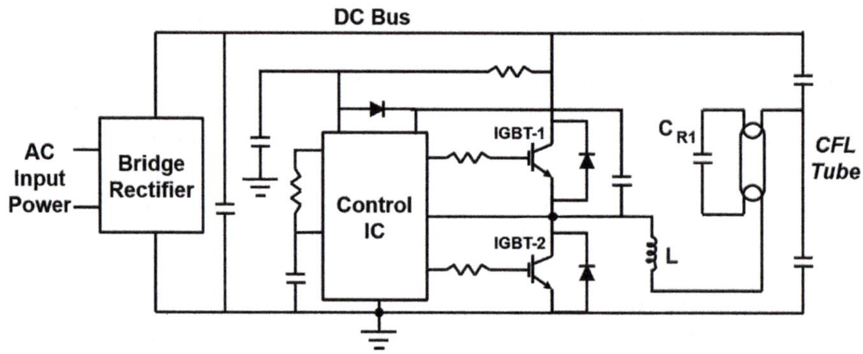

FIGURE 11.5

Typical half-bridge circuit topology used to operate compact fluorescent lamps (CFLs).

$$f_{R1} = \frac{1}{2\pi\sqrt{L \cdot C_{R1}}} \tag{11.1}$$

The impedance of the resonant circuit is:

$$Z_S = R + j2\pi fL + \frac{1}{j2\pi fC_{R1}} \tag{11.2}$$

At low frequencies, the impedance of the capacitor is dominant while at high frequencies, that of the inductor becomes dominant. At the resonant frequency, the impedance becomes equal to the resistance value.

During the preheating step for the CFL filament, the control circuit operates the resonant circuit at a high frequency located above the resonant frequency, as indicated by point A in Fig. 11.6. This produces a high current to heat the filament rapidly. In the second step, a high voltage is generated across the tube to produce an arc discharge. This is achieved by moving the operating frequency close to resonance, as indicated by point B in the figure. The voltage across the tube is equal to the voltage across the resonant capacitor C_{R1}:

$$V_{\text{Tube}} = Q \cdot V_{\text{DCBus}} \tag{11.3}$$

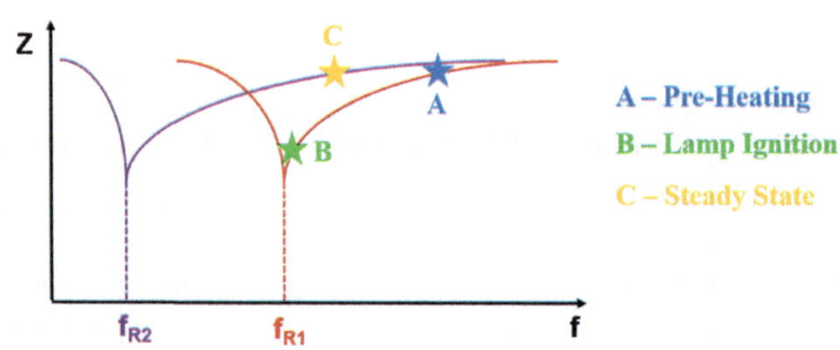

FIGURE 11.6

Operating points for the resonant circuit in the compact fluorescent lamp.

where Q is the quality factor of the resonant circuit and V_{DCBus} is the DC bus voltage (300 V for the case of 220 V AC input power). The quality factor can be designed to be more than 10 to generate the desired high voltage to ignite the gas discharge. For continuous operation of the CFL, the resonant frequency of the circuit is reduced by adding additional capacitance (not shown in Fig. 11.5) to a new value f_{R2}. The control circuit then operates the circuit at point C shown in Fig. 11.6. Because this frequency is well above the new resonant frequency, the impedance is dominated by the inductor, which determines the steady-state current level. The impedance of the fluorescent tube is low after producing the arc in the gas. A typical steady-state operating frequency for the CFL ballast is 40 kHz.

It is worth pointing out that a reverse conducting diode must be placed across each of the IGBTs to carry the inductor current in the reverse direction. The integration of the fly-back diode into the IGBT structure (see Fig. 2.7) provides an optimal solution for the CFL application because this reduces the tail current of the IGBT as well during turn-off. This is important for reducing the energy loss per cycle and allowing the IGBT to operate at a relatively high frequency (40 kHz) without exceeding a junction temperature of 175°C in spite of an ambient temperature of 90°C in the lamp base.

The resonant circuit current and voltage waveforms for the IGBT are illustrated in Fig. 11.7 during steady-state operation of the CFL. The maximum collector current for the IGBT is 0.4 A with a maximum collector voltage of 300 V. The IGBTs are turned off with zero voltage switching (ZVS) conditions. This greatly reduces the switching losses, allowing the IGBTs to operate at the relatively high circuit frequency (40 kHz).

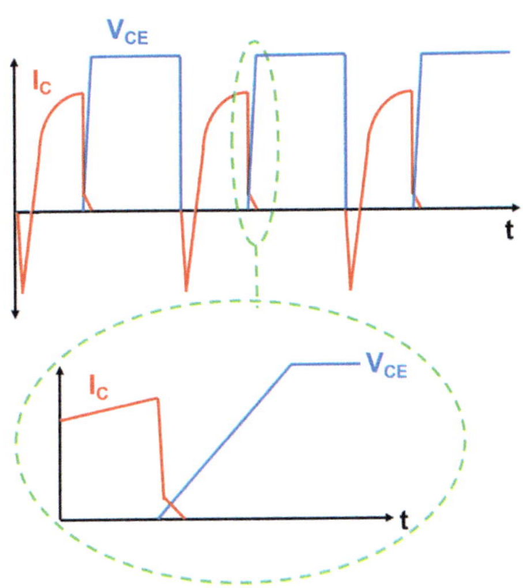

FIGURE 11.7

Compact fluorescent lamp resonant circuit waveforms.

11.2.3 Power Transistor Comparison

In 1996, Motorola published an application note (AN1576) that compared the available transistor options [20] for the half-bridge series resonant voltage-fed topology discussed above. They examined the power bipolar transistor, the power MOSFET, and the PowerLux IGBT. Their conclusions are summarized in Table 11.4. The bipolar power transistor has a long storage time that is strongly dependent on the reverse base drive current. This limits its high-frequency operating capability. The device also requires a very high open-base breakdown voltage. The power MOSFET has excellent gate drive and switching performance suitable for high-frequency applications. However, at the high breakdown voltage required for the CFL circuit, its on-resistance is large, leading to a large die size and cost. The PowerLux IGBT design provides the simplicity of gate drive with acceptable power loss at the CFL operating frequency.

The relative performances of the three types of transistors are compared in Table 11.5. The high on-resistance of the power MOSFET requires a 30% larger die size when compared with the bipolar transistor. Both these transistors must be housed in a large package (DPAK) with a footprint of 6.5 × 6.5 mm. In contrast, the PowerLux IGBT has a die size that is 60% smaller than the bipolar transistor and 70% smaller than that of the power MOSFET. When combined with the simplicity of the gate drive, the IGBT offers the lowest cost solution for the CFL application.

Table 11.4 Comparison of Power Transistor Types

Transistor (1996)	BV(CEO)	V(sat)	FBSOA	RBSOA	Storage Time	Fall Time
Bipolar	High 1600 V	Low	Second BV	V(BE) bias dependent	High	Medium
MOSFET	Medium 600 V	High	Square	Square	Small	Very fast
IGBT PowerLux	Medium 600 V	Medium	Second BV	Gate bias dependent	Small	Fast

IGBT, *insulated gate bipolar transistor.*

Table 11.5 Comparison of Power Transistor Performance

Transistor (1996)	Relative Die Size	Package	Tj MAX	Drive	Relative Cost	Dimming
Bipolor	1.00	DPAK	175°C	Complex	Medium	Complex
MOSFET	1.30	DPAK	175°C	Very easy	High	Easy
IGBT PowerLux	0.40	SOT223	175°C	Very easy	Low	Easy

IGBT, *insulated gate bipolar transistor.*

In addition, its smaller footprint is important because of the limited space for the printed circuit board (PCB) located in the base of the CFL. The diameter of the PCB is restricted to less than 4 cm. Due to its smaller size, the IGBT chip can be housed in the SOT223 package with a size of 6.5 × 3.5 mm, reducing the footprint on the PCB by 50%. This early work established the IGBT as the lowest cost technology for the development of CFLs for consumer residential use.

11.2.4 Self-Resonant Ballast Topology

The self-resonant half-bridge topology is another option for delivering power to the tube in the CFLs [20]. This is a low-cost CFL ballast option because it eliminates the control or drive IC. Instead, a transformer (T_1) is used to provide positive feedback drive to the two IGBTs, as shown in Fig. 11.8. The gate drive network (R, D, C) is included to assist the dynamic response of the IGBTs. A Zener diode (Z_1, Z_2) is also included between the gate and emitter of each IGBT to limit the gate voltage. Some of the disadvantages of the self-resonant ballast are transformer saturation, no protection if the lamp fails to strike, and no capability to detect open filaments.

11.2.5 Power Factor Correction

Regulations do not require power factor correction (PFC) and reduction of third harmonic distortion (THD) in the case of low-power lamps (below 25 W). Low-cost lamp ballasts for these low-power levels have a power factor of about 0.5 and 100% TDH. However, European regulations require CFLs to have a power factor of at least 0.94 with a maximum THD of 25%.

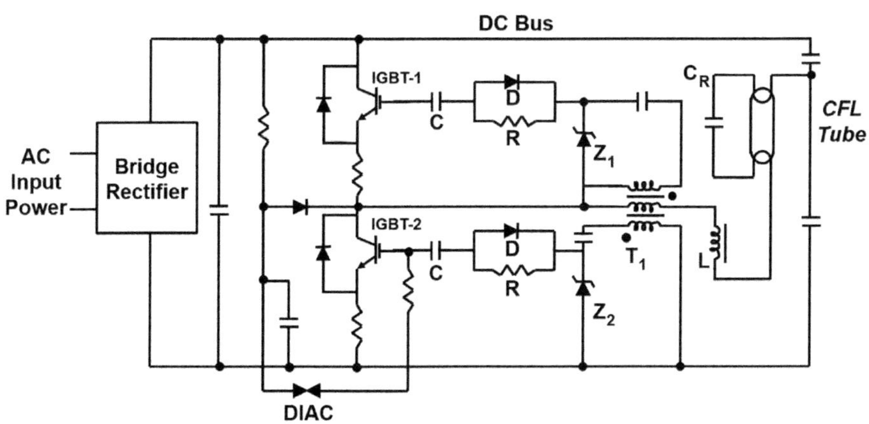

FIGURE 11.8

Self-resonant circuit topology for compact fluorescent lamps. *IGBT*, insulated gate bipolar transistor.

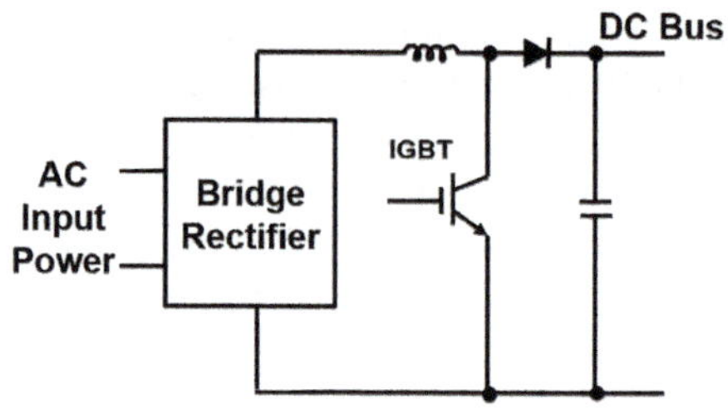

FIGURE 11.9

Power factor correction for compact fluorescent lamps. *IGBT*, insulated gate bipolar transistor.

Commercial CFL ballasts utilize small DC link capacitors due to space and cost constraints. This leads to large ripple currents. A simple low-cost PFC circuit for CFLs is shown in Fig. 11.9. This boost topology is the most popular approach [21,22]. The PFC circuit draws sinusoidal input current, which is in phase with the input AC source voltage. It regulates the DC bus voltage, allowing more stable operation of the CFL.

11.2.6 Discrete IGBT Designs for CFLs

Many companies have developed optimized IGBT products specifically for lamp ballast applications due to the large market for these devices. One example is the "LightMOS" IGBT product available from Infineon [23]. In this device structure, a reverse conducting diode has been integrated into the IGBT structure, as shown in Fig. 11.10. This is achieved in the LightMOS IGBT by starting with a high-resistivity N-type

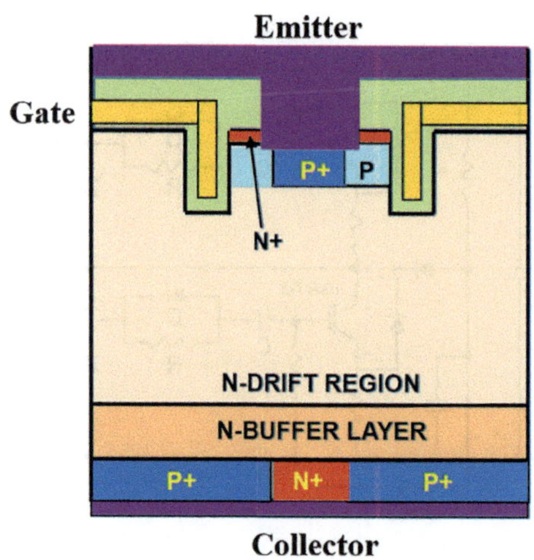

FIGURE 11.10

Insulated gate bipolar transistor with built-in reverse conducting diode for compact fluorescent lamps.

wafer that is first processed to create the upper structure containing the gate, the P-base, and N^+-emitter regions. A trench gate structure is utilized to increase channel density. The wafer thickness is reduced by grinding and polishing to about 70 μm. The N^+ and P^+ regions shown in the figure on the back of the wafer are created by masking and ion implantation of appropriate dopants. The elimination of a separate reverse conducting diode saves space in the CFL ballast and reduces the cost of the additional diode package, die attachment, and wire bonding.

One of the problems encountered when the reverse conducting diode is integrated into the IGBT is a snapback in the on-state characteristics [24]. This occurs because the injecting P-N junction is shorted by the ohmic current path through the cathode region of the diode. For the IGBT, this produces a MOSFET current path at low current levels, as illustrated in Fig. 11.11. Once a sufficient forward bias is applied, the P-N junction begins to inject minority carriers into the drift region, producing conductivity modulation. This reduces the resistance of the drift region, causing a snapback in the *i-v* characteristics, as shown in the figure. The snapback can be reduced by proper scaling of the N^+ region formed on the back of the IGBT chip, as discussed in Chapter 2 [23].

An important advantage of the IGBT for CFL application is that its on-state voltage drop remains relatively constant with increasing temperature [11]. In contrast, the on-state voltage drop of the power MOSFET (including the CoolMOS devices) increases by a factor of two from 25°C to 125°C. The authors state [23]: "*Summarizing, the LightMOS (IGBT) offers an excellent price performance ratio for this (lamp ballast) extreme cost-sensitive segment of applications.*"

The optimization of the reverse recovery behavior of the reverse conducting diode integrated into the IGBT structure must be performed while taking into account impact on the IGBT characteristics [25]. The use of nonuniform lifetime control with helium irradiation and the adjustment of the doping of the upper P^+

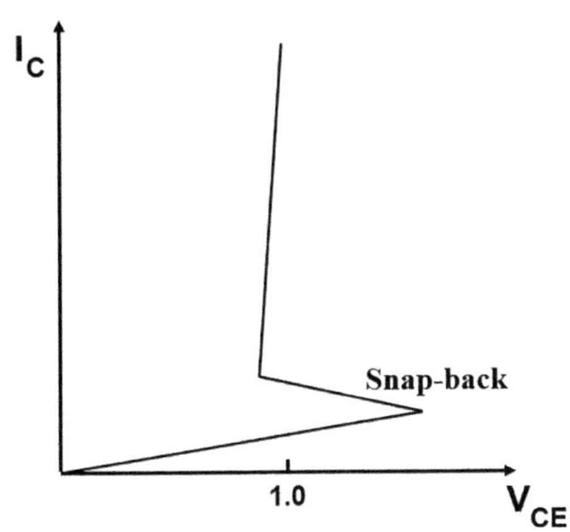

FIGURE 11.11

On-state characteristics of the insulated gate bipolar transistor with built-in reverse conducting diode for compact fluorescent lamps.

region in the IGBT cell structure have been explored. This P^+ region is usually doped very heavily to suppress latch-up in the IGBT. However, in the RC-IGBT structure, the upper P^+ region serves as the anode of the reverse conducting diode. The injection level and profile of the stored charge in the diode can be improved by reducing the doping concentration of the upper P^+ region.

11.2.7 Integrated IGBT Designs for CFLs

The integration of the control circuit and the IGBT power devices is attractive for reducing the space occupied by the ballast in the CFL. The integration of the IGBT devices with control circuits requires bringing all three terminals of the device to the upper surface. This can be achieved using a lateral device structure, as shown in Fig. 11.12. Supporting a high voltage in a lateral device structure can be done with the Reduced Surface Field (RESURF) principle first proposed using junction isolation [26]. Although this approach is suitable for integration of power MOSFETs, it is not suitable for the integration of a bipolar device such as the IGBT due to strong injection of current into the substrate, which produces interaction between adjacent devices on the chip.

The problem was solved by using the RESURF principle with dielectric isolation [27]. The dielectric layer between the active region of the lateral IGBT and the substrate prevents injection of carriers into the substrate and allows isolation of multiple IGBTs on the same monolithic chip. The first lateral IGBTs fabricated on dielectrically isolated wafers using the RESURF principle were reported in 1992 [28]. Since then, many reports on power-integrated circuits have been published with lateral

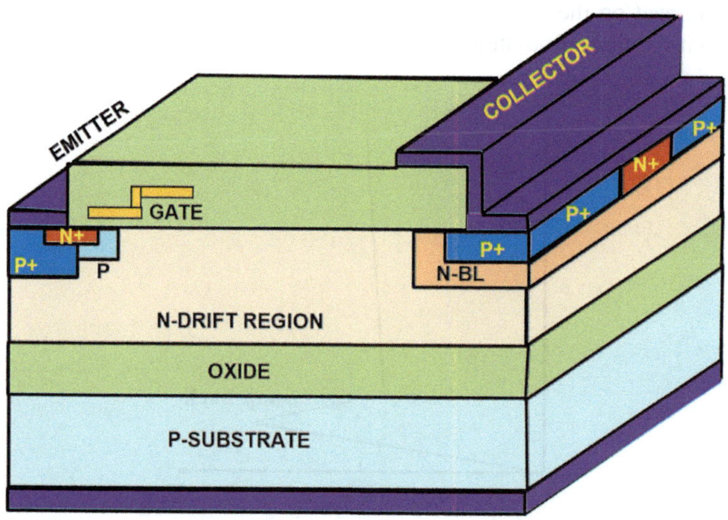

FIGURE 11.12

Lateral IGBT with built-in reverse conducting diode for compact fluorescent lamps.

IGBTs as power devices. The development of power integrated circuits (ICs) for off-line applications such as lamp ballasts and half-bridge inverters for motor drives was reported by Siemens [29] in 1995. They successfully integrated lateral IGBTs and fly-back diodes on a monolithic chip with 600 V blocking capability and several amperes of current-handling capability.

The lateral IGBT monolithically integrated in a power IC chip must be capable of switching at a high frequency for the CFL ballast application. The design of the structure must be engineered without resorting to lifetime control because electron irradiation would alter the performance of other devices integrated on the chip. Lateral IGBTs capable of operating at 20 kHz have been produced by using the structure shown in Fig. 11.13. Here, a lightly doped P-region is included on the collector side to reduce the injection efficiency of the device in the on-state. Devices with an on-state voltage drop of 3.2 V at 3.0 A of collector current have been integrated with diodes and control elements [30]. A shallow n-type ion implant is also used under the gate electrode to reduce the junction field effect transistor (JFET) resistance.

Lateral IGBTs with high blocking voltage capability have also been developed for use in plasma display panel drives [31–33]. This technology can also be utilized for building monolithic chips for CFL ballasts. A CFL ballast IC using power MOSFETs has been developed with junction isolation [34]. The development of a CFL ballast housed within a single package is motivated by space constraints. The space available for the ballast is less than 3 mm^2 because the diameter of the PCB is less than 6 cm for a 20 W lamp. The control IC in the SiP package performs the preheating mode, the ignition mode, the running mode, the dead-time control mode, and the shutdown mode. The size of the power devices can be reduced by replacing the lateral RESURF power MOSFETs with dielectrically isolated RESURF IGBTs. This has been demonstrated using triple trench gate lateral IGBTs [35]. Three trench gate electrodes are used per emitter region to reduce the on-state voltage drop of the

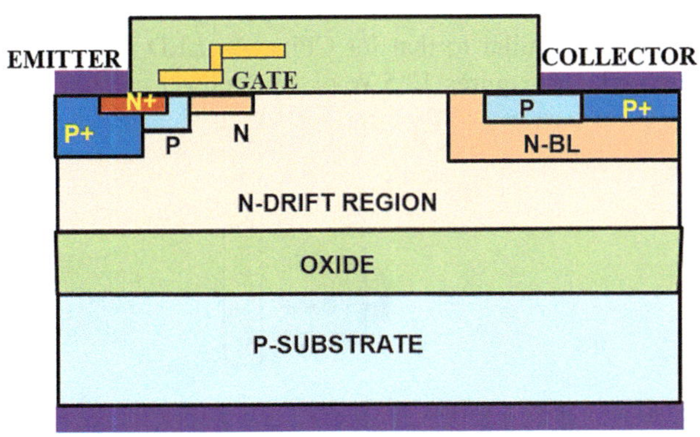

FIGURE 11.13

Improved lateral insulated gate bipolar transistor for high-frequency operation.

lateral IGBTs. In addition, a deep trench process is used for the isolation of adjacent IGBTs and the fly-back diodes with 0.8 μm technology.

11.3 Light-Emitting Diodes

During the last 5 years, LED lamps have become available at reasonable prices for residential and commercial lighting. The market for LED lighting has grown to $51 billion in 2020 and occupies about 50% of lighting sales. The LED is basically a P-N junction diode fabricated from direct large energy bandgap semiconductor material, such as gallium nitride [36]. Current flows through the diode with the injection of minority carriers on both sides of the junction when the P-N junction is forward biased. These carriers recombine, releasing energy in the form of photons. The energy bandgap of the semiconductor must be in the range of 3 eV to obtain the emission of visible light. A larger amount of light can be derived from the LED by increasing the current flow through the diode. Although an LED lamp is far more efficient than an incandescent lamp, a considerable fraction of the power is still converted into heat. The light output from the LED decreases when the junction temperature increases. Consequently, each LED is operated using about 0.35 A of current as a compromise between light output per LED and thermal management inside the lamp.

The typical housing for the LED, shown in Fig. 11.14, is designed to resemble that for the incandescent lamp with an identical screw-in base to be compatible with fixtures in homes and offices. Multiple LEDs are placed inside a lamp to generate the desired light output, as shown in the figure on the right side.

The efficiency of the LED is described by its efficacy—the ratio of output light to input electrical power—is units of lumens per watt. The efficacy of a packaged LED lamp can be substantially less than the diodes due to losses in the driver and optics. The DOE is targeting a packaged LED efficacy of 266 lm/W but products in the market can range from 10 to 120 lm/W [37]. Most products have an efficacy of 40–80 lm/W, which is similar to that for CFLs. An LED lamp that can replace a 60 W incandescent lamp requires 12.5 W of input power compared with 15 W for the CFL.

FIGURE 11.14

Light-emitting diode lamp.

LEDs require a constant DC current flow through the junction to maintain a constant light output. The lamp must contain an electronic ballast that converts the input AC power to a regulated DC current through a string of LEDs. The efficiency of the driver is typically 85%.

11.3.1 LED Driver

As in the case of CFL lamps, the driver for the LEDs must fit within the small space in the base of the lamp. A driver must convert the available 120 or 220 V AC power into a lower DC voltage to drive the LEDs. A resonant circuit class-E topology using a piezoelectric transformer [38] has been demonstrated for driving LEDs. Piezoelectric transformers are much smaller in size and weight when compared with conventional inductive transformers. The piezoelectric transformer operates using acoustical coupling rather than magnetic coupling used in the conventional transformer [39]. The input voltage applied across the bar of piezoelectric material produces a mechanical vibration at the resonant frequency of the material, which is typically between 0.1 and 1 MHz. The vibration is transferred to the secondary side of the bar, creating a voltage on the secondary side [40]. The equivalent circuit for the piezoelectric transformer is shown in Fig. 11.15. Because the piezoelectric transformer operates with electromechanical energy transfer, the EMI is reduced when compared with conventional transformers.

The driver circuit for LEDs using the piezoelectric transformer is a class-E resonant converter, as shown in Fig. 11.15. The ZVS of the IGBT is used to reduce switching power loss. This topology has the advantages of needing only one power switch and no high side driver (as required for the half-bridge topology). Because an AC signal is produced at the secondary side of the transformer, a set of back-to-back LEDs can be used to produce the desired light output. Regulation of the LED current can be performed using an auxiliary tap on the piezoelectric transformer to change the pulse width modulation (PWM) gate drive for the IGBT.

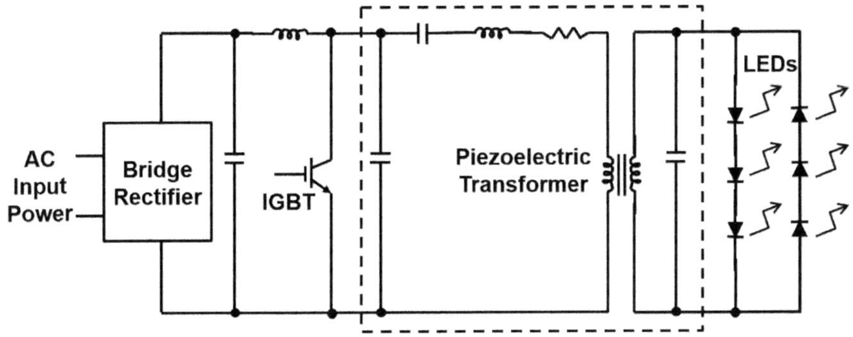

FIGURE 11.15

Light-emitting diode driver using piezoelectric transformer. *IGBT*, insulated gate bipolar transistor.

11.3.2 Conventional LED Driver

The more commonly used LED driver circuit is shown in Fig. 11.16 [41]. A bridge rectifier is used to create a DC bus voltage across capacitor C_B. The power MOSFET is switched to control the current flowing through the primary of the transformer and regulate the current delivered to the LEDs on the secondary side. The diode and capacitor on the secondary side create a DC voltage across the series string of LEDs. The author states: "*...the power losses of the MOSFET account for 82–97% of the total driver losses over the input voltage range. ... Because the MOSFET operates in linear mode at a low switching frequency, the MOSFET power losses are approximately equal to its conduction loss...*" High-voltage power MOSFETs have a large on-resistance, degrading the driver efficiency. Replacing them with IGBTs would allow the reduction of power losses and the power device footprint in the small space within the lamp housing.

The control IC in Fig. 11.16 provides many functions beyond the gate drive signals to the power device. One example of a commercial control chip is the Fairchild FL7733A product (now manufactured by On Semiconductor) [42] for use with low- to mid-power (5–60 W) LED lamps. Its features include: (a) LED short/open protection; (b) output diode short protection; (c) sensing resistor short/open protection; (d) overvoltage protection; (e) undervoltage lock-out; and (f) overtemperature protection. The control IC maintains constant current for the LEDs within ±3% under all operating conditions, such as input AC voltages ranging from 80 to 308 V rms. It includes PFC to maintain a power factor of >0.9 with third harmonic distortion of <10%.

11.3.3 Multiple Series/Parallel LED Driver

An inverted buck converter circuit has been demonstrated for controlling a string of LEDs connected in series, as shown in Fig. 11.17 [43]. The input AC voltage is rectified using the diode bridge and fed to an inductor L_B connected in series with the LEDs. Multiple power MOSFET switches (T_1 ...T_4) are used to provide an alternate path for the current through each LED string whenever the MOSFET is turned on.

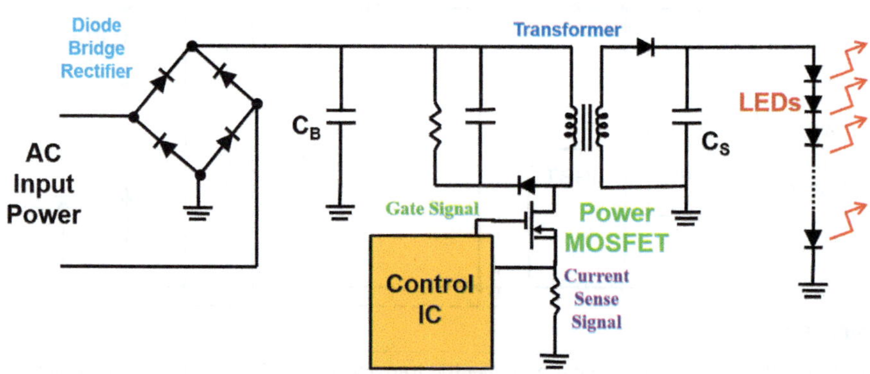

FIGURE 11.16

Control circuit for driving a string of light-emitting diodes.

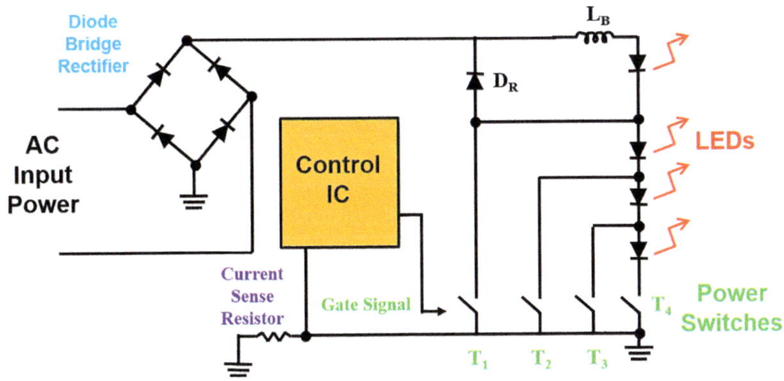

FIGURE 11.17

Control circuit for driving series/parallel combination of light-emitting diodes.

The inductor current circulates through the diode D_R when the MOSFETs are turned off. The circuit was tested with 20 V across each LED segment. The total voltage drop across all four LED segments was 80 V. The control circuit can be similar to one described in the previous section. The measured efficiency was 94% with a power factor of 0.92. This approach requires many power MOSFETs that can occupy significant space on the board.

A more elegant approach for implementation of the driver for multiple LEDs in series and parallel strings employs lateral IGBTs (LIGBTs) that can be integrated on a single chip [44]. The authors state: "*LIGBTs offer much higher current densities (5–10x), significantly lower leakage currents, (and) lower parasitic device capacitances and gate charge compared to conventional vertical MOSFETs commonly used in LED drivers.*" The IGBTs were implemented using the structure shown in Fig. 11.18 with an N-type drift region implanted into a P-type substrate. A P-top layer was implanted on top of the N-drift layer for achieving RESURF charge balance. Each LIGBT is self-isolated, allowing the integration of multiple high-voltage (800 V) IGBTs on the same chip to make a compact solution. Solder bump technology was used to make connections to the LIGBTs.

The driver must be fitted into the limited space inside the GU-10 housing of the LED lamp. This is difficult to achieve with multiple discrete vertical MOSFETs. The LIGBT technology discussed above provides a compact solution. The solder bump reliability has been modeled for this configuration [45]. The solder bump underfill material was found to be important to withstand high voltages and achieve reliability.

11.3.4 Conducted EMI

One of the issues with LED lighting is the conducted EMI generated by the driver. The EMI has been linked to the high dI/dt and dV/dt produced by the fast switching of power MOSFETs. The EMI must be brought within the limits set by the CISPR 15 standard for lighting equipment. This was achieved by increasing the gate drive

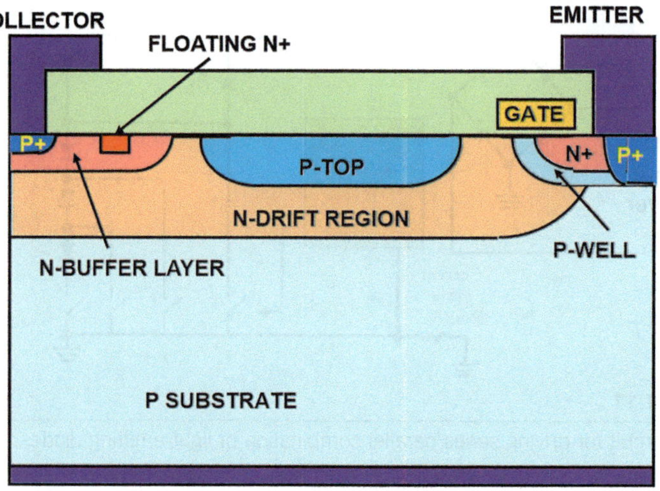

FIGURE 11.18

Lateral IGBT structure designed for driving series/parallel combination of light-emitting diodes.

resistance for the power MOSFETs with the previously discussed Fairchild FL7733 controller [46]. However, it resulted in a loss of efficiency by about 1%. Better performance is possible by using an IGBT here due to the lower dI/dt for the MOS bipolar device.

11.4 Strobe Flash Light

The digital still camera, shown in Fig. 11.19(a), employs an embedded flash unit to illuminate the scene being photographed when the ambient light is insufficient. Flip cell phones have also incorporated a flash capability in them, as shown in Fig. 11.19 (b), for the convenience of customers. Most of these applications were served using a xenon flash tube to generate the bright light required to achieve the desired color temperature. More recent iPhones have changed to LED-based flash for providing lighting to take photographs under low ambient light.

Xenon gas was most commonly used in a flash tube due its high efficiency ($\sim$50%) for conversion of electrical energy into light [47]. The emission of light from the xenon gas is produced by ionization of the gas using a high-voltage discharge. Once the discharge occurs, electrons flow from the cathode toward the anode, colliding with the xenon atoms and producing more ionization. When the xenon atoms recover the electron, the energy is converted into radiation from the gas. Xenon gas emits light with spectral lines in the ultraviolet, blue, green, and infrared regions. The bright light required for photographic applications is produced with a continuum emission called "gray body radiation." The production of gray body radiation requires current densities of above 2500 A/cm^2.

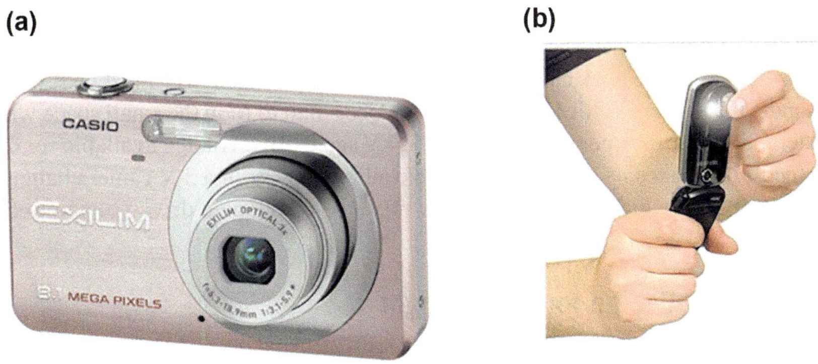

FIGURE 11.19

(a) Digital still camera; (b) cell phone.

In order to initiate the discharge of the xenon gas, a three-electrode configuration is used in the flash. A voltage is applied between the anode and cathode electrodes to keep the lamp below its self-flash threshold. To initiate the discharge, a trigger pulse is applied to a third electrode located between the anode and cathode electrodes but outside the glass tube. This creates a high electrostatic field inside the tube sufficient to discharge the gas.

11.4.1 Strobe Flash Circuit

The delivery of the current and voltage pulse required to trigger the discharge of the xenon tube is a challenge due to the high voltage and current requirements. The circuit topology commonly used for the flash in digital still cameras is shown in Fig. 11.20 [48–50]. To deliver the needed energy to the xenon tube, the main capacitor (C_M) with a value of 330 μF is usually charged to 300 V, resulting in 15 J of stored energy [48]. The discharge of the xenon tube is initiated with a much larger

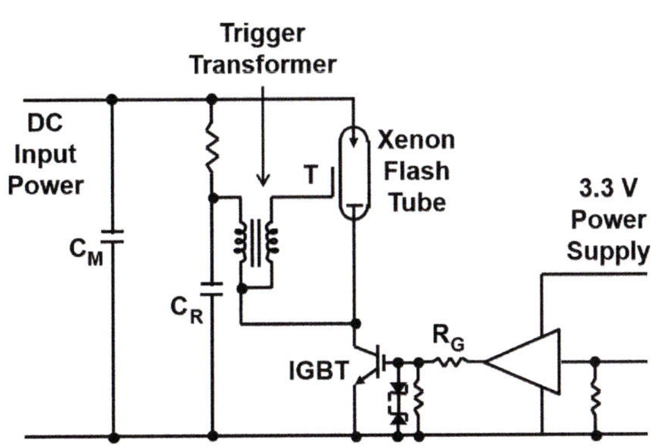

FIGURE 11.20

Camera flash circuit. *IGBT*, insulated gate bipolar transistor.

voltage (4–5 kV) applied to the trigger electrode (T) generated via a trigger transformer. The 130–200 A current pulse flows through the xenon tube via the IGBT during the flash. The peak current-handling capability of the IGBT must be sufficient to handle this current while keeping the chip size small to fit in the small space within modern digital still cameras. The low (2.5–4 V) gate drive voltage available for driving the IGBT makes its design even more challenging. With a 3 V camera battery, it is necessary to include a boost circuit to raise the IGBT gate drive voltage to 5 V.

11.4.2 IGBT Design for Strobe Light

Due to the large market for IGBTs in digital still cameras and cell phones, many companies have developed IGBT products specifically engineered for the strobe flash application. Two examples are the Toshiba GT10G131 IGBT with 200 A peak collector current and the Fairchild Semiconductor FGR15N40A IGBT with 150 A peak collector current. Both products have a collector blocking voltage capability of 400 V.

The optimization of the IGBT structure for the strobe flash application is unique. Unlike industrial motor drives where IGBTs operate with an on-state current density of 30–200 A/cm^2, the IGBTs for the strobe flash are designed to handle a very high peak current density of more than 3000 A/cm^2. This allows handling the large peak collector current during the flash with a small chip size that fits within the small space in a camera with a low cost.

Because the IGBT designed for the camera flash tube operates with a very low frequency, the turn-off times for the devices can be long. Typical products have collector turn-off and fall times of about 2 μs. To achieve very high peak collector current-handling capability, IGBTs are fabricated using trench gate structures that produce large channel density. Product design trends for IGBTs for the strobe flash market are shown in Fig. 11.21 [51]. The peak collector current-handling capability

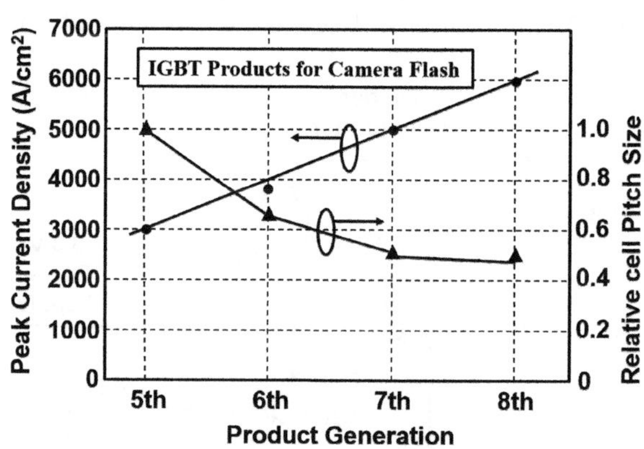

FIGURE 11.21

Insulated gate bipolar transistor product trends for camera flash.

has been increased from 3000 to 6000 A/cm^2 over three generations of technology, primarily by reducing the cell pitch to increase the channel density. A thinner gate oxide must be used for these IGBTs to reduce the channel resistance and obtain the large transconductance required to produce a high collector saturation current density in spite of the low available gate drive voltage. The cell structure must also be designed to suppress current-induced latch-up of the parasitic thyristor [52]. Zener diode clamps are included between the gate and emitter of the IGBT for immunity against electrostatic discharge, as shown in Fig. 11.20.

11.4.3 Professional Flash

An adage of professional photographers is "*Good lighting makes the difference between a good and a great photograph.*" Achieving good lighting requires control over the color temperature, illumination distribution, and the duration of the exposure. The ability to control the light generated from a professional xenon flash has been greatly improved by using electronic control with IGBTs. Many high-end flash units available from major manufacturers, such as Nikon, Canon, Pentax, and Olympus, mention the use of IGBT circuits, including Yongnuo Professional Flash [53] shown in Fig. 11.22(a). A typical professional flash set-up with a reflecting umbrella is shown in Fig. 11.22(b).

The circuit for driving the professional flash unit is similar to that already discussed in Section 11.4.1 with IGBTs optimized to handle the high peak collector

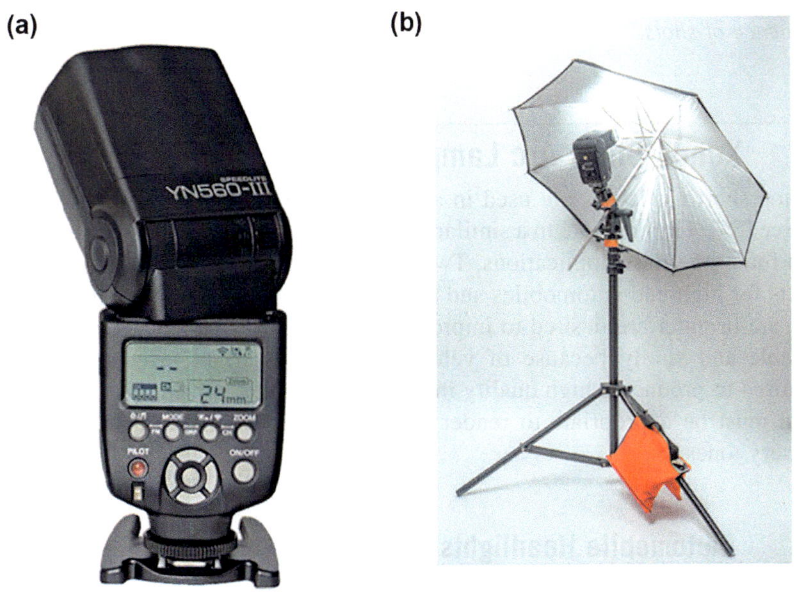

FIGURE 11.22

(a) Professional flash unit; (b) professional flash setup.

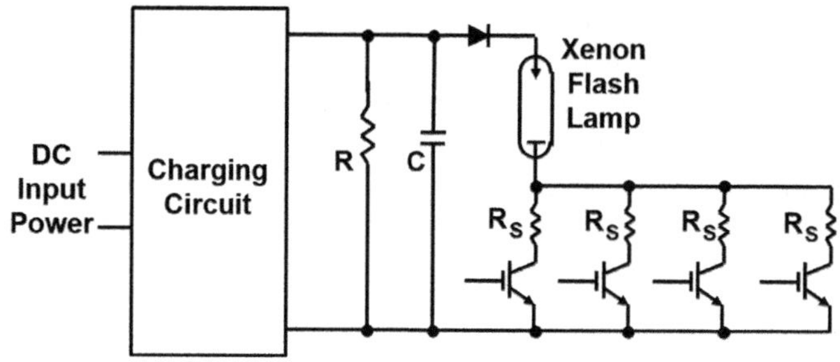

FIGURE 11.23

Insulated gate bipolar transistor control circuit for professional flash units.

currents for a short duration. The light intensity from a xenon tube has been found to be proportional to the current flowing through the lamp [54]. One method for controlling the magnitude of the current is changing the value of a resistor in series with the IGBT. This can be achieved by using multiple IGBTs in parallel with a resistor in series with each one, as shown in Fig. 11.23. An example of an advanced integrated studio flash system is the Einstein model E640 flash unit [55]. The manufacturer states: "*Einstein is a powerful, all-digital, self-contained flash unit using multiple IGBT devices to allow a 9 f-stop power variability range... An advantage of the IGBT flash control is that, at reduced power settings, the flash capacitors are not completely drained when the unit is fired, making it possible to perform a rapid sequence of shots.*"

11.5 Xenon Short Arc Lamps

Xenon short arc lamps are used in applications that require a bright white light source [56]. They operate in a similar manner to that described in the previous section for strobe flash applications. Two well-recognized examples are in the headlights for high-end automobiles and in cinema projectors. Automobile headlights that are brighter are desired to improve visibility and safety. The lamps must be reliable and sturdy because of vehicle vibrations. Bright projector lamps are required to produce a high quality image on the theater screen. Their color spectrum must be appropriate to render a true image of realistic facial and natural scenery tones.

11.5.1 Automobile Headlights

In the automobile headlights shown in Fig. 11.24(a), the gas in the tube consists of xenon and a mixture of metal halides. The xenon gas is only used during the initial period of 20–30 s before the metal halides are vaporized by the heat from the xenon

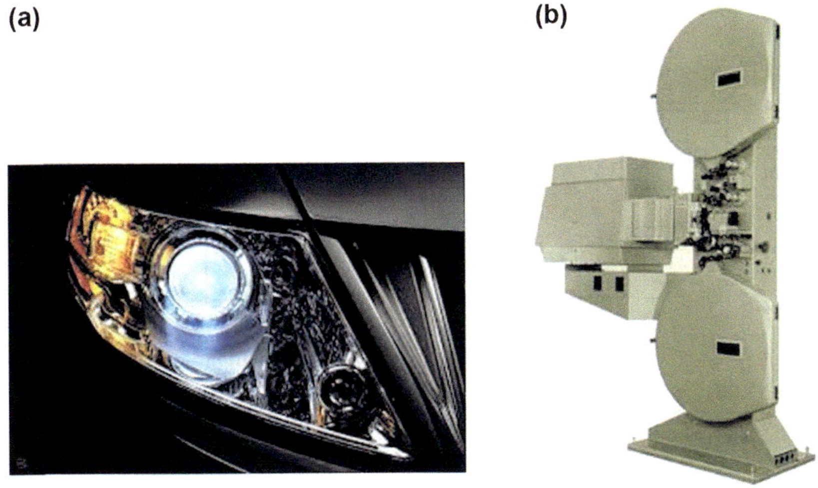

FIGURE 11.24

(a) Xenon automobile headlight; (b) cinema projector.

discharge. A very high voltage (20–50 kV) is required to ignite the xenon gas, after which the tube operates at 18 V and 25 A. The electronic ballast controls these functions and maintains a stable operating point for the lamp after the start-up phase. The main advantage of the xenon headlight is that it produces 2–3 times the light output compared with halogen bulbs and consumes only half as much power. The color of the light from the xenon headlight is also closer to daylight. Sixty percent of automobile accidents occur at night. Brighter automobile lights improve safely by the illumination of debris, animals, cyclists, and pedestrians at a greater distance and clarity to reduce reaction time for the driver [57]. The drive circuits for the xenon headlights are similar to those discussed in Section 11.4.

11.5.2 Movie Theater Projectors

All the movie projectors, such the one shown in Fig. 11.24(b), used in theaters (including modern digital projection systems) utilize xenon short arc lamps because of their brightness and color spectrum [56,58]. Their power ratings range from 900 W to 12 kW. The IMAX projectors use single xenon lamps with ratings of 15 kW.

A PFC circuit [59] has been developed using IGBTs for xenon short arc lamps, as shown in Fig. 11.25. The first stage performs the PFC. The second stage, a buck converter circuit, regulates the power delivered to the xenon short arc lamp. The PFC stage is operated at a switching frequency of 2 kHz, allowing any harmonics to be removed by a line filter and obtaining a power factor of 0.98. The second stage regulates the lamp voltage to 75 V for lamp ignition.

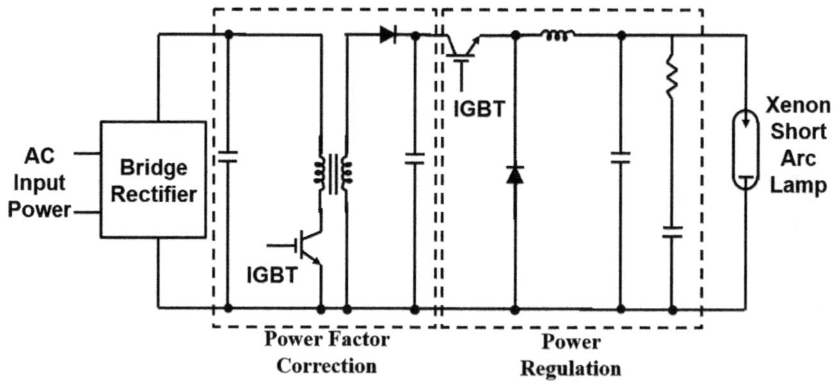

FIGURE 11.25

Movie theater projector power circuit. *IGBT*, insulated gate bipolar transistor.

11.6 Stroboscopic Imaging

A strobe light produces pulses of light at regularly spaced intervals in time [60]. A xenon flash lamp is commonly used for this application to produce a color temperature of 5600 K for a daylight ambient. Early work with strobe lights by H.E. Edgerton allowed examination of the movement of people, animals, and even bullets in flight.

Significant improvement in strobe speed has been derived by replacing silicon-controlled rectifier (SCR) (thyristor) control of the xenon tube with an IGBT control circuit [61]. The circuit shown in Fig. 11.26 uses a Toshiba MG400Q1US11 IGBT with a blocking voltage of 1200 V and a collector current of 400 A. The capacitors

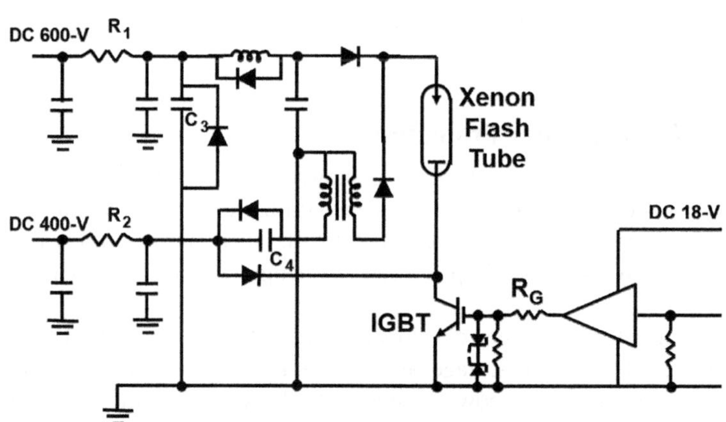

FIGURE 11.26

High-speed strobe circuit. *IGBT*, insulated gate bipolar transistor.

C_3 and C_4 must be sufficiently large (0.22 μF) to store enough energy to trigger the discharge of the xenon tube. To obtain high-intensity flashes at higher speeds, the resistors R_1 and R_2 are reduced when compared with the SCR circuit. The circuit is able to deliver 5000 flashes per second with a flash duration of 100 μs. The IGBT-based strobe has allowed observation of the flagellum of a single cilium from sea urchin larvae.

11.7 Dimmable Luminaries

Adjustable spotlights like the one shown in Fig. 11.27(a) are commonly used in exhibitions and museums to showcase artwork. These devices require electronic dimming capability. The luminaries are made with a dimmable electronic transformer using IGBT technology [62,63]. They can be located in acoustically sensitive spaces such as live theaters, schools, churches, and auditoriums. The IGBT dimming eliminates the mechanical hum or buzz observed with SCR dimmers. These dimmers shown in Fig. 11.27(b) use convection cooling and have an energy efficiency of 98%.

11.8 Rapid Thermal Annealing

Rapid thermal processing has become commonly used by the semiconductor industry for annealing ion-implanted layers [64]. A typical single wafer rapid thermal processing unit is shown in Fig. 11.28(a). The short duration of this process, when compared with annealing in a conventional furnace, allows retaining shallow junction depths by avoiding the diffusion of the dopants. As transistor geometries become

(a) (b)

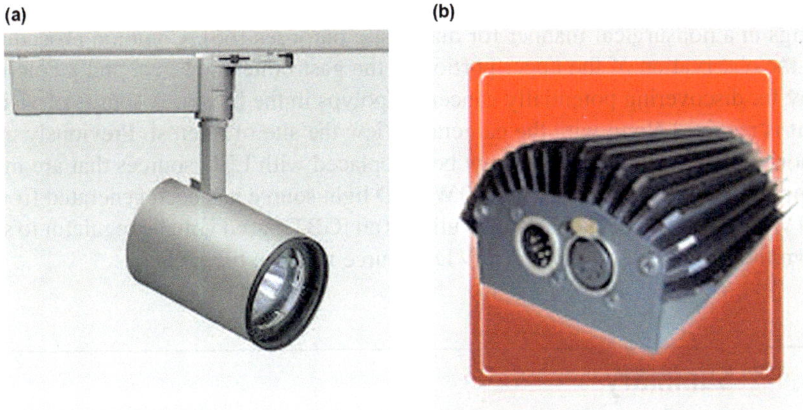

FIGURE 11.27

(a) Dimmable luminary; (b) insulated gate bipolar transistor dimmer.

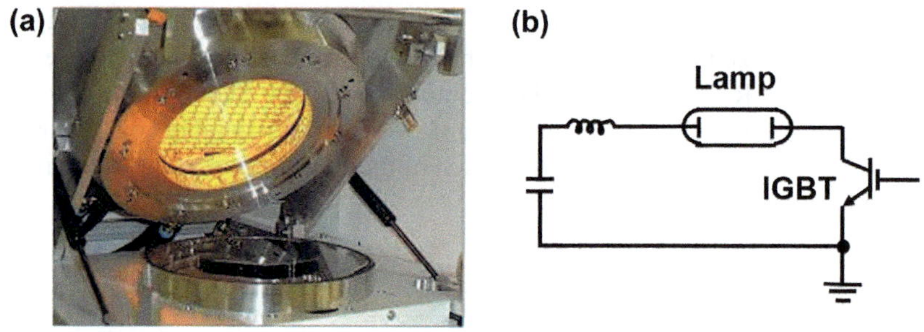

FIGURE 11.28

(a) Rapid thermal annealing furnace; (b) lamp control circuit. *IGBT*, insulated gate bipolar transistor.

smaller and smaller according to Moore's law, the annealing duration has to decrease to retain a smaller size for the junctions.

A subsecond annealing (SSA) technology has been demonstrated by using IGBTs to control the duration of the radiation produced from a xenon flash lamp source [65]. The basic LCR circuit used for controlling the radiation pulse is shown in Fig. 11.28 (b). The discharge of the capacitor produces radiation from the lamp. The flash can be turned on by gating the IGBT into its on-state, but it can also be interrupted by turning off the IGBT. SSA pulses with duration as small as 1.4 ms have been achieved. The damage from boron ion implants using BF2 has been shown to recover with no diffusion of the dopant.

11.9 LED-Based Endoscopy

Medical endoscopes are used to observe organs and cavities in animals and human beings in a nonsurgical manner for diagnostic purposes [66]. Common procedures are the observation of the upper portion of the gastrointestinal tract and a colonoscopy for discovering potentially cancerous polyps in the bowel. A source of visible light is required along with the camera to view the site of interest. Previously used xenon and halogen lamps have now been replaced with LED sources that are more compact and need less power. A 100 W LED light source has been generated from a 220 V source [67]. The power source utilizes an IGBT-based voltage regulator to step down the 200 V AC source to a 12 V DC source to drive the LEDs.

11.10 Summary

The introduction of the CFL has had a major impact on reducing the energy consumption in the lighting sector of the economy. The IGBT enabled the introduction of the CFL because a small IGBT chip can handle a large amount of power and its gate drive circuit can be integrated, making the control electronics compact. The economic and environmental impact of this technology will be discussed in a later chapter of the book. Commercial and residential lighting with LEDs also benefits from using IGBTs to reduce the size of the control circuit.

The availability of the IGBT has also allowed the miniaturization of digital still cameras by reducing the space taken by the electronics for the flash tube. Modern flash units for professional cameras are advertised with IGBT circuits, making these devices a recognized brand enhancer.

The diversity of IGBT applications for lighting includes dimmable luminaries for museums and auditoriums. It has even found use in rapid thermal processing equipment used for its own fabrication. IGBT-based power supplies are utilized for medical endoscopy as well.

References

[1] World Governments Adopting Bright Idea. www.worldwatch.org/node/4941.
[2] How Much Electricity Is Used for Lighting in the United States? www.eia.gov/tools/faqs/faq.cfm?id=99&t=3.
[3] Today in Energy, 2012. www.eia.gov/todayinenergy/detail.cfm?id=6210.
[4] How Is Electricity Used in U.S. Homes? www.eia.gov/tools/faqs/faq.cfm?id=96&t=3.
[5] Incandescent Light Bulb. en.wikipedia.org/wiki/Incandescent_light_bulb.
[6] Bans of Incandescent Light Bulbs. en.wikipedia.org/wiki/Bans_of_Incandescent_light_bulbs.
[7] M.D. Bloomer, W.J. Laughton, D.L. Watrous, TRIAD Low-Voltage Converter: Final Report, 1986. General Electric Technical Report 86CRD011.
[8] How Does a Halogen Light Bulb Work? http://home.howstuffworks.com/question151.htm.
[9] Halogen Lamp. en.wikipedia.org/wiki/Halogen_lamp.
[10] Parabolic Aluminized Reflector Light. en.wikipedia.org/wiki/Parabolic_aluminized_reflector_light.
[11] B.J. Baliga, Temperature behavior of insulated gate transistor characteristics, Solid State Electron. 28 (1984) 289–297.
[12] Compact Fluorescent Lamp. en.wikipedia.org/wiki/Compact_fluorescent_lamp.
[13] Let there be light, Time Magazine Article, 2008, pp. 68–70.
[14] Strong Growth in Compact Fluorescent Bulbs Reduces Electricity Demand, Worldwatch Institute, 2011. http://www.worldwatch.org/node/5920.
[15] L.B. Vestel, As C.F.L. Sales Fall, More Incentives Urged, N.Y. Times, 2009. http:/green.blogs.nytimes.com/2009/09/28/as-cfl-sales-fall-more-incentives-urged/.
[16] Widespread CFL Use Could Slash Global Lighting Demand for Electricity in Half, 2008. http://www.facilitiesnet.com/lighting/article/Widespread-CFL-Use-Could-Slash-Globa-Lighting-Demand-For-Electricity-By-Half.htm.
[17] India Lights Future with Efficient Lamps and Carbon Credits, Worldwatch Institute, 2011. http://www.worldwatch.org/node/6438.
[18] How CFL Bulbs Work. http://science.howstuffworks.com/environmental/green-tech/sustainable/cfl-bulb2.html.

[19] Gases that Emit Light. http://scifun.chem.wisc.edu/chemweek/gasemit/gasemt.html.

[20] M. Bairanzade, Reduce Compact Fluorescent Cost with Motorola PowerLux IGBT, 1996. Motorola Semiconductor Application Note AN1576.

[21] T. Ribarich, The Top Five Global Lighting Technologies, Power Electronics Technology Magazine, 2004, pp. 56–62.

[22] R.B. Arroyo, et al., Power-quality model based on IGBT dynamic conduction for non-polluting lighting applications, in: IEEE International Conference on Electrical Engineering, Computer Science, and Automation Control, 2012, pp. 1–5.

[23] E. Griebl, L. Lorenz, M. Purschel, LightMOS a new power semiconductor concept dedicated for lamp ballast application, in: IEEE Industry Applications Conference, 2003, pp. 768–772.

[24] B.J. Baliga, Fundamentals of Power Semiconductor Devices, second ed., Springer-Science, New York, 2019, pp. 1024–1032. Section 9.17.

[25] H. Ruthing, et al., 600-V reverse conducting (RC) IGBT for drives applications in ultra-thin wafer technology, in: IEEE International Symposium on Power Semiconductor Devices and ICs, 2007, pp. 89–92.

[26] J.A. Appels, H.M.J. Vaes, High voltage thin layer devices (RESURF devices), in: IEEE International Electron Devices Meeting, Abstract 10.1, 1979, pp. 238–241.

[27] Y.S. Huang, B.J. Baliga, Extension of RESURF principle to dielectrically isolated power devices, in: IEEE International Symposium on Power Semiconductor Devices and ICs, Paper 2.1, 1991, pp. 27–30.

[28] Y.S. Huang, et al., Comparison of DI and JI lateral IGBTs, in: IEEE International Symposium on Power Semiconductor Devices and ICs, Paper 3.1, 1992, pp. 40–43.

[29] M. Stoisiek, et al., A dielectrically isolated high voltage IC technology for off-line applications, in: IEEE International Symposium on Power Semiconductor Devices and ICs, Paper 8.35, 1995, pp. 325–329.

[30] A. Nakagawa, et al., Improvement in lateral IGBT design for 500V 3A one chip inverter ICs, in: IEEE International Symposium on Power Semiconductor Devices and ICs, 1999, pp. 321–324.

[31] H. Sumida, et al., A high performance plasma display panel driver IC using SOI, in: IEEE International Symposium on Power Semiconductor Devices and ICs, Paper 7.1, 1998, pp. 137–140.

[32] H. Sumida, et al., 250-V class lateral SOI devices for driving HDTV PDPs, in: IEEE International Symposium on Power Semiconductor Devices and ICs, 2007, pp. 229–232.

[33] J. Sakano, et al., Large current capability 270V lateral IGBT with multi-emitter, in: IEEE International Symposium on Power Semiconductor Devices and ICs, Paper 4.4, 2010, pp. 83–86.

[34] J.T. Hwang, et al., 550V SiP compact fluorescent lamp ballast IC, in: IEEE European Solid State Circuits Conference, 2006, pp. 576–579.

[35] S.E. Berberrich, et al., Triple trench gate IGBTs, in: IEEE International Symposium on Power Semiconductor Devices and ICs, 2005, pp. 1–4.

[36] Light-emitting Diode. en.wikipedia.org/wiki/Light-emitting_diode.

[37] Energy Efficiency of LEDs, U.S. Department of Energy, 2013. Fact sheet number PNNL-SA-14206 www.energy.gov.

[38] F.E. Bisogno, et al., A line power-supply for LED lighting using piezoelectric transformers in class-E topology, in: IEEE Power Electronics and Motion Control Conference, vol. 2, 2006, pp. 1–5.

[39] Piezoelectricity. en.wikipedia.org/wiki/Piezoelectricity.

[40] Comparing Magnetic and Piezoelectric Transformer Approaches in CCFL Applications. www.ti.com/sc/analogapps.

[41] H. Zhang, Developing highly reliable LED luminaires for high temperature applications using AC-direct driving LED technology, in: IEEE Applied Power Electronics Conference, 2018, pp. 3466–3470.

[42] Fairchild LED Control IC FL7733A Datasheet, 2022.

[43] J. Baek, S. Chae, Off-line buck LED driver for series connected LED segments, IEEE Trans. Ind. Electron. (2017) 7852–7861.

[44] A.M. Aliyu, et al., Chip-on-Board Assembly of 800 V L-IGBTs for High-Performance Ultra-Compact LED Drivers, 2017, pp. 431–434.

[45] C. Bailey, et al., Mechanical modelling of high power lateral IGBT for LED driver applications, in: IEEE Electronic Components and Technology Conference, 2018, pp. 1375–1381.

[46] Y. Chia, et al., Effects of MOSFET Gate Driving on Conducted Emissions in a Flyback LED Driver, 2019, pp. 262–266.

[47] Flashtube. en.wikipedia.org/wiki/Flashtube.

[48] Xenon Flash Light IGBT Application Note. www.onsemi.com.

[49] Discrete IGBT Product Guide Section 5.3 Strobe Flash Applications. www.semicon.toshiba.co.jp/eng.

[50] IGBT Application Note AN9006 for Camera Strobe. www.fairchildsemi.com.

[51] S. Umekawa, et al., New discrete IGBT development for consumer use—application specific advanced discrete IGBTs with optimized chip design, in: IEEE International Power Electronics Conference, 2010, pp. 790–795.

[52] B.J. Baliga, Fundamentals of Power Semiconductor Devices, second ed., Springer-Science, New York, 2019, pp. 939–968. Section 9.10.

[53] Amazon. https://www.amazon.com/Yongnuo-Professional-Speedlight-Flashlight-Olympus/.

[54] R. Gelagaev, et al., Multi-objective optimization of a flash lamp drive, in: IEEE European Power Electronics and Applications Conference, 2013, pp. 1–8.

[55] The Einstein E640 Flash Unit. http://www.paulcbuff.com/e640.php.

[56] Xenon Arc Lamp. en.wikipedia.org/wiki/Xenon_arc_lamp.

[57] Xenon—How and Why? www.autolamps-online.com.

[58] USHIO Xenon Short Arc. www.ushio.eu/xenonshortarc.html.

[59] F. Yan, PFC electronic ballast for xenon short arc lamps, in: IEEE International Conference on Power Electronics and Drive Systems, 1999, pp. 5010–5015.

[60] Strobe Light. en.wikipedia.org/wiki/Strobe_light.

[61] M.O. Miyake, et al., Improvement of time and space resolution of stroboscopic micrography using high power xenon flash, Rev. Sci. Instrum. 69 (1) (1998) 325–326.

[62] Bak Pak Individual IGBT Dimmer. www.lightolier.com.

[63] Bak Pak Individual IGBT Intelligent Dimmer. www.etdimming.com.

[64] Rapid Thermal Processing. en.wikipedia.org/wiki/Rapid_thermal_processinge.

[65] H. Kiyama, et al., Flexibly controllable sub-second flash lamp annealing, in: IEEE International Symposium on Junction Technology, 2008, pp. 206–208.

[66] Endoscopy. https://en.wikipedia.org/wiki/Endoscopy.

[67] H. Wehby, et al., 100-watt MATLAB/Simulink model LED based light source for medical endoscopy, in: IEEE Conference on Emerging Smart Technologies and Applications, 2021, pp. P1–P6.

CHAPTER

IGBT Applications: Consumer 12

The availability of electricity in homes has enriched our lives with the ability to provide 24 h of light per day, as discussed in the previous chapter. This revolutionary impact on society has been followed by conveniences that an average citizen of any developed nation would not be willing to live without. The ability to refrigerate our food has prolonged its shelf life and made more dietary choices available to us. The commonplace use of washing machines and dryers in homes has eliminated the drudgery of manually cleaning clothes. The cooking of food with microwave energy makes reheating food extremely convenient and has added another category of processed food at our disposal. The removal of dust with electric vacuum cleaners allows improving our environment while reducing problems of allergens. The quality of our entertainment while at home has been enhanced with brighter and larger color television (TV) sets.

The distribution of electricity that is used to perform the wide variety of tasks in the home [1] is shown in Fig. 12.1. More than 50% of the energy is used for heating/cooling of the home and for refrigeration and cooking. More efficient ways of performing these tasks are desired to reduce electrical demands and costs for consumers. The availability of the insulated gate bipolar transistor (IGBT) has had a major impact in this respect.

The shipment of home appliances [2] over a period of 10 years from 1999 to 2008 is provided in Fig. 12.2. Robust sales of refrigerators and air conditioning units can be observed over this time frame. The use of these appliances has resulted in an average of 125 motors used in a typical American home with a daily residential energy consumption of more than 12 kWh of electricity per person [3]. The importance of reducing energy consumption with more efficient appliances has been recognized by the passage of minimum efficiency standards for air conditioning (seasonal energy efficiency ratio (SEER) ratings) and for appliances (Energy Star ratings). In 2006, 31% of refrigerator sales and 38% of clothes washer sales met the Energy Star rating. The deployment of variable-speed motor drives constitutes an important component of the strategy for increasing the efficiency of home appliances.

The various appliances in the typical residence can be classified into two broad categories, namely those with motors (or rotating loads) and those without motors (or static loads) [4]. They are shown ranked in order of energy requirements in Fig. 12.3. The loads that make use of IGBTs for power control are highlighted with a red box. All the rotating loads are driven with IGBT-based power electronics. The static loads with higher energy requirements are also driven using IGBT-based power electronics.

The IGBT Device. https://doi.org/10.1016/B978-0-323-99912-0.00001-5

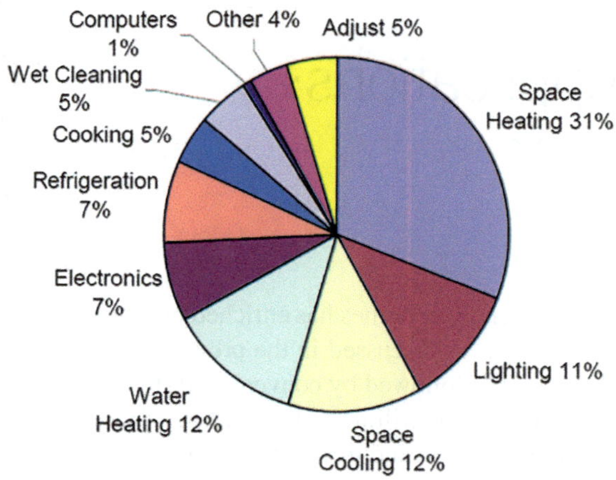

FIGURE 12.1

Energy use in the home.

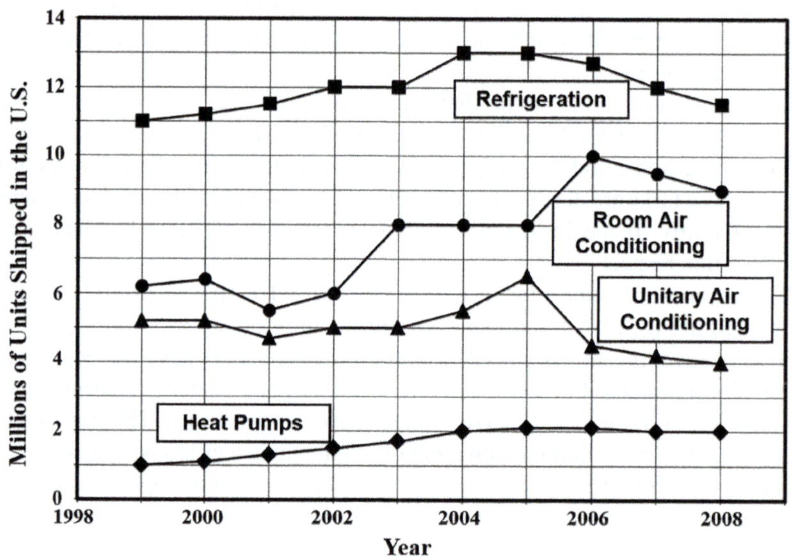

FIGURE 12.2

US appliance annual shipments.

This chapter describes the reliance of home appliances on the control of electricity with IGBTs. Their pervasive influence on improving the quality of life for people is not readily apparent because the power electronics are embedded into a larger device (such as a refrigerator or washing machine) and hidden from the eyes of the consumer. The IGBT is an enabling technology essential for building a smart home where all the appliances can be programmed not only in advance but via the Internet.

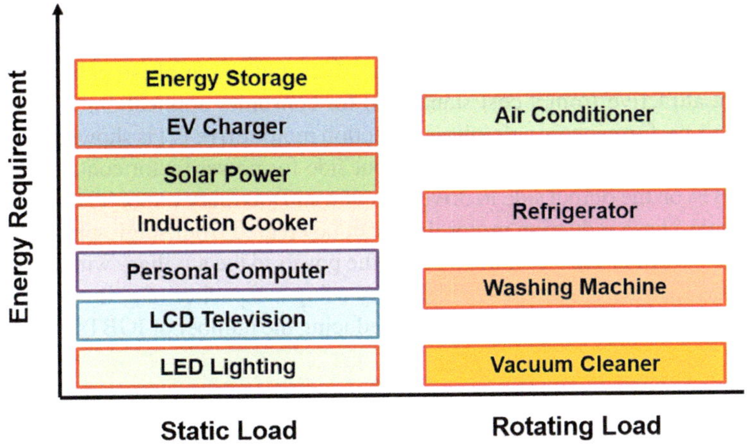

FIGURE 12.3

Classification of loads within a typical residence.

12.1 Large Appliances

Appliances used in homes are generally classified into the categories of major and small appliances [5]. Major appliances commonly found in homes include the air conditioning heat pump unit, kitchen cooking range, washing machine, microwave oven, and refrigerator. They are sometimes referred to as white goods because of their typical color, although stainless steel finishes have become increasingly popular. In the United States, more than 90% of the market is served by five companies: General Electric (GE), Whirlpool, Electrolux, Maytag, and Thermador. European appliance makers include Bosch and Miele while in Asia, appliances are produced by LG Electronics, Samsung, Mitsubishi Electric, and Toshiba Corporation.

These relatively large appliances are major consumers of electrical energy in a home and must be powered directly from the AC power line. The large peak voltages of the AC power line (170 V in the United States and 310 V in Europe) require power devices with high blocking voltage ratings in the control circuits. The IGBT is ideally suited for this type of application.

12.1.1 Air Conditioners (Heat Pumps)

As discussed in the introductory chapter, the IGBT was first developed at the General Electric Company for creating an adjustable-speed motor drive for air conditioning applications. This work was described in Chapter 10.

Toshiba states [6]: "*Air conditioners were the first major appliance to use inverter control. Key semiconductor devices include the insulated gate bipolar transistors. Power factor correction (integrated circuits) or IGBTs are used in the power supply circuit to keep harmonics in the input current below the (International Electrotechnical Commission, IEC) limit.*" This is consistent with the introduction of adjustable-speed drives using IGBTs by GE in the early 1980s for heat pumps, like the one shown in Fig. 12.4, built by Trane and Carrier. Industrial applications commonly utilize three-phase induction motor drives, but single-phase induction motors are more attractive from a cost standpoint for consumer applications.

A high-performance single-phase induction motor drive [7] is shown in Fig. 12.5. This approach uses two IGBTs on the input side for power factor control (PFC) and four IGBTs on the output side to drive the main and auxiliary motor windings. A low-cost, single-phase induction motor drive can be created from single-phase AC input power, as shown in Fig. 12.6. In this case, the power to the auxiliary winding is-phase shifted with respect to the main winding by using a capacitor (C_R) in series with the auxiliary winding. This approach allows reducing the number of IGBTs on the output

FIGURE 12.4

Typical home air conditioning unit.

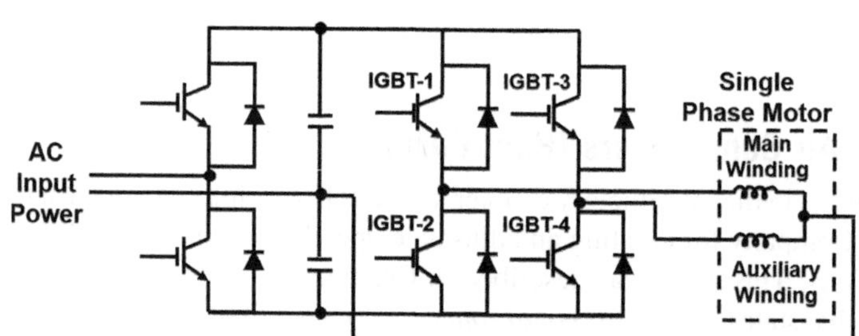

FIGURE 12.5

High-performance, single-phase induction motor drive with two-phase inverter. *IGBT*, insulated gate bipolar transistor.

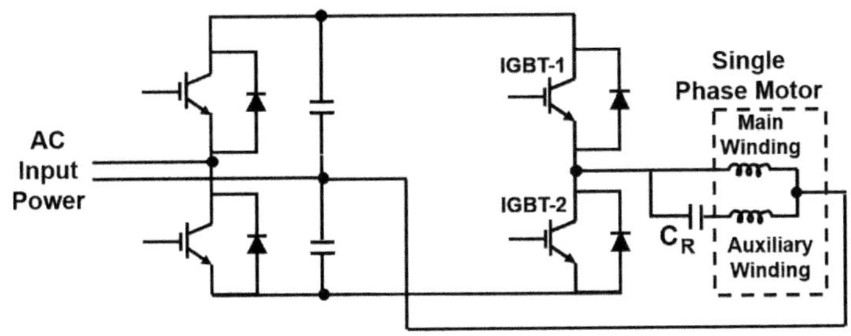

FIGURE 12.6

Low-cost, single-phase induction motor drive. *IGBT*, insulated gate bipolar transistor.

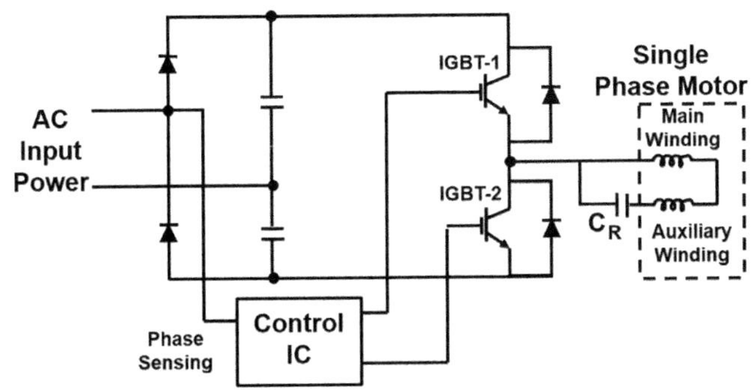

FIGURE 12.7

Reduced cost, single-phase induction motor drive.

side. The IGBTs used for PFC are included in this topology. An even lower-cost single-phase induction motor drive [8] is shown in Fig. 12.7. Here, the IGBTs on the input side are eliminated to reduce the cost.

The term electronically commutated motor (ECM) is used by the heating, ventilation, and air conditioning industries. The ECM consists of a permanent magnet (PM) motor and a variable speed drive for fans and blowers [9]. The 1/3 to 1 hp. motors are controlled by H-bridge modules with IGBTs, as illustrated in Fig. 12.8. The controller for the IGBTs can use digital signal processors or application-specific microcontrollers depending on cost/performance trade-offs. This approach allows for the continuous delivery of airflow at a low rate into the residential space to produce minimum variations of the ambient temperature and less acoustical noise.

12.1.2 Refrigerators

Refrigeration is the process by which heat is removed from one location and delivered to another location [10]. It has had an enormous impact on society not only in terms of lifestyle but also agricultural growth and settlement of populations.

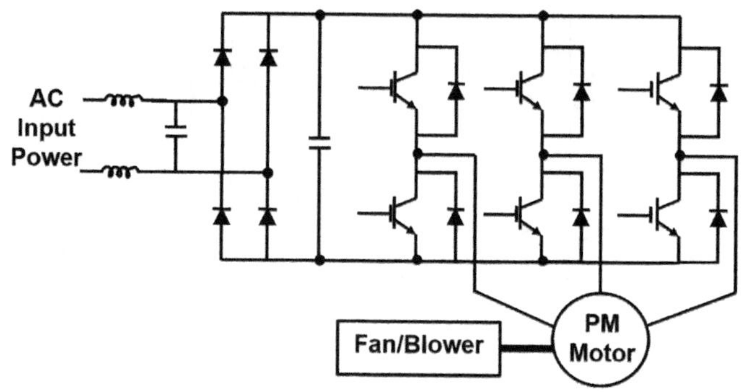

FIGURE 12.8

Electronically commutated motor for heating, ventilation, and air conditioning fan/blower drive. *PM*, permanent magnet.

The benefits of cooling food for preservation have been recognized for a long time. In the early part of the 19th century, this was achieved by using ice harvested from frozen lakes in the north, then transporting it to cities. Contamination of ice by pollution and sewage led to serious health hazards. Exploration of means to cool spaces and items using electricity can be traced to experiments conducted in 1755 by William Cullen and in 1758 by Benjamin Franklin. The first commercially viable vapor compression refrigeration system was produced by James Harrison in the 1850s. The refrigeration of food began to play an important role by the 1890s. Refrigeration in farms reduces spoilage and allows food preservation for transport to the supermarket for consumption. This replaced the consumption of locally grown, seasonally available produce with a much richer variety of produce available in modern supermarkets. GE produced the first refrigerators for home use in 1927. The popularity of refrigerators and freezers like those shown in Fig. 12.9 for residential use has grown exponentially since that time. They consume a significant fraction of electric power in our homes, as shown in Fig. 12.1.

The reverse-Rankine vapor compression cycle is the most commonly used method for refrigeration systems in homes and supermarkets. The operating principle for the vapor compression cycle is schematically illustrated in Fig. 12.10. A gas, originally Freon and now R134a, is liquefied using a compressor and a condenser. The gas is compressed to a high-pressure superheated vapor by the compressor. When the gas flows through the condenser coils, it is liquefied by cooling produced by exposure to the air in the room. The gas leaves the condenser at slightly above room temperature and high pressure. The liquefied gas is forced through a small pinhole-sized constriction in the expansion valve. The sudden reduction in pressure results in cooling the mixture of gas and liquid that exits the expansion value. When the mixture of gas and liquid flows through the evaporator, it is converted back into the vapor-phase, absorbing heat from the warm air surrounding the evaporator. The resulting cold air can be utilized for preserving food in the refrigerator at low temperatures (~40°F) and for making ice from water.

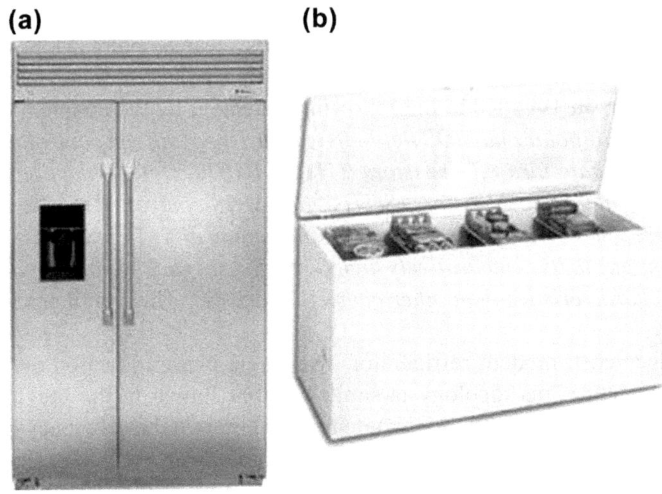

FIGURE 12.9

(a) Refrigerator; (b) freezer.

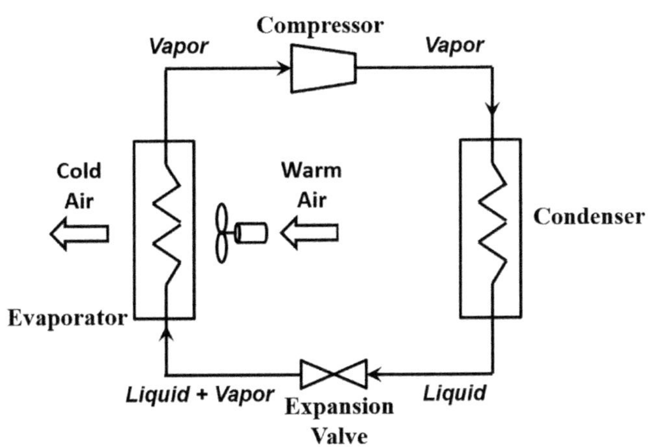

FIGURE 12.10

Vapor compression cycle used for refrigeration.

The external electrical energy required to drive the process is performed by an inverter that drives the compressor motor. Toshiba states [6]: "*(Because) refrigerators work continuously, it is important to reduce their power consumption. The compressor driver consists of IGBTs. Power factor correction with IGBTs is used in the power supply circuit to keep harmonics in the input current below the IEC limit.*"

Hermetically sealed units for household use have increased the space available for food storage [11]. At that time (1952), single-phase four-pole, 60-Hz,

1725 rpm motors were usually used in refrigerators. By 1994, compressors were operated with three-phase induction motors using pulse-width-modulated, variable-speed inverters. A comparison of existing power semiconductor devices for the inverters in 1994 led to the following statement by the authors [12]: "*These devices include bipolar and MOS field-effect transistors, silicon controlled rectifiers (SCRs), triacs, gate turn-off thyristors (GTOs), IGBTs, static induction transistors (ISITs), static induction thyristors (SITHs), and MOS-controlled thyristors (MCTs). Among these, the IGBT converter is the most common drive in medium power range applications due to its comparatively high switching speed, low noise, and better performance characteristics than other competing drives.*" This continues to hold true even today.

To reduce cost, modern refrigerators use single-phase induction motors with a starting capacitor. This topology is similar to that shown in the previous section for air conditioning in homes. Many attempts to improve the efficiency of the compressor used for refrigeration have been proposed over the years. One example shown in Fig. 12.11 is the use of a brushless DC motor with variable speed drive [13], which allows energy-efficient control without a shaft position sensor. Other advantages are fewer components to reduce the cost, and less electrical and acoustical noise. The compressor operates at speeds ranging between 1500 and 5000 rpm with a continuous speed of about 2500 rpm. The drive efficiency was found to be only weakly dependent on compressor speed for the topology shown in the figure. A total energy savings of 40% was reported with quiet operation and temperature control within 0.1°C.

The refrigerator with a compressor that utilizes a conventional single-phase induction motor with a starting capacitor creates high in-rush currents and low power factor [14]. Simple on/off or "bang–bang" control for temperature regulation creates constant in-rush currents that produce flicker problems for other appliances connected to the AC power source in the home. These problems can be overcome using

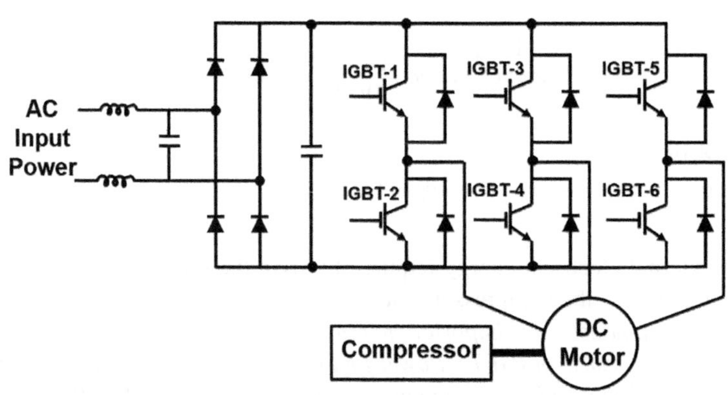

FIGURE 12.11

Pulse-width modulation voltage source inverter drive for brushless DC motor. *IGBT*, insulated gate bipolar transistor.

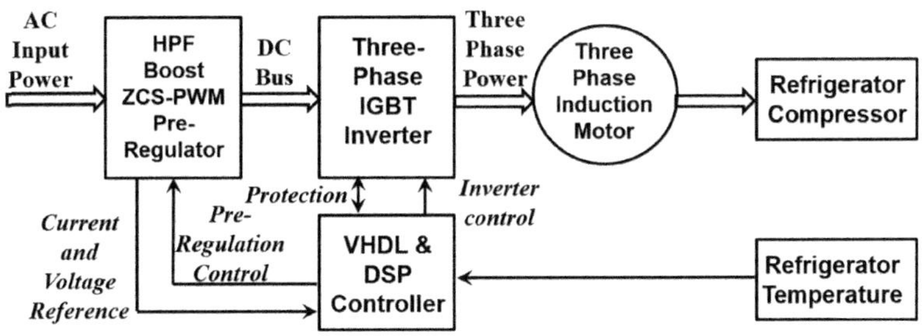

FIGURE 12.12

Variable-speed inverter drive for three-phase induction motor. *IGBT*, insulated gate bipolar transistor; *PWM*, pulse-width modulation; *ZCS*, zero current switching; *VHDL*, VHSIC hardware description language; *DSP*, digital signal processor.

a three-phase 0.5 hp. induction motor driven with an IGBT inverter, as shown in Fig. 12.12. A high power factor (HPF) preregulator is used at the input to prevent in-rush currents and achieve a good power factor. The input current ripple is reduced with an interleaved boost rectifier. Zero-current turn-on and zero-current–zero-voltage turn-off for IGBTs, and zero-current turn-on and zero-current turn-off of diodes are used to reduce switching losses and electromagnetic interference (EMI).

The implementation of the SEER rating for air conditioning and the Energy Star rating for appliances has increased the adoption of variable-speed drives. In order to reduce energy consumption, brushless PM motors are being used driven with IGBT inverters [9]. Evaporators for frost-free refrigerators now use brushless DC motors in place of conventional single-phase induction motors because the efficiency of the motor and drive electronics is improved by three times.

Mitsubishi Electric Corporation has reported improved efficiency for refrigerators by replacing the induction motor with a brushless DC motor and driving it with sine wave current rather than previously used trapezoidal wave current [15]. The sine wave current decreases copper losses due to resistance in the motor windings and iron core losses due to reduced harmonics. An improvement in efficiency by 4% was obtained with acoustical noise reduction by 17 dB. The technology has been implemented in Type S, G, and A series Mitsubishi refrigerators.

Refrigerator compressors usually operate at low speeds requiring low output power from the inverter to the PM motor. A better efficiency is obtained with square wave drive when compared to a sine wave drive. However, the sine wave drive provides reduced torque ripples and starting performance. An improvement in overall efficiency by 2% has been reported by using dual-mode operation [16].

12.1.3 Washing Machine

Maintaining good hygiene requires regular laundering of clothes to remove dust and body odors. The manual cleaning of clothes is still practiced around the world but the convenience of performing this task using machines has been embraced by all developed countries. The laundering process involves soaking the clothes in water, beating and scrubbing the clothes to release the dirt, and then rinsing with clean water to separate the dirt from the clothes. Performing these tasks manually is time consuming and laborious. Mechanical washing machines were introduced in the mid-1780s with hand cranking [17]. Additional operations by hand were required to remove most of the water from the clothes by wringing. The electric wringer/mangle machine became available in the early part of the 20th century with the washing and wringing (mangling) operations performed as separate tasks. The combination of the washing and wringing functions created the automatic washing machine. The availability of low-cost motors enabled the introduction of the spin-drying concept. The first top-load washing machine, shown in Fig. 12.13(a), was commercialized immediately after World War II. The front-load washing machine, shown in Fig. 12.13(b), was also introduced at that time but has only become popular in recent years. Early washing machines used electromechanical timers with motors running at constant speed. Since the 1990s, modern washing machines have relied on microcontrollers with sophisticated drives for the motors that move the clothes. This allows the implementation of predefined programs for laundering different types of clothing with selected temperature and washing duration.

The growing popularity of washing machines in developed countries is evident from the large number of units manufactured each year [17]; 2005 data are shown in Fig. 12.14. The top-load machine shown in Fig. 12.13(a) is the most popular version in Australia, Canada, and the United States. This design contains a perforated basket within a concentric water-retaining tub. The axis of rotation for the top load design is vertical. A large rotating screw is placed along the axis of rotation. The clothes are

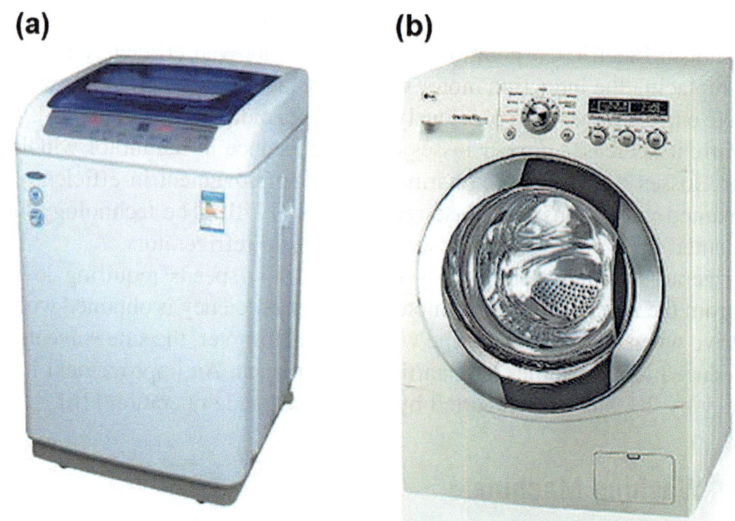

FIGURE 12.13

Washing machines: (a) top load; (b) front load.

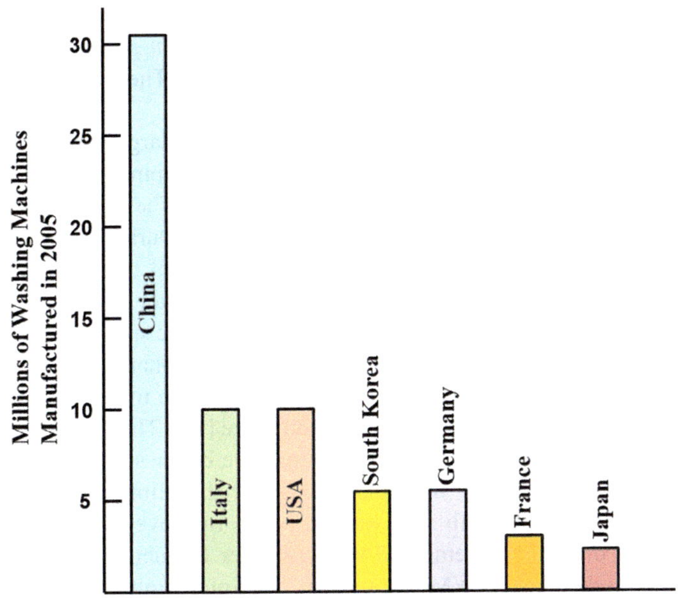

FIGURE 12.14

Washing machine production in 2005.

suspended in water filled into the water-retaining tub. An agitator rotates the screw in the center of the tub to push the water outward, up the sides of the perforated basket, and then back down. The agitator rotation must be periodically reversed to avoid spinning the water around the basket. Originally, top-load washers contained a gearbox to achieve the desired motion of the agitator, making them cumbersome and unreliable. Modern top-load washers have replaced the gearbox with electronic inverters that can provide versatile control over the motion of the screw. A rapid spinning cycle is used at the end of the wash cycle to remove the water from the clothes by centrifugal action.

Front-load machines, shown in Fig. 12.13(b), are popular in Europe and have been introduced as high-end machines in the United States. This design places the perforated basket and water-retaining tub along a horizontal axis. The agitation of the clothes is performed by the repeated rotation of the tub in both directions. The clothes are lifted and dropped using paddles on the inside walls of the basket. This process uses less water than top-load washers because the clothes do not have to be suspended in the water. The amount of detergent required is consequently also reduced. As in the case of the top-load models, a rapid spinning cycle is used at the end of the wash cycle to remove the water from the clothes by centrifugal action.

Toshiba states [6]: "*The latest washing machines use inverter control for both washing and spin drying. Inverter control helps reduce wash/spin noise and vibration and enables a washing machine to adjust the amount of water and motor torque to suit the washload. IGBTs are used for motor drive, and microcontrollers for overall control. Power factor correction with IGBTs is used in the power supply circuit to keep harmonics in the input current below the IEC limit.*" The Toshiba IGBT-based washing machine drive circuit is shown in Fig. 12.15.

The motor in the washing machine must operate over a large speed range. During the wash cycle, the drum speed is maintained at about 50 rpm while during the spin cycle, it must be increased to more than 1000 rpm [18]. The torque speed requirements for the washing machine are shown in Fig. 12.16. During the washing cycle, a large torque must be generated by the motor because of the presence of water, which adds considerable weight inside the drum. The water is pumped out before the spin cycle. Consequently, the torque required during the spin cycle is low. When the drum transitions between these speeds of rotation, a resonance usually occurs that must be controlled by the drive. The three-phase induction motor is well suited for this application due to its light weight and speed range [19]. They are simpler to manufacture, are robust in construction, and do not have brushes that can wear out.

The direct drive washing machine allows for the elimination of gears. An IGBT inverter is utilized with sine wave drive to improve the noise level [20]. The sine wave drive is implemented with a carrier frequency of 16 kHz for the pulse-width modulation (PWM) waveform. A ripple voltage is generated on the DC bus even with the sine wave drive. This can be corrected, as shown in Fig. 12.17, by monitoring the DC bus voltage and correcting for the fluctuations with the control circuit. With these measures, the noise level has been reduced by 29 dB during the wash cycle (making it equivalent to the level in a suburban area) and by 40 dB during the spin cycle (making it equivalent to the level in a quiet

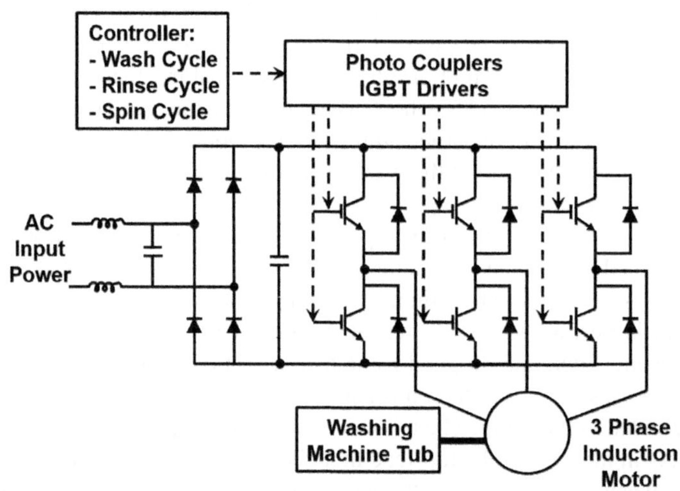

FIGURE 12.15

Washing machine drive with three-phase induction motor. *IGBT*, insulated gate bipolar transistor.

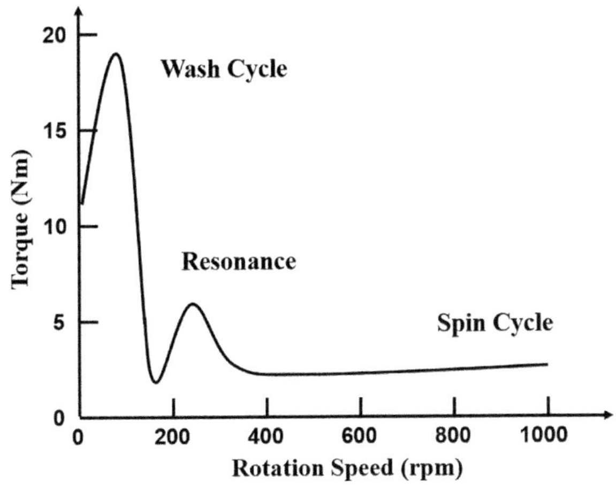

FIGURE 12.16

Torque speed requirements for a top-load washing machine.

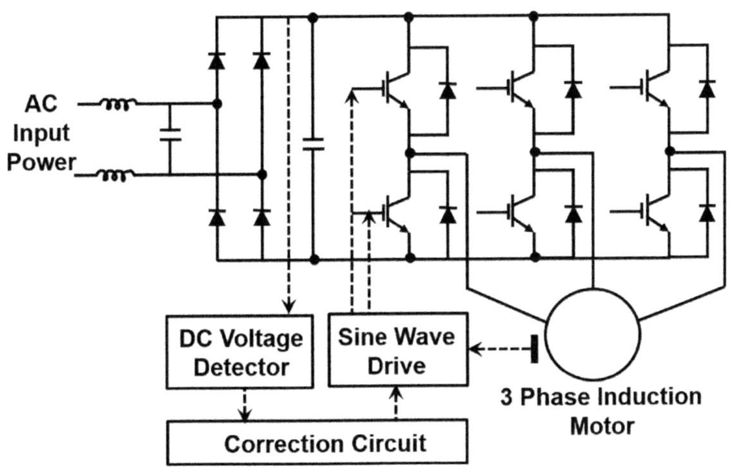

FIGURE 12.17

Sine wave drive with DC voltage correction for direct drive washing machine.

park!). A resistor-less electromagnetic braking method is also implemented to reduce the size and cost of the inverter.

Investigations of ways to improve the efficiency of washing machine drives have been proposed in more recent years for the conservation of energy use in the home. A comparison between single-phase induction motors and three-phase induction motors has been performed [21] for top-load washing machines. The single-phase induction motor operated at an efficiency of only 18.1% during the wash cycle and 16.7% during the spin cycle. In contrast, the three-phase induction motor operated at an efficiency of 46% during the wash cycle and 51% during the spin cycle. An impressive 2.5 times improvement in efficiency is gained by using three-phase induction motors for washing machines.

Commercial laundry washing machines have also shifted to three-phase induction motors controlled by IGBT inverters [22]. Variable frequency drives allow a large range of speed, detection of unbalanced loads, and improved power factor and in-rush currents. The cost of operation is improved and performance can be tailored for unique loads such as heavy cotton towels in hotels and resorts.

A comparison between using the IGBT and bipolar transistors for washing machine drives was performed in 1997 [22]. It was demonstrated that the IGBT drive produced a superior torque–speed characteristic, less harmonics in the current, and "virtual elimination" of motor noise in the audible frequency range. Since then, washing machine drives have all been served by IGBTs.

12.1.4 Microwave Oven

Today, the microwave oven is considered an indispensable appliance in the kitchens of people in the developed world. It provides the convenience of rapidly reheating food and cooking vegetables. Modern microwave ovens are capable of stewing, frying, baking, and steaming food as well as making popcorn. Many supermarket products have been configured for "cooking" in the microwave oven for the benefit of consumers with limited time for meal preparation. Microwave ovens were originally made as countertop appliances like the one shown in Fig. 12.18(a). However, they have become popular as over-the-range units, as shown in Fig. 12.18(b), often with built-in exhaust fans to save counter space in kitchens. Although first introduced for the residential market in the 1960s, their adoption did not begin until the 1970s due to

(a) (b)

FIGURE 12.18

Microwave oven: (a) countertop; (b) over the range.

the high initial cost [23]. Twenty-five percent of US households owned a microwave oven by 1986; almost all households have one today.

Heating food in a microwave oven uses the presence of water molecules in the food. Water molecules are electrical dipoles. When subjected to microwave radiation, the water molecules rotate in alternating directions in response to the electrical field of the microwave radiation. The rotating water molecules impart energy to surrounding molecules in the food, increasing their temperature. Microwave ovens utilize radiation at 2.45 GHz corresponding to a wavelength of 12.2 cm. The radiation is able to penetrate most food items placed in the chamber, resulting in uniform heating. With this approach, the food is usually not heated to above 100°C and the cookware is not heated. This is considered an advantage of microwave heating from the point of view of safety when compared with heating food in a conventional oven with resistive elements.

The microwaves generated within the cooking cavity (chamber) in a microwave oven are produced using a magnetron, which was first developed during World War II for radar applications. The cavity magnetron is a high-powered vacuum tube that generates microwaves using the interaction of a stream of electrons with a magnetic field. A typical oven that generates 700 W of microwave power requires about 1100 W of input power.

The *i-v* characteristics of a typical magnetron are shown in Fig. 12.19. The anode to cathode voltage of the magnetron must be increased above 3.5 kV before significant current begins to flow. Microwave energy can be generated only with sufficient current flow to provide energy to the electrons within the magnetron cavity [24]. The magnetron power supply must be capable of producing a voltage of at least 3.5 kV.

The first power supply used for magnetrons in microwave ovens was the ferroresonant design shown in Fig. 12.20. The input 110 and 220 V AC power sources in

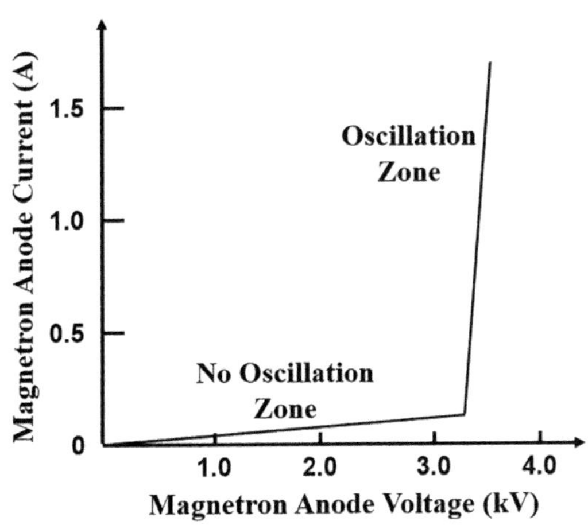

FIGURE 12.19

Current–voltage characteristics of a typical magnetron.

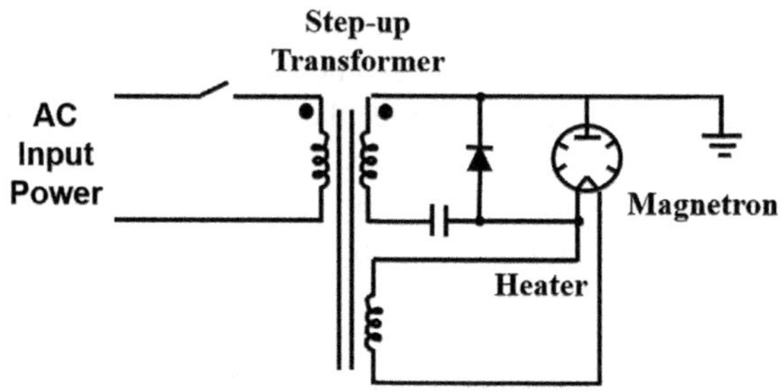

FIGURE 12.20

Ferro-resonant power supply for the magnetron in a microwave oven.

the United States and Europe were stepped up to the high voltages required to operate the magnetron by using a transformer. A second winding was used on the secondary side to provide power for the magnetron heater. The transformer had a large size and weight because of operation at a line frequency of 50 or 60 Hz. This made the microwave oven bulky and unwieldy for the consumer.

A significant improvement in the size and weight of microwave ovens has been achieved by replacing the ferro-resonant topology with an IGBT inverter power supply. The magnetron power supply [25] for modern microwave ovens is shown in Fig. 12.21. Using the IGBT-based inverter circuit, the weight of the transformer could be reduced by more than 10 times from 4000 g in the ferro-resonant converter

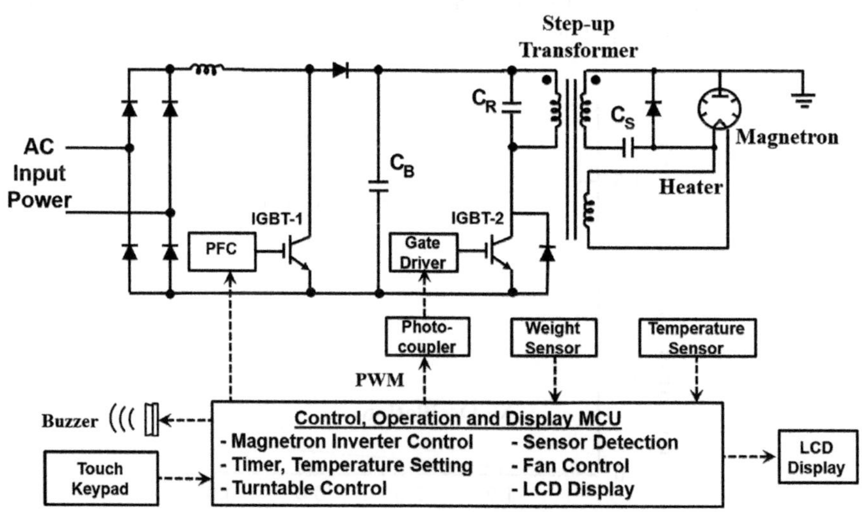

FIGURE 12.21

Insulated gate bipolar transistor inverter power supply for the magnetron in a microwave oven. *PFC*, power factor correction; *PWM*, pulse-width modulation.

to 380 g. In addition, a new turbo-up mode of operation could be introduced using the IGBT inverter to shorten the cooking time by 30%.

After the AC input voltage is rectified, the voltage across the DC bus capacitor (C_B) is maintained using a PFC circuit with IGBT-1 to reduce harmonics on the power line. The second IGBT-2 is connected in series with the primary winding of the step-up transformer. A capacitor (C_R) is connected across the primary winding of the transformer to create a resonant circuit. The current in the transformer primary is controlled by switching IGBT-2 using a PWM signal from the microprocessor. The size and weight of the transformer can be reduced, as mentioned above, by operating the PWM signal at a high frequency (33 kHz).

The current flowing through the primary winding and the collector voltage of the IGBT is shown in Fig. 12.22 for the magnetron power supply circuit when the input power level is 1.4 kW. The current in the primary winding increases linearly with time when the IGBT-2 is turned on to charge the inductance. The maximum collector current for the IGBT is 70 A, as shown by the waveform with the red line in Fig. 12.22. When the IGBT-2 is turned off, the capacitor is charged and discharged in the resonant circuit. A resonant voltage of 700 V must be supported by the IGBT-2, as shown by the waveform with the blue line in Fig. 12.22. The resonant voltage on the primary side is stepped up on the secondary side by the transformer and impressed on capacitor C_S. When the IGBT-2 is turned on again, the peak secondary voltage plus the capacitor voltage is applied to the anode of the magnetron, doubling the magnitude. This voltage is designed to be sufficient to produce current flow in the magnetron, generating microwave radiation. The output power delivered by the magnetron can be controlled by adjusting the on-time for IGBT-2.

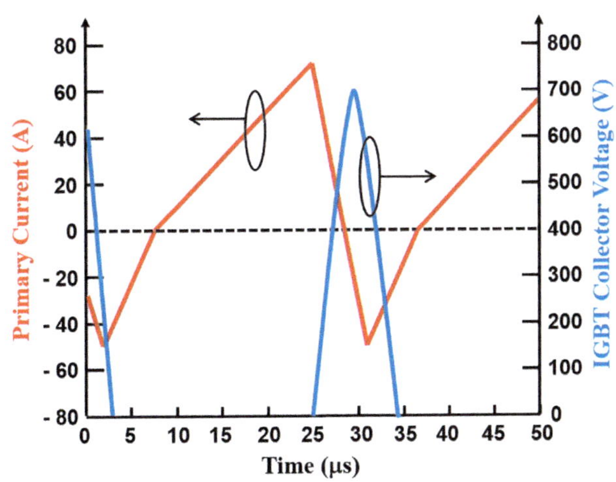

FIGURE 12.22

Power supply waveforms in the microwave oven power supply.

Based on the voltage and current waveforms in the power supply circuit, the IGBT-2 specifications are a blocking voltage of 900 V, a collector current of 70 A, and a turn-off time of 0.25 μs. The optimization of the trade-off curve between on-state and switching losses discussed in previous chapters is essential to minimize the power losses in the IGBT, even at an operating frequency of 33 kHz. One of the advantages of the IGBT inverter power supply is that the output power can be increased from 600 to 700 W for 5 min at the beginning of the cooking period [25]. This allows the temperature of the food to increase more rapidly, reducing the total cooking time by 30%. The weight of the transformer is reduced from 4000 g for the ferro-resonant case to 380 g for the IGBT inverter case. This has made modern tabletop microwave ovens much easier for consumers to move around and for the installation of over-the-range units as well.

Various improvements in the IGBT inverter power supply designs for the magnetron have been proposed over the years to increase the efficiency and reduce the stress on the magnetron. One example is the active voltage-clamped, ZVS–PWM (zero voltage switching–pulse-width modulation) quasiresonant inverter [26] shown in Fig. 12.23. A full wave voltage doubling is achieved on the secondary side by using two capacitors (C_1 and C_2) with rectifiers. In addition, the peak voltage on the magnetron is clamped by using the capacitor C_C and IGBT-2. This keeps the magnetron current less than 1.2 A to enhance its lifetime. Another advantage reported with this approach is that the heater temperature does not increase as much with increasing microwave output power.

A half-bridge IGBT inverter has been developed for the magnetron power supply [20], as shown in Fig. 12.24. This topology has the advantage of reducing the blocking voltage requirements of the IGBT in half when compared with the topology shown in Fig. 12.21. The blocking voltage of the IGBT could be reduced to 430 V, allowing a reduction of the on-state voltage from 2.3 to 1.8 V while maintaining the same turn-off time. A full wave voltage doubling is achieved on the secondary side by using two capacitors (C_1 and C_2) with rectifiers. The operating frequency of

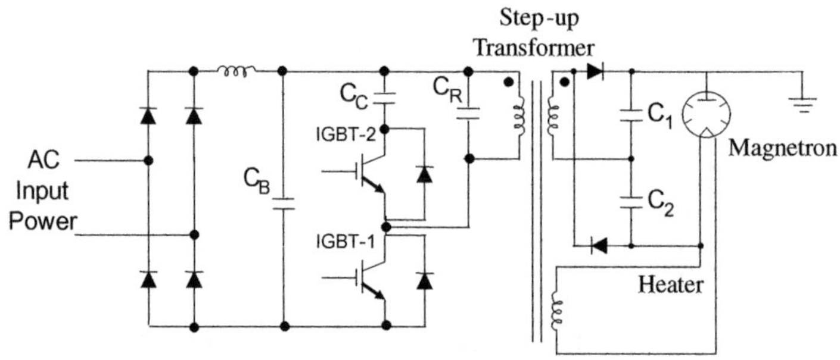

FIGURE 12.23

Active voltage clamped insulated gate bipolar transistor inverter power supply for the magnetron in a microwave oven.

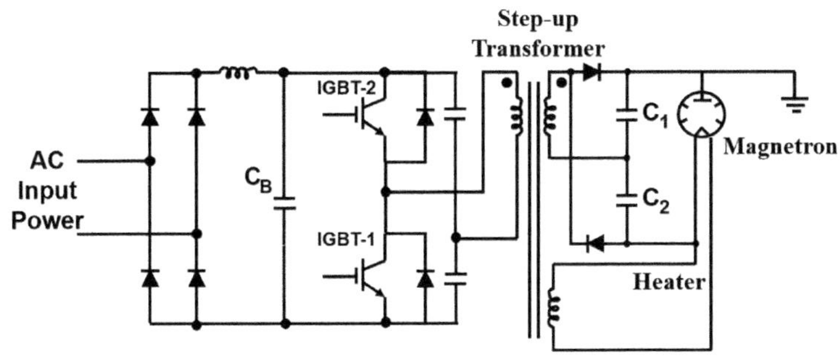

FIGURE 12.24

Half-bridge insulated gate bipolar transistor inverter power supply for the magnetron in a microwave oven.

the half-bridge topology could be increased to between 39 and 53 kHz, allowing a reduction in the inverter size by 20% (from 1395 cm^2 to 1165 cm^2).

Homes in the United States and Japan use 120 V rms power while in the Europe and China, 220 V rms is utilized. A universal microwave oven design is convenient from the production point of view to serve the worldwide market. The single-ended push-pull high-frequency inverter [24] shown in Fig. 12.25 has been developed with this goal in mind. ZVS with PWM control is used in this topology. A capacitor, C_S, is placed in series with the transformer primary to equalize the current in the 220 V rms market to that in the 120 V rms market.

A single chip controller [27] for the quasiresonant ZVS inverter using a single IGBT has been developed to reduce the cost. The microcontroller functions have been merged with the IGBT drive circuit, as shown in Fig. 12.26. The input power factor is maintained close to unity by controlling the on-time of the single IGBT. The topology operates in open loop when the microwave oven is first started to generate the high voltage across the magnetron. After this, current feedback is used to maintain a good power factor during operation of the microwave oven.

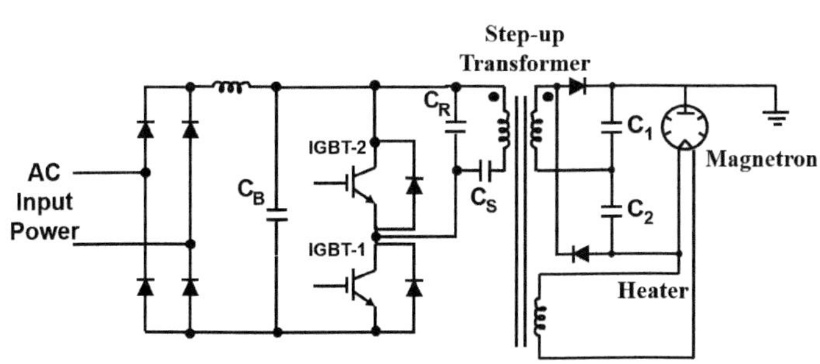

FIGURE 12.25

Single-ended push-pull insulated gate bipolar transistor inverter power supply for the magnetron in a microwave oven.

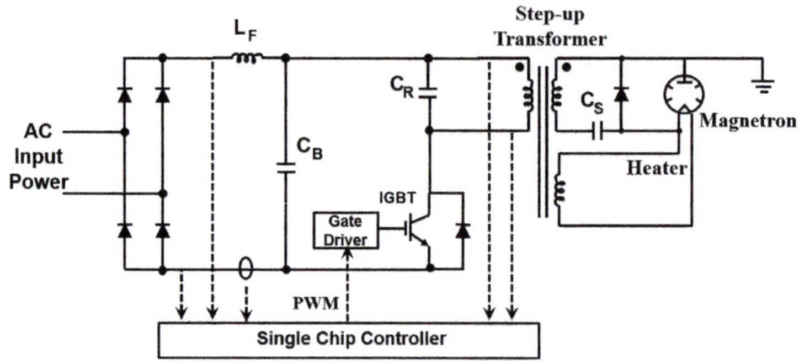

FIGURE 12.26

Single-chip controller for insulated gate bipolar transistor inverter power supply for the magnetron in a microwave oven.

12.1.5 Induction Cooktop Range

Induction cooking, using the range shown in Fig. 12.27(a), is accomplished by heating a vessel via power transferred with a high-frequency electrical transformer that consists of a primary coil and the cooking vessel as a secondary winding. The coil has a spiral shape on a horizontal plane, as shown in Fig. 12.27(b), to increase the surface area for generating the magnetic flux that is coupled into the vessel. The primary coil is designed with many windings while the cooking vessel behaves as a single winding that is short circuited. A high current is induced in the cooking vessel due to the large ratio of primary to secondary turns. This current flow through the vessel produces heating due to the resistance within the vessel material [28]. The primary coil is fabricated from litz wire, which is a bundle of a large number of insulated wires in parallel. This reduces the resistance in the primary coil due to mitigation of the skin

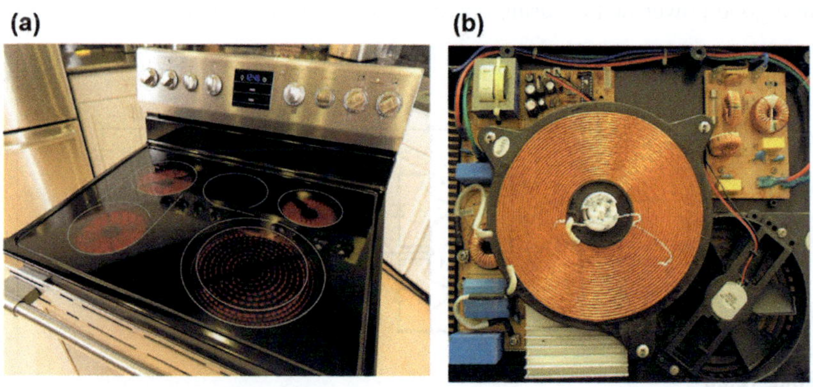

FIGURE 12.27

(a) Induction cooktop range; (b) induction range coil.

depth effect. The coil has a spiral shape on a horizontal plane to increase the surface area for generating the magnetic flux that is coupled into the vessel. The magnetic coupling is enhanced by placing rectangular ferrite bars at equal intervals around the winding. The cooking vessel must be made of ferrous material such as cast iron or stainless steel to induce current in the pot. Aluminum pots do not work well but aluminum cladding of stainless steel can be used to improve the heat distribution in the pot.

Induction cooktops are favored by consumers because they are safer than electric or gas cooktops. The heat is transferred only to the pots, making the surrounding temperature lower. They are easy to clean because of a flat ceramic surface between the coil and the pot. However, the pots must have a flat surface to obtain a strong magnetic coupling between the coil and the pot. The energy transfer efficiency of the induction cooktop is 84% when compared with 74% for a range with a resistive burner and only 40% for a gas range.

Interest in the development of commercial induction cooktop ranges dates back to the 1970s. To avoid unpleasant acoustic noise, the operating frequency of the induction current must exceed 20 kHz. The performance and cost of induction cooktops have been limited by the power semiconductor devices available to generate the high-frequency current [29]. The other issue was skin depth for the current flow in the cooking vessel. The skin depth (d) of the current induced in the vessel is given by:

$$d = 3160\frac{\sqrt{\rho}}{\mu f} \tag{12.1}$$

where ρ is the resistivity, μ is the permeability, and f is the operating frequency. The typical frequency used in induction cooktops is 24 kHz. In this case, the skin depth is 0.004 in for stainless steel, 0.022 in for aluminum, and 0.017 in for copper. For this reason, the cooking vessels made of stainless steel are recommended for induction cooktops.

An early version of the induction cooktop utilized self-excited, shunt-fed series-commutated SCRs [30]. The output power for heating the cooking vessel was controlled by switching a multiplicity of small capacitors. These units did not become popular products.

With the availability of IGBTs, several circuit topologies for induction heating were developed: the half-bridge, the series resonant converter, the single-ended converter, and the full-bridge converter. The half-bridge and full-bridge converters are used for powering multiple burners on a cooktop range while the single-ended topology is used for smaller tabletop units and rice cookers. The converters for cooktop ranges must be able to generate relatively high power levels of 2.5–3 kW per heating element. The single-ended topology that is more suitable for lower power levels will be discussed later in the chapter in the section on small appliances where its smaller size and cost is attractive.

The impact of improved power semiconductor devices on induction cooktops has been reviewed recently [31]. The first generation of induction cooktops developed in 1974 made use of thyristors, which required using a current-source resonant circuit.

The large switching losses in the power devices resulted in a weight of 23 kg and a volume of 48 L for the induction cooktop. With the availability of power bipolar transistors in the early 1980s, the voltage resonant circuit topology could be developed due to the ability to turn-off the switch. The drawback was the large number of components needed to implement the base drive circuit for the bipolar power transistor. With the availability of IGBTs with high-frequency switching performance, the drive circuit could be "drastically" reduced in size, making the inverter much smaller in size and cost. The authors state: "*IGBT has made (a) breakthrough with drastic advancements in semiconductor process technology and its breakthrough has made a great contribution to (the) improvement of efficiency in induction heating cooking heater (IHCHs). For example, in the case of (the) IH cooker, the weight and volume (are) reduced to 2.3 kg and 7 L.... It is no exaggeration that the inverter for consumer IH cooking has evolved with the progress of the power semiconductor devices.*"

The half-bridge topology used for modern induction cooktop ranges [20] is shown in Fig. 12.28. Efforts to develop this topology for induction cooking applications date back to the 1990s [32]. This half-bridge topology is used for twin burners with 3 and 2 kW power levels, respectively, operated from a 200 V rms AC power source. The IGBT-1 and IGBT-2 in the H-bridge are alternately switched on and off at 21 kHz to produce the high-frequency current in the heating coil. The on-time for IGBT-2 is held constant and the on-time of IGBT-1 is varied to control the output power.

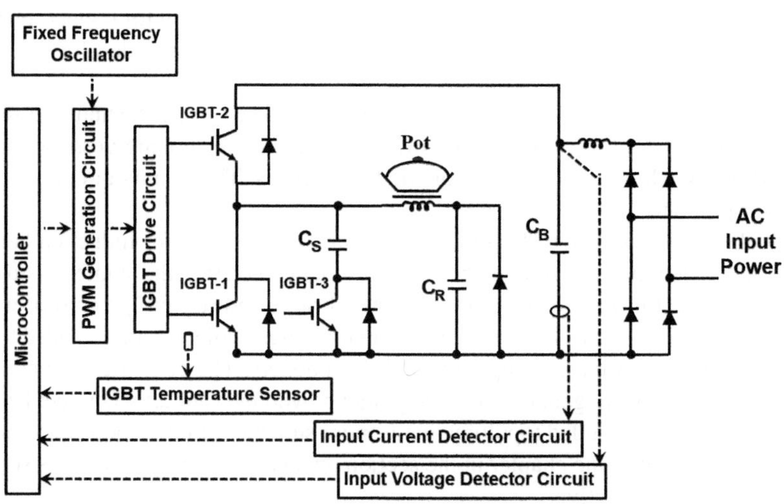

FIGURE 12.28

Half-bridge inverter topology for induction cooktop ranges. *IGBT*, insulated gate bipolar transistor.

Soft switching of the IGBTs is used with an active snubber capacitor (C_S) controlled using IGBT-3. This allows reducing the dV/dt of the IGBT voltage waveform, as shown in Fig. 12.29, with a sinusoidal current generated in the coil. A high output power of 3 kW is achieved under these operating conditions. For induction heating at low output power levels (down to 120 W), the IGBT-3 is turned off to isolate the snubber capacitor (C_S).

A quasiresonant half-bridge IGBT inverter [33] has also been proposed and demonstrated for induction cooktops to heat iron and steel pans. The circuit topology for this approach is shown in Fig. 12.30. Diode D_{1c} is required for allowing zero-voltage soft switching with different types of metal vessels placed on the induction burner. Capacitor C_{1a} is a high-frequency plastic film capacitor used to achieve quasiresonant operation. Capacitor C_{1b} is a high-frequency capacitor used to reduce the peak voltage impressed across IGBT-1 from 1200 to 500 V. Consequently, IGBT-1 has a voltage rating of 600 V and a current rating of 80 A while IGBT-2 has a voltage rating of 900 V and a current rating of 65 A. This allows improving the power loss trade-off curve for IGBT-1.

The buck-boost full-bridge inverter shown in Fig. 12.31 can heat all pans, which includes not only iron and stainless steel pans but also aluminum and copper pans [4]. Aluminum and copper pans are nonmagnetic materials with low resistance in contrast to iron and stainless steel pans. For Al and Cu pans, the frequency of eddy currents generated in the pan is increased to raise the skin resistance. The magnetic field is also increased by raising the current and number of turns in the primary coil in the burner. The DC link voltage must be smoothed to avoid buzzing noise and power factor correction is required. Instead of the pulse frequency modulation used for Fe and SS pans, pulse amplitude modulation is utilized for Al and Cu pans. IGBTs

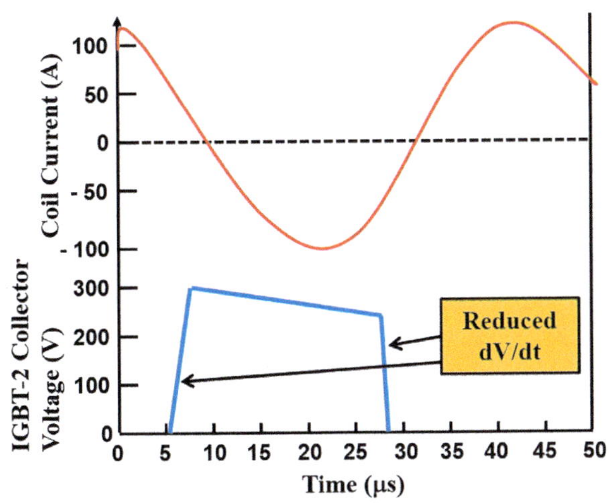

FIGURE 12.29

Half-bridge inverter waveforms. *IGBT*, insulated gate bipolar transistor.

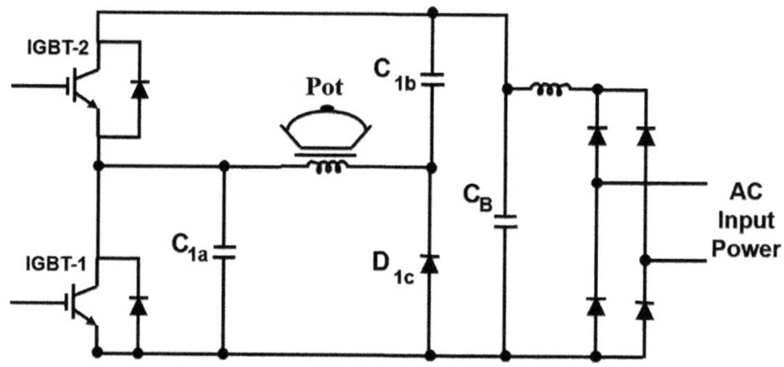

FIGURE 12.30

Quasiresonant half-bridge inverter topology for induction cooktop ranges. IGBT, insulated gate bipolar transistor.

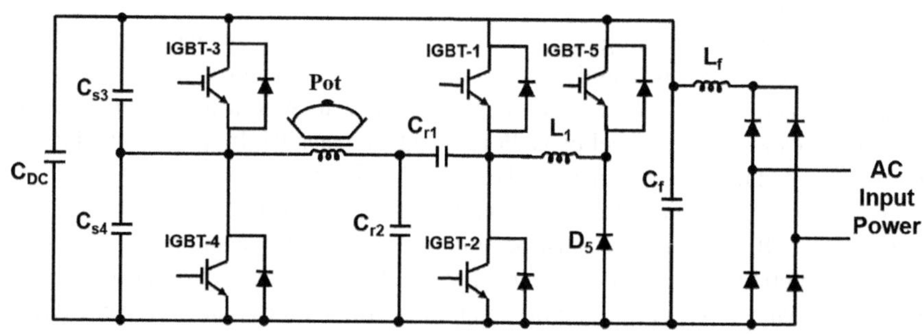

FIGURE 12.31

Full-bridge resonant topology for all-metal pan induction cooktop ranges.

with 600 V blocking voltage ratings are commonly used for these half-bridge and full-bridge inverters.

The reverse blocking IGBT enables the development of cycloconverters or matrix converters. The first IGBTs were fabricated with a symmetric voltage blocking capability of 600 V [34]. The focus of the power semiconductor industry quickly shifted to building IGBTs with asymmetric blocking capability due to interest in variable-speed drive applications. Recently, IGBTs with reverse blocking capability have been released as products by some companies, such as IXYS Corporation and the Fuji Electric Company. An inverter for the induction cooktop has been proposed and developed using these symmetric blocking IGBTs [35]. The cycloconverter topology has the advantages of directly converting the low-frequency AC input power to the high-frequency AC power fed to the induction coil without using a DC power bus. This eliminates the bulky and costly electrolytic capacitor (C_B) needed for the previous topologies shown in this section. The reverse blocking IGBTs are configured in antiparallel, as shown in Fig. 12.32. The voltage drop across

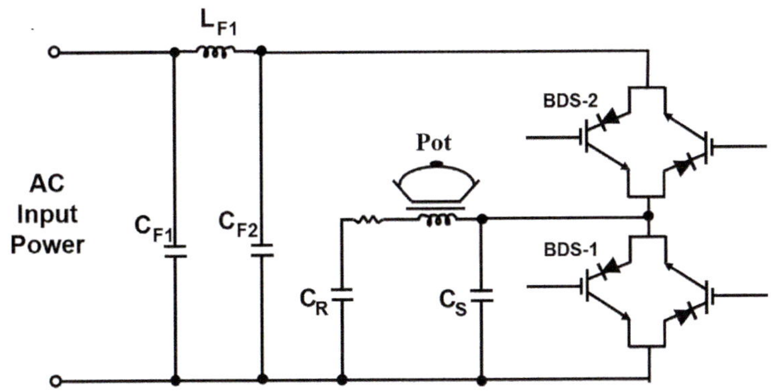

FIGURE 12.32

Half-bridge inverter topology for induction cooktop ranges with reverse blocking IGBTs.

the reverse blocking IGBT is smaller than if a reverse blocking diode was placed in series with the asymmetric IGBT structure. The 1200 V reverse blocking IGBT has an on-state voltage drop of 2.3 V at an on-state current of 55 A. The cost of this converter is also reduced by elimination of the bridge rectifier at the input.

12.1.6 Dishwasher

Manual cleaning of dishes and utensils relies on physical scrubbing by hand with a sponge containing detergent to aid in the removal of grease and food particles. The process is tedious and time consuming with deleterious impact on the skin. Dishwashers are electromechanical devices used in homes (see Fig. 12.33(a)) and commercial establishments (see Fig. 12.33(b)) such as hotels and restaurants. The modern dishwasher relies on spraying hot water between 55°C and 75°C onto dishes.

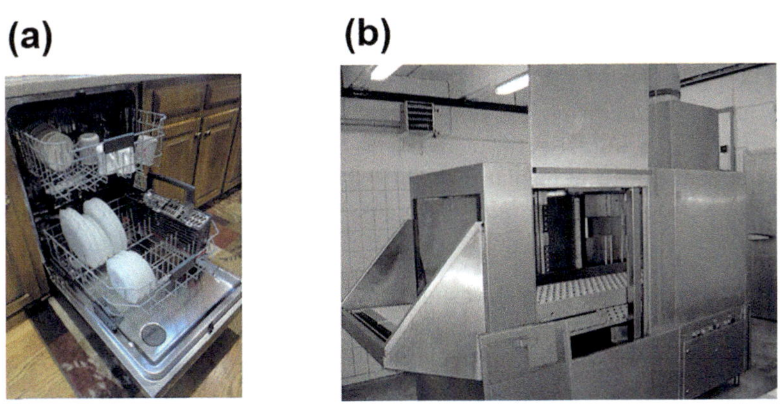

FIGURE 12.33

(a) Household dishwasher; (b) commercial dishwasher.

This concept was invented and patented by Josephine Cochrane in 1887. The basic configuration of a drop-down front-door panel with wire racks to house the dishes and a rotating water spray arm is still employed in modern dishwashers.

Dishwashers began to be sold as domestic appliances during the post-World War II boom [36]. They became widely adopted in homes in the United States and Europe in the 1970s with more than 75% of homes using them by 2012. The cleaning cycle consists of three basic steps: (1) the wash cycle to blast the dishes with water and detergent; (2) the rinse cycle to remove the dirt and food particles from the dishes, and (3) the heat cycle to dry the dishes. Modern dishwashers make use of microprocessor control with sensors to monitor the quantity of dishes and the amount of dirt to optimize the cleaning cycle. Large commercial dishwashers used in hotels and restaurants do not utilize the drying cycle. Instead the final water rinse is performed at a higher temperature to allow drying in open air.

The evolution of dishwashers has been recently reviewed [37]. The article points out that the cleaning effect is caused by three physical actions: the mechanical impact of the water spray on the dishes and pans, the chemical action of the detergent, and the thermal action of the water. The water spray is heated to 50°C during energy-saving wash cycles, requiring about 1 KWh of energy consumption. During the drying phase, a water rinse at 65°C is used that also consumes a significant amount of energy.

An important change in dishwasher technology has occurred recently to achieve improved efficiency. As mentioned earlier, a large amount of electrical energy is required for heating the water in the dishwasher and for running the centrifugal pump used for water circulation. In the 2000s, the induction motor was replaced with the brushless motor. The authors state [35]: "*The use of these motors was made possible by the introduction of adequately sophisticated electronics, capable to generate and handle waveforms able to control the three-phase brushless motors.*" The energy savings derived from making this change are shown in Table 12.1. A nearly 40% reduction in power is enabled by using the brushless motor driven using IGBT inverters.

An small low-loss intelligent molded module nano (SLLIMM) has been developed [38] in an NDIP 26L transfer molded package for providing 100 W drive to the water pumps in dishwashers. The module incorporates six very fast-switching 600 V IGBTs, six ultrafast soft-recovery fly-back diodes, and a high-voltage integrated circuit (HVIC) in bipolar CMOS (BCD) technology within a compact package. The authors state: "*Despite the higher IGBT losses at turn-off, the overall*

Table 12.1 Evolution of Dishwasher Motors

Motor Type	Power Used	Motor Efficiency	Hydraulic Efficiency	Absorbed Power
Induction	25 W	53%	52%	90 W
Brushless	25 W	75%	52%	65 W

Table 12.2 Comparison of IGBT With Power MOSFET for Dishwasher Application

Device	Breakdown Voltage	V (On) @ 1 A	E (On)	E (Off)	E (Total)
IGBT	600 V	2.6 V	33 μJ	25 μJ	58 μJ
MOSFET	500 V	1.4 V	138 μJ	8 μJ	146 μJ

IGBT, *insulated gate bipolar transistor.*

dynamic losses of the IGBT are significantly lower than that of the MOSFET because of the copackaged diode's excellent energy recovery, whose impact in the turn-on phase is far lighter than the body diode of the MOSFET." The losses measured for the IGBT and MOSFET can be compared in Table 12.2 when operating a brushless DC motor to run a dishwasher water pump at 3200 rpm. The lower power losses obtained with the IGBT solution allow operation of the module at higher ambient temperatures without a heat sink.

More recently, an intelligent power module (IPM) has been developed with 600 V, 50 A capability for consumer and industrial motor drives [39]. The IPM contains six field-stop, high density trench gate IGBTs and six fly-back diodes. The low-voltage IC and high-voltage level-shifting IC for driving the IGBTs are packaged with the IGBTs in a DBC package (see Fig. 6.12).

12.2 Small Appliances

Appliances used in homes are generally classified into the categories of major and small appliances [5]. Small appliances (sometimes referred to as brown goods) are portable units that can be placed on kitchen countertops to perform household tasks [40]. Examples are toasters, mixers, and food processors. Recently, single-burner induction plates and rice cookers have been added to kitchens, especially in Asia.

12.2.1 Portable Induction Cooktop and Rice Cooker

The portable induction cooktop has the same advantages previously described for the induction cooktop range in Section 12.1.5. The advantage of the portable or tabletop unit is the convenience of moving it to selected locations on the countertop in a kitchen or on the patio of the home. An example of the portable induction cooktop is shown in Fig. 12.34(a). It is manufactured under popular brand names such as Fagor, Hamilton Beach, NuWave, and Waring.

Rice is a staple food for billions of people around the world, especially in Asian countries with large populations such as China, India, and Japan [41]. The annual per capita rice consumption is between 100 and 150 kg in Southeast Asia while that in China, South Korea, and Japan is about 50 kg. The Asian Rice Foundation states [42]: "*Rice is arguably the world's most important food. It is the second most widely*

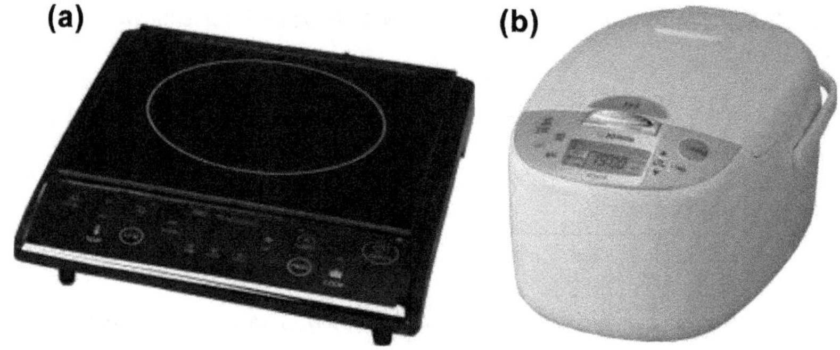

FIGURE 12.34

(a) Portable induction cooktop; (b) rice cooker.

cultivated cereal in the world, after wheat, and is a staple food of (more than) half the world's population. In much of Asia, rice is so central to the culture that the word is almost synonymous with food. In Chinese the line in the Lord's Prayer is translated as 'Give us this day our daily rice,' and a Japanese proverb states that 'A meal without rice is no meal.' With (more than) half the population of the world eating rice two or three times a day, it is not surprising that there are many arguments over how it should be cooked. Indeed rice can be cooked in a variety of ways, including boiling, baking, roasting, frying, and pressure cooking. Cooking rice in an automatic rice cooker is becoming very popular, as it ensures consistent results and cooking instruction is much simpler to follow."

The induction-heated rice cooker, shown in Fig. 12.34(b), is quite popular in Asian countries. Its manufacturers include Sanyo and Zojirushi. Rice is found to taste better in an induction-heated cooker compared with the conventional resistive heated rice cooker. A high induction power is required to bring the water in the pot to a boil quickly to produce better-tasting rice [20]. The design of the coil in the rice cooker is different than that for the induction cooktop because it has to conform to the shape of the vessel, as shown in Fig. 12.35, to maximize energy transfer [43]. The vessel is made of stainless steel to achieve strong inductive coupling to the coil. It is clad on the inside with aluminum to improve the heat distribution within the vessel.

Due to the smaller power requirements for the portable single burner cooktop units and the rice cooker, the single-ended inverter topology, shown in Fig. 12.36, is commonly used [44]. Only one IGBT is required for this circuit topology, reducing the cost. The typical current and voltage waveforms for the IGBT are shown in Fig. 12.37 together with the current in the coil. The current in the IGBT increases linearly with time from 30 to 40 μs to charge the coil when it is turned on by the gate drive circuit. When the IGBT is turned off, its current rapidly drops to zero and the inductor transfers energy to the resonant capacitor C_R. The voltage across the IGBT increases at a rate of about 100 V/μs and increases to 900 V during the turn-off. The IGBT must have a blocking voltage rating of 1200 V to satisfy this circuit operating condition. The power loss in the IGBT is reduced by the zero-voltage turn-off

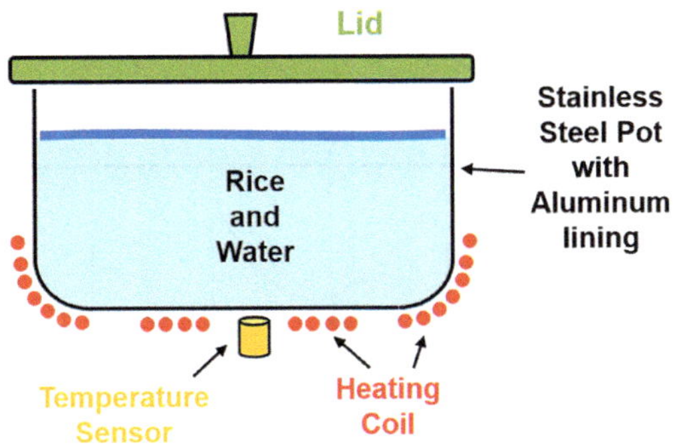

FIGURE 12.35

Rice cooker vessel design.

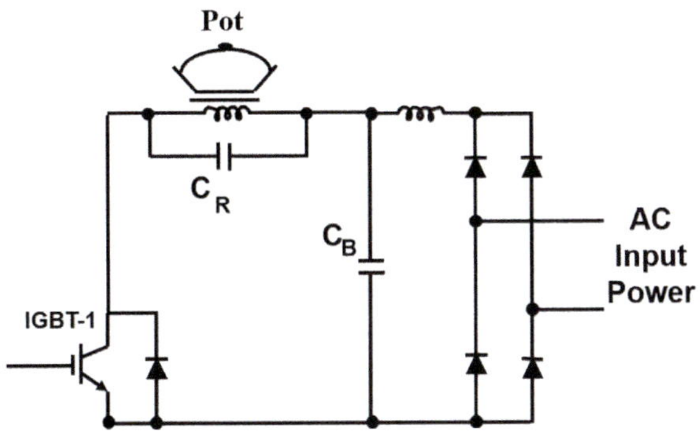

FIGURE 12.36

Single-ended inverter topology for portable induction cooktops and rice cookers. *IGBT*, insulated gate bipolar transistor.

process in the resonant circuit. Note that the voltage across the IGBT reverses at 30 μs, which forward biases the diode across the IGBT (the small on-state voltage (~2 V) across the diode is not visible in Fig. 12.37). The power delivered to the vessel can be regulated by changing the on-time of the IGBT.

12.2.2 Food Processors (Blenders, Juice Makers, Mixers)

In Western countries, it is now commonplace to have a variety of gadgets on the kitchen counter for the preparation of food in different ways. The term food processor can be broadly defined to include the preparation of food using various speeds and form factors. The gadgets shown in Fig. 12.38 include: (a) the blender, (b) the juicer, (c) the mixer, and (d) the food processor. The blender [45] is used to process food in liquefied form with a rotating metal blade run by a motor located at the bottom of the jar. Blenders are commonly used to make milk shakes and

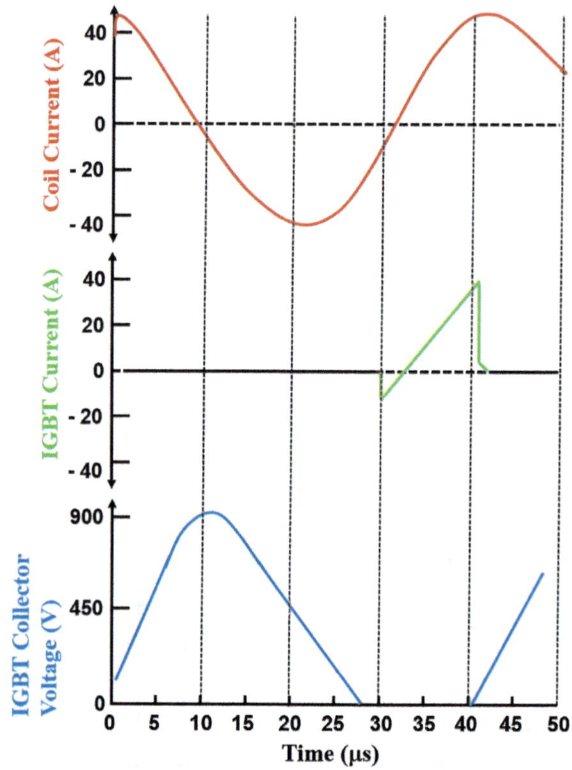

FIGURE 12.37

Single-ended inverter waveforms. *IGBT*, insulated gate bipolar transistor.

FIGURE 12.38

Food processors. (a) blender; (b) juicer; (c) mixer; (d) food processor.

alcoholic drinks. The blender has an operating speed between 6500 and 37,000 rpm. The juicer [46] is used to separate the juice from the pulp in fruits and vegetables. The juicer operates at speeds between 1500 and 6500 rpm. The mixer [47] contains a pair of beaters that allows stirring or whisking food in a bowl. It can also be accessorized with a dough hook to knead dough for making bread. The mixer's operating speed ranges between 50 and 1500 rpm. In contrast to blenders, food processors [48] use interchangeable blades and disks instead of a fixed blade and their bowls are wider and shorter for the solid or semisolid foods usually worked in a food processor. They make use of little or no liquid during operation. The use of a food processor reduces the large amount of time spent in the kitchen with chopping, shredding, and mixing of ingredients. Manufacturers of food processors include Cuisinart, KitchenAid, and Hamilton Beach.

By 1992, the permanent magnet DC motor driven by an IGBT-controlled chopper circuit was replacing the AC universal motor [49] because of increased efficiency, reduced motor noise, and improved speed control. The power circuit used for operation of the food processor is shown in Fig. 12.39. The AC input power is first rectified by a diode bridge to create a DC bus with a capacitor C_B. The PM DC motor used in the food processor is then driven by an IGBT-controlled chopper. The high input impedance of the IGBT allows its control with a microprocessor that can be programmed to perform various functions. This creates a compact, low-cost design suitable for the consumer market. Home appliances were an early adopter of the IGBT technology because it enhanced the simplicity and flexibility of design while deriving more functionality.

A modern food processor uses the brushless DC motor because of enhanced efficiency and speed response [50]. The high speed is achieved with stability by using IPMs, as shown in the block diagram of Fig. 12.40. Three Hall sensors are located at

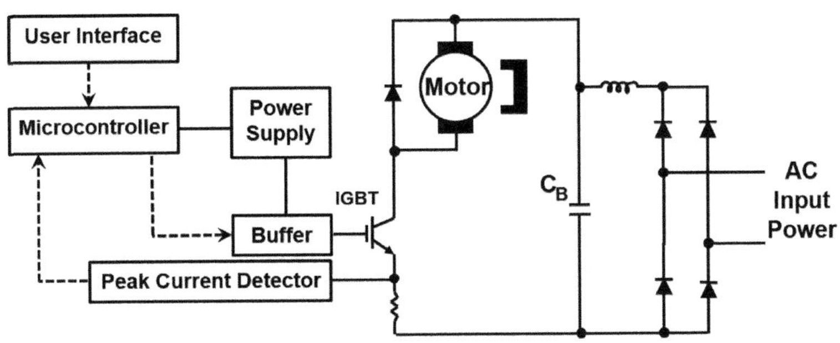

FIGURE 12.39

Insulated gate bipolar transistor control circuit for food processors.

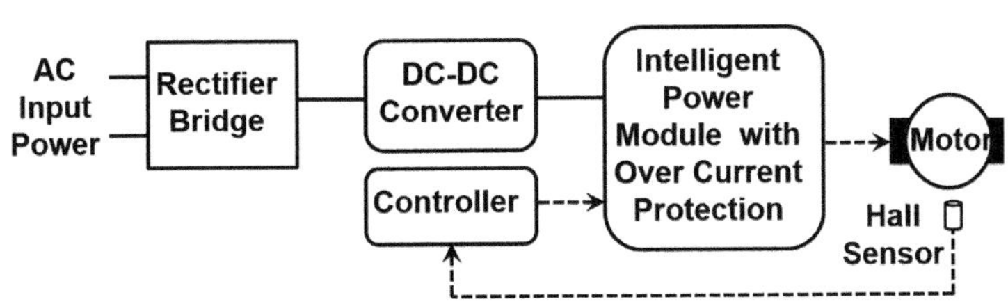

FIGURE 12.40

Food processor with brushless DC motor.

the motor to provide information to the microcontroller regarding motor current and speed. This configuration allows rapid changes in speed, fast braking, and other functions.

12.2.3 Vacuum Cleaners

Vacuum cleaners [51] remove dust and dirt from surfaces by using the suction principle. The dirt is collected in a bag or canister for subsequent disposal. Although the concept of electric vacuum cleaners was first developed in the early 1900s, they did not become popular until after World War II. They are especially attractive for consumers in Western countries who prefer to have carpeted rooms. Vacuum cleaners come in two basic formats. An example of the canister models popular in Europe is shown in Fig. 12.41(a). The wheeled canister containing the motor and dust collection bin is located separately from the vacuum head. This provides good mobility and maneuverability during cleaning. The upright model shown in Fig. 12.41(b) is popular in the United States and the United Kingdom. The cleaning head containing a beater brush is attached to the handle and dust collection bin. Two motors are commonly used in both models—one for generating the suction and the second for driving the beater brush. The high-speed vacuum cleaner (cyclone design) is shown in Fig. 12.41(c). This design employs centrifugal force to remove dust and particles.

An ultrahigh-speed vacuum cleaner has been achieved by using the switched reluctance machine [52]. Replacing the universal motors is desirable because of problems with wear in the brushes that preclude operating at very high speeds. The suction power of vacuum cleaners has been increased by 30 W every year [53] by increasing the motor speed; it reached more than 500 W in 2001 with the use of switched reluctance motors.

The sensored 1600 W, 100,000 rpm 4–2 switched reluctance motor is now used in several Dyson vacuum cleaners [52]. A block diagram of the drive circuit is shown in Fig. 12.42. An optical sensor is used to monitor the rotor position for feedback to the controller. IGBTs in an H-bridge configuration are used to provide drive to the

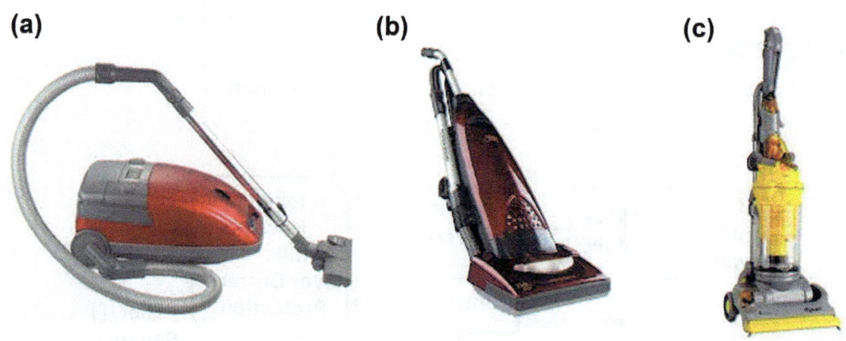

FIGURE 12.41

Vacuum cleaners: (a) canister, (b) upright, and (c) cyclone.

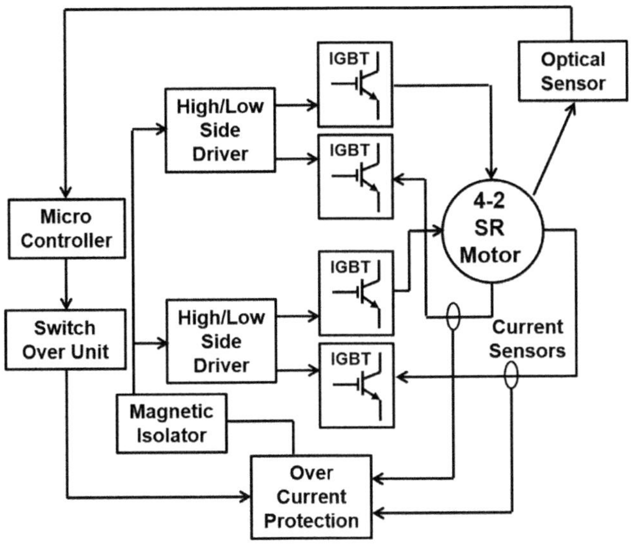

FIGURE 12.42

Insulated gate bipolar transistor control circuit for the vacuum cleaner.

motor. High- and low-side gate drive is provided, as previously discussed, for adjustable speed drives. An open-loop program is used to start the motor until it reaches a speed of 60,000 rpm. A closed-loop control is then performed to accelerate the motor to the rated speed. A small 10 μF DC-link capacitor is sufficient, allowing cost and size reduction of the drive electronics.

12.3 Television

The word television [54] is derived from the Greek term *tele* for far and the Latin term *visio* for sight. It allows telecommunication by sending images from a high-power central transmitter (or increasingly by cable today) and receiving them in homes using TV sets. TV broadcasts began in Schenectady, NY, in 1928 but did not become popular in homes until 1935. By 2009, at least one TV set was owned by 78% of the households around the world. The TV sets shown in Fig. 12.43(a) were originally made using the cathode ray tube (CRT) shown in Fig. 12.43(b).

In the CRT display, electrons are emitted from a heated filament located at the cathode end [55]. The electrons are focused into a narrow beam and accelerated toward the screen, which serves as the anode electrode. The screen is coated with a phosphor that is excited by the energy imparted from the incident electrons and emits visible light from the screen to the viewer. The electron beam is rastered across the screen by using a deflection circuit to sweep it in the horizontal direction and then

(a) (b)

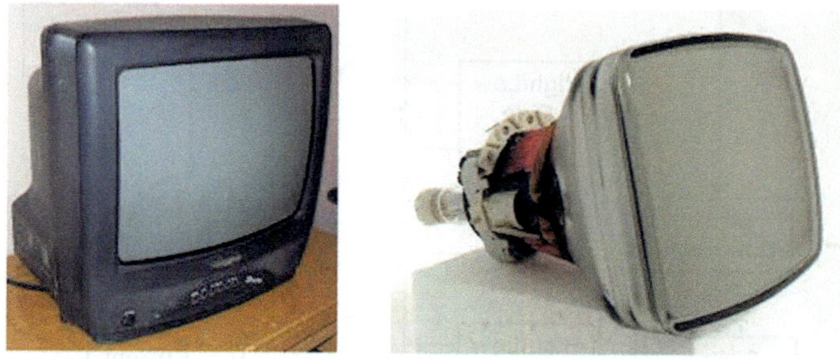

FIGURE 12.43

(a) Television set, (b) cathode ray tube.

(a) (b)

FIGURE 12.44

(a) Plasma display TV, (b) LCD TV.

step it periodically in the vertical direction. The pattern has to be repeated sufficiently quickly to create an image that the human eye observes without flicker. In a color TV set, three independent electron beams are used to generate the red, green, and blue colors with the screen composed of corresponding red, green, and blue dots.

The CRT display remained dominant until the 1990s when flat-panel displays emerged. The primary flat-panel displays are the plasma TV shown in Fig. 12.44(a) and the LCD TV shown in Fig. 12.44(b). These models displaced the CRT-based TVs because of a slim form factor and much-reduced weight [56]. Their picture quality has also superseded that achievable with CRT displays. In 2006, 42 million LCD TVs and 8 million plasma TV sets were manufactured. In 2012, LCD TVs comprised 87% of the worldwide shipment of TV sets while plasma TVs and CRT TVs had 5.7% and 6.9% of the market share, respectively [54].

12.3.1 TV Sets With CRTs

The CRT display requires rapid rastering of the electron beam to generate proper screen refresh rates. The horizontal deflection of the electron beam across the wider screens of larger TVs requires a high voltage (20 kV) at a frequency of 100 kHz [57].

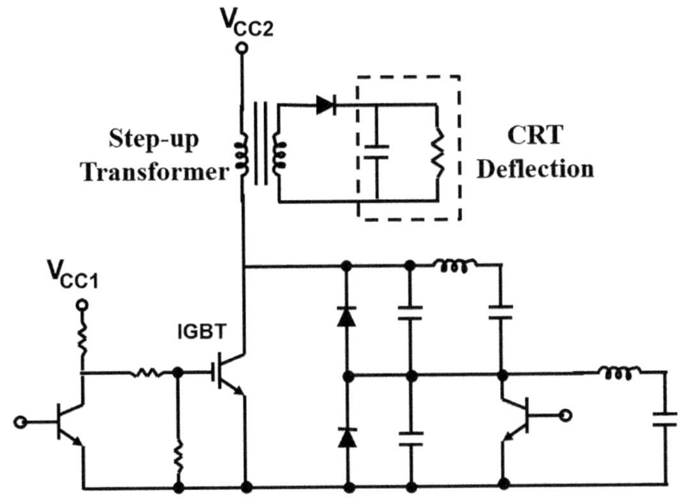

FIGURE 12.45

Cathode ray tube deflection circuit using an insulated gate bipolar transistor driver.

This function is performed by using a step-up transformer, as shown in Fig. 12.45. Despite a high turns ratio for the transformer, the power switch in series with the primary winding must withstand at least 900 V during each switching event. The bipolar power transistor was used to perform the switching function in the CRT deflection circuit [58] until the availability of high-voltage IGBTs with reduced switching losses that were optimized for this application. With reference to bipolar junction transistors, the authors state: "*Long storage times (6–10 μs) can pose circuit operating problems and the slow switching performance of these devices, although acceptable for operation at normal TV horizontal line scanning rates, cause thermal problems when used at higher line rates, with shorter retrace times.*" Attempts to improve this situation by using complex base drive circuits add to the cost [55].

Specially optimized IGBTs with high switching speed (100 kHz) and a blocking voltage capability of 1500 V were developed in the early 1990s for use in CRT deflection circuits [59,60]. The current and voltage waveforms for the CRT deflection circuit are shown in Fig. 12.46. The current in the primary of the step-up transformer increases linearly with time when the IGBT is turned on. When the IGBT is turned off, its voltage increases to a peak of 1200 V due to the fly-back converter operating in the resonant mode. The IGBT turn-off occurs with ZVS conditions, leading to low losses. This is important to enable operation of the IGBT at a high frequency of 100 kHz.

12.3.2 Plasma TV Sets

Plasma TV sets are an important segment of modern high-definition digital televisions that have displaced the previous vacuum tube models. Plasma displays derive their name because of utilizing millions of small cells containing electrically charged

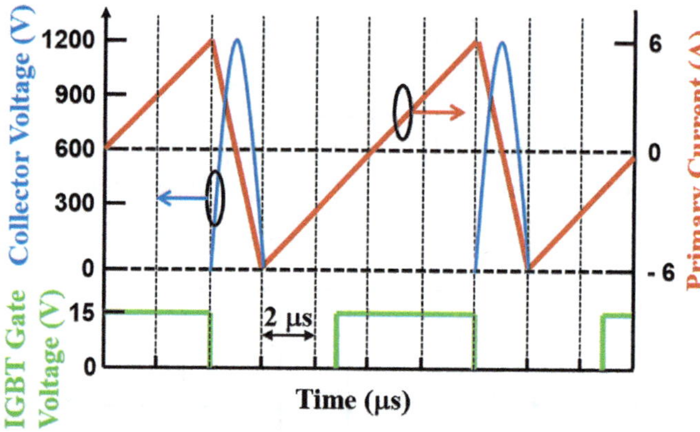

FIGURE 12.46

Cathode ray tube deflection circuit waveforms for the insulated gate bipolar transistor.

ionized gases [61]. Each of the cells operates in a manner similar to fluorescent lamps. The cell contains a gas whose temperature reaches at least 1200°C during gas discharge. The gas is composed of a noble gas (rarefied neon or xenon) plus a small amount of mercury. The gas discharge is produced by application of a high voltage at each pixel. The discharge results in the flow of energetic electrons that collide with mercury atoms, imparting energy to them. Ultraviolet radiation is produced when the mercury atom returns to its equilibrium state. This radiation is absorbed by the phosphor coating in each of the cells. The coating is designed to emit red, green, or blue light when excited by the ultraviolet radiation.

The arrangement of the pixels inside the plasma TV sets is illustrated in Fig. 12.47. Millions of pixels with red, green, and blue phosphor coatings are distributed throughout the large TV panel. The high voltage required to produce the discharge of the gas and produce the plasma is delivered to a specific pixel by using X/Y sustain electrodes and the address electrodes oriented orthogonal to each other on opposite sides of the cell. The X and Y electrodes are covered with dielectric and magnesium oxide layers. This approach allows addressing 0.25 million cells using a 400 by 640 matrix of wires when compared with 256,000 wires required for direct addressing of each cell. In spite of this simplification, the cost, reliability, and performance of the plasma TV set is governed by the drive electronics [54].

Some of the attractive features of plasma TVs are high brightness, wide color range, and excellent viewing angle. They can be produced with very large sizes (up to 150 in diagonally). The "dark-room" black level for plasma TVs is superior to that for LCDs. However, the typical power consumption for a plasma TV is 400 W for a 50-in screen due to the need to produce the gas discharge using high voltages. To reduce power consumption, it is necessary to utilize energy recovery circuits. Panasonic states that plasma display panels (PDPs) will consume only half the power of their previous series of plasma sets to achieve the same overall brightness for a given display size.

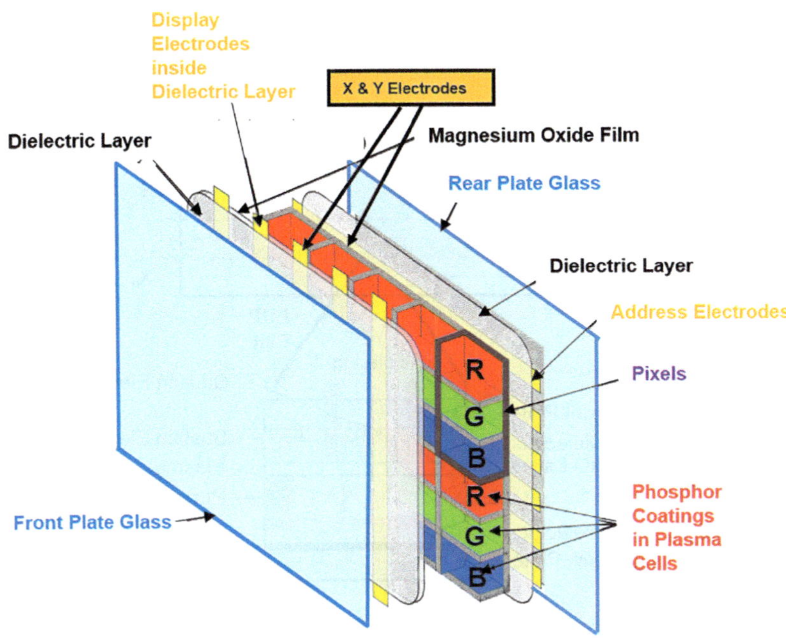

FIGURE 12.47

Plasma television panel pixel configuration.

The architecture of the PDP is illustrated in Fig. 12.48. It consists of an array of X and Y sustain electrodes across the surface of the panel. The electrodes are made of transparent conductive electrodes. An address electrode is placed on the opposite side of the cells. The gas is contained within the cells with a phosphor coating. When a high voltage is applied between the X and Y sustain electrodes, a high electric field is produced in the gas, resulting in a discharge that stimulates light emission from the phosphor [62].

A full-bridge IGBT circuit configuration shown in Fig. 12.49 is usually used to convert the DC supply voltage to high-voltage and high-frequency square wave pulses by the sustaining circuit. The electrodes with the intervening dielectric and MgO layers form a capacitor that must be charged and discharged rapidly during each cycle. For an 8-in diagonal PDP, the capacitance is about 1 nF and the current flow through the IGBTs 1–4 in the H-bridge is 2 A [62]. With increasing panel size, this current can exceed 100 A [63]. The voltage between the X and Y sustain electrodes is 400 V, requiring high blocking voltage ratings for the power devices.

The energy used for charging and discharging the capacitance ($2 \times C_P \times V_s^2$) was dissipated in the parasitic resistance of circuits in older-generation plasma panels [64]. This energy is recovered using the energy recovery circuit, shown in Fig. 12.49, in modern plasma panels to reduce the energy consumption. The energy recovery circuits utilize four additional IGBTs labeled IGBT 5–8. Two energy recovery capacitors (C_1 and C_2) and two energy recovery inductors (L_1 and L_2) are utilized

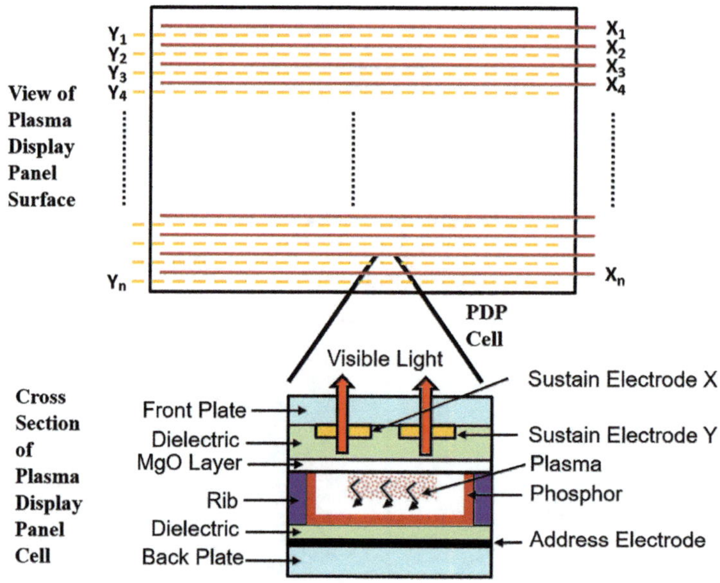

FIGURE 12.48

Plasma display panel architecture with cell structure.

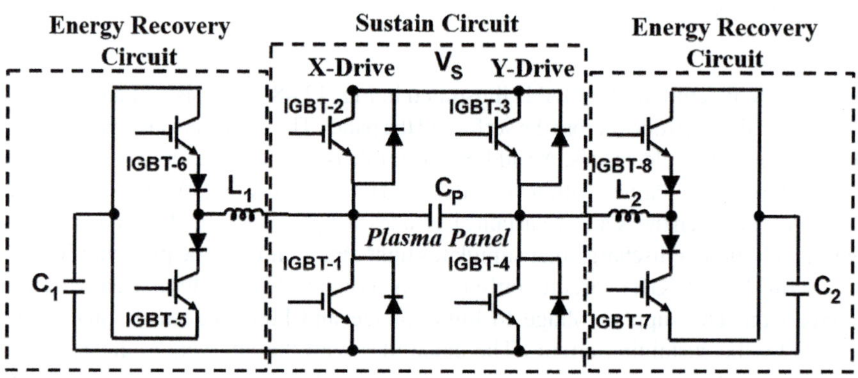

FIGURE 12.49

Plasma TV drive circuit with energy recovery. *IGBT*, insulated gate bipolar transistor.

here. The authors state: "*(Because) all power switches can be turned-off under zero current switching operation and have no current tailing problem, IGBTs can be used in the proposed ERC successfully*." A 42-in PDP with a C_P of 80 nF was controlled with a supply voltage of 165 V at a sustain electrode switching frequency of 200 kHz. The current flow through the IGBTs in the energy recovery circuit has a peak value of 40 A. The replacement of power MOSFETs with IGBTs resulted in reducing power loss from 74 to 48 W, an improvement of 50%.

For large-screen plasma TV sets, it is preferable to integrate multiple output stages with high voltage blocking capability for the scan driver and the address driver functions. This was first performed using lateral power MOSFETs in a junction isolation technology. However, it has been found that using dielectric isolation technology is more cost effective because it allows the use of lateral IGBTs that have improved current-handling capability [65]. The power devices had a blocking voltage of 200 V and a current handling capability of 0.4 A. It was found that the chip size could be reduced by 30% by using IGBTs in a dielectrically isolated substrate.

The demand for higher voltage and current-handling capability for the integrated power devices for high-definition TV (HDTVs) sets has been met by further innovations in the lateral IGBT design for the dielectrically isolated technology [66]. A shorted channel length was achieved to improve the saturation current while making the short circuit safe operating area longer by adding a hole bypass layer.

A multiemitter lateral IGBT structure has been demonstrated in dielectrically isolated technology for plasma panel drives [67]. These devices exhibit a high saturation current for driving the capacitive load in the scan mode and operate at high current density in the light emission discharge mode with a low on-state voltage drop.

12.3.3 Preregulator Circuit

Using a preregulation circuit in place of a switch mode power supply reduces cost by the elimination of inductances and the use of a smaller number of power devices [68]. This approach is suitable for creating the DC power supply bus for TV sets from the available AC power source. A single chip preregulator has been achieved by using junction-isolated lateral IGBTs as power devices to deliver a load current of up to 1 A from an AC source voltage of up to 260 V.

A block diagram of the preregulator circuit is shown in Fig. 12.50. A diode bridge rectifier is used at the front end with the DC output voltage taken across the filter capacitor C_F. The IGBT in the power IC is connected in series with the filter capacitor. The DC output voltage is continually monitored using the "differential voltage

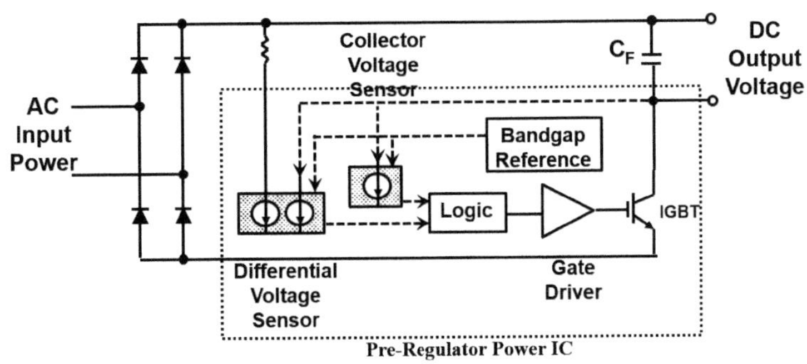

FIGURE 12.50

Preregulator circuit for television sets. *IGBT*, insulated gate bipolar transistor.

sensor." The IGBT is switched off whenever the output voltage exceeds the desired preset value. The IGBT is turned back on at the next input cycle to recharge the filter capacitor. The ripple current is determined by the size of the filter capacitor and the load current [69]. The "collector voltage sensor" monitors the IGBT collector voltage to determine when to safely turn it on.

Lateral IGBTs in junction isolation technology are suitable for this application for many reasons. The switching frequency is very low (input AC power line frequency), the power device current is very high (up to 10 A), and the blocking voltage of the power device must be high (more than 500 V). Under these operating conditions, the lateral IGBT has a much smaller active area than a lateral power MOSFET. The slow switching of the IGBT reduces ringing, allowing meeting Federal Communications Commission (FCC) requirements for EMI. The power dissipation in the IGBT is less than 3 W at an output DC power of 150 W, leading to a high efficiency of 98% for the preregulator circuit.

12.4 IGBT Design Optimization for Consumer Applications

The consumer sector offers many high-volume market opportunities for manufacturers of the IGBT. They have been developing products optimized for each of the applications discussed in the previous sections to remain competitive. This has helped in reducing power losses in the devices and bringing down the cost of the products. The IGBTs optimized for consumer applications operate at much larger collector current density than those used in the industrial sector [63], as shown in Fig. 12.51. The collector current density for the strobe applications discussed in

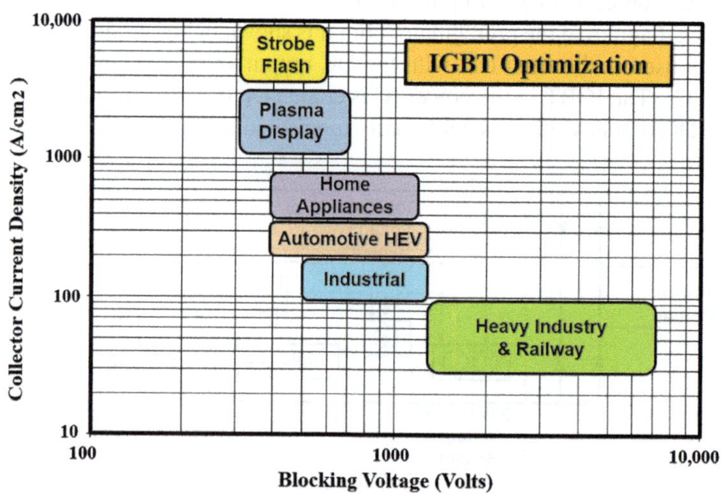

FIGURE 12.51

Insulated gate bipolar transistor optimization for consumer applications.

the chapter on the lighting sector is the largest (4000–9000 A/cm^2) due to the short duty cycle and low operating frequency, and the requirement to minimize the chip size to fit the small space in digital still cameras. The collector current density for IGBTs in the consumer sector is in the range of 1000–3000 A/cm^2 for PDPs and 400–600 A/cm^2 for home appliances such as induction cooktops and microwave ovens. The collector current density for IGBTs designed for industrial applications is lower because of their greater blocking voltage capability, which results a larger on-state voltage drop. This holds true for IGBTs used in heavy industry and traction (railway) applications due to their very high voltage operating levels.

12.4.1 IGBT Optimization for Motor Drives

Optimized IGBT structures are required for the control of motors in a variety of consumer appliances discussed in this chapter. This includes air conditioners, refrigerators, and washing machines. The main goal of the optimization is to improve the motor drive efficiency by reducing the on-state voltage drop and switching losses. This has to be done by performing a trade-off between these power loss mechanisms, as discussed in Chapter 3. The reverse recovery behavior of the fly-back diode in these circuits is very important because the reverse recovery current of P-i-N rectifiers contributes to significant losses in the diode and the IGBT [70].

The reverse conduction IGBT structure shown in Fig. 12.52 has been developed by Infineon for use in motor drives for refrigerators, washing machines, and air conditioners [71]. These devices are made with the asymmetric IGBT structure containing a buffer or field stop layer. The area occupied by the N$^+$ region on the collector side can be adjusted with the proper choice of lithographic dimensions when doing

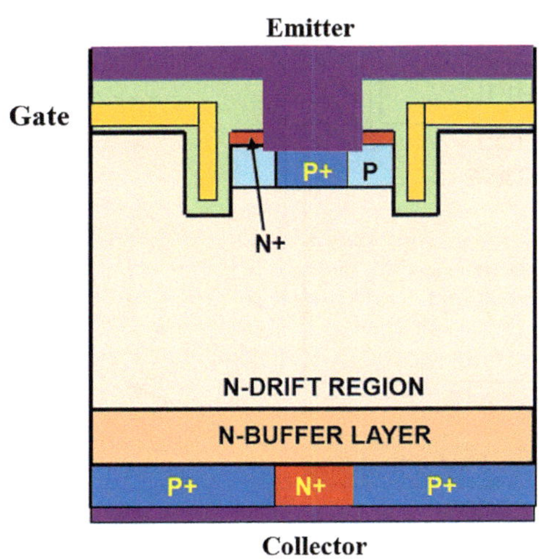

FIGURE 12.52

Reverse conduction insulated gate bipolar transistor structure with integrated fly-back diode.

N^+ and P^+ diffusions. The on-state voltage drop for the integrated diode in the reverse conduction mode is very close to that for the IGBT in the forward conduction mode. The reverse recovery behavior of the integrated diode matches that for previous external fly-back diodes. The reverse conducting IGBT eliminates the fly-back diode, reducing the space occupied within the package. Six reverse conducting IGBTs can be copackaged with the drive, protection, and control circuits within a single CIPOS module with 600 V, 15 A capability.

The development of IPMs containing multiple IGBTs, fly-back diodes, and the drive plus control circuits has been of interest for a long time due to the advantages of a compact footprint [72]. Progress in the optimization of IGBTs for home appliance motor drives has been recently reviewed by Mitsubishi Electric [73] for a period of six generations. The sixth-generation IGBT structure in 2010 has a smaller cell pitch achieved by reducing the dimensions of the trench and mesa widths by using VLSI process tools, as shown in Fig. 12.53. Using the optimized IGBT cell design, the on-state power loss was reduced by a factor of three while also reducing the switching loss by a similar amount.

Fairchild Semiconductor makes use of a non-punch-through or symmetric IGBT structure in their technology for home appliances [74]. Two optimized IGBT structures are used for home appliances—a design for low-frequency operation in air conditioners and refrigerators and a design for high-frequency operation in washing machines. The design for low-frequency operation has a reduced on-state voltage

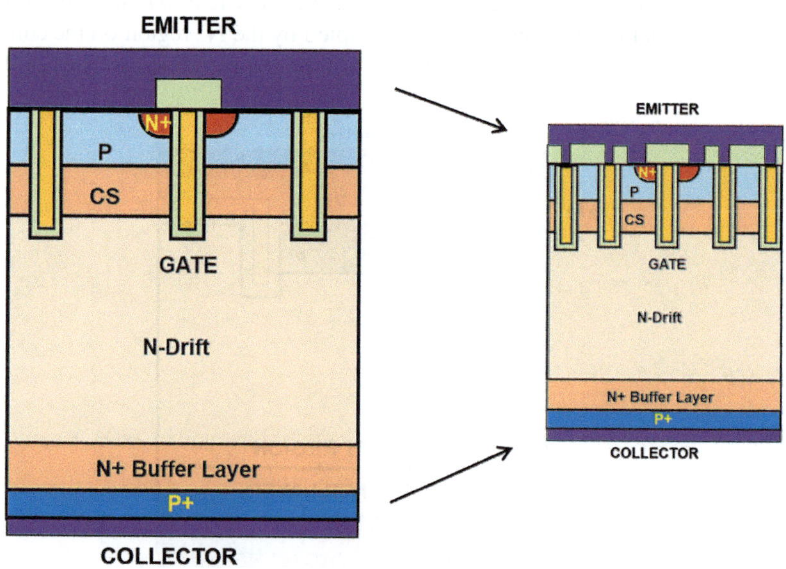

FIGURE 12.53

Scaling of insulated gate bipolar transistor structures using very large scale integrated (VLSI) technology.

drop of 1.7 V while that for high-frequency operation has an on-state voltage drop of 2.1 V. These devices are copackaged with fast recovery fly-back diodes and control circuits in a micro-mini DIP SPM package.

Toshiba offers a variety of 600 and 1200 V IGBTs optimized for low- and high-frequency home appliance use [75]. The IGBTs are also available with copackaged fly-back diodes.

12.4.2 IGBT Optimization for Induction Cooking

As discussed in Sections 12.1.5 and 12.2.1, induction cooking requires power device operation at relatively high frequencies (20 kHz). The switching loss for the IGBT can be reduced in this application by using soft switching with zero-voltage turnoff. The H-bridge topology used for induction cooktops requires IGBTs with a blocking voltage of 600 V while the single-ended topology used for portable induction plates and rice cookers requires devices with a blocking voltage capability of 1200 V.

The optimization of the IGBT structures made by Toshiba Semiconductors for induction heating home appliances has been charted by Matsushita Electric [31]. The evolution in IGBT performance over five generations (until 2007) is shown in Fig. 12.54. A significant improvement in the trade-off curve between on-state voltage drop and turn-off time for the IGBT is evident. Toshiba offers a lineup of IGBT products [75] for the AC 100 V and AC 200 V rice cookers and cooktops on their

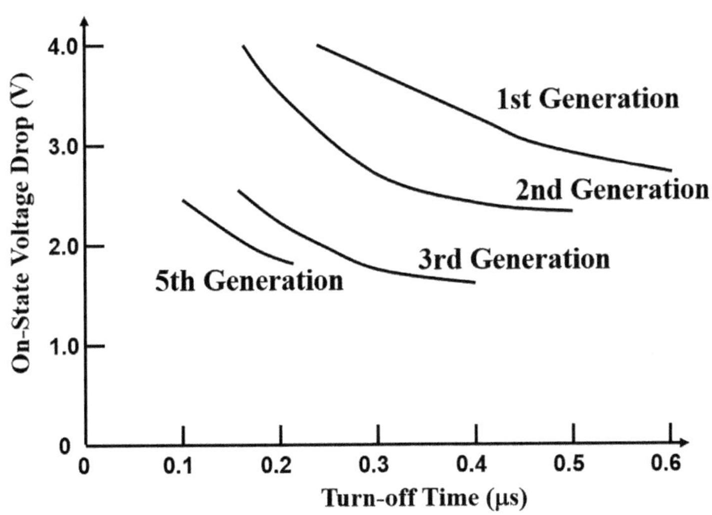

FIGURE 12.54

Toshiba semiconductor insulated gate bipolar transistor optimization for H-bridge induction heating applications.

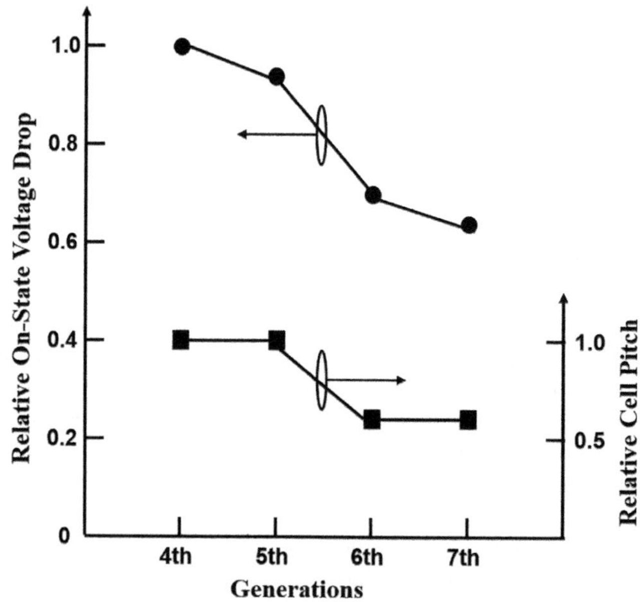

FIGURE 12.55

Optimizing insulated gate bipolar transistor cell design for H-bridge induction heating applications.

website. Some of the products include an integrated fly-back diode, as shown in Fig. 12.52. The improved performance is derived by improving the carrier distribution near the emitter, by reducing the injection efficiency of the collector P^+ region, and by reducing the cell pitch [63], as shown in Fig. 12.55. The power loss in the IGBT was reduced by 15% with these improvements in design.

Matsushita has developed IGBTs suitable for the single-ended induction cooking topology used in portable induction plates and rice cookers [76]. These devices have a larger blocking voltage than required for the H-bridge induction heating topology. The improvement in the trade-off curves from generation 1–3 shown in Fig. 12.56 was achieved by reducing the cell pitch for the planar gate structure. The improved trade-off curve for fourth-generation IGBTs was achieved by using the trench gate structure.

The reverse conducting IGBT structure shown in Fig. 12.52 for motor control application has also been refined by Infineon for use in half-bridge and single-ended induction heating applications [77]. For the H-bridge topology, the devices have a blocking voltage of 600 V while 1200 V devices are needed for the single-ended typology. The integration of the reverse conducting diode into the IGBT structure reduces component and package costs. An IGBT with a 1200 V blocking voltage capability requires a wafer thickness of 120 μm with the trench stop (or asymmetric) structure.

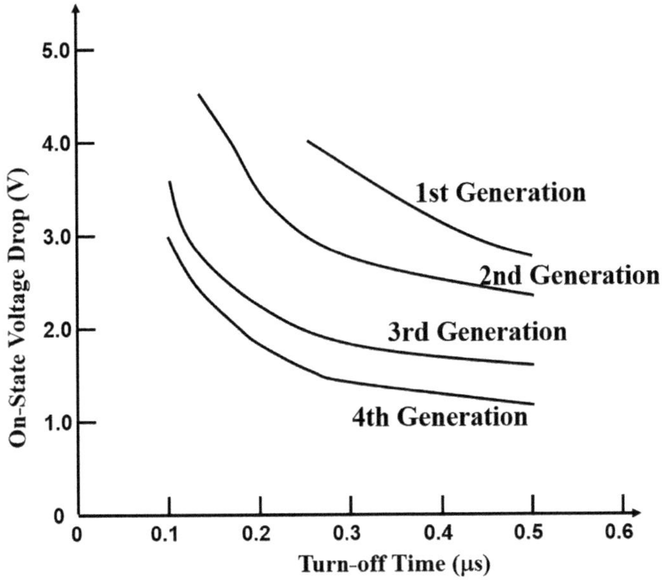

FIGURE 12.56

Matsushita electric insulated gate bipolar transistor optimization for single-ended induction heating applications.

12.4.3 IGBT Optimization for TV Sets

The CRT for TV sets requires IGBT structures with a blocking voltage capability of 1500 V and the ability to operate at 100 kHz. The planar IGBT structure has been optimized to deliver this performance [59]. A self-aligned process was created to reduce the cell size from 20 μm to only 5 μm. The trade-off curve between on-state voltage drop and turn-off switching loss is improved when compared to the conventional structure, as shown in Fig. 12.57. The self-aligned DMOS (SA-DMOS) IGBT has a power loss of one-third that of a power MOSFET over the scan frequency range of 30–100 kHz.

The IGBTs used in the sustain circuits for the PDP drives must handle very high currents, as shown in Fig. 12.58. As shown previously in Fig. 12.51, these IGBTs are designed to work at a current density of more than 1000 A/cm [2] at the peak collector current. The power loss during turn-on of the IGBT must be reduced to achieve greater efficiency for the plasma panel drive. The turn-on energy loss in the IGBT has been reduced by reducing the on-state voltage drop through optimization of the collector diffusion profiles [63]. The turn-on energy loss with very high peak current density was reduced by 20%, as shown in Fig. 12.59, by shrinking the cell size by 40% to increase the channel density.

Due to the large market size for plasma display TV sets, many companies have IGBT products that are specifically engineered for this application with high peak

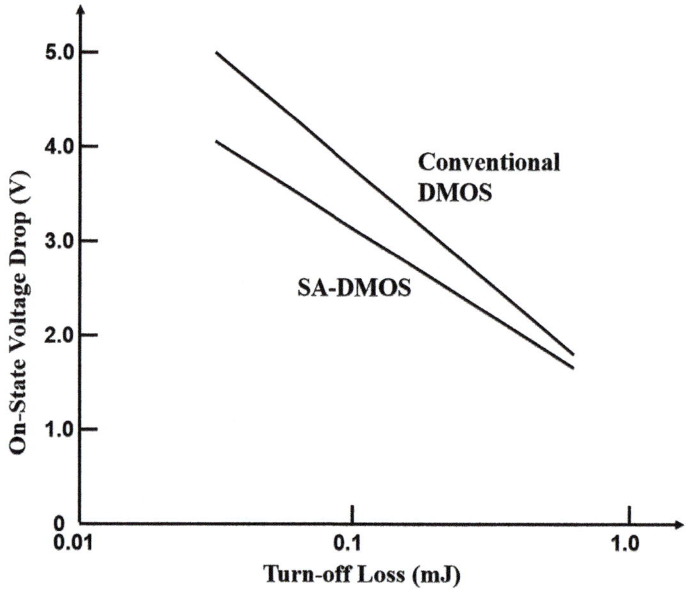

FIGURE 12.57

Hitachi SA-DMOS insulated gate bipolar transistor structure for cathode ray tube deflection circuit.

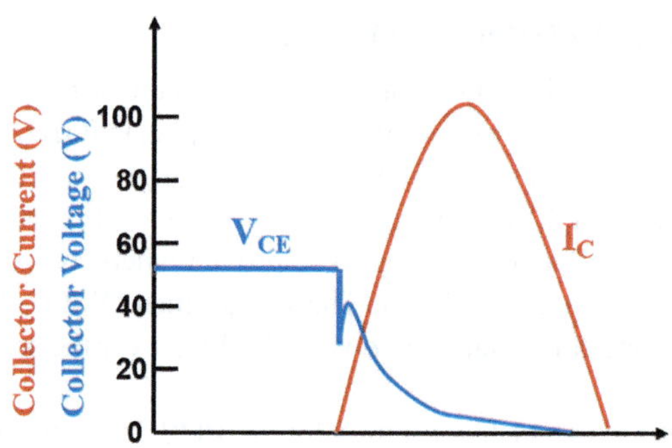

FIGURE 12.58

Current and voltage waveforms for insulated gate bipolar transistor in the plasma display panel.

collector current-handling capability. Two examples are the IGBT from International Rectifier (IRGP4050) with a blocking voltage of 250 V and a continuous current of 104 A. The device can deliver more than 200 A at a gate bias of 15 V without current saturation or latch-up. Similarly, IXYS has a plasma display power IGBT product (IXGQ90N33TCD4) with a blocking voltage of 330 V and a peak current

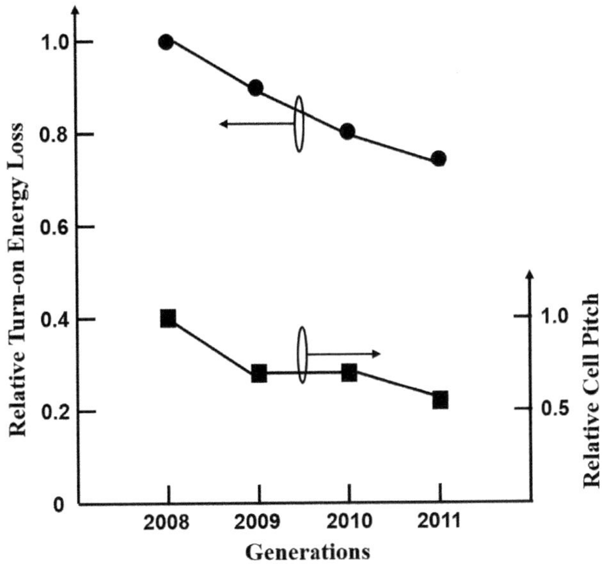

FIGURE 12.59

Toshiba plasma display drive insulated gate bipolar transistor design evolution.

rating of 360 A; and Toshiba has 300, 400, and 600 V IGBTs with peak currents of 120 A for PDP applications [74].

12.4.4 IGBT Optimization for Power Factor Correction

PFC is desirable for home air conditioners to reduce reactive power and EMI noise. Operation at up to 20 kHz leads to high dV/dt noise to the ground wire as well as conducted and radiated EMI [78]. The authors demonstrate that optimization of the blocking voltage from 710 V down to 650 V by reducing the thickness of the N-drift region results in reducing the energy loss during turn-off to one-third.

12.5 Summary

It is demonstrated with the examples shown in this chapter that the IGBT is extensively used for the control of power in home appliances used by consumers. It has become an essential component that the enables efficient and cost-effective operation of major appliances such as air conditioners, refrigerators, microwave ovens, induction cooktops, dishwashers, and washing machines. The IGBT is also required to obtain versatile performance from small appliances on kitchen countertops such as food processors, mixers, blenders, and induction rice cookers. It has enabled the development of cyclone model vacuum cleaners with very high motor speed to improve the removal of dirt from floors and carpets in homes. Plasma TV sets derive their high display quality by using IGBT-based sustain drive circuits and improved efficiency by using IGBT-based energy recovery circuits. Before this technology was available, the IGBT-based deflection circuit drives improved the performance of CRT-based TV sets. It can be confidently stated that the quality of life for consumers has been profoundly impacted by the development of the IGBT and its application to household appliances.

The evolution of the home appliance industry has been reviewed over the years. A review [79] published in 2003 stated that the US home appliance industry produced 325 million products in 2002 comprising 7.7 million clothes washers, 11.2 million refrigerators, 6.1 million dishwashers, 10.7 million microwave ovens, and 15 million ranges. The major appliance industry produced 80 million units in Europe, 65 million units in Japan, and 49 million units in Latin America in 2002. Nearly a billion fractional hp. motors were utilized for these appliances, requiring power electronic controls. In the home, 61% of the energy was used for comfort conditioning, 25% for refrigerators and freezers, and 5% for dishwashers, dryers, and clothes washers. Operating appliances with improved efficiency is an enormous opportunity to reduce the consumption of fossil fuels for the generation of electricity and consequently carbon dioxide emissions.

The energy utilization profile for a variety of appliances has been recently studied [80]. The authors report a growth in worldwide sales volume of microwave ovens from 40 million in 2006 to 70 million units in 2011. A total of 73 million refrigerators were sold in 2011. When compared with numbers reported in 2002, a growth rate of more than 15% is observed with no end in sight due to expanded worldwide prosperity. The impact of adding many new appliances in the kitchen can severely load the power grid. The electric stove consumes 1000 W of power and is used for about 30 min each day at around 5–6 pm. The microwave oven uses 800 W of power for about 6 min each day. The refrigerator consumes 100 W of electricity for 12 min and then stops for 24 min, repeating this cycle ~40 times per day. The freezer has a similar energy use profile. The heaviest power load occurs at 5–6 pm during the day. The authors predicted that the power grid in California will be overtaken by the demand from these appliances by 2018. Improving the efficiency of home appliances was paramount to avoid and re-extensive utilization of IGBTs in power electronics.

Another trend that will impact the consumer is the creation of a home network system. Various methodologies for implementing such a smart home environment have been proposed over the last decade such as using the UPnP middleware [81]. A home network system enables the automatic configuration of appliances that join or leave the network. The appliances communicate with each other and a central home server that monitors the entire system to optimize energy use. The home network can interact with the utility to reduce energy consumption during peak demand periods to smooth the loading and reduce bills for the consumer and investments in energy generation capacity by the utility [82]. Smart washing machines that can sense the load and optimize energy and water use will provide the greatest benefit. Smart refrigerator sales exceeded $2.5 billion in 2018 and are projected to reach $7.5

billion in 2026 [83]. These trends cannot be realized without advancements in power electronics and the IGBT is poised to continue to play a major role in enabling this progress into the foreseeable future.

References

[1] A.D. Little, Opportunities for Energy Savings in Residential and Commercial Sectors with High Efficiency Electric Motors, U.S. Department of Energy, 1999. Contract No. DE-AC01-90CE23821. December 1.

[2] US Appliance Industry Factory Unit Shipment Statistics. www.ApplianceMagazine.com.

[3] US Department of Energy, Annual Energy Review 2008, Energy Information Administration, 2009.

[4] A. Kanouda, et al., Expectations of next-generation power devices for home and consumer appliances, in: IEEE Int. Power Electronics Conf, 2014, pp. 2058–2063.

[5] Home Appliance. en.wikipedia.org/wiki/Home_appliance.

[6] Air Conditioners. www.semicon.toshiba.co.jp/eng/application/homeappliance/index.html.

[7] M.B.R. Correa, et al., Single-phase induction motor drives systems, in: IEEE Appl. Power Electron. Conf, vol. 1, 1999, pp. 403–409.

[8] M. Chomat, T.A. Lipo, Adjustable speed drive with single-phase induction machine for HVAC applications, in: IEEE Power Electron. Spec. Conf, vol. 3, 2001, pp. 1446–1451.

[9] D.M. Ionel, High efficiency variable speed electric motor drive technologies for energy savings in the US residential sector, in: IEEE International Conference on Optimization of Electrical and Electronic Equipment, 2010, pp. 1403–1414.

[10] Refrigeration. en.wikipedia.org/wiki/Refrigeration.

[11] L.C. Packer, Application of motors to household refrigeration compressors, Trans. Am. Inst. Electr. Eng. 71 (1952) 70–77.

[12] S.A. Tassou, T.Q. Qureshi, Performance of a variable speed inverter/motor drive for refrigeration applications, IEEE Comp. Control Eng. J. 5 (1994) 193–199.

[13] C.B. Rasmussen, E. Ritchie, Variable speed brushless DC motor drive for household refrigerator compressor, in: IEEE International Conference on Electrical Machines and Drives, 1997, pp. 128–132.

[14] C.A. Canesin, et al., Variable speed refrigeration system with HPF input rectifier stage, in: IEEE International Symposium on Power Electronics, Electrical Drives, Automation, and Motion, 2008, pp. 848–853.

[15] M. Yabe, K. Sakanobe, M. Kawakubo, High efficient motor drive technology for refrigerator, in: IEEE Conference on Environmentally Conscious Design and Inverse Manufacturing, Paper 3C-1-3S, 2005, pp. 708–709.

[16] K.-W. Lee, S. Park, S. Jeong, A seamless transition control of sensorless PMSM compressor drives for improving efficiency based on a dual-mode operation, IEEE Trans. Power Electron. 30 (2015) 1446–1456.

[17] Washing Machine. en.wikipedia.org/wiki/Washing_machine.

[18] K. Harmer, P.H. Mellor, D. Howe, An energy efficient brushless drive system for domestic washing machine, in: IEEE International Conference on Power Electronics and Variable Speed Drives, 1994, pp. 514–519.

[19] H.S. Rajamani, R.A. Mc Mahon, Induction motor drives for domestic appliances, IEEE Ind. Appl. Mag. 3 (3) (1997) 21–26.

[20] T. Tanaka, Environment friendly revolution in home appliances, in: IEEE International Symposium on Power Semiconductor Devices and ICs, Paper P.6, 2001, pp. 91–95.

[21] J.S. Moghani, M. Heidari, High efficient low cost induction motor drive for residential applications, in: IEEE International Symposium on Power Electronics, Electrical Drives, Automation, and Motion, 2006, pp. 1399–1402.

[22] D.G. Kokalj, Variable frequency drives for commercial laundry machines, IEEE Ind. Appl. Mag. 3 (3) (1997) 27–30.

[23] Microwave Oven. en.wikipedia.org/wiki/Microwave_oven.

[24] M. Ishitobi, et al., Pulse width and pulse frequency modulation pattern controlled ZVS inverter type AC-DC power converter with lowered utility AC grid side harmonic current components for magnetron drive, in: IEEE Power Electron. Spec. Conf, 2002, pp. 2062–2067.

[25] H. Kako, T. Nakagawa, R. Narita, Development of compact inverter power supply for microwave oven, IEEE Trans. Consum. Electron. 37 (1991) 611–616.

[26] T. Matsushige, et al., Voltage clamped soft switching PWM inverter-type DC-DC converter for microwave oven and its utility AC side harmonics evaluations, in: IEEE Power Electron. Motion Control Conf, vol. 1, 2000, pp. 147–152.

[27] Y.-J. Woo, et al., One-chip class-E inverter controller for driving a magnetron, IEEE Trans. Ind. Electron. 56 (2009) 400–407.

[28] I. Cooking. en.wikipedia.org/wiki/Induction_cooking.

[29] W.C. Moreland, The induction range: its performance and its development problems, IEEE Trans. Ind. Appl. IA-9 (1973) 81–85.

[30] P.H. Peters, A portable cool-surface induction cooking appliance, IEEE Trans. Ind. Appl. IA-10 (1974) 814–822.

[31] K. Yasui, et al., Latest developments of soft-switching pulse modulated high frequency conversion systems for consumer induction heating power appliances, in: IEEE Power Conv. Conf, 2007, pp. 1139–1146.

[32] H.W. Koertzen, J.S. van Wyk, J.A. Ferreira, Design of the half-bridge series resonant converter for induction cooking, in: IEEE Power Electron. Spec. Conf, vol. 2, 1995, pp. 729–735.

[33] S.P. Wang, et al., A constant frequency variable power regulated ZVS–PWM load resonant inverter for induction heating appliance, in: IEEE Int. Symp. Ind. Electron, vol. 2, 1997, p. 5670571.

[34] B.J. Baliga, et al., The insulated gate rectifier: a new power switching devices, in: Abstract 10.6 IEEE Int. Electron Devices Meeting, 1982, pp. 264–267.

[35] H. Sugimura, et al., Direct AC-AC resonant converter using one chip reverse blocking IGBT-based bidirectional switches for HF induction heaters, in: IEEE Int. Symp. Ind. Electron, 2008, pp. 406–412.

[36] Dishwasher. en.wikipedia.org/wiki/Dishwasher.

[37] F. Rosa, et al., Dishwasher history and its role in modern design, in: IEEE Hist. Electro-Technol. Conf, 2012, pp. 1–6.

[38] B. Rubino, C. Parisi, S. Buonomo, Potential of new SLLIMM-nano intelligent molded module for low power home appliance motor drives, in: IEEE Int. Power Electron. Motion Control Conf, 2012. pp. LS6d.1-1–LS6d.1-7.

[39] B. Jin, et al., A new generation of 600V intelligent power module for consumer and industrial motor drive applications, in: PCIM Europe Conf, 2017, pp. 930–934.
[40] Small Appliance. en.wikipedia.org/wiki/Small_appliance.
[41] V. Smil, Feeding the World: How Much More Rice Do We Need? www.vaclavsmil.com.
[42] Let's Eat!. www.asiarice.org.
[43] I. Hirota, et al., Performance evaluations of single ended quasi-load resonant inverter incorporating advanced 2nd generation IGBT for soft-switching, in: IEEE International Conference on Industrial Electronics, Control, Instrumentation, and Automation, vol. 1, 1992, pp. 223–228.
[44] IGBT Power Losses in Induction Heating Applications, 2011. On-Semiconductor Application Note AND9064/D. October http:/onsemi.com.
[45] Blender. en.wikipedia.org/wiki/Blender.
[46] Juicer. en.wikipedia.org/wiki/Juicer.
[47] Mixer (Cooking). en.wikipedia.org/wiki/Mixer_(cooking).
[48] Food Processor. en.wikipedia.org/wiki/Food_processor.
[49] T. Castagnet, J. Nicolai, Digital drive for home appliance DC motor, in: IEE Colloquium on Variable Speed Drives and Motion Control, 1992, pp. 6/1–6/4.
[50] Z.-S. Ho, et al., Implementation of food processor application using brushless DC motor control, in: IEEE Power Electronics and Drive Systems, 2011, pp. 272–277.
[51] Vacuum Cleaner. en.wikipedia.org/wiki/Vacuum_cleaner.
[52] C.J. Bateman, et al., Sensorless operation of an ultra-high-speed switched reluctance machine, IEEE Trans. Ind. Appl. 46 (2010) 2329–2337.
[53] J.-Y. Lim, et al., Single-phase switched reluctance motor for vacuum cleaner, in: IEEE Int. Symp. Ind. Electron, vol. 2, 2001, pp. 1393–1400.
[54] Television. en.wikipedia.org/wiki/Television.
[55] A.R. Kmetz, Flat-panel displays, IEEE Aerosp. Electron. Syst. Mag. 2 (8) (1987) 19–24.
[56] P. O'Donovan, Goodbye CRT, IEEE Spect. 43 (2006) 38–42.
[57] I. Oh, A new base driving technique of a high voltage BJT for the horizontal deflection output using a CRT, in: IEEE Asia-Pacific Conference on ASICs, 1999, pp. 67–74.
[58] P.S. Wilson, Power transistor developments for color CRT deflection, IEEE Trans. Consum. Electron. CE-32 (1986) 264–267.
[59] M. Mori, et al., High switching speed 1500 V IGBT for CRT deflection circuit, in: IEEE International Symposium on Power Semiconductor Devices and ICs, 1991, pp. 237–241.
[60] Y. Seki, A 1500 V IGBT with a self-aligned DMOS structure for high resolution CRT operated up to 100 kHz, in: IEEE International Symposium on Power Semiconductor Devices and ICs, 1991, pp. 215–219.
[61] Plasma Display. en.wikipedia.org/wiki/Plasma_display.
[62] H.-B. Hsu, et al., Regenerative power electronics driver for plasma display panel in sustain-mode operation, IEEE Trans. Ind. Electron. 47 (2000) 1118–1125.
[63] S. Umekawa, et al., New discrete IGBT development for consumer use, in: IEEE Int. Power Electron. Conf, 2010, pp. 790–795.
[64] S.-K. Han, et al., IGBT-based cost-effective energy-recovery circuit for plasma display panel, IEEE Trans. Ind. Electron. 53 (2006) 1546–1554.
[65] H. Sumida, et al., A high performance plasma display panel driver IC using SOI, in: IEEE International Symposium on Power Semiconductor Devices and ICs, Paper 7.1, 1998, pp. 137–140.

[66] H. Sumida, et al., 250-V class lateral SOI devices for driving HDTV PDPs, in: IEEE International Symposium on Power Semiconductor Devices and ICs, 2007, pp. 229–232.
[67] J. Sakano, et al., Large current capability 270 V lateral IGBT with multi-emitter, in: IEEE International Symposium on Power Semiconductor Devices and ICs, Paper 4–4, 2010, pp. 83–86.
[68] S.L. Wong, N. Majid, A single-chip pre-regulator circuit using LIGBT and current mode sensing, in: IEEE International Symposium on Power Semiconductor Devices and ICs, Paper 8.3, 1994, pp. 385–389.
[69] A.R. Hambley, Electrical Engineering, Chapter 10, fifth ed., Prentice Hall, 2011, pp. 492–494.
[70] B.J. Baliga, Power semiconductor devices for variable frequency drives, Proc. IEEE 82 (1994) 1112–1122.
[71] J. Song, et al., A new intelligent power module with reverse conducting IGBTs for up to 2.5 kW motor drives, in: IEEE Int. Power Electron. Conf, 2010, pp. 156–158.
[72] G. Mujumdar, et al., Novel intelligent power modules for low-power inverters, in: IEEE Power Electron. Spec. Conf, vol. 2, 1998, pp. 1173–1179.
[73] G. Majumdar, Power module technology for home power electronics, in: IEEE Int. Power Electron. Conf, 2010, pp. 773–777.
[74] T.-S. Kwon, et al., Development of new smart power module for home appliances motor drive applications, in: IEEE International Electric Machines and Drives Conference, 2011, pp. 95–100.
[75] Toshiba Product Guide: Discrete IGBTs. http://www.semicon.toshiba.co.jp.
[76] H. Terai, et al., Comparative performance evaluations of IGBTs and MCT in single-ended quasi-resonant zero voltage soft switching inverter, in: IEEE Power Electron. Spec. Conf, vol. 4, 2001, pp. 2178–2182.
[77] S. Voss, O. Hellmund, W. Frank, New IGBT concepts for consumer power applications, in: IEEE Industry Applications Conference, 2007, pp. 1038–1043.
[78] C.M. Yun, B.S. Suh, T.H. Kim, Optimization of IGBT in power factor correction module for home appliance, in: IEEE International Symposium on Power Semiconductor Devices and ICs, 2002, pp. 181–184.
[79] J.C. Moreira, Evolution and future of power electronic applications in home appliances, in: IEEE Int. Conf. Ind. Electron, vol. 3, 2003, pp. 3023–3024.
[80] Y.W. Leung, et al., The energy profile study of electric kitchen utensils for residential smart kitchen, in: IEEE International Conference on Advances in Power System Control, Operation, and Management, 2012, pp. 1–5.
[81] D.-S. Kim, J.-M. Lee, W.H. Kwon, Design and implementation of home network systems using UPnP middleware for networked appliances, IEEE Trans. Consum. Electron. 48 (2002) 963–972.
[82] A. Grogan, Smart appliances, Eng. Technol. Mag. 7 (6) (2012) 44–45.
[83] Smart Refrigerator Market Size. https://www.gminsights.com/industry-analysis/smart-refrigerator-market.

CHAPTER

IGBT Applications: Medical

13

Advances in medicine have been revolutionary over the last century, providing huge improvements in the quality of care for patients. The ability to observe trauma within the human body using noninvasive tools has given healthcare providers unprecedented assistance during diagnosis and treatment. These capabilities allow avoiding unnecessary surgical intervention and allow for more precise decision-making when performing required surgical operations.

The first noninvasive instrument for observing the interior of the human body was developed using x-rays discovered by Roentgen in 1895. An x-ray is electromagnetic radiation with energy between 100 eV and 100 keV corresponding to a wavelength of 0.1–100 Å [1]. X-rays with energies above 5–10 keV are referred to as "hard x-rays" while those with lower energies are referred to as "soft x-rays." Hard x-rays have the ability to penetrate soft tissue in the human body, making them suitable for medical imaging and for airport security scanners.

X-ray computed tomography (CT) is a more recent and sophisticated technology developed in the 1970s to provide two-dimensional sectional images of the human body with high resolution. In 2007 in the United States, 72 million CT scans were performed.

Magnetic resonance imaging (MRI) was developed in the 1970s based on radio frequency (RF) excitation of the protons in human tissue. A strong and stable magnetic field must be generated to obtain high-quality images.

Ultrasonography is an imaging approach that makes use of sound waves at above the human hearing range. The echo of the sound wave from organs within the human body can be used to create images. It is commonly used during pregnancies to observe the development of the fetus in the womb.

Implanted defibrillators have allowed stabilization of the heart's rhythm. In the 1980s, compact portable defibrillators with simple operating instructions were introduced. These defibrillators have been widely deployed in buildings, airplanes, and emergency vehicles to resuscitate victims of cardiac arrest. The American Medical Association estimates that more than 100,000 lives are saved in the United States every year due to this technology.

The worldwide market for the above medical equipment was about $300 billion in 2012, with the United States as the largest market share at 38% [2]. The worldwide market for medical equipment grew to $400 billion in 2017 and is expected to reach $500 billion in 2023. More than 6500 companies serve the medical market in the United States. The subsectors of the medical equipment market include:

The IGBT Device. https://doi.org/10.1016/B978-0-323-99912-0.00023-4

(1) electromedical equipment such as pacemakers, MRI scanners, and ultrasound machines; (2) irradiation equipment such as x-ray machines and CT scanners; and (3) dental equipment. The top medical equipment companies based on revenue are Medtronic, Johnson and Johnson, Siemens AG, General Electric Company, and Philips Electronics [3].

All these sophisticated medical imaging systems rely power supplies enabled by insulated gate bipolar transistors (IGBTs). The availability of the IGBT has been critical to making stable power sources that are essential to obtaining high-quality images. The compact size and smaller weight of IGBT-enabled power sources have important impacts on the design and functionality of medical equipment. This chapter discusses the principles behind the above types of medical equipment and the circuit topologies that are used for each of these machines.

13.1 X-Ray Machine

Some typical x-ray equipment widely used by the medical industry is shown in Fig. 13.1. Fig. 13.1(a) is an x-ray machine for imaging any part of the human body with the patient reclining on the bed. Fig. 13.1(b) shows a portable x-ray machine that can be moved to the patient's location for diagnostics in military, sports, or disaster response situations. Fig. 13.1(c) is an example of an x-ray unit used by dentists for prophylactic examination of teeth and gums to detect cavities and fractures.

The x-rays required in medical equipment are generated using the Coolidge tube developed by William Coolidge at the General Electric Company in 1913. This configuration is still used today. The basic principle of operation of the Coolidge x-ray tube is illustrated in Fig. 13.2. X-rays are generated by first producing electrons from a filament that serves as the cathode electrode. The electrons are then accelerated to a high energy by using a high-voltage power supply, as shown in Fig. 13.2. They then

(a) (b) (c)

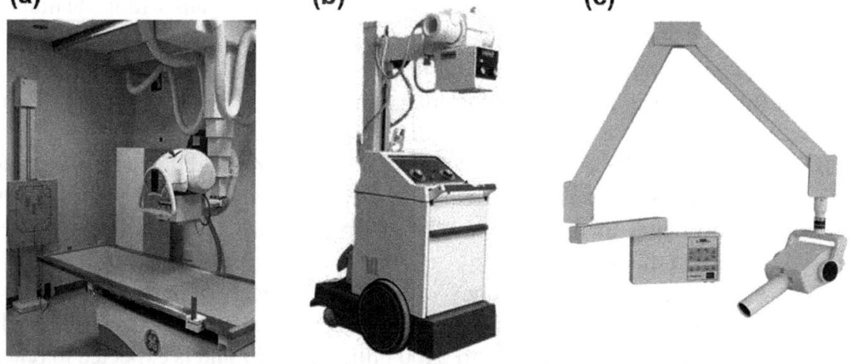

FIGURE 13.1

X-ray units: (a) full body; (b) portable; (c) dental.

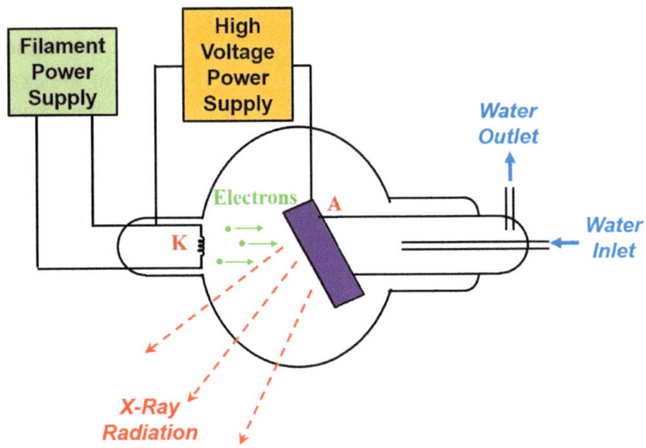

FIGURE 13.2

Principles of x-ray generation.

bombard the anode electrode, which is made using tungsten or tungsten with 5% rhenium to resist cracking [1]. The energetic electrons give up their energy to the target, producing the x-ray spectrum shown in Fig. 13.3, using two basic mechanisms: (1) x-ray fluorescence and (2) Bremsstrahlung. In x-ray fluorescence, the energetic electrons displace an orbital electron out of the inner shell of the target atom. X-ray radiation is produced when electrons from higher-energy orbitals fill the vacancies created in the inner orbitals. This x-ray emission has a specific wavelength corresponding to the energy difference between the orbitals of the target atoms. In the case of tungsten targets, the x-rays are emitted at energies of 59.3 and 67.2 keV as long as the incident electrons are accelerated using a voltage larger than 70 kV. This produces the hard x-rays that can be used for imaging. Bremsstrahlung is radiation

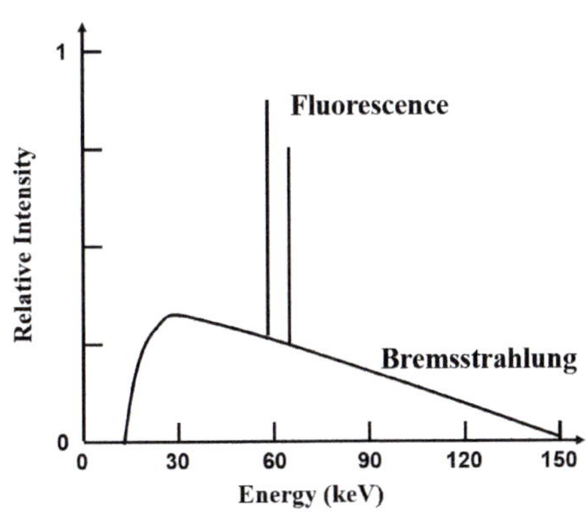

FIGURE 13.3

X-ray spectrum generated by the Coolidge tube using 150 kV source.

produced when the energetic electrons are scattered by the nucleus. It has a continuous spectrum, as shown in Fig. 13.3, with zero intensity at the highest available energy corresponding to the power supply voltage (150 keV in the example).

Unfortunately, only about 1% of the input energy generates hard x-rays for imaging purposes [4]. Most of the energy is converted into heat, resulting in a temperature of 2500°C at the tungsten electrode where the electron beam is incident. In the Coolidge tube, the heat is removed using water cooling, as shown in Fig. 13.2, to allow stable operation of the x-ray source. In addition, aluminum filters are used at the exit port for the x-rays to absorb the undesirable soft x-rays. This reduces the radiation exposure incurred by patients.

The first imaging of bone structure was reported by Roentgen soon after his discovery of x-rays. Since then, this tool has been widely used around the world with more than 5 billion images taken by 2010. Early images were prepared by using photographic plates, but modern machines create digital images using x-ray detectors. Radiographs can be used to detect the pathology of the skeletal system because of the high density of bones when compared with surrounding tissue. X-rays of the chest allow detection of lung diseases such as pneumonia and lung cancer. Abdominal x-rays allow imaging of intestinal obstruction and gall bladder/kidney stones. Dental x-rays are routinely performed for imaging of cavities in teeth. X-ray mammograms are used for the early detection of breast cancer. X-ray radiation at higher intensities is utilized for the management of cancer.

The quality of the x-ray images is impacted by fluctuations in the output of the x-ray tube, which can appear as noise or artifacts in the image. It is critical to precisely regulate the voltage supplied to the x-ray tube to avoid these problems and generate high-quality images. As the demands of the medical community increase for higher-resolution images for enhanced diagnosis and patient outcomes, it has become increasingly important to improve the precision and regulation of the x-ray tube voltage with faster response time and reduced ripple currents. At the same time, it is desirable to make the power supply more efficient as well as smaller in size and weight for managing the equipment. These features have been made possible by the availability of IGBT modules that can operate at high frequencies.

13.1.1 Series–Parallel Resonant Power Supply

The x-ray tube must be operated at a very high voltage (up to 150 kV) to generate the desired hard x-ray radiation. The x-ray power supply must deliver the high voltage for a very short duration of time. The output voltage for the power supply must reach its steady-state value within 10 ms to prevent noise and defects in the image. The power delivered by the x-ray supply ranges from 20 to 150 kW.

A series–parallel resonant converter has been designed to satisfy the power delivery requirements for the x-ray tube [5]. The output voltage of the power supply is controlled by altering the switching frequency and duty cycle. A circuit diagram for the power supply is shown in Fig. 13.4. The input AC power is rectified to create a DC bus voltage across capacitor C_B. The four IGBTs in the pulse-width modulated

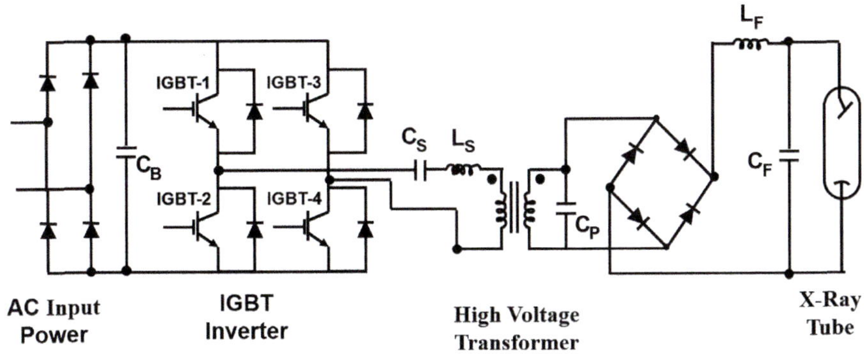

FIGURE 13.4

X-ray tube resonant converter power supply. *IGBT*, insulated gate bipolar transistor.

(PWM) inverter are operated under zero-voltage switching conditions to reduce switching losses. The resonant circuit is formed by C_S and L_S. The high-voltage transformer has a turns ratio of 400–1000 to deliver the high voltage required for the x-ray tube. The voltage across the x-ray tube can be controlled from 60 to 100 kV with a risetime of 1 ms. The inductor L_F and capacitor C_F provide filtration of ripple currents.

The IGBT structure must be optimized for x-ray power supply applications to reduce the power dissipation and improve the efficiency. The waveforms for the IGBT collector voltage and current in the series resonant circuit are illustrated in Fig. 13.5 [6]. The zero-voltage switching produces a high *dI/dt* for the IGBT during the turn-on period t_1. This results in a forward voltage overshoot that increases the power losses. An analytical model for the forward recovery in IGBTs has been

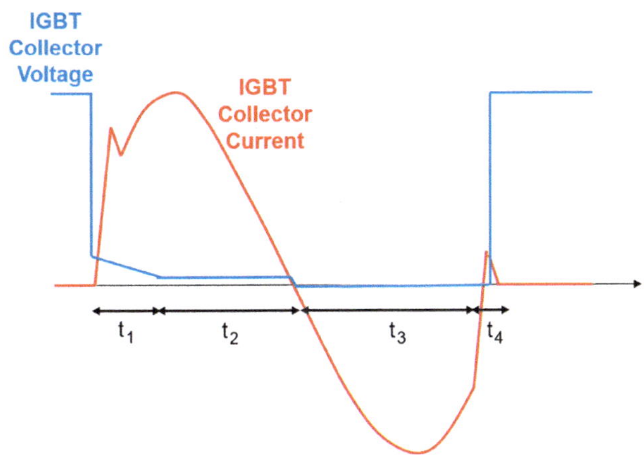

FIGURE 13.5

Insulated gate bipolar transistor waveforms in x-ray tube power supply.

provided in the textbook [7]. The forward voltage overshoot is due to the time taken for modulation of the conductivity of the high resistivity drift region in IGBT structures. It has been found that the forward recovery losses are a major contributor to the total power loss in the x-ray power supply circuits [6]. Mitsubishi has developed their sixth-generation IGBTs with trench gate and collector stop (also called asymmetric and field stop) structures to reduce power losses in high-frequency circuits for this application.

13.1.2 Dual-Mode Power Supply

As discussed in the previous section, the voltage across the x-ray tube must be increased at a rapid rate to the operating value when generating the images. This voltage change must be accomplished without a voltage overshoot or ringing. Several dual-mode power supply topologies have been proposed and demonstrated to achieve these characteristics [8,9].

One of the dual-mode power supply topologies is shown in Fig. 13.6. It consists of a phase-shifted PWM series resonant inverter for driving the input of the high-voltage, high-frequency transformer. The output of the transformer is rectified using a multistage voltage amplifier. The resonant frequency for the circuit is chosen using capacitor C_S together with the leakage inductance of the transformer L_S. The PWM switching frequency for the IGBTs is chosen to be larger than the resonant frequency. The voltage ripple and dynamic response of the circuit improve with increasing operating frequency. IGBT-1 and IGBT-2 are operated with a fixed 50% duty cycle (reduced slightly to prevent shoot-through currents). IGBT-3 and IGBT-4 are operated similarly but phase-shifted relative to IGBT-1 and IGBT-2. With a zero phase shift, the input voltage and the two output voltages of the transformer are at their maximum and equal to the DC bus voltage (V_{DC}). When the phase shift is increased to 180 degrees, the voltage applied to the primary side of the transformer is reduced to zero. Consequently, the voltage developed at the secondary side of the transformer

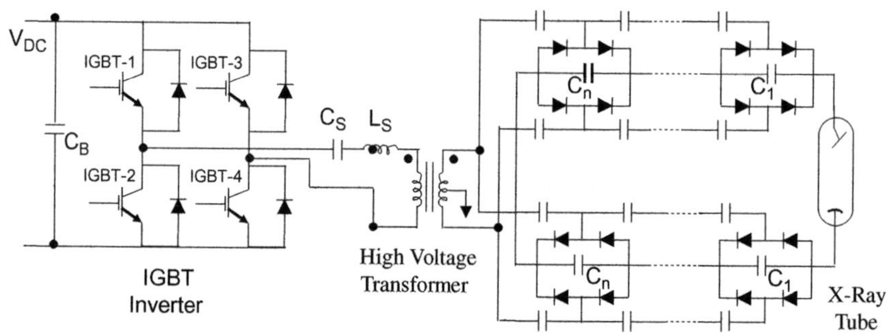

FIGURE 13.6

Dual-mode x-ray tube power supply. *IGBT*, insulated gate bipolar transistor.

can be adjusted from zero to V_{DC} based on controlling the phase shift for the gate drive signals to IGBT-3 and IGBT-4.

The rectified DC voltage produced by the voltage multiplier stages is given by:

$$V_o = 2nV_{in}(\max) - \frac{I_o}{f_s C}\left[\frac{n^3}{24} + \frac{n^2}{8} + \frac{n}{3}\right] \tag{13.1}$$

where V_{in} is the input voltage to the transformer ranging from zero to V_{DC}, I_o is the current delivered to the x-ray tube, f_s is the PWM switching frequency, C is the value of all the capacitors in the voltage multiplier circuits ($C_1, \cdots, C_n$), and n is the number of multiplier stages.

The dynamic response of the circuit is improved by maintaining a maximum voltage of V_{DC} at the primary side of the transformer at the beginning of the pulse transient. When the output voltage approaches 90% of the desired target voltage across the x-ray tube, the phase shift is increased to a value required to obtain the final target voltage. The time taken for the output voltage to reach its steady-state value has been found to be less than 1 ms using this approach with a switching frequency of 39 KHz.

13.2 Computed Tomography

A significant advancement in x-ray radiography was achieved in the 1970s with the development of x-ray computed tomography, commonly referred to as a CT scanner. The word "tomography" is derived from the Greek word *tomos* for slice and *graphein* for writing. Tomographic images are two-dimensional virtual slices through the human body that can be combined to generate a three-dimensional image. A typical scanner, the GE Prospeed model, is shown in Fig. 13.7(a). The patient lies on the table in a horizontal position. The x-ray tube and detectors are located in the gantry with the circular cavity for insertion of the body during the scanning, as illustrated in Fig. 13.7(b). The gantry is rotated around the patient to obtain information along

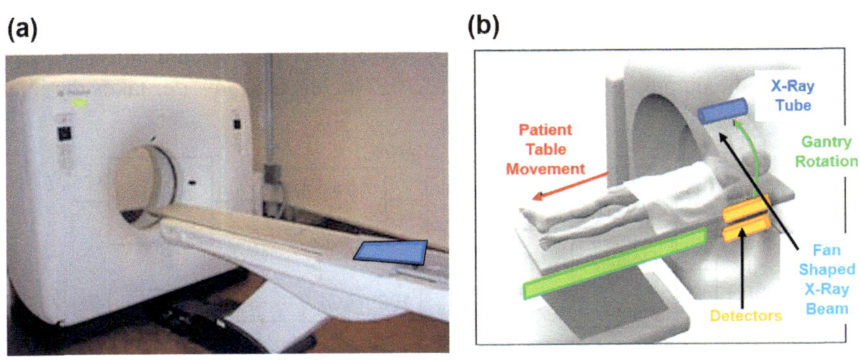

FIGURE 13.7

(a) CT scanner, (b) CT operation.

various radii to generate the image for a single slice orthogonal to the body. The patient table is then repositioned as indicated in the figure to generate a slice along the length of the body.

The use of CT scans by the medical community has gone up drastically during the last two decades, with 72 million scans performed in the United States in 2007 [10]. CT scans of the cranium are used to detect tumors, hemorrhages, and bone trauma. CT scans of the lungs allow the detection of emphysema and fibrosis. CT scans of the heart enable imaging of coronary arteries to detect blockage. CT scans of extremities are used to diagnose complex fractures and dislocations. Most hospitals in the United States have CT scanners for use by the emergency room and outpatients. Some well-known brands for CT scanners include the GE Lightspeed/Prospeed, Siemens Somatom, Phillips Brilliance, and Toshiba Aquilion models. IGBT products especially tailored for CT scanners include the IGBT Inverter 6509454 for the Siemens SOMATOM AR.STAR CT scanner [11], the IGBT drive unit 53159006 for the Philips Tomoscan SR 7000 CT scanner [12], and the IGBT CT unit assembly 5017456402 for the Shimadzu SCT-5000T CT scanner [13].

13.2.1 PWM-Resonant Converter Power Supply

A block diagram for the power supply for the CT scanner circa 1995 is shown in Fig. 13.8 [14]. The input AC power source is rectified to create a DC bus voltage for the PWM converter. The hard-switched PWM converter operating at 50 kHz steps the voltage down and increases the current. The output of the PWM converter is fed to the resonant inverter. The resonant inverter is soft-switched at 50 kHz to generate a sinusoidal high-frequency input voltage to the high-voltage, high-frequency transformers. The output of the transformers is rectified using a delta-wye configuration.

The circuit diagram for the PWM converter is shown in Fig. 13.9. It makes use of a single IGBT operating at a frequency of 50 kHz. The output of the PWM converter is fed to the resonant inverter. The output of the PWM converter is determined by the duty cycle used for driving the IGBT.

The resonant inverter consists of two IGBTs in a totem-pole configuration with a Pi filter at the output, as shown in Fig. 13.10. The resonant inverter takes the input

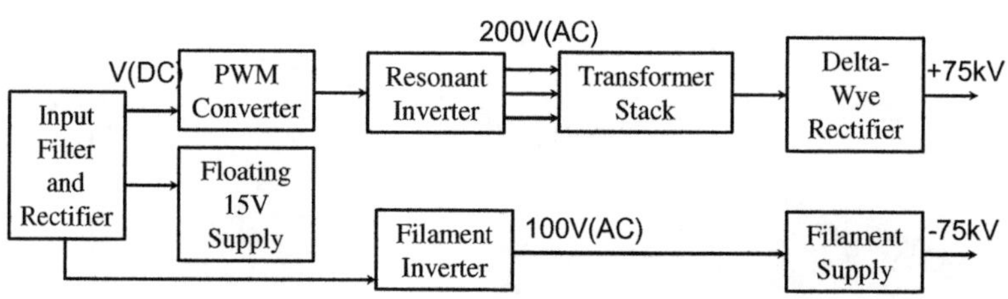

FIGURE 13.8

Pulse width modulation resonant converter power supply for CT scanner.

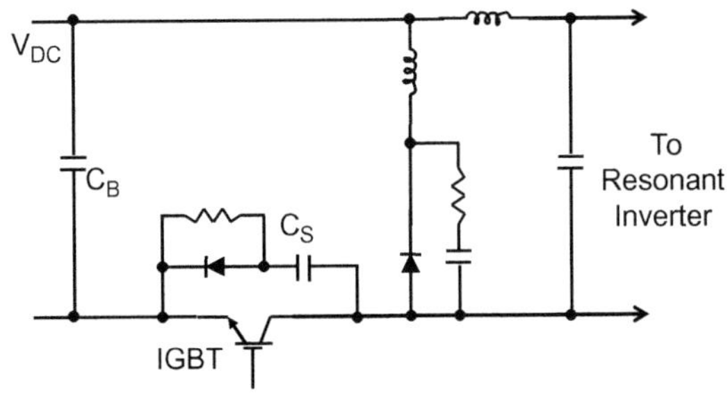

FIGURE 13.9

Pulse width modulation converter circuit for CT scanner. *IGBT*, insulated gate bipolar transistor.

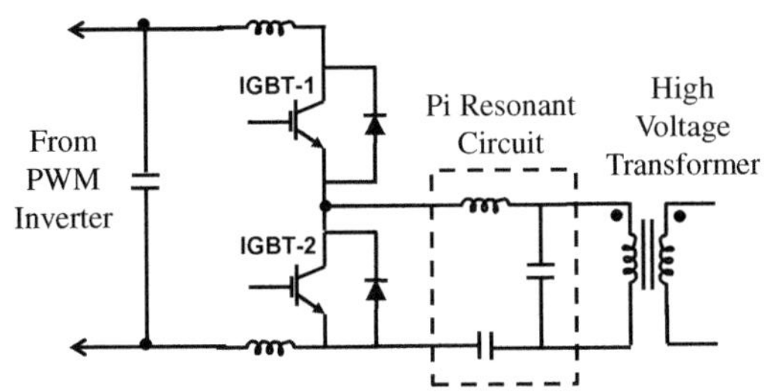

FIGURE 13.10

Resonant inverter with Pi resonant circuit for CT scanner. *IGBT*, insulated gate bipolar transistor.

from the PWM converter and produces high-frequency sinusoidal input to the high-frequency, high-voltage transformers. The resonant inverter is soft-switched at close to the resonant frequency (about 50 kHz). The 600 V rated IGBTs are operated as zero voltage switches to reduce switching power loss and improve the efficiency. The Pi resonant circuit produces the desired sinusoidal input to the transformer input windings.

Multiple sets of the above PWM converter/resonant inverter stages are used with multiple transformers to reduce the ripple current by phase shifting them with respect to each other. A 150 kV anode-to-cathode voltage is delivered to the x-ray tube in the CT scanner with a 3.5% ripple at 600 kHz with this approach.

13.2.2 Resonant Inverter Power Supply in Rotating Gantry

By 2000, the power supply for the CT scanner was designed with an IGBT-based resonant inverter mounted within the movable gantry [15]. The input AC power source was rectified and filtered to create a DC bus on the stationary part of the gantry. The DC power was then transferred to the rotating part of the gantry by using slip

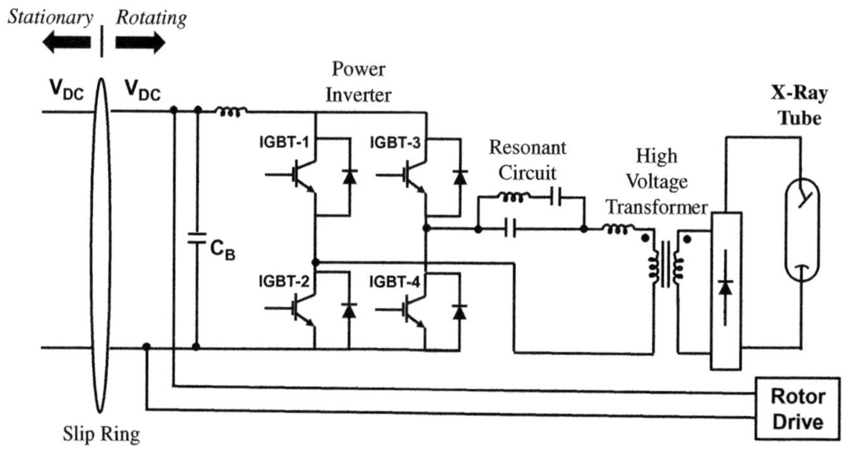

FIGURE 13.11

Inverter inside rotating gantry of CT scanner. *IGBT*, insulated gate bipolar transistor.

rings with brushes. The power electronics housed in the rotating part of the gantry included the x-ray power inverter and resonant circuit, the high-frequency high-voltage transformer, the high-voltage rectifiers, the x-ray tube, and its rotor drive.

The CT x-ray tube power supply is shown in Fig. 13.11. The slip ring for transferring DC power to the IGBT inverter stage is on the left side. The four IGBTs are used together with a series resonant circuit to produce the desired high-frequency input power for the high-frequency high-voltage transformer. Resonant circuit operation allows reducing the switching losses in the IGBTs while allowing a high operating frequency to reduce the size and weight of the transformer. The output of the transformer is rectified and fed to the x-ray tube to generate the x-ray radiation for the CT images. The rotor drive for the x-ray tube also utilized IGBTs.

13.2.3 Resonant Inverter Power Supply in Stationary Gantry

The weight and volume of the rotating gantry are increased by the location of the IGBT inverter inside it. In 2006, this problem was addressed by locating the IGBT inverter in the stationary part of the gantry and transferring high-frequency AC power via a contactless rotary transformer [15]. The contactless rotary transformer eliminates the brushes used in the previous DC slip ring configuration. The wear of the brushes was a reliability problem for CT scanners.

The new CT scanner power supply configuration is shown in Fig. 13.12. The main high-power (~150 kW) IGBT inverter shown at the top of the figure is used to provide power to the x-ray tube. The input AC power is first rectified to create a DC bus for the IGBT power inverter stage located on the stationary part of the CT gantry. The IGBT inverter is a series resonant circuit used to produce high-frequency AC power at the input of the rotary transformer. The output of the rotary

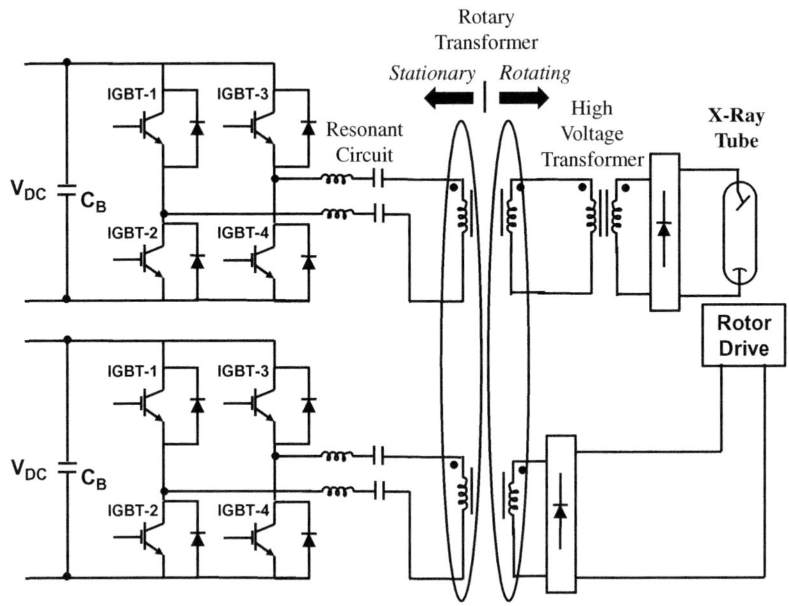

FIGURE 13.12

Inverter outside rotating gantry of CT scanner. *IGBT*, insulated gate bipolar transistor.

transformer is fed to the high-frequency, high-voltage transformer whose output is rectified and filtered to reduce the ripple voltage. This voltage is then used to drive the x-ray tube in the CT scanner.

A lower-power (~5 kW) auxiliary IGBT inverter, shown at the bottom of Fig. 13.12, is also incorporated in this design. The output of the auxiliary inverter is used to provide power to the x-ray tube rotary drive and other electronic components housed inside the rotating part of the gantry.

The design of the rotary transformer is illustrated by its cross-section in Fig. 13.13. It consists of two opposing platters within which the transformer windings are housed. The E-shaped magnetic cores are used to confine the magnetic field and enhance the coupling between the primary and secondary sides of the transformer. The windings of the outer cores are used for the auxiliary power delivery and the windings of the inner core are used to deliver power to the x-ray tube.

13.3 Magnetic Resonance Imaging

The CT scanner discussed in the previous section makes use of x-ray radiation, which can be harmful to the human body with the accumulation of exposure. An imaging modality that does not utilize such ionizing radiation is MRI, which has grown in importance since the 1980s. An example of the MRI scanner is shown in

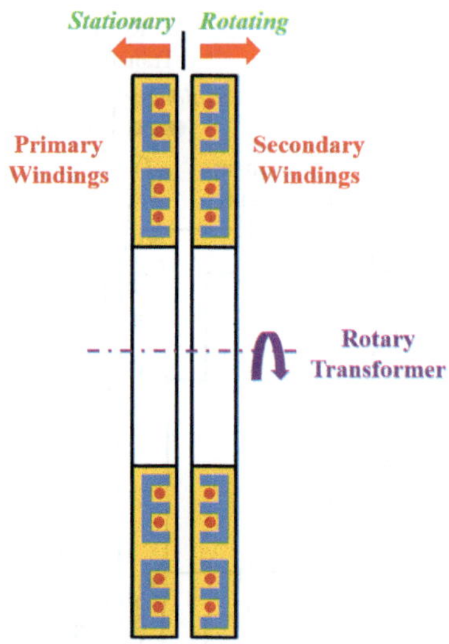

FIGURE 13.13

Rotary transformer design for the CT scanner.

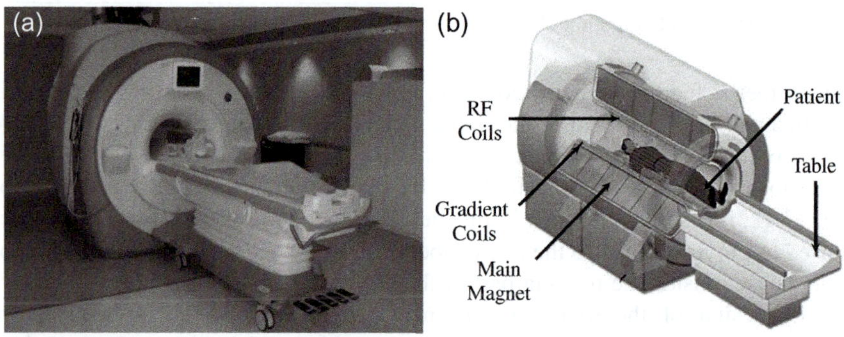

FIGURE 13.14

(a) MRI scanner, (b) MRI components.

Fig. 13.14(a). Its main components required to produce the images are illustrated in Fig. 13.14(b). The MRI imaging technique relies on the precession of the hydrogen nucleus (or protons) [16]. Hydrogen atoms are prevalent in human tissue within the water and fat molecules. In the natural state, the spin of the hydrogen nuclei is randomly oriented. When a very high magnetic field (typically 1–2 T) is applied along the axis of the MRI machine using the main magnet, the hydrogen nuclei line up along the axis with half the atoms oriented in each direction. However, a few of them are not equally matched. A radio frequency (RF) pulse is applied, tuned to the precession of the hydrogen nucleus to achieve resonance (Larmour frequency), to the area of the body where the imaging is to be performed. The unmatched hydrogen nuclei absorb the RF energy, changing their spin direction into a plane transverse to the main magnetic field. The transverse magnetization is detected using an RF receiver coil. When the RF input energy is stopped, the excited hydrogen nuclei return to equilibrium while releasing energy that can be picked up by detectors. The relaxation of the proton spin occurs via spin–lattice relaxation and spin–spin relaxation mechanisms. This information is used to generate an image called a "slice," which can be a few millimeters in thickness.

Three gradient magnets are also located along the axis of the MRI scanner together with the main magnet, as shown in the figure. They are turned on and off rapidly to vary the magnetic field in a local region of the body where the slice image is to be generated. The slice-selection gradient is produced by the z-axis gradient coil to excite protons only in a slice that is a few millimeters thick. The x and y gradient coils are used to apply a magnetic field to create controlled variations in the frequency and phase for the proton spin. This approach allows producing images along any orientation without moving the patient—a significant advantage of MRI machines. The application of rapid pulses of high currents to the gradient magnets produces the hammering sound during the MRI procedure.

Due to many opportunities for medical diagnosis created with MRI machines, they have become prevalent tools around the world in medical facilities. It is estimated that more than 25,000 MRI scanners were in use worldwide in 2015 [17]. This has grown to 50,000 machines in use globally in 2020 with 5000 units being manufactured per year. MRI scanners are used for the detection of various cancers, imaging vascular diseases, observing lesions of the liver and pancreas, and even changes in neural activities within the brain. The global market for MRI machines was $5.5 billion in 2010 with 80% of the sales occurring in the United States, Europe, and Japan [18]. It reached $8 billion in 2020 and is projected to exceed $11.5 billion in 2027 [19]. The leading manufacturers for MRI machines are GE Healthcare, Siemens, Philips, Hitachi, and Toshiba. More than 30 million MRI scans were performed in 2010. Of these, 29% were used for the brain, head, and neck; 25% were used for the spine; and 24% were used for extremities. MRI products are subdivided as low-field systems (magnetic field less than 0.5 T), medium-field systems (magnetic field of 0.5–1.0 T), high-field systems (magnetic field of 1.5 T), and very high-field systems (magnetic field of 3.0 T). High-field and very high-field systems dominate the market. Very high-field MRI machines can be used for whole body scans.

13.3.1 Two-Paralleled Four-Quadrant DC Chopper Power Amplifier

To produce MRI images rapidly (every 100 ms), the gradient coils must be driven with power amplifiers that deliver a rapid increase in the current and maintain it at a stable value with a small ripple current. A pair of four-quadrant DC chopper circuits was reported to accomplish this in 1999 [20] by Hitachi Medical Corporation. The two-paralleled four-quadrant DC chopper topology is shown in Fig. 13.15. Eight

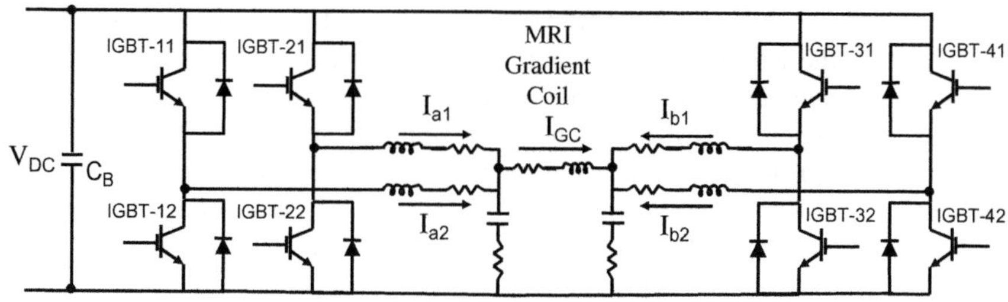

FIGURE 13.15

Pulse width modulation resonant converter power supply for the MRI scanner. *IGBT*, insulated gate bipolar transistor.

IGBTs are used to produce the desired current waveform in the gradient coil. The IGBTs are switched on and off such that the voltage across the L–R–C filter circuits alternates between V_{DC} and zero. The current (I_{GC}) in the gradient coil is the sum of the currents I_{a1} and I_{a2} or currents I_{b1} and I_{b2} in the filter circuits. The ripple current in the gradient coil is reduced because the ripple currents in the filters are mutually canceled out.

The gradient coil inductance is 200 μH and its current is 150 A in each direction. By operating the IGBT at a switching frequency of 20 kHz, a current rise and fall time of 0.5 μs can be achieved for the 300 A transient. A power loss of 2 kW is observed for the IGBTs due to the turn-off tail currents when delivering 125 kW of power to the gradient coil.

13.3.2 Four-Paralleled Full-Bridge Power Amplifier

The demand to improve the imaging speed and obtain better image quality requires increasing the response speed for the power amplifiers to drive the gradient coil magnets. To achieve the enhanced performance, Hitachi Medical Systems reported the development of a four-paralleled full-bridge topology for the power amplifier in 1999 [21]. This approach is able to deliver any current waveform with a peak current of 400 A and an output voltage of 600 V. The basic principle of the approach is illustrated in the block diagram shown in Fig. 13.16. This topology contains a main current amplifier at the top of the figure and a slave current amplifier at the bottom of the figure. Each has three multiple-bridge PWM amplifiers. The AC–DC converters produce the 650 V DC voltage from the 200 V AC input power source and feed it to the PWM output circuits. The output circuits of the main and slave power amplifiers are combined to drive the X, Y, and Z channels of the gradient coil.

The four-paralleled full-bridge power amplifier circuit is shown in Fig. 13.17. It employs 16 IGBTs with ratings of 1200 V and 300 A. In addition, a power factor correction active AC–DC converter is used containing six additional IGBTs with ratings of 1200 V and 400 A to rectify the AC input power source and create a stable DC bus for the power amplifier. The IGBT switches S-11, S-12, S-13, and S-14 comprise the first H-bridge network driving the gradient coil current; the IGBT switches S-21,

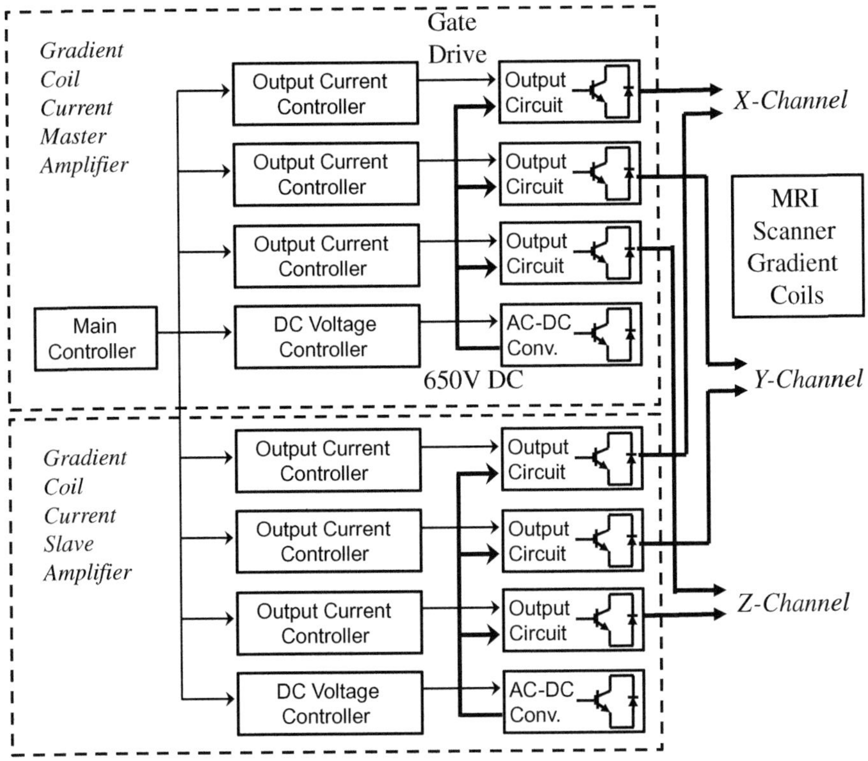

FIGURE 13.16

Divided gradient field generation topology for the MRI scanner.

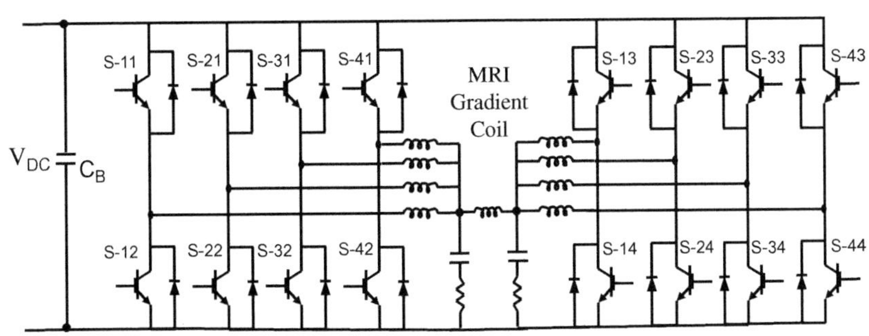

FIGURE 13.17

Four-paralleled full-bridge power supply for MRI scanner.

S-22, S-23, and S-24 comprise the second H-bridge network driving the gradient coil current; the IGBT switches S-31, S-32, S-33, and S-34 comprise the third H-bridge network driving the gradient coil current; and the IGBT switches S-41, S-42, S-43, and S-44 comprise the fourth H-bridge network driving the gradient coil current.

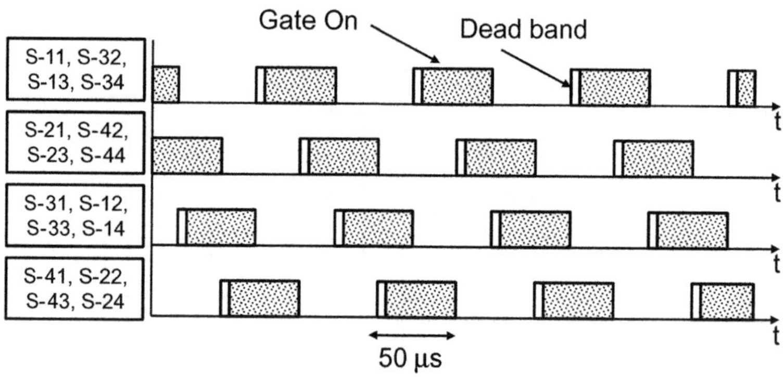

FIGURE 13.18

Gate control signals for 16 insulated gate bipolar transistors in the MRI scanner power amplifier.

The control signals for gating the IGBTs in the four-paralleled full-bridge circuit are shown in Fig. 13.18. To prevent shoot-through currents that produce destructive heating in the devices, the IGBTs in totem poles (such as S-11 and S12, etc.) are switched with a dead band, as indicated in the figure. The IGBTs in the second H-bridge (e.g., S-21) are gated with a 90-degree phase shift from the corresponding IGBT in the first H-bridge (e.g., S-11). A similar 90-degree phase shift is applied to the IGBTs in the third and fourth H-bridges. This approach significantly reduces the output current ripple. The output current supplied to the gradient coil of the MRI scanner can be adjusted up to 200 A with a current risetime of 200 μs. It has been found that an output current ripple at frequencies below 200 Hz impacts the quality of the MRI images. The approach described here keeps ripple noise less than 5 mA between 10 and 500 Hz.

13.3.3 Stacked Three-Bridge Power Amplifier

There has been a continuous demand for improving the quality and speed for producing MRI images over the years. This can be achieved by increasing the current driving the gradient coils, which requires increasing the drive voltage across the gradient coil as well. By 2004, MRI systems with the gradient coils driven with several hundred amperes using more than 1500 V were feasible. The drive also needs to have a fast response time and to create only small ripple currents in the gradient coil.

General Electric Healthcare has reported a stacked three-bridge approach to achieve an improved power amplifier for driving gradient coils that meets the increasing demands for MRI scanners [22]. This approach makes use of three full IGBT bridge circuits that are stacked, as shown in Fig. 13.19, to provide a drive voltage that is the sum of the output voltage for each of them. Two of the full-bridge inverters have a high DC input voltage of 800 V and are operated at a switching

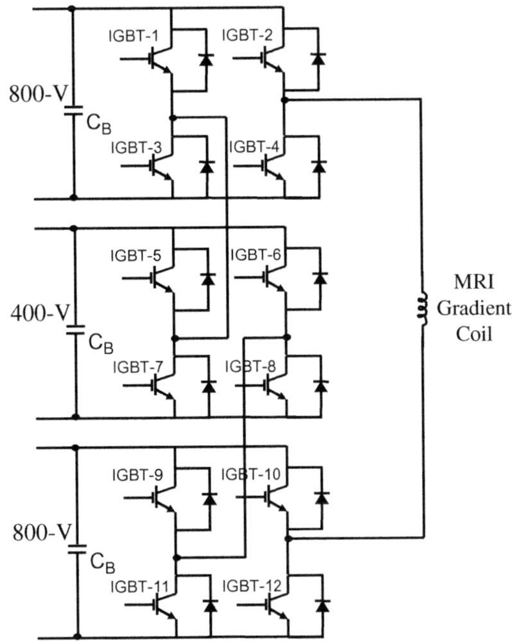

FIGURE 13.19

Stacked three-bridge power amplifier for the MRI scanner. *IGBT*, insulated gate bipolar transistor.

frequency of 31.25 kHz. The third half-bridge inverter with IGBT-5, IGBT-6, IGBT-7, and IGBT-8 has a lower DC input voltage of 400 V and is operated at a switching frequency of 62.5 kHz. Its smaller DC voltage allows using IGBTs with lower blocking voltage capability that have smaller switching losses required to accommodate operating at the higher switching frequency. The higher operating frequency of this inverter provides a faster response for the power amplifier.

The three inverter bridges are driven with interleaved gate control signals to reduce the ripple current in the coil and increase its frequency. Feed-forward controls are used for the two high-voltage inverters and feedback is used for the low-voltage inverter. The two high-voltage inverters are not pulse width modulated when the drive voltage is less than 400 V, allowing operation in a free-wheeling mode with no output voltage. For an output voltage ranging up to 1200 V, one of the high-voltage inverters is switched on while an output voltage up to 2000 V is generated by switching on both the high-voltage inverters. This approach was able to produce a trapezoidal gradient coil current changing from plus 50 A to minus 50 A with a slew rate of 0.3 A/μs. The flat top of the current had a duration of 1 ms. The method allows creating any arbitrary wave shape for the gradient coil current.

13.3.4 Multioutput Phase-Shifted Power Amplifier

By 2007, improvements in MRI scanner images were achieved by driving the gradient coil with currents of more than 600 A using a drive voltage of more than 1600 V for each of the three axes. Although the gradient coil is an almost purely reactive (inductive) load, the power amplifier must be capable of generating about 1 MVA of power per axis. The power losses in the gradient coil and the inverters add up to 60 kW per axis. Three power amplifiers are required for driving the three independent gradient coils for the x, y, and z axes.

The approach described in the previous section is based on each inverter comprised of multiple isolated sources—two 800 V sources and one 400 V source. This had evolved to two 600 V and two 400 V sources by 2007 [23]. The topology for one power amplifier for one of the gradient coils is shown in Fig. 13.20. Each of the gradient coils for the x, y, and z axes required two 600 V and two 400 V power supplies for a total of 12 isolated voltage sources.

A less complex topology for the power supplies was reported in 2007 [23]. This approach makes use of a three-leg IGBT bridge circuit with phase shift PWM control of the devices. One leg is used as a common element with each of the other legs to form the H-bridge circuit for supplying current to the transformers for each of 600 V and 400 V power supplies, as shown in Fig. 13.21.

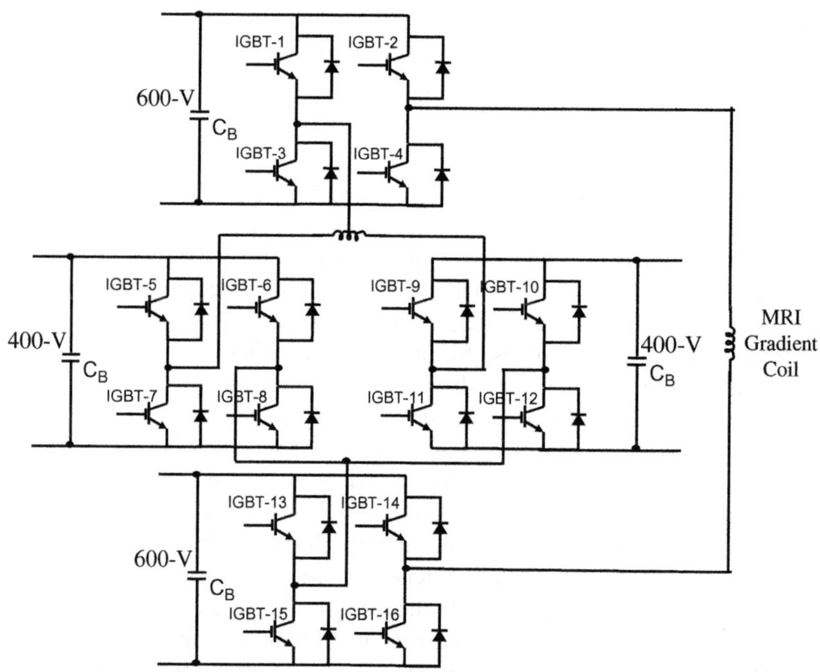

FIGURE 13.20

Multioutput phase-shifted power amplifier for the MRI scanner. *IGBT*, insulated gate bipolar transistor.

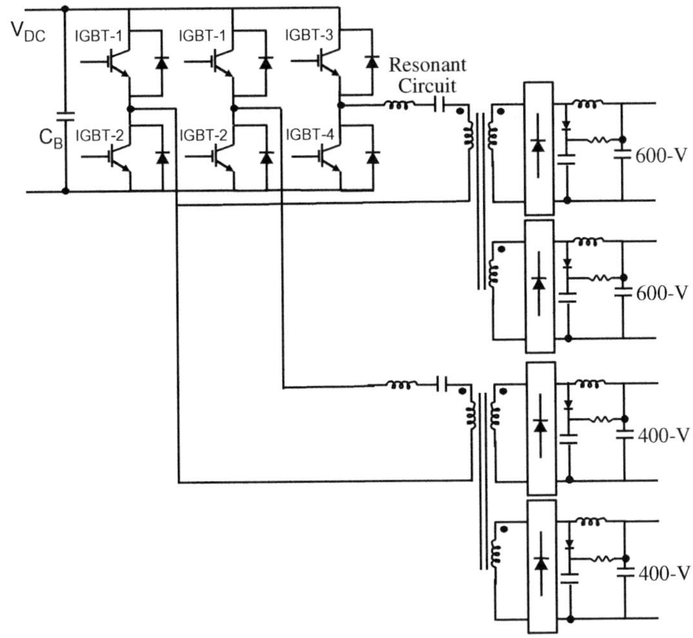

FIGURE 13.21

Three-leg power supply for one axis of the gradient coil in the MRI scanner. *IGBT*, insulated gate bipolar transistor.

In the simplified power supply, all three IGBT legs are operated at close to 50% duty cycle with the outputs controlled using phase shift between the common leg and the leg corresponding to each output voltage. The resonant circuits are used to achieve zero voltage switching. For further savings in space and weight, the common leg is also shared between the x-axis, y-axis, and z-axis but a better modular design is achieved with separate common legs for each axis of the gradient coil. The IGBTs are switched at 20 kHz as a good trade-off between the switching losses in the IGBT and the size of the output filter.

Because the gradient coil is almost purely inductive, significant energy is exchanged between it and the capacitor at the output of each of the power supplies. This capacitor must be sufficiently large to handle the power. The presence of the inductance in series with the primary of the transformer, in conjunction with the LC filter on the output, produces large voltage ringing across the secondary-side rectifiers. This is damped by using the passive resistor-capacitor-diode (RCD) clamp to reduce the maximum voltage rating for the rectifier diodes.

This power supply design has been tested while driving the gradient coil in MRI scanners. The voltage transients are less than 20 V for each of the output voltages. The gradient coil current can be supplied with a plus 400 A and a minus 400 A with a ramp rate of 1.8 A/μs.

13.3.5 Series Voltage Compensated Power Supply

As already mentioned, there has been a continual increase in the operating power level for gradient coil power amplifiers due to demands by radiologists for higher-gradient magnet strength to obtain better signal-to-noise ratio in the MRI images. An approach to simplify the topology for the multiple power supplies required for the gradient power amplifiers was proposed in 2009 [24]. This approach makes use of an AC preregulator to compensate for fluctuations in the input AC power source and stabilize the DC output voltage from the multiple power supplies.

A schematic diagram for the series compensated power supply is shown in Fig. 13.22. A series transformer is added to the input AC source before feeding the power to the multiwinding transformer. The secondary side voltage of the multiwinding transformer is sensed and compared to a reference voltage. The error signal is fed to the IGBT inverter via an IGBT active filter. The IGBT inverter generates the compensating AC signal that is injected into the series transformer to correct for the AC input source fluctuation. A voltage variation form—15% to +10% in the input voltage—can be compensated with this approach. The approach does not compensate for load variations. This is performed separately in the power amplifiers for the gradient coils.

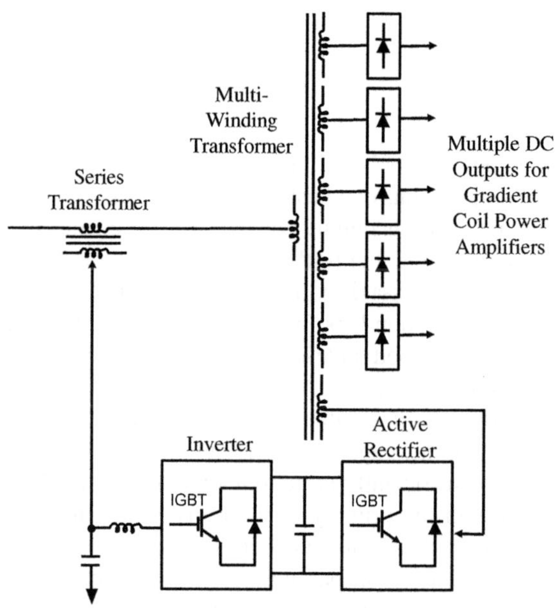

FIGURE 13.22

Series-compensated power supply for the gradient coil amplifier in the MRI scanner. *IGBT*, insulated gate bipolar transistor.

13.3.6 Supercapacitor Energy Storage Power Supply

By 2014, MRI systems made use of superconductivity-based main magnets with a magnetic field strength of 1.5 T with a uniformity of better than 10 ppm over a 30 cm sphere for a whole body scanner [25]. These large machines installed in hospitals require power of more than 1 MVA. It is desirable to manufacture smaller portable MRI scanners for use by emergency services and by the military. These machines are operated using mobile generator sets at power levels of about 10 kVA. The MRI gradient coils require very high current pulses with a trapezoidal shape consisting of a pulse duration of a few milliseconds long with a typical risetime of 200 μs. The pulses are irregular and widely spaced. The energy for this type of pulse can be supplied using supercapacitors.

A schematic illustration of the supercapacitor-based power supply for the gradient coils of the portable MRI scanner is shown in Fig. 13.23. The supercapacitor makes use of porous electrodes placed in a highly conductive electrotype. The electrolyte enters the pores of the electrodes. This creates a very large surface area with a very small effective dielectric thickness, resulting in large capacitance values in a compact space. The supercapacitors are designed to operate in the 150–300 V range to be compatible with the voltage required to drive the gradient coils in portable MRI scanners.

The constant current charger used for charging the supercapacitor bank is shown in Fig. 13.24. It makes use of a single IGBT to regulate the voltage across the

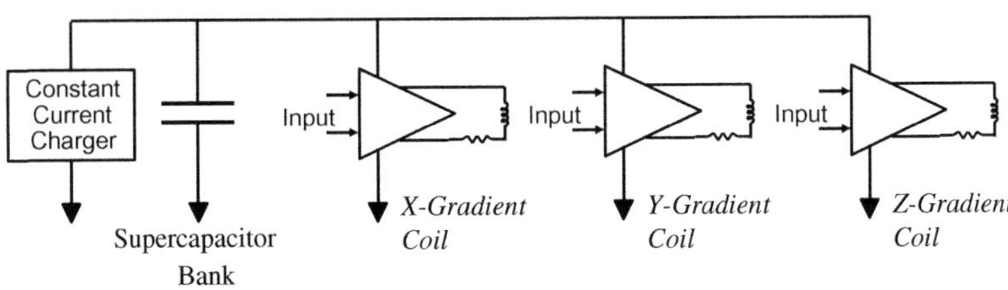

FIGURE 13.23

Supercapacitor-based power supply for the gradient coil amplifiers in the MRI scanner.

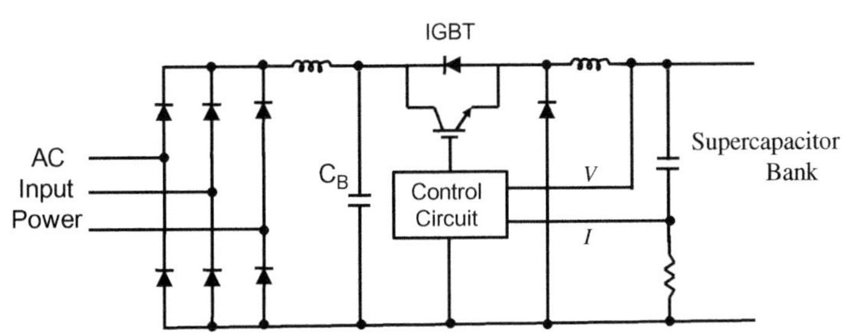

FIGURE 13.24

Charger for supercapacitor. *IGBT*, insulated gate bipolar transistor.

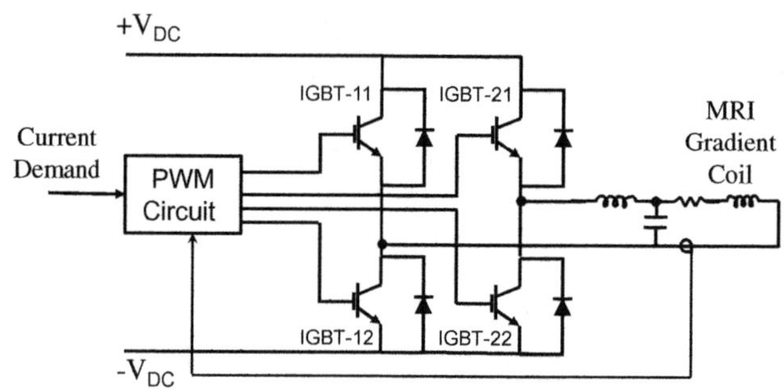

FIGURE 13.25

PWM gradient coil amplifier. *IGBT*, insulated gate bipolar transistor.

supercapacitor bank and the current (I) flowing through it. These parameters are monitored by the control circuit used to drive the IGBT. The input power for the charger is derived from the available three-phase AC power source.

A gradient coil amplifier using the PWM switch-mode design is shown in Fig. 13.25. It makes uses of four IGBTs in an H-bridge configuration for driving the current into the gradient coil. This topology provides high linearity and a response bandwidth of 10 kHz.

The portable MRI system without energy storage has a rating of 110 kVA, mostly for driving the three gradient coils. This power can be reduced by a factor of 10 by using the above energy storage with supercapacitor banks. The supercapacitor bank has a weight of 120 kg and occupies 0.42 m^3 of space.

13.4 Medical Ultrasonography

Ultrasonography or diagnostic sonography is widely used by the medical community for imaging internal human organs such as tendons, joints, and muscles [26]. It is commonly used by obstetricians for observing the development of the fetus during pregnancy. Sound waves at frequencies above human hearing ability are referred to as ultrasound. They can range from 20 kHz to several hundred MHz. The ultrasound wave is generated by a transducer. The ultrasound wave reflects off surfaces where changes in density occur. The reflected signals collected by detectors provide information of the depth of the reflection from the time taken for the wave to return to the transducer. By using an array of detectors, it is possible to generate an image of the organ at a particular plane.

As examples, ultrasound is used by anesthesiologists for guiding needles to place the anesthetic at selected locations; by cardiologists to detect dilation of the heart and operation of ventricles; by gastroenterologists for observation of obstructions in the bladder or kidney; by otolaryngologists for visualizing and detecting cancer of the thyroid and salivary glands; by neurologists for the assessment of blood flow in arteries; and by urologists for the diagnosis of testicular cancer. Two examples of modern

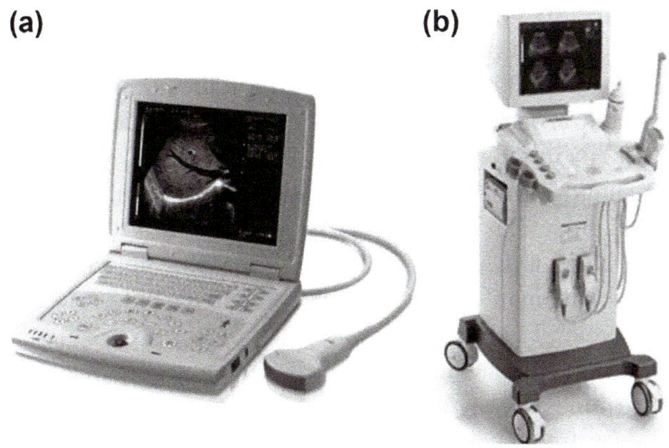

FIGURE 13.26

Ultrasound scanners. (A) Table top unit and (B) mobile unit.

ultrasound scanners are shown in Fig. 13.26. The global market for ultrasound systems was $4.6 billion in 2012 with 33% of the share in the United States [27]. The leading suppliers of ultrasound equipment are GE Healthcare, Philips Healthcare, Siemens Healthcare, Toshiba Medical Systems Corporation, and Hitachi Medical Corporation. The demand for ultrasound scans is expected to grow due to the aging of the worldwide population and the prevalence of chronic diseases. The market for diagnostic ultrasound equipment will reach $7 billion by 2026. In addition, the market for portable diagnostic instruments is expected to reach $86 billion in 2016 [28].

13.4.1 Ultrasonography Principles

The basic concept of the ultrasound scanner is based on creating a sound wave in the human body using a transducer. The transducer is placed in contact with the body using a gel to maximize transfer of the sound energy pulse into the body. The pulse travels through the body, as illustrated in Fig. 13.27, until it encounters a change in tissue density that produces a reflected wave. The reflected wave returns to the probe,

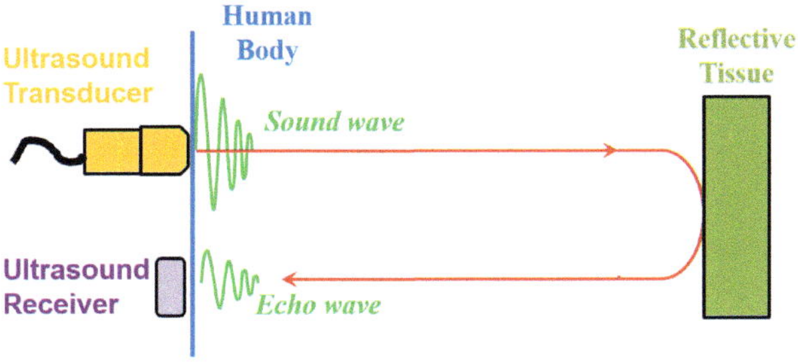

FIGURE 13.27

Principles of ultrasound scanners.

which contains detectors for sensing the ultrasound echo signal. The time taken for the pulse to return to the probe provides an accurate measure of the distance of the reflective tissue because ultrasound waves travel in the human body at the speed of 154,000 cm/s. An ultrasound echo from a reflection at a distance of 10 cm inside the body will take 130 μs to return to the probe. The ultrasound pulses are usually separated by several hundred microseconds to allow for this time. For sonography imaging, the ultrasound pulses have a width of 2–3 cycles, as illustrated in the figure, with a time duration of about 4 μs. The typical duty cycle for sonography is about 2%. For Doppler imaging to obtain information on movement, the pulses are 5–20 cycles long and have a duty cycle of about 5% [29].

The resolution of the ultrasound image is dependent on the frequency of the sound wave, which determines its wavelength. For sonography, the typical frequency range is 2–18 MHz. Lower frequencies allow imaging to larger depths in the body but have less resolution. Higher frequencies enable better resolution but the imaging depth is reduced because these frequencies are absorbed at a shorter distance in the body. Frequencies of 7–18 MHz are used for imaging muscles, tendons, breasts, and testes. Frequencies of 1–6 MHz are employed for deeper organs such as the liver and kidneys. Higher-energy ultrasound systems are used for therapeutic applications. Examples are physical therapy, cancer treatment of cysts and tumors, breaking up kidney stones (lithotripsy), and cataracts (phacoemulsification). Doppler ultrasonography can be applied to observe blood flow and muscle movement.

13.4.2 Pulsed Power Supply

The ultrasonic sound wave in the body is generated using a transducer. The transducer is a piezoelectric ceramic structure with a capacitive electrical equivalent circuit. The sound wave is created by the application of a high-voltage pulse to the piezoelectric element with a short duration of about 1 μs. The voltages required can range from 1000 to 2000 V, making IGBTs the best option for controlling the pulse. Historically, negative voltage pulses have been used for the transducer in ultrasound systems [30].

A circuit for providing the negative voltage pulse to the ultrasound transducer is shown in Fig. 13.28 [30]. The high voltage applied to the transducer is delivered

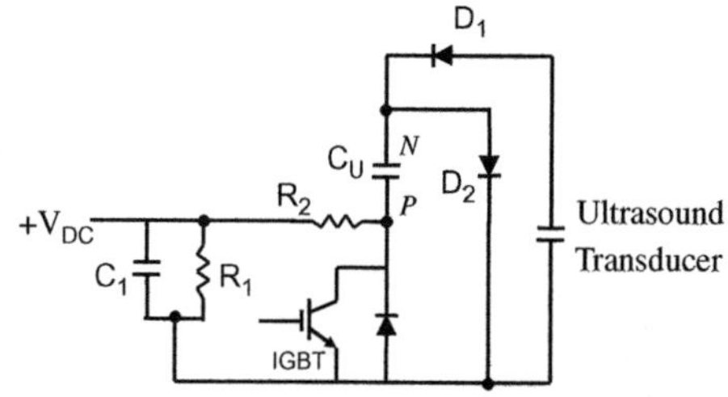

FIGURE 13.28

Nondoubling voltage pulser for ultrasound transducer. *IGBT*, insulated gate bipolar transistor.

using the capacitor C_U. When the IGBT is maintained in its off-state, the positive end of the capacitor (P) is charged to the DC supply voltage (V_{DC}) via resistor R_2 and diode D_2, which is forward biased. The transducer voltage pulse is delivered by turning on the IGBT, which pulls the potential at the positive end of the capacitor to ground. Because the voltage across the capacitor cannot change instantaneously, the potential at its negative end (N) jumps to a value of $-V_{DC}$. This voltage is applied to the transducer via diode D_1, which is forward biased. The voltage applied to the transducer during the pulse duration is approximately equal to V_{DC} if capacitor C_U is chosen with a much larger value than the equivalent capacitance of the transducer. During this time, diode D_2 becomes reverse biased and must withstand the DC supply voltage. The DC supply voltage can range from 100 to 1200 V. The pulse currents can range up to 100 A. The IGBT must be rated at a blocking voltage of 1200 V and a collector current of 120 A to meet the circuit operating requirements. The resistor R_2 is chosen so that capacitor C_U is rapidly charged while at the same time limiting the current through the IGBT when it is turned on. When the IGBT is switched off again, the voltage across the transducer returns to zero. The voltage waveform for the transducer obtained by using the above circuit is shown in Fig. 13.29. The voltage rises to its steady-state value of 900 V in 60 ns with a slew rate of 12 V/ns. After the IGBT is switched off, the voltage decays to zero in about 0.5 μs.

A larger negative voltage pulse can be applied to the ultrasound transducer by using the voltage doubling pulser circuit shown in Fig. 13.30 [30]. When IGBT-1 is in the off-state, capacitor C_{U1} is charged to the DC supply voltage V_{DC} via series resistor R_2 and diodes D_2 and D_3 with a positive voltage at terminal $P1$. Similarly, when IGBT-2 is in the off-state, capacitor C_{U2} is charged to the DC supply voltage V_{DC} via series resistor R_3 and diode D_4 with a positive voltage at terminal $P2$. These diodes are forward biased during this charging operation. As in the previous

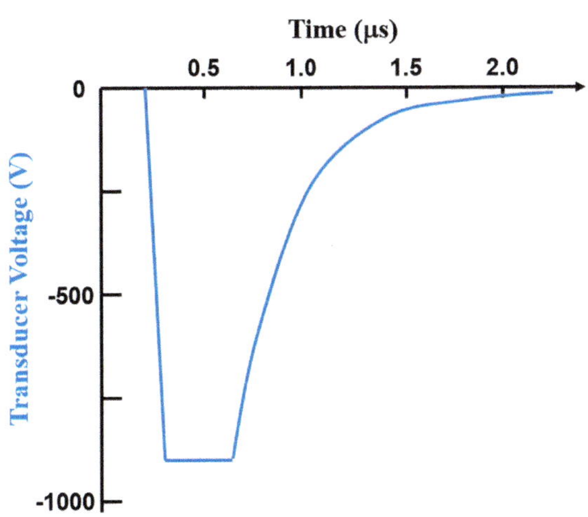

FIGURE 13.29

Transducer voltage waveform for the nondoubling voltage pulser.

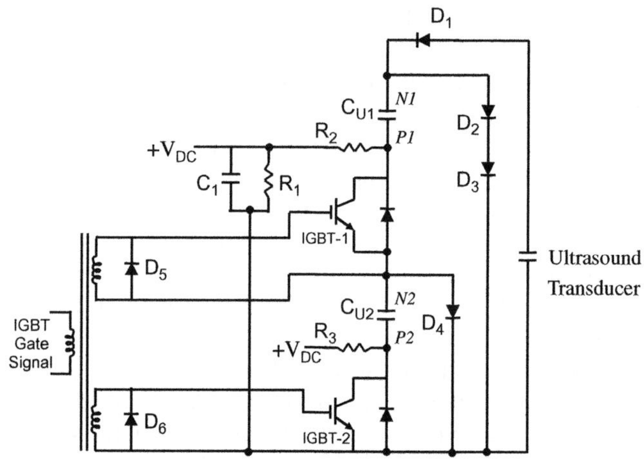

FIGURE 13.30

Doubling voltage pulser for ultrasound transducer. *IGBT*, insulated gate bipolar transistor.

nondoubling circuit, the capacitors C_{U1} and C_{U2} are chosen with much larger values than the capacitance of the transducer to produce a rectangular-shaped voltage pulse. Both IGBTs are turned on simultaneously using the transformer coupled gate drive signal. IGBT-2 pulls the terminal $P2$ of capacitor C_{U2} to ground, making the potential at its terminal $N2$ equal to $-V_{DC}$. At the same time, IGBT-1 pulls terminal $P1$ of the capacitor to the potential of terminal $N2$ of capacitor C_{U2}. This rapidly increases the potential at terminal $N1$ of capacitor C_{U1} to $-(2^*V_{DC})$. During this operation, diodes D_2, D_3, and D_4 become reverse biased. The voltage at terminal $N1$ of capacitor C_{U1} is applied to the transducer via forward biased diode D_1. Two diodes are used in series with terminal $N1$ to ground because its potential reaches twice that at terminal $N2$. A negative voltage pulse of 1800 V in magnitude is generated by this circuit on the transducer. The voltage rises to its steady-state value of 1800 V in 100 ns with a slew rate of 14 V/ns. After the IGBT is switched off, the voltage decays to zero in about 0.5 μs.

The authors state [30]: "*The nondoubling pulse was tested by replacing the IGBT with a MOSFET rated at 1000 V. The rise time was comparable but the peak voltage was about 20% less than an IGBT operating at 1000 V due to the significantly larger internal resistance of the MOSFET. Thus, higher output voltages can be obtained with the IGBT.*"

13.5 Defibrillators

Every year, nearly 424,000 people experience the onset of sudden cardiac arrest (SCA) in the United States, leading to death for nine of 10 victims [31]. It is the leading cause of death in the developed world. SCA is the sudden stopping of the pumping of blood by the heart, usually related to ventricular fibrillation. The lack of blood flow deprives vital organs such as the brain of oxygen, causing the victim to collapse. During the SCA episode, the regular beating of the heart (systole and diastole) stops occurring. The victims of SCA can be revived by the immediate use of cardiopulmonary resuscitation (CPR) and treatment with defibrillators. The chances of revival diminish by 7–10% for every minute of delay before the defibrillator is used. Because the average response time for paramedics to respond to an emergency call exceeds 10 min, it is important to have defibrillators widely distributed in society to save lives.

The development of the low-cost and portable automatic external defibrillator (AED) and its widespread deployment in offices, public entertainment and sports venues, airplanes, cruise ships, and shopping centers has been estimated to save 50,000–100,000 lives each year in the United States [32]. An example of the AED unit is shown in Fig. 13.31(a). The low cost ($1000 to $2000) of these defibrillators, due to advanced technology including IGBT switches, has been an enabling factor in their widespread deployment. The leading manufacturers of portable AEDs are HP HeartStart (formerly HeartStream), Philips, and Medtronic [33].

The heart pumps blood through the body using two stages [34,35]. In the first stage, the right and left atrium contract pumping blood into the right and left ventricles. In the second stage, the right and left ventricles contract to pump the blood out of the heart. The heart muscle is comprised of two types of cells: conducting cells that transmit electrical signals and muscle cells that produce contractions. The heartbeat begins by an electrical signal from the sinoatrial (SA) node [36]. The signal is distributed rapidly through the wall of the atria and triggers contraction of the upper heart chamber (atria). The atrioventricular (AV) node then sends an electrical signal to contract the lower heart chambers (ventricles) after a delay of 0.1 s. This sequence is repeated for each subsequent regular heartbeat.

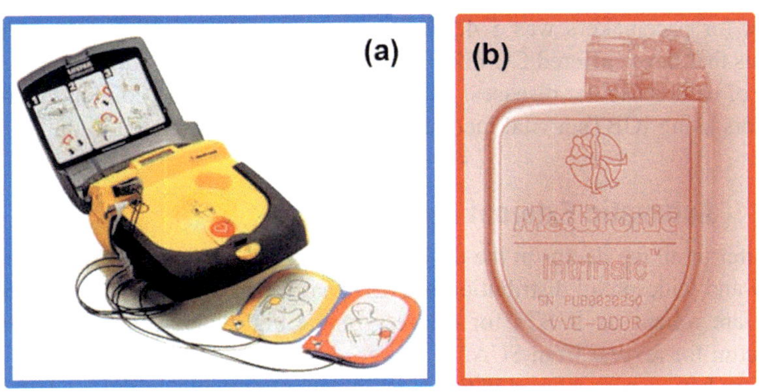

FIGURE 13.31

(a) Automatic external defibrillator; (b) implantable cardioverter defibrillator.

Sudden cardiac arrest is accompanied by ventricular fibrillation [37] or atrial fibrillation [38] with uncoordinated contractions of the cardiac muscle, which makes them quiver rather than pump blood. It was discovered in 1947 by Claude Beck that a regular heartbeat can be restored by the application of a pulse of electrical energy [39]. Early defibrillators were used during surgery to recover the heartbeat of patients with open chest cavities. Subsequently, Bernard Lown and Barouh Berkovits found that a high-energy jolt of electricity administered using paddles pressed against the chest could be used to revise victims of ventricular fibrillation. The DC or uniphasic Lown-Berkovits waveform required a large amount of energy (360 J), making the defibrillator large and heavy [40].

Biphasic defibrillation makes use of energy pulses with alternate voltage direction in a 10 ms period. This approach was originally developed for implantable cardioverter defibrillators (ICDs). Biphasic defibrillation requires much smaller energy levels (<200 J), decreasing the risk of burns and myocardial damage [39]. It has also allowed the development of AEDs that are the size of a laptop. The implanted and external defibrillators utilize IGBTs for regulating the voltage pulse transmitted to the heart.

ICDs or pacemakers are recommended for people suffering from cardiac dysrhythmia caused by blockage and myocardial infarction [41]. This device combines the detection of cardiac arrhythmia and the application of short electrical impulses to correct it. Advances in semiconductor technology have greatly reduced the size and efficiency of the ICDs, improving the implantation procedure. Patients with ICDs can resume normal physical activity, including sports, resulting in a great enhancement to the quality of life. The ICD operates by charging a capacitor from a low-voltage battery in the unit and discharging it through the muscles of the heart when an abnormally high heart rate (tachyarrhythmia) is detected. The therapeutic shock of 750 V and current of 15 A is controlled using IGBTs. The first implantable defibrillator in a human patient was demonstrated in 1980 at Sinai Hospital, Baltimore. More than 1 million pacemakers were implanted in the world in 2009 with the largest number (225,000) in the United States [42]. An example of the ICD made by the Medtronic company is shown in Fig. 13.31(b). It is the size of a microcassette. The global market for pacemakers was $ 4.5billion in 2020 with 2 million units sold [43]. Patients that have suffered COVID-19 infection have been found to have high incidence of cardiovascular diseases. The coronavirus outbreak in 2020 has made defibrillators using IGBTs a vital medical aid for saving lives related to the pandemic.

13.5.1 Automatic External Defibrillators

A quarter of all the deaths in the developed world are attributed to SCA. The availability and widespread distribution of AEDs has saved thousands of lives. Modern AEDs are reliant on IGBTs for delivery of the high-voltage biphasic voltage to the victim for revival from SCA.

There has been an interest in the development of portable defibrillators for many decades. In 1962, the development of a portable defibrillator was reported using capacitor discharge [44] for use in a hospital setting. The Hopkins a-c defibrillator placed in service in 1957 had a weight of more than 200 pounds. The authors state: "*Although it may be easily moved about the hospital, this defibrillator is not, by any stretch of the imagination, portable.*" An improved portable defibrillator was constructed in 1959 by using a set of capacitors charged to 2200 V. One set was used to provide a positive pulse and a second set a negative pulse for the biphasic voltage waveform. The switching of the capacitors was achieved using a rotary switch due to the lack of suitable semiconductor switch technology at that time. This design had a weight of about 45 pounds. The rotary switch produced arcing at the contacts, making this defibrillator not suitable for use in an operating room due to the presence of explosive gases.

Modern AEDs are much more sophisticated in design and far smaller in weight and size due to advancements in semiconductor technology [45]. The units available from the leading manufacturers listed above have a weight of just 3.5 pounds and the size of a small laptop computer. They provide simple audio instructions to allow someone in the vicinity of the SCA victim to provide lifesaving assistance. The unit prompts the placement of two electrodes with adhesive surfaces on the victim's chest. The AED has a built-in electrocardiograph that detects and analyzes the victim's heartbeat. It then determines whether to administer a shock if ventricular fibrillation is occurring. The bystander is asked to simply push a button to initiate the biphasic voltage waveform. More than 90% of victims are revived if the shock is provided within the first minute of the onset of SCA.

The basic biphasic voltage waveform that is used in modern AEDs is shown in Fig. 13.32. The first phase of the voltage waveform charges the membranes of cells in the heart muscle; the second phase of the waveform with reversed voltage returns the cell membranes to the original state [45]. The voltage applied to the electrodes is about 2000 V. This is accomplished using an H-bridge circuit with four IGBTs that control the direction of the voltage to the electrodes from a single charged capacitor, as shown in Fig. 13.33.

The EKG system initially monitors the victim's heartbeat to detect arrhythmia with a heartbeat greater than 150 per minute and then initiates the shock sequence. The defibrillator capacitor C_D is first charged to about 2000 V. In modern AEDs, a

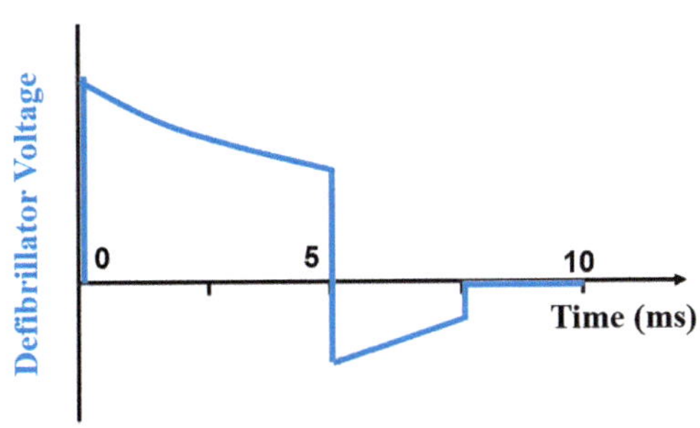

FIGURE 13.32

Biphasic voltage waveform used in modern defibrillators.

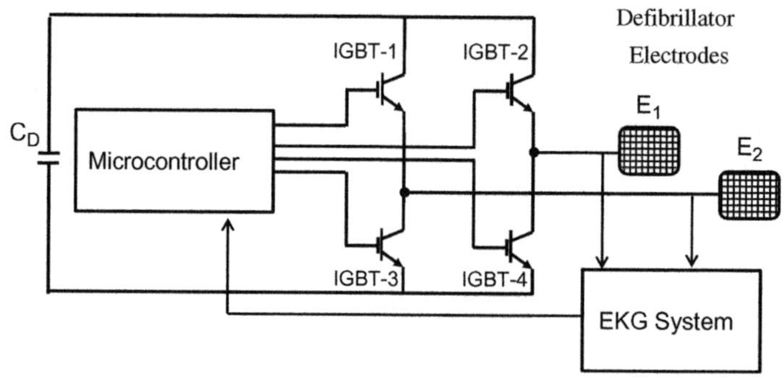

FIGURE 13.33

Insulated gate bipolar transistor H-bridge circuit used for generating the biphasic voltage waveform in automatic external defibrillators.

bank of lightweight aluminum electrolytic capacitors is connected in series to achieve a small weight and size. Compact IGBT switches are then used to direct the capacitor voltage to the electrodes. The positive voltage for the first phase is applied to electrode E_1 by turning on IGBT-2 to connect it to the positive end of capacitor C_D, and electrode E_2 is connected to the negative end of capacitor C_D by turning on IGBT-3 while the other IGBTs remain in the off-state. After a time duration of 5 ms, as shown in Fig. 13.32, the voltage at electrode E_1 is reversed by turning on IGBT-4 to connect it to the negative end of capacitor C_D, and electrode E_2 is connected to the positive end of capacitor C_D by turning on IGBT-1. The voltage during the second phase is smaller because the capacitor C_D is partially discharged during the first phase. The EKG system then rechecks the patient's heartbeat to ensure that the heart has returned to its regular rhythm.

13.5.2 Energy Generation and Pulse Forming in AEDs

A fly-back topology for charging the capacitor in the AED has been reported [46]. The circuit, shown in Fig. 13.34, charges the defibrillator capacitor C_D over multiple cycles of switching the IGBT. The AEDs are powered using lithium-ion batteries with a DC voltage of 12–16 V. To achieve a final charge voltage of 2000 V across the capacitor C_D, the turns ratio for the transformer becomes large. To avoid this, a boost converter (not shown) is used to increase the voltage to 200 V (V_{DC}) for the fly-back converter. The charging process takes 6–15 s [46]. The energy stored in the capacitor ranges from 30 to 260 J.

13.5.3 Implantable Cardioverter Defibrillator

As mentioned earlier, ICDs or pacemakers have become lifesaving innovations for people suffering from cardiac dysrhythmia. The device combines monitoring cardiac arrhythmia and the application of short electrical impulses when needed to correct

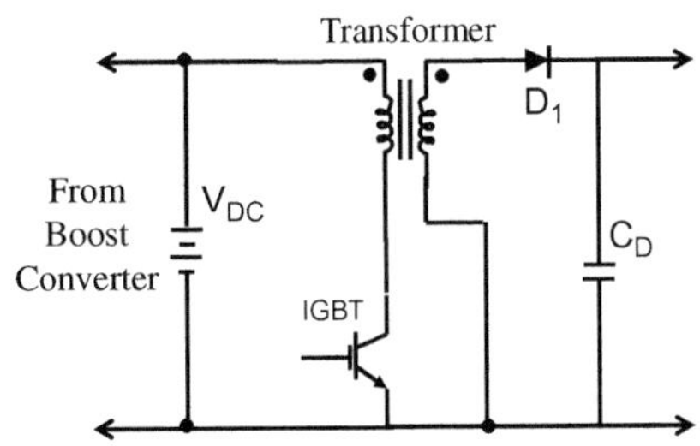

FIGURE 13.34

Fly-back topology for charging capacitor in defibrillator. *IGBT*, insulated gate bipolar transistor.

any problems. Modern ICDs monitor and store a record of the patient's heart signals, providing diagnostic data to physicians. Advances in semiconductor technology have reduced the volume of ICDs to less than 70 cc. Because the ICD is placed subcutaneously in the human body, only a small average current of 20 μA must be drawn from the battery inside the ICD [47]. State-of-the-art ICDs use a single transvenous lead with multiple electrodes located in the right ventricle for defibrillation.

When the ICD detects abnormally fast heart rates above 150 beats per minute (called tachyarrhythmia), the output capacitor is charged to 750 V. The capacitor is discharged into the heart muscle by using embedded electrodes to shock the heart back into its normal rate of about 72 beats per minute. The therapeutic shock is delivered by the ICD every 5 months on average. The availability of the ICD has reduced the mortality rate for people who have survived a previous SCA event to only 2% in 1 year.

The authors state: "*The therapeutic pulse that is delivered by the ICD is 750 V with a leading edge current of 15 A and (a) pulse termination current spike of 210 A.*" This energy must be derived from a 3.2 V lithium silver vanadium oxide battery that delivers an average current of only 20 μA to achieve a lifetime of 4–5 years. However, during the capacitor charging operation, the battery must supply 2 A for several seconds. The conversion from the battery voltage of 3.2 V to the 750 V across the capacitor is performed using the fly-back converter topology discussed in the previous section. The circuit is operated at high frequency (30–60 kHz) to reduce the size and weight of the transformer. A typical capacitor energy of 35 J can be obtained in 8 s using a 10 W DC–DC converter. The energy storage capacitor C_D consists of two aluminum electrolytic capacitors in series.

For application of the biphasic waveform shown in Fig. 13.32, the authors state: "*...(IGBTs) are used in the output bridge configurations capable of polarity switching.*" The peak currents delivered by the IGBTs are about 40 A. The output voltage across the capacitor is monitored during the discharge and the pulse is reversed when it reaches 40% of the initial voltage. The IGBT chips of 1 cm^2 in size used in the ICD have a voltage rating of 1200 V and a peak surge current handling capability of 55 A [48].

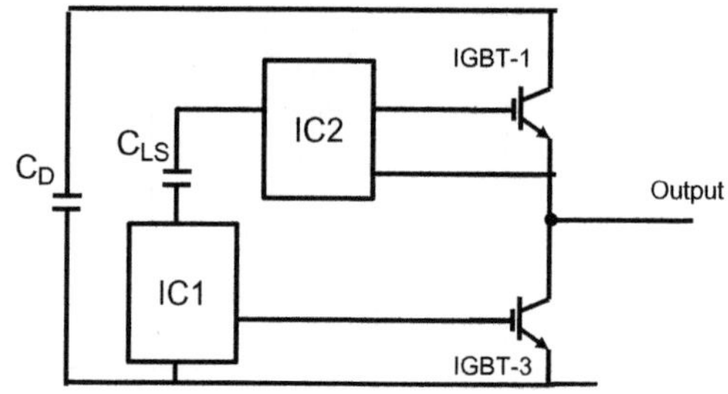

FIGURE 13.35

Level-shift circuit for driving insulated gate bipolar transistors in implantable cardioverter defibrillators.

The H-bridge circuit requires level shifting of the gate signal for IGBT-1 and IGBT-2 in Fig. 13.33. An innovative method to achieve this function is used in the ICDs, as illustrated in Fig. 13.35 [48]. The high-side level shifting is performed using capacitive coupling with capacitor C_{LS}. A high-frequency oscillator in IC1 drives the charge transfer via capacitor C_{LS} to IC2. A similar configuration is used for driving IGBT-2 in the H-bridge circuit. An H-bridge configuration for ICDs has also been described with multiple channels [49].

13.5.4 Cardioverter Defibrillator for Surgery

The onset of sudden arrhythmias is not uncommon during surgical operations. A low-energy cardioverter defibrillator powered by batteries is required in the operating rooms of hospitals. Defibrillators can be designed to operate with external application, as described in Section 13.5.1, or internally using the epicardial and endocardial configurations. Among these options, the surgeons prefer the epicardial approach because it requires less energy [50]. The biphasic truncated exponential waveform was selected for the design with a 3 ms positive voltage followed by a 1 ms pause and a 3 ms negative voltage application. Two 680 μF electrolytic capacitors were used in series with a combined charging voltage of 850 V. This allows discharging about 60 J of energy into the heart. The H-bridge utilized four STGH100N160S IGBTs. The system can charge the capacitor to 850 V in 15 s. The biphasic waveform generated by this circuit has a positive voltage of 220 V for 3 ms and a negative voltage of 180 V for 3 ms.

13.6 Medical Synchrotron

The medical synchrotron, such the one from Grenoble shown in Fig. 13.36, is used to generate light ion or proton beams for use in cancer therapy. Particle therapy for cancer consists of bombarding the tumor with high-energy ions. The damage to the DNA in the cancer cells causes their death because they are unable to repair the damage [51].

FIGURE 13.36

Synchrotron particle accelerator.

Protons and heavier ions exhibit the dose distribution illustrated in Fig. 13.37. It can be seen that dose increases with depth from the surface and has a sharp peak called the Bragg peak. The depth of the peak is determined by the particle energy (150 MeV in the example shown in Fig. 13.37). This allows targeting the particle bombardment to the tumor inside the human body while reducing the radiation dose received by healthy tissue around the tumor. This is better than radiation with x-rays whose dose decreases exponentially down from the surface.

In 2009, 28 synchrotron facilities were available around the world for cancer treatment and more than 70,000 patients have been treated with particle therapy [51]. This has grown to 60–70 synchrotron facilities around the world in 2020. More than 5000 patients have been treated using carbon ions because their higher density of ionization is more effective in destroying the tumor cells. These machines contain

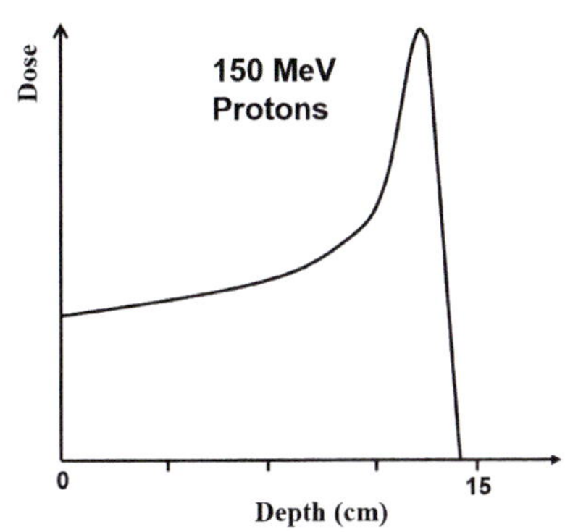

FIGURE 13.37

Ion dose distribution.

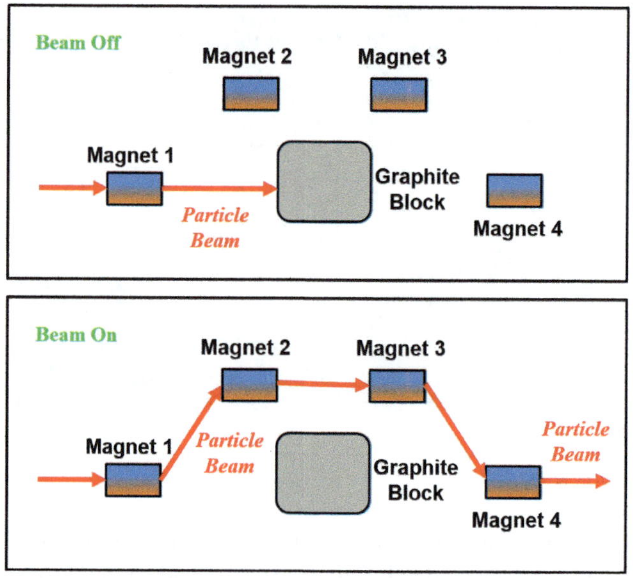

FIGURE 13.38

Ion particle deflection system in a synchrotron particle accelerator.

a linear accelerator (LINAC) to accelerate particles to an energy of 7 MeV. The particles are then injected into the synchrotron ring, which increases their energy from 60 to 250 MeV for protons and 120 to 400 MeV for carbon ions [52]. During treatment of cancer tumors, the particle beam must be scanned over its entire volume to be effective. This rastering is performed similar to an electron beam in a cathode ray tube. Powerful magnets in the path of the beam are used to deflect the particles over the tumor voxel. Once the proper dose has been delivered to a voxel, the beam must be moved to a new line. During this time, the beam must be blanked by using a graphite block, as illustrated in Fig. 13.38. This is accomplished by turning on several magnets that deflect the particle beam, allowing it to go around the graphite block and get delivered to the target. An IGBT-based power supply is used for driving the powerful magnets.

13.6.1 CNAO Magnet Coil Power Supply

The power supply topology used for driving the beam magnets in the Centro Nazionale di Adroterapia Oncologica (CNAO) in Pavia, Italy, [52] is shown in Fig. 13.39. It contains two DC power supplies formed by rectifying the output of two secondary windings of a transformer. One of these creates a high-voltage (2.9 kV) DC source labeled V_{HV}. The second of these creates a low voltage (60 V) DC source labeled V_{LV}. The R–L circuit represents the coils of all four beam magnets that are driven simultaneously by the IGBTs in the circuit. The inductance

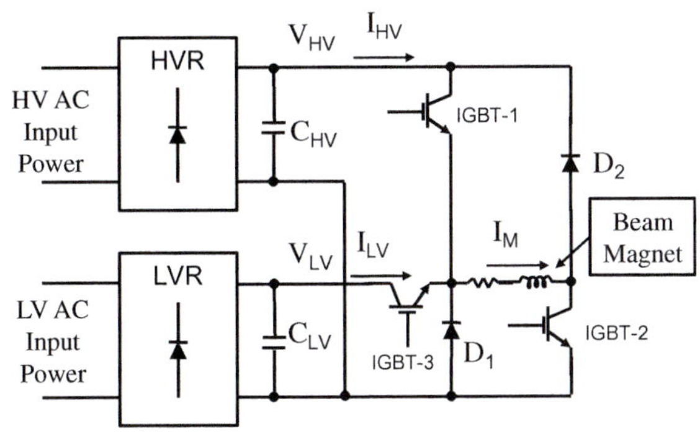

FIGURE 13.39

CNAO power supply for driving beam chopper magnets. *IGBT*, insulated gate bipolar transistor.

of the coils is about 1 mH. When all the IGBTs are in the off-state, no current flows through the magnet coils, resulting in no deflection of beam, which is then truncated into the graphite block. To initiate the particle beam treatment, IGBT-1 and IGBT-2 are turned on, resulting in the voltage from the high-voltage rectifier (HVR) supply (V_{HV}) applied to the beam magnets. The current in the magnet coils (I_M) and from the HV supply (I_{HV}) ramps up as determined by the inductance of the coils and the magnitude of V_{HV}, as shown in Fig. 13.40. When the current reaches the desired level (660 A in the example), IGBT-1 is turned off. The coil current

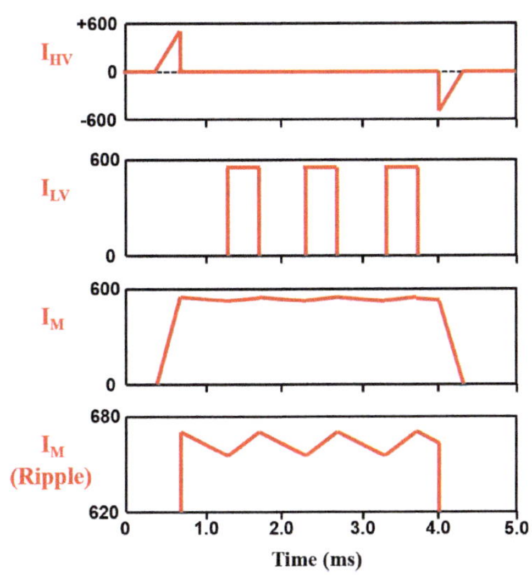

FIGURE 13.40

CNAO power supply waveforms.

circulates through forward biased diode D_1 and IGBT-2. The current in the coil decreases during this time. The second power supply low-voltage rectifier (LVR) is used to maintain a relatively constant current in the magnet coils with only a small ripple. To achieve this, IGBT-3 is switched on and off with a fixed duty cycle resulting in pulses of current, as shown in Fig. 13.40, from the low-voltage supply. The period and on-time for IGBT-3 can be adjusted to obtain any desired magnitude for the ripple current. In the example, the ripple current has a peak-to-peak magnitude of 20 A, as shown at the bottom of Fig. 13.40. When the radiation is to be stopped to blank the particle beam, all the IGBTs are turned off, preventing deflection of the particle beam. During the ramp-down of the magnet coil current, energy is partially returned to the high-voltage supply capacitor (C_{HV}). The power supply is designed to deliver nearly 2 MW of power to the magnet coils.

In this topology, IGBTs with 6.5 kV blocking voltage capability are used to withstand the high voltage and to provide sufficient reliability despite the synchrotron being operated continuously day and night. Because IGBT-1 carries current only during the ramp-up and ramp-down of the coil current, its continuous current rating can be 400 A. However, because IGBT-2 conducts current throughout the pulse duration, two 400 A rated IGBTs are required in parallel for this part of the circuit. IGBT-3 must not only have the same voltage and current ratings but also its switching losses should be low. A larger switching frequency for IGBT-3 reduces the ripple current but increases the power losses in the device.

13.6.2 GUNMA Magnet Coil Power Supply

The synchrotron used for cancer treatment at the Gunma University Heavy Ion Medical Center in Gunma, Japan, makes use of a different IGBT-based circuit topology for driving the deflection magnets for the particles [53]. The specifications for the power supply are a maximum output current of 2220 A and a maximum output voltage of 1460 V for a maximum output power of 3.2 MW. It delivers the current to the magnet coils with a tracking error of only 1×10^{-4} and a ripple of 1×10^{-5}. This is achieved by using an IGBT chopper circuit with four IGBTs rated at 1700 V and 400 A connected in parallel for each leg of the H-bridge.

The topology of the two-quadrant IGBT chopper is shown in Fig. 13.41. Each quadrant is provided power by a separate three-phase transformer. For the first quadrant chopper, the IGBTs of the three choppers that are working in parallel are gated with a phase of 0 degrees, 90 degrees, and 180 degrees. For the second quadrant chopper, the IGBTs of the three choppers that are working in parallel are gated with a phase of 45 degrees, 135 degrees, and 225 degrees. The two chopper blocks are connected in series for driving the current in the beam magnets.

Eighteen beam magnets are connected in series with the power supply. The switching frequency for the IGBTs in each of the choppers is 1.953 kHz. The phase-shifted quadruplex operation results in a combined frequency of 15.624 kHz. The common-mode ripples are canceled out with this topology. The

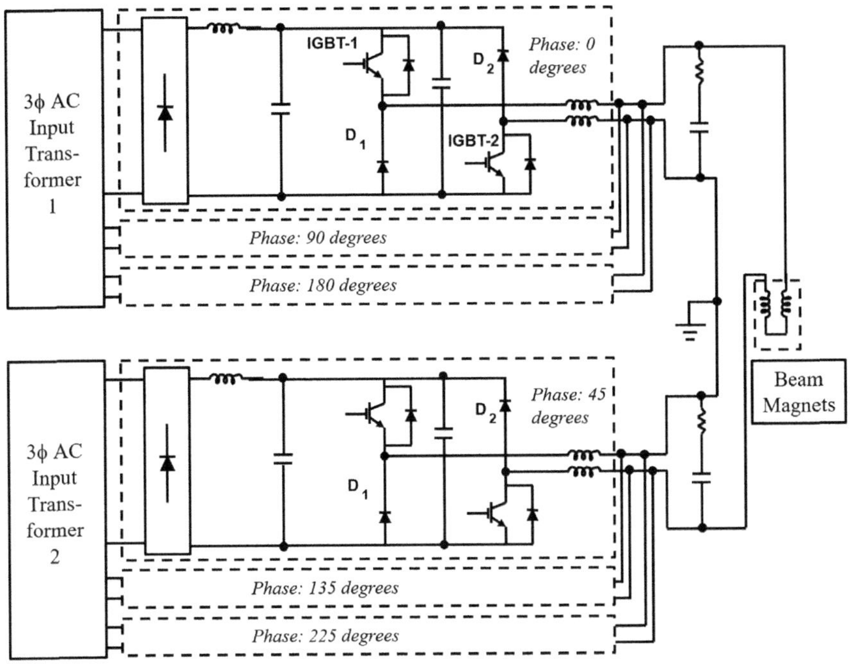

FIGURE 13.41

GUNMA power supply for driving beam chopper magnets. *IGBT*, insulated gate bipolar transistor.

R–C filter at the output of the power supply is designed to attenuate the 15.624 kHz ripple by 28 dB. A trapezoidal current with a flat top of 2220 A is delivered by this power supply with a slew rate of nearly 4000 A/s, that is, a current rise and fall time of 675 ms. The 3 MW IGBT chopper-based power supply has been in use for cancer treatment since March 2010.

13.7 Medical Lasers

Lasers are used by the medical community for a large variety of procedures. They include surgeries performed using a surgical scalpel such as the one shown in Fig. 13.42(a); the dental procedures using a laser scalpel shown in Fig. 13.42(b); and the LASIK eye surgery with the system shown in Fig. 13.42(c). Laser surgery employs a laser to make incisions in human tissue instead of the traditional steel scalpels used by surgeons [54]. A laser beam of the appropriate wavelength is absorbed by the water present in soft tissue, allowing its vaporization with sufficient laser intensity. Carbon dioxide, Er:YAG, excimer, argon, Nd:YAG, copper(II) bromide, and holmium lasers are employed for specific applications due to their different

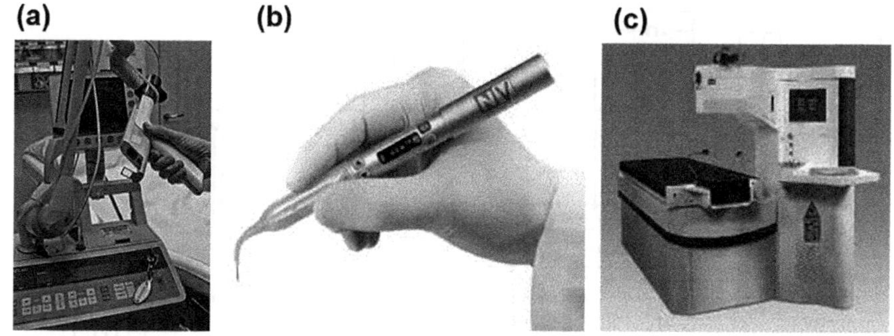

FIGURE 13.42

(a) Laser surgical scalpel; (b) laser dental scalpel; (c) LASIK eye surgical system.

wavelengths. Laser surgery offers many advantages [55]: (1) the laser energy reduces bleeding by sealing small blood vessels; (2) the laser reduces postoperative pain by sealing the ends of nerves; (3) it reduces infections because no sterilization is required; (4) surgeons get better visualization of the procedure because there is no blockage from the scalpel; and (5) optical fibers can be used to perform internal operations without opening the body.

Examples of the use of lasers in the medical field include the following:

(1) The LASIK eye surgery procedure used to correct myopia, hypermetropia, and astigmatism. More than 28 million LASIK operations had been performed around the world as of 2011 with 11 million of them in the United States [56]. 700,000 LASIK surgeries were performed in 2020. The excimer or exciplex laser is most commonly used for eye surgery [57]. It operates in the ultraviolet region at a wavelength of 172 nm, which is effectively absorbed by human tissue. The excimer laser removes tissue by disruption of the molecular bonds and discharges them into the air rather can causing burning. This leaves the surrounding tissue intact, allowing greater precision in the surgery especially for the eye.

(2) Glaucoma treatment [58] by laser trabeculoplasty, which burns areas of the trabecular mesh near the base of the iris to increase fluid flow, and by iridotomy to burn a hole in the iris near its base.

(3) Cataract treatment [59] by capsulotomy where a tiny incision is made in the crystalline lens of the eye to allow light rays to reach the retina.

(4) Dental surgeries and procedures [60] to cure gingivitis, bone cutting, and tooth drilling without local anesthesia. Nd:YAG lasers with wavelengths of 800–900 nm are the best for dental procedures because of absorption by red tissues.

(5) Dental tooth decay removal [61] using Er:YAG lasers operating at 2940 nm has been found to be effective without the need for a local anesthetic. The laser procedure eliminates the risk of microfractures produced by dental drills.

(6) Photorejuvenation [62] to remove wrinkles and scars in the skin. The Er:YAG laser operating at 2940 nm and the CO_2 laser operating at 10,600 nm have been found to be the most effective for this procedure.
(7) Lithotripsy [63] for the removal of kidney stones and gallstones by laser shock waves to break up the obstruction. The holmium laser operating at 2100 nm has been found to be the most effective for this procedure.
(8) Prostatectomy [64] for the removal of a part or the entire prostate gland in case of cancer with reduced blood loss. The holmium–Nd:YAG laser has been found to be the best for this procedure. It can also relieve benign prostatic hyperplasia symptoms.
(9) Transoral microsurgery [65] for the removal of small- and medium-sized tumors in the mouth or voicebox.
(10) Endovenous laser treatment [66] for the removal of varicose veins.
(11) Laser hair removal [67] using pulsed laser light to destroy hair follicles.
(12) Laser neck and back surgery [68] to relieve pain.

Lasers have even become popular tools for the removal of tattoos [69]. The tattoo pigments are removed by using Q-switched lasers. Black and dark-colored ink can be more readily removed with this technique.

The laser operates by stimulated emission achieved by creating a population inversion where the lasing medium has large numbers of particles in one excited state when compared to other particles in a ground state [70]. The most common method for producing the population inversion is by laser pumping using a flash lamp. The basic flash lamp configuration most often used in laser pumping has the elliptical cross-section illustrated in Fig. 13.43. The flash lamp is located at one of the foci of the ellipse while the laser medium is located at the other focal point. The lamp and lasing medium are water-cooled due to a large loss in energy transfer from the flash lamp to the laser medium. Xenon flash lamps are extensively used due to their high efficiency, but krypton flash lamps provide greater energy transfer efficiency for Nd:YAG lasers.

The flash or arc lamp requires a high voltage to first strike the arc with a triggering pulse. This is followed by a boost voltage to heat the gas to a plasma state. After this, a current control phase begins to stabilize the lamp current. The currents typically range from 10 to 50 A. Power supplies for the flash lamp employ IGBTs for controlling the voltage and current delivered to the flash lamp.

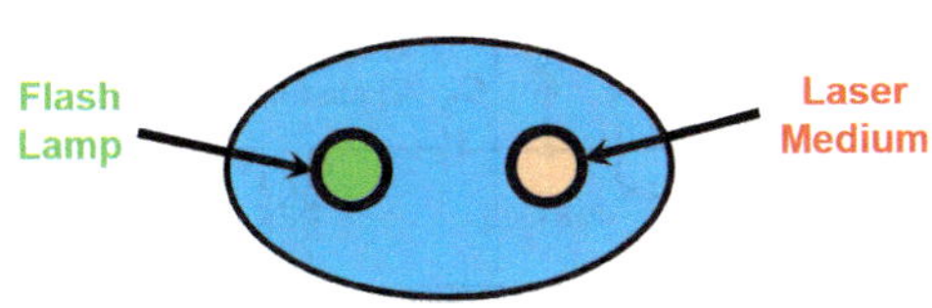

FIGURE 13.43

Laser pumping arrangement using flash lamp.

13.7.1 Pulse Compression Network Power Supply

Lasers used for surgery and other medical procedures require power supplies to pump the laser. Flash lamps are located at one end of the laser to pump the laser. The gas in the flash lamp must be ionized to generate the flash. The ionization is produced using a high-voltage pulse. A pulse compression network for triggering the lamp discharge [71] is shown in Fig. 13.44. The energy stored in the capacitor C_S is applied to the lamp by using the IGBT. An impedance matching network is used to control the discharge duration to 230 μs for an Nd:YAG laser. This power supply has been used for the Nd:YAG, argon, krypton, and carbon dioxide lasers used in dermatology. A pulse of several hundred volts is first delivered and converted at the laser head to higher voltage using a 20:1 trigger transformer. The energy storage capacitor C_S is then discharged into the simmering lamp using the pulse compression network.

Excimer lasers are used for photorefractive keratectomy and LASIK surgeries for correcting eyesight. An excitation pulse of 45 kV is generated using the pulse compression network shown in Fig. 13.44 [71]. This approach is also used for the erbium-YAG lasers for cataract surgery.

13.7.2 Capacitor Discharge Power Supply

Until the availability of the IGBT, the optical pumping of carbon dioxide lasers for medical applications was performed by using vacuum tubes and hydrogen thyratrons. The commercialization of IGBT modules having high-voltage and current-handling capability has replaced these technologies with a solid-state alternative with more precise control [72]. The new power supply topology, shown in Fig. 13.45, is based on the controlled discharge of energy stored in a capacitor into the flash lamp via an IGBT.

When the IGBT is in the off-state, the capacitor C_S is charged to the supply voltage (1000 V) via charging resistor R_1 with a low current flow in the transformer

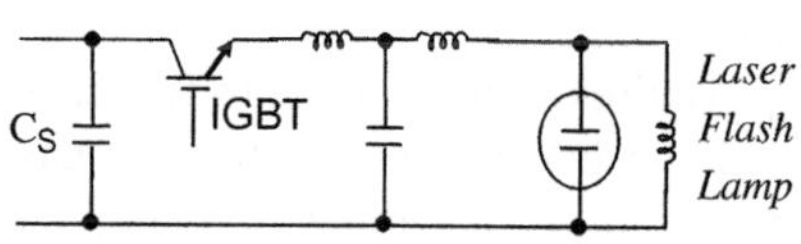

FIGURE 13.44

Pulse compression network for laser discharge. *IGBT*, insulated gate bipolar transistor.

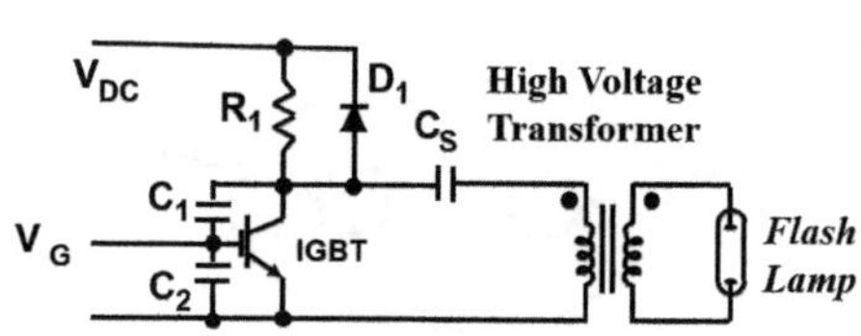

FIGURE 13.45

Pulse power supply for flash lamp used to pump lasers. *IGBT*, insulated gate bipolar transistor.

primary. When the IGBT is turned on with a positive gate bias pulse, it allows high current to flow from the capacitor into the primary of the transformer with a large dI/dt. The supply voltage is supported across resistor R_1 during this time. The secondary winding increases the voltage to 20–30 kV, producing the desired discharge of the flash lamp. Discharge pulses of a few microseconds are achievable with this scheme. The current in the primary of the transformer is about 100 A. Consequently, the IGBT used for this power supply has a voltage rating of 1200 V and a peak current rating of 100 A (Toshiba MG400Q1US51 product).

The inductive reactance of the primary coil of the transformer can produce a high ringing voltage across the IGBT that can cause its destructive failure. Diode D_1 is used to prevent this from happening by clamping the collector voltage to one diode drop above the V_{DC} (1000 V). The capacitors C_1 and C_2 are added to augment the internal IGBT capacitances and suppress fast voltage spikes at its gate electrode. It increases the rise and fall times for the gate voltage and limits the pulse width of the primary current.

The pulse power supply is able to deliver peak power of 1 kW with a pulse width of 10 μs, 3 kW with a pulse width of 4 μs, and 5 kW with a pulse width of 0.35 μs. Efficient laser operation was observed with a pulse length of 0.3–30 μs at a repetition rate of 100 Hz to a few kHz. A peak carbon dioxide laser output power of 300 mW was obtained with the IGBT-based power supply compared with only 50 mW with the previous thyratron design.

13.7.3 Series–Parallel Transformer Power Supply

Another approach to supplying the high voltage to the laser flash lamp has been developed for the CuBr-laser used in dermatology [73]. In this topology, multiple capacitor discharge power supplies are connected in parallel to drive the primary side of the transformer [74]. The secondary side outputs of the transformers are connected in series to generate the desired high voltage to drive the flash lamp in the laser. The circuit diagram is shown in Fig. 13.46.

Each of the IGBT pulse power supplies contains a storage capacitor (C_1, C_2, ⋯, C_n). The capacitor is charged up to the supply voltage V_{DC} (500 V) via diode D_1, D_2, ⋯, D_n. A single inductor L_1 allows for resonant charging of the capacitors. The voltage across each capacitor depends on the on-time for the IGBT because this decides the energy accumulated in the inductor L_1. The total voltage across the flash lamp is the sum of all the secondary voltages. Its value is given by $[n^*V_C^*T_n]$ where n is the number of IGBT cells operated in series, V_C is the voltage stored in each of the capacitors, and T_n is the turns ratio of the transformers.

13.8 Sterilization and Disinfection

The worldwide outbreak of COVID-19 in 2020 produced a major dislocation of society with the death of more than a million people as well as a massive impact on the economies of countries. In addition to vaccines that can suppress the spread of the

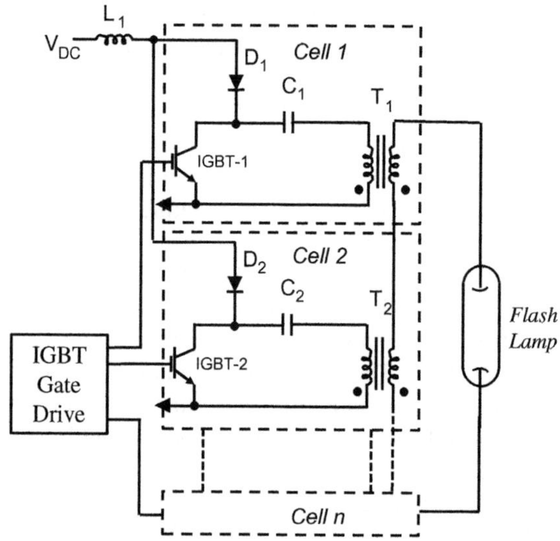

FIGURE 13.46

Series–parallel IGBT pulse power supply for flash lamp used to pump lasers. IGBT, insulated gate bipolar transistor.

virus, it is necessary to deploy technology for the disinfection of public spaces such as hospitals, offices, and mass transit systems. Germicidal ultraviolet (UV) radiation can be used to sterilize air we breathe, surfaces we touch, and personal protective equipment (PPE) needed by medical personnel when interacting with infected patients [75]. Germicidal UV radiation has a short wavelength of 280–100 nm, which is required to damage the DNA and RNA of viruses and germs. Because this radiation is harmful to humans, the radiation must be conducted in isolation in spaces such as the inside of airplanes and trains, or by directing the radiation to the ceilings of hospitals with air circulation. Robots with UV radiation emanating from lamps have been programmed to roam through an airplane to decontaminate all the exposed surfaces.

Germicidal UV disinfection equipment is built using mercury arc lamps to generate the radiation. An arc is induced in a quartz tube containing mercury that is vaporized by the heat. Electrons that are transported through the tube gain energy due to the high applied voltage. They interact with the mercury atoms to excite electrons into higher orbital shells. The relaxation of electrons back to lower orbital levels generates the desired UV radiation. The decontamination of surfaces from SARS-CoV-2 using UV radiating from LED lamps was also used during the 2020 pandemic [76].

The generation of UV radiation from the mercury arc lamp can be performed with a continuous wave or a pulsed wave. The pulsed method provides short bursts of UV radiation of higher energy. A full-bridge IGBT-based driver is shown in Fig. 13.47

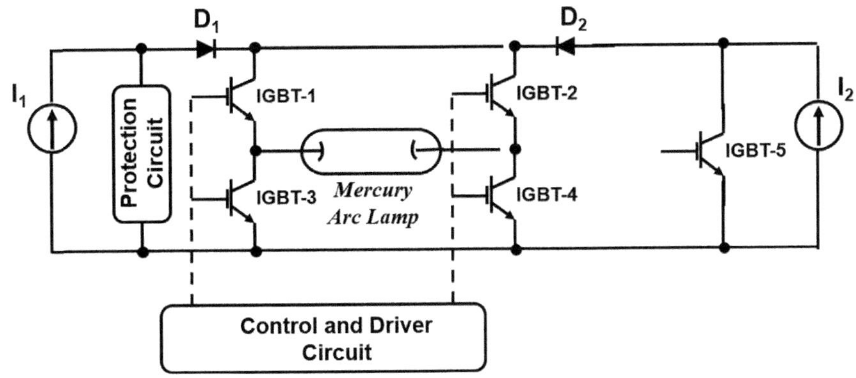

FIGURE 13.47

IGBT drive circuit for mercury flash lamp to generate UV-V radiation to disinfect surfaces.

for the mercury arc lamp [77]. Square-wave operation is achieved using IGBT 1–4. Pulse power operation is achieved with IGBT-5. The IGBTs are rated at 1200 V.

A mercury-free plasma (MFP) lamp for the generation of UV light has been reported using a single-phase AC-DC converter with unity power factor [78]. Two IGBTs are used on the primary side of a high-voltage transformer to produce a 5 kV pulse on the secondary side to discharge the lamp with 2.5 A delivered for generating UV radiation. The pulse repetition frequency is 25 kHz with a pulse width of 2 μs.

13.9 IGBT Design for Medical Applications

Suppliers of IGBTs for life critical applications such as implanted defibrillators must receive approval from the Food and Drug Administration for use in the United States. Several companies, such as IXYS and Microsemi, sell IGBT products with this qualification stated on their websites. Microsemi product MED427 is specifically targeted for ICDs. It contains a half-bridge multichip module with IGBTs in a miniature size suitable for ICDs. The IGBT chips, mounted using ball grid arrays, have a maximum voltage rating of 1000 V and can deliver 56 A of peak current. Two of these products can be combined to create the H-bridge ICD circuit shown in Fig. 13.33. A single IGBT is also available from Microsemi in die form as product MSAGA11F120D. It has a maximum blocking voltage of 1200 V and a peak current of 55 A.

IGBT replacement modules are available for various CT scanners, such as the part Gantry JEDI IGBT module, and for MRI scanners, such as the Mitsubishi Electric IGBT module NX series. The Mitsubishi Electric fifth-generation technology uses the carrier stored trench bipolar transistor (CSTBT) structure.

The Mitsubishi IGBT CSTBT structure optimized for use in x-ray and MRI machines [79] is shown in Fig. 13.48. A trench gate region is used in this design to increase the channel density and eliminate the junction field effect transistor

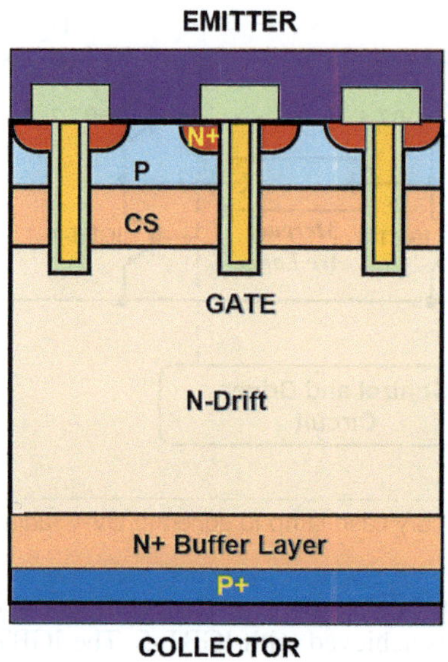

FIGURE 13.48

Insulated gate bipolar transistor optimized for MRI applications.

(JFET) resistance so that the on-state voltage drop can be reduced. The structure contains an N-buffer layer with the N-drift region optimized for non-punch-through operation under typical collector supply voltage conditions but with punch-through during the blocking voltage conditions. A carrier-storage (CS) layer is used to improve the carrier distribution near the emitter to reduce the on-state voltage drop. The trade-off curve between on-state voltage drop and the switching energy loss (given in mJ per ampere of collector current) for this chip design is provided in Fig. 13.49. The IGBTs for the medical applications are optimized using lifetime control to reduce the switching losses when compared with those optimized for industrial applications. They operate at a higher on-state voltage to bring down the energy loss per cycle, as shown in the figure.

13.10 Summary

Numerous applications of the IGBT in the medical sector have been discussed in this chapter. The examples demonstrate that the IGBT has taken a critical role in power delivery for major medical diagnostic tools such as CT scanners, MRI imagers, ultrasound machines, and lasers. The small size and weight of the power supplies for CT and MRI scanners are crucial to maintaining a reasonable size for these machines. These features are enabled by using IGBTs. In addition, IGBTs are used to precisely regulate the output of the power supplies and reduce their ripple current for generating improved images for examination by radiologists and surgeons. The

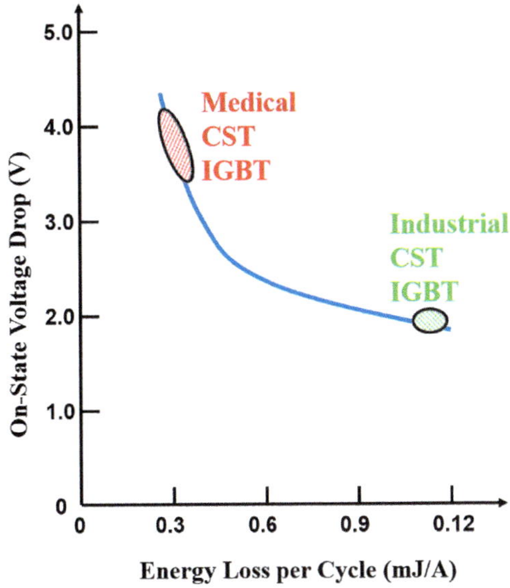

FIGURE 13.49

Mitsubishi Electric insulated gate bipolar transistor optimization for medical applications. *CST-IGBT*, carrier stored trench insulated gate bipolar transistor.

availability of IGBTs allowed creating portable defibrillators that are saving thousands of lives each year. IGBT-based power supplies are used for generating ultraviolet C radiation to disinfect surfaces to mitigate pandemics created by viruses. Another medical application enabled by IGBTs that was discussed in Chapter 11 is for endoscopy [80,81]. It is likely that IGBTs will continue to serve the medical profession far into the future.

References

[1] X-ray. en.wikipedia.org/wiki/X-ray.
[2] The US Medical Device Industry. http://selectusa.commerce.gov/industry-snapshots/medical-device-industry-united-states.
[3] Top 40 Medical Device Companies. http://www.mddionline.com/article/top-40-medical-device-companies.
[4] X-ray Tube. en.wikipedia.org/wiki/X-ray_tube.
[5] S.-S. Liang, Y.-Y. Tzou, DS control of a resonant switching high-voltage power supply for X-ray generators, in: IEEE International Conference on Power Electronics and Drive Systems, vol. 2, 2001, pp. 522–526.
[6] G. Majumdar, Advanced IGBT technologies for HF operation, in: IEEE European Power Electronics Conference, Keynote Speech, 2009.
[7] B.J. Baliga, Fundamentals of Power Semiconductor Devices, 2nd ed., Springer-Science, New York, 2019, pp. 876–883. (Chapter 9), Section 9.7.1.

[8] S. Iqbal, G.K. Singh, R. Besar, A dual-mode input voltage modulation control scheme for voltage multiplier based X-ray power supply, IEEE Trans. Power Electron. 23 (2008) 1003–1008.
[9] S. Iqbal, R. Besar, C. Venkataseshaiah, A novel control scheme for voltage multiplier based X-ray power supply, in: IEEE International Conference on Power and Energy, 2008, pp. 1456–1460.
[10] X-ray Computed Tomography. en.wikipedia.org/wiki/X-ray_computed_tomography.
[11] http://www.medwow.com/igbt-inverter-6509454/ct-scanner/593602739.item.
[12] http://www.medwow.com/igbt-dr-unit-53159006/ct-scanner/575721247.item.
[13] http://www.medwow.com/igbt-ct-unit-assy-left-5017456402/ct-scanner/226095562.item.
[14] J. Kociecki, T. Resnick, High Voltage Power Supply for X-ray Tubes, US Patent 5602897, Issued February 11, 1997.
[15] J.S. Katcha, et al., Multichannel Contactless Power Transfer System for a Computed Tomography System, US Patent 7054411, Issued May 30, 2006.
[16] How MRI Works. http://science.howstuffworks.com/mri1.htm.
[17] Magnetic Resonance Imaging. en.wikipedia.org/wiki/Magnetic_resonance_imaging.
[18] http://www.magnetica.com/page/innovation/todays-mri-market.
[19] https://www.mordorintelligence.com/industry-reports/global-mri-market-industry.
[20] S. Watanabe, et al., Analysis on a PWM power conversion amplifier with IGBT macro model to generate gradient magnetic fields in MRI systems, in: IEEE International Conference on Power Electronics and Drive Systems, vol. 1, 1999, pp. 127–132.
[21] H. Takano, S. Watanabe, M. Nakaoka, Multiple-bridge PWM current-regulated power amplifier for magnetic resonance imaging system and its feasible digital control implementation, in: IEEE Industrial Electronics Society Conference, vol. 2, 1999, pp. 785–790.
[22] J. Sabate, et al., High-Power High-Fidelity Switching Amplifier Driving Gradient Coils for MRI Systems, vol. 1, 2004, pp. 261–266.
[23] J. Sabate, Y. Liu, M. Wiza, Power supply with independently regulated multiple outputs, in: IEEE European Conference on Power Electronics and Applications, 2007, pp. 1–8.
[24] P. Zhu, Y. Liu, J. Sabate, Multi-output power supply with series voltage compensation capability for magnetic resonance imaging system, in: IEEE Power Electronics and Motion Control Conference, 2009, pp. 993–997.
[25] M. Ristic, et al., Supercapacitor energy storage for magnetic resonance imaging systems, IEEE Trans. Ind. Electron. 61 (2014) 4255–4264.
[26] Medical Ultrasonography. en.wikipedia.org/wiki/Medical_ultrasonography.
[27] Ultrasound Systems Market to 2019. http://www.marketresearch.com/GBI-Research-v3759/Ultrasound-Systems-Technological.html.
[28] https://www.globenewswire.com/news-release/2019/12/11/1959425/0/en/Ultrasound-Equipment-Market-Size-to-Reach-US-10-Bn-by-2026.html.
[29] M. Helguera, An Introduction to Ultrasound. http://222.cis.rit.edu/research/ultrasound/ultrasoundintro/ultrasoundintro.html.
[30] P.M. Gammel, G.R. Harris, IGBT-based kilovoltage pulsers for ultrasound measurement applications, IEEE Trans. Ultrason. Ferroelectr. Freq. Control 50 (2003) 1722–1728.
[31] Sudden Cardiac Arrest: A Healthcare Crisis. http://www.sca-aware.org/about-sca.
[32] A. Gonzallez, Top Defibrillators Makers in the region, Vying to Dominate the Market. http://seattletimes.com/html/businesstechnology/2004006644_defibrillators.

[33] T. Morva, Defibrillator Manufacturers. http://ezinearticles.com/?Defibrillator-Manufacturers&id=428920.
[34] Heart. en.wikipedia.org/wiki/Heart.
[35] F. Pratama, M. Haryanti, Y. Dewanto, The design defibrillators based on AT89C51 microcontroller, in: IEEE International Conference on Electrical Engineering and Informatics, 2011, pp. A2.1–A2.7.
[36] Electrical System of the Heart. http://www.webmd.com/heart/tc/electrical-system-of-the-heart.
[37] Ventricular Fibrillation. en.wikipedia.org/wiki/Ventricular_fibrillation.
[38] Atrial Fibrillation. en.wikipedia.org/wiki/Atrial_fibrillation.
[39] Defibrillation. en.wikipedia.org/wiki/Defibrillation.
[40] N. Thongpance, T. Kaewgun, R. Deepankaew, Design and construction of the low-cost defibrillator analyzer, in: IEEE International Conference on Biomedical Engineering, 2013, pp. 1–4.
[41] Implantable Cardioverter-Defibrillator. en.wikipedia.org/wiki/Implantable_cardioverter_defibrillator.
[42] The 11th World Survey of Cardiac Pacing and Implantable Cardioverter-Defibrillators, 2009. http://www.ncbi.nlm.nih.gov/pubmed/21707667.
[43] https://www.statista.com/statistics/731763/cardiac-pacemaker-units-sales-worldwide/.
[44] W.B. Kouwenhoven, G.G. Knickerbocker, The development of a portable defibrillator, AIEE Trans. Power Appar. Syst. Part III (1962) 428–431.
[45] M.W. Kroll, K. Kroll, B. Gilman, Idiot-proofing the defibrillator, IEEE Spectr. 45 (2008) 40–45.
[46] J. Ramon, et al., Energy generation and discharge for a semiautomatic defibrillator EBT, in: IEEE Electronics, Robotics, and Automotive Mechanics Conference, 2006, pp. 220–225.
[47] J.A. Warren, et al., Implantable cardioverter defibrillators, Proc. IEEE 84 (1996) 468–479.
[48] F. Lynch, Current Design Trends in Medical Electronics. www.mdci.com.
[49] D.J. Dosdall, D.E. Rothe, J.D. Sweeney, Programmable arbitrary waveform generator for internal defibrillation research, in: IEEE International Conference on Engineering in Medicine and Biology Systems, 2004, pp. 3971–3974.
[50] Y. Zhou, et al., A special cardioverter-defibrillator in surgery, in: IEEE Bioinformatics and Biomedical Engineering Conference, 2007, pp. 1061–1064.
[51] Particle Therapy. en.wikipedia.org/wiki/Particle_therapy.
[52] E. Dallago, et al., The power supply for the beam chopper magnets of a medical synchrotron, in: IEEE Annual Conference on Industrial Electronics, 2008, pp. 1016–1020.
[53] C. Yamazaki, et al., Development of a high-precision power Supply for bending electromagnets of a heavy ion medical accelerator. Paper WeP1-098, in: IEEE International Conference on Power Electronics, 2011, pp. 3013–3016.
[54] Laser Surgery. en.wikipedia.org/wiki/Laser_surgery.
[55] R. Ariele, The Laser Adventure, Section 9.2.1 Lasers in Medical Surgery. http://www.um.es/LEQ/laser.htm.
[56] LASIK. en.wikipedia.org/wiki/LASIK.
[57] Excimer Laser. en.wikipedia.org/wiki/Excimer_laser.
[58] Glaucoma Surgery. en.wikipedia.org/wiki/Glaucoma_surgery.
[59] Capsulotomy. en.wikipedia.org/wiki/Capsulotomy.

[60] Dental Laser. en.wikipedia.org/wiki/Dental_laser.
[61] Er:YAG Laser. en.wikipedia.org/wiki/ER:YAG.laser.
[62] Photorejuvenation. en.wikipedia.org/wiki/Photorejuvenation.
[63] Lithotripsy. en.wikipedia.org/wiki/Lithotripsy.
[64] Prostatectomy. en.wikipedia.org/wiki/Prostatectomy.
[65] Transoral Laser Microsurgery. en.wikipedia.org/wiki/Transoral_laser_microsurgery.
[66] Endovenous Laser Treatment. en.wikipedia.org/wiki/Endovenous_laser_treatment.
[67] Laser Hair Removal. en.wikipedia.org/wiki/Laser_hair_removal.
[68] Laser Spine Institute, Minimally Invasive Spine Surgery. http://www.laserspineinstitute.com/.
[69] Tattoo Removal. en.wikipedia.org/wiki/Tattoo_removal.
[70] Laser Pumping. en.wikipedia.org/wiki/Laser_pumping.
[71] M. Iberler, K. Frank, Small-size pulsed lasers in medical applications, in: IEEE Power Modulator Symp, 2002, pp. 608–611.
[72] A. Bertolini, et al., Solid-state power supply for gas lasers, Rev. Sci. Instrum. 75 (2004) 2686–2691.
[73] Laser in Dermatology. http://www.dermanetnz.org/procedures/lasers.html.
[74] S.N. Torgaev, M.V. Trigub, F.A. Gubarev, Studying of solid-state power supply unit of CuBr-laser, in: IEEE International Conference on Micro/Nanotechnologies and Electron Devices, 2011, pp. 411–414.
[75] M. Anderson, The ultra-violet offense, IEEE Spectr. (2020) 51–55.
[76] F. Juarez-Leon, et al., Design and implementation of a germicidal UVC-LED lamp, IEEE Access (2020) 196951–196962.
[77] M. Mustapha, et al., Effect of a pulsed power supply on the ultra violet radiation and electrical characteristics of low pressure mercury discharge, in: IEEE Int. Design and Test Workshop, 2008, pp. P1–P6.
[78] S. Ram, et al., Simulation and analysis of pulse power supply for mercury-free plasma UV-lamp, in: IEEE India Int. Conf. Power Electronics, 2018, pp. P1–P5.
[79] J. Yamada, Y. Yu, Y. Ishimura, Low turn-off switching energy 1200-V IGBT module, in: IEEE Industrial Applications Conference, 2002, pp. 2165–2169.
[80] Endoscopy. https://en.wikipedia.org/wiki/Endoscopy.
[81] H. Wehby, et al., 100-watt MATLAB/Simulink model LED based light source for medical endoscopy, in: IEEE Conf. Emerging Smart Technologies and Applications, 2021, pp. P1–P6.

CHAPTER

IGBT Applications: Defense

14

The defense or military industry is one of the largest sectors of the economy for most countries. The defense industry is involved in the development and production of equipment needed by the air force, navy, army, and other services [1]. The United States Department of Defense (DoD) has an annual budget of more than $700 billion. Its budgetary resources for fiscal year 2010 were $1.2 trillion [2]. It is also a huge consumer of electricity with almost 30,000 gigawatt hours used annually.

The global expenditures by the military around the world were more than $1.5 trillion in 2010 [3]. The spending exceeded $2 trillion in 2021 with 38% spent by the United States, 14% spent by China, and 16.7% spent by non-US NATO countries in Europe [4]. The top 100 largest arms producers had sales of $315 billion in 2006, which increased to $531 billion in 2020 [5]. Arms sales include airplanes, missiles, aircraft carriers, submarines, armored vehicles, etc. The leading arms exporters are Russia, the United States, China, France, and the United Kingdom, according to the Stockholm International Peace Research Institute. The top US defense companies include Lockheed Martin, Boeing, BAE Systems, Northrup-Grumman, Raytheon, and General Dynamics.

Insulated gate bipolar transistors (IGBTs) are extensively used today by the defense industry for power delivery and management, propulsion systems, and weapon systems. The defense sector of the United States embraced IGBT technology rather reluctantly. A consortium group was formed in 1986 to coordinate funding of power semiconductor technology in the United States between the defense agencies and the Electric Power Research Institute (EPRI) [6]. Its stated goal was: "*To restore the US leadership in the power semiconductor field through coordinated investments in research*." The authors state: "*The MOS-controlled thyristor (MCT) was exclusively chosen by the consortium for funding*." The primary basis for this decision was the low on-state voltage drop of the MCT because of operation as a latched-up four layer structure. At that time, it was well known that thyristor structures, such as the gate turn-off thyristor (GTO), require bulky snubbers and do not provide the current control available in transistor structures. The MCT was also found to lack the features available in the IGBT in terms of the forward and reverse biased safe operating area. These disadvantages offset any benefit from the lower on-state voltage drop of the MCT when compared with the IGBT. Despite tens of millions of dollars in investments by EPRI and DoD, the commercialization of the MCT was not successful because of technical difficulties in manufacturing the device.

The IGBT Device. https://doi.org/10.1016/B978-0-323-99912-0.00015-5

Meanwhile, the superior attributes of the IGBT from a systems perspective as well as its compatibility with power MOSFET manufacturing allowed its rapid commercialization for consumer and industrial applications without support by DoD and EPRI. With widespread commercial adoption of the IGBT in other sectors of the economy and the establishment of a strong worldwide production base for IGBTs by the semiconductor industry, it eventually became impossible for the defense sector to not take advantage of this technology. In 2003, the members of the consortium group concluded [7]: "*Given the market strength of IGBTs, improvements continue to be made in terms of ratings and packaging...further strengthening their market position against thyristor devices. Transistors (IGBTs) generally have superior switching performance, in terms of faster switching and low gate drive power.*" In 2004, members of the consortium group concluded [8]: "*The IGBT has progressed to become a device of primary choice, replacing the GTO, a thyristor device, in a wide range of low- and medium-power applications going up to several megawatts and even competing up to a few hundred megawatts....It has fast switching capability and controlled di/dt during turn-on. These are a major advantage for small- to medium-sized converters and even high-power converters...However, the MOS turn-off thyristor (MTO) has not yet been commercially introduced.*"

The greatest emphasis for migration to solid-state devices in military applications was generated by the US Navy through the power electronic building block (PEBB) program. The PEBB was defined as a universal power processor that converts any input electrical power to the desired voltage, current, and frequency output [9]. Although the concept is device agnostic, it initially targeted the use of MCTs. However, it was eventually implemented using IGBTs as the power devices within the PEBB. After discussion of the PEBB concept, this chapter will cover the use of IGBTs in the Navy for ships, aircraft carriers, and submarines; in the Air Force for fighter jets; in the Army for hybrid electric vehicles (HEVs); in electromagnetic launchers for projectiles; and in the missile defense network.

14.1 Power Electronic Building Blocks

The PEBB concept was proposed and promoted by the US Navy in 1995. It was considered a crucial technology for the Navy with implications for the entire power electronics industry [10]. The PEBB concept is illustrated in Fig. 14.1. It contains a power processing unit in the middle with sensing and control blocks shown on three sides. The PEBB is a plug-and-play architecture that could be used in applications with all its operational settings automatically configured. The end application would recognize that the PEBB is plugged in, identify the manufacturer, and automatically operate it. Each power processing unit would manage its own safe operating limits. Standard interfaces and protocols would be employed to allow use in a wide range of applications. These features would lower the cost for defense and commercial use.

The PEBB concept is revolutionary because past practices in power electronics were based on custom designs for each application, requiring significant engineering

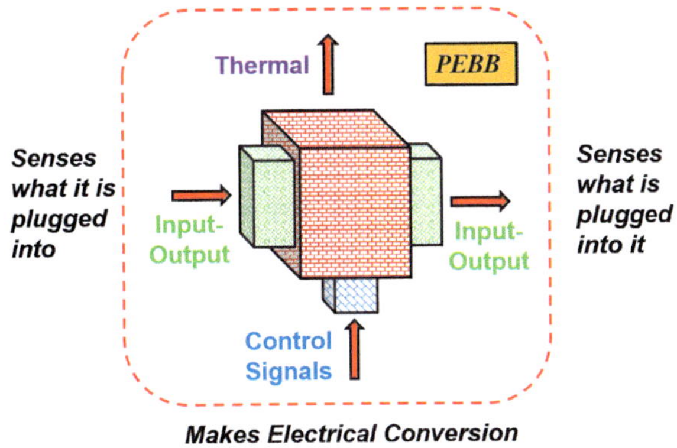

FIGURE 14.1

Power electronic building block concept.

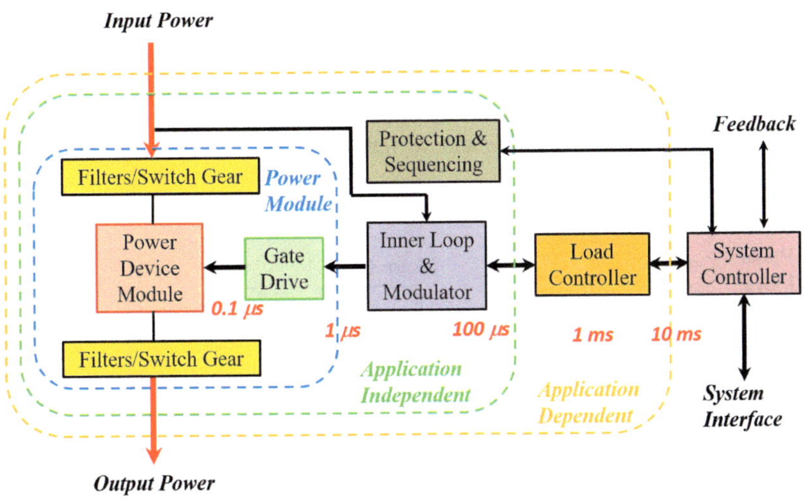

FIGURE 14.2

Hierarchical control of power electronic building block.

expertise, design effort, and development time. Time to market can be shortened using the PEBB concept and new semiconductor technologies can replace older ones within the power processing unit without reengineering the entire system. The hierarchical architecture of the PEBB concept is illustrated in Fig. 14.2. The power module contains the power switches, filters, and control circuits. It can also be partitioned into cells. At power levels of less than 100 kW, the cell can be an entire three-phase inverter. At power levels of 100 kW to 1 MW, the cell can be a phase leg. Sufficient intelligence with control circuits is included within each PEBB to allow them to be "snapped together" to serve a variety of applications. The control architecture is hierarchical: The gate driver controls the time domain of less than 10 μs; the circuit topology controller controls the time domain between 10 μs and 1 ms; the load controller controls the time domain between 1 ms and 1 s; and the system controller regulates operation beyond 1 s. The output of the PEBB is software configured and controlled based on the needs of the end users. This allows manufacturing high-volume products that can serve a variety of applications with minimal engineering effort.

A naval ship power system implemented using the PEBB concept is illustrated in Fig. 14.3 [9]. The ship is powered from a variety of sources to maintain a high redundancy and reduce vulnerability to faults and attacks. The main power distribution system utilizes a DC bus, which simplifies the ship builder's complexity and the end user's interface management. DC power distribution systems have become practical with the availability of the IGBT capable of switching in the microsecond regime, allowing accurate monitoring and control of energy flow. The PEBBs in the system control all the energy sources and loads. The sources include prime movers (turbines and diesel engines), fuel cells, batteries, and fly wheel storage units. The loads include the propellers for locomotion, auxiliary electronics, and pulse power weapons. High-frequency transformers are used for the DC to DC and AC to DC conversion to reduce their size and weight. Efficient power conversion requires IGBTs with low on-state and switching losses.

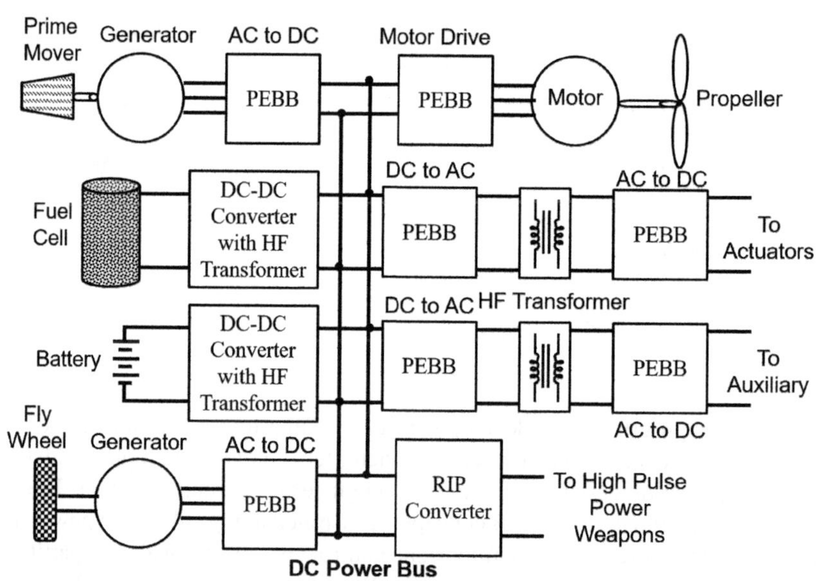

FIGURE 14.3

Naval-ship power system using power electronic building blocks.

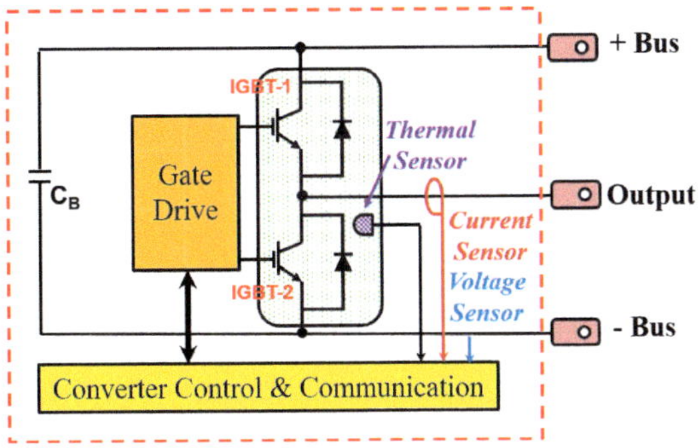

FIGURE 14.4

Universal power converter PM1000 power electronic building block design.

The Navy has funded the development of PEBB technology and encouraged manufacturers to release commercial products. In 2005, a PEBB product called LoPak5 was made available [11] by ABB with 1200 V/1700 V rated IGBTs. The maximum output for the PowerPak3 unit containing the LoPak5 IGBT modules was 600 kVA with water cooling. In 2006, a universal power converter PM1000 IGBT-based PEBB module was reported with 0.5 MVA power handling capability [12]. The configuration of this module is shown in Fig. 14.4. In addition to the IGBTs, it contains the gate drivers and controls. Thermal, current, and voltage sensors are included for rapid local protection of the devices.

Before the development of the PEBBs using IGBTs, a 60 Hz AC power distribution system was being used in naval vessels [13]. A major problem with this approach was the size and weight of the transformers. Transformers were used at each load in the marine system for galvanic isolation. They were of the same size and weight as the propulsion motors at the point of use. The use of IGBT-based PEBBs to make DC to AC converters to generate high-frequency AC power enabled the use of high-frequency transformers, which are much smaller in size and weight.

14.1.1 PEBB-1, PEBB-2, and PEBB-3

The first phase of PEBB demonstration was a five-port device whose schematic drawing is shown in Fig. 14.5. The five ports consist of the two (+ and −) input power ports and the three (A, B, C) output power ports. In addition, the PEBB-1 contains a communication bus, an analog bus, and a digital bus. It could be externally programmed to perform the following functions: (1) DC to AC inverter; (2) AC to DC converter; (3) DC to AC motor controller; (4) AC to DC boost converter; and (5) DC to DC boost converter [14].

The PEBB-2 developed in the second phase defined the packaging requirements, including not only thermal issues, but also electromagnetic interference (EMI), interconnection, sensors, and passive elements. It incorporated faster switching devices

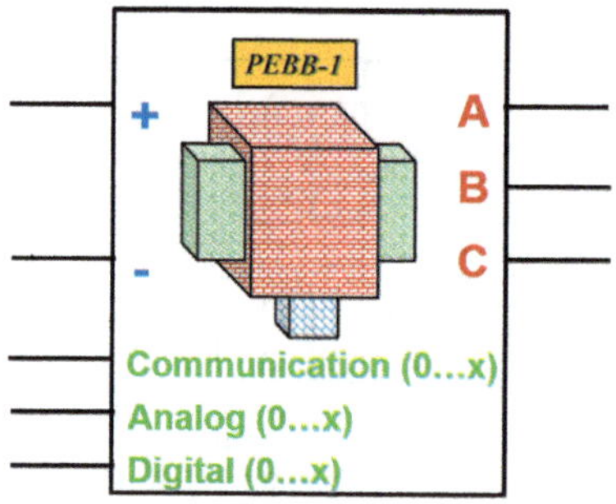

FIGURE 14.5

Five-port power electronic building block-1 demonstration.

operating at higher power levels than for PEBB-1. The PEBB-3 effort was focused on improving the form, fit, and function. It utilized two-sided cooling of the semiconductors, even faster switching devices, and distributed control.

14.1.2 Naval Frequency Changers

Radars in naval ships require pulse power exhibiting a nonlinear load to the power supply. Naval vessels use frequency changers for power modulator applications such as radar systems [14]. They convert 60Hz distributed power to 400Hz. The frequency changer isolates the radar load from the power distribution to the rest of the ship. Electrical filters constitute two-thirds of the size and three-quarters of the weight in shipboard equipment. The size and weight can be significantly reduced by using a higher operating frequency. However, this requires devices that can switch at higher frequencies with low losses. Zero-current and zero-voltage resonant circuit topologies also allow reducing switching losses, enabling higher operating frequencies. A PEBB-based frequency changer was able to reduce the filter size by 40% while cutting costs by 35–40% as well. A power density goal of 1.8 MW/m^3 was achieved by 1998.

14.1.3 Shunt Active Power Filter

An active filter with compact size and low weight has been designed using the PEBB concept for shipboard use in naval vessels [15]. Conventional passive filters offer poor performance for naval vessels due to not being adaptive and working over a narrow band of frequencies. An active filter is adaptive and has the flexibility of updating. The active filter contains a power electronic converter, a real-time controller, and sensors. It was implemented using a commercially available PEBB from American Superconductor (PM-1000) with a high-power density of 130 W/cubic inch. The PEBB contains three intelligent IGBT modules with ratings of 1200V and 180 A. A current-sensing loop was added to the module to create the active filter.

14.1.4 Three-Level ANPC-VSC PEBB

Advances to PEBB capability have been possible with the availability of IGBT modules capable of handling higher power levels. One example is the three-level neutral-point clamped voltage source converter (3L-NPC-VSC) that is widely used for high-power, medium-voltage (MV) applications with the multilevel VSC topology. This topology is popular for high-power, medium-voltage industrial drives for rolling mills, fans, pumps, conveyors, marine propulsion systems, and mining [16]. The 3L-NPC-VSC topology is illustrated in Fig. 14.6 with 12 IGBTs. Each phase leg is designed to deliver 1000 A rms, with $2 \times 2.8\,\text{kV}$ on the DC link side and 3.3 kV rms between the three AC phases. Trench gate 4.5 kV, 1.5 kA IGBT modules built by Hitachi were used. Ultrasoft fast recovery diodes were used as fly-back diodes across each IGBT. The advantages of creating a PEBB stated by the authors are: (a) improved reliability and manufacturability; (b) robust, fast power connectors; and (c) reduced commissioning time due to standardization. The distribution of energy loss was 35% IGBT turn-on loss, 38% IGBT turn-off loss, and 27% diode reverse recovery loss.

14.1.5 PEBB for More-Electric Aircraft

Electric propulsion drives for more-electric aircraft require power converters that are lightweight and highly efficient. Standardization using the PEBB concept improves reliability, cost, and maintenance. A 100 kW PEBB design has been created using a

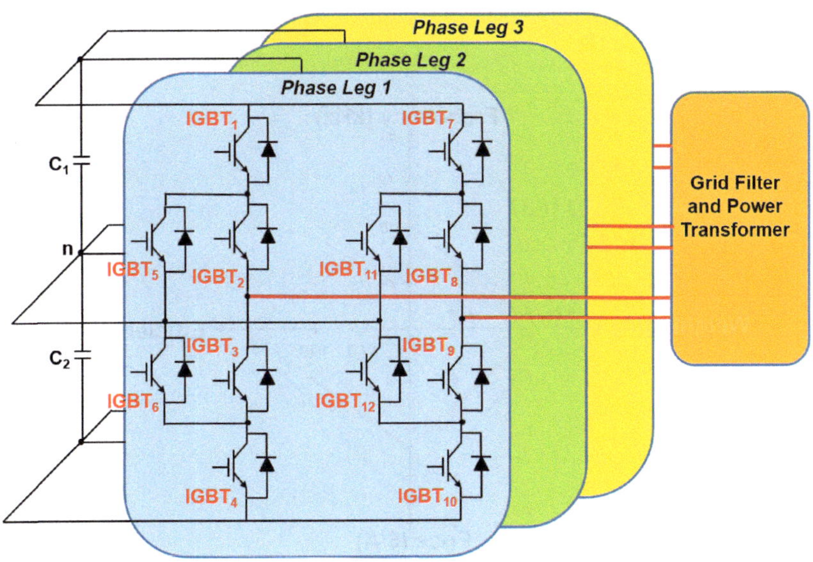

FIGURE 14.6

The three-level neutral-point clamped voltage source converter.

hybrid switch. The hybrid switch consists of an Si IGBT in parallel with an SiC power MOSFET. In the hybrid switch concept, the on-state losses are reduced by using the Si IGBT while the switching losses are reduced by using the SiC power MOSFET. This requires that the SiC MOSFET be turned-on and turn-off before the Si IGBT (see Chapter 10, Fig. 10.56). The three-level t-type topology (3L-T^2-NPC) is shown in Fig. 14.7, implemented using hybrid switches [17]. Each hybrid switch has a fly-back diode as well. The advantage of building this topology with the hybrid switch can be observed on the "radar" diagram shown in Fig. 14.8. The lowest cost is achieved using the pure Si IGBT approach but its efficiency is limited by switching losses to about 97%. The all SiC MOSFET approach increases the efficiency to 99% due to reduced switching loss but its cost is much larger. The hybrid switch provides the same high efficiency of 99% with a cost reduction of 30%. The resulting lightweight PEBB design with the hybrid switch has a specific power

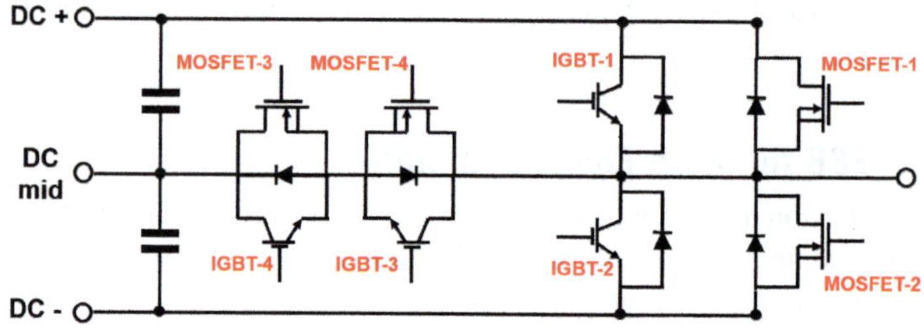

FIGURE 14.7

The three-level t-type topology implemented with hybrid Si IGBT and SiC power MOSFET devices.

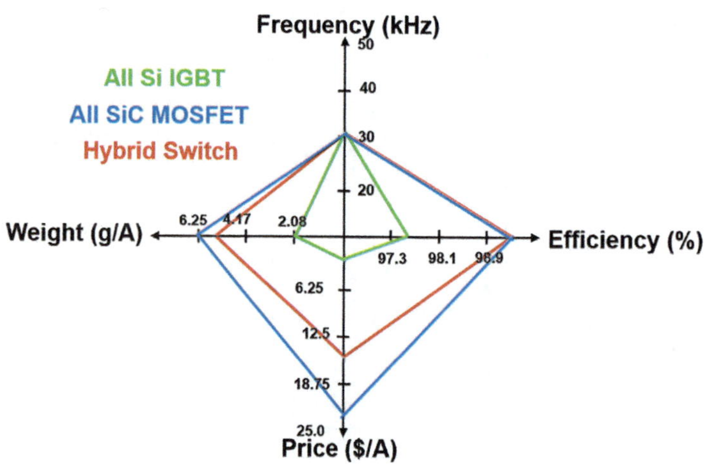

FIGURE 14.8

Radar diagram comparing performance of three switch options.

density of 27.2 kW/kg and a volumetric power density of 309 W/in^3 (18.8 W/cm^3). This is beneficial to the more-electric aircraft application.

14.2 The Electric Warship

Frigates are warships of various sizes, with a typical vessel shown in Fig. 14.9. Warships of the frigate capacity are used for antisubmarine warfare to protect merchant ship convoys, provide escort for aircraft carriers, and support amphibious assault forces [18]. These warships operate at a maximum nautical speed of 30 knots but are normally cruising at a speed of 18 knots [19] (for landlubbers: 1 knot equals 1.15 mph). The propulsion system must move a 4000 ton vessel at these speeds.

By 1998, an electrical propulsion system using IGBTs was already considered favorably by the British Royal Navy [19]. The existing mechanical power transmission system had eight separate prime movers. This could be reduced to four with an integrated full electric propulsion (IFEP) system. The projected cost reduction with IFEP was 37% for antisubmarine warfare and 30% for carrier escort. The authors compared various power semiconductor devices available and under development at that time. The GTO was found to be unattractive even though it had adequate power ratings because of the cumbersome gate drive. In contrast, the IGBT was regarded favorably due to the simplicity of its gate drive and its larger operating frequency. The authors state: "*Their major advantage is that they are voltage driven and hence the control circuits are much simpler with commensurate reduction in weight and volume. The IGBT can also be switched at frequencies greater than 10 times that of the GTO, in the order of 20 kHz. This has the effect of reducing the amount of filtering required and gives further reduction in overall weight and volume.*"

The "total electric ship" has also been of interest to the United States Navy [20]. The notional navy littoral combat ship requires a total propulsion and service power of 15,000–20,000 kW. The naval ships require a power density of 10 times that used

FIGURE 14.9

The frigate warship.

for cruise ships. In addition, the electric warship requires a power density of 20 times that for a cruise ship when pulse power requirements are taken into consideration.

14.2.1 Propulsion Drive Options

The migration to electric propulsion for warships requires the determination of the best circuit topology and power semiconductor technology to meet the needs of the variable speed motor drives. A careful analysis of the available options has been performed by American Superconductor [21], which was tasked with the development of superconductor-based motors and generators for shipboard applications. The three topologies evaluated were: (1) the cycloconverter; (2) the series-connected low-voltage inverters; and (3) multilevel inverters.

Cycloconverter: The cycloconverter has the advantage of performing direct AC to AC power conversion. It makes use of self-commutated thyristors with reverse blocking capability. This approach was ruled out due to poor power quality causing flicker noise, control problems with motor current distortion, significant motor voltage ripple at the line frequency, and low power factor for the distribution system. The torque ripple increased from 1% at 100% speed to 25% at 50% speed.

Series-connected low-voltage inverters: In 2005, 1200 V IGBTs were available from several manufacturers and had been proven to be reliable high-performance devices with efficient switching at 4–8 kHz for commercial motor drives [21]. Using this technology, it is possible to configure a motor drive for ship propulsion from a 9 kV DC bus by placing multiple IGBT modules in series, as illustrated in Fig. 14.10. Each DC-DC converter provides an isolated 800 V bus to the DC-AC inverter stages containing the 1200 V-rated IGBTs in the standard H-bridge configuration. The output of the inverters is connected in a wye configuration for driving

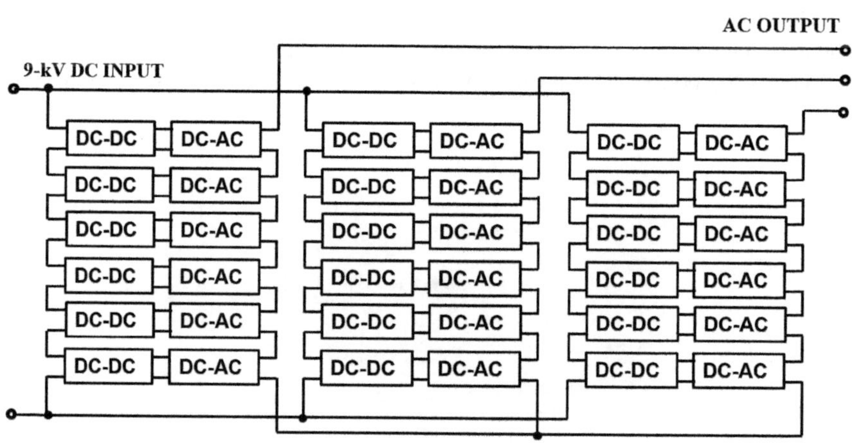

FIGURE 14.10

Series connected low-voltage inverter drive.

the motor. A 6 kV motor can be operated using a nine-phase inverter arrangement with 126,300 kW inverter modules. The 600 kW DC-DC converters make use of two high-frequency transformers driven by a quasiresonant H-bridge inverter with IGBTs. The DC-AC inverter uses 300 kW IGBT modules operating at a pulse width modulation (PWM) carrier frequency of 8 kHz.

This approach has the advantages of using standard commercially available IGBTs to create a common building block for all inverters in naval vessels. Their high switching frequency reduces motor noise. However, it does not allow regenerative operation. The motor current has a third harmonic distortion of only 0.6% and the torque ripple at full speed and load is only 0.05% of the average torque independent of motor speed and load. This is the best performance among the topologies. It is able to deliver an efficiency of 96.8%.

Multilevel inverter: The diode-clamped topology, shown in Fig. 14.11, was evaluated for naval ship propulsion. 6.5 kV IGBTs could be used for operation from a 3.6 kV DC bus. PWM control of the IGBTs with a carrier frequency of 1 kHz results in an IGBT switching frequency of 0.4 kHz, reducing switching losses. The IGBTs were compared with insulated gate controlled thyristors (IGCTs) for this application. The authors concluded: "*IGCTs typically have switching losses of more than 20 J per switching cycle while IGBTs typically have losses less than 10 J. IGBTs require a relatively small gate drive circuit and no snubbers while IGCT circuits*

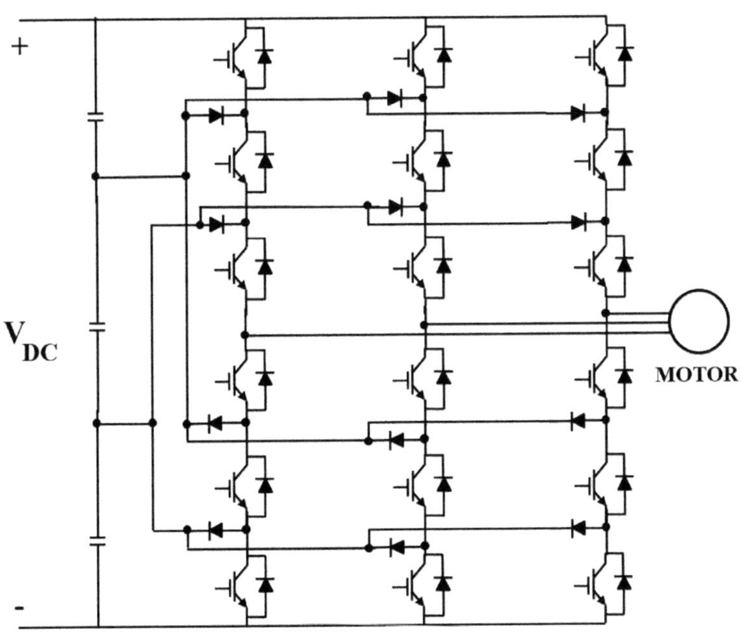

FIGURE 14.11

Multilevel inverter drive.

typically incorporate a very large built-in gate driver and require snubbers and other components to limit di/dt and dv/dt....There is a significant increase in weight and size for the IGCT design, so it was ruled out." The isolated IGBT module was selected for this application rather than the hockey puck module because it is available from multiple manufacturers. The torque ripple with the IGBT multilevel inverter drive was found to be 0.42% at full speed and increased to 2% at 50% of full speed. The efficiency of this topology exceeded 97% for this range of speeds.

The three motor drive topologies for naval vessel propulsion motor drives are compared in Table 14.1. The cycloconverters have the best efficiency and smallest weight and size. However, they operate with a poor power factor and produce severe torque ripple, which makes quiet operation impossible, an important consideration for naval applications. The series-connected low-voltage inverter creates the best waveform for the motor current but has a large weight and size. Based on all the factors of relevance to a naval vessel (efficiency, size, and acoustic noise), the diode-clamped multilevel inverters using 6.5 kV IGBT modules offer the best overall performance.

14.2.2 Naval Shipboard Power Distribution

Naval ships will utilize DC power distribution in the future. Inverters are required for ship services at the point of load. A 500 kVA three-phase 60 Hz AC power source operating from the DC bus, called the ship service inverter module, was reported in 1995 [22]. It uses 1200 V, 300 A IGBT modules to create three-phase AC power using four sets of IGBT half bridges. A power density of 3 kW/kg and 3.6 W/cm^3 was achieved. When operated in the hard-switched mode at an output power of 267 kW (corresponding to an inductive load with power factor of 0.8), an overall conversion efficiency of 97.1% was observed. The total harmonic distortion at full load was 2.4% with a PWM ripple of 0.1%.

A novel snubber circuit was proposed to reduce losses in the IGBTs. The snubber consists of a diode and capacitor in series connected in parallel with the collector-emitter of the IGBTs. A MOSFET fly-back converter was also used to return the

Table 14.1 Comparison of Ship Propulsion Motor Drive Topologies

Drive Topology	Cycloconverter	Multilevel Inverters	Series Connected Low Voltage
Total weight (kg)	3850	9820	13,700
Magnetic weight (kg)	3060	2405	6311
Size (cubic meters)	7.6	11.13	21.4
Efficiency (%)	99.3	98.7	97.6

energy stored in the snubber capacitor to the DC bus. The efficiency was found to increase by 0.7% with no change in harmonic distortion using this approach.

A Z-source-based isolated DC-DC building block for mission critical applications such as electric ships, aircraft, and subsea oil rigs has been designed [23]. An IGBT is used at the input of the Z-source impedance network to achieve the fault-tolerance feature.

14.2.3 Solid-State Transfer Switch

Mission-critical naval operations require uninterrupted power delivery to loads such as radar, weapon systems, and communication resources. Uninterruptible power supplies are prone to critical component failures and maintenance problems. An alternative approach to robust power delivery in naval vessels is the static automatic transfer switch (SABT) [24]. SABT is an automatic two-pole nonpolarized switch that allows rapid transfer of the load from one power source to another. The SABT was first developed using reverse parallel connection of thyristors for each phase of the AC source. The SABT power transfer time is designed to be much less than the criteria established by the computer industry for "ride through" during a power fluctuation or interruption.

The Computer and Business Equipment Manufacturers Association curve published as the IEEE Standard 446-1987 is shown in Fig. 14.12. The shipboard power system is designed to fall within the acceptable range for the power source voltage and time duration of the interruption. Based on these criteria, the SABT must be capable of detecting the change in power supply voltage and transferring to another power source within one-quarter of the 60 Hz power cycle. The SABT is typically designed to react in less than 250 μs to a sag in power source voltage to 90% of its nominal value with 70% of nominal value indicating a failure of the source. The SABT usually operates from a 450 V AC, three-phase, three-wire, 60 Hz

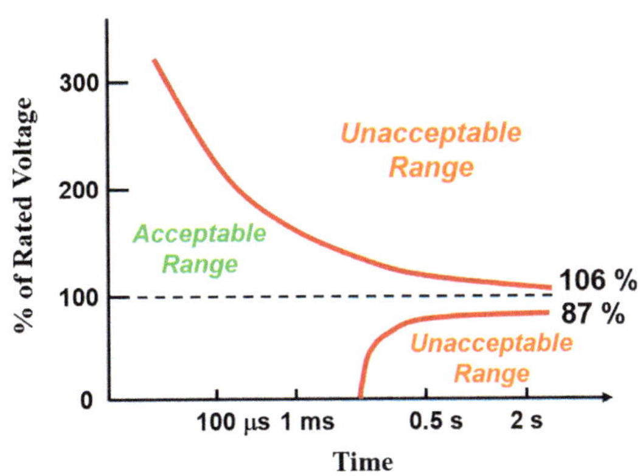

FIGURE 14.12

Computer and Business Equipment Manufacturers Association power interruption curve.

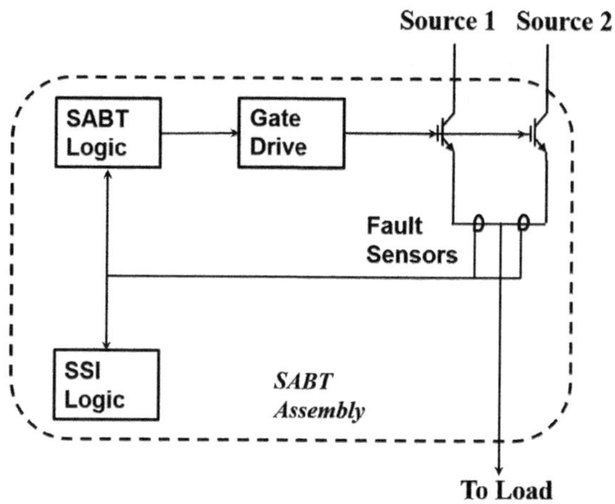

FIGURE 14.13

Static automatic transfer switch block diagram.

power source or a 120 V AC, three-phase, three-wire, 60 Hz power source. They are also designed to sense overcurrent conditions as well as frequency and phase deviation between multiple power sources. The operation must be conducted with a "break-before-make" sequence to avoid connection of the two power sources involved in the switching.

A dual-function switch (DFS) has been developed using IGBTs for Navy shipboard use. A block diagram of the DFS with SABT implemented using IGBTs is shown in Fig. 14.13. It was successfully used for load currents of up to 100 A. The selected IGBTs had an on-state voltage drop of 1.5 V (including a series diode for reverse blocking) with a turn-off time of 40 μs for a load current of 50 A. The power loss in the IGBT when carrying the load current of 50 A is 80 W. Although the on-state voltage drop for the IGBT option is twice that for a thyristor, the authors state: "*Test results show that the transfer time for DFS with IGBT is only 10% of SABT with SCR, close to ideal seamless transfer operation. ...Implementation of current limit control technique to the DFS (with IGBT) will meet both source transfer and fault current interruption requirement.*"

14.2.4 Solid-State Circuit Breakers

Circuit breakers are required in power delivery networks to interrupt high current flow during faults that might result in a fire hazard. Circuit breakers are designed to detect a fault condition where the current increases to a certain threshold above normal operating conditions and then interrupt the current flow rapidly [25]. Circuit breakers can be reset after the fault is removed from the system, unlike a fuse that would have to be replaced. Mechanical circuit breakers, like the ones in a residential garage, rely on a mechanical switch that is opened when the fault condition is detected. Although widely deployed in power distribution networks, their disadvantages include a slow response to the fault resulting in high current levels during the current interruption and the creation of an arc across the mechanical contact. The arc results in significant wear of the contacts in the mechanical circuit breaker and can be hazardous for environments with explosive gases. This is an important consideration in navy shipboard applications.

Solid-state circuit breakers rely upon turning off a semiconductor device connected in series with the load in the power distribution network. The on-state voltage drop for the solid-state device is an important consideration because it produces continuous power losses in the power distribution system. For AC power systems, the devices must also exhibit reverse blocking capability. Alternately, a diode can be connected in series with the switch but this increases the on-state power loss. Thyristors were initially used in solid-state circuit breakers for AC power distribution systems due to their reverse blocking capability and low on-state voltage drop. Disadvantages of the thyristors include current interruption only after zero crossing occurs (which can be as long as 8 ms after a fault is detected in the 60 Hz power distribution system) and the inability to limit the current during the fault. Solid-state circuit breakers utilizing IGBTs have the advantages of very rapid (<20 μs) current interruption after detection of the fault and the ability to limit the fault current.

The three-phase power distribution system is shown in Fig. 14.14 with IGBT-based circuit breakers [26]. The circuit breaker is constructed using two IGBTs

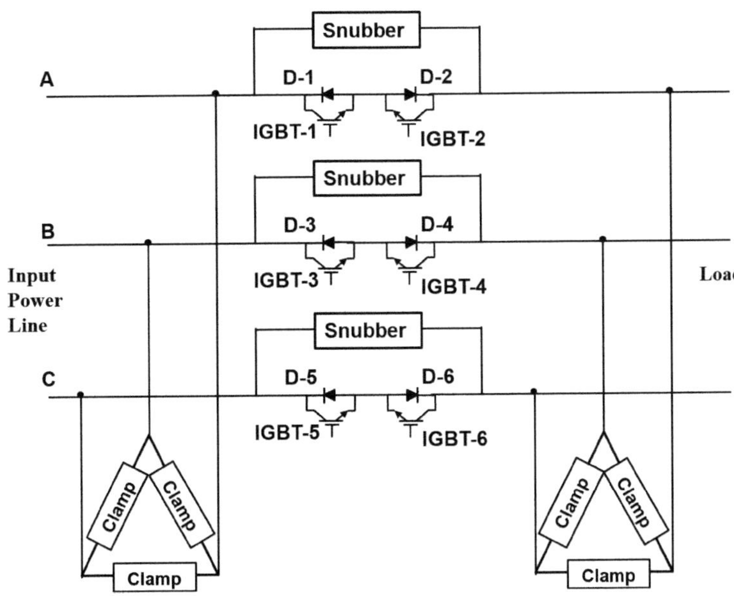

FIGURE 14.14

AC solid-state circuit breaker with insulated gate bipolar transistors.

and two diodes for each phase leg. All the IGBTs are maintained in their on-state during normal (no fault) operating conditions. During the positive voltage excursion for phase A, current flows via IGBT-1 and its series diode D-2 to the load. Consequently, current can be interrupted during the positive half–cycle for phase A by turning off IGBT-1. Similarly, during the negative voltage excursion for phase A, current flows via IGBT-2 and its series diode D-1 to the load. Consequently, current can be interrupted during the negative half-cycle for phase A by turning off IGBT-2. This applies to other phase legs as well. With recent development of IGBTs with reverse blocking capability, the series diode can be eliminated, resulting in reduced power losses. The snubbers across the IGBTs include a metal oxide varistors (MOV) to absorb the energy in the fault current.

The IGBT-based circuit breaker for a DC power distribution system [26] is shown in Fig. 14.15. Two IGBTs are used here in the current path between the DC power source and the load. Both IGBTs are turned off together during a fault condition but either device can clear the fault from the DC power supply. This configuration has been tested using 4500 V, 900 A IGBT modules to create a solid-state circuit breaker for protecting a propulsion drive inverter in a naval ship application.

The relative merits of the IGBT and the IGCT have been compared for Navy ship power distribution systems [27]. The IGCT is a GTO structure with copackaged gate turn-off circuitry. It has larger current-handling capability than IGBTs but cannot limit the load current during a fault. Although six times slower in switching speed, the turn-off time for the IGCT is at least 900 times faster than the mechanical circuit breaker. It is therefore recommended for use at medium current upstream locations in the power distribution system. The IGBT is appropriate for lower current downstream circuit breakers where their very high-speed response reduces the fault current levels.

14.2.5 Direct Conversion System

AC electric propulsion is commonly used for driving the asynchronous electric motors in electric ship propulsion. The voltage inverter with intermediate DC link is used with PWM switching of IGBTs. The option of direct conversion based on the cycloconverter topology is attractive due to the smaller number of components,

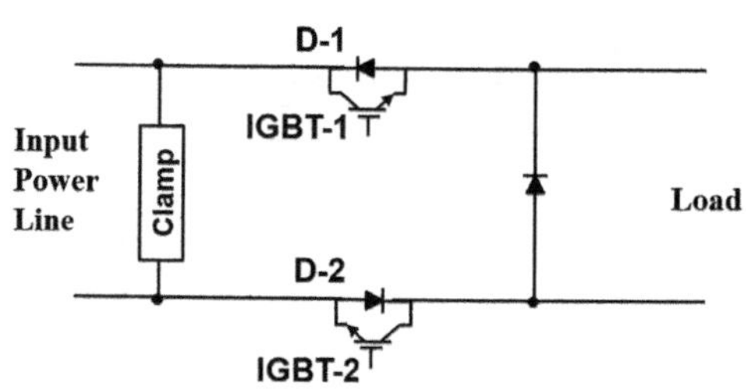

FIGURE 14.15

DC solid-state circuit breaker with insulated gate bipolar transistors.

which reduces the size and weight. The usual implementation with six sets of three-phase reversing switches requires 72 IGBTs. The number of IGBTs can be reduced in half by using series-connected reverse bridges to achieve size and weight reduction [28].

14.2.6 Hybrid ANPC H-Bridge System

The Si IGBT/SiC power MOSFET hybrid switch was discussed in Section 14.1.5 for electric aircraft. In that approach, the Si IGBT and SiC power MOSFET were connected in parallel. Another hybrid approach using Si IGBTs and SiC power MOSFETs has been applied to ship propulsion [29]. The active-neutral-point-clamped (ANPC) topology shown in Fig. 14.16 has two half-bridges—one formed using the SiC power MOSFETs and the other using Si IGBTs. This creates a three-level ANPC circuit. The Si IGBTs are operated at the fundamental frequency while the SiC power MOSFETs are switches at the carrier frequency (2 kHz). This allows taking advantage of the low conduction losses of the Si IGBT and low switching of the SiC power MOSFETs. Simulations of a three-phase 5 kV(DC) 10 MW converter were performed to validate performance.

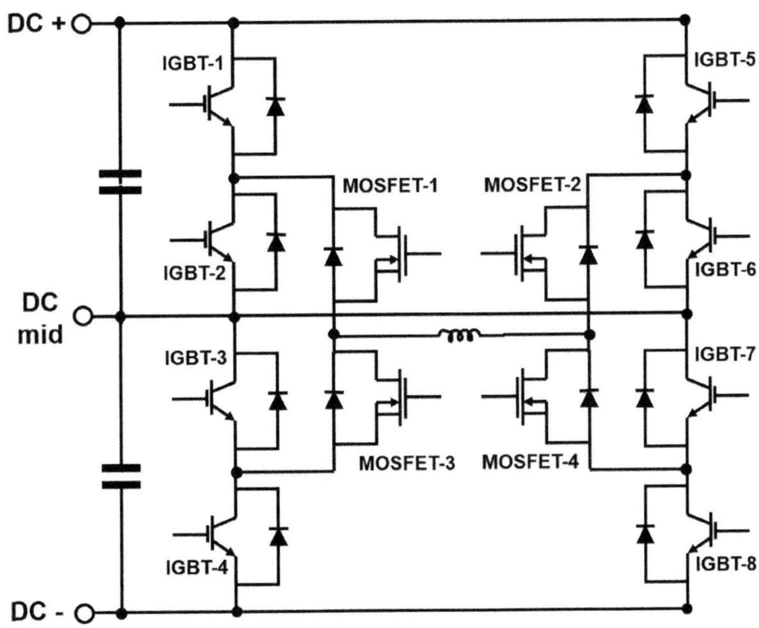

FIGURE 14.16

The active-neutral-point-clamped topology built using Si IGBT and SiC power MOSFET devices.

14.3 Aircraft Carriers

Aircraft carriers allow nations to deploy air power around the globe far removed from airfields on land. The aircraft carrier requires a full length flight deck and storage facilities for the aircraft that it can launch and recover [30]. The nuclear-powered USS Nimitz (CVN-68) aircraft carrier [31] is shown in Fig. 14.17 with numerous aircraft on its flight deck. The aircraft can also be housed below the flight deck and brought to it using elevators.

The navy has decided to replace hydraulic actuators with electromechanical actuators to reduce weight, space, and maintenance. Applications in an aircraft carrier include [32] weapons elevators, aircraft elevators, hangar doors, rudder mechanisms, and propulsion systems. Propulsion systems were discussed in the previous section. The weapons elevators have a load capacity of 42,000 pounds and must move at 2 ft per second even when the sea is rough. The aircraft elevator has a load capacity of 500,000 pounds and must also be moved at 2 ft per second. The aircraft carrier rudder requires a 500 hp. motor running at 3000 rpm with high torque. The authors state: "*The goal is to drive everything that moves on the aircraft carrier with torque-dense electromechanical actuators...removing 500 personnel, reducing ship weight by 1,400,000 pounds, freeing up to 60,000 square feet of floor space, reducing average maintenance of the affected systems by 2.7x, reducing their complexity by 2.2x, and reducing their power consumption by 2.7x.*" These electromechanical actuators are designed using the power electronic building blocks discussed in Section 14.2.

The Navy plans to utilize high-power IGBT switching arrays to serve shipboard power needs [33]. The authors state: "*Within the last 10 years, advances in energy storage, solid-state switching, computer technology, and material science have made electromagnetic power conversion a viable alternative to conventional methods being used today in aircraft launch assist, elevators, motion damping, actuators, and power supplies. The key enabling technology is the IGBT and the controls that*

FIGURE 14.17

The USS Nimitz aircraft carrier.

can rapidly switch them simultaneously." In addition, railguns are under development for launching projectiles at hypersonic velocities during defensive and offensive tactical maneuvers.

14.3.1 Railgun Projectile Launcher

Railguns are projectile launchers that use electromagnetic energy to propel an object to great distances [34]. An example of a railgun projectile launcher under development by the US Navy is shown in Fig. 14.18. It is considered the next game-changing weapon technology. General Atomics demonstrated the 3 MegaJoule Blitzer railgun to the US Navy in 2017 [35]. Rear Admiral Bryant Fuller, the Navy's Chief Engineer, has stated [36]: "*The electromagnetic railgun represents an incredible new offensive capability for the US Navy. This capability will allow us to effectively counter a wide range of threats at a relatively low cost while keeping our ships and sailors safer by removing the need to carry as many high-explosive weapons.*" The railgun has a range of 100 miles without using explosive warheads, making it cheaper and safer. Each 18 in. projectile will cost $25,000 compared with $0.5–1.5 million for conventional missiles [37].

The railgun consists of parallel conducting rails, as shown in Fig. 14.19, with windings designed to generate a strong magnetic field in the middle of the tube containing the projectile. A power source with a power switch in series with the stationary winding is used to generate a pulse of current through it when the switch is closed. This creates the high magnetic field along the rails that is coupled to the moving winding on the projectile or a sabot, which can drop away after the projectile is launched. The Lorentz force applied to the projectile can accelerate it to extremely high velocities (3500 m/s (about Mach 10) or hypervelocity). These high velocities enable railguns to have a greater range with less time to reach the target than conventional guns that utilize explosives. This is advantageous for defending against aircraft and missile moving at "merely" supersonic speeds. A huge

FIGURE 14.18

A railgun projectile launcher.

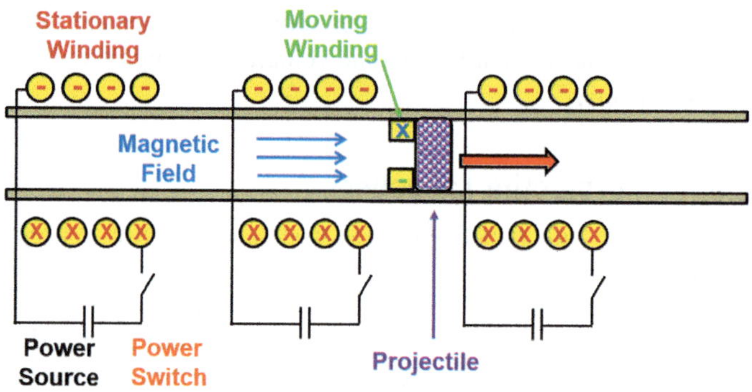

FIGURE 14.19

Principles of the railgun launcher.

challenge in building a railgun is the extremely high pulse currents (millions of amperes) with enormous energy expenditure (more than 30 MJ). These current and power levels are feasible on nuclear power naval vessels.

The biggest challenge in making a practical railgun for use in a shipboard environment is also the size and weight of the power supply. In the railgun, energy is stored in a capacitor and discharged through the stationary winding [38]. The rail launcher must provide 64 MJ of kinetic energy to the projectile. The conversion efficiency between the electrical stored energy and the kinetic energy is 40%. At this efficiency, the rail launcher becomes smaller in size and weight than the conventional gun and magazine it is going to replace. The pulse length for the electromagnetic launch is about 9 ms. With a recharge duration of 4 s, it is possible to achieve 12 rounds per minute with the railgun.

14.3.2 Aircraft Launchers

The shipboard rail launcher for the aircraft on the carrier can also be electrically powered. The USS Gerald R. Ford aircraft carrier plans to use railgun technology to launch fighter jets instead of the steam catapults used in previous aircraft carriers [39]. The electromagnetic aircraft launch assist system uses the same principles as the railgun [33]. The requirements are a peak current of 50 kA delivered using 10 kV peak voltage to achieve a 5 ms slew rate. The IGBTs are switched at submicrosecond speed equivalent to operation at 1–50 kHz. A densely packed array of 4.5 kV IGBTs operating in series and parallel has been proposed for this application. In 2010, the US Navy reported the successful launch of an aircraft on land using a railgun [40]. The electromagnetic aircraft launcher was a linear induction motor capable of accelerating the 100,000 pound aircraft to 240 miles per hour in a space of 300 ft. Compared with the steam catapult, it is considered to be much smaller in size, more efficient, and simpler to maintain. A total of 8000 launch/recovery cycles were completed on the USS Gerald R. Ford aircraft carrier through April 2021 [41].

14.4 Nuclear and Diesel-Electric Submarines

Submarines are submersible watercraft manned by a crew [42]. They were first deployed during World War I. The most powerful modern submarines make use of nuclear power. The advantages of a nuclear submarine [43] are independence of air for propulsion and a large stored nuclear energy source for operation at high speed for long time intervals. The largest submarines built for the US Navy are 18 Ohio Class nuclear-powered submarines (example shown in Fig. 14.20) [44]: 14 armed with ballistic missiles and four with guided missiles. Submarines must often rely on being stealthy to avoid detection and attack by the enemy. A moving submarine produces a noise signature above the ocean's natural noise, which can reveal its location and direction of movement. A primary source of acoustic noise is caused by the rotation of its propellers. Another source of noise is generated by its power plant and motor drive. The noise level increases rapidly with increasing speed of the submarine [45], as shown in Fig. 14.21. Diesel-electric submarines have the advantage of operating in the electric mode using power stored in batteries to reduce their noise signature dramatically [46], as shown in the figure, as the "ultraquiet mode."

14.4.1 Quiet Electric Drive

A high-efficiency quiet electric drive [47] technology has been described for reducing the noise signature of submarines. It has been found that the harmonic current in the AC power bus of submarines originating from the motor drives can produce noise from the generators. A hybrid cascaded voltage source multilevel inverter topology fed by a synchronous rectifier enables reducing the harmonic currents in the input AC power bus.

A block diagram for the submarine quiet electric drive is shown in Fig. 14.22. A three-phase phase-shift transformer is used to supply power to the synchronous

FIGURE 14.20

The USS Ohio Class nuclear submarine.

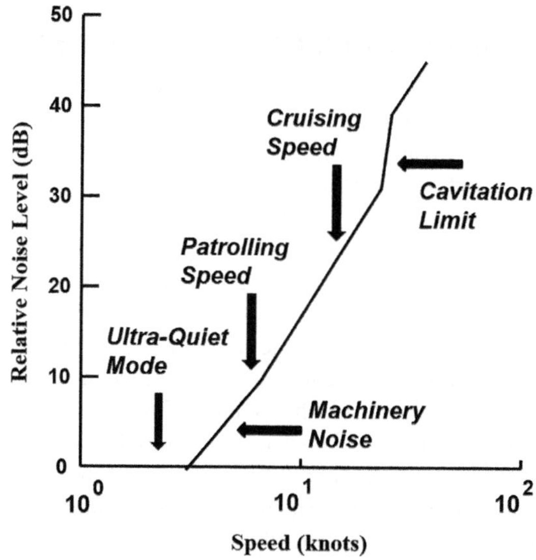

FIGURE 14.21

Noise signature of submarines.

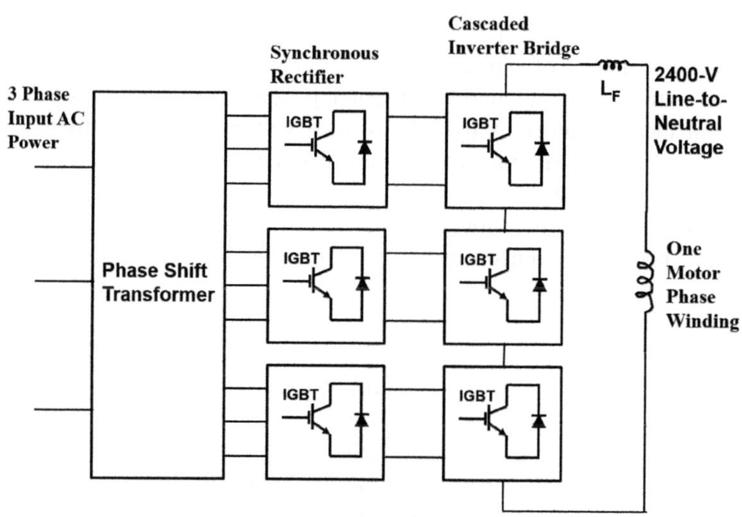

FIGURE 14.22

One phase of the quiet drive module.

rectifier. The transformer isolates the motor drive from the ship's AC power source. The synchronous rectifier circuit is similar to an inverter circuit with the power angle chosen to produce power flow to the inverter. The IGBTs in the synchronous rectifier operate at a PWM frequency of 10 kHz. The synchronous rectifiers allow minimizing the harmonics introduced into the AC power source while at the same time regulating the DC output voltage. Three inverter stages are connected in series (cascaded) to deliver 2400 V line to neutral AC voltage to one phase winding of the motor.

The sinusoidal AC voltage waveform is delivered to one phase winding of the motor using the cascaded inverter shown in Fig. 14.23. It contains three stages of IGBT H-bridge circuits that are connected in series to generate the 2400 V AC output. Two of the IGBT inverter stages use 1400 V DC buses while the third makes use of a 700 V DC bus. The high-voltage stages operate with a quasisquare wave at a low frequency while the third low voltage stage operates in the PWM mode at a high switching frequency. This reduces overall switching losses because the IGBTs in the PWM stage need lower blocking voltage rating. The full power efficiency of this motor drive is 94% with an efficiency of 76.4% at 10% load.

14.4.2 IGBT Power Cycling

The crew of a submarine must be assured of reliable operation of all onboard electrical systems as well as each of the components. The failure of IGBTs in the submarine motor drive could leave the crew stranded underwater, an obviously

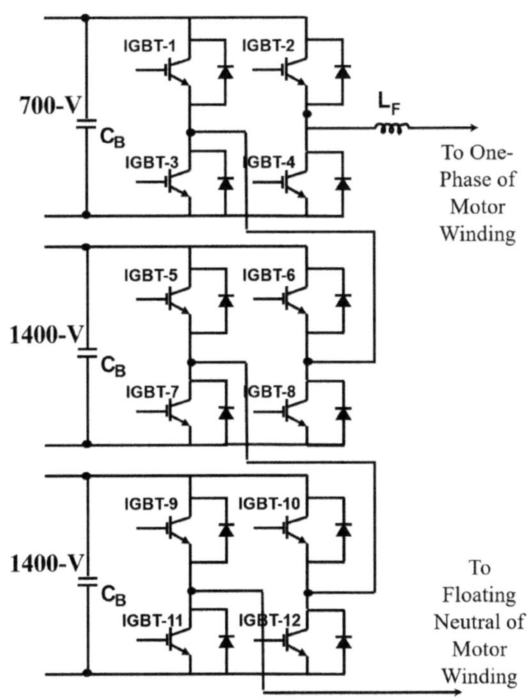

FIGURE 14.23

Cascaded inverter topology. *IGBT*, insulated gate bipolar transistor.

untenable option. In 2000, the British Royal Navy performed thermal cycling tests on IGBT modules to determine their reliability for submarine applications [48]. The rise in junction temperature for the IGBT under nominal working conditions was found to be well below the 125°C suggested by manufacturers. Repetitive thermal cycling was performed from 50 to 65°C to examine failure from wirebonds and up to 75°C for solder failure (see Section 6.7). More than 110,000 electrical cycles were applied to four IGBT modules. This corresponds to the total number of switching events for all the IGBTs on five submarines during a 25-year lifespan. No IGBT failures were observed. The authors concluded: "*A high degree of confidence can be taken that the in-service thermal cycling duty will not cause any degradation in performance characteristics of the IGBT power switches.*"

14.5 Army Vehicles

The US military is the world's biggest consumer of fossil fuels. It is extremely costly for the army to provide fuel to its vehicles in remote theaters of war such as encountered recently in Afghanistan and Iraq. In addition, the US government operates more than 600,000 vehicles. DoD has committed to purchasing 92,400 electric vehicles (EVs) for noncombat operations by 2020 [49]. These vehicles will be derived from the hybrid EV (HEV) and EV platforms being pursued by all the major automotive manufacturers (see Chapter 9) using IGBTs to drive the motors.

HEVs are also of interest to the army for combat use. BAE Systems has developed a hybrid-electric tracked armored vehicle [50]—a 15 ton M113 vehicle shown in Fig. 14.24(a). It has a 600 hp. electric motor compared with a 275 hp. motor in the previous M113 vehicle. Similar HEVs are being developed by the British, French, and Swedish militaries. The vehicle requires delivery of 100 kW of power to the electric motor. The Bradley Fighting Vehicle shown in Fig. 14.24(b) will also be replaced by an HEV according to the DoD [51]. The new Ground Combat Infantry Fighting Vehicle [52] is a gas–electric hybrid that operates in the EV mode to reduce

(a) (b)

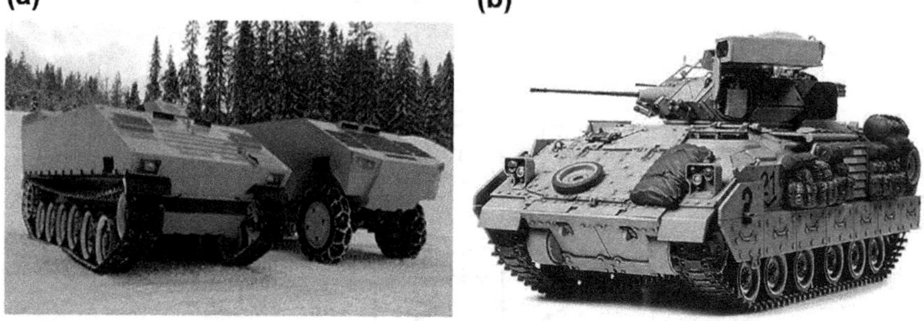

FIGURE 14.24

(a) M113 hybrid electric transporter; (b) Bradley Fighting Vehicle.

noise. The electric motor provides faster acceleration than the Bradley Fighting Vehicle. According to BAE, the hybrid system not only saves on fuel but reduces weight by 4 tons, improving maneuverability. The fewer moving parts in the HEV make it more reliable, reducing maintenance cost and making it easier to service [53].

14.5.1 Bidirectional DC–DC Converter

The new HEVs being developed by the US Army will require bidirectional DC–DC power converters (BDC) [54]. The BDC controls power flow between the high-voltage (600 V) propulsion DC bus and the lower-voltage (300 V) battery bank. The BDC is operated in the boost mode when additional power is needed for vehicle acceleration and in the buck mode for recharging the battery bank. The nonisolated bidirectional converter circuit is shown in Fig. 14.25. Boost mode operation is performed by switching IGBT-2 and using antiparallel diode D-1. Buck mode operation is performed by switching IGBT-1 and using antiparallel diode D-2.

The operation of the BDC topology is discussed using the waveforms shown in Fig. 14.26. In the boost mode of operation, the gate signal (V_{GE}) is applied to IGBT-2. With turn-on of IGBT-2 at time t_0, the battery bank voltage (300 V) is applied across the inductor (L), producing an increase in its current (I_L), which also flows through IGBT-2 as collector current I_C. When IGBT-2 is switched off at time t_1, the voltage across IGBT-2 increases until it is clamped by the antiparallel diode D-1 to the propulsion bus high voltage (600 V). Because the voltage polarity across the inductor is reversed, its current decreases until time t_2. During this time, the inductor current flows through diode D-1, as shown by waveform I_D. No current flows during the time from t_2 to t_3. The voltage across IGBT-2 oscillates at time t_2 to t_3 due to energy transfer between the output capacitance of IGBT-2 and the inductor. At time t_3, IGBT-2 is turned on again in the boost mode.

Similarly, in the buck mode of operation, the gate signal (V_{GE}) is applied to IGBT-1. With turn-on of IGBT-1 at time t_0, the voltage difference (300 V) between

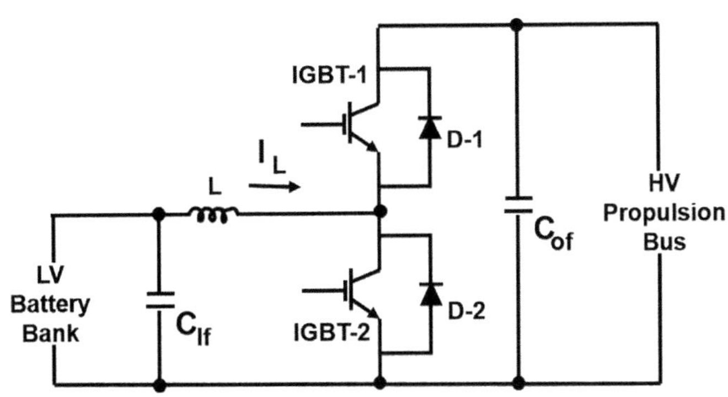

FIGURE 14.25

Nonisolated bidirectional converter topology. *IGBT*, insulated gate bipolar transistor.

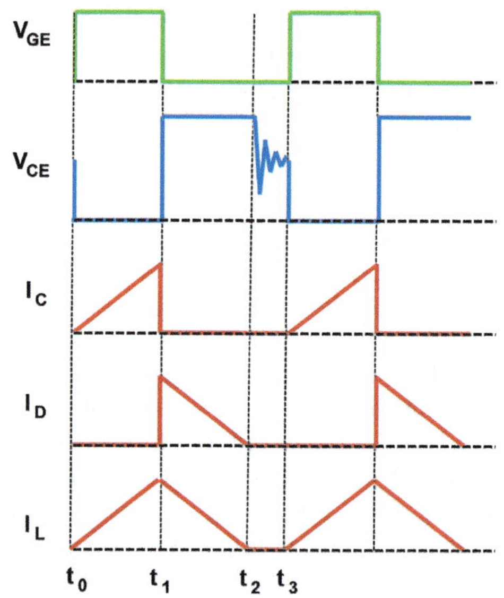

FIGURE 14.26

Circuit waveforms for bidirectional converter.

the propulsion bus high voltage (600 V) and battery bank voltage (300 V) is applied across the inductor (L), producing an increase in its current (I_L) in the opposite direction to the boost mode, which also flows through IGBT-1 as collector current I_C. When IGBT-1 is switched off at time t_1, the voltage across IGBT-1 increases until it is clamped by the antiparallel diode D-2 to the propulsion bus high voltage (600 V). The voltage polarity across the inductor is reversed to the battery bank voltage (300 V) and its current decreases until time t_2. During this time, the inductor current flows through the diode D-2 as shown by waveform I_D. No current flows during the time from t_2 to t_3. The voltage across IGBT-1 oscillates at time t_2 to t_3 due to energy transfer between the output capacitance of IGBT-1 and the inductor. At time t_3, IGBT-1 is turned on again in the buck mode.

The BDC topology needs only two IGBTs and smaller inductors, giving it higher power density. A switching frequency of 20 kHz is chosen to reduce the size of the passive elements while keeping the IGBT switching losses acceptable. The IGBTs must handle a large collector current at turn-off requiring 300–400 A modules. Modules with larger on-state voltage drop and lower switching losses are preferable for this topology. A volumetric power density of 2.7 kW/L was achieved for a 90 kW BDC. The IGBTs operated at a junction temperature of 100°C, even with 80°C coolant temperature. By using microchannel liquid cooling for the IGBTs to reduce the thermal resistance, the BDC could be operated at 100 kW with the 80°C coolant increasing the volumetric power density to 4 kW/L.

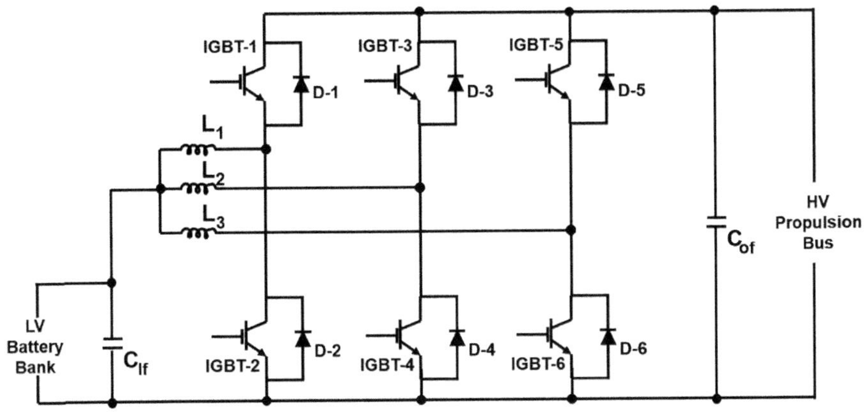

FIGURE 14.27

Three-phase bidirectional converter topology. *IGBT*, insulated gate bipolar transistor.

A three-phase version of the bidirectional converter topology is shown in Fig. 14.27. Its operation is similar to that for the single inductor circuit shown in Fig. 14.25. Six IGBTs with fly-back diodes are used here with a single input side capacitor (C_{if}) and a single output side capacitor (C_{of}). The waveforms for the inductor currents can be controlled, as described in the context of Fig. 14.26.

14.6 Air Force Jets

The Air Force has relied on high-performance aircraft throughout history to gain superiority and control over the battlefield. Since World War II, fighter jets have relied on the use of large numbers of electromechanical relays, wiring harnesses, and pin/socket connectors [55]. The aircraft electrical system is powered from a 400 Hz three-phase AC system. The US Air Force relies on jets such as the F-16 Fighting Falcon shown in Fig. 14.28 for combat missions [56]. Other jet fighters include the F-15 Eagle and F-22 Raptor.

The Air Force began investigation of the more-electric airplane (MEA) in the 1980s because of the following technological drivers [55]: (1) movement to digital control systems that cannot tolerate power interruptions; (2) fly-by-wire option; (3) communication over multiplexed data buses; and (4) increasing number of electrical power loads on the aircraft. The major impetus to migration to all-electric operation is to replace hydraulic systems with electrical actuators and controls for use in gear retraction, nosewheel steering, braking, and actuating flaps and primary flight surfaces. To be competitive, the MEA has to surpass the light, reliable, hybrid hydraulic system previously used in aircraft. In addition, the Air Force wants high combat availability and low maintenance costs for its aircraft. To obtain a more aerodynamic surface, the new generation of aircraft requires thin, hinge-line actuators on

FIGURE 14.28

F-16 Falcon fighter jet.

the primary flight surfaces. The power electronics must not only have high reliability but also self-diagnostic capability to allow repair or replacement with minimal tools and skills without large ground gear that is expensive to transport, house, and defend on frontlines. In 1984, NASA estimated fuel savings of more than 20% by replacing hydraulics with all-electric drives by eliminating engine bleed and electric flight control systems [57].

The aircraft used by the Air Force in the 1990s were powered by multiple secondary power sources: hydraulic, pneumatic, electrical, and mechanical [58]. The secondary mechanical power was extracted from the main engine's drive shaft and distributed to a gearbox to drive lubrication pumps, fuel pumps, hydraulic pumps, and electrical generators. Hydraulic power was used for flight control actuators, landing gear brakes, utility actuators, avionics, and weapon systems. In the MEA, electric motor-driven actuators would replace the hydraulically driven actuators; electric motor-driven pumps would replace the gearbox-driven fuel and lubrication pumps; and an electric motor-driven compressor would replace the pneumatically driven compressor for environmental control. In 1993, the author stated: "*The Air Force has focused its resources in this thrust area on the development of the MOS controlled thyristor (MCT) switching device and the MCT driver.*" This is consistent with the mission of the power semiconductor research coordination consortium [6]. However, the Air Force eventually abandoned the MCT in favor of the IGBT. In 2003, the same author stated [59]: "*Key to the advancements in power electronics are the advent of silicon-based metal oxide semiconductor gated power devices, such as field effect transistors and insulated gate bipolar transistors. These power semiconductor devices offer substantial increases in power densities due to the significantly smaller drive power requirements for control.*"

A twin-engine MEA fighter jet would be operated from a 250 kW starter/generator using a 270 V DC bus [59]. The power devices required for the starter/generator

Table 14.2 Electrical Loads in a Generic More Electric Aircraft Fighter Jet

Aircraft System	Maximum Continuous Total Power (kW)	Number of Motor Drives	Largest Motor Drive (kW)
Flight controls	80	28	50
ECES system	40	10	10
Fuel management system	35	10	9
Pneumatic system	30	2	15
Landing system	30	20	5
Other systems	20	10	1

have ratings of more than 1000 V and 1000 A. This source would power all the MEA motor drive loads listed in Table 14.2. The flight controls are the most numerous loads with the highest power demands. They represent mission-critical loads whose failure can endanger the aircraft and pilot. The highest continuous maximum power is 80 kW. The typical motor drive is a three-phase inverter containing IGBTs with fly-back diodes. Because the motor ratings range from fractional hp. to 100 hp, the IGBTs must have a blocking voltage of 600 V and a current-handling capability of ten to hundreds of amperes.

In 1998, the Air Force reported demonstration activities of the MEA concept [60]. The authors stated: "*Generator power control units, inverters, converters, and motor controllers consist of state-of-the-art silicon-based power semiconductor switching devices, including IGBTs.*" The demonstrations were performed on the C-141 Electric Starlifter whose ailerons were "electrified." A total of 500 h of "near-flawless flight" and "100% confident" reports from pilots were reported. The technology was being applied to the electric stabilator for the F/A-18 aircraft and the electric flight control actuation for the AFTI/F-16 Falcon jet shown in Fig. 14.28.

14.6.1 Electrical Power Distribution Architecture

By 2007, the MEA concept had matured to an electrical distribution system that performs all the functions of the previous hydraulic/pneumatic system [61]. The hydraulic/pneumatic power distribution system can be contrasted with the new electrical distribution system using Figs. 14.29 and 14.30. In the legacy power distribution system shown in Fig. 14.29, the main engine power is converted to either hydraulic or pneumatic power before being utilized to operate the various high-power loads. The electrical power is used only for auxiliary loads. In the new MEA system, the main engine power is converted mostly into electrical power to serve high- and low-power loads.

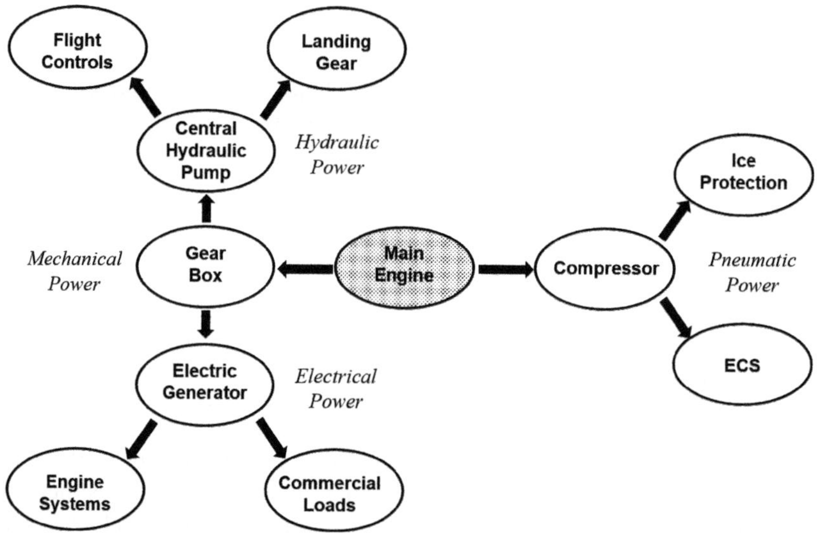

FIGURE 14.29

Conventional power distribution system. Environmental control system.

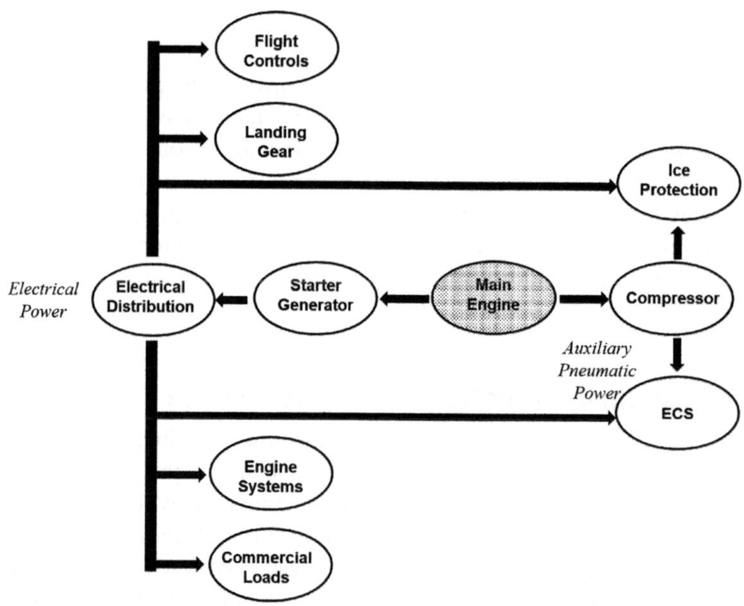

FIGURE 14.30

More electric airplane power distribution system.

14.6.2 Portable Railgun

In Section 14.3.1, the railgun being developed for naval shipboard applications was discussed. A small, lower-velocity railgun is also of interest to the Air Force for use on aircraft. In this railgun, a 15 mF capacitor needs to be charged to about 2 kV within 200 ms and then discharged into the railgun to launch the projectile. IGBTs are used not only to deliver the current pulse to the railgun, as discussed in Section 14.3.1, but to also charge the large capacitor within 200 ms. The charging time for the capacitor limits the firing rate for the projectiles. The 200 ms time for charging allows achieving three rounds per second from the railgun.

For a portable railgun suitable for deployment in aircraft, the capacitor charging circuit must be compact and lightweight. The development of a suitable power supply for charging the capacitor in the railgun to store a large amount of charge has been under investigation since 2007 [62]. The specifications call for charging a 15 mF capacitor bank to a stored energy of 30 kJ within 200 ms. The circuit topology for this application is illustrated in Fig. 14.31. In an aircraft environment, the capacitor (C_S) must be charged from a 200 V battery bank on the airplane. A DC–DC converter with four IGBTs in an H-bridge configuration is used with a high-frequency transformer to generate 2 kV across the capacitor bank to achieve 30 kJ of stored energy. Turning on IGBT-1 and IGBT-4 delivers current into the primary of the transformer in one direction while turning on IGBT-2 and IGBT-3 delivers current into the primary of the transformer in the opposite direction. The transformer has a turns ratio of 1:10 to produce the 2 kV voltage across the capacitor (C_S). By switching the IGBTs at a high frequency (up to 10 kHz), the size and weight of the transformer can be reduced. The IGBTs have a voltage rating of 600 A and 1200 V. An 8.4 mF capacitor was successfully charged to 2 kV using this circuit topology in 300 ms.

In 2008, the authors reported charging a 16.6 mF bank of capacitors to 2 kV in 660 ms [63]. An average power of 50 kW was used here. In 2012, the authors had made progress toward achieving the goal of three rounds per second (RPS) by adding a second module to the charging circuit [64]. Compared with one RPS in 2008, they were able to achieve two RPS with this enhancement.

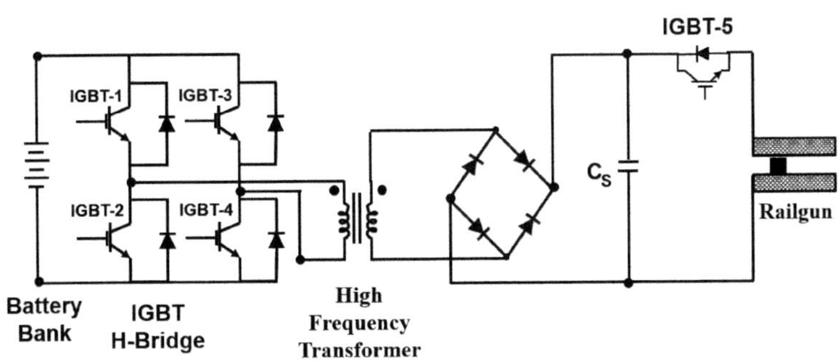

FIGURE 14.31

Capacitor charging topology for capacitor in small portable railgun. IGBT, insulated gate bipolar transistor.

14.7 Missile Defense

Protection of the civilian population and military targets from hostile missile attacks is a necessity of modern warfare. The threat can extend to nonwartime missile launches from terrorist organizations as well. The technology used for the detection of a missile launch, tracking its trajectory, and the destruction of the missile before it reaches the intended target is referred to as missile defense [65]. The hostile missile is destroyed by the use of an interceptor missile like the one shown in Fig. 14.32(a). The strategy for protection against hostile missiles is dependent on the type of missile under consideration: (1) strategic missile defense aims at the removal of threats from long-range intercontinental ballistic missiles that travel at about 16,000 mph; (2) theater missile defense aims at the elimination of threats from medium-range missiles that travel at 6700 mph; (3) tactical missile defense is aimed at the destruction of incoming short-range (up to 50 miles) tactical ballistic missiles that travel at less than 3400 mph. Many countries (United States, Russia, France, India, and Israel) have developed and deployed missile defense systems.

Protection against missile threats requires the detection of an incoming missile and its trajectory. This is performed by using radar systems like that shown in Fig. 14.32(b). Radar was originally developed during World War II and since then been applied to civilian use as well [66]. The radar works by emitting a pulse of radio waves or microwaves scattered by objects in their path. Radio or microwaves are reflected from electrically conductive surfaces (metallic objects) that form the outer shell of aircraft and missiles. These reflected waves are detected by sensors located usually at the location of the system emitting the radar pulse but can be at another location as well. The signal received from the missile is inversely proportional to the fourth power of the distance between the pulse transmitter/receiver and the object. A shift in the frequency of the radio waves or microwaves due to the motion of the object (the Doppler effect) can be used to monitor the velocity of incoming missiles.

Ballistic missile early warning systems make use of UHF frequencies ranging from 300 to 1000 MHz. They have a very long range of transmission through the

(a) (b)

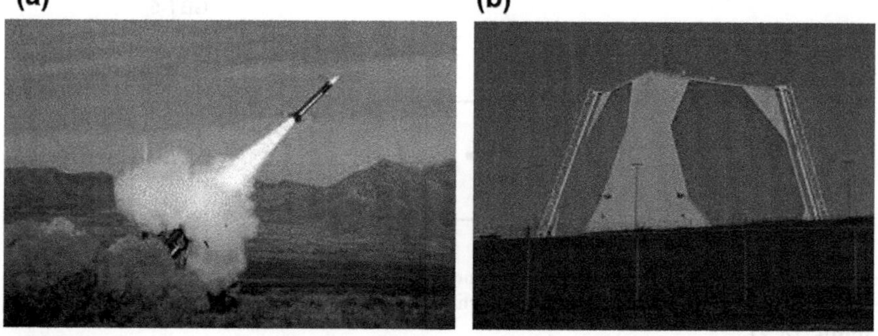

FIGURE 14.32

(a) Antiballistic missile; (b) missile defense radar.

Earth's atmosphere. Pulses of energetic microwaves are generated using a magnetron driven from a pulsed high-voltage power supply. Another approach for the generation of microwave radiation is the mirror radar. In this concept, a "plasma mirror" is formed using a fast rise time high-voltage pulse to a vacuum chamber containing a gas. Because the plasma has no inertia, it can be steered easily for rapidly scanning the skies against missile threats.

14.7.1 Radar Transmitter

The radio waves or microwaves for the radar beam are generated by using a vacuum electronic device (VED), which is either a crossfield device such as a magnetron or a linear beam device such as a Klystron or traveling wave tube (TWT) [67]. The VED emits radio waves or microwaves when stimulated using a high-voltage pulse. The types of VED used for various applications are shown in Table 14.3 with the corresponding frequency band. It can be seen that some technologies are more suitable for shipboard use, some for ground radar, and some for airborne use. The Klystron and TWT VEDs require driving the grid while for the magnetron, it is necessary to drive the cathode voltage at a larger voltage level.

14.7.2 Klystron Radar Power Supply

A block diagram for driving the VED [67] is shown in Fig. 14.33. The Klystron VED requires a pulsed power supply for controlling the grid and a high-voltage power supply to deliver the high voltage to the collector. The circuit topology for driving the Klystron grid [67] is shown in Fig. 14.34. It consists of IGBT-1 for connecting the grid to the on-voltage DC power supply (V_{on} bias) and IGBT-2 for connecting

Table 14.3 Radar Systems Used for Various Applications

Frequency Band	Application	VED	Type	Beam Voltage (kV)	Beam Current (A)
L	Ship	Klystron	Mod-anode	−43	25
S	Ship	CC TWT	Grid	−45	18
C	Ground	CC TWT	Grid	−53	17
C	Ground	CFA	Cathode	30	35
X	Air	Magnetron	Cathode	−23	27
X	Air	Helix TWT	Grid	−11	0.7
W	Various	EIK	Focus anode	−16	0.6

VED, *vacuum electronic device.*

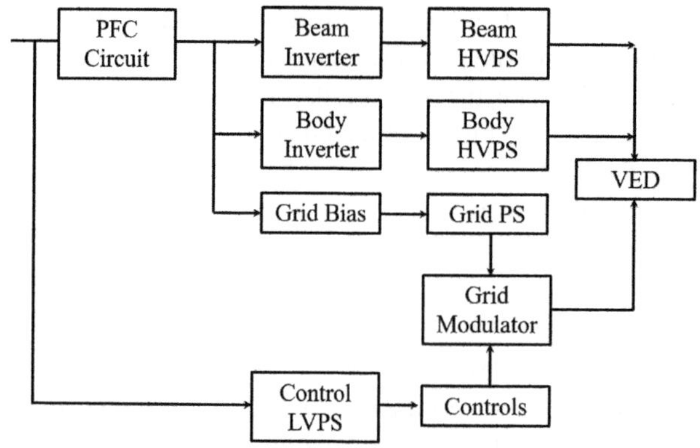

FIGURE 14.33

Block diagram for pulsed radar system. *PFC*, power factor correction; *LVPS*, low-voltage power supply; *HVPS*, high-voltage power supply; *VED*, vacuum electronic device.

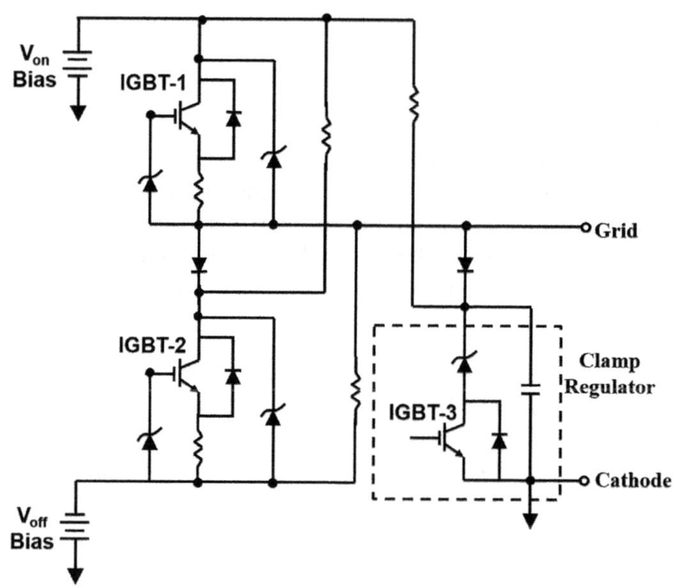

FIGURE 14.34

Circuit topology for Klystron grid power supply. *IGBT*, insulated gate bipolar transistor.

the grid to the off-voltage DC power supply (V_{off} bias). Because both the IGBTs must hold off 2000 V, a series stack of IGBTs was recommended in 2006 but a single IGBT device can now perform this task. The timing of the gate signals must be offset to avoid shoot-through current by shorting the two power supplies. The IGBTs have protective clamps to safeguard the gate and collector side. The authors state

[67]: "*In modern radar transmitters, modulator designs use stacked IGBTs for high current and low repetition rate applications and use stacked MOSFETs for low current and high repetition rate applications.*"

14.7.3 Doppler Radar Pulse Power Supply

The introduction of solid-state switches to the radar systems was being considered by 1998 [68] because of the availability of IGBTs. Until this time, ground-based radar systems were using thyratrons in radar modulators with peak powers of 700 kW. The motivation for replacing the thyratron is the extension of mean time to failure by at least a factor of two. The solid-state approach also eliminates the filament/heater power supply and the high-voltage trigger circuit of the thyratron, reducing the complexity. Thyratrons exhibit increasing delay and jitter as they age, contributing to the phase noise in the radar system.

The circuit topology for driving the magnetron is shown in Fig. 14.35. The magnetron requires a pulse of 27 kV in magnitude with a maximum magnetron current of 27 A. A high-voltage transformer is used to generate the high voltage on its secondary side. A pulse-forming network is used on the primary side to develop the voltage pulse triggered by turning on and off the IGBT. Several pulse-forming networks can be switched into the circuit using the pulse width switch. The pulse-forming network 1 generates a long pulse width of 2 μs at a repetition rate of 500 Hz. The pulse-forming network two generates a short pulse width of 0.8 μs at a repetition rate of 1200 Hz.

Ten IGBTs are connected in series because the power supply voltage is 11 kV. They must handle a peak current of 1200 A with a rise time of 150 ns. The average power dissipation in the IGBTs is 150 W. The authors state: "*Although the rate of rise of current is higher for the thyratron than for the solid-state switch, its jitter is much better. The difference in rise time does not affect the performance of the transmitter.*"

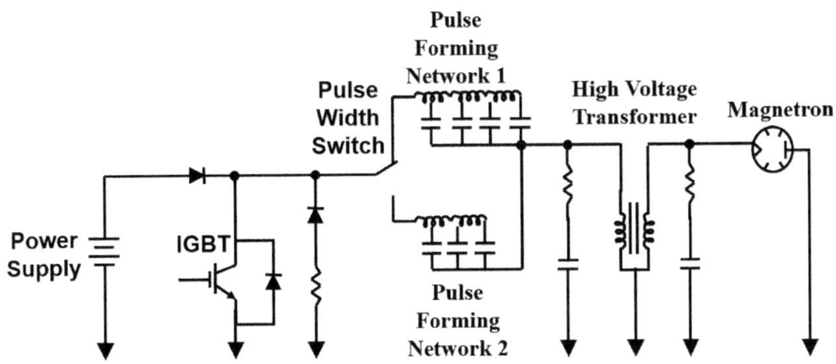

FIGURE 14.35

Circuit diagram for magnetron power supply. *IGBT*, insulated gate bipolar transistor.

14.7.4 Agile Mirror Radar

It is advantageous to be able to rapidly steer the radar along the azimuth and altitude. A plasma sheet offers an inertia-free method to steer the microwave beam in the radar [69]. A 60 × 60 cm plasma sheet can be used to direct X-band microwave beams. The plasma mirror is formed by using a high-voltage, variable-width pulse with fast rise time and high repetition frequency. The "crossatron" tube was used until 1996 as the primary switching element for development of the agile mirror radar. For the solid-state version, the authors state [69]: "*The IGBT was chosen for this application due to its high current capacity and high frequency capability at voltages of 1.2–1.6 kV.*"

The principles of the plasma mirror concept are illustrated in Fig. 14.36. The beam from a 10 GHz source is steered by using the plasma mirror. The 60 × 60 cm plasma mirror is generated inside a vacuum chamber at a pressure of 100–200 mTorr. It is magnetically confined to a thickness of 1 cm by using the Helmholtz field coils that create a 200 G axial field. The plasma is created using a high-voltage (8 kV) pulse. The pulse modulator must create square waves with voltages of up to 8 kV and deliver currents of 30 A in magnitude. Ten 500 μs wide pulses are required for every second. To steer the beam, an array of cathodes is pulsed for each orientation of the beam. Consequently, multiple modulators are needed, making the solid-state solution with IGBTs more attractive than the crossatron. The authors state [69]: "*As a less-expensive, less-complex alternative to the crossatron, IGBTs were chosen to serve as the primary switching element for the cathode array experiments. IGBTs have high current capability with low ohmic losses and can be gated at high frequency using a low voltage.*" At least 10 IGBTs were required in series to generate the desired 8 kV pulse based on available commercial IGBT ratings of only up to 1.6 kV in 1996.

The circuit topology for operating the IGBTs in series is shown in Fig. 14.37. Ten IGBTs are connected in series to deliver the high-voltage pulses to the plasma mirror

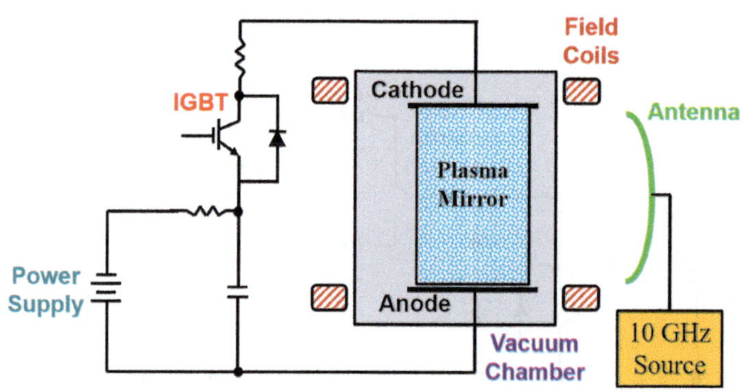

FIGURE 14.36

Plasma mirror steering topology. *IGBT*, insulated gate bipolar transistor.

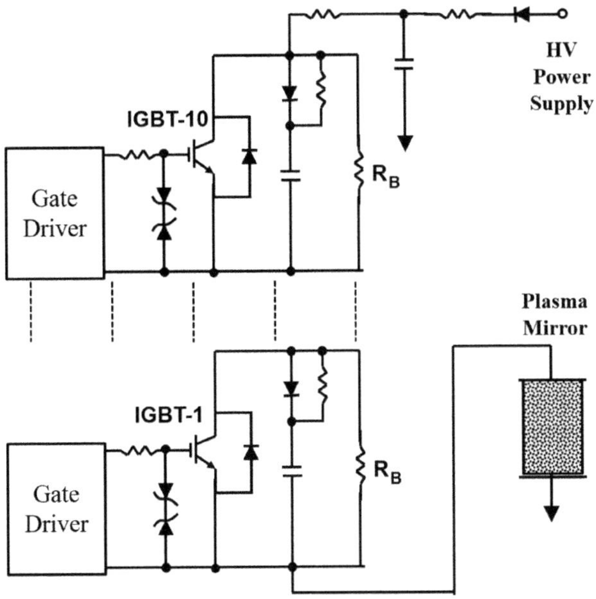

FIGURE 14.37

Plasma mirror steering topology. *IGBT*, insulated gate bipolar transistor.

for an 8 kV high-voltage power supply. It is important that the voltage is equally shared by the IGBTs. In steady-state operating conditions, the voltage sharing is determined by the leakage current of the IGBTs, which can defer from device to device. A parallel balancing resistor (R_B) is used to make sure the voltage is shared equally by the IGBTs, independent of their leakage currents. The value for the balancing resistor can be computed using:

$$R_B = \frac{(0.9nV_{CES} - V_S)}{[\Delta I_{CES}(n-1)]} \tag{14.1}$$

where V_{CES} is the voltage across each IGBT (e.g., 1.2 kV), V_S is the high-voltage supply voltage (8 kV), ΔI_{CES} is the leakage current for each IGBT (e.g., 10 mA), and n is the number of devices in series (e.g., 10). For the indicated values, the balancing resistor value is found to be 10 kΩ. The supply voltage must also be shared equally by the IGBTs during the switching transients. This was accomplished by synchronizing the gate drive to within 5 ns for all the IGBTs using a Fuji Electric gate driver (EXB840) with optical isolation.

14.7.5 Ground-Based Radar for Theater Missile Defense

The ground-based radar system for theater missile defense systems requires a megawatt-rated inverter [70]. The inverter provides three-phase, 60 Hz, AC power at 4.16 kV with continuous power of 1 MW. A transformerless, three-level neutral

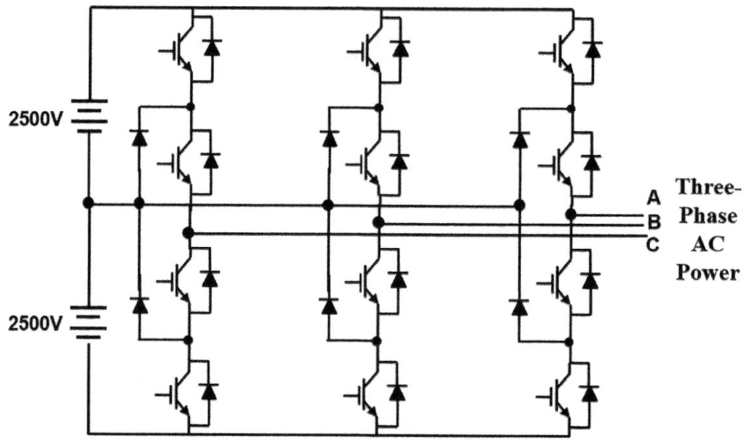

FIGURE 14.38

Neutral point clamped converter inverter for theater missile defense ground radar.

point clamped converter (NPCC) has been designed to satisfy these requirements. The basic topology for the NPCC inverter is shown in Fig. 14.38. Six IR modules and three SEMIKRON modules were used with IGBTs rated at 600 V and 165 A. This inverter was able to deliver 100 kW of power. The diodes are used to clamp the midpoint voltage, leading to proper voltage distribution among the four IGBTs within each phase leg.

14.8 IGBTs for Defense Applications

The ratings of IGBTs required for a variety of military applications have been discussed in the previous sections. In many cases, the devices are required for pulse-powered applications. For defense applications, an important criterion is the reliability of the device. Another consideration is higher junction operating temperature for the IGBT because this can reduce the heat sink size and weight in portable applications.

14.8.1 Pulse Power Capability

IGBT modules with 2.5 kV, 2000 A manufactured by ABB have been evaluated for such applications [71]. This module was originally developed for traction applications. The module was evaluated for generating short current pulses from a capacitive energy source as a closing switch and from an inductive source as an opening switch. The IGBT module consists of six submodules each containing six 1.5 cm^2-sized IGBTs and three diodes. The 36 total IGBT chips have a continuous rated current of 2000 A. With a high-voltage capacitor as the source, it was found that the IGBT module was capable of generating a 12 kA current pulse with a pulse width of at least 5 μs. The turn-on and turn-off times for the voltage transient were measured at 910 and 680 ns, respectively. With an inductive energy source, the module was able to generate a 4.8 kA pulse with a rise time of 10 μs.

14.8.2 **Reliability**

The well-known bathtub curve for the reliability analysis of electronic components, such as the IGBT, is illustrated in Fig. 14.39. It consists of three parts. The portion with the constant failure rate is taken as the lifespan of the component. Components that undergo the high early failure rates can be screened out by the manufacturer before distribution. The high failure rate due to wear out is a function of the materials and processes used to make the device. The mean time between failures (MTBF) is defined as the inverse of the failure rate (λ).

In the case of IGBT modules, the main failure modes are: (1) breakdown and leakage of the die insulation, (2) failure of wirebonds to the die, and (3) failure of the solder used for die attachment [72]. Thermal cycling of the module leads to liftoff of the aluminum wirebonds. This is manifested as an increase in the on-state voltage drop of the IGBT due to an increase in series resistance and degradation of current uniformity in the die. A large amount of load cycling in pulse power applications leads to wear out of the solder joints in the module. This is manifested by an increase in the thermal resistance for the die, which eventually leads to failure of the die due to a high junction temperature.

The failure rate model for the IGBT is given by:

$$\lambda_{IGBT} = \lambda_{b\_IGBT} \cdot \pi_{Tj} \cdot \pi_A \cdot \pi_S \cdot \pi_E \tag{14.2}$$

where λ_{IGBT} is the number of failures in 10^6 h, λ_{b_IGBT} is the base failure rate (0.133 for medium-power IGBT modules), and π_{Tj} is the junction temperature (T_j in °C) factor, which can be calculated using:

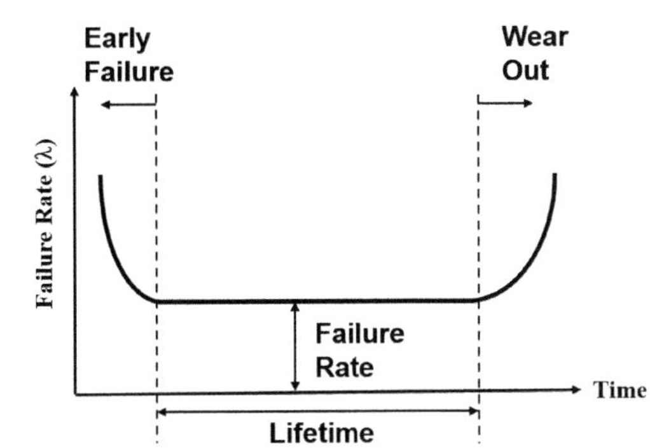

FIGURE 14.39

Reliability bathtub curve.

$$\pi_{Tj} = \exp\left[-2114\left(\frac{1}{T_j + 273} - \frac{1}{298}\right)\right] \quad (14.3)$$

π_A is the application type factor (0.7 for switching applications) and π_S is the voltage stress factor, which can be calculated using:

$$\pi_S = 4 \cdot 5 \cdot 10^{-3} \cdot e^{3.1 V_{DC}/V_{Rated}} \quad (14.4)$$

where V_{DC} is the DC link voltage, V_{Rated} is the device blocking voltage rating, and π_E is the environmental factor, which is 13 for airborne inhabited aircraft.

The lifetime of IGBT modules is determined by the lifetime for the wirebonds and the solder joints. The number of power cycles to failure of wirebonds (N_{fWB}) is given by:

$$N_{fWB} = k_{fTj} \cdot (\Delta T_j)^{k}_{\Delta T_j} \cdot \exp\left(\frac{k_{Tj}}{T_j + 273}\right) \quad (14.5)$$

where k_{fTj} is 9830, $k\Delta_{Tj}$ is -5, and k_{Tj} is 1124. The number of power cycles to failure of the solder joint (N_{fSJ}) is given by:

$$N_{fSJ} = k_{fTc} \cdot (\Delta T_c)^{k}_{\Delta T_c} \cdot \exp\left(\frac{k_{Tc}}{T_c + 273}\right) \quad (14.6)$$

where k_{fTc} is 9830, $k\Delta_{Tc}$ is -5, and k_{Tj} is 337.

When computing the failure rates and MTBF for power converters, it is necessary to take into account other components such as capacitors. It is also dependent on the PWM modulation scheme and frequency of operation as well. This analysis has been performed for the case of the more electric aircraft with six-pack IGBT EUPEC FS450R12KE3 modules [72]. The impact of changing the PWM control mode from the low-loss D-PWM to the continuous sinusoidal S-PWM at 20kHz is given in Table 14.4. It can be seen that the D-PWM control mode leads to a longer MTBF and lifetime. A similar analysis has been performed regarding the impact of increasing the PWM switching frequency from 20 to 30 kHz. The results given in Table 14.5 indicate a reduction in the MTBF and lifetime due to larger switching losses in the IGBT module at the higher frequency. The impact of using a single versus two IGBT modules in parallel has also been examined. The results shown in Table 14.6 demonstrate an increase in the lifetime because of reduced electrical and thermal stress for the IGBTs, but the failure rate for the converter increases due to the larger number of components.

Table 14.4 IGBT Reliability Analysis: Control Mode

IGBT Module	D-PWM	S-PWM
Failure rate (per million hours)	1.45	2.12
MTBF (h)	690,000	472,000
Lifetime (h)	458,000	12,000

IGBT, *insulated gate bipolar transistor;* MTBF, *mean time between failures.*

Table 14.5 IGBT Reliability Analysis: Switching Frequency

IGBT Module	20 kHz	30 kHz
Failure rate (per million hours)	1.45	1.81
MTBF (h)	690,000	552,000
Lifetime (h)	458,000	37,000

IGBT, *insulated gate bipolar transistor;* MTBF, *mean time between failures.*

Table 14.6 IGBT Reliability Analysis: Single Versus Two Modules in Parallel

IGBT Module	Single	Two in Parallel
Failure rate (per million hours)	1.45	2.52
MTBF (h)	690,000	397,000
Lifetime (h)	458,000	10,323,000

IGBT, *insulated gate bipolar transistor;* MTBF, *mean time between failures.*

Table 14.7 Converter Reliability Analysis

	IGBT Module	Capacitors	Inductor	Power Stage
Failure rate (per million hours)	1.45	1.07	0.012	2.532
MTBF (h)	690,000	937,000	83,333,000	395,000
Lifetime (h)	458,000	125,000	2,153,000	125,000

IGBT, *insulated gate bipolar transistor;* MTBF, *mean time between failures.*

The reliability prediction for the converter requires inclusion of the capacitors and inductor in the analysis. The results are given in Table 14.7 for the case of a design with eight film capacitors, one six-pack IGBT module, and a three-phase inductor [72]. It can be concluded that the IGBT and capacitor are the main components limiting the reliability.

14.9 Summary

The IGBT was shunned by the DoD during the early years of its development in favor of the MCT despite evidence of the benefits of a wide forward and reverse safe operating area for the IGBT. The IGBT became rapidly commercialized and widely adopted for commercial and industrial applications while the MCT was not successfully manufactured. Consequently, US military systems eventually made use of the commercially available IGBTs to build high-performance systems. The IGBT has now been embraced and widely utilized by the US Navy, Army, and Air Force. It has also become an important component in the theater missile defense network.

References

[1] Defense Industry. en.wikipedia.org/wiki/Defense_industry.
[2] United States Department of Defense. en.wikipedia.org/wiki/United_States_Department_of_Defense.
[3] Arms Industry. en.wikipedia.org/wiki/Arms_industry.
[4] Share of World Military Expenditures. https://www.axios.com/countries-highest-military-spending-us-china-48102c49-7dd7-4215-a24b-1d31947cf373.
[5] A. Marksteiner, et al., The SIPRI Top 100 Arms-Producing and Military Services Companies, 2020. SIPRI Factsheet 2021, 2021.
[6] N.G. Hingorani, et al., Research coordination for power semiconductor technology, Proc. IEEE 77 (1989) 1136–1388.
[7] N. Hingorani, Power electronics building block concepts panel discussion, in: IEEE Power Engineering Society General Meeting, vol. 3, 2003, pp. 1339–1343.
[8] T. Ericsen, N. Hingorani, Y. Khersonsky, Power electronics and future marine electrical systems, in: IEEE Petroleum and Chemical Industry Technical Conference, 2004, pp. 163–171.
[9] T. Ericsson, The second electronic revolution, in: IEEE Petroleum and Chemical Industry Conference, 2009, pp. 1–10.
[10] T. Ericsen, Power electronic building blocks – a systematic approach to power electronics, in: IEEE Power Engineering Society Summer Meeting, vol. 2, 2000, pp. 1216–1218.
[11] T. Ericsen, Y. Khersonsky, P.K. Steimer, PEBB concept applications in high power electronics converters, in: IEEE Power Electronics Specialists Conference, 2005, pp. 2284–2289.
[12] T. Ericsen, N. Hingorani, Y. Khersonsky, PEBB – power electronics building blocks – from concept to reality, in: IEEE Petroleum and Chemical Industry Conference, 2006, pp. 1–7.
[13] T. Ericsen, The ship power electronic revolution: issue and answers, in: IEEE Petroleum and Chemical Industry Conference, 2008, pp. 1–11.
[14] T. Ericsen, Power electronics building blocks and potential power modulator applications, in: IEEE Power Modulator Symposium, 1998, pp. 12–15.
[15] Q. Huang, K. Borisov, H.L. Ginn, PEBB-based shunt active power filter for shipboard-power systems, in: IEEE Electric Ship Technologies Symposium, 2005, pp. 393–399.
[16] A. Mayor, et al., Commutation behavior analysis of a dual 3L-ANPC-VSC phase-leg PEBB using 4.5-kV and 1.5-kA HV-IGBT modules, IEEE Trans. Power Electron. 34 (2019) 1125–1141.
[17] A. Deshpande, et al., Design of a high-efficiency, high specific-power three-level T-type power electronics building block for aircraft electric-propulsion drives, IEEE J. Emerg. Sel. Top. Power Electron. 8 (2020) 407–416.
[18] Frigate. en.wikipedia.org/wiki/Frigate.
[19] D.S. Parker, C.G. Hodge, The electric warship, Power Eng. J. 12 (1) (1998) 5–13.
[20] T. Ericsen, Engineering 'Total electric ship', in: IEEE Petroleum and Chemical Industry Conference, 2007, pp. 1–6.
[21] D. Gritter, S.S. Kalsi, N. Henderson, Variable speed electric drive options for electric ships, in: IEEE Electric Ship Technologies Symposium, 2005, pp. 347–354.
[22] A.R. Millner, et al., A high power density DC-AC inverter with staggered regenerative recovery circuit, in: IEEE Applied Power Electronics Conference, vol. 2, 1995, pp. 973–976.

[23] H. Sayed, H.S. Krishnamoorthy, Fault-tolerant Z-source-based isolated DC-DC building blocks for Mission-critical DC distribution applications, in: IEEE Annual Conf. Industrial Electronics Society, 2021, pp. P1–P6.
[24] J. Commerton, M. Zahzah, Y. Khersonsky, Solid-state transfer switches and current interruptors for mission-critical shipboard power systems, in: IEEE Electric Ship Technologies Symposium, 2005, pp. 298–305.
[25] Circuit Breaker. en.wikipedia.org/wiki/Circuit_breaker.
[26] S. Krstic, et al., Circuit breaker technologies for advanced ship power systems, in: IEEE Electric Ship Technologies Symposium, 2007, pp. 201–205.
[27] R.F. Schmerda, et al., IGCTs vs IGBTs for circuit breakers in advanced ship electrical systems, in: IEEE Electric Ship Technologies Symposium, 2009, pp. 400–405.
[28] E. Koptjaev, E. Popkov, Direct conversion electric propulsion system, in: IEEE International Conference on Industrial Engineering Applications and Manufacturing, 2018, pp. P1–P5.
[29] X. Xu, et al., Modulation and control of an ANPC/H-bridge hybrid inverter for ship electric propulsion drives, in: IEEE International Conference on Electrical Machines and Systems, 2019, pp. P1–P6.
[30] Aircraft Carrier. en.wikipedia.org/wiki/Aircraft_carrier.
[31] USS Nimitz (CVN-68). en.wikipedia.org/wiki/USS_Nimitz_(CVN-68).
[32] D. Tesar, Electro-mechanical actuators for the Navy's ships, in: IEEE Electric Ship Technologies Symposium, 2005, pp. 387–391.
[33] D.A. Fink, et al., High-voltage IGBT switching arrays, IEEE Trans. Magn. 45 (2009) 282–287.
[34] Railgun. en.wikipedia.org/wiki/Railgun.
[35] https://www.ga.com/general-atomics-successfully-tests-railgun-hypersonic-projectiles.
[36] Navy to Deploy Electromagnetic Railgun Aboard JHSV. www.navy.mil/submit/dispaly.asp?story_id=80055.
[37] A. McDuffee, Navy's New Railgun Can Hurl a Shell over 5000 Mph, 2014. www.wired.com/2014/04/electromagnetic-railgun-launcher/.
[38] L.N. Domaschk, et al., Coordination of large pulsed loads on future electric ships, IEEE Trans. Magn. 43 (2007) 450–455.
[39] Railguns to Launch Planes from Aircraft Carriers. http://forum.worldofwarships.com/index.php?/topic/1801-railguns.
[40] E. Ackerman, Electromagnetic Railgun Launches Fighter Jet for the First Time, 2010. www.dvice.com/archives/2010/12/electromagnetic_1.php.
[41] Electromagnetic Aircraft Launch System. http://en.wikipedia.org/wiki/Electromagntic_aircraft_launch_system.
[42] Submarine. en.wikipedia.org/wiki/Submarine.
[43] Nuclear Submarine. en.wikipedia.org/wiki/Nuclear_submarine.
[44] Ohio-class Submarine. en.wikipedia.org/wiki/Ohio-class_submarine.
[45] E. Miasnikov, What is Known about the Character of Noise Created by Submarines. http://www.armscontrol.ru/subs/snf/snf03221.htm.
[46] G. Jean, Diesel-Electric Submarines, the U.S. Navy's Latest Annoyance, 2008. http://www.nationaldefensemagazine.org/archive/2008/April/Pages/.
[47] Y. Khersonsky, R. Lee, C. Mak, High efficiency quiet electric drive, in: IEEE Power Electronics, Machines, and Drives, vol. 2, 2004, pp. 568–573.
[48] S. Large, P. Walker, Rapid cycle testing of high current IGBT power switch modules, in: IEEE Power Electronics and Variable Speed Drives, 2000, pp. 235–240.

[49] N. Choudhury, US Military to Spend $2.4 Billion on Electric Vehicles by 2020, 2013. http://www.rtcc.org/2013/11/07/us-military-to-spend.
[50] Hybrid Electric Drives (HED) for Armored Fighting Vehicles. http://defense-update.com/features/du-3-05/feature-HED-afv.htm.
[51] Is this Hybrid Tank the Future of American Warfare. http://www.fastcoexist.com/1681938/is-this-hybrid-tank.
[52] GCV Infantry Fighting Vehicle. en.wikipedia.org/wiki/GCV_Infantry_Fighting_Vehicle.
[53] Auto BAE to Offer Hybrid Electric Vehicle to Replace Army's Bradley Armored Personnel Carrier. http://www.dailytech.com/BAE+to+Offer+Hybrid+Electric+Vehicle.
[54] D.P. Uriuoli, C.W. Tipton, Development of a 90-kW bi-directional DC-DC converter for power dense applications, in: IEEE Applied Power Electronics Conference, 2006, pp. 1375–1378.
[55] J.D. Engelland, The evolving revolutionary all-electric airplane, IEEE Trans. Aerosp. Electron. Syst. AES-20 (1984) 217–220.
[56] List of Active United States Military Aircraft. en.wikipedia.org/wiki/List_of_active_United_States_military_aircraft.
[57] C.R. Spitzer, The all-electric aircraft: a systems view and proposed NASA research programs, IEEE Trans. Aerosp. Electron. Syst. AES-20 (1984) 261–265.
[58] J. Weimer, Electrical power technology for the more electric aircraft, in: IEEE Digital Avionics Systems Conference, 1993, pp. 445–450.
[59] J. Weimer, The role of electric machines and drives in the more electric aircraft, in: IEEE International Electric Machines and Drives Conference, vol. 1, 2003, pp. 11–15.
[60] J.S. Cloyd, Status of the United States air force's more electric aircraft initiative, IEEE Aerosp. Electron. Syst. Mag. 13 (2) (1998) 17–22.
[61] J.A. Rosero, et al., Moving towards a more electric aircraft, IEEE Aerosp. Electron. Syst. Mag. 22 (2) (2007) 3–9.
[62] R. Allen, J. Neri, A battery powered 200-kW rapid capacitor charger for a portable railgun in burst mode operation at 3 RPS, in: IEEE Pulsed Power Conference, 2007, pp. 1500–1504.
[63] R. Allen, et al., Development of a 150-kW, battery powered, rapid capacitor charger for a small railgun in burst mode operation at 3 RPS, in: IEEE International Power Modulator and High Voltage Conference, 2008, pp. 106–108.
[64] R. Allen, et al., Progress towards a self-contained rapid capacitor charger for a small railgun in burst mode operation at 3 RPS, in: IEEE International Power Modulator and High Voltage Conference, 2012, pp. 218–220.
[65] Missile Defense. en.wikipedia.org/wiki/Missile_defense.
[66] Radar. en.wikipedia.org/wiki/Radar.
[67] D. Okula, Series stacked switches for radar transmitters, in: IEEE International Power Modulator Symposium, 2006, pp. 256–259.
[68] F. Gekat, D. Ruhl, J. Gosch, An IGBT-based switch as replacement for thyratrons in Doppler radar transmitters, in: IEEE International Power Modulator Symposium, 1998, pp. 110–113.
[69] M.C. Myers, et al., Power modulator for broadband agile mirror radar utilizing semiconductor switching, in: IEEE International Power Modulator Symposium, 1996, pp. 118–121.

[70] M. Giesselmann, B. Crittenden, J. Fonseca, Design and test of a high power inverter for a ground based radar system for theater missile defense, in: IEEE International Pulsed Power Conference, vol. 2, 1997, pp. 1542–1547.
[71] S. Scharnholz, et al., Investigation of IGBT-devices for pulsed power applications, in: IEEE International Pulsed Power Conference, vol. 1, 2003, pp. 349–352.
[72] R. Burgos, et al., Reliability-oriented-design of three-phase power converters for aircraft applications, IEEE Trans. Aerosp. Electron. Syst. 48 (2012) 1249–1263.

CHAPTER

IGBT Applications: Renewable Energy

15

The need for the global economy to shift from using fossil fuels for supplying our electrical energy demands to renewable energy sources has become widely accepted and increasingly urgent due to the acceleration of global warming. The proliferation of intense and large wildfires in the United States and Australia has been linked to droughts caused by global warming. The melting of glaciers and ice shelfs in the Artic and Antarctic regions is producing a rise in sea levels that is inundating islands in the Pacific Ocean.

There has been a surge in activity around the world to invest in energy generation from renewable sources. Renewable energy is available from sources that are constantly replenished within the timescale of humans [1]. Examples are hydropower, direct sunlight, wind, and ocean waves that are fueled by the radiation from the Sun. An additional resource is geothermal energy fueled by the heat stored in the Earth's core and by natural radioactivity.

Although hydropower has been used for a long time for assisting human activity, it is relevant here from the point of view of generating electricity. Large hydroelectric stations, such as the Hoover Dam, were traditionally of interest. Small hydroelectric generation capability has recently become of interest in developing countries. Hydroelectric power comprises 16% of the electricity generated in the world.

The fastest growth segments for producing electricity from renewable resources are photovoltaic cells and windmills. Wind power has an annual growth rate of 30% with more than 280 GW of power being generated in 2012. The wind power generation capacity has grown to 743 GW in 2021 [2]. The potential for the generation of electrical power using wind energy is 40 times the present worldwide electricity demand.

It is estimated that the radiation received by the Earth from the Sun can generate 122 PW (petawatts $PW = 10^{15}$ W) of power, which is about 10,000 times the 13 TW (terawatts $TW = 10^{12}$ W) of power presently being consumed [3]. The capacity for generating electricity with photovoltaics has been growing at the rate of more than 30% with a worldwide installed capacity of 139 GW in 2013, which has grown to 629 GW in 2019. Worldwide, 183 GW of solar power was installed in 2021 [4]. Wind and solar power generation has exceeded 10% of global electricity demand for the first time in 2021 [5].

The impact of the "Green Revolution" has been highlighted by mainstream media such as an article in Time Magazine in 2014 [6], which states: "*The prices for photovoltaic panels have dropped by 80% in the last 5 years. A new solar-power*

The IGBT Device. https://doi.org/10.1016/B978-0-323-99912-0.00020-9

system is now installed on an American roof every three or four minutes." Saudi Arabia is investing $100 billion in solar electricity generation and China is becoming the world's largest market for wind and solar energy with 208 GW installed by 2020.

In all the renewable electricity generation systems, it is essential to use an inverter to convert the power being generated, either as variable frequency AC, for example in windmills, or as DC in solar cells, to a well-regulated 60 Hz AC power that can be delivered and distributed in the energy grid. This requires highly efficient inverters based on insulated gate bipolar transistors (IGBTs) as switching elements. Consequently, the entire renewable energy sector of the economy relies on IGBTs to be a successful resource for society [7].

This chapter describes all the major approaches for generating electricity from renewable energy sources. It will begin with hydropower because this resource was established well before other approaches. It will then delve into solar and wind power, which are eliciting a lot of interest in the power electronics community. A discussion of wave power and geothermal power is included for completeness. Energy storage is becoming increasingly important to complement the generation of electricity from solar and wind power because their output varies during the day and is dependent on the availability of sunshine or a stiff wind. This effort is described in Chapter 18.

15.1 Hydroelectric Power

One of the earliest hydropower electricity-generating dams was set up in 1892 at Niagara Falls between the border of the United States and Canada [8]. The power station shown in Fig. 15.1(a) was the largest hydroelectric power facility in the world when installed and still provides 2.4 GW of power within New York state [9]. In 1936, the Hoover Dam, shown in Fig. 15.1(b) with the power station visible at the bottom of the image, was built with the ability to produce 1.345 GW of power [10]. The water turbines used to generate electricity at Hoover Dam are shown in Fig. 15.1(c). This was followed by construction of the 6.8 GW Grand Coulee

(a) (b) (c)

FIGURE 15.1

(a) Niagara Falls dam; (b) Hoover dam; (c) turbines.

Dam in 1942, which remains the largest hydroelectric generator in the United States. These hydroelectric plants have been dwarfed by the Three Gorges Dam built in China in 2008 with 22.5 GW of generating capacity.

Hydropower is considered an indirect means of utilizing solar energy incident on the Earth. The solar radiation produces evaporation of water from oceans, which is precipitated on land to create rivers and lakes that can be dammed for generating electricity. In 2010, 3427 TWh of electricity was provided by hydroelectric power, comprising 16% of global electricity production. By 2020, the worldwide electricity production by hydropower had grown to 1330 GW, with the largest share produced by China (28%), the second largest by Brazil (8%), and the next largest by the United States (7.1%) [11]. The Asia–Pacific region accounts for 32% of global hydropower production. In the United States, 6.4% of the electricity is supplied by more than 2000 hydroelectric power plants. Globally, the largest hydroelectric facilities are the Three Gorges Dam in China (22.5 GW), the Itaipu Dam in Brazil (14 GW), and the Guri Dam in Venezuela (10.2 GW). Small hydroelectric plants that produce up to 10 MW of power have been growing at the rate of 28% and provide 85 GW of power. Among these, 65 GW was installed in China, 3.5 GW in Japan, 3 GW in the United States, and 2 GW in India. In addition, micro hydroelectric plants that generate up to 100 kW and pico hydroelectric plants that produce under 5 kW are popular in developing countries for rural areas.

Hydroelectricity is considered a flexible energy source that can be rapidly ramped up and ramped down based on consumer demand because hydroturbines can be started up in a few minutes. Another advantage of hydroelectricity is the elimination of fuel costs associated with coal- and gas-powered electricity generation. The average cost for electricity from a hydroelectric plant is $ 0.05 per kWh, about half the average cost of electricity production in the United States [10]. Hydroelectric plants have a long lifespan with some still in service after 100 years (with upgraded equipment). The major disadvantage of hydroelectric plants is the disruption of the local ecosystem both upstream and downstream by damming the flow of water from a river.

15.1.1 Large Power Plants

In the case of large hydroelectric power plants, the electricity is distributed over high-voltage DC (HVDC) transmission lines shown in Fig. 15.2 from the point of power generation to the point of use, such as industrial plants or cities. This has traditionally been performed using high-power thyristors with a blocking voltage of 8 kV and 2 kA current-handling capability with current source controllers (CSC). A new generation of HVDC systems based on voltage source control (VSC) has become recently feasible with the availability of IGBTs with high blocking voltage and current-handling capability. The classical HVDC system using thyristors and CSC is contrasted with the new VSC-based HVDC system using IGBTs in Fig. 15.3.

The traditional HVDC system for power transmission and distribution (T&D) shown at the top of Fig. 15.3 makes use of thyristors with line commutation to control current flow [12]. The DC current flow is unidirectional as determined by the DC

FIGURE 15.2

HVDC power transmission.

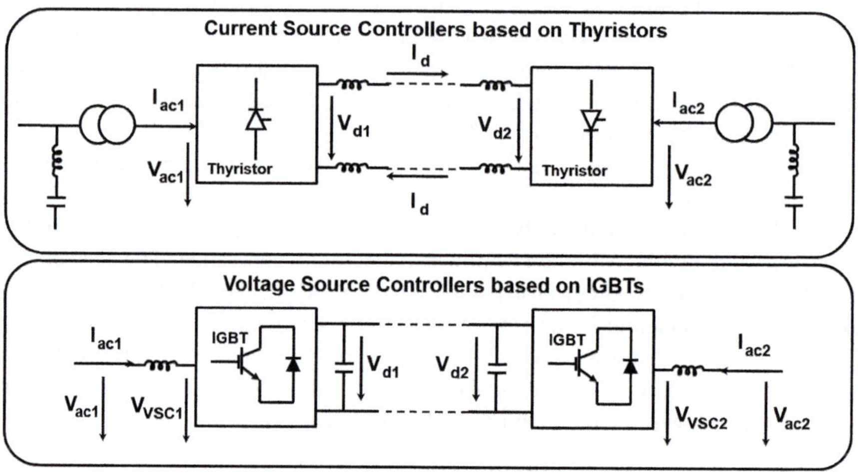

FIGURE 15.3

CSC *(top)* and VSC *(bottom)*-based HVDC transmission systems.

side equilibrium voltage of the two converters. The valve requires reactive power generation using the shunt connected filter shown in the figure. The new HVDC system based on VSC is shown in the bottom of Fig. 15.3. In this case, the AC side voltage can be independently determined with the current decided by the series inductance shown in the figure. Active and reactive power can be controlled on both ends of the network. Harmonic generation is low in this approach because of high switching frequencies for the IGBTs.

The authors state [12]: "*Today, the IGBT is the preferred choice for VSC-based converters in T&D applications because of the following characteristics: (1) low-power control because it is an MOS-controlled device. This is advantageous when operating at very high-voltage levels (several 100 kV); (2) transistor action, which enables precise control of the device in a manner that is not possible with latching alternatives. For instance, the converter can be turned off even in short circuit conditions; (3) high switching speed, (which) makes high switching frequency feasible.*" A 300 MW HVDC VSC-based system using IGBTs was commissioned in 2002 with three-level converter topology.

One GW of power transmission has been achieved using the latest VSC-HVDC implemented at +/− 320 kV with a transmission line current of 1550 A by using existing IGBT modules rated at 3.3 kV [13]. The next-generation 3 GW VSC-HVDC transmission system operating at +/− 500 kV with 3000 A line currents will be implemented using press-pack IGBT modules. The use of reverse-conducting (RC) IGBTs is beneficial to integrate the IGBT with the fly-back diode to achieve space savings.

The new VSC-HVDC installations utilize IGBTs in press packs (previously shown in Chapter 6). The IGBT cell design is the 4.5 kV trench gate structure with field stop technology (previously shown in Fig. 9.13). The devices can handle 3000 A of current with a voltage drop of only 2.2 V [14]. The fly-back diodes are separately packaged. These devices will be implemented in 800 kV 8 GW power stations with the VSC-HVDC system used for long-distance power transmission within China. The VSC-HVDC power transmission with IGBTs is also being utilized for delivering power from off-shore wind farms to the mainland [15]. These systems also require circuit breakers for protection during faults. An ultrahigh-voltage gate driving approach has been proposed for trench gate IGBTs (IEGTs) to obtain very high saturation currents during the fault [16]. A gate drive voltage of 30 V is used, even though the device manufacturer recommends 15 V, to raise the saturation current. Up to 32,000 A of current was turned off using a gate bias of 30 V for an IGBT rated at 3000 A by the manufacturer.

15.1.2 Small Power Plants

Small hydropower electricity-generating systems have become of interest in developing countries, such as India, for delivering electricity to remote and rural areas. India has the potential to generate 10 GW of power from small hydroelectric plants. Plants that generate more than 100 kW can be grid connected while lower power units can serve a local community. The small micro and pico power plants supply mainly the needs for lighting and fans where a moderate power quality delivered at minimal cost is adequate.

The input power is constant in the case of power generated from flowing water (or a small waterfall) using a turbine. A load side controller is then required to account for variable consumer demands, as shown in Fig. 15.4 [17]. A three-phase self-excited induction generator (SEIG) converts the mechanical energy from the water driven turbine to electrical output. A three-phase induction motor with three-phase balanced capacitors connected to the stator windings serves as a cost-effective SEIG. The load dump circuit is used to compensate for variable consumer demands for

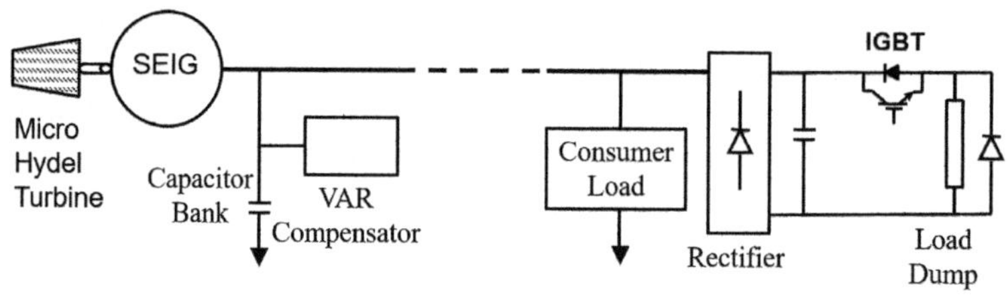

FIGURE 15.4

Micro hydropower system.

power. The load controller has a rectifier stage followed by an IGBT used as a chopper switch to regulate power flow to the load dump resistor. The heat generated in the load dump resistor can be used for heating water or the home. A 6 kW unit was deployed in Kerala State in South India in 1998 as a demonstrator, with more units to follow [17].

15.1.3 Decoupled Voltage and Frequency Controller

In the case of micro and pico hydroelectric power generation, the energy source is an isolated asynchronous generator, as discussed in the previous section. It has a low cost and compact size while also being affordable and reliable. However, it suffers from poor voltage and frequency regulation. This problem can be addressed using a decoupled voltage and frequency controlled converter (DVFC) [18]. The DVFC combines a static compensator (STATCOM) and electronic load controller (ELC). The ELC was discussed in the previous section and serves to compensate for variable consumer demands for electricity while being supplied with a constant power from the generator. The DVFC circuit topology is shown in Fig. 15.5. The STATCOM on the upper right side uses six IGBTs to deliver reactive power to the AC power line using energy stored in the DC bus capacitor C_B and the filter inductors. The combination of the STATCOM and the ELC enables regulation of both the voltage and frequency of the power line.

15.1.4 Auxiliary Generation Units

For reliable operation with good continuity of service, the hydroelectric power plant must operate without interruption of its auxiliary services. Even short interruption of auxiliary services can result in a miss-trip of the generator [19]. The auxiliary generator is mechanically connected to the main generator of the hydroelectric power plant. The inductive impedance of the auxiliary generator can be compensated using a parallel capacitor that is switched using antiparallel IGBTs, as shown in Fig. 15.6. The IGBT controlled series capacitor is a flexible alternating current transmission system (FACTS) device that provides voltage support, dampens oscillations, and reduces losses. Pulse width modulation (PWM) control of the IGBTs has been shown to provide rapid adjustment of the active power.

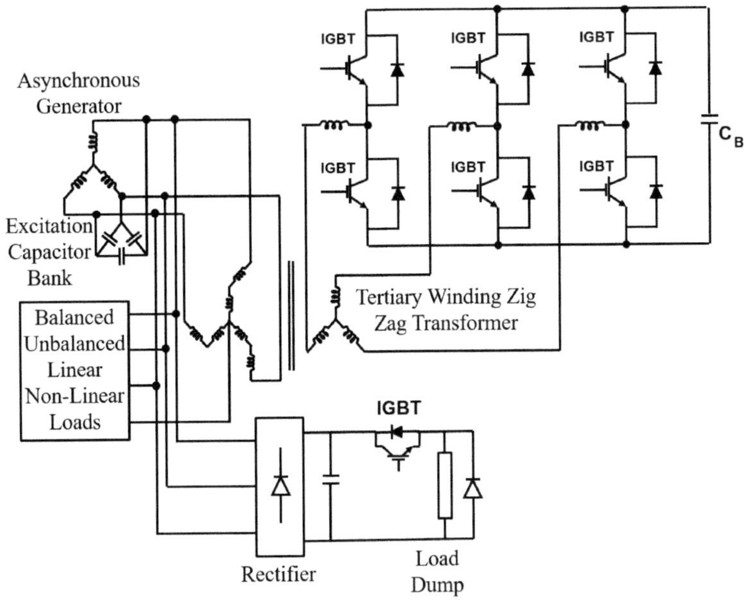

FIGURE 15.5

Decoupled voltage and frequency controller for pico hydropower application.

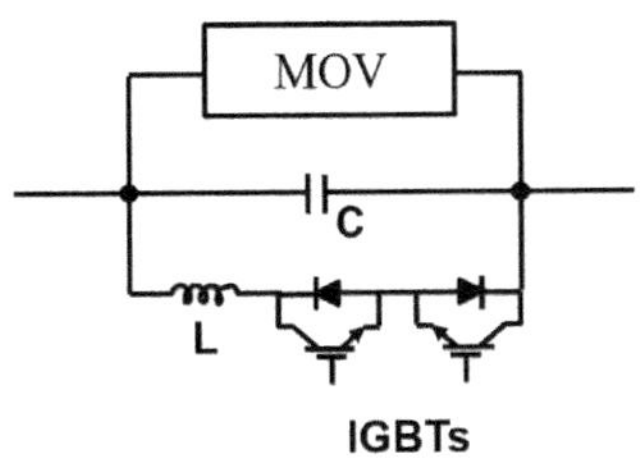

FIGURE 15.6

Auxiliary services regulation using IGBT controlled series capacitor.

15.2 Photovoltaic Power

It is estimated that the solar radiation received by the Earth can be used to generate 122 PW of electric power, which is 7000 times the 18 TW of power that humans consumed in 2021. Among all the renewable energy options, solar power has the highest power density of 170 W/m^2. Photovoltaic (PV) power refers to the generation of electric power by directly converting solar radiation into electric current. This is performed using solar cells that contain semiconductor P-N junctions [20], as illustrated in Fig. 15.7. Sunlight is absorbed by the semiconductor, producing electron–hole pairs or excitons in the depletion region and the neutral regions of

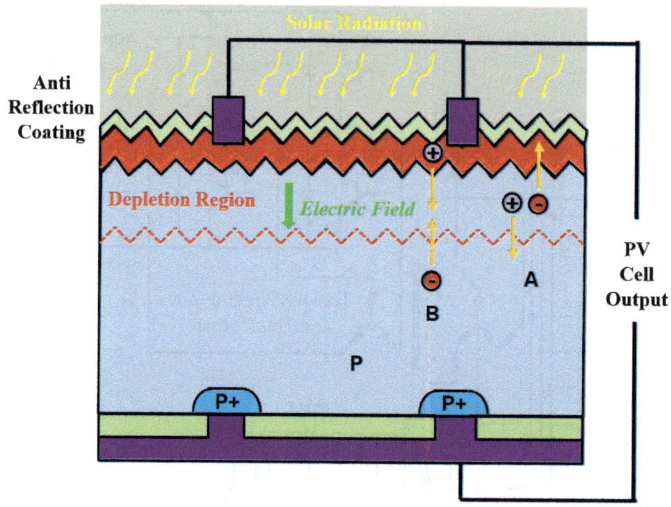

FIGURE 15.7

Principles of solar cell operation.

the semiconductor. The electron–hole pairs are separated by the electric field within the depletion region at the P-N junction before they can recombine (process A). Electrons generated in the P-region within one diffusion length of the depletion region boundary are also collected (process B). Similarly, holes generated in the N^+-region within one diffusion length of the depletion region boundary are collected. The separated carriers flow through the external circuit, producing current flow.

The efficiency of the solar cell is a combination of the reflectance efficiency, the thermodynamic efficiency, the charge carrier separation efficiency, and the current conduction efficiency. The reflectance efficiency is enhanced by corrugating the upper surface of the semiconductor and by the application of antireflection coatings to maximize the penetration of the radiation. The thermodynamic efficiency is dependent on matching the bandgap of the semiconductor to the spectral energy distribution of the sunlight. A high charge carrier separation efficiency is achieved by the strong electric field within P-N junctions in semiconductors. A good current conduction efficiency is derived by making low-resistance ohmic contacts to the semiconductor.

Most commercial photovoltaic panels installed for electric power generation by utilities and on homes rely on silicon as the semiconductor material due to its abundance and low cost when compared with alternatives. Typical large-scale and small-scale photovoltaic installations are shown in Fig. 15.8(a) and (b). The efficiency of solar cells made from silicon is reaching the theoretical power efficiency of 33.7%, called the Shockley–Queisser limit [20]. Solar cells that make use of polycrystalline silicon are popular despite their lower efficiency due to the much lower cost of production. The average efficiency for solar cells used in photovoltaic installations ranges between 12% and 18%.

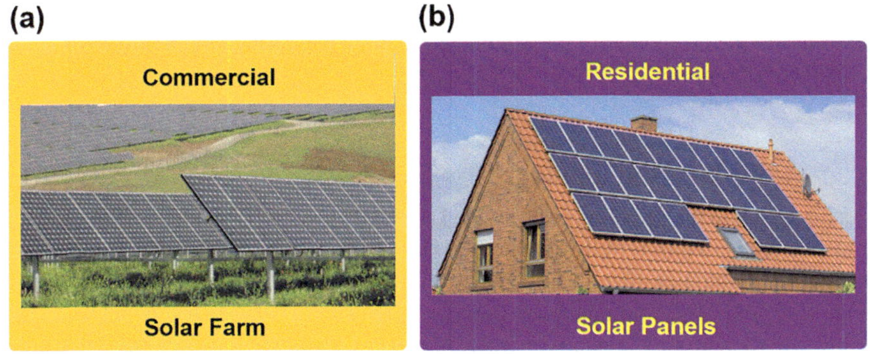

FIGURE 15.8

(a) Large photovoltaic generation plant; (b) home photovoltaic panel.

Grid parity is defined as the point at which the cost of photovoltaic electricity generation becomes equal to that of traditional power generation using coal and natural gas. In islands such as Hawaii, grid parity has already been reached due to the high cost of transporting fossil fuels to the middle of the Pacific Ocean. Grid parity was reached in Spain by 2013. Many countries, such as Germany, are subsidizing the installation of photovoltaic power generation to promote greater security from vulnerable fossil fuel sources delivered from outside the country, leading to a growth rate of 58% for PV in recent years [21]. Electricity delivery from photovoltaics grew from 3.7 TWh (0.02% of total electricity production) in 2005 to 125 TWh (0.54% of total electricity production) in 2013. It has risen to 720 TWh in 2019, comprising 3% of the world's electricity production, and 1000 TWh in 2021 [22]. The world's largest photovoltaic power stations are the Bhadla Solar Park in India (2.25 GW), the Huanghe Hainan Solar Park in China (2.2 GW), the Shakti Sthala Solar Project in India (2 GW), the Benban Solar Park in Egypt (1.65 GW), and the Tengger Desert Park in China (1.55 GW) [23]. A major advantage of photovoltaic power generation is the absence of moving parts, making them very reliable for long-term operation.

Among the various renewable energy options for generating electricity, photovoltaics have the advantage of allowing distributed energy generation by the installation of solar panels on the rooftops of residential and commercial buildings. This creates a unique ability to move the point of generation of the electric power close to the point of use. It eliminates the expensive infrastructure for transmission and distribution of electricity that is associated with fossil fuel power generation. The energy loss in the power distribution network is also eliminated.

Solar cells produce DC output voltages and currents when converting sunlight into electricity. For photovoltaic electric power to be utilized for large-scale power generation and for residential applications, it is essential to convert the generated DC power into the well-regulated 60 Hz (or 50 Hz in some countries) AC power required to operate appliances in homes and equipment in factories. Consequently, the success of photovoltaic technology depends not only on the efficiency and cost of the solar panels but also on the efficiency, cost, and size of the power electronic inverters used to translate the DC power to the regulated AC power. These inverters are built using IGBT-based circuits.

15.2.1 PV Inverter Topologies

As discussed in the previous section, the DC power produced by photovoltaic panels must be converted to a well-regulated 60 Hz (or 50 Hz) AC power. The basic power converter topologies that can be employed to perform this function are shown in Fig. 15.9 [24]. The first approach shown in Fig. 15.9(a) makes use of a DC-to-DC converter to increase the relatively low solar panel DC voltage to a higher value suitable to generate the desired peak AC voltage at the output. The high DC voltage is then converted to the regulated 60 Hz single- or three-phase AC output power with an inverter DC-to-AC stage. This topology does not provide the desired galvanic isolation needed to comply with safety regulations.

In the second topology shown in Fig. 15.9(b), a high-frequency transformer is used to provide the galvanic isolation. This requires conversion of the photovoltaic panel DC voltage to a high-frequency AC voltage fed to the primary side of the transformer. The output of the transformer is rectified to form a DC bus across the capacitor. The desired regulated 60 Hz single- or three-phase AC output power is then generated with an inverter DC-to-AC stage. This approach has the advantage of reducing the size and weight of the transformer by using the high operating frequency.

In the third topology shown in Fig. 15.9(c), a DC-to-AC inverter is directly connected to the DC output of the photovoltaic panel to generate 60 Hz AC power. The AC output voltage is then amplified to the desired level by using a low-frequency

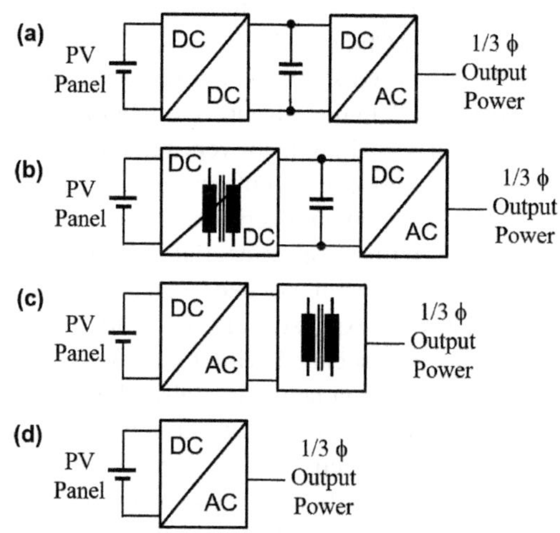

FIGURE 15.9

Power converter topologies for photovoltaic applications.

transformer to generate the desired regulated 60 Hz single- or three-phase AC output power. Although it is the simplest topology, the converter is bulky due to the size and weight of the low-frequency 60 Hz transformer.

In the fourth topology shown in Fig. 15.9(d), the DC output power from the photovoltaic panel is converted to the desired regulated 60 Hz single- or three-phase AC output power in a single stage. Although the topology is simple in form, it requires photovoltaic panels that deliver a higher output DC voltage than the peak AC output voltage, which creates serious safety issues. It can be implemented using the current source inverter (CSI) topology.

Transformerless converters are dominant in the European market because safety class II does not require galvanic isolation there [25]. This reduces not only the cost and weight but eliminates about 2% of losses in the transformer. About 50% of the PV converters in Japan make use of transformerless inverters while most converters in the United States employ transformers.

The solar power inverters for residential use operate at 360–600 V at power levels ranging from 3 to 25 kW. Centralized inverters used for solar farms to generate 0.5–3.5 MW operate at 460–1500 V [26].

15.2.2 HERIC PV Inverter

The HERIC inverter topology has been commercially successful for photovoltaic installations [25]. An H-bridge circuit with four IGBT is used to create the AC power output, as shown in Fig. 15.10. Two additional IGBTs (IGBT-5 and IGBT-6) form the freewheeling path on the output side. IGBT-5 is turned on for positive inductor current while IGBT-6 is turned on for negative inductor current. During the freewheeling mode, IGBTs 1 through 4 are turned off, isolating the PV panel. This topology requires a PV input voltage always greater than the peak AC output voltage, typically 360 VDC for a 230 VAC grid. The HERIC topology implemented with 600 V and 1200 V IGBTs is able to achieve an efficiency of 95% at low (10%) output power and 97% efficiency at full (100%) output power. The single-phase inverter can generate up to 11 kW of AC power.

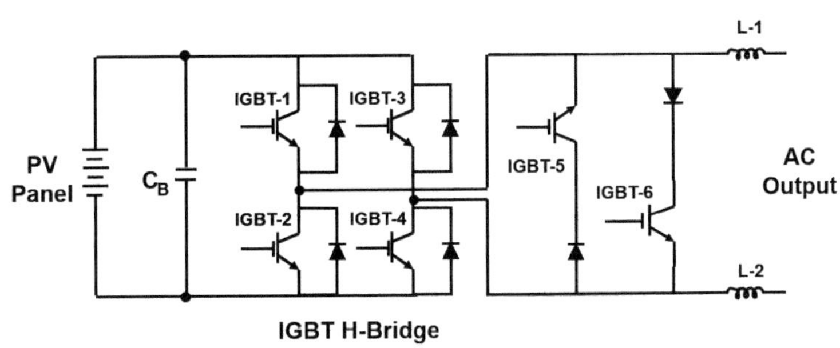

FIGURE 15.10

The HERIC topology for PV panels.

15.2.3 Three-Phase PV Inverter

For operation at larger power levels of PV output, a three-phase topology is required. Six IGBTs are utilized to generate the three-phase AC output waveforms. The capacitors C_1 and C_2 are used to generate a "quiet DC rail," which forms the neutral wire for the AC output connection, as shown in Fig. 15.11. Efficiencies ranging up to 97% have been achieved using this topology [25]. AC output power levels of 30 kW can be achieved using the three-phase inverter topology. Many other variations in the topology can also be configured [25].

15.2.4 Nonisolated Interactive PV Inverter

Due to limited access to fossil fuel resources, the generation of electricity using solar radiation has been of interest in Japan for a long time. The AC power supply using PV panels is provided with a 100 V single-phase three-wire system [27], as shown in Fig. 15.12. This AC power is generated using a PWM H-bridge inverter with four IGBTs. To generate the desired 100 V AC output, the DC voltage from the PV panels has to be boosted using the boost chopper with IGBT-5. The boost chopper increases the voltage across the capacitors C_1 and C_2 to 400 V.

15.2.5 Nonisolated Buck-Boost PV Inverter

Another transformerless inverter topology [28] developed in Japan for residential photovoltaic applications is shown in Fig. 15.13. In this topology, one buck-boost chopper is formed using the PV panel 1, capacitor C_1, IGBT-1, diode D_1, IGBT-2, and inductor L_1. This chopper is used to generate the positive half-cycle of AC output power. A second buck-boost chopper is formed using the PV panel 2,

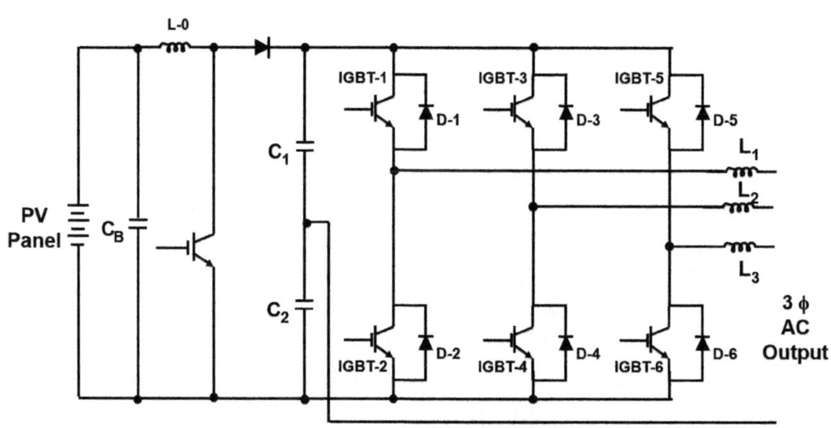

FIGURE 15.11

Three-phase inverter for PV panels.

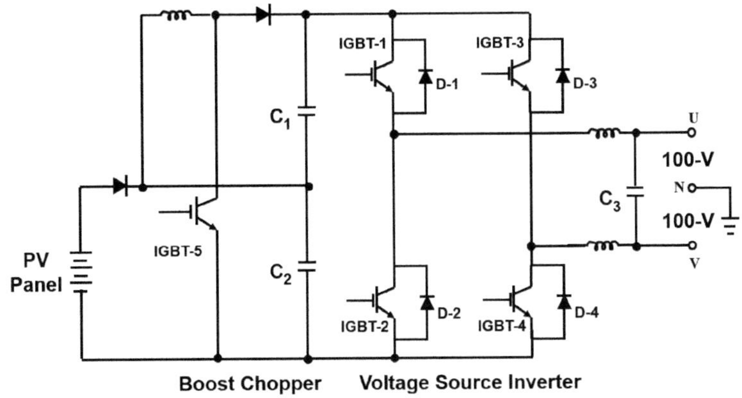

FIGURE 15.12

Nonisolated interactive inverter for Japanese residential PV panels.

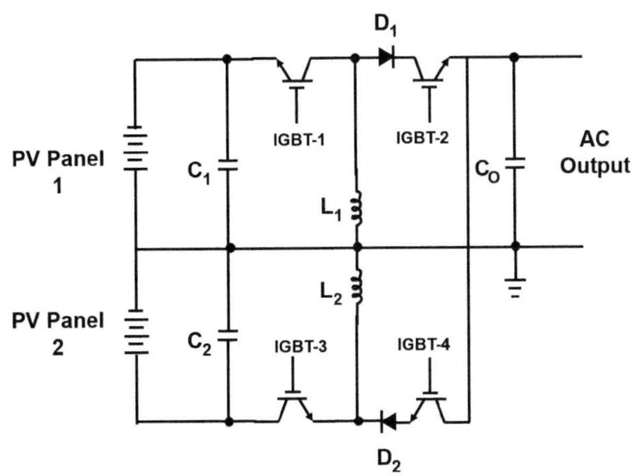

FIGURE 15.13

Nonisolated buck-boost chopper for Japanese residential PV panels.

capacitor C_2, IGBT-3, diode D_2, IGBT-4, and inductor L_2. This chopper is used to generate the negative half-cycle of AC output power.

During operation of this circuit, energy from the PV panel is used to store energy in the inductor by turning on IGBT-1, keeping all other IGBTs in the off-state. During this time, energy is delivered from capacitor C_O to the grid in the positive direction. When IGBT-2 is turned on with all other IGBTs in the off-state, the stored energy in the inductor L_1 and in capacitor C_O is supplied to the grid in the positive direction. This is repeated using the second buck-boost chopper to supply energy to the grid in the negative direction. The IGBTs were switched at 9.2 kHz in this topology. An efficiency of 96% has been achieved using a capacitor C_O value of 500 μF.

15.2.6 Maximum Power Point Tracking Circuit for PV Inverter

The extraction of the greatest amount of power from a solar cell requires operation at the maximum power point. A DSP controlled maximum power point tracking (MPPT) system [29] has been implemented using IGBTs, as shown in Fig. 15.14. The PV panel consists of 16 arrays, each capable of generating 200 W of power with an open circuit voltage of 32.9 V and a short circuit current of 8.2 A. The 16 arrays are connected in series to produce a DC voltage of 526.4 V. The DSP monitors the input voltage and current to the MPPT circuit and the output voltage to calculate the available power from the PV arrays.

PV arrays operating at their maximum power point can sometimes overload the grid. The MPPT control can also be used to curtail the power delivered by the PV array to match the needs of the grid or end user. A methodology has been proposed and implemented to achieve this goal [30].

15.2.7 Current Source PV Inverter

The replacement of the voltage source inverters (VSI) discussed in the previous sections with a current source inverter (CSI) has been proposed to reduce cost, size, and weight [27]. The CSI topology is shown in Fig. 15.15. Four IGBTs were used in 1998 in this topology with series diodes to provide reverse blocking capability. Recently, reverse blocking IGBTs have become commercially available, eliminating the need for the series diodes, which reduces the power losses.

In the CSI topology, IGBT-1 is turned on during the entire positive half cycle of the output voltage and IGBT-2 is turned on during the entire negative half cycle of the output voltage. Consequently, devices with low on-state voltage drop should be chosen in these cases. The IGBT-3 and IGBT-4 are operated using PWM control at a frequency of several tens of kilohertz. Consequently, devices with low switching

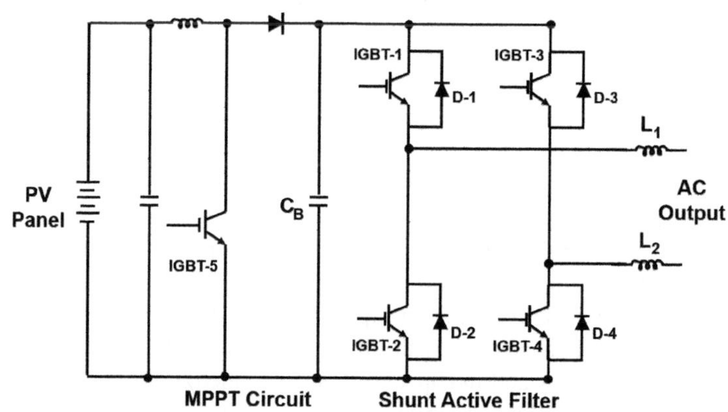

FIGURE 15.14

Maximum power point tracking circuit for residential PV panels.

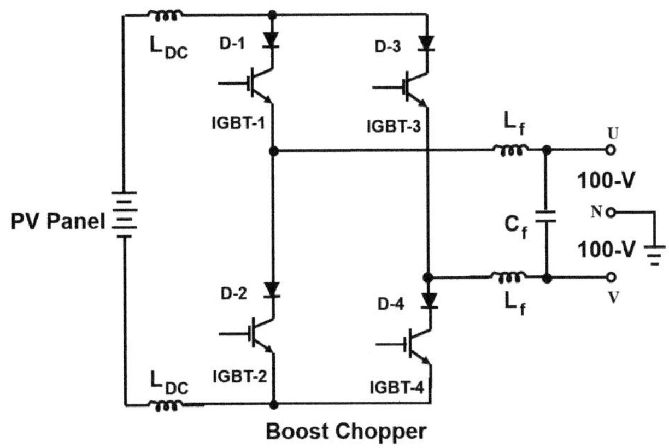

FIGURE 15.15

Current source inverter topology for residential PV panels.

losses should be chosen in these cases. If the inductance L_{DC} is large, its ripple current becomes small, resulting in a low distortion of the output current. Higher harmonic currents can be filtered out using the low pass filter comprising C_f and L_f.

However, it is desirable to reduce the size of the inductance L_{DC} to 10 mH to achieve small size and weight. This results in a large ripple on the output current and significant harmonics at the output if the conventional PWM approach is utilized. The ripple current and harmonics can be greatly reduced by employing a novel PWM scheme, called pulse area modulation, which makes the output current pulse amplitude gradually increase while its width gradually decreases. This method has been demonstrated to achieve reduced ripple in the output current with low harmonic content.

15.2.8 Three-Phase Current Source PV Inverter

A three-phase implementation of the CSI topology [24] is shown in Fig. 15.16. Once again, IGBTs with reverse blocking capability are required for this approach. Using the recently available IGBTs with reverse blocking capability, the number of devices required for the CSI topology becomes less than that for the VSI topology due to the elimination of diodes.

The reduction in power losses achieved by replacing the asymmetric blocking IGBT and series diode with the reverse blocking IGBT has been analyzed for the case of the BP 275 solar PV panel [24]. A summary is provided in Table 15.1. The conduction losses are reduced by a factor of 1.66 times. However, the switching losses for the reverse blocking IGBTs are much greater than for the asymmetric IGBTs [31]. The total power loss is only slightly reduced with the use of reverse blocking IGBTs, but this approach eliminates the series diodes and reduces component count. The total power loss is 2.74% of the output power, making the overall efficiency very good.

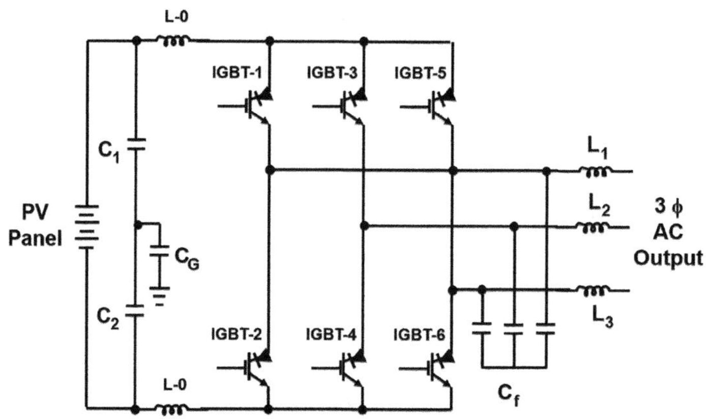

FIGURE 15.16

Three-phase current source inverter topology for residential PV panels.

Table 15.1 Power Losses in Current Source Inverter Topology for Residential PV Panels

Power Loss	IGBT With Series Diode	Reverse Blocking IGBT
Conduction (W)	386	232
Switching (W)	76	209
Total (W)	462	440

15.2.9 Commercial PV Converter

A 1 MW commercial photovoltaic installation has been constructed in Japan and Europe using the CSI topology [32]. "Mega solar" systems require increasing the PV panel DC output voltage to 1000 V. The inverters for such high voltages must be either two-level or NPC. The advanced neutral-point converter topology is shown in Fig. 15.17. Fuji Electric has made an IGBT module (4MB1300VG-120R-50) especially for this application. It contains two asymmetric IGBTs (IGBT-1 and IGBT-2) with fly-back diodes and two reverse blocking IGBTs (IGBT-3 and IGBT-4) connected in antiparallel. Three of these modules can be used to produce three-phase AC output power, as shown in the figure. IGBT-1 and IGBT-2 have a blocking voltage capability of 1200 V while IGBT-3 and IGBT-4 have a symmetric blocking voltage capability of 600 V. This topology results in conduction losses similar to the two-level inverter with lower switching losses. The efficiency of the inverter ranges from 97% to 98.5% over the entire output range. Four 250 kW power inverters are used to achieve the desired 1 MW output power capability. This creates a redundancy that leads to high reliability.

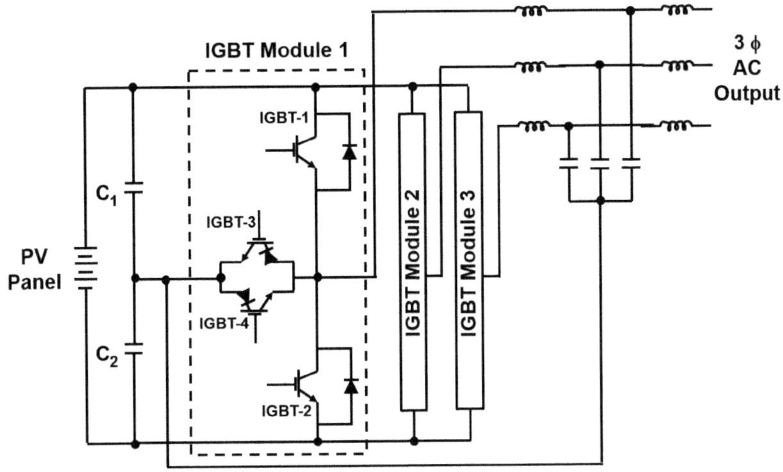

FIGURE 15.17

Three-phase current source inverter topology for residential PV panels.

15.2.10 NPC2 Topology for Solar Farm

Advances in IGBT technology are enabling higher power delivery from solar farms. The availability of IGBT modules with 1800 A, 2.3 kV capability based on seventh-generation micropattern trench and submicron mesa width allows creating 1500 V DC solar inverters to meet cosmic radiation failure specifications [33]. It is beneficial to reduce the number of IGBT devices in the inverter to bring down the cost. A comparison between the conventional NPC-1 topology illustrated in Fig. 15.18(a) and an improved NPC-2 topology illustrated in Fig. 15.18(b) shows a reduction in IGBT device count from six to four. IGBT-1 and IGBT-2 are modules with ratings of 1800 A and 2.3 kV while IGBT-3 and IGBT-4 are modules with ratings of 2400 A and 1.2 kV. The authors' analysis indicates that an output current of 1750 A can be delivered at 690 V (1.2 MW) with total device power losses of 6 kW (0.5%).

15.2.11 Dual-Source Multilevel Inverter for Residential Solar Power

Continual efforts are under way to bring down the cost for domestic or residential solar power generation to increase penetration into the home community. One novel approach is the dual-source multilevel inverter topology [34]. The topology with nine devices can generate output AC power with nine levels and 13 levels from symmetric and asymmetric input DC PV sources. The DC-MLI topology was reported to achieve an efficiency of 92% at 1 kW output, an increase of 3% compared with previous methods. The reduction in the number of components brought down the cost from \$170 to \$125.

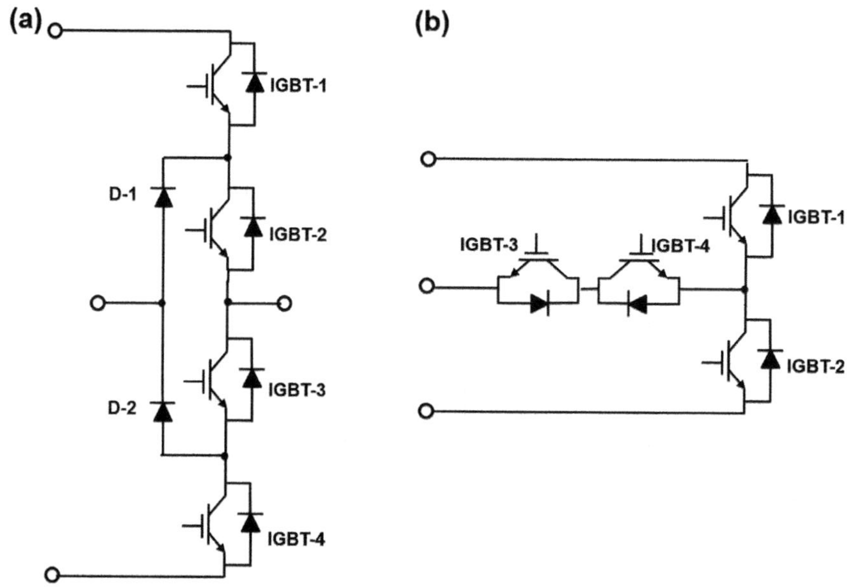

FIGURE 15.18

(a) Conventional NPC-1 topology; (b) improved NPC-2 topology.

15.2.12 PV Energy Storage

Fuel cells based on hydrogen are attractive means for the storage and transportation of energy. In a fuel cell, electricity is generated by the ionization of hydrogen at the anode and transport into an electrolyte [35]. A continuous supply of hydrogen is required to sustain the operation of the fuel cell and generate electricity. The hydrogen used in the fuel cells can be conveniently stored and transported. Fuel cells can then be used in stationary applications such as remote buildings and for electric vehicles. No fuel-cell electric vehicles are commercially available for sale as yet. A hundred buses were used in Japan as of August 2011. A total of 4000 fuel cell-based forklift trucks were in use in the United States in 2013.

Hydrogen can be generated by the electrolysis of water. Photovoltaic systems have been proposed for providing electricity for the electrolysis process. A precise match between the electricity generated by the PV panels and the electrolysis process is required. High operating voltages are favored for reducing losses in parasitic resistances, but the construction is simpler if lower voltages are used [36]. This application requires a high-efficiency DC-to-DC converter because of the different operating voltages of the PV panel and the electrolysis process.

The buck converter circuit developed for producing hydrogen by electrolysis is shown in Fig. 15.19. The circuitry for reducing switching losses in the IGBTs by using zero-voltage switching is indicated within the dashed lines. The diode on the output side is connected to a center tap for the buck inductor. This topology

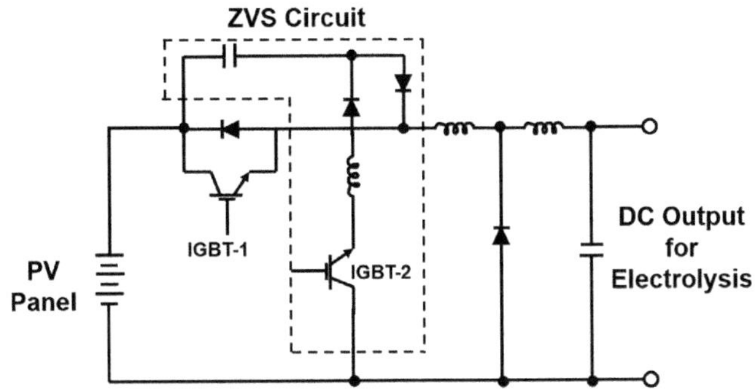

FIGURE 15.19

DC–DC converter for hydrogen generation in fuel cell applications.

has been constructed for operation with a PV panel input voltage ranging from 150 to 400 V and a DC output voltage of 30–40 V for 250 A of electrolysis current. An efficiency of 92% was achieved using 100 A IGBT modules at the PHOEBUS solar hydrogen plant in Julich, Germany [36].

Energy storage can also be accomplished using supercapacitors. Supercapacitors are based on the double-layer capacitance principle where a Helmholtz double layer is formed at the conductive electrode placed in an electrolyte, providing a charge separation of only a few angstroms [37]. They have a large capacitance per unit volume and a large power density. They can tolerate more charge and discharge cycles than batteries.

The power generated using photovoltaic panels can be stored in supercapacitors for use at a later time. A DC-DC converter has been developed to translate the DC voltage generated by the PV panel to the small DC voltage required to charge the supercapacitor [38]. The circuit topology is shown in Fig. 15.20. A diode clamped resonant snubber is used to suppress oscillations in the voltage across IGBT-2.

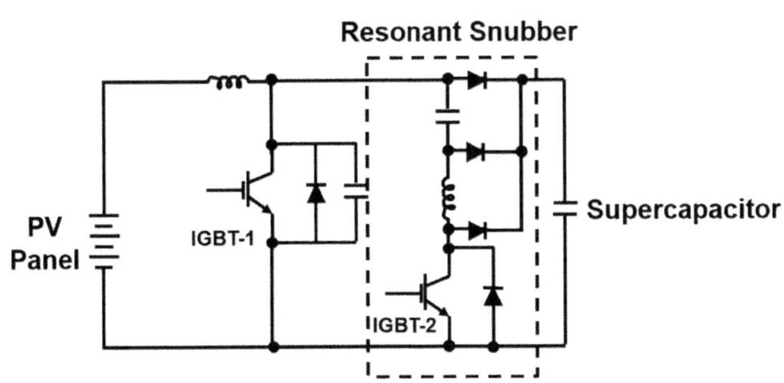

FIGURE 15.20

DC–DC converter for charging supercapacitors.

15.2.13 IGBTs for PV Applications

The large potential market for inverters used in photovoltaic installations has motivated many manufacturers to optimize IGBTs especially for this use. The high switching frequencies of the PWM circuits discussed in the previous sections require optimization of the IGBT to reduce switching losses. Reverse blocking IGBTs have also been developed for implementation of the CSI topology.

The optimization of the collector-stop IGBT technology has been discussed by Mitsubishi [39]. The IGBT chip optimized for the intelligent power modules (IPMs) used for residential PV installations has a low switching loss with a moderate on-state voltage drop. These IGBTs with 600 V blocking voltage ratings have equal on-state and switching losses in the PV application.

IGBTs with reverse blocking capability are required for the CSI topology. Although a reverse blocking capability is inherent in the basic IGBT structure, in the past this has been sacrificed by using the asymmetric structure to optimize IGBTs for motor control applications. The reverse blocking IGBT structure does not contain a buffer layer, as shown in Fig. 15.21. An edge termination must also be devised that can avoid reach-through breakdown and support the large reverse voltage. Without such a termination, a high leakage current is produced by the sawing of the wafers to separate out individual devices. This problem is solved by using thin silicon wafers and performing a deep P^+ diffusion at the edges of the chips [40], as shown in Fig. 15.21. The scribe lanes for the chips are placed within the deep P^+ diffusion. IGBTs capable of supporting 600 V in both directions have been achieved using this technique. Using these reverse blocking IGBTs, a neutral-point converter with 51% lower power losses has been constructed compared with a two-level converter. A 1 MW solar plant has also been installed using the reverse blocking IGBTs developed by Fuji Electric in the inverters [41].

The deep P^+ region used in the reverse blocking IGBT structure shown in Fig. 15.21 requires a prolonged high temperature diffusion cycle. This can be circumvented by forming a V-groove cut from the back of the silicon wafers. A P^+ isolation diffusion must be performed on the surface of this V-shaped cut, as shown in

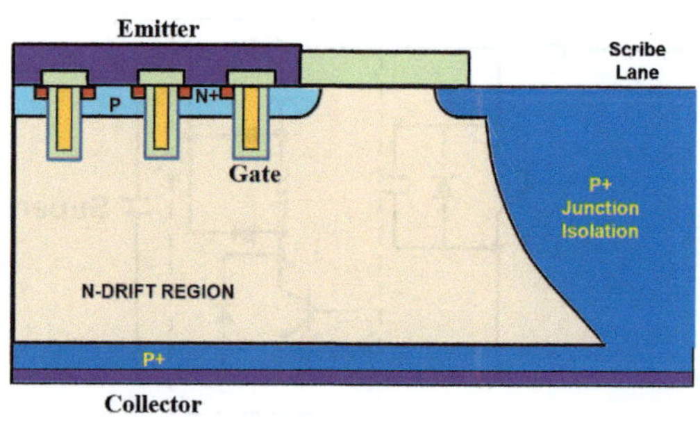

FIGURE 15.21

Reverse blocking IGBT structure for CSI applications.

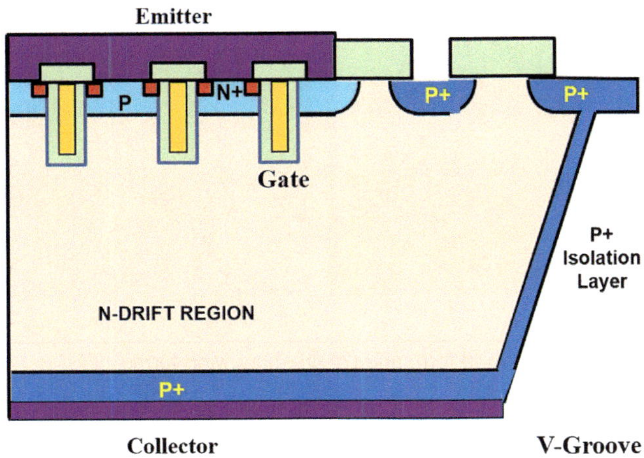

FIGURE 15.22

V-groove-based reverse blocking IGBT structure for CSI applications.

Fig. 15.22. IGBTs with symmetric 1200 V blocking voltage capability have been reported using this approach by Fuji [42].

15.3 Wind Power

The largest renewable resource for the generation of electricity after solar power is wind power. Solar radiation on the Earth generates the movement of air due to uneven temperature distribution. The energy in the wind resulting from the movement of air can be captured by using large turbines. The first windmill for the generation of electricity was built in 1887 [43]. Small wind power generators have been in use in farms and rural areas throughout the 20th century. The interest in harnessing wind power for the delivery of electricity into the grid has become of interest since the turn of the 21st century.

Large wind farms for generating electricity that can be fed into the power grid have been deployed, first on land as shown in Fig. 15.23(a), and then at sea as shown in Fig. 15.23(b). Off-shore wind farms have stronger and more constant wind gusts when compared with those on land but are more expensive to deploy and maintain. Land-based wind power turbines are often placed near the tops of mountains and in land leased from farmers. This allows dual use of the land for agriculture and power generation. According to the US Department of Energy, more than 900 GW of electric power could be generated using off-shore wind power to serve the entire needs of the country. The cost of generating electricity with wind power is expected to decrease by 17 35% by 2035 [44]. Thc output capacity per on-shore windmill will increase from 2.5 MW in 2019 to 5.5 MW in 2035 by increasing the hub height from 89 to 130 m and increasing the rotor diameter from 120 to 174 m. The output capacity

FIGURE 15.23

Wind farms: (a) terrestrial wind form and (b) off-shore wind form.

per off-shore windmill will also increase from 6.0 MW in 2019 to17 MW in 2035 by increasing the hub height from 103 to 151 m and increasing the rotor diameter from 150 to 250 m. The cost of the generated electricity will drop from \$0.035 to \$0.030 for the on-shore case and from \$0.08 to \$0.05 for the off-shore case by 2035.

The electricity generation capacity using wind power has grown rapidly since the beginning of this millennium [43,45]. With growing concerns regarding the depletion of fossil fuel resources and the environmental impact of using these sources, investment in wind power accelerated rapidly in 2000. Wind power generation capacity has been doubling every 3 years, as shown in Fig. 15.24 [43]. The installed wind power generation capacity had increased to 282 GW by the end of 2012 and

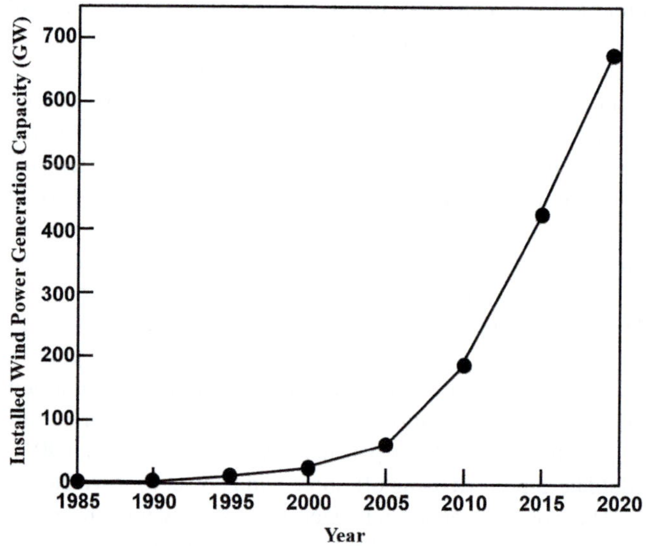

FIGURE 15.24

Installed worldwide wind power generation capacity.

reached 730 GW in 2020. Of this, 200 GW has been installed in China, 118 GW in the United States, 62 GW in Germany, 38 GW in India, and 27 GW in Spain [46]. The largest on-shore wind farms are the 20 GW Gansu Wind Farm in China, the 1.6 GW Alta Wind Farm in the United States, and the 1.6 GW Jaisalmer Wind Farm in India. The largest off-shore wind farm is the 1.2 GW Hornsea One plant in the United Kingdom.

In a typical wind power generator, the energy produced by the rotation of the blades is used to generate three-phase AC power. However, the frequency of the generated power varies widely due to changes in the wind speed. To deliver constant frequency power (50 Hz or 60 Hz) to the power grid with a well-regulated voltage, it is necessary to first rectify the generated power to produce a DC voltage across a capacitor, and then use IGBTs in an inverter stage to generate the desired AC output power at constant frequency and voltage. Before the availability of IGBTs, inverters were made using thyristors. Replacement of thyristors with IGBTs enables a reduction of the size and weight of the power electronics, allowing their location in the gantry crane of the windmill. With IGBTs, PWM is possible, allowing control of real and reactive power. EDN magazine states [47]: "*Such devices handle kiloamp-level currents at several kilovolts, and progress within this (IGBT) technology has made the biggest single contribution to wind-turbine advances over the past several decades.*" It is now well established that the IGBT is the most suitable power semiconductor device for wind power generation applications [48]. The IGBT-based inverters eliminate the need for the heavy and unreliable gearbox that had to be located in the gantry. A good discussion of the advantages of the IGBT-based PWM inverters can be found in many papers in the literature.

15.3.1 Wind Power Generator Configurations

The wind power generator configuration has evolved over the years [49]. In the early phase of wind power deployment, a fixed speed power generator concept was proposed and implemented. In this "Danish concept" shown in Fig. 15.25, a

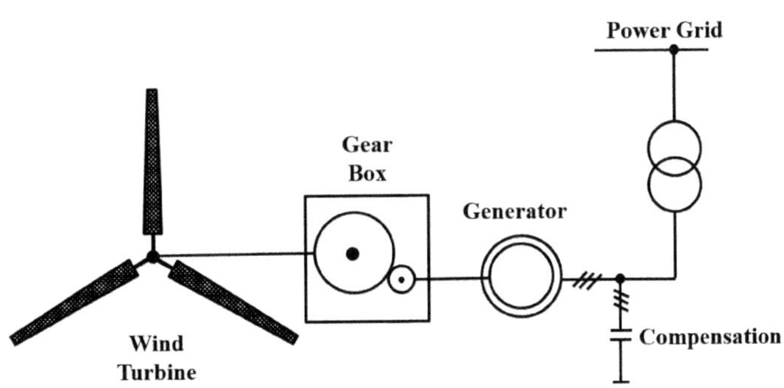

FIGURE 15.25

Fixed-speed wind power generator.

squirrel-cage induction motor is directly connected to the power grid. The wind turbine rotor is coupled via a fixed ratio gearbox to the generator shaft. The operation of this fixed-speed wind turbine configuration is controlled with mechanical subcircuits. The response time for the mechanical control is tens of milliseconds. The electrical output fluctuates whenever a gust of wind occurs. Stable operation is possible only with a stiff power grid and the mechanical components have to absorb large stresses that degrade reliability.

The modern wind power generation concept using adjustable speed generators is illustrated in Fig. 15.26. The adjustable frequency output of the generator produced by the variation in the rotational speed of the wind turbine is first converted to DC and an inverter stage is then used to generate the well-regulated 50/60 Hz AC power output fed into the grid. This approach, made possible using IGBTs in the AC-DC and DC-AC converter stages, allows simple cost-effective pitch control, reduces mechanical stresses, improves power quality, eliminates flicker, and increases overall efficiency [50].

The configuration shown in Fig. 15.26 has the disadvantages of requiring expensive power converters with the same ratings as the output power being delivered by the generator and output filters also rated at the output power as well. These disadvantages can be overcome using the doubly fed induction generator shown in Fig. 15.27. Here, the stator windings of the generator are directly connected to the grid while a four-quadrant AC-to-AC converter is connected to the rotor windings of the generator. Compared with the direct in-line configuration of Fig. 15.26, the doubly fed approach reduces inverter costs because its rating is reduced to 25% of the generator. This also applies to the cost and ratings for the filter components. An efficiency improvement of 3% is also obtained with this configuration.

The power delivered by individual wind turbines has been increasing over the years. The average size of the on-shore wind turbine was 1.8 MW in 2012 while

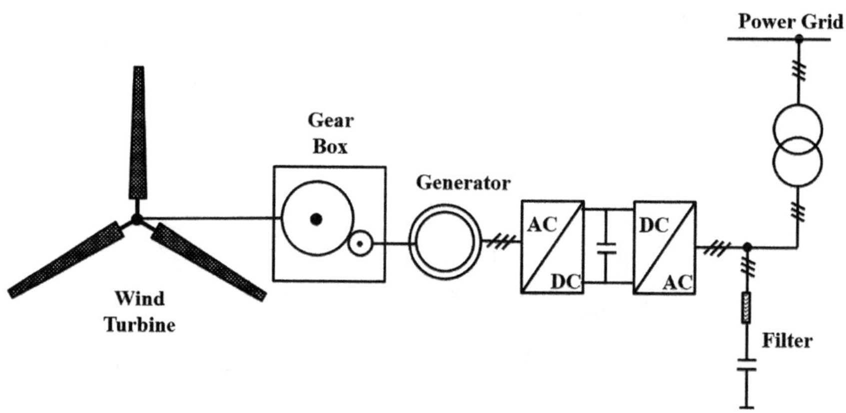

FIGURE 15.26

Adjustable-speed wind power generator.

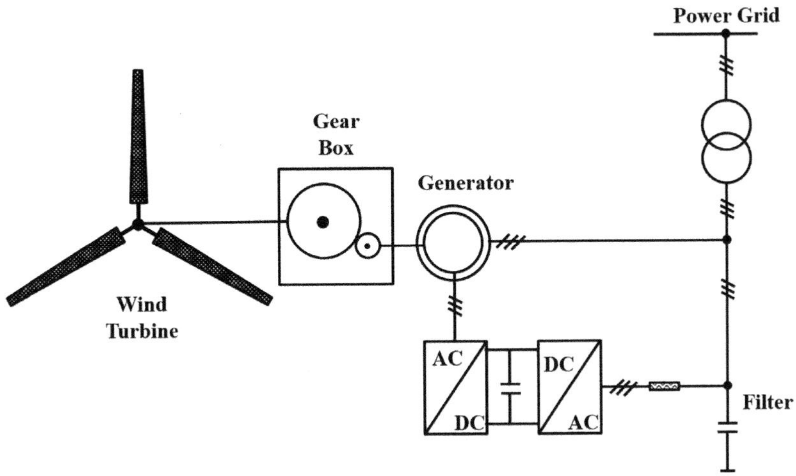

FIGURE 15.27

Doubly fed induction wind power generator.

the average off-shore wind turbine had a size of 4 MW [50]. This has grown to 2.5 MW and 6 MW for on-shore and off-shore cases by 2019 and is expected to grow to 5.5 MW and 17 MW, respectively, by 2035. A multicell converter topology has been developed to allow handling these larger power ratings. The multicell back-to-back converter approach is illustrated in Fig. 15.28. It allows using IGBT modules with lower power ratings to achieve the desired large output power (>3 MW).

15.3.2 Basic Converter Topology

The typical converter topology used for the wind generator is shown in Fig. 15.29. The AC-to-DC rectification can be performed using diodes because the DC voltage on the bus remains approximately constant [51]. An active front end bridge using IGBTs is not required here. The VSI has six IGBTs with fly-back diodes and an LC filter to remove the harmonics. The IGBTs are controlled using the PWM technique. This basic converter topology applies to all the wind generation configurations shown in the previous section. The types of IGBTs used for this converter are discussed in a later section.

15.3.3 Off-Shore Wind Power Installations

The off-shore wind power generation systems require the transmission of electrical power from the off-shore generation site to the mainland, where it can be delivered to the power grid. The transmission of electrical power can be accomplished by using either an AC line or a DC line. AC power lines have been found to be more cost-effective for shorter distances of transmission while DC power transmission (HVDC)

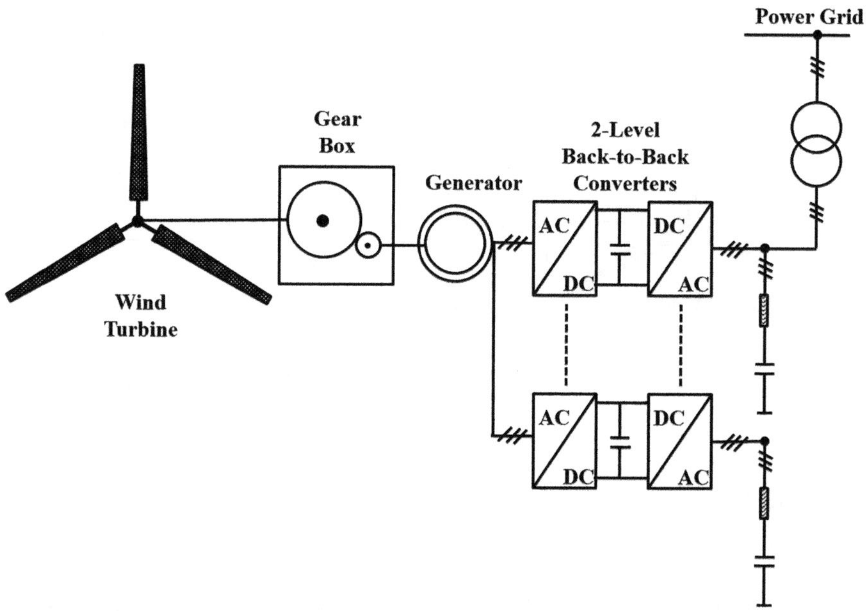

FIGURE 15.28

Multicell parallel BTB converter-based wind power generator.

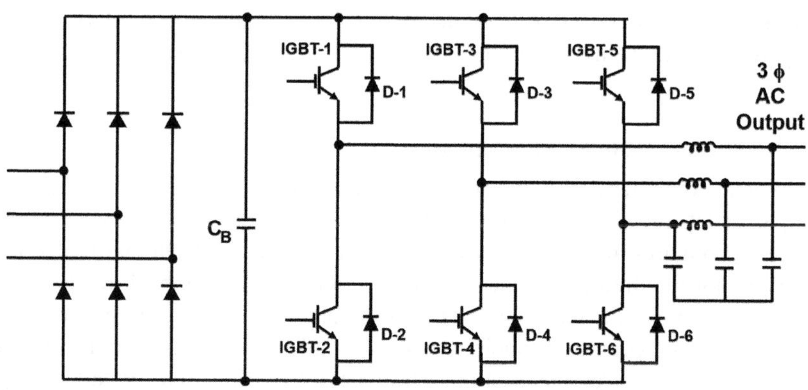

FIGURE 15.29

Basic converter topology for wind turbines.

is favored for long distances. The typical crossover point is at a distance of 100 km. Today, most off-shore wind power stations have been installed at short distances from the shore. In the future, they will be located at longer distances, making HVDC a more cost-effective approach. In this case, the generated AC power needs to be converted to a high DC voltage rather than regulated 50/60 Hz AC power, as discussed in previous sections.

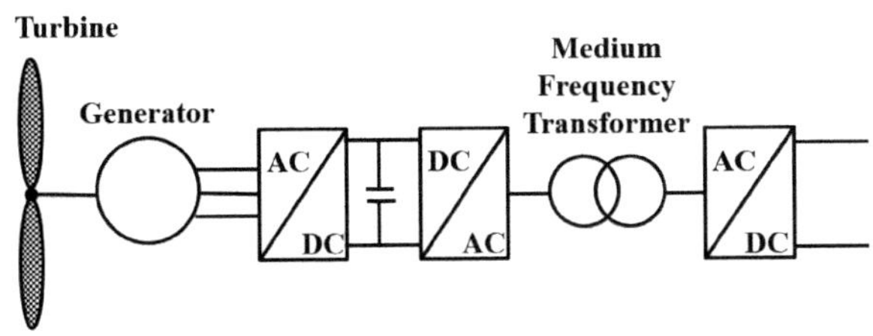

FIGURE 15.30

Off-shore wind turbine with HVDC output.

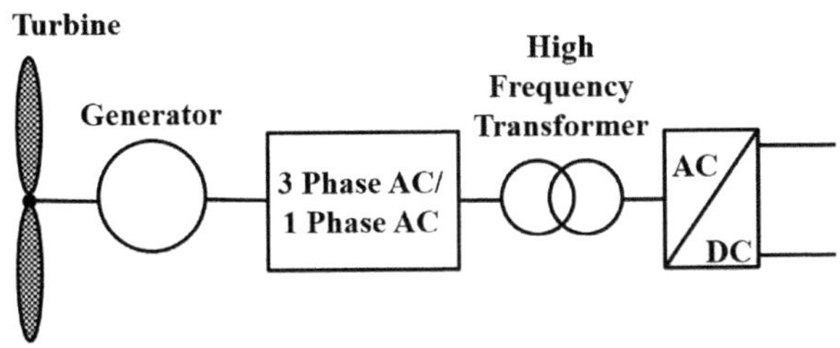

FIGURE 15.31

Alternate off-shore wind turbine with HVDC output.

Two configurations for creating the HVDC power transmission from an off-shore wind farm are shown in Figs. 15.30 and 15.31 [52]. The first approach makes use of adjustable speed generation, as previously shown in Fig. 15.26. The AC voltage is stepped up using a medium-frequency transformer and then a rectifier to obtain the desired HVDC output for transmitting the power. In the second approach shown in Fig. 15.31, the generated power is stepped up using a high-frequency transformer and then rectified to obtain the desired HVDC output for transmitting the power. The entire converter is housed in the nacelle of the wind turbine, eliminating a substation. The output from each of the turbines is then combined in series, as shown in Fig. 15.32, to create the HVDC for transmission of power to the mainland.

The three-phase AC power generated by the turbine must be converted to a single-phase, high-frequency AC output that is provided to the primary of the high-frequency transformer. A bidirectional AC link between the generator and the transformer has been constructed using reverse blocking IGBTs, as shown in Fig. 15.33 [53]. Twelve IGBTs in a back-to-back connection are used in this topology to allow bidirectional current flow. Each leg (e.g., IGBT-1, IGBT-2, IGBT-3, and IGBT-4) is connected to one phase of the generator. The generator delivers a low-frequency sinusoidal waveform. The IGBTs are switched at a high frequency to generate a square wave voltage with a high frequency at the primary of the transformer. The reverse blocking IGBTs can be formed by using asymmetric IGBTs with a series diode to provide the reverse blocking capability. However, the use of

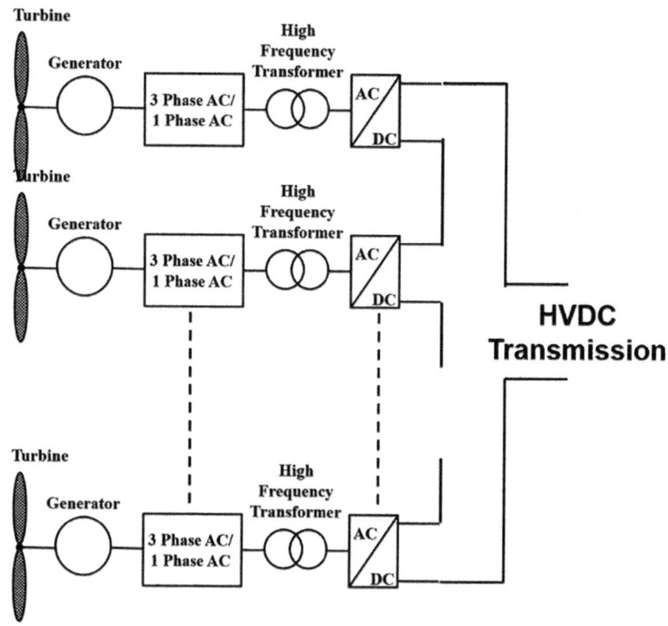

FIGURE 15.32

HVDC transmission from an off-shore wind farm.

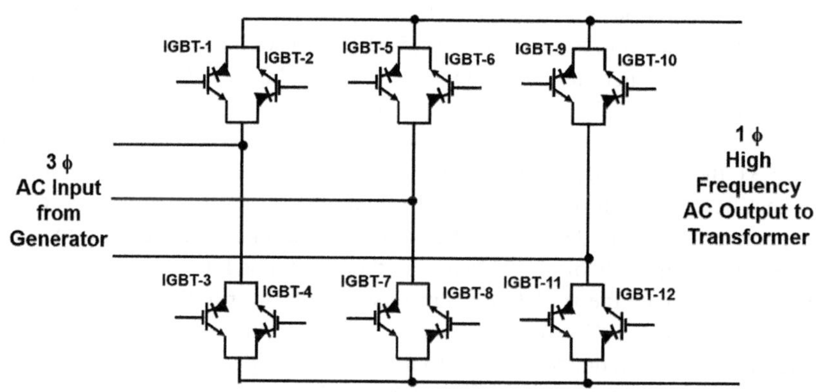

FIGURE 15.33

Bidirectional AC link between the generator and the transformer.

recently commercially available reverse (or symmetric) blocking IGBTs has reduced losses, improving the efficiency by 1.9% [53].

In the past with wind park power levels of 3 MW, the output of each off-shore wind turbine has been delivered at 690 V, which is then combined in series, as shown in Fig. 15.32. The trend to increase the output power from a wind park to above

10 MW will require increasing the output voltage from each wind turbine to greater than 3 kV. Reverse blocking IGBTs with the capability to withstand 5 kV must be developed to meet this requirement.

15.3.4 Chinese off-Shore Wind Power Installation

China has been investing in large wind power generation capacity. The installed wind power capacity reached 20 GW in 2010 and grew to 200 GW by 2020. The largest off-shore wind power plant in China is the Donghai Bridge off-shore wind farm [53]. The wind farm is located 10 km south of Shanghai near the Shanghai East Sea Bridge. It contains 34 wind turbines with the capability to generate 100 MW of power to serve the needs of 20 million Shanghai residents. A submarine cable is utilized to transfer power from the wind farm to the mainland. An AC power transmission system is used for the Donghai Bridge wind farm with a 110 kV busbar. One important consideration in the design is the impact of typhoons that occur regularly every year. The large wind speed during a typhoon can produce huge voltage fluctuations on the grid. All the turbines in the wind farm must be shut down during a typhoon to avoid this problem.

15.3.5 European off-Shore Wind Power Installation

The deployment of off-shore wind power generation has been increasing rapidly in Europe to move away from fossil fuel-based electricity production. Wind farm installations in Europe include [54]: (1) the 5 MW AREVA WIND M5000 plant using medium-voltage (3.1 kV) full converter (FC) topology; (2) the 5 MW REPOWER power plant using a doubly fed induction generator with a low-voltage (690 V) converter; (3) the 3.6 MW SIEMENS SWT plant using an FC asynchronous induction generator with a 690 V converter; (4) the 6 MW ALSTHOM Haliade 150 plant using a direct drive operating at 900 V; and (5) the 3 MW VESTAS V90 plant using an FC generator. These plants must satisfy the grid code: total harmonic distortion below 3%, line frequency excursion of less than 5%, and a power factor of better than 0.9. Because off-shore power plants are expensive to maintain and repair, the use of electrolytic capacitors is avoided. The design includes redundancy and fault monitoring with remote diagnostics and configurability.

The low-voltage full converter (LV-FC) topology makes use of two, three, or four power conversion lines (PCLs) working in parallel [55]. Each PCL consists of an AC-to-DC converter and a DC-to-AC inverter, as shown previously in Fig. 15.26. Each PCL makes use of power electronic building blocks (PEBBs) or basic power modules (BPMs) with 1.7 kV PrimePACK IGBTs made by Infineon. The BPMs can be removed easily using sliding rails for maintenance. The LV-FC topology has a high power density of 0.21 m^3 per MW of power [55]. The low-voltage doubly fed induction generator topology makes use of two PCLs working in parallel. A three-level neutral-point-clamped (NPC) BPM is used here with 1.7 kV PrimePack IGBTs. A maximum power density of 0.3 m^3 per MW of power has been achieved with this approach. The medium-voltage full converter topology makes use of two PCLs connected to two isolated stator windings in the generator. A PEBB with four IGBTs and two diodes is used to produce an output voltage of 3.3 kV or 4.16 kV [55]. The overall best choice is the DFIG system with two PCLs [55]. It provides the highest efficiency with the smallest cost, volume, and weight.

15.3.6 Standalone Wind Power Installation

The generation of electricity using wind power is attractive in remote areas for supporting small communities and farms. These standalone power sources are not connected to the power grid. They must regulate not only the power flow but also the voltage and frequency on the AC output power line. The gearbox must be eliminated to achieve a reliable wind power source. This favors the direct drive permanent magnet synchronous generator [55]. The standalone wind power station consists of the wind turbine, the directly gearless driven permanent magnet synchronous generator, a three-phase switch-mode rectifier with single IGBT control for maximum power extraction, and a PWM VSI implemented using IGBTs to regulate the AC output voltage and frequency.

Variable speed operation of the generator is required to be able to extract the maximum available power from a fluctuating wind source. A switch-mode rectifier with a single IGBT switch (IGBT-1) is used for MPPT, as illustrated in Fig. 15.34. The maximum power point shifts with increasing wind speed, as illustrated in Fig. 15.35. The turbine torque and the generator torque match when the optimum turbine speed

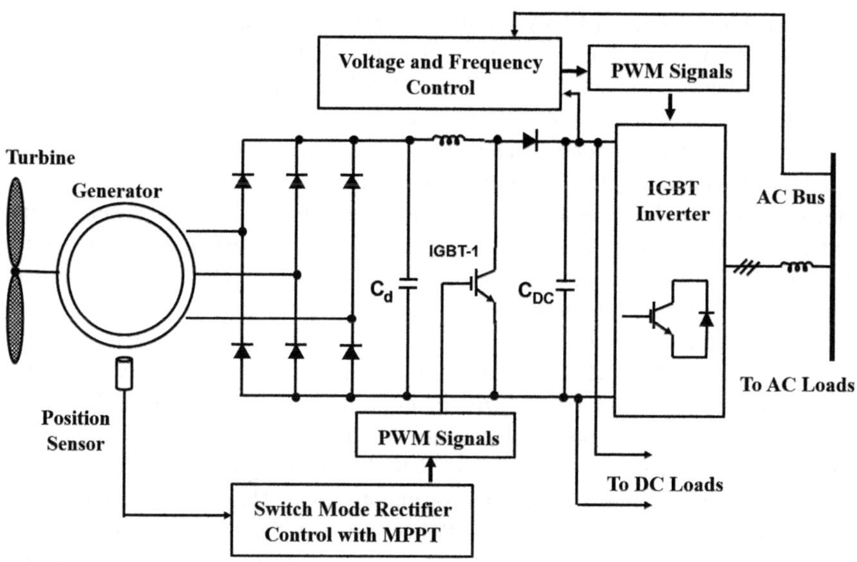

FIGURE 15.34

Standalone wind power generation.

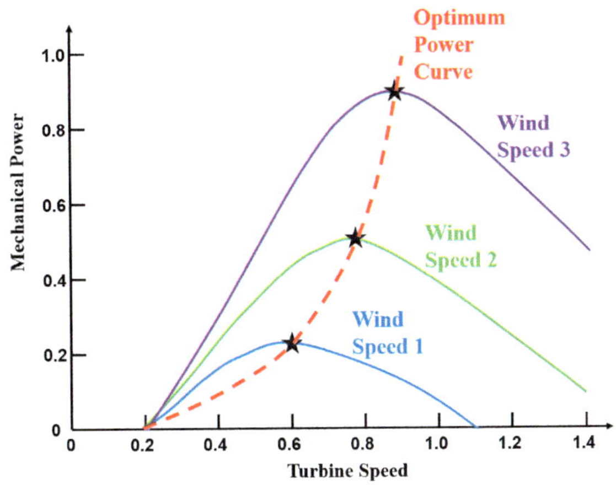

FIGURE 15.35

Maximum power point tracking.

is reached, as shown in this figure. The standalone configuration can also provide power to DC loads, as shown in Fig. 15.34. Because excess generated power cannot be fed into a grid, it must be diverted to energy storage (in a DC load) or be dissipated in a resistor.

15.3.7 STATCOM for Reactive Power Compensation

The transmission of power from wind power plants requires support from static VAR compensators (STATCOM) to supply reactive power to stabilize the line voltage. An IGBT-based STATCOM has been investigated for the Muhuri Dam wind power station in Bangladesh [56], a 125 kV grid network operating at 900 kW. It was able to reduce the average voltage fluctuation of 21% to only 0.7%.

15.3.8 IGBTs for Wind Power Applications

Three voltage ratings are important for IGBTs used in wind power installations [57]: (1) the maximum DC voltage for operation within its safe operating area; (2) the blocking voltage rating required to tolerate voltage spikes during turn-off; and (3) the maximum DC voltage based on cosmic radiation failure.

In a two-level inverter, each IGBT must withstand the full DC bus voltage. Based on this, the maximum DC-link voltage for cosmic ray failure of IGBTs is provided in Table 15.2. It also gives the preferred repetitive blocking voltage rating for the IGBTs. In a three-level inverter, each IGBT must withstand half the DC bus voltage. Based on this, the maximum DC link voltage for cosmic ray failure of IGBTs is provided in Table 15.3. It also gives the preferred repetitive blocking voltage rating for the IGBTs.

Table 15.2 Preferred Blocking Voltages for IGBTs Used in Two-Level Inverters

Nominal RMS Line Voltage (V)	Nominal DC-Link Voltage for Cosmic Ray Rating (V)	Preferred Repetitive Blocking Voltage Rating (V)
400	620	1200
750	900	1700
690	1070	1700
1500	1800	3300
1700	2800	4500
3000	3600	6000
3300	4000	6500

Table 15.3 Preferred Blocking Voltages for IGBTs Used in Three-Level Inverters

Nominal RMS Line Voltage (V)	Nominal DC-Link Voltage for Cosmic Ray Rating (V)	Preferred Repetitive Blocking Voltage Rating (V)
2300	1900	3300
3300	2700	4500
4160	3400	5500
6000	4900	8000
6600	5400	8500
6900	5600	9000
7200	5900	9500

The HiPak IGBT module made by ABB is used in low- and medium-voltage wind energy converters [58]. It is available with power ratings of 1700 V/2400 A and 6500 V/750 A. Wind power installations with 2.5 MW generation capability have been installed using these devices in a DFIG topology. The soft-punch-through (SPT) IGBT structure with a planar gate cell has been shown to provide a good combination of on-state and switching losses. The reverse conducting IGBT with an integrated fly-back diode has also been developed for use in wind power stations. This improves the power density by elimination of the external fly-back power rectifiers.

Infineon has reported the optimization of IGBTs and power modules for use in the wind power generation market [58]. Their preferred IGBT structure has a trench gate architecture, as shown in Fig. 15.36, with an SPT design. The trench gate structure was found to have on on-state voltage drop of 3.7 V, which is 1.6 V less than that for the planar gate design. In addition, a new edge termination design using a variation of lateral doping (or junction termination extension) was used to shrink the edge termination width by 25%. This allowed increasing the active area by 20% for the same die size.

Multimegawatt wind power generation must be performed using higher DC output voltages. The case of a 5 MW wind power installation using a10 kV DC

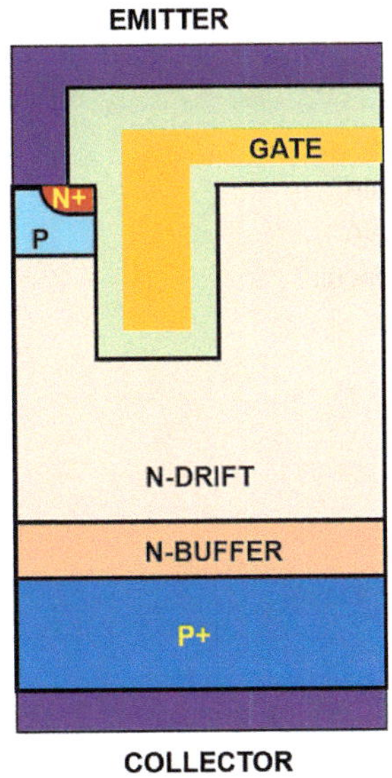

FIGURE 15.36

Infineon trench gate IGBT with field stop layer.

bus voltage has been analyzed from the point of view of using 1700 V, 3300 V, or 6500 V IGBT modules [59]. More modules must be connected in series in the case of devices with the lower blocking voltages. This increases the on-state voltage drop and power loss. However, it was found that the switching losses are dominant, even for a switching frequency of 500 Hz. Because lower blocking voltage devices have smaller switching losses, the 1.7 kV IGBT module-based system was found to produce the highest conversion efficiency of 98%.

A comparison between the IGBT module and the IGBT press pack is shown in Table 15.4 [51]. The IGBT module technology has been in use for traction applications for a long time. The IGBT press pack technology offers better thermal properties, leading to higher current ratings and better reliability. It is estimated that more than 250,000 IGBT modules are required to serve just the China market for wind power [26].

The revenue generated from a wind farm is adversely impacted by breakdown and repair costs. The high failure rate components in the direct-drive wind turbine topology shown in Fig. 15.31 have been identified as IGBT modules, converter control units, and the cooling system [60]. The authors propose a strategy combining wind energy abandonment and reactive power compensation to achieve a reduction in cost for repairs.

The revenue can also be improved by using state-of-the-art 4.5 kV IGBT structures. One example is the Hitachi HiGT IGBT structure illustrated in Fig. 15.37. It contains floating P^+ regions surrounding the trench gate portion. The transparent emitter concept discussed in Chapter 3 is employed here with a thin P^+ emitter.

Table 15.4 Comparison of IGBTs Used in Wind Power Applications

Parameter	IGBT Module	IGBT Press Pack
Power density	Low	High
Reliability	Moderate	High
Cost	Moderate	High
Failure mode	Open circuit	Short circuit
Easy maintenance	+	–
Heat sink installation	+	–
Snubber requirement	–	–
Thermal resistance	Large	Small
Switching loss	Low	Moderate
Conduction loss	Moderate	Moderate
Gate driver	Moderate	Moderate
Manufacturers	Infineon, Semikron, Mitsubishi, ABB, Fuji	Westcode, ABB
Medium voltage ratings	3.3/4.5/6.5 kV	2.5/4.5 kV
Maximum current ratings	1.5/1.2/0.75 kA	2.3/2.4 kA

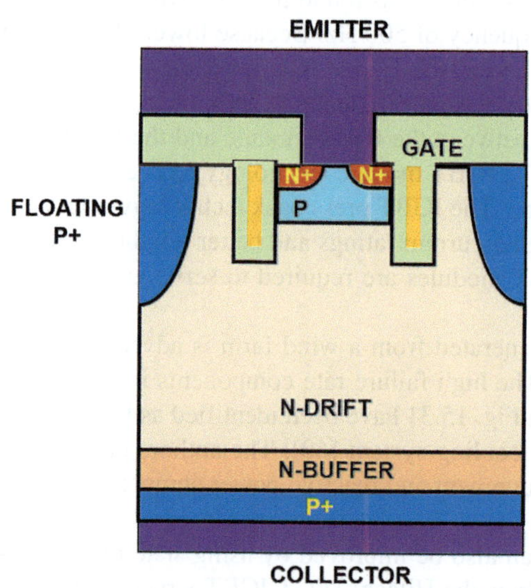

FIGURE 15.37

Hitachi HiGT IGBT structure.

This structure enabled reducing switching losses by 40% for the same on-state voltage drop of 4 V. The module was able to handle 1500 A instead of 1200 A for previous technology to enhance the power delivered from the wind turbine [61].

15.4 Wave Power

Like wind power, wave power is an indirect method of harvesting some of the energy radiation from the Sun that is captured by the Earth. The uneven absorption of energy received by the Earth from the Sun generates wind over bodies of water. The passage of wind over the water produces waves, as illustrated in Fig. 15.38. Energy is transferred from the wind to the waves when they propagate at a slower speed than the wind velocity [62]. Waves also generate oscillatory motion at the surface of the water. Wave power is proportional to the period of the ripples and the square of the height of the ripples. It is estimated that 2 TW of power could be harnessed from wave power around the globe [63].

The kinetic energy in the waves can be converted to electrical energy by various methods: (1) point absorber buoys that float on the water's surface. The rise and fall of the buoys due to the waves can be used to drive hydraulic pumps, which generate electricity; (2) surface attenuators with multiple flexible segments. The flexing movement of the segments due to the waves is used to drive hydraulic pumps, which generate electricity; (3) oscillating water column where the water enters an air chamber to produce compressed air. The compressed air drives a turbine generating electricity; (4) an overtopping device where the crests of waves fill a reservoir. The potential energy in the reservoir is used to run turbines and generate electricity; and (5) an oscillating wave surge converter with one end of a structure fixed to the sea bed. The movement of the other end of the structure provides kinetic energy that can be converted into electricity.

Many wave energy extraction projects have been deployed around the world over the last three decades. The Islay LIMPET on-shore wave power generation platform was set up in 1991 in Scotland. It makes use of an oscillating water column to generate 500 kW of electrical power. The Ocean Swell Powered Renewable EnergY (OSPREY) wave energy machine was set up in 1995 in Scotland. It makes use of

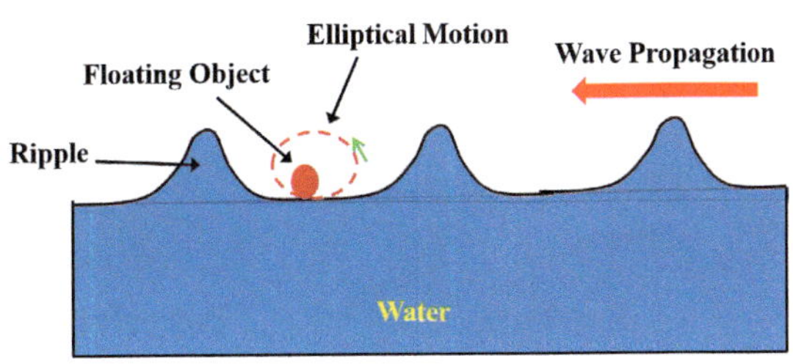

FIGURE 15.38

Wave propagation.

oscillating water columns to generate 2 MW of energy. The Oceanlinx near-shore wave power generator was set up in 1997 in Australia. It also made use of the oscillating water column method to generate 1 MW of electricity. Although the original installation sank in extreme seas, a 1 MW installation is under way. The PowerBuoy wave power generator was set up in the United States in 1997. It makes use of buoys to generate electricity. The Pelamis Wave Energy Converter was set up in 1998 in Scotland. It makes use of the surface following attenuator method to generate electricity. The Wave Star wave power generator was set up in Denmark in 2000. It makes use of the multipoint absorber to generate electricity. The Wave Dragon power generator was set up in Denmark in 2003. It makes use of the overtopping method to generate electricity. The Anaconda Wave Energy Converter was set up in the United Kingdom in 2008. It makes use of the surface following attenuator to generate electricity. The AWS-III wave power generator was set up in Scotland in 2010. It makes use of a floating toroidal vessel to generate electricity. The Agucadourda Wave Farm was set up in Portugal in 2008. It makes use of the Pelamis generators to produce 2 MW of electricity. As an example, images of the Pelamis and PowerBuoy generators are shown in Fig. 15.39.

15.4.1 Osprey Wave Energy

The OSPREY wave energy power generation system, shown in Fig. 15.40, was installed in Scotland in 1997 [64]. It was the world's first commercial-scale wave-powered electricity generation installation with the ability to deliver 2 MW of power to 1000 homes. The OSPREY is installed on the sea bed floor, as illustrated in Fig. 15.41, with most of the structures submerged below the sea level at a depth of 14.5 m. The generators are located above the water level.

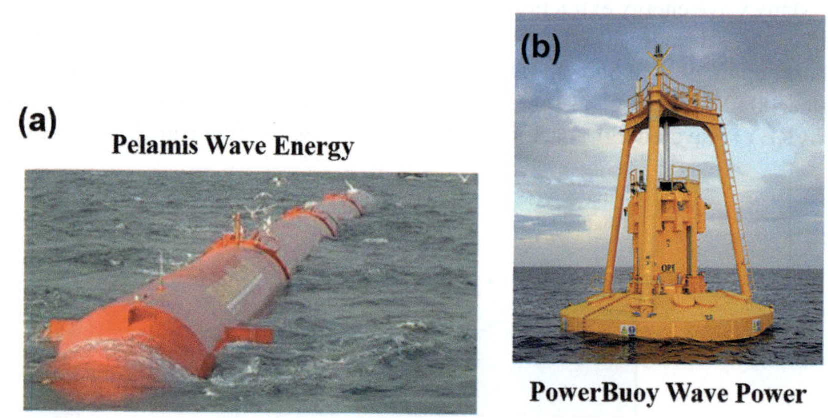

FIGURE 15.39

Wave power generators: (a) Pelamis; (b) PowerBuoy.

FIGURE 15.40

Osprey wave energy installation.

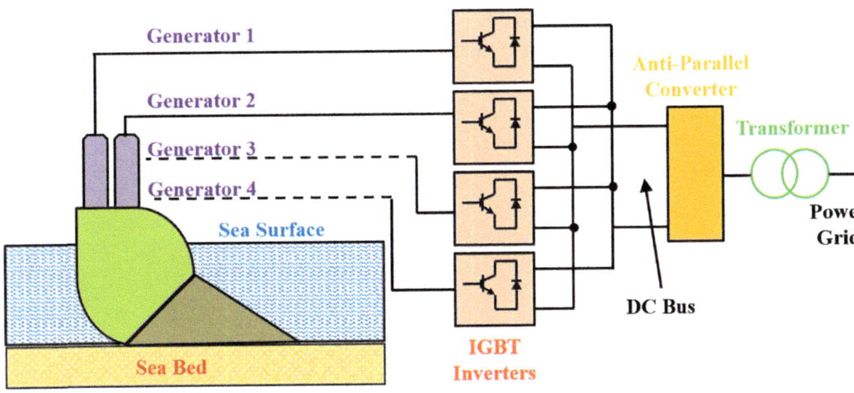

FIGURE 15.41

The Osprey wind energy generator design.

The OSPREY system operates by extracting energy from the natural rise and fall of the sea water level near the shoreline. The changes in water level in the structure push out and suck in air through a Wells turbine located in the generators. The Wells turbine is designed to rotate in the same direction for either direction of air flow through its chamber. The turbine is coupled to an electric generator to produce power. The energy delivered by the generator is highly oscillatory, requiring IGBT-based converters to change the variable frequency AC input power to a DC bus. An antiparallel converter is used to translate the DC power to regulated

50 Hz AC power. Four 500 kW generators and IGBT converters are used to deliver 2 MW of power to the power grid.

The OSPREY design allows each generator to operate independently to derive the highest efficiency; maximum energy extraction for any given sea state by optimizing the generator speed; maintaining a low generator speed at low sea levels to reduce losses; and absorbing 150% of rated torque for 60 s in case of large pulses of waves. The authors state [65]: "*IGBT-based inverters offer the most advanced control and highest performance.*"

15.4.2 Wave Dragon Energy

The Wave Dragon energy generator is based on the overtopping principal. The water in the ocean waves is focused by baffles to gain in height on a ramp, as illustrated in Fig. 15.42. The cresting of the wave over the top of the ramp fills a reservoir. The water is drained from the reservoir through a turbine to create mechanical energy. The turbine drives an electric generator to produce electrical power [65]. The Wave Dragon installation contains 16 turbines that can be switched on- and off-shore using a gate to regulate the flow of water and maintain a desired level in the reservoir.

The 16 hydroturbine generators in the Wave Dragon installation are individually rated at 380 kW. The generator produces a variable frequency 690 V AC output due to fluctuation in the water flow from the ocean waves. This is converted into well-regulated 50 Hz 690 V AC power by using an IGBT-based voltage source converter, as shown in Fig. 15.43 [66]. The output of the converter is then fed to the power grid

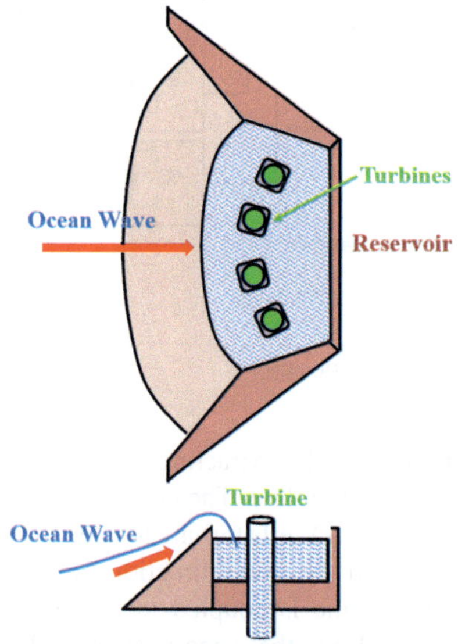

FIGURE 15.42

The Wave Dragon energy generator design.

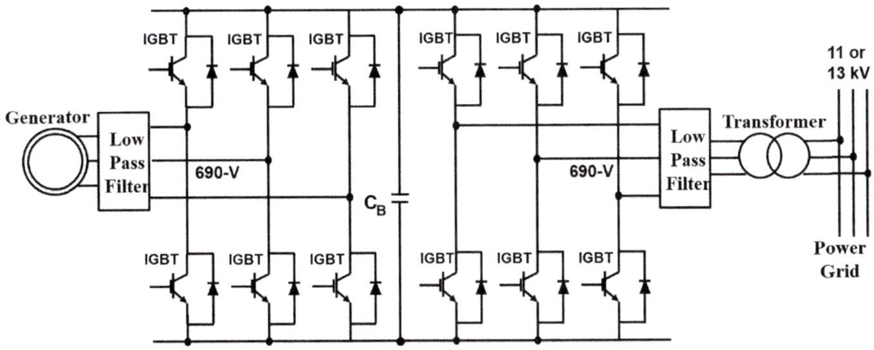

FIGURE 15.43

AC/DC/AC converter using IGBTs.

by boosting the voltage to either 11 or 13 kV by using a transformer. This configuration has the advantages of rapid regulation of active and reactive power, smaller harmonic currents, and a high power factor.

15.4.3 Bolt Buoy Energy

The Fred Olsen Bolt wave energy generator, shown in Fig. 15.44, was launched in 2004 [66] outside Risor in Norway. The principle of operation is illustrated in Fig. 15.45. The power unit is tightly moored to the sea bed using a winch and drum system. Movements of the absorber produce rotation of the drum that is fed to a generator via a gearbox. It produced 3.36 MWh of electricity by 2014. In this approach, the output power fluctuates over a large range with many periods when the energy output is zero.

Bolt Bouy Wave Energy

FIGURE 15.44

Fred Olsen's WEC Bolt buoy energy generation.

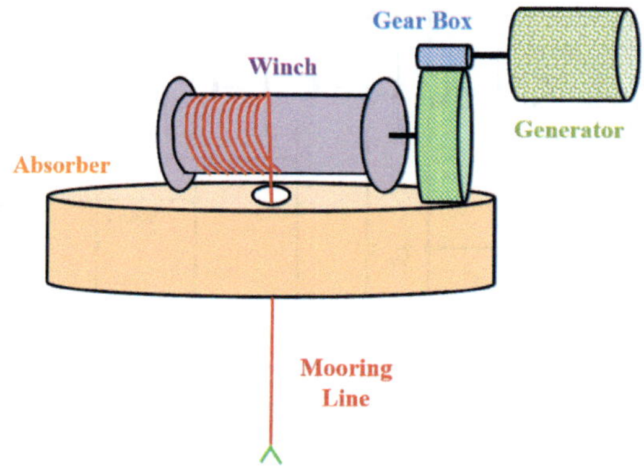

FIGURE 15.45

The Fred Olsen Bolt buoy energy generator design.

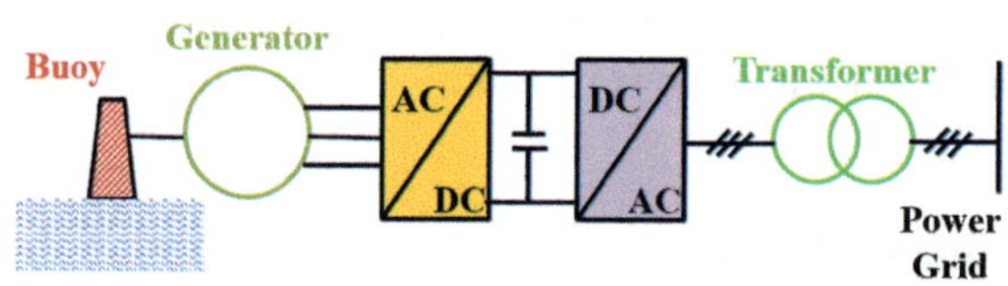

FIGURE 15.46

The Bolt buoy power topology.

The fluctuating output from the generator is converted into a DC bus using IGBTs and then an IGBT inverter is used, as shown in Fig. 15.46, to produce the well-regulated 50 Hz AC output power that can be delivered to the power grid. The large power fluctuations in this energy-generation method produce significant thermal cycling stress on the IGBTs. The wave-to-wave movement occurs in periods of 4–10 s. The IGBT modules must be qualified for reliable operation without detachments of wire bonds and dies. The IGBT module must be oversized to provide enough margin for long-term operation under these conditions [67].

A similar conclusion was derived for the Système Electrique Autonome de Récupération de l'Energie des Vagues (SEAREV) direct wave energy converter [67]. The location of wave energy installations in the ocean at energetically attractive sites provides only few windows of opportunity to perform maintenance. The current rating for the IGBT modules had to be increased by 30% to achieve a 99% reliability rate after 20 years of operation.

15.4.4 Optimum Damping Strategy

The output power obtained from a wave power generating system has large fluctuations due to ambient changes. This produces large thermal stresses to the

IGBT modules used for the power conversion. The output energy that can be derived from the wave power generator can be maximized by using a different optimization than achieved with maximum power output [68]. The authors created a thermal model for the IGBT module, containing IGBTs and fly-back diodes, to analyze the temperature excursions of the internal devices. Based on a temperature fluctuation of 30°K for a peak power of 50 kW and average power of 1 KW, the Coffin-Manson model was used for a 1500 s irregular wave cycle to predict the lifetime of the IGBT modules. It was found that a converter input impedance (also damping impedance) of 10 Ω resulted in maximizing the lifetime. The optimum maximum energy damping impedance produced an output energy of 60 MWh, which was 38.5% larger than the 44 MWh obtained with optimization for maximum output power.

15.4.5 Oscillating Water Column

The oscillating water column wave generator has been used for a 300 kW power generation plant at Mutriki, Spain. Pressure is generated by waves inside a chamber to produce airflow through a Wells turbine. The turbine is connected to an electric generator to deliver electrical power to the grid. Ocean waves are highly irregular in nature with a variation of airflow within a few seconds. The fault-tolerant IGBT-based converter topology shown in Fig. 15.43 was utilized. This allows achieving robust speed control with fast dynamic response, which is vital for the generating plant. The topology in the figure was found to be tolerant to a failure of two IGBT modules [69].

15.5 Tidal Power

Tidal power is fundamentally different from wave power because tides are generated due to gravitational pull of the moon and the Sun [70] rather than energy delivered by the Sun to the Earth via solar radiation, which fuels the production of waves. The first large power generation installation using tidal energy was the Rance Tidal Power Station set up in France in 1966 with an installed capacity of 240 MW. Other large tidal electricity generation installations are the 20 MW Annapolis Royal Generation Station set up in Nova Scotia, Canada, in 1984; the 3 MW Jiangxia Tidal Power Station set up in Hanzhou China in 1985; the 254 MW Sihwa Lake Tidal Power Plant set up in South Korea in 2011; and the 50 MW Tidal Power Station set up in Gujarat India in 2012. An advantage of tidal power when compared with wind or solar power is its consistency due to the regular orbit of the moon around the Earth. Tides create energy loss that slows down Earth's rotation, causing the period of Earth's rotation to increase from 21.9 to 24 h during the last 620 million years. Extraction of tidal energy has a negligible impact on this phenomenon. The 1.4 MW SeaGen tidal energy generation station is shown in Fig. 15.47. It takes advantage of 400 million

FIGURE 15.47

SeaGen tidal energy generation station.

gallons of water that flows in and out of Strangford Lough in the United Kingdom twice a day due to the tides [71].

The fluctuating energy produced by the tidal energy generator can be delivered to the grid using the topology previously shown in Fig. 15.46 where the variable AC power is first rectified to create a DC bus. The regulated energy to the grid is then delivered using PWM control of an IGBT inverter stage. A demonstration of the "stream power" technology has been reported by using power electronic loads [72].

A tidal power generator combined with an off-shore wind turbine power generator has been analyzed [73]. An IGBT-based bidirectional converter was used for both generators. An experimental model with a 1.5kW servo tidal motor and a 3.0kW servo wind motor was built and analyzed. Good performance with stable operation was demonstrated.

In the case of tidal power generation, the IGBT modules can be cooled by utilizing the natural convection of water. Although not as efficient as forced water cooling used in terrestrial power electronics, the convection-based system is compatible with seabed power electronics for tidal wave generation. The case study of a 100 kVA generator has been performed with thermal analysis of IGBT modules in enclosures placed on the seabed [74]. The thermal modeling indicated that the spreading resistance of the mounting plate and cabinet walls made significant contributions. Despite the lower cooling efficiency, the passive cooling approach has advantages of lower cost, simplicity of design, and greater reliability.

15.6 Geothermal Power

Energy stored in the Earth's crust can be extracted to generate electricity. Twenty percent of the geothermal energy is the consequence of the original process for the formation of the planet and 80% is due to the radioactive decay of elements in the past [75]. 15 GW of geothermal electric power generation capacity has been installed in 24 countries. An example of a geothermal plant in Iceland is shown in Fig. 15.48. Geothermal power generation has been cost-effective in places near tectonic plate boundaries. The United States has the largest amount (3.7 GW) of electricity production from geothermal energy, followed by the Philippines (1.9 GW), and Indonesia (1.2 GW). Iceland has the highest (30%) national percentage of electricity in the world produced using this resource. Geothermal energy generation has the advantages of producing electricity 24 h per day at any time of the year independent of the climate. An overview of geothermal power generation suggests that abandoned oil and gas wells could be utilized as low-cost options for extracting geothermal energy [76].

15.6.1 Power Generation Architecture

A schematic diagram for the typical geothermal energy generation installation [77] is shown in Fig. 15.49. The hot fluid (steam or superheated water) from the magma pool-heated geothermal reservoir located underground is pumped up and circulated around a loop at the Earth's surface. It heats an organic fluid using the heat exchanger. The organic vapor runs through a turbine, producing mechanical energy that is then converted to electrical energy by the generator. The output of the generator is rectified to form a DC bus and an IGBT-based inverter is then used to create

FIGURE 15.48

Iceland geothermal plant.

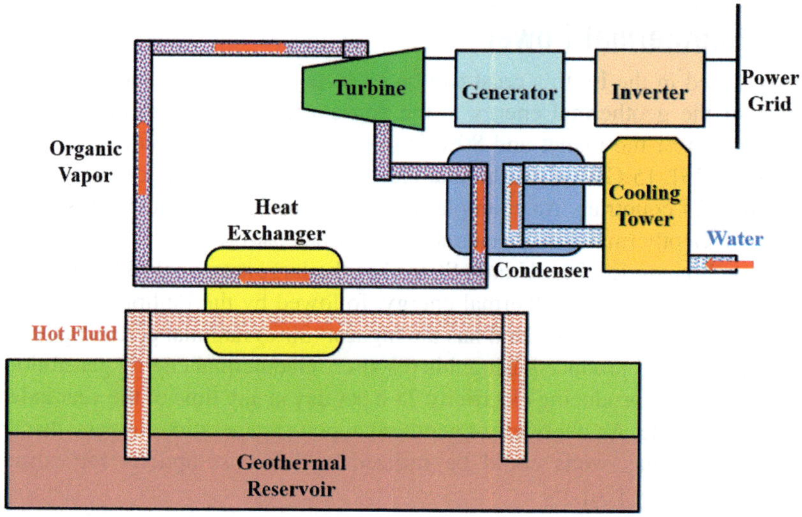

FIGURE 15.49

Geothermal energy generation station.

the well-regulated 60 Hz AC power fed into the power grid. A two-pole induction generator is used to obtain the 480 V 60 Hz output power. The IGBT inverter is operated at a carrier frequency of 2 kHz.

Several manufacturers have optimized IGBTs for use in geothermal installations. The 140 MW Nga Awa Purua geothermal power plant was constructed by Fuji Electric in New Zealand [78]. Infineon has also recognized this as a significant market opportunity for IGBTs [79].

15.7 Summary

Only a few years ago, generating electricity from solar and wind power was mocked as "hippie power" [80]. A reversal of public opinion on renewable energy and a change in the economics for delivering energy from clean sources has resulted in a turnaround of investments by venture capitalists. Green electricity is no longer considered avant garde. The market for commercial solar generation is expected to grow from $50 billion in 2019 to $200 billion in 2026. Annual investments in renewable energy sources grew from $130 billion in 2008 to $214 billion in 2013 [1].

Significant progress has also been made since 2013 for wind power generation. In 2019, China added 10,300 wind turbines to introduce 25.7 GW of grid connected wind power and achieve a total wind power generation capacity of 210 GW [26]. The number of IGBT modules annually required for wind power is expected to exceed 200,000 with ratings of 1700 V and 450–2400 A. The market for the IGBT in China is $500 million for wind power use. The global solar power installation reached 100 GW in 2019 with a total installed capacity of 500 GW [26]. The IGBTs for solar power can have relatively lower power ratings, allowing packaging them in TO packages with 600–1200 V capability at 100 A for residential use, and in modules rated at 1200–1700 V with 450–1800 A capability for commercial use. The IGBT market in China for solar power is $225 million.

The cost of electricity depends on a number of factors, including cost of generation, initial capital investment cost, and maintenance cost. The levelized cost of energy (LCOE) is the price that must be charged for electricity from a particular source to break even over the lifetime of the investment. The LEC is usually determined over a period of 20–40 years in units of cents per kWh. The levelized cost of electricity in cents/kWh generated from various sources has been estimated, as shown in Table 15.5, for power plants entering service in 2019 [81]. It can be seen that the costs for generating electricity using solar and wind power are becoming competitive with conventional coal and nuclear power but above that for natural gas. This does not factor the impact on the environment due to carbon emissions or storage of nuclear waste. The high capital cost for installation of off-shore wind power is a detriment to its commercialization. A more recent study reported in June 2021 [82] states: "*(a) LCOE of photovoltaics ranges from 3.12 to 11.01 €Cent /kWh; (b) the LCOE of on-shore wind turbines ranges between 3.94 and 8.29 €Cent /kWh; (c) Off-shore wind turbines are significantly more expensive at 7.23 to 12.13 €Cent/kWh; (d) Forecasts show that in 2040, the LCOE will be between 3.58 and 6.77 €Cent €Cent/kWh for small rooftop systems and between 1.92 and 3.51 €Cent €Cent/kWh for ground-mounted PV systems. As of 2024, the LCOE is expected to be below 10*

Table 15.5 Levelized Cost in Cents per kWh for Electricity Generation in 2019

Power Source	Capital Cost	Fixed O&M	Variable O&M	Transmission Investment Cost	Total System Cost
Conventional coal	6.00	0.42	3.03	0.12	9.56
Coal-gasification	7.61	0.69	3.17	0.12	11.59
Natural gas fired	1.43	0.17	4.91	0.12	6.63
Advanced nuclear	7.14	1.18	1.18	0.11	9.61
Geothermal	3.42	1.22	0.00	0.14	4.79
Biomass	4.74	1.45	3.95	0.12	10.26
Wind	6.41	1.30	0.00	0.32	8.03
Off-shore wind	17.54	2.28	0.00	0.58	20.41
Solar PV	11.45	1.14	0.00	0.41	13.00
Hydro	7.20	0.41	0.64	0.20	8.45

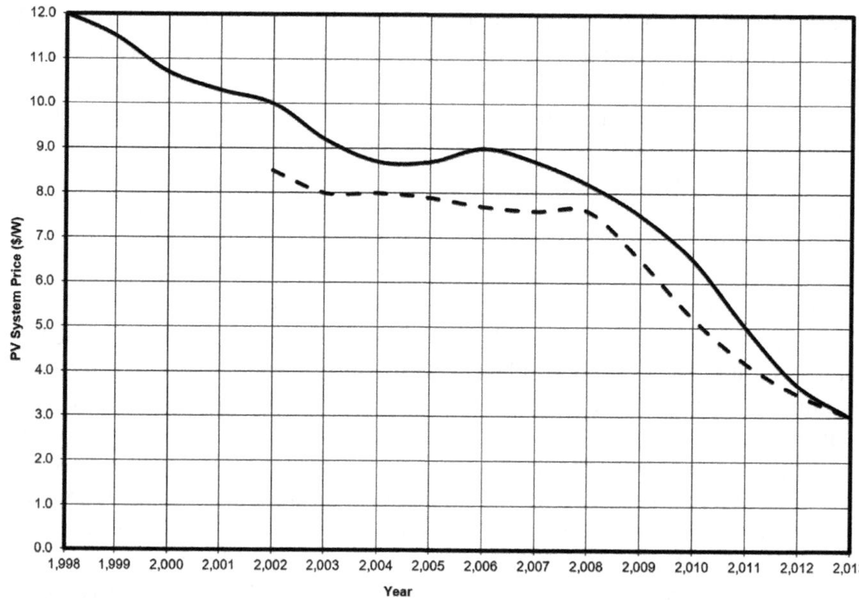

FIGURE 15.50

Decline in solar PV system prices.

€Cent €Cent/kWh for all PV systems." These numbers show that renewable sources for generating electricity can now outperform fossil fuel-based electricity generation.

The installed system cost for solar PV panels has been steadily declining over the years [83], as shown in Fig. 15.50. The residential prices are for systems of 5 kW in size while commercial prices are for systems of 100 kW in size. The price per watt has dropped from $12 in 1998 to about $3 in 2013. The sharp decline in prices for PV panels in 2009 is due to the huge growth in solar panel production in China. The prices reported for utility-scale solar PV plants of more than 2 MW in size was close to $2 per watt.

The cost of electricity generated by any renewable energy source must reach "grid parity" before it becomes commercially viable. Grid parity is reached when the cost of generating electricity becomes equal or less than power already available from the grid without any subsidies [20]. The cost of electricity generated using solar PV has been rapidly declining [84], as shown in Fig. 15.51. By 2013, it had reached 10 cents per kWh, which is the average cost for generating electricity in the United States. It has dropped to 6–8 cents per kWh for utility-scale generation and 8–10 cents per kWh for residential generation in 2021. Grid parity was reached much earlier in Hawaii due to the high cost of generating electricity from fossil fuels there because they must be imported to the islands. Grid parity was also reached earlier in Japan and Germany.

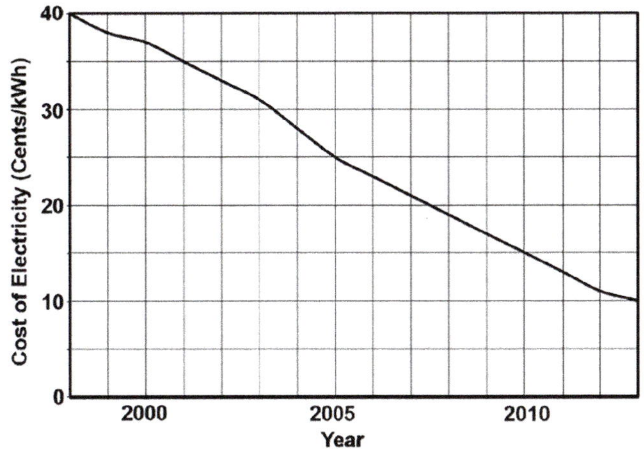

FIGURE 15.51

Decline in solar PV-generated electricity cost.

The cost of the IGBT inverter has remained only a small fraction of the total cost of the solar PV system [85]. This can be observed in Fig. 15.52, which includes the cost of the PV panels, the hardware for installation, and soft costs (related to permits and installation labor). Although the cost of the solar PV panel has dropped significantly, the inverter cost has remained below 10% of the total cost of the solar PV generation system. The cost of the inverter has also been correlated to the installed capacity of PV

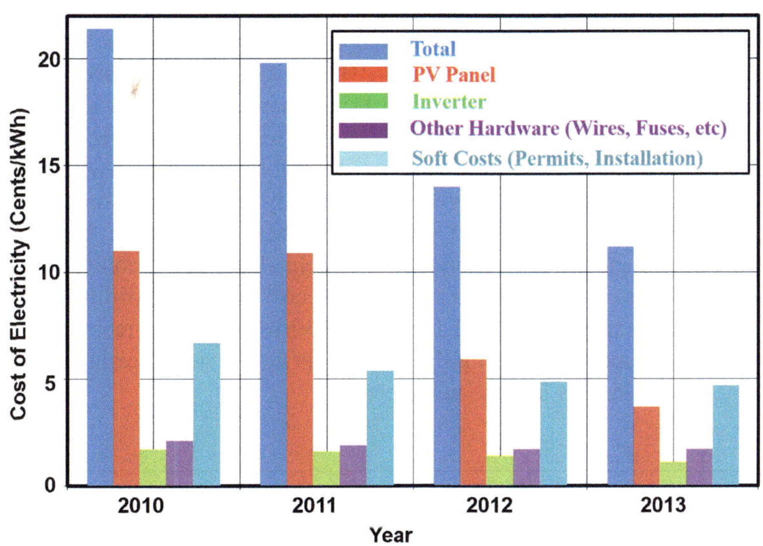

FIGURE 15.52

Inverter cost trends in solar PV utility installations.

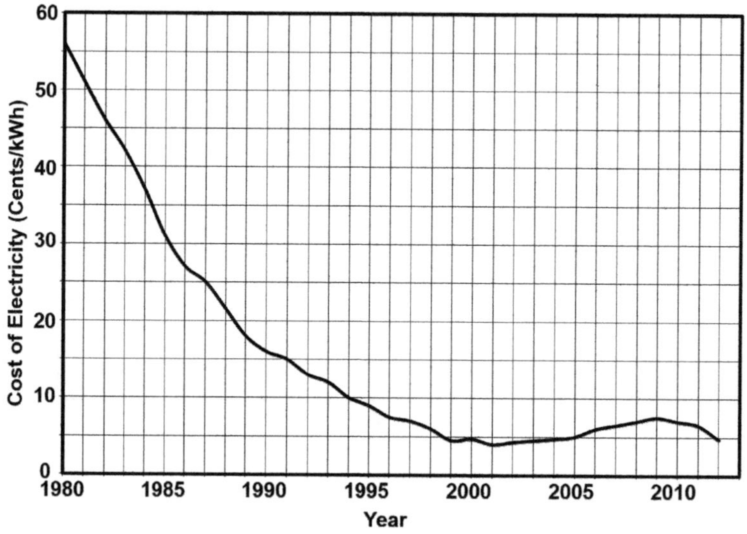

FIGURE 15.53

Decline in wind power-generated electricity cost.

systems as a learning curve [25]. As the cumulative power generated from solar power grew from 300 MW in 1991 to 15 GW in 2010, the cost of the inverter decreased from 1 Euro per watt to 0.25 Euros per watt. The world market for solar PV inverters in 2011 reached \$6.5 billion with 10% of this in the United States [86].

The cost of wind-generated electricity has dropped even faster than for the case of solar power [87]. This trend is shown since 1980 in Fig. 15.53. The cost of wind power electricity fell below 10 cents per kWh by 1995 and has remained there since that time, making this technology cost competitive. It has now reached 5 cents per kWh, making the technology cheaper than the 10 cents per kWh for coal power generation.

In conclusion, solar and wind power have already become cost competitive with energy generated using fossil fuel sources. This has laid a strong foundation for future growth in energy generation by these technologies. The ability to generate electrical power from renewable sources at a low cost makes the worldwide commitment to discontinue use of fossil fuels by 2030 realistic. All renewable energy generation relies upon the use of IGBTs for power conversion, creating a very large market for these devices into the next decade.

References

[1] Renewable Energy. en.wikipedia.org/wiki/Renewable_energy.
[2] Global Wind Report, 2021. https://gwec.net/global-wind-report-2021/.
[3] Photovoltaics. en.wikipedia.org/wiki/Photovoltaics.

[4] Global PV, Installation Estimate and Forecast, 2022, January https://www.pv-magazine.com/2022/02/01/.
[5] Wind and Solar Generated 10% of Global Electricity in 2021 – A Wolds First. https://www.weforum.org/agenda/2022/04/wind-solar-electricity-global-energy/.
[6] The green revolution, Time Mag. (2014) 42–45. June 16.
[7] B.J. Baliga, Enabling green and renewable energy solutions with power semiconductor technology, in: Virtual Global Renewable Energy Researchers Meet, May 7–8, 2021, Australia, 2021.
[8] Niagara Falls Hydroelectric Generating Plants. en.wikipedia.org/Niagara_Falls_hydro electric_generating_plants.
[9] Niagara Power Plant. www.nypa.gov/dacilities/Niagara.htm.
[10] Hydroelectricity. en.wikipedia.org/wiki/Hydroelectricity.
[11] https://www.hydropower.org/status-report.
[12] R. Chokhawala, B. Danielsson, L. Angquist, Power semiconductors in transmission and distribution applications, in: IEEE International Symposium on Power Semiconductor Devices and ICs, 2001, pp. 3–10. Paper P1.
[13] H. Wang, K. Ma, IGBT technology for future high-power VSC-HVDC applications, in: IEEE International Conference on AC and DC Power Transmission, 2016, pp. P1–P6.
[14] H. Wang, et al., A new press pack IGBT for high reliable applications with short circuit failure mode, in: CPSS Transaction on Power Electronics and Applications, 2021, pp. 107–114.
[15] M.A. Faruque, J. Zhang, Study of DC short circuit fault & protection of VSC-HVDC based off-shore DFIG wind farm, in: IEEE International Renewable Energy Conference, 2021, pp. P1–P5.
[16] J. Wang, et al., An IGBT current boosting method by using ultrahigh driving-voltage in HVDC circuit breaker applications, IEEE Trans. Power Electron. 37 (2022) 8146–8156.
[17] S.S. Murthy, R. Jose, B. Singh, A practical load controller for stand alone small hydro systems using self excited induction generator, in: IEEE International Conference on Power Electronic Drives and Energy Systems for Industrial Growth, vol. 1, 1998, pp. 359–364.
[18] G.K. Kasal, B. Singh, VSC with zig-zag transformer based decoupled controller for a pico hydro power generation, in: IEEE India Conference, 2006, pp. 441–446.
[19] M.C.R. Paz, et al., Active power regulation of auxiliary generation units in hydroelectric power plants using power electronic switching, in: IEEE Transmission and Distribution Latin American Conference, 2012, pp. 1–7.
[20] Solar Cell. en.wikipedia.org/wiki/Solar_cell.
[21] Solar Power. en.wikipedia.org/wiki/Solar_power.
[22] Global Energy Review, Renewables, 2021. https://www.iea.org/reports/global-energy-review-2021/renewables.
[23] https://www.nsenergybusiness.com/features/largest-solar-power-plants/.
[24] C. Klumpner, A new single-state current source inverter for photovoltaic and fuel cell applications using reverse blocking IGBTs, in: IEEE Power Electronics Specialists Conference, 2007, pp. 1683–1689.
[25] B. Burger, D. Kranzer, Extreme high efficiency PV-power converters, in: IEEE European Conference on Power Electronics and Applications, 2009, pp. 1–13.
[26] Y. Liu, et al., Development review and market research of IGBT devices for PV inverter and wind power converter, in: IEEE China International Forum on Solid-State Lighting, 2020, pp. 30–33.

[27] K. Hirachi, et al., Pulse area modulation control implementation for single-phase current source-fed inverter for solar photovoltaic power conditioner, in: IEEE International Conference on Power Electronic Drives and Energy Systems for Industrial Growth, vol. 2, 1998, pp. 677–682.
[28] N. Kasa, et al., A transformer-less inverter using buck-boost type chopper circuit for photovoltaic power system, in: IEEE International Conference on Power Electronic Drives and Energy Systems for Industrial Growth, vol. 2, 1999, pp. 653–658.
[29] P. Neves, et al., Single-phase shunt active filter interfacing renewable energy sources with the power grid, in: IEEE Industrial Electronics Conference, 2009, pp. 3264–3269.
[30] Y. Yang, et al., A hybrid power control concept for OV inverters with reduced thermal loading, IEEE Trans. Power Electron. 29 (2014) 6271–6275.
[31] B.J. Baliga, Fundamentals of Power Semiconductor Devices, Second Edition, Springer-Science, 2019.
[32] K. Fujii, et al., 1-MW advanced T-type NPC converters for solar power generation system, in: IEEE European Power Electronics and Applications Conference, 2013, pp. 1–10.
[33] X. Hao, K. Ma, Z. Jia, 7th generation IGBT technology enables highest power density stack in solar application, in: IEEE International Power Electronics and Motion Control Conference, 2020, pp. 2056–2060.
[34] P. Ponnuswamy, et al., A new multilevel inverter topology with reduced power components for domestic solar PV applications, IEEE Access (2020) 187483–187497.
[35] Fuel Cell. en.wikipedia.org/wiki/Fuel_cell.
[36] G. Scheible, H. Solmecke, D. Hackstein, Low cost soft switching DC-DC converter with autotransformer for photovoltaic hydrogen systems, in: IEEE International Conference on Industrial Electronics, Control, and Instrumentation, vol. 2, 1997, pp. 780–785.
[37] Supercapacitor. en.wikipedia.org/wiki/Supercapacitor.
[38] K. Ogura, et al., Non-isolated ZVS-PWM boost chopper-fed DC-DC converter with auxiliary edge resonant snubber, in: IEEE Industrial Electronics Society Conference, vol. 3, 2003, pp. 2356–2362.
[39] G. Majumdar, Power module technology for home power electronics, in: IEEE International Power Electronics Conference, 2010, pp. 773–777.
[40] Z. Minghui, K. Komatsu, Three-phase advanced neutral-point-clamped IGBT module with reverse blocking IGBTs, in: IEEE International Power Electronics and Motion Control Conference, 2012, pp. 229–232.
[41] S. Chen, et al., T-type 3-level IGBT power module using authentic reverse blocking IGBT (RB-IGBT) for renewable energy applications, in: IEEE International Future Energy Electronics Conference, 2013, pp. 229–234.
[42] S. Igarashi, et al., Advanced three-level converter with newly developed 1200V reverse blocking IGBTs, in: IEEE European Power Electronics and Applications Conference, 2011, pp. 1–7.
[43] Wind Power. en.wikipedia.org/wiki/Wind_power.
[44] https://www.renewableenergyworld.com/wind-power/wind-power-experts-expect-wind-energy-costs-to-decline-up-to-35-by-2035/#gref.
[45] World Cumulative Installed Power Capacity. www.earthpolicy.org.
[46] . https://en.wikipedia.org/wiki/Wind_power_by_country.
[47] D. Marsh, Wind chimes in, EDN Mag. (2004) 38–47. December 7.
[48] D. Morrison, Wind Mill Power Component Development. Power Electronics Technology, 2006, p. 8. July.

[49] S. Muller, et al., Adjustable speed generators for wind turbines based on doubly-fed induction machines and 4-quadrant IGBT converters linked to the rotor, in: IEEE Industry Applications Society Meeting, vol. 4, 2000, pp. 2249–2254.
[50] F. Blaabjerg, K. Ma, Future on power electronics for wind turbine systems, IEEE J. Emerg. Sel. Top. Power Electron. 1 (2013) 139–152.
[51] A. Jamehbozorg, W. Gao, A new controller design for a synchronous generator-based variable-speed wind turbine, in: IEEE North American Power Symposium, 2010, pp. 1–6.
[52] A.B. Mogstad, et al., A power conversion system for offshore wind parks, in: IEEE Annual Conference on Industrial Electronics, 2008, pp. 2106–2112.
[53] Y. Miao, The impact of large-scale offshore wind farm on the power system, in: IEEE China International Conference on Electricity Distribution, 2010, pp. 1–5.
[54] J. Chivite-Zabalza, et al., Comparison of power conversion topologies for a multi-megawatt off-shore wind turbine, based on commercial power electronic building blocks, in: IEEE Industrial Electronics Society Conference, 2013, pp. 5242–5247.
[55] M.E. Haque, K.M. Muttaqi, M. Negnevitsky, Control of a stand alone variable speed wind turbine with a permanent magnet synchronous generator, in: IEEE Conference on Conversion and Delivery of Electrical Energy in the 21st Century, 2008, pp. 1–9.
[56] K. Mamun, et al., Design and simulation of a reactive power compensator using STATCOM for 'Muhuri dam' wind power plant, in: IEEE International Conference on Energy and Power Engineering, 2019, pp. P1–P5.
[57] B. Backlund, et al., Topologies, voltage ratings, and state of the art high power semiconductor devices for medium voltage wind energy conversion, in: IEEE Power Electronics and Machines for Wind Applications, 2009, pp. 1–6.
[58] J.G. Bauer, T. Duetemeyer, L. Lorenz, New IGBT development for traction drive and wind power, in: IEEE International Power Electronics Conference, 2010, pp. 768–772.
[59] P. Roshanfekr, et al., Selecting IGBT module for a high voltage 5 MW wind turbine PMSG-equipped generating system, in: IEEE Power Electronics and Machines in Wind Applications Conference, 2012, pp. 1–6.
[60] Y. Chen, et al., A health-oriented power control strategy of direct drive wind turbine, IEEE Trans. Power Delivery (2022), https://doi.org/10.1109/TPWRD.2021.3084216.
[61] A. Jokin, et al., Output power increase of a 3-level converter using state-of-the-art 4.5kV IGBT modules, in: European Conference on Power Electronics and Applications, 2016, pp. P1–P9.
[62] Wave Power. en.wikipedia.org/wiki/Wave_power.
[63] J. Cruz, et al., Green Energy and Technology, Ocean Wave Energy, Springer-Science, 2008.
[64] J.F. Childs, The role of converters and their control in the recovery of wave energy, in: IEE Colloquium on Power Electronics for Renewable Energy, 1997, pp. 3/1–3/7.
[65] Z. Zhou, et al., Permanent magnet generator control and electrical system configuration for wave dragon MW wave energy take-off system, in: IEEE International Symposium on Industrial Electronics, 2008, pp. 1580–1585.
[66] J. Sjolte, et al., Reliability analysis of IGBT inverter for wave energy converter with focus on thermal cycling, in: IEEE International Conference on Ecological Vehicles and Renewable Energies, 2014, pp. 1–7.
[67] T. Kovaltchouk, et al., Influence of IGBT current rating on the thermal cycling lifetime of a power electronic active filter in a direct wave energy converter, in: IEEE European Power Electronics and Applications Conference, 2013, pp. 1–10.

[68] C. Chen, L. Zuo, Damping selection strategy for maximum energy on wave energy power converters, in: IEEE Energy Conversion Congress and Exposition, 2019, pp. 4988–4993.

[69] M. Zarie, et al., Three-phase four-switch converter for SPMS generators based on model predictive current control for wave energy applications, IEEE Trans. Power Electron. 35 (2020) 289–302.

[70] Tide. en.wikipedia.org/wiki/Tide.

[71] SeaGen. en.wikipedia.org/wiki/SeaGen.

[72] S. Wang, et al., Design and implementation of power electronic load using a test tidal current energy generator sets, in: IEEE International Conference on Fuzzy Systems, 2014, pp. 354–358.

[73] M. Rahman, S. Oka, Y. Shirai, Hybrid power generation system using offshore-wind turbine and tidal turbine for power fluctuation compensation (HOT-PC), IEEE Trans. Sustain. Energy 1 (2010) 92–98.

[74] F. Wani, et al., A study on passive cooling in subsea power electronics, IEEE Access 6 (2018) 67543–67554.

[75] Geothermal Energy. en.wikipedia.org/wiki/Geothermal_energy.

[76] H. Kulasekara, V. Senulabdeen, A review of geothermal energy for future power generation, in: International Conference on Advances in Electrical Engineering, 2019, pp. 223–228.

[77] S. Dahal, et al., Modeling and simulation of the interface between geothermal power plant based on organic Rankin cycle and the electric grid, GRC Trans. 34 (2010) 1011–1015.

[78] https://www.fujielectric.com/products/thermal_power_generation/.

[79] http://www.infineon.com/cms/en/product/applications/Renewables/Geothermal.

[80] M. Grunwald, Wall street goes green, Time Mag. (2014). September.

[81] Cost of Electricity by Source. en.wikipedia.org/wiki/Cost_of_electricty_by_source.

[82] C. Kost, Study: Levelized Cost of Electricity-Renewable Energy Technologies, https://www.ise.fraunhofer.de/en/publications/studies/cost-of-electricity.html.

[83] D. Feldman, et al., Photovoltaic (PV) Pricing Trends: Historical, Recent, and Near-Term Projections. U.S. Department of Energy Report DOE/GO-102012-3839, 2012. November.

[84] S. Combs, n.d. US Solar Market Trajectory, Texas State Energy Conservation Office.

[85] D. Wood, n.d. Falling Price of Utility-Scale Solar Photovoltaic Projects, National Renewable Energy Labs.

[86] L.F. Casey, et al., Power devices for grid connections, in: IEEE International Symposium on Power Semiconductor Devices and ICs, 2012, pp. 1–7.

[87] http://futurepowernow.com/category/wind-turbine-power/.

CHAPTER

IGBT Applications: Power Transmission 16

Traditionally, electric power has been generated in large-capacity generation plants utilizing hydroelectric power, fossil fuels such as coal, or nuclear power. These sources for electrical energy are located remotely from the point of use by human populations concentrated in cities and by factories for manufacturing goods. It is necessary to efficiently transfer the electrical energy between the source and the customer. A major component for power losses in the transmission system is the ohmic resistance of electrical wires. Although high currents can be delivered through a power line by increasing the diameter of the conductor, this becomes problematic due to the increased cost and weight of the transmission cable. Power transmission has therefore relied upon using very high voltages in excess of 100 kV to maintain good efficiency.

Two basic approaches to power transmission are the use of either DC or AC voltages. AC power waveforms have the advantage of allowing transformation of the magnitude of the voltage by using inexpensive transformers. However, this leads to complications due to the reactive power in the cables. The transmission of electrical power over long distances is therefore usually performed using high-voltage DC (HVDC) networks.

Over the last 50 years, the thyristor has been extensively used for HVDC transmission systems. Thyristors can be manufactured with deep P–N junctions formed using aluminum and gallium diffusions with excellent uniformity across large-diameter silicon wafers [1]. The application of neutron transmutation doping has enabled the production of high-resistivity n-type silicon wafers with excellent uniformity. These technologies have allowed increasing the ratings of silicon thyristors to more than 8 kV and 1 kA. Light-triggered thyristors have been developed allowing gate control with optical fibers so that a large number of thyristors can be connected in series to hold off the extremely large HVDC transmission voltages. The first commercial HVDC transmission system with thyristors was installed in 1970 at Gotland, Sweden, with 50 kV lines [2]. The first HVDC system using light-triggered thyristors was installed in 1988 in the Konti–Skan project between Denmark and Sweden operating at 250 kV [3].

However, in the past 10 years, there has been a growing interest in replacing thyristor valves with insulated gate bipolar transistor (IGBT) valves for HVDC systems. The availability of IGBTs allows building voltage source converters (VSCs) with pulse-width modulation (PWM). The high commutation frequency allows

The IGBT Device. https://doi.org/10.1016/B978-0-323-99912-0.00026-X

compensation of rapid changes in the voltage on the AC side. This approach is referred to as HVDC light because it eliminates the need for converter transformers [2].

16.1 HVDC Transmission

Electricity generated from large power plants is distributed over HVDC transmission lines from the point of power generation to the point of use, such as factories or urban areas. Large towers like those shown in Fig. 16.1 are used to keep the very high voltages applied to the lines far above the ground. HVDC power transmission has advantages over a high-voltage AC (HVAC) system because it avoids high charge and discharge currents due to cable capacitance (particularly in underwater transmission) and the need for synchronization of frequency and phase between the source and the load. In addition, it allows power transfer between networks operating at 50 Hz and 60 Hz. The longest HVDC power link was 2000 km between Xiangjiaba Dam and Shanghai in China until 2013 when the 2375 km Rio Madeira link was built in Brazil to serve the citizens of the Sao Paulo area [4].

During the 1970s, thyristor-based HVDC systems replaced those previously implemented using mercury arc valves. Thyristor valves are known as line-commutated converter (LCC) HVDC valves. During the last 10 years, the availability of IGBTs with large blocking voltages and modules with high current-handling capability enabled extending the VSCs developed for motor control applications to the HVDC transmission system. The ABB group refers to these VSC-HVDC valves as HVDC light; Siemens calls them the HVDC PLUS for power link universal system; and Alsthom's product is named HVDC MaxSine. With multilevel converters, the harmonic filtering, which occupied nearly half the LCC installation, can be

FIGURE 16.1

High-voltage DC power transmission lines.

reduced or eliminated [4]. It is projected that the VSC–HVDC approach will completely displace the previous LCC HVDC approach, even at the highest voltage and power levels.

The transmission system contains switching valves at each end of the network. The valves have traditionally been built using high-power thyristors with a blocking voltage of 8 kV and a 2 kA current-handling capability with current source converters (CSC). A new generation of HVDC systems based on VSCs has become recently feasible with the availability of IGBTs with larger blocking voltage and current handling capability. The classical CSC HVDC system using thyristors can be contrasted with the new VSC-based HVDC system using IGBTs in Fig. 16.2 [5].

The traditional HVDC system shown at the top of Fig. 16.2 makes use of thyristors with line commutation to control current flow. The DC current flow is unidirectional as determined by the DC-side equilibrium voltage of the two converters. The valve requires reactive power generation using the shunt connected filter shown in the figure. The new HVDC system based on VSC is shown at the bottom of Fig. 16.2. In this case, the AC side voltage can be independently determined with the current decided by the series inductance shown in the figure. Active and reactive power can be controlled on both ends of the network. Harmonic generation is low in this approach because of the high switching frequencies for the IGBTs.

The authors state [5]: "*Today, the IGBT is the preferred choice for VSC-based converters in T&D applications because of the following characteristics: (1) Low-power control because it is an MOS-controlled device. This is advantageous when operating at very high voltage levels (several 100 kV); (2) Transistor action, which enables precise control of the device in a manner that is not possible with latching alternatives. For instance, the converter can be turned off even in short*

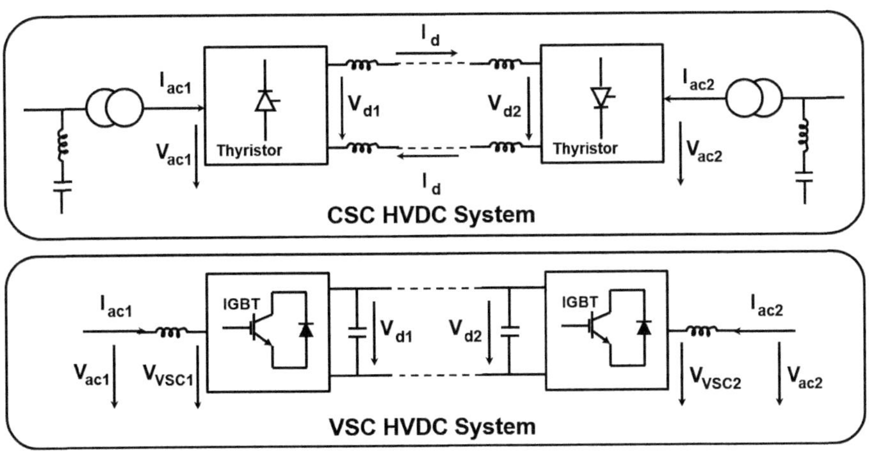

FIGURE 16.2

Current source converter *(top)* and voltage source converter *(bottom)* high-voltage DC transmission systems.

circuit conditions; (3) High switching speed, (which makes) high switching frequency feasible." A 300 MW VSC–HVDC-based transmission and distribution (T&D) system using IGBTs was commissioned in 2002 with three-level converter topology.

16.2 HVDC Components

Power electronics play an increasing role in HVDC systems. Many subsystems are required to support the HVDC network [6]:

1. The static compensator (STATCOM) is used for voltage stabilization and load balancing. It corrects for both leading and lagging power factors.
2. The unified power flow controller (UPFC) enables optimal utilization of power lines by controlling power flow.
3. The dynamic voltage restorer (DVR) detects a drop in the line voltage during a cycle and restores the rest of the waveform by using energy in storage elements such as batteries.
4. Transfer switches move loads to alternate lines to maintain power quality
5. Static breakers interrupt currents produced by faults within one cycle.
6. Interties allow connection of asynchronous power sources.
7. Active filters remove harmonic distortion in power networks.

The implementation of power electronics is strongly impacted by the available power semiconductor device technology. The system demands on the power devices include a low device cost, rugged operation without snubbers, highly reliable operation despite power cycling, high voltage and current ratings, fast switching speed, and modularity. These features are available in modern IGBT modules.

16.3 HVDC Trends

The HVDC transmission voltage has been increasing over the years, as shown in Fig. 16.3, to meet the larger power transmission needs of countries [7]. HVDC transmission began by using mercury arc rectifiers. The transmission was first conducted using 50–100 kV power lines. The replacement of mercury arc rectifiers with silicon thyristors marked a major step in improvement of the transmission system. The thyristor-based HVDC networks began in 1970 with 400 kV transmission lines. A large number of thyristors had to be connected in series to hold off such high voltages because the blocking voltage capability for each device was about 3 kV at that time. The blocking voltage capability was continually increased such that 10 kV devices are available today [1]. This enabled increasing the transmission line voltage as well. The area of each thyristor was also increased, using larger wafer diameters to obtain larger current handling capability.

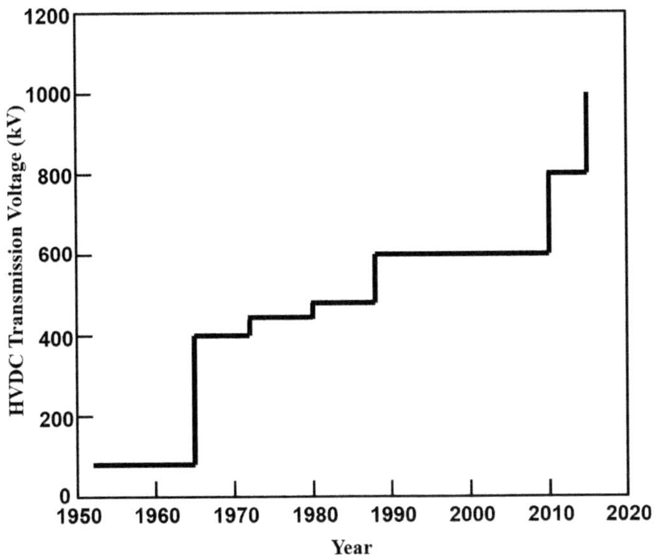

FIGURE 16.3

Growth of high-voltage DC transmission voltage.

In the figure, the increase in transmission line voltage to 500 kV occurred with the Cahora Bassa installation in Mozambique [8] in 1980. The next jump in HVDC transmission voltage to 600 kV occurred with the building of the Itaipu Dam [9] at the border between Brazil and Paraguay in 1987. The Rio Madeira tributary [10] of the Amazon in Brazil was dammed to produce electricity delivered over a 600 kV transmission line in 1987. Power transmission of 12 GW at 1100 kV was achieved in China in 2019 over a distance of 3300 km for the large Changji-Guquan hydroelectric project [11].

HVDC interconnections can be classified into four types [12]: (1) Back-to-back configuration with the converters and valves in the same building to maintain a short DC busbar; (2) cable transmission configuration where an underground or submarine cable is needed; (3) line transmission with overhead power lines between converters separated by a long distance; and (4) multiterminal configuration with more than three converters interconnected using cables or overhead lines. The VSC-HVDC system using IGBT converters is being increasingly adopted because they generate much fewer harmonics than prior LCC converters based on thyristors. The authors state [12]: "*The use of VSC technology will become predominant in the future in the other end of the ratings up to 1500 MW HVDC systems, especially cable schemes (allowing the use of lower cost XLPE cable).*"

For power levels of a few 100 MW, the VSC-HVDC topology with IGBTs is gaining favor. The ability to turn-off current flowing in the IGBT allows building self-commutated converters. The authors state [6]: "*Several candidate high-power devices are available such as GTOs, IGCTs, IEGTs, etc., but the most commonly used device is the IGBT.*" The IGBT has eclipsed the gate turn-off thyristor (GTO) and the insulated gate controlled thyristor (IGCT). The injection enhanced IGBT (IEGT) is an IGBT with deep trenches. The limited blocking voltage capability of IGBTs has led to a variety of topologies to serve HVDC installations.

16.3.1 Gratz Bridge

The VSC using IGBTs was originally developed for motor drive applications. To apply this approach to the very high voltages encountered in HVDC transmission systems, it is necessary to place a large number of IGBTs in series in each leg of the two-level converter topology. This is illustrated in Fig. 16.4 with a three-phase Gratz configuration [6]. The topology requires synchronized gate drives for all the IGBTs that are in series. To ensure that the voltage is shared equally among the IGBTs in series, large and heavy passive snubber circuits are required. To reduce harmonic losses, the IGBTs must be switched many times per cycle with PWM leading to significant power losses. The fast transients of the IGBTs create a large dV/dt because of the high bus voltage leading to EMC noise problems. This topology has been used to build a modular multilevel converter without passive filters [13]. It was able to tolerate severe fault conditions such as a short circuit on the DC side.

Multilevel converters allow reduction of the harmonics. The multilevel converter, called a "full chain link," shown in Fig. 16.5 has many independent converter cells connected in series. Each converter cell has four IGBTs with fly-back diodes and a DC link capacitor. By creating tens or even hundreds of discrete output voltage levels, almost perfectly sinusoidal voltage and current waveforms can be generated to reduce the harmonics without the need for filters [6].

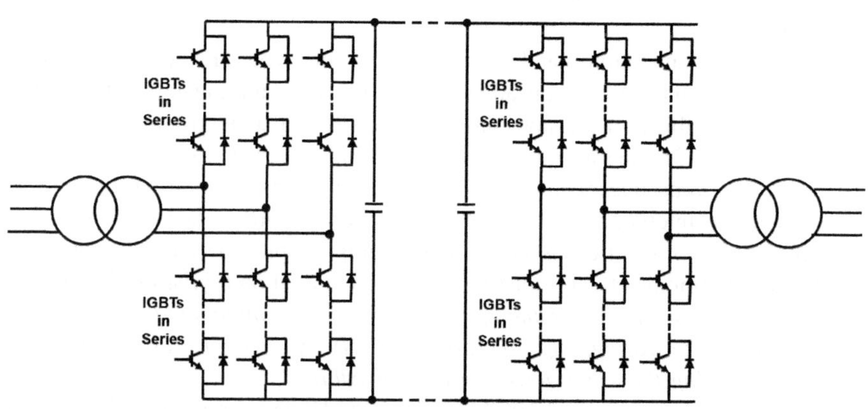

FIGURE 16.4

Two-level, three-phase Gratz bridge converter.

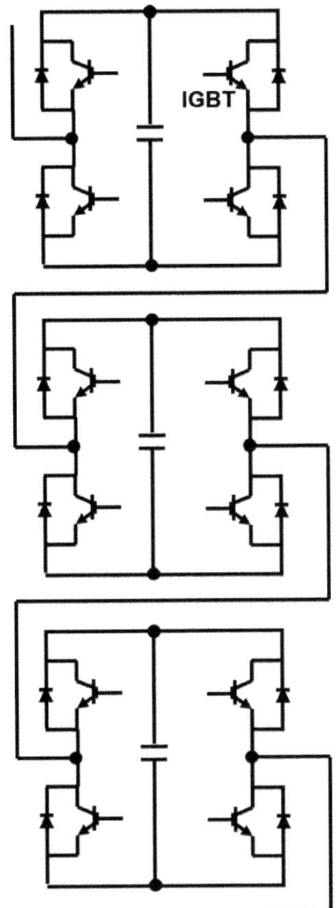

FIGURE 16.5

Multilevel chain link converter.

16.3.2 CSC-Based HVDC

The basic configuration for the CSC HVDC topology shown in Fig. 16.2 was implemented using thyristors. The availability of IGBTs with reverse blocking capability has encouraged the development of CSC HVDC converters using IGBTs in place of thyristors. This approach is part of the HVDC light technology. It allows independent control of reactive power as well as the current and voltage [14]. The CSC contains a rectifier bridge and an inverter bridge, as illustrated in Fig. 16.6, together with AC filters.

A detailed review of many different topologies for the VSC and CSC has been provided by Nami et al. [15]. The authors also contrasted the advantages and disadvantages of the CSC and VSC, as shown in Table 16.1. The authors conclude: "*It is clear that the development of power semiconductor devices and new demands and*

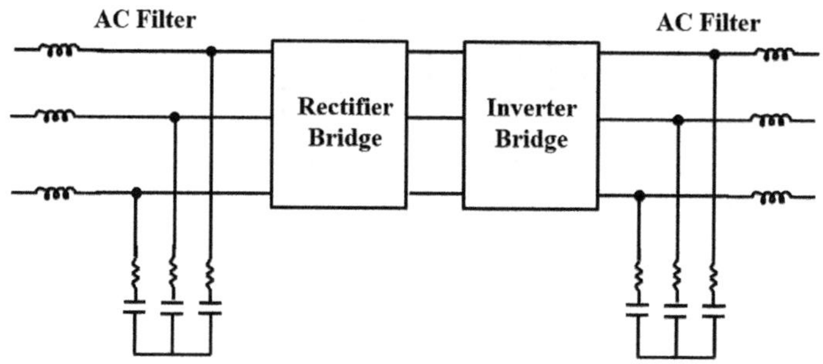

FIGURE 16.6

CSC high-voltage DC converter.

Table 16.1 CSC versus VSC HVDC Converters

	VSC Converters	CSC Converters
Advantages	• PQ controllability • Voltage scaling by series connection of cells • Well-defined voltage across semiconductors • Nearly pure sine wave output voltage	• PQ controllability using self-commutation devices • Current scaling by paralleling cells • Short-circuit fault tolerance • Nearly pure sine wave output current
Challenges	• Large energy storage in capacitors • Protection against internal and external DC short circuit faults	• High loss due to inductors and RB devices • Large footprint for inductors • Protection against open circuit faults and overvoltage across devices • Limited voltage scaling capability

regulations will drive the future of MMC technology." Modular multilevel converters (MMC) are the most suitable choice for VSC-based HVDC power transmission.

16.3.3 Static Synchronous Compensator

STATCOM is used in an HVDC system to provide reactive power compensation. It makes use of the VSC to synthesize a controllable sinusoidal voltage at the fundamental frequency. When the generated voltage by the STATCOM exceeds that of the AC system, it behaves like a shunt capacitor generating reactive power. When the generated voltage by the STATCOM falls below that of the AC system, it behaves like a shunt inductor absorbing reactive power. VSC-HVDC systems make use of two- or three-level PWM converters and modular multilevel series-connected chain link converters.

The two- and three-level converters are designed in many configurations: (1) a single converter using a single IGBT in series within a valve; (2) multiple converters operating in parallel with an IGBT in series within a valve; (3) a single converter with multiple IGBTs in series within a valve; (4) multiple converters connected in parallel with multiple IGBTs in series within a valve with PWM modulation; and (5) multiple converters connected in parallel with multiple IGBTs in series within a valve without PWM modulation.

For serving large power levels, the STATCOMs are built using cascaded H-bridges to form the chain link circuit shown in Fig. 16.5 in each phase to produce the sinusoidal voltage. This approach is particularly suited to unbalanced loads such as electric arc furnaces and traction load balancers [6].

16.3.4 Self-Powered IGBT Switch

Gate drive signals must be provided to each IGBT connected in series to operate at the very high HVDC voltage. The high-voltage isolation barrier between the series-connected IGBTs cannot be breached by the gate drive signal. A self-power IGBT switch configuration (called a director switch) has been developed to address this issue [16]. Each IGBT switch is built using the configuration shown in Fig. 16.7. The IGBT has a large resistor R_B in parallel to balance voltage sharing between series-connected devices in spite of differences in leakage currents of each of the devices. The charge on capacitor C_S is used to provide power to the gate drive circuit for the IGBT. The capacitor has a loading resistor R_L (100 kΩ) across it to avoid a drift in the voltage across the capacitor. The voltage across the capacitor is fed to a DC-to-DC converter to change the 2500 V across the IGBT to 15 V needed for the gate drive circuit. The crowbar circuit is used to keep the voltage across the capacitor within its safe operating area during any hard switch turn-off events. The communications and diagnostic functions are performed by the field-programmable gate

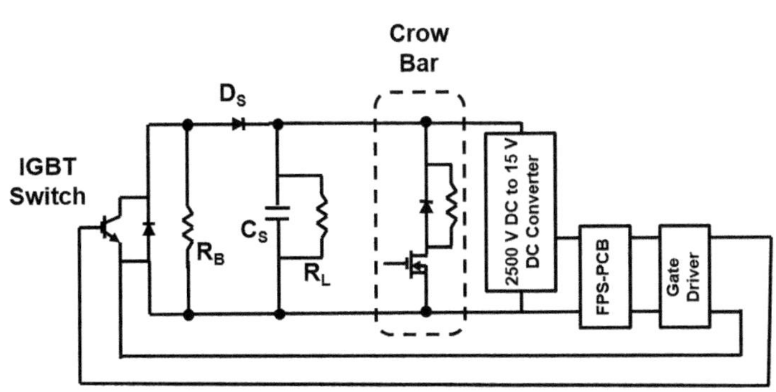

FIGURE 16.7

Director switch configuration for series connected IGBTs in a high-voltage AC transmission network with self-powered gate drive.

array (FPGA) on the FPS-PCB. The operation of seven 2.5 kV IGBT press-pack devices in series has been experimentally demonstrated.

The turn-off delays between the director switch IGBTs in the above topology can be adjusted using an algorithm to maintain voltage balance between the IGBTs during switching [17]. The diode DS and capacitor Cs also behave as a snubber to obtain a balance in the voltage across the IGBTs during the switching events.

16.3.5 Hockey-Puck Press-Pack IGBT Design

The transition to IGBT-based VSC-HVDC power transmission systems has been accelerating due to the advantages of black-start capability, elimination of filters, and modular scalable architecture. Commercially available IGBT modules for industrial applications have been utilized for HVDC systems up to 2 GW. For larger power levels, the press-pack IGBT modules are preferred [18]. Soft switching freewheeling diodes have been manufactured with >80 kA surge current capability. They can handle >10 MW of turn-off power with a dI/dt of >15 kA/μs. A 4.5 kV, 3 kA press-pack IGBT with a 125 mm pole-piece diameter contains 36 chips. This is a hockey puck-shaped module design similar to that used previously for thyristors in legacy HVDC systems rather than the rectangular module design commonly used for industrial applications. A hermetically sealed housing provides enhanced reliability and field lifetime. The trench gate structure shown in Fig. 12.52 with field stop design is used. A low on-state voltage drop for the IGBT is desirable here due to the low switching frequency of 50–150 Hz.

16.3.6 IGBT Ratings for VSC-HVDC

The use of IGBTs for VSC-HVDC systems requires relating the transmission line currents to the IGBT-rated current [19]. The authors also compare the option of using IGCT devices instead of IGBTs. Although IGCTs offer lower on-state losses, they have many disadvantages such as a minimum operating temperature of 100°C, a maximum turn-on dI/dt of 1 kA/μs, a series inductor to reduce the turn-on slew rate, and resistor, capacitor, diode (RCD) snubber to mitigate voltage overshoots. These elements increase the cost and volume of the device. In addition, IGCTs do not offer short circuit overcurrent limiting capability, which is a safety concern. These issues make the IGBT a favorable choice.

A one-to-one correlation between the IGBT current rating and the transmission line current was observed by the authors. This allows implementation of VSC-HVDC systems of hundreds of MWs with transmission line currents of 1500 A. A 3 GW VSC-HVDC system is possible by using IGBT modules described in the previous section with 3000 A capability. A 6.5 kV IGBT module has been constructed using the reverse conducting IGBT (RC-IGBT) structure with 1000 A capability [19]. The integration of the fly-back diode into the IGBT structure improved performance and space.

16.4 AC Power Transmission

Most of the power generated from large power plants is delivered to users via AC power transmission lines. The basic power transmission arrangement is illustrated in Fig. 16.8 [20]. The voltage of the AC power from the generator (ranging from 2.3 to 30 kV) is first increased to the transmission line AC voltage level using a step-up transformer. The transmission line voltages have been increasing over the years [21], as shown in Fig. 16.9, to handle larger power levels without huge increases in the current in the lines. The 110kV line (number 1) was first installed in 1911 at Lauchhammer-Riesa in Germany. This was followed by a 220kV line (number 2) installed in 1929 at Brauweiler–Hoheneck in Germany. In 1932, the AC transmission line voltage was increased to 287 kV (number 3) at Boulder Dam in the United States followed by a 380kV line (number 4) installed in 1952 at Harspranget–Hallsberg in Sweden. The voltage was increased to 735 kV in 1965 for the Manicouagan line (number 5) in Montreal, Canada. A 1200kV line (number 6) was installed in 1985 at Ekibastuz–Kochetav in the Soviet Union. A 100 mile, 1000 MW AC power transmission line operating at 765 kV has about 1% losses compared to 4% for a 345kV line.

The very high voltage at the AC transmission line must be first stepped down with a transformer for utilization of power by end users. This is first performed at the sub-transmission level with AC voltages ranging from 20 kV to 69 kV. This voltage level is suitable for power delivery to a large factory for manufacturing consumer and industrial products. In order to serve the primary customers, the AC voltage is stepped down to the 4–13 kV level at a substation. Secondary customers such as homes and offices are provided power at 120 V or 240 V using a distribution transformer on the power pole. The production price for electricity ranges from 1 to 2.5

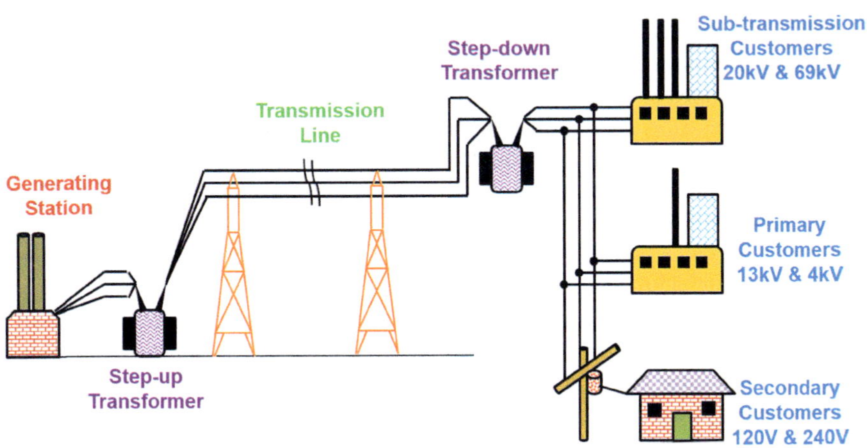

FIGURE 16.8

High-voltage AC transmission network.

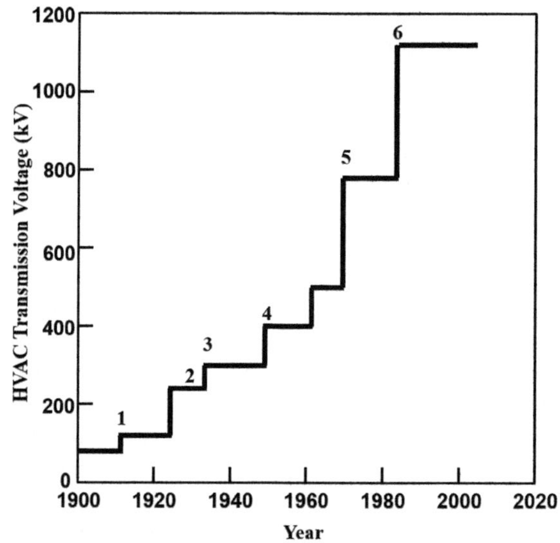

FIGURE 16.9

Growth of high-voltage AC transmission line voltage.

cents per kWh while the long-distance transmission cost ranges from 0.5 to 2 cents per kWh. The average residential price for electricity in the United States is 13.7 cents per kWh, but this is much larger (3–5 times) in Europe and Japan.

16.4.1 Facts

The concept of flexible AC transmission systems (FACTS) was proposed to provide stability and increased power-handling capacity by the injection of reactive power using power electronically controlled elements connected either in series or in parallel with the lines [22]. Static volt amperes reactive (VAR) compensators (SVCs) and STATCOMs are utilized for providing the reactive energy [23]. The parameters that determine power flow along the transmission line are the magnitude and phase angle of the voltages at the ends of the line and the impedance of the line. These parameters must be dynamically controlled to ensure stable operation, even in the presence of disturbances. High-speed FACTS devices damp down power oscillations and stabilize the end-point voltages in the transmission network.

The two basic FACTS approaches for stabilization of the AC power transmission network [23] are shown in Fig. 16.10. The first approach is shunt compensators where the capacitor banks and inductors are connected in a shunt configuration to inject reactive power. The second approach makes use of capacitor banks connected in series with the network line that can be switched in and out by using power devices.

Until the turn of the 21st century, the silicon thyristor was the only semiconductor device capable of handling the power levels required for FACTS devices. In 1995,

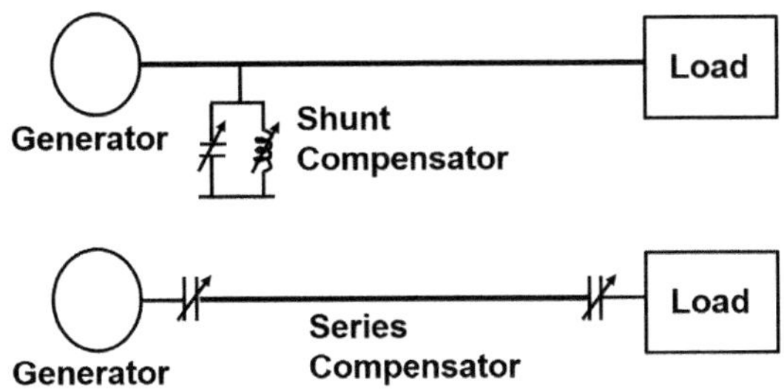

FIGURE 16.10

Basic FACTS device options.

the authors stated [23]: "*One fundamental limitation to the use of the conventional thyristor is that the device can only be turned-off at the next current zero crossing. Successful development of turn-on, turn-off devices, which have a high unit power capability, high switching speeds (greater than say 1 kHz), and low unit losses, will see a new phase of advanced FACTS devices (that) can greatly reduce equipment size by eliminating the need for large capacitor banks and inductors, and provide harmonic current cancellation using PWM techniques.*" The situation indeed changed after the availability of devices that can both turn-on and turn-off current such as GTOs at first and then IGBTs.

16.4.2 Static VAR Compensator (SVC)

The SVC is used to adjust the reactive power in the AC power transmission network. Its basic topology, illustrated in Fig. 16.11, consists of inductors and capacitors that can be connected to the network using power devices. The figure illustrates this by utilizing thyristors [24], but IGBTs are being increasingly used for this application. In addition, fixed capacitors are also used to balance the presence of inductive loads,

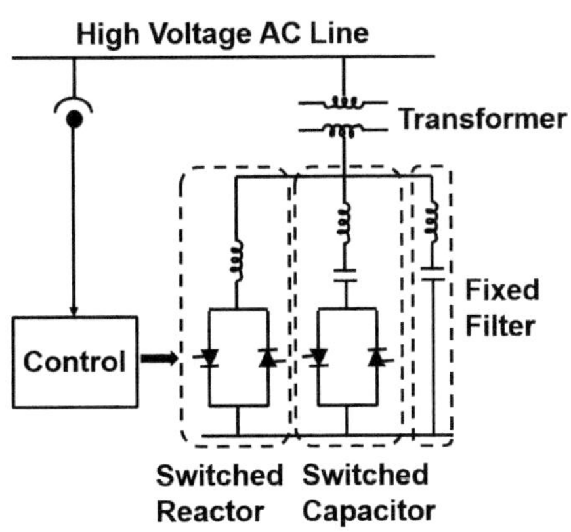

FIGURE 16.11

Basic static VAR compensator.

such as motors, on the network. The fixed capacitors are tuned using series inductors for filtering harmonics. A transformer is used to connect these reactances to the network so that the SVC components do not have to be designed for operation at very-high network voltages.

16.4.3 Static Synchronous Compensator

The STATCOM generates a current with controlled amplitude and phase angle that can be fed into the AC power transmission network. The principle of the STATCOM is illustrated in Fig. 16.12 [24]. A VSC is used to generate a sinusoidal voltage V_{VSC} at one end of the reactor connected to the AC power line. The current I_{VSC} that flows through the reactance is determined by the difference in the magnitude and phase angle of the AC line voltage V_L and the voltage created by the VSC. In practice, the current generated is either leading or lagging the line voltage by about 90 degrees [24].

In comparison with the SVC, the STATCOM can provide greater power under low line-voltage conditions. STATCOMs implemented using GTOs encountered problems because of difficulty with sharing of voltages equally among the series-connected GTOs [21]. The high switching losses in GTOs did not allow PWM control to generate good sinusoidal voltages from the STATCOM. The availability of the IGBT modules with sufficiently high voltage and current ratings enabled high switching frequencies during PWM operation with good sinusoidal waveforms [21].

Modern STATCOMs make use of VSCs with IGBTs [21]. The authors state: "*The latest compensation devices are based on VSC using first GTOs and now IGBTs.*" The IGBTs are configured, as shown in Fig. 16.5, allowing sharing of the voltage. This reduces the voltage rating of the IGBT as well as that for the capacitors. A sinusoidal waveform is achieved by switching the IGBTs at a high PWM modulation frequency in the kHz range.

An IGBT-based VSC has been analyzed for reactive power compensation [25]. A cost-effective 200 kVArms distributed STATCOM was demonstrated.

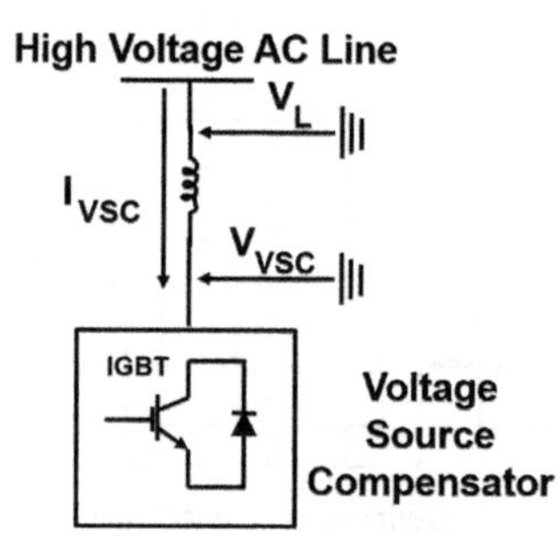

FIGURE 16.12

Static compensator principle.

16.4.4 SVC Light

The SVC light concept was developed by ABB and put into operation in 1999 [26]. Steel mills that utilize large electric arc furnaces create flickers in the power line voltage. Voltage dips or fluctuations produce disturbances in power delivered to adjacent communities. These problems can be overcome using SVCs. The SVC light concept is a VSC with IGBTs as switches. It provides a high-efficiency method to prevent line voltage fluctuations due to its high dynamic response. The SVC light technology was first installed in 1999 at Uddeholm Tooling in Sweden. The steel mill has a 37.8 MVA electric arc furnace and a 7.7 MVA ladle furnace. The VSC was operated from a 10.5 kV bus. The installation was upgraded in 2000 for connection to 20 kV power lines and since 2002 to 33 kV power lines.

The three-level VSC configuration using IGBTs is shown in Fig. 16.13. The capacitor bank acts like a DC voltage source. The converter outputs (a, b, and c) can be connected by the IGBTs to the positive pole, the neutral, or the negative pole. This allows the production of a variable AC output voltage using PWM control of the IGBTs [26]. A sinusoidal AC output waveform can be generated without harmonic filtering, leading to a compact and robust design because no bulky transformers are required. ABB continues to market this technology around the world [27]. An SVC has been installed outside the city of Oslo, Norway, to correct for voltage fluctuations on a 420 kV power transmission line [28]. It has also been employed to stabilize the power grid to allow fluctuating power delivery from wind farms in the McCamey area of Texas in the United States.

16.4.5 SVC and STATCOM in China

There has been a rapid growth in the power systems industry in China since the 1980s. It has progressed from 17 GW of installed generation capacity in 1990 to 320 GW by the end of 2000, the second highest in the world [29]. The power-generation capacity in China grew to 2200 GW by 2021. The addition of power generation in China creates problems with power quality and reliability that must be solved by using SVC and STATCOM devices. The Three Gorges Dam Hydroelectric Power Station requires a 1-GVAR (giga VAR) STATCOM and 1.2 GVAR of SVC capability.

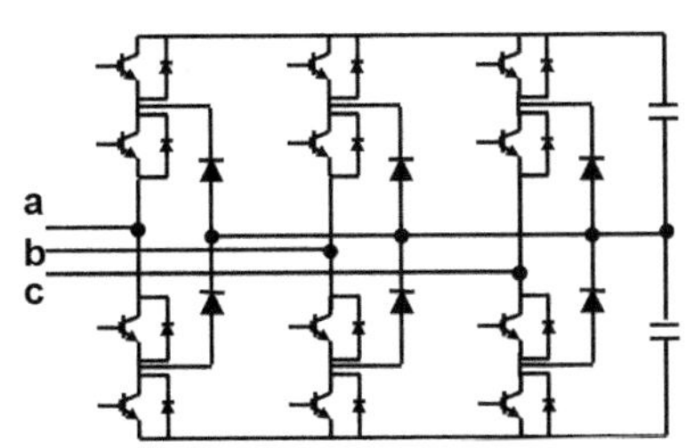

FIGURE 16.13

Three-level insulated gate bipolar transistor-based voltage source control topology.

The STATCOM approach used in China is a cascaded multilevel converter like that shown in Fig. 16.5. This approach allows generating a sinusoidal waveform without bulky transformers. The IGBT structure named the injection enhanced gate transistor (IEGT) by Toshiba with ratings of 4500 V/1500 A is the most promising technology [29].

16.4.6 Urban STATCOM Design

Sometimes, residential areas develop close to power plants, creating environmental and power quality issues. One such example is the Holly Power Plant located in Austin, Texas, in the United States [30]. Although this power plant was planned for phase out in 1995, energy shortages required its continued operation. The plant required upgrading to address the safety concerns of local residents due to prior incidents of fire. A comparison was made between the SVC and STATCOM approaches to improve power quality while keeping the footprint of the solution small to ameliorate aesthetic concerns. A STATCOM installation using VSCs with IGBTs was chosen as the optimal solution.

The basic line diagram for the STATCOM is shown in Fig. 16.14. The IGBT valve must be capable of operating at the 32 kV AC bus voltage. A series connection of 32 IGBTs was required due to their limited voltage rating. A three-level, neutral-point clamping inverter bridge topology was selected with PWM at a base frequency of 1260 Hz. This relatively high switching frequency allowed a reduction of the size of the filter components with ratings of 5–10 MVAR (mega VAR). The compact STATCOM footprint of 20 × 30 m allowed installation inside a two-story building, making it aesthetically acceptable.

An interesting aspect of the Holly STATCOM design is enhanced reliability with built-in redundant IGBTs in each valve. Because IGBTs fail in a shorted mode, the

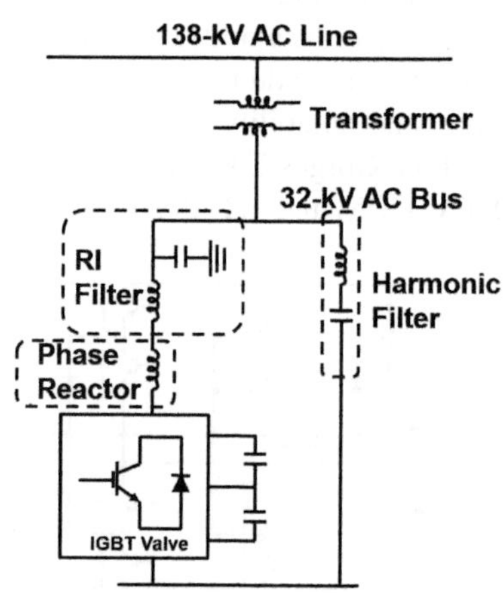

FIGURE 16.14

Austin Holly urban static compensator installation.

valve can continue to operate even if two of the three valves are no longer functional. The damaged valves can be repaired during routine scheduled maintenance without interruption of power delivery.

16.4.7 STATCOM Stability Analysis

Fluctuations in the DC transmission voltage can occur due to the connection of unbalanced loads. The loads, such as electric trains and arc furnaces, produce a ripple at about 100 Hz. In China, the current imbalance is 15% for a rural distribution system, 16% for a coal mine distribution system, 40% due to electric arc furnaces, and 55% due to electric railways. The magnitude of the ripple voltage for a STATCOM for 10–35 kV HVDC transmission lines with series-connected IGBT modules has been analyzed for the worst case situation of 55% imbalance in load current [31]. The ripple voltage could be reduced to 0.4 kV for a 10 kV DC transmission line.

16.5 HVDC Back-to-Back Converter

With recent deregulation of the electric power industry, it has become important to manage power flow between the three major power networks in the United States–namely, the Western Interconnection, the Eastern Interconnection, and the Texas Interconnection. The intertie between these networks is performed at the Tres Amigas HVDC Super Station located in Clovis, New Mexico [32]. This location is also favored due to a burgeoning wind-power generation capacity in the region.

The back-to-back HVDC converter is comprised of HVDC valves with short internal interconnection buses housed within the same building. Due to the high power capacity of 5 GW of transmission power, the installation requires six back-to-back converters. The HVDC valves are implemented using IGBTs in a multilevel converter topology. The authors state: "*Full-scale (11 MW) IGBT converter testing continues to show remarkable reliability and fault-free performance of the valve module design.*"

16.6 Off-Shore Power Transmission

Power transmission must be performed from power sources located on land to loads that are located off-shore and from power sources located off-shore to land-based power grids. For example, power needs to be delivered to off-shore oil platforms for running the oil pumps located on rigs in the sea. The oil rigs are often located at a significant distance (10–100 km) from the shore. In the case of wind farms located away from the shoreline, because of both higher wind strength and for aesthetic reasons, the generated power must be transferred to the power grid located on-shore.

16.6.1 Oil Rig Power Transmission

Oil platforms are often located off-shore for the extraction of oil from under the sea bed. The oil extraction requires electric submarine pumps (ESPs). As an example, the off-shore oil platforms in Saudi Arabia operate 1742 ESPs, which demand more than 500 MW of power [33]. This power is delivered using a submerged cable. It was found that the most efficient approach was an HVDC transmission system with a line voltage of 230 kV.

The HVDC system requires conversion of AC power to a DC transmission voltage. This is performed using VSCs with IGBTs, as shown in Fig. 16.15. PWM is employed in the converter to control the active and reactive power independently. The analysis of the Saudi Arabian off-shore power delivery system showed that HVDC transmission for lines of up to 100 km was at least three times less expensive than the HVAC alternative.

16.6.2 Wind Farm Power Transmission

In many countries, such as Germany, there are fewer and fewer places left on land for wind farms. The installation of wind farms off-shore in the North Sea has seen rapid growth as the future for this renewable energy generation resource. Due to the relatively long distance for transmission of the generated power from off-shore wind farms, the HVDC transmission approach is favored [34,35] over HVAC for distances beyond 80 km. Losses in the HVDC line have been found to be about 3% per 1000 km of line length and the losses in the converter are 1.5% of the transmitted power.

Thyristor-based HVDC transmission has a long history. However, it has the disadvantages of needing reactive power and filtering to achieve good power quality. The thyristor inverter requires reactive power to achieve commutation. This power

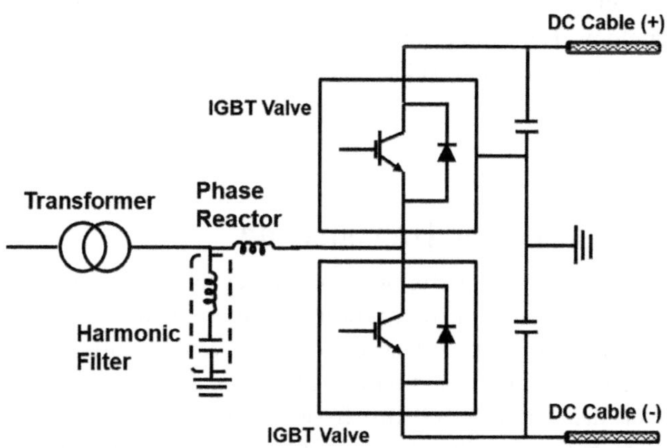

FIGURE 16.15

Off-shore power transmission with high-voltage DC.

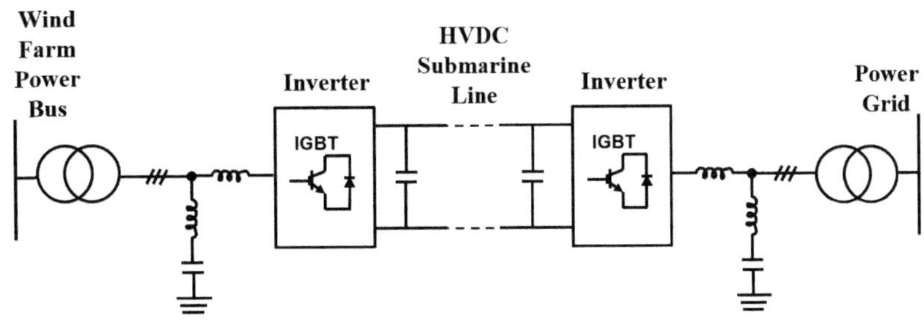

FIGURE 16.16

Wind farm power transmission using the insulated gate bipolar transistor–high-voltage DC.

must be supplied by a reactive power source such as a synchronous generator [34]. The VSC makes use of the turn-off capability of IGBTs to greatly improve the HVDC system. It is located on both the generation side and the delivery side of the HVDC transmission cable, as illustrated in Fig. 16.16. PWM of the IGBT at relatively high frequencies allows synthesis of the desired AC waveforms and reduces the filter size. The IGBT units have either the two-level configuration shown in Fig. 16.4 or the three-level configuration shown in Fig. 16.13. The IGBT-HVDC system allows control of power flow in both directions, which is beneficial during installation and lack of wind at the off-shore wind farm. It prevents the propagation of faults and improves the network stability, resulting is better power quality.

As the generated wind power output increases, it becomes necessary to use a hybrid approach with thyristor and IGBT inverters because IGBT inverters can operate at only up to the 100 MW level. In the hybrid approach shown in Fig. 16.17, the thyristor rectifier and inverter handle the high power transmission and the IGBT portion is disconnected by the circuit breaker. The IGBT inverters can still serve as an

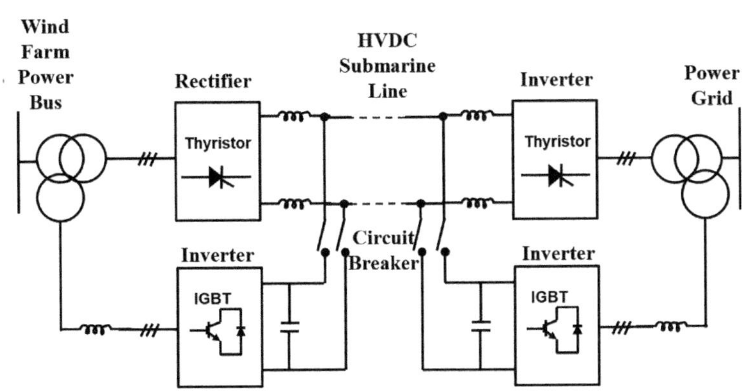

FIGURE 16.17

Wind farm power transmission using the hybrid thyristor/insulated gate bipolar transistor–high-voltage DC.

active filter supplying reactive power to the thyristor units. During start-up of the wind power installation, the IGBT-HVDC is utilized for powering auxiliary equipment such as cooling systems, oiling systems, and controls. This is also made use of during periods of no wind.

16.7 Premium Quality Power Park

Large cities make use of underground cables for their subtransmission and distribution systems [36]. The authors state: "*With the advent of IGBT VSCs, multiterminal high-voltage direct current (HVDC) has the possibility of being an attractive alternative to AC transmission in city centers where underground cable transmission is preferred for safety and environmental reasons.*" Any interruption in power delivery to banks and financial institutions is prohibitively expensive due to reliance on electronic computation and communication. The cost of downtime is several million dollars per hour for Internet-based brokerage and credit card operations. Such businesses can be preferably located at premium quality power parks that guarantee high-power quality.

Power quality issues include (1) voltage dips; (2) voltage swells; (3) voltage transients; (4) harmonic currents; (5) flickers; (6) voltage imbalance; (7) frequency variance; (8) transient power interruption; and (9) power outage. These issues can be dealt with using VSC multiterminal HVDC (VSC–M–HVDC). The VSC–M–HVDC is designed with 100% of the transmission power so that it can handle not only transients, but supply power during power outages [36]. It is possible to insulate the premium quality power park customers from disruption of power delivery from the utility supplier, equipment failure from neighboring users, and harmonics generated from office electronics, lamp ballasts, and arc welders injected by local communities.

A five-terminal VSC-HVDC system [36] is illustrated in Fig. 16.18 as an example. The VSCs are connected in parallel at their DC terminals by creating a ring of underground DC cables. Each VSC has the IGBT topology shown in Fig. 16.19 to generate the DC voltage (V_{DC}) using sinusoidal PWM control. A fault in any one of the independent AC power supplies can be isolated using the IGBTs (acting as circuit breakers) within the VSC connected in series with the source. The IGBTs can be turned on again after the fault has been cleared. The multiterminal VSC–HVDC network creates a reliable DC power grid to which multiple energy generation sources can be connected so that all the loads on the network are immune to faults in any individual source.

16.8 IGBT Designs for Power Transmission

As previously pointed out, IGBTs are the preferred semiconductor devices for VSCs used in transmission and distribution networks. Their low gate-drive power derived from the MOS gate structure is advantageous when operating multiple devices in

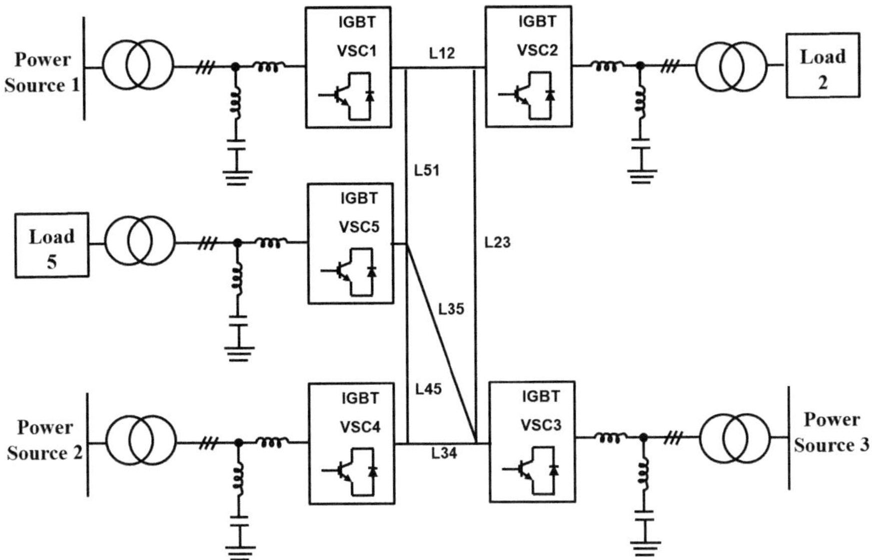

FIGURE 16.18

Multiterminal voltage source control–high-voltage DC system.

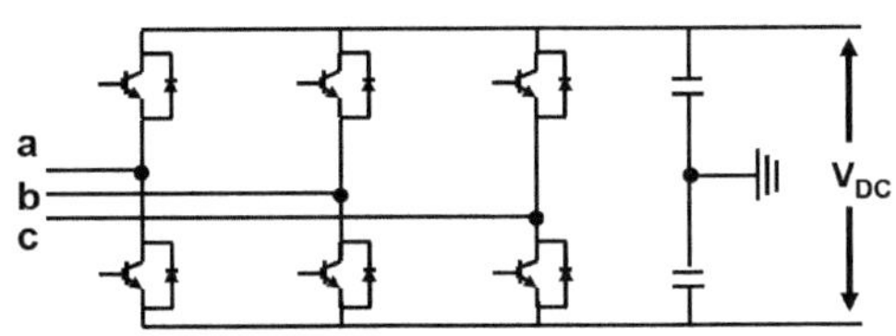

FIGURE 16.19

VSC with insulated gate bipolar transistors.

series for very high voltage (several 100 kV) lines. Their transistor characteristics provide the fast current turn-off capability not possible with thyristors. Their high switching speed enables operation at high PWM frequencies to reduce filter size to remove harmonics. The main drawback for IGBTs is the larger converter losses when compared with thyristor-based HVDC systems [5]. However, by using three-level converters to reduce switching losses, a 300-MW IGBT-based VSC-HVDC system was deployed in 2002.

Continual progress with increasing the voltage ratings of IGBTs has been made while trading off on-state losses, switching losses, and the safe operating area. In addition, the placement of hundreds of IGBTs in series required proper voltage sharing between them. This is achieved by reducing the spread in the characteristics of manufactured devices and matching them within each HVDC valve. Due to operation at high voltages over long periods of time, the IGBT design must take into consideration cosmic particle radiation. The IGBTs used in HVDC systems utilize the press-pack package design. This package is the most effective for handling the high current levels and for heat removal from the IGBT chips [5].

Mitsubishi Electric has discussed the critical design issues for IGBT modules used in VSCs for HVDC applications [37]. An IGBT module with chips capable of blocking 3.3 kV has been developed using a fine-patterned DMOS-IGBT structure (R-series) shown in Fig. 16.20. The device structure was optimized using an N-type carrier storage region located adjacent to the P-base diffusion at the top of the chip. This carrier storage region provides the additional carriers for current flow in the MOSFET portion of the IGBT structure to reduce its on-state voltage drop. It must be optimized in terms of doping and thickness to avoid reduction of the breakdown voltage and safe operating area. A thinner N-drift region with the optimized N-buffer layer is used to produce soft-punch-through of the electric field. This aids in reducing the on-state voltage drop and the stored charge that must be removed during turn-off of the IGBT. The device was found to have a reduced on-state voltage drop of 2.8 V compared with 3.6 V for the conventional IGBT structure. The on-state voltage drop has a strong positive temperature coefficient to enable paralleling of multiple chips.

The IGBT module construction is improved by increasing the number of wire bonds on the chips by 200%. This provides enhanced reliability from wire bond lift-off failures. An aluminum nitride substrate is used in the module for better expansion match to the silicon chips and superior heat removal. Eighteen IGBT chips are placed in parallel within each IGBT module to achieve a 1500 A current rating. For

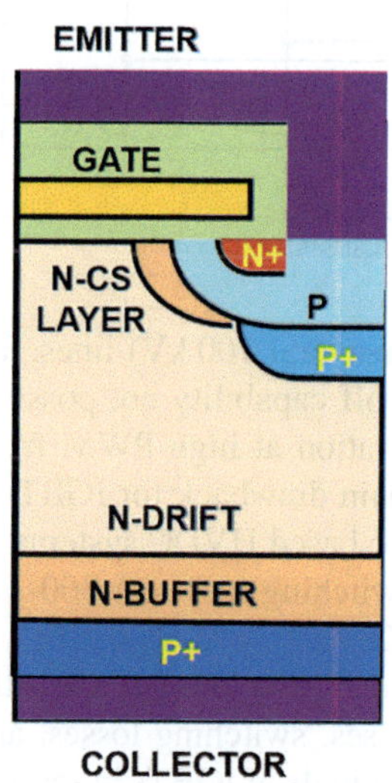

FIGURE 16.20

Insulated gate bipolar transistor design for volume source control in high-voltage DC application.

immunity against cosmic radiation-induced failure, the maximum allowed DC link voltage for the 3.3 kV IGBT module must be set at 2.5 kV. The module has an excellent reverse biased safe operating area, demonstrated by turning off 6000 A at this DC link voltage. Its short circuit safe operating area is 20 μs at a peak short circuit current of 8600 A.

16.9 Summary

The IGBT was originally developed for relatively low-power applications in the consumer and industrial sectors. During the last 30 years, its power ratings have continually grown by increasing its voltage rating to 6.5 kV. The ability to make multichip modules, which take advantage of the excellent paralleling capability of IGBTs in terms of current sharing, has enabled the current-handling capability to reach at least 1500 A. These developments are now enabling a displacement of thyristors in HVDC and HVAC systems. The IGBT-based VSC provides substantial advantages in improving the power quality, balancing reactive power, and correcting for faults and transients. The penetration of this technology is anticipated to grow and be important to the integration of large amounts of power generated using renewable energy sources [6]. The authors state: "*Power electronics have, for many decades, been a niche application in T&D systems. However, the situation is changing and power electronic systems are increasingly being seen not just as a niche application but central to the whole operating principles of the grid.*" The key enabling technology for power electronics in T&D systems has been identified as the IGBT.

References

[1] B.J. Baliga, Fundamentals of Power Semiconductor Devices, second ed., Springer-Science, New York, 2019.

[2] V.F. Lescale, Modern HVDC: state of the art and development trends, in: IEEE International Conference on Power System Technology, 1998, pp. 446–450.

[3] Konti-Skan. en.wikipedia.org/wiki/Konti-Skan.

[4] High-Voltage Direct Current. en.wikipedia.org/wiki/High_voltage_direct_current.

[5] R. Chokhawala, B. Danielsson, L. Angquist, Power semiconductors in transmission and distribution applications, in: IEEE International Symposium on Power Semiconductor Devices and ICs, 2001, pp. 3–10. Paper P1.

[6] E.I. Carroll, Power electronics for very high power applications, in: IEEE International Conference on Power Electronics and Variable Speed Drives, 1998, pp. 218–223.

[7] C.C. Davidson, G. de Preville, The future of high power electronics in transmission and distribution power systems, in: IEEE Conference on Power Electronics and Applications, 2009, pp. 1–14.

[8] Cahora-Bassa Dam. en.wikipedia.org/wiki/Cahora_Bassa_Dam.

[9] Itaipu Dam. en.wikipedia.org/wiki/Itaipu_Dam.

[10] Madeira River. en.wikipedia.org/wiki/Madeira_River.

[11] 12,000 Megawatts of Electricity over 3,000 Kilometers at 1.1 Million Volts, Setting New World Records on Voltage Level, Transmission Capacity and Distance. https://www.hitachiabb-powergrids.com/us/en/references/hvdc/changji-guquan-uhvdc-link.

[12] N.M. Kirby, HVDC system solutions, in: IEEE Transmission and Distribution Conference, 2012, pp. 1–3.

[13] S. Allebrod, R. Hamerski, R. Marquardt, New transformerless, scalable modular multilevel converters for HVDC transmission, in: IEEE Power Electronics Specialists Conference, 2008, pp. 174–179.

[14] N. Stretch, et al., A current-sourced converter-based HVDC light transmission system, in: IEEE International Symposium on Industrial Electronics, 2006, pp. 2001–2006.

[15] A. Nami, et al., Modular multilevel converters for HVDC applications: review on converter cells and functionalities, IEEE Trans. Power Electron. 30 (2015) 18–36.

[16] F. Chivite-Zabalza, et al., Resistive balance compensation of a self-powered high voltage switch using series connected IGBTs for HVDC applications, in: European Conference on Power Electronics and Applications, 2019, pp. P1–P10.

[17] P. Briff, et al., Turn-OFF delay compensation of series-connected IGBTs for HVDC applications, IEEE Trans. Power Electron. 35 (2020) 11294–11298.

[18] H. Wang, et al., The new high reliable press pack IGBT for modular multilevel converter in VSC-HVDC applications, in: IEEE International Conference on HVDC, 2020, pp. 14–19.

[19] H. Wang, K. Ma, IGBT technology for future high-power VSC-HVDC applications, in: IET International Conference on AC and DC Power Transmission, 2016, pp. P1–P6.

[20] Electric Power Transmission. en.wikipedia.org/wiki/Electric_power_transmission.

[21] H.K. Heinz, F. Schettler, Historical overview on dynamic reactive power compensation solutions from the begin of AC power transmission towards present applications, in: IEEE Power Systems Conference and Exposition, 2009, pp. 1–7.

[22] Flexible AC Transmission System. https:/en.wikipedia.org/wiki/Flexible_AC_transmission_system.

[23] P. Moore, P. Ashmole, Flexible AC transmission systems, Power Eng. J. 9 (1995) 282–286.

[24] H.K. Tyll, F. Schettler, Power system problems solved by FACTS devices, in: IEEE Power Systems Conference and Exposition, 2009, pp. 1–5.

[25] B. Panigrahy, et al., Design and modeling of a filter assisted 6-pulse VSC-IGBT based D-Statcom for reactive power compensation, in: Asian Conference on Innovation in Technology (ASIANCON), 2021, pp. 1–7.

[26] R. Grunbaum, Voltage source converters for maintaining of power quality and stability in power distribution, in: IEEE European Conference on Power Electronics and Applications, 2005, p. P10.

[27] G. Bopparaju, VSC based FACTS and HVDC: ABB experience, in: IEEE Chennai and Dr. MGR International Conference on Sustainable Energy and Intelligent Systems, 2011, pp. 1–2.

[28] R. Grunbaum, FACTS to enhance availability and stability of AC power transmission, in: IEEE Bucharest Power Tech Conference, 2009, pp. 1–8.

[29] Q. Yu, et al., Overview of STATCOM technologies, in: IEEE International Conference on Electric Utility Deregulation, Restructuring, and Power Technologies, 2004, pp. 647–652.

[30] E.M. John, A. Oskoui, A. Petersson, Using a STATCOM to retire urban generation, in: IEEE Power Systems Conference and Exposition, vol. 2, 2004, pp. 693–698.
[31] X. Zhang, et al., Analysis of DC voltage fluctuation for series-connected STATCOM with IGBT Presspack modules compensation of unbalanced loads, in: IEEE China International Youth Conference in Electrical Engineering, 2020, pp. 1–6.
[32] M.A. Reynolds, Tres amigas super station–large scale application of VSC back-to-back technology, in: IEEE Transmission and Distribution Conference and Exposition, 2012, pp. 1–5.
[33] Z.E. Al-Haiki, A.N. Shaikh-Nasser, Power transmission to distant offshore facilities, IEEE Trans. Ind. Appl. 47 (2011) 1180–1183.
[34] T. Peter, H. Raffel, B. Orlik, Parallel operation of thyristor and IGBT-based HVDC, in: IEEE European Conference on Power Electronics and Applications, 2007, pp. 1–10.
[35] T. Volker, et al., New HVDC concept for power transmission from off-shore wind farms, in: IEEE Wind Energy, 2008, pp. 1–6.
[36] W. Lu, B.-T. Ooi, Premium quality power park based on multi-terminal HVDC, IEEE Trans. Power Delivery 20 (2005) 978–983.
[37] X. Gong, A 3.3 kV IGBT module and application in modular multilevel converter for HVDC, in: IEEE International Symposium on Industrial Electronics, 2012, pp. 1944–1949.

CHAPTER

IGBT Applications: Financial 17

It is not obvious that a power semiconductor device like the insulated-gate bipolar transistor (IGBT) would have an impact on the financial sector of the economy. However, modern financial institutions such as banks and investment firms are completely reliant on high-speed transactions and large-scale computations of big datasets. Even a brief interruption of power delivery can have a major impact on this sector of the economy [1]. The estimated cost associated with power outages is shown in Table 17.1 for various industries. It ranges from $40,000 per hour to $6.5 million per hour. The cost of a break in operations is on the order of millions of dollars per hour for Internet-based credit card and brokerage institutions [2].

The cost for loss of power in data centers is so prohibitive that they are being located as close as possible to "digital quality" or "always-perfect, always-on" facilities such as premium quality power parks (PQPPs). In many cases, the utilities are asked to provide power transmission directly to the data center, which serves as a load of tens of megawatts, without going through a distribution substation [2]. As an alternative, many customers sensitive to power outages rely on local backup energy sources such as flywheel storage devices to provide power for up to 30 s.

Uninterruptible power supplies (UPS) are also used to handle short fluctuations in power line voltages. In addition to financial institutions, they can become extremely important components of an electrical power delivery system for medical facilities, such as hospitals and trauma centers, where lives can be threatened by the loss of equipment functionality. The IGBT is an essential technology for manufacturing cost-effective and efficient UPS systems.

17.1 Power Quality Equipment

Due to the high loss of revenue from power outages or even brief interruptions, the financial institutions often rely on many types of equipment to provide immunity from loss of power. The basic equipment sets are UPS, standby generators, and alternative ties, as shown in Table 17.2 [3]. Alternative ties represent power supplied to the enterprises from multiple power utilities. These methods to suppress the likelihood of loss of power are employed by not only financial institutions, data processing centers, and telecommunication centers but also by refineries and hospitals. Even paper mills and semiconductor chip fabrication faculties make use of UPS and standby generators to avoid costly interruption of production lines.

The IGBT Device. https://doi.org/10.1016/B978-0-323-99912-0.00017-9

Table 17.1 Economic Impact of Power Outages

Industry	Average Costs of Power Interruption ($/h)
Cellular communications	41,000
Telephone ticket sales	72,000
Airline reservations	90,000
Credit card operations	2,580,000
Brokerage operations	6,480,000

Table 17.2 Power Quality Equipment Types

Customer	UPS	Standby Generator	Alternative Tie
Financial institution	★	★	★
Data processing center	★	★	★
Telecommunication center	★	★	★
Refinery	★	★	★
Hospital	★	★	★
Paper mill	★	★	
Semiconductor plant	★	★	

17.2 Power Reliability and Quality

Power being delivered to the consumer by utility suppliers is judged using two criteria, namely, power delivery reliability and power delivery quality. Problems with power delivery reliability are illustrated in Fig. 17.1, which displays the voltage waveform for the utility AC power at the consumer site [3]. The first problem is a power outage, where the AC voltage completely collapses, for instance, when the utility power line is broken by a tree falling on the lines. The power outage interruption is defined to occur for a period of more than 1 min [4]. It is obvious that this would stop all the electrical equipment at the customer's site, such as computers and servers, from continuing to function. The second problem is a sag in the AC voltage. The voltage dip is defined to be in the range of 0.1–0.9 p.u. of the rms input voltage for a time period of more than half the period of the AC input voltage and less than 1 min [4]. Voltage sags are caused by the start-up of large motors connected to the line. If the available AC voltage on the power line falls below the specifications for the electrical equipment, it will trigger a shutdown of the equipment, leading to the interruption of operations at the customer's location. Similarly, an unusually high AC voltage on the line, called a swell, created by power network instabilities can trigger shutdown of equipment to protect the internal electronics. The voltage swell is defined as an increase in the rms voltage of the AC input power in the range of 1.1–1.8 p.u. for a time period of more than half the period of the AC input voltage and less than 1 min [4]. Voltage swells are produced by load switching events in the power network.

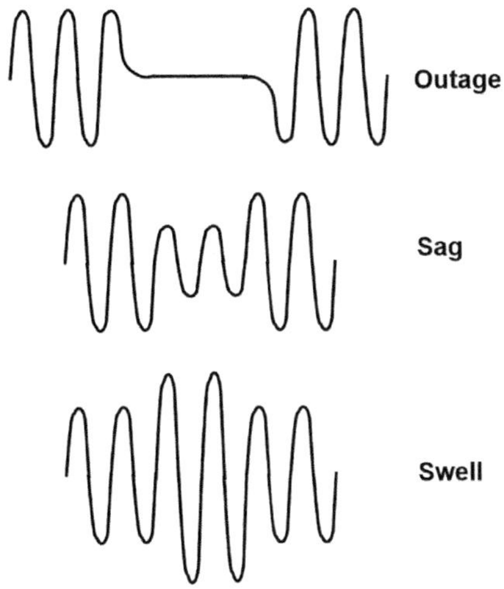

FIGURE 17.1

Power delivery reliability issues.

Power quality issues are illustrated in Fig. 17.2 [3]. Harmonics are injected into the AC power line by large electronic loads, such as inverters for motor drives and by arc furnaces and welders. They are present for many cycles of the AC voltage waveform. In addition, impulse voltages can be created by lightning strikes on the power line that propagate to the customer's site. Other power quality issues [4] are voltage imbalance between phases in a three-phase power source both in terms of deviations of the voltage and the appropriate phase angle; frequency deviation outside the nominal AC source frequency by more than 0.1%; and flicker produced by small voltage fluctuations that impact light produced by tungsten filament lamps.

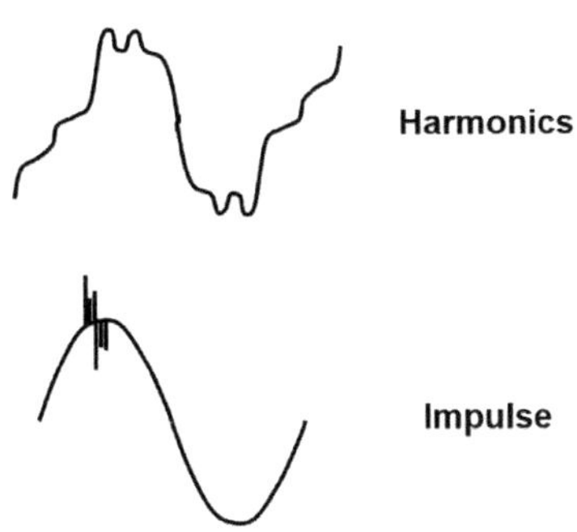

FIGURE 17.2

Power quality issues.

17.3 Dynamic Voltage Restorer

The dynamic voltage restorer (DVR) detects and compensates for sags in the voltage of the AC power source so that the loads are insulated from these power reliability issues. The operating principle for the DVR is illustrated in Fig. 17.3 [3]. The DVR consists of DC power sources, an IGBT converter, and an injection transformer that is connected in series with the power line and the sensitive load. The DC power sources that can be used are batteries, supercapacitors, superconducting magnetic storage units, and flywheels. A sag in the input AC power line voltage may propagate through the power network due to a fault in one of the distribution feeder lines. The DVR detects the sag and generates AC power from the DC power source by using the IGBT converter. The generated power is fed to the line by the transformer to correct the sag so that the sensitive load receives highly reliable AC input power.

Siemens has published several papers that describe their Siemens power conditioner (SIPCON) DVR technology. A 300kW system was described in 1999 for use with sensitive loads such as industrial plants and railway lines [5]. By creating modular DVRs, Siemens was able to scale them to applications with loads of up to 50 MVA by 2000 [6]. The 2 MVA DVR was used to provide reliable power for a large paper mill in Scotland. It provided protection against voltage sags at power levels up to 47 MW to stop disruptions in paper production. It was also used in Phoenix, Arizona, for protection of a semiconductor power plant at a power level of 17 MVA where any sag in the power line voltage stopped the exhaust fans from

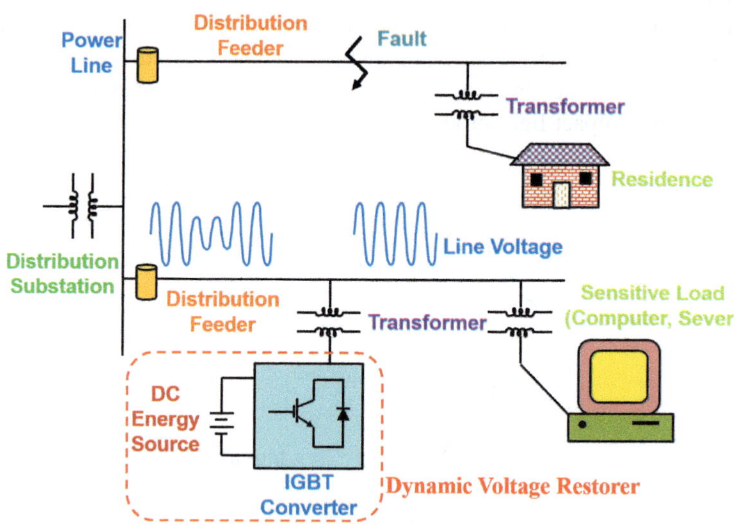

FIGURE 17.3

Dynamic voltage restorer operating principle.

functioning, requiring evacuation of the semiconductor facility with large losses in productivity. In addition to voltage sags, the SIPCON system can be utilized for harmonic filtering [7]. The authors state: "*The SIPCON system is composed of two main elements. The fundamental component of the system is a pulse width modulation (PWM) converter based on IGBT technology. No additional control hardware is necessary.*"

The DVR technologies described above can also be applied to compensating for voltage sags in power delivery systems to financial institutions, credit card services, and banks.

17.4 Uninterruptible Power Supplies

UPS has become indispensable for providing highly reliable power to sensitive loads such as data centers and financial institutions. A small UPS unit suitable for an office computer is shown in Fig. 17.4(a). Large UPS systems suitable for data centers and telecommunication centers are shown in Fig. 17.4(b). The UPS unit contains a DC power source that can provide power to the load for a brief time frame whenever there is an outage in the utility power delivery. It is capable of instantaneous protection of computers and servers to avoid large losses in revenue, as discussed in Table 17.1. The ratings of UPS equipment can range from 200 VA for a single computer to more than 10 MW for large data centers [8]. In the case of financial institutions, the online double-conversion UPS is the most suitable configuration [8]. The batteries within this UPS are connected by the IGBT-based inverter to the AC power line without any transfer switches, allowing rapid power restoration. The batteries are recharged from the utility power source once the disruption has been corrected. In a class-1 UPS system, the output voltage is maintained within ±30% under all circumstances.

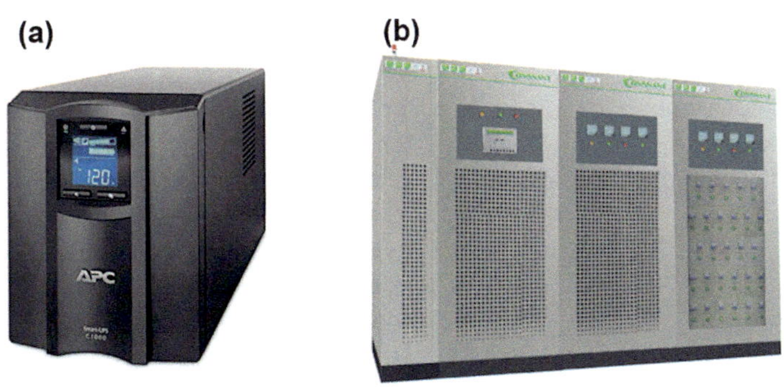

FIGURE 17.4

Uninterruptible power supplies: (a) office use; (b) data center use.

17.4.1 Fuji Electric 200 kVA UPS

In 1990, Fuji Electric reported the development of a 200kVA UPS using IGBTs for use by financial institutions [9]. The goal of this UPS is to provide power to computers that are sensitive to power source voltage fluctuations and prevent halting operations. The authors state [9]: "*As compared with the conventional-type UPS using bipolar junction transistors, this UPS utilizing IGBT provides an equivalent efficiency in spite of having an approximately 10 times higher switching frequency. As a result, the new UPS has realized a compact, lightweight, low acoustic noise and high-performance design, resulting in a reduced input harmonic current (less than 5% third harmonic distorsion (THD)), and also a reduced output voltage distortion (less than 8% THD) under nonlinear load.*" The specifications for this 200kVA UPS are provided in Table 17.3. These results demonstrate that IGBTs became the chosen power device technology for UPS systems by 1990.

The topology for the Fuji Electric 200kVA UPS is shown in Fig. 17.5. Two 100 kVA units are configured in parallel here [9]. Each unit consists of a converter with a high power factor, a PWM inverter, an inverter transformer, and AC filters. The converters make use of IGBT modules operated at 8 kHz to achieve a high input power factor to reduce harmonics introduced into the input power line. The inverters are

Table 17.3 Fuji Electric Insulated Gate Bipolar Transistor-Based 200 kVA Uninterruptible Power Supplies From 1990

Item	Specification
Input voltage	200 V ± 10%
Input frequency	50 or 60 Hz ±5%
Input phases/wires	3 phases/3 wires
Capacity	220 kVA; 160 kW
Input power factor	Over 95%
THD input current	Under 5%
Output voltage	200 V
Voltage accuracy	±1.5%
Output frequency	50 or 60 Hz
Output frequency accuracy	±0.1%
Output phases/wires	3 phases/3 wires
Load power factor	0.7 lagging to ~1.0
Transient output voltage fluctuation	(a) 100% sudden load change ±8%
	(b) 10% input voltage sudden change ±5%
	(c) Major power interruption and recovery ±5%
	(d) UPS bypass switching ±8%
Response time	100 ms
Output waveform THD	100% linear load—under 5%
	100% nonlinear load—under 8%
Voltage imbalance between phases	100% unbalanced load—±3%

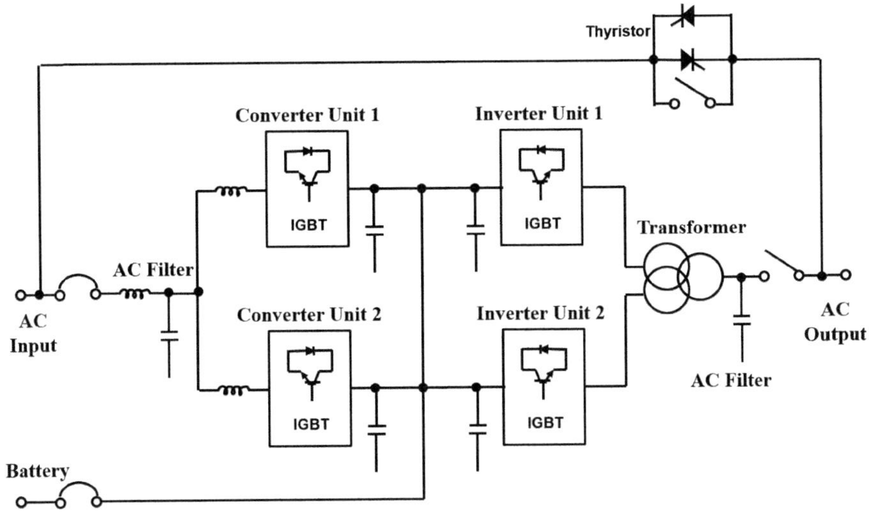

FIGURE 17.5

Configuration of the Fuji Electric uninterruptible power supplies.

also built using IGBT modules operated at a PWM carrier frequency of 8 kHz. This allows delivering AC sinusoidal output waveforms with low distortion using small AC filters. A bypass path with thyristors is included to handle the situations when the UPS may fail and need servicing.

IGBT modules with ratings of 600 V and 150 A were used for this UPS application. To reach the 200 kVA UPS capability, six IGBT modules were connected in parallel to create the IGBT stack shown in Fig. 17.6. Each stack includes the gate

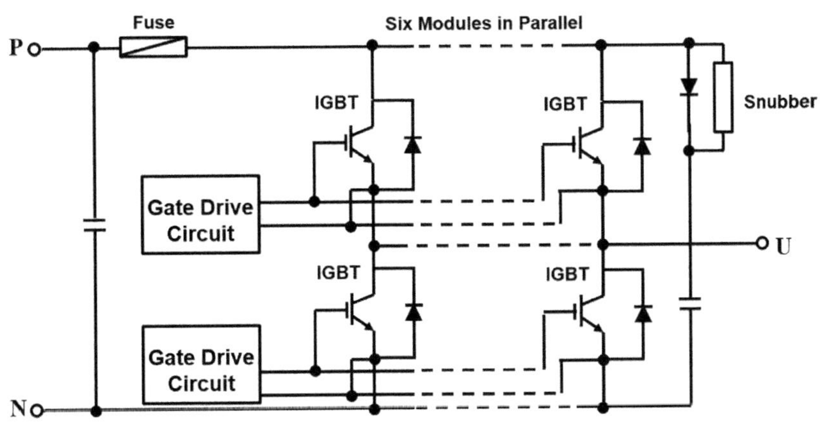

FIGURE 17.6

Insulated gate bipolar transistor stack for the Fuji Electric uninterruptible power supplies.

drive circuits, the electrolytic DC bus capacitors, and fuses. The snubber circuit was sufficiently small to fit directly on top of each IGBT module.

The UPS configuration described above with IGBTs was able to achieve an efficiency of 90% over a range of 20–100% of the total output power. It delivered a compact, lightweight UPS unit with low acoustic noise for use by customers in the financial sector.

17.4.2 Fujikura 10 kVA UPS

In 1991, Fujikura described a 10 kVA UPS that can operate from a 200 V input source and deliver either 100 V or 200 V output [10]. They implemented the UPS using a fully digital control with a digital signal processor. The power circuit makes use of 1000 V, 150 A IGBTs in the configuration shown in Fig. 17.7. As in the case of the previous UPS in Section 17.4.1, the input 200 V AC voltage is first converted into a 450 V DC link. The backup batteries are connected to this DC link. The converter operates by sensing the input AC voltage and switching the IGBTs (IGBT-1 and IGBT-2) in the rectifier stage to maintain a stable DC link voltage with sinusoidal input current drawn from the AC source. The THD of the input current was maintained at 5.3% with a power factor of 0.98.

The IGBTs in the inverter stage were switched at a PWM carrier frequency of 14.7 kHz. The output voltage was sensed and digitally controlled to achieve a THD of 3% at full load of 8 kW. The UPS is capable of riding through a loss of input power for 2.5 cycles. The authors state [10]: "*The advantage of using IGBTs, which are devices for high power and high speed switching, is low power loss at high speed switching. The overall conversion efficiency is 87% or more.*" This small, lightweight UPS was capable of operating under battery backup for more than 5 min.

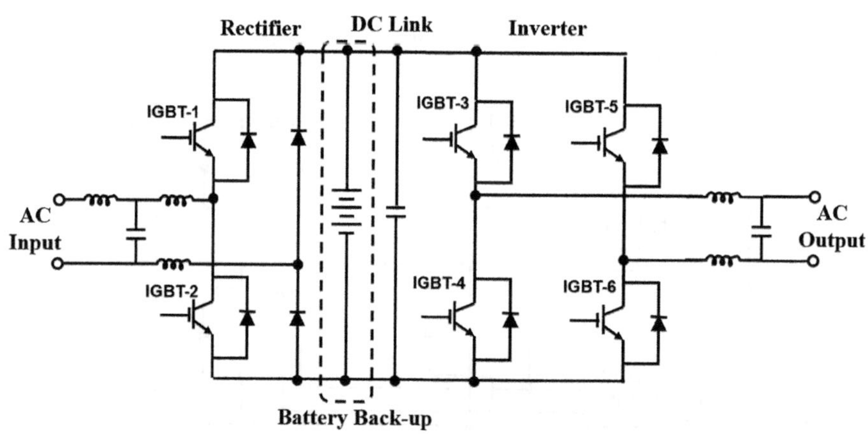

FIGURE 17.7

Insulated gate bipolar transistor power circuit for Fujikura uninterruptible power supplies.

17.4.3 Toshiba 500 kVA UPS

In 1991, Toshiba reported the development of a UPS unit using IGBTs with a capacity of 500 kVA [11] with the specifications given in Table 17.4. It was designed to operate with nonlinear loads such as the input rectifiers in the power supplies for computers and servers. The 1200 V, 300 A IGBTs were switched at 7 kHz in the inverter stage. Each inverter bridge made use of six IGBTs in parallel. The authors state: "*The IGBTs can operate at (a) much higher switching frequency than can bipolar transistors or gate turn-off thyristors (GTO). That realizes one-fourth capacity of an AC filter compared with the conventional one. Consequently, the prototype UPS is 40% smaller in volume and has 5 dB less audible noise than our conventional UPS.*" Toshiba already had dozens of these UPS units working in the field by 1991.

17.4.4 Yuasa Corporation 3 kVA UPS

In 1995, the Yuasa Corporation reported the development of a commercial UPS unit with a 3 kVA rating for application in computer and telecommunication systems [12]. The conventional UPS topology shown in Fig. 17.5 requires an isolation transformer at the output. This transformer has a large size and weight because it operates at the low (50 or 60 Hz) power line frequency. An approach to eliminate this transformer has been developed with a common bus line between the input and output, as shown in Fig. 17.8. This is a simple topology with a half-bridge switch-mode rectifier implemented with IGBTs and a half-bridge sine-wave PWM inverter, also implemented using IGBTs. Its cost is low because of using only two IGBT modules and it delivers high efficiency. However, the topology requires a DC link voltage of 340 V to enable delivering 100 V rms output AC waveforms. The large number of storage batteries that must be connected in series to achieve this voltage adds to the cost and degrades the reliability.

A UPS topology that circumvents the above issues is shown in Fig. 17.9 [12]. It includes a bidirectional two-quadrant chopper circuit implemented using two additional IGBTs (IGBT-5 and IGBT-6). When the AC input power is interrupted, the circuit acts as a step-up chopper, switching the IGBTs at 20 kHz to maintain the DC link voltage at 340 V in a stable manner. The energy for this function is derived

Table 17.4 Toshiba Insulated Gate Bipolar Transistor-Based 500 kVA Uninterruptible Power Supplies From 1991

Item	Specification
Input voltage	415 V +10%/−15%
Input frequency	50 Hz ±10%
Input phases	3 phases
Capacity	500 kVA
Output voltage	415 V
Output voltage regulation	±0.6%

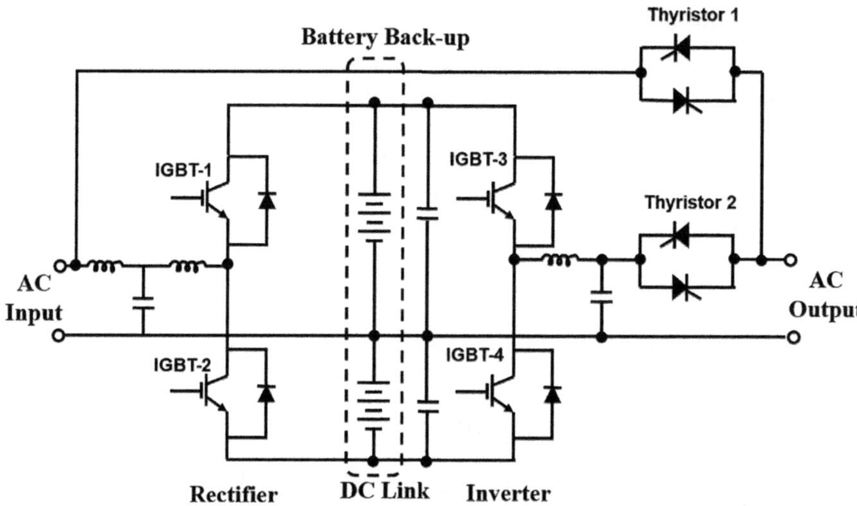

FIGURE 17.8

Yuasa Corporation insulated gate bipolar transistor power circuit for uninterruptible power supplies.

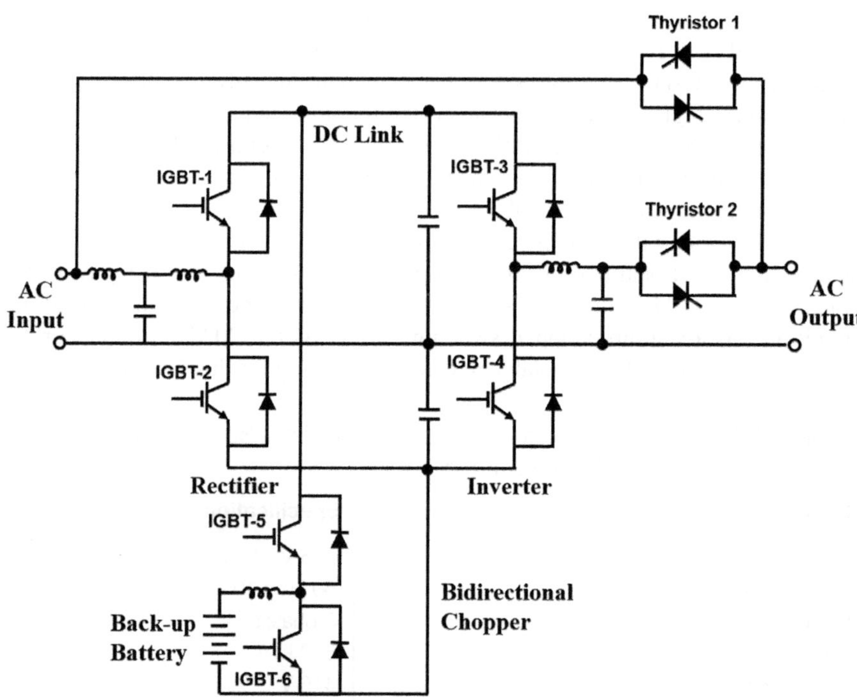

FIGURE 17.9

Two-quadrant Yuasa Corporation insulated gate bipolar transistor power circuit for uninterruptible power supplies.

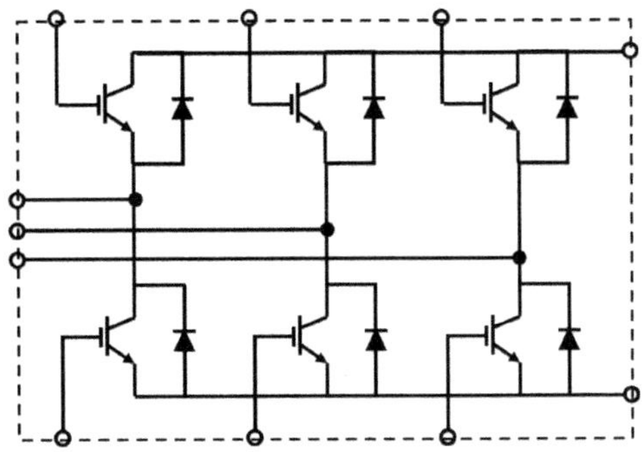

FIGURE 17.10

Six element insulated gate bipolar transistor module with pinouts for uninterruptible power supply application.

from a smaller number of batteries connected in series in the chopper leg. When the AC input power is restored, the circuit is operated as a step-down chopper, switching the IGBTs at 20 kHz to provide energy to charge the back-up batteries. The entire UPS topology described above can be implemented using a single six-cell IGBT module. The device layout and pinouts of the IGBT module are illustrated in Fig. 17.10.

The specifications for the Yuasa Corporation UPS unit are given in Table 17.5. In addition to its high-performance capability, the unit has a small size of 220 cm × 560 cm × 580 cm and a light weight of 75 kg.

Table 17.5 Yuasa Corporation Insulated Gate Bipolar Transistor-Based 3 kVA Uninterruptible Power Supplies From 1995

Item	Specification
Input voltage	100 V 50/60 Hz
Input power factor	99% at full load
Input current THD	3% at full load
Output voltage	100 V 50/60 Hz
Output voltage regulation	±2%
Output frequency regulation	±0.1%
Output voltage distortion	3.4% for nonlinear load
AC/AC efficiency	89.1% at full load
Battery type	Sealed lead acid
Backup time	10 min
Audible noise	Less than 10 phon

17.4.5 Daikin UPS

In 2006, Daikin reported a UPS developed with bidirectional IGBTs [13]. The bidirectional IGBT was created by connecting two IGBTs with flywheel diodes in anti-parallel with each other, as shown in Fig. 17.11. For each direction, the current flows through one forward biased diode and one IGBT in series. The IGBTs were operated using PWM signals at a carrier frequency of 16 kHz. The input AC chopper controls the input power factor and line harmonics. It produces a DC link that makes use of supercapacitors to reduce the size and weight of the UPS unit.

17.4.6 Single-Stage UPS Topology

The UPS units discussed in the previous sections make use of two stages—the first one to convert the input AC power to a DC bus to which the battery back-up is connected and a second stage to derive the regulated AC output. A UPS topology was reported in 2007 that performs this function with a single stage [14]. This topology makes use of nine IGBTs, as shown in Fig. 17.12. The input AC power is used to generate a DC power bus across the capacitor using six IGBTs (IGBT-1, IGBT-2, IGBT-4, IGBT-5, IGBT-7, and IGBT-8), which serve as a rectifier stage. The back-up battery source is also connected to the DC bus to keep it charged. Power is delivered to the three-phase AC load from the DC bus by generating an AC output

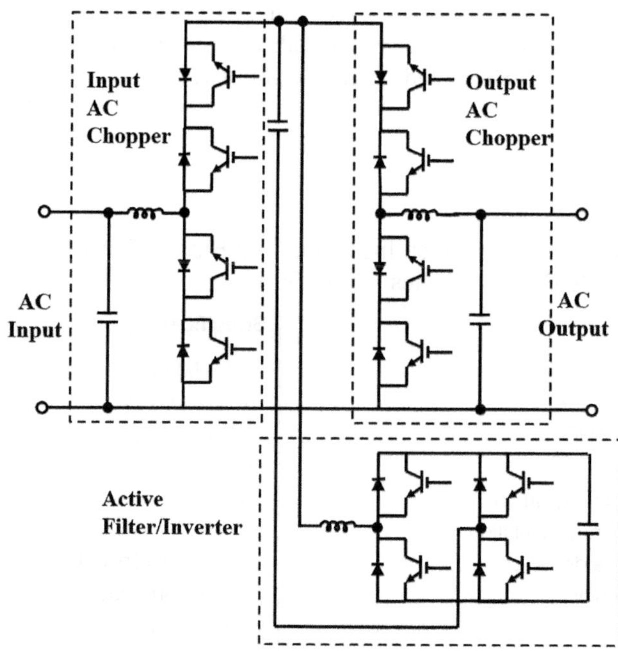

FIGURE 17.11

Single-phase uninterruptible power supplies with AC chopper and active filter.

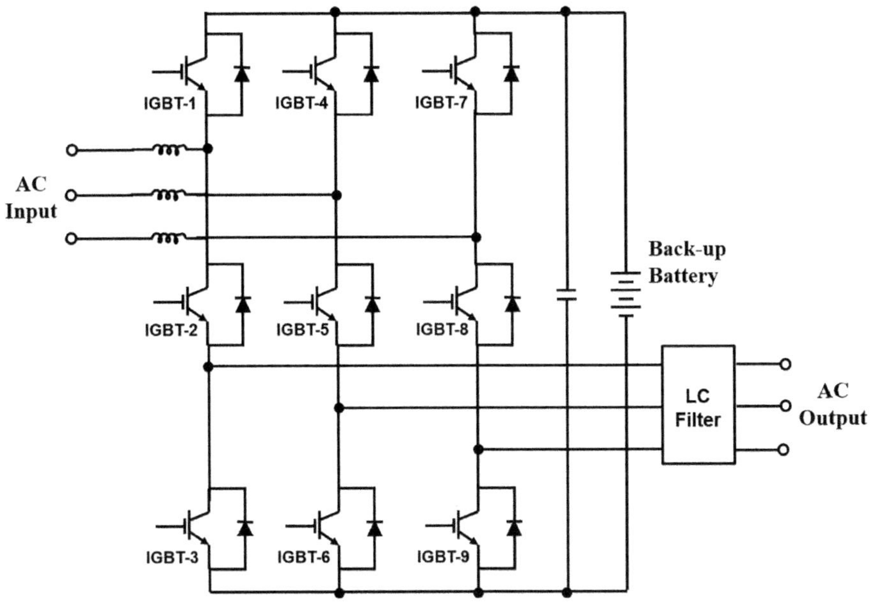

FIGURE 17.12

Single-stage uninterruptible power supplies with nine insulated gate bipolar transistors.

using six IGBTs (IGBT-2, IGBT-3, IGBT-5, IGBT-6, IGBT-8, and IGBT-9), which serve as the inverter stage. Consequently, IGBT-2, IGBT-5, and IGBT-8 are used in both the rectifier and inverter stages, which reduces the cost. A 5 kVA UPS unit has been demonstrated with the IGBTs operating at a switching frequency of 3240 Hz. Sinusoidal input currents and output voltages are obtained with a unity power factor.

17.4.7 Transformerless 300 kVA UPS

A large-capacity UPS system is required with the capability of delivering more than 300 kVA of power for large financial data centers (as well as telecommunication centers and hospitals). Two basic types of configurations have been developed for this application [15]. The offline configuration shown in Fig. 17.13(a) provides power to the computers (load) directly from the utility AC power during normal operation. A static switch is used to utilize battery back-up power when the input AC power has problems. The switching can be performed within 10 ms, allowing the uninterrupted operation of computers. A bidirectional inverter is used to charge the batteries and for delivering AC power to the load during power failure. The offline UPS configuration addresses power outages and protects against sags, dips, or surges in the input AC voltage.

The online UPS shown in Fig. 17.13(b) provides the greater electrical performance and reliability needed for large data centers. It consists of a fully rated rectifier

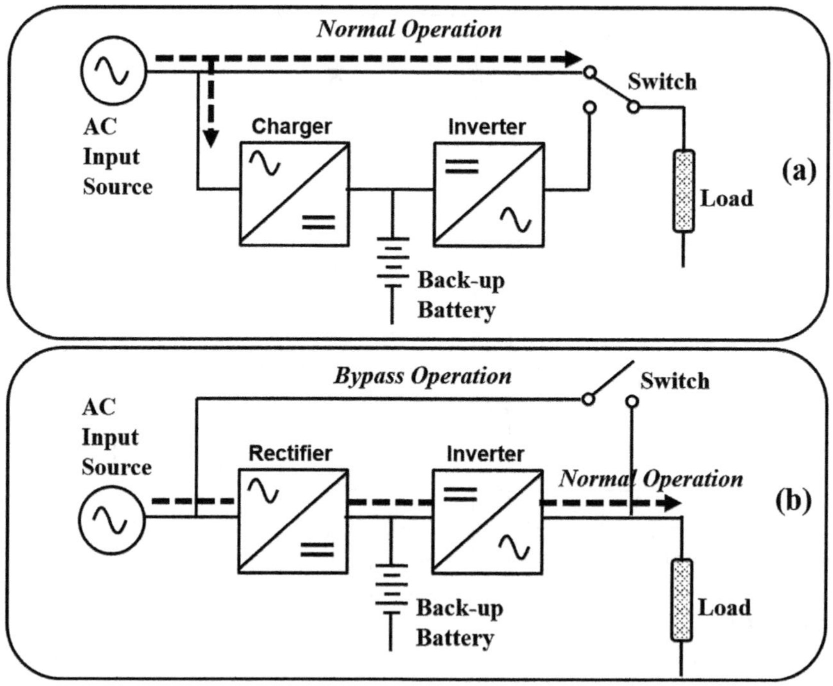

FIGURE 17.13

Large uninterruptible power supplies configurations: (a) offline; (b) online.

and inverter stage to deliver power to the computers (load) via a DC bus to which the back-up battery is connected. In addition to the functions listed above for the offline UPS, it can protect against transients such as voltage spikes, harmonic distortion, and frequency variations at the input AC source.

The online UPS is the most commonly used approach for large financial data centers, banks, and hospitals [15]. The authors state: "*With advancement of the IGBT, which uses high frequency PWM, the UPS started adapting the IGBT inverters in the early to mid-1990s.*" The thyristor-based rectifier stage was replaced with IGBT-based rectifiers because they could eliminate the input isolation transformer, reducing the size and weight of the UPS unit. The transformerless UPS is dominating the data center applications for the power range of 100–300 kVA. The authors state: "*The IGBT is the key driver for development of the transformerless UPS.*"

A transformer-less three-phase output inverter stage is shown in Fig. 17.14 [15]. It utilizes two capacitors across the DC intermediate bus to create a neutral wire. The elimination of the transformer greatly reduces the size and weight of the UPS unit. However, a larger DC bus voltage of 800 V is required to generate the desired 480 V rms AC output line-to-line voltage.

The installation cost for the transformerless UPS is 30% less than that of the previous designs with isolation transformers. A size reduction of 50–60% is also

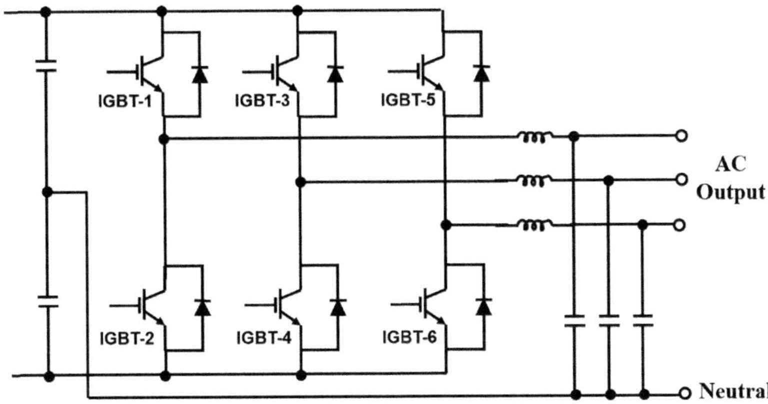

FIGURE 17.14

Transformer-less uninterruptible power supplies output stage.

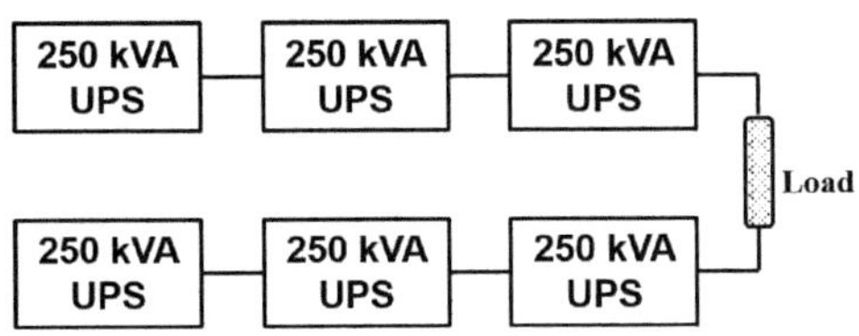

FIGURE 17.15

750 kVA uninterruptible power supplies configuration.

achieved. This is an advantage in large data centers where the cost per square foot of raised floors is $600–900 per year. The transformerless UPS units using IGBTs operated at high switching speed have an input power factor of close to unity when compared with only 0.8 for previous transformer-based UPS units. In addition, the noise from the UPS is greatly reduced and the efficiency is enhanced by about 1%. A 750kVA UPS can be derived using multiple 250kVA transformerless units, as shown in Fig. 17.15 [15]. Here, six 250kVA UPS units are used to create redundancy and achieve a more reliable configuration.

17.4.8 UPS With Static Transfer Switch

The passive standby UPS topology is shown in Fig. 17.16. It operates in two modes: the normal operation mode where the power is provided to the load directly from the AC power source and the battery mode where the power is supplied to the load from a battery back-up whenever the AC power source is compromised. The transfer of operations between these modes is accomplished by the static transfer switch. The static transfer switch must be capable of handling current flow in both directions when it is in the closed position and hold off the AC peak voltage in both directions

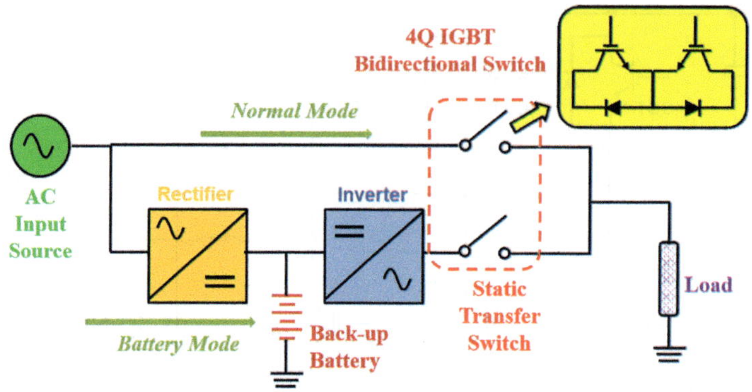

FIGURE 17.16

Static power transfer switch for a passive standby uninterruptible power supplies system.

whenever it is in the off position [24]. The poles of the switch were previously constructed using back-to-back triacs. They offer a bulky solution with a long transfer time of about 10 ms. In contract, the pole can be formed as shown in Fig. 17.16 with IGBTs and their fly-back diodes. The fly-back diodes provide the reverse blocking capability for each of the asymmetric IGBT devices. In this case, the on-state current flows via one of the IGBTs and one of the diodes in series. The IGBTs are connected with a common-emitter terminal to allow driving the devices with a single reference. The static transfer switch with IGBTs has a much faster transfer time of 0.27 ms compared to the thyristor case. One shortcoming of this design with IGBTs is the high switch on-state voltage drop of the sum of the IGBT on-state voltage drop and the diode on-state voltage drop. The current supplied to the load most of the time in the normal operating mode flows through the transfer switch. There is continuous power dissipation in the IGBT-based switch during this power flow, which is wasteful.

Recently, a monolithic bidirectional field effect transistor (BiDFET) has been invented using silicon carbide junction barrier Schottky FETs (JBSFETs) [25]. The characteristics of a 1.2 kV, 20 A BiDFET fabricated at a six-inch SiC foundry were reported with an on-state resistance of only 50 mΩ [26]. This creates a pole for the static transfer switch with an on-state voltage drop of 1 V, which is a factor of 5 smaller than the silicon IGBT approach with the series diode. The SiC JBSFETs (MOSFETs with integrated JBS fly-back diodes) can also switch much faster, reducing the transfer time for the static transfer switch.

17.4.9 Three-Phase Four-Wire Hybrid Frequency Parallel UPS

A three-phase, four-wire UPS design has been shown to provide high performance using IGBTs instead of the more expensive SiC power MOSFET option [30]. The hybrid-frequency parallel UPS (HbFPUPS) showed reduced output voltage distortion and improved dynamic response. Experimental results on a 10 kVA validated the performance.

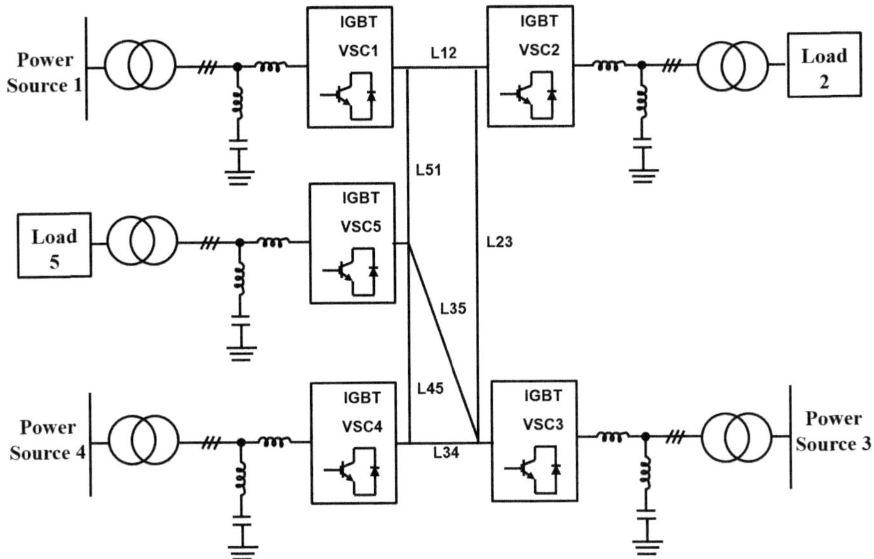

FIGURE 17.17

Power quality park multiterminal power system.

17.5 Premium Quality Power Park

As discussed above, any interruption in power delivery to banks and financial institutions is prohibitively expensive due to reliance on electronic computation and communication. One option is for such businesses to be located at PQPPs that guarantee high power quality. Power quality issues can be dealt with using a voltage source converter (VSC) multiterminal high voltage direct current (HVDC) (VSC-M-HVDC). The VSC-M-HVDC is designed with 100% of the load power so that it can handle not only transients but supply power during power outages. It is able to insulate the PPQP customers from disruption of power delivery from the utility supplier, equipment failure from neighboring users, and harmonics generated from office electronics, lamp ballasts, and arc welders injected by local communities.

As an example, a five-terminal VSC-HVDC system was considered [16], as shown in Fig. 17.17. The VSCs are connected in parallel at their DC terminals by creating a ring of underground DC cables. Each VSC has the IGBT topology shown in Fig. 17.18 to generate the DC voltage (V_{DC}) using sinusoidal PWM control. A fault in any one of the independent AC power supplies can be isolated using the IGBTs (acting as circuit breakers) within the VSC connected in series with the source. The IGBTs can be turned on again after the fault has been cleared. The multiterminal VSC-HVDC network creates a reliable DC power grid to which multiple energy-generation sources can be connected so that all the loads on the network are immune to faults in any individual source.

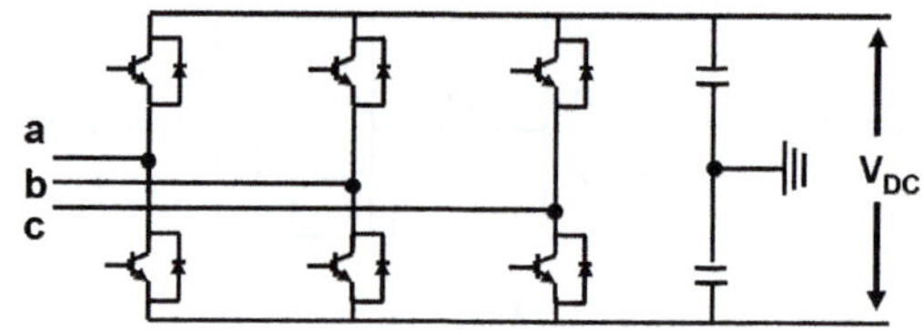

FIGURE 17.18

Voltage source converter using insulated gate bipolar transistors.

17.6 IGBT Designs for UPS

The IGBTs used in UPS units must be optimized in terms of the trade-off between on-state conduction and switching losses. In addition, they must have good ruggedness to achieve high reliability in this application. For the UPS application, the switching frequency is higher than for motor control, requiring optimization of the IGBT to reduce switching losses. Both the planar and trench gate IGBT structures are options and the drift layer can be configured with a buffer layer to form the punch-through (PT) case or without a buffer layer to create the non-punch-through (NPT) case. The on-state voltage drop can be reduced for the planar structure by decreasing the junction depth of the P-base region to reduce the JFET effect [17].

The optimization of the IGBT for various applications has been discussed by Fairchild Semiconductor [18]. A cross-section of their NPT IGBT structure is shown in Fig. 17.19. The junction depth of the P-base region was reduced by a factor of 2–3 and a self-aligned P^+ region [17] was formed to suppress latch-up of the internal

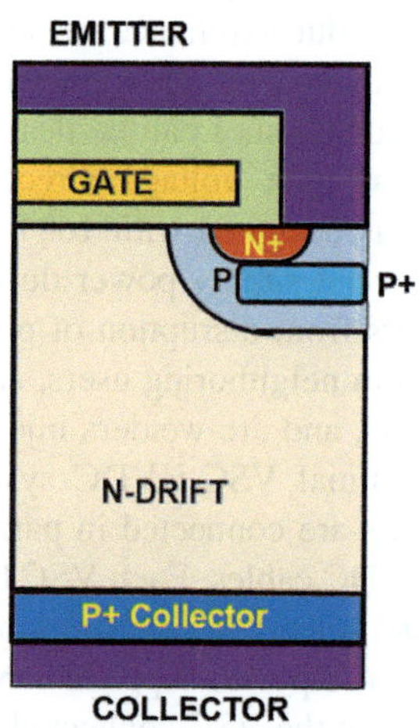

FIGURE 17.19

Non-punch-through insulated gate bipolar transistor design for uninterruptible power supplies applications.

thyristor. A linear cell topology was chosen with improved lithography design rules to shrink the cell size. The doping and thickness of the drift region was optimized to reduce on-state voltage drop and switching losses. The lifetime profile in the drift region was tailored to reduce the tail current and switching losses.

The resulting trade-off curve between on-state voltage drop and turn-off switching energy loss for the planar-gate IGBT structure is provided in Fig. 17.20. The trade-off curve for the trench gate IGBT structure with the same drift region and lifetime profile is also shown in the figure. The trench gate structure allows reducing the on-state voltage drop by about 0.4 V. However, the short circuit safe operating area (SCSOA) and switching time for the trench gate device are poor. The SCSOA for the planar-gate device is more than 10 μs while that for the trench gate device is less than 2 μs. The authors conclude that the planar gate NPT IGBT is the most suitable structure for UPS applications while the trench gate NPT IGBT structure is more suitable for resonant converters like those used for induction cooking.

However, it has been recently shown that the SCSOA of trench gate IGBTs with low on-state voltage drop can be improved beyond 10 μs by using the Baliga short circuit improvement concept (BaSIC) [27]. The BaSIC concept makes use of a low-voltage Si power MOSFET connected to the emitter terminal of the IGBT. A DC gate bias is applied to the Si power MOSFET to maintain it in the on-state during normal operation of the IGBT. The Si MOSFET is chosen to exhibit a low on-resistance in the on-state to minimize the net on-state voltage drop of the composite transistor consisting of the IGBT and Si power MOSFET. The Si power MOSFET is chosen so that

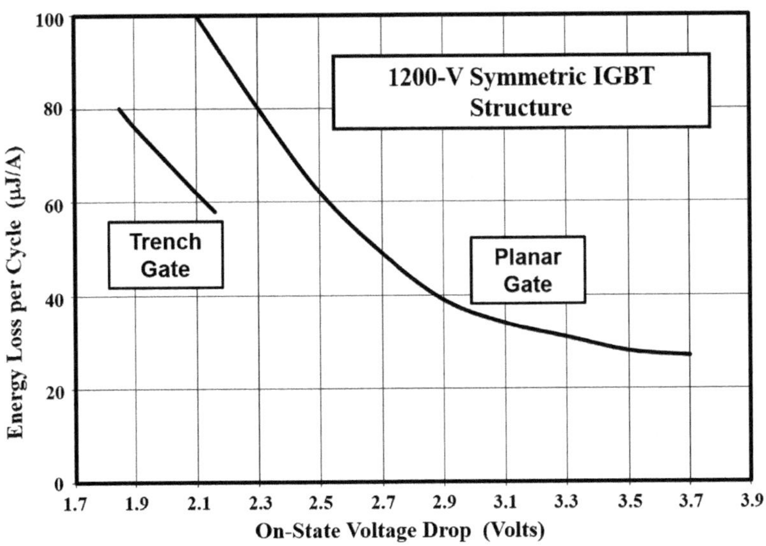

FIGURE 17.20

Trade-off curve for various insulated gate bipolar transistor chip designs for uninterruptible power supplies applications.

it limits the current flow during the short circuit event at a level much below the saturation current of the IGBT. Consequently, the IGBT current during the short circuit event is also reduced, resulting in a large short circuit withstand time. The BaSIC topology has been experimentally demonstrated for the case of 1.2 kV trench gate IGBTs with low on-state voltage drop [28].

17.7 IGBT UPS Failure Modes

A boost converter is used at the front end of UPS systems and other applications. The failure modes for IGBTs in the boost converter are: (a) IGBT short; (b) IGBT open; (c) IGBT blast; (d) high leakage current; (e) device burn-out; and (f) loss of gate control. These modes have been related to the internal failure causes [29].

17.8 Summary

The financial sector requires high-quality power delivery to protect its computers from interruptions in operation due to breaks in utility service or even dips in the voltage supplied from the mains. The very high economic losses incurred by even short interruptions of financial transactions have mandated protection of computer systems in data centers using UPS and even locating them within PQPPs.

Many companies have developed and marketed UPS units for use by the financial sector. For example, Liebert/Emerson has developed Series 7400 UPS units with ratings ranging from 10 to 60 kVA. Their brochure states [19]: "*The system's advanced true-online, double-conversion topology features a microprocessor-based controlled IGBT inverter.*" Toshiba is marketing their G8000MM and G9000 series of UPS units with "*true-online, double-conversion IGBT PWM, unity power factor output*" to data centers and financial institutions [20]. Their brochure states: "*Technologically advanced control of fast switching IGBT delivers 94% efficiency*," reducing air-conditioning costs and offering a rapid return on investment. The G9000 series is a transformerless UPS topology that is lighter and has a smaller footprint. Many Chinese [21] and Indian [22,23] companies market their UPS units highlighting IGBT-based designs.

All the modern UPS units make use of IGBTs for conversion of the input AC power to a stable DC power bus and then an IGBT-based inverter to create stable AC output power for either single- or three-phase loads. The IGBT-based units have been able to achieve an efficiency of more than 90% to minimize the internal cooling needs and their impact on the air-conditioning systems. The financial sector is therefore reliant on IGBT technology for smooth operation without costly interruptions in transactions.

References

[1] W. Lu, B.-T. Ooi, Premium quality power park based on multi-terminal HVDC, IEEE Trans. Power Deliv. 20 (2005) 978–983.

[2] S. Rahman, Power for the internet, IEEE Comput. Appl. Power 14 (4) (2001) 8–10.
[3] N.G. Hingorani, Introducing custom power, IEEE Spectr. 32 (1995) 41–48.
[4] J. Stones, A. Collinson, Power quality, Power Eng. J. 15 (2) (2001) 58–64.
[5] I. Papic, et al., 300 kW battery energy storage system using an IGBT converter, IEEE Power Eng. Soc. Summer Meet. 2 (1999) 1214–1218.
[6] N.H. Woodley, Field experience with dynamic voltage restorer (DVR-MV) systems, IEEE Power Eng. Soc. Winter Meet. 4 (2000) 2864–2871.
[7] X. Lei, D. Retzmann, M. Weinhold, Improvement of power quality with advanced power electronic equipment, in: IEEE International Conference on Electric Utility Deregulation and Restructuring, and Power Technologies, 2000, pp. 437–442.
[8] Uninterruptible power supply. en.wikipedia.org/wiki/Uninterruptible_power_supply.
[9] M. Yatsuo, et al., Three-phase 200 kVA UPS with IGBT consisting of high power factor converter and instantaneous waveform controlled by HF PWM inverter, IEEE Ind. Electron. Soc. Meet. 2 (1990) 1057–1062.
[10] I. Kubo, et al., A fully digital controlled UPS using IGBTs, IEEE Ind. Appl. Soc. Meet. 1 (1991) 1042–1046.
[11] H. Ohahima, K. Kawakami, Large capacity 3-phase UPS with IGBT PWM inverter, in: IEEE Power Electronics Specialists Conference, 1991, pp. 117–122.
[12] K. Hirachi, et al., Cost-effective practical developments of high-performance and multi-functional UPS with new system configurations and their specific control implementations, in: IEEE Power Electronics Specialists Conference, vol. 1, 1995, pp. 480–485.
[13] I. Ando, H. Haga, K. Ohishi, Development of single phase UPS having AC chopper and active filter ability, in: IEEE International Conference on Industrial Technology, 2006, pp. 1498–1503.
[14] C. Liu, et al., A novel nine-switch PWM rectifier-inverter topology for three-phase UPS applications, in: IEEE European Conference on Power Electronics and Applications, 2007, pp. 1–10.
[15] F. Al Dubaikel, Comparison between transformer-based versus transformer-less UPS systems, in: IEEE Symposium on Industrial Electronics and Applications, 2011, pp. 167–172.
[16] W. Lu, B.-T. Ooi, Premium quality power park based on multi-terminal HVDC, IEEE Trans. Power Deliv. 20 (2005) 978–983.
[17] B.J. Baliga, Fundamentals of Power Semiconductor Devices, 2nd, Springer-Science, New York, 2019 (Chapter 9).
[18] P.M. Shenoy, S. Shekhawat, B. Brockway, Application specific 1200 V planar and trench IGBTs, in: IEEE Applied Power Electronics Conference, 2006, pp. 160–164.
[19] Maximum protection for mission critical network applications. www.emersonnetworkpower.co.in.
[20] http://www.p-s-s.com/toshiba.htm.
[21] Shenzhen ZLPower Electronics, Uninterruptible Power Supply – UPS 1KVA-20KVA, with IGBT (T1k-t3/1 20KL). http://zlpower.en.made-in-china.com/product.
[22] Akhilesh Power Technologies, Smart, Mega, and Giga UPS Units with IGBT Design. http://Akhileshpower.com/Ups.htm.
[23] Best Power Equipments India, MF Series Online UPS 1KVA – 20 KVA, Technology IGBT2. www.bpee.com.
[24] M. Rodrigues, N. Silva, F. Goncalves, Static transfer switch applied to single-phase uninterruptible power supply, in: IEEE Brazilian Power Electronics Conference, 2019, pp. P1–P8.

[25] B.J. Baliga, Monolithically-Integrated AC Switch having JBSFETs therein with Commonly-Connected Drain and Cathode Electrodes, 2020. U.S. Patent # 10,804,393, Issued October 13.

[26] K. Han, et al., Monolithic 4-terminal 1.2 kV/20 A 4H-SiC Bi-Directional Field Effect Transistor (BiDFET) with integrated JBS diodes, in: IEEE Int. Symp. Power Semiconductor Devices and ICs (2020), 2020, pp. 242–245.

[27] Baliga, B.J. n.d. Patent filed and pending.

[28] A. Kanale, B.J. Baliga, User-programmable short-circuit capability enhancement for 1.2 kV Si IGBTs using a 40 V Si enhancement-mode MOSFET connected in series with the emitter, in: IEEE Applied Power Electronics Conference, 2020, pp. 680–685.

[29] R.B. Darla, A frontend PFC boost converter failure analysis for power supply and UPS applications, in: International Conference for Convergence in Technology, 2021, pp. 1–5.

[30] T.F. Wu, et al., 3f4W hybrid frequency parallel uninterruptable power supply for reducing voltage distortion and improving dynamic response, IEEE J. Emerg. Sel. Top. Power Electron. (2022) 906–918.

CHAPTER

IGBT Applications: Energy Storage 18

Large energy-storage capability has become increasingly important for many power delivery systems. One example is in the case of photovoltaic and wind energy generation power stations where the generated power fluctuates depending on the availability of sunlight and wind. Large energy-storage capability colocated with these renewable energy sources allows gathering the excess generated energy when it cannot be delivered into the grid. This energy is then available for delivery during intervals when the energy-generation capability is compromised by weather conditions. Another example is in the case of large datacenters where interruption of power can have severe economic consequences. Localized energy-storage capability produces assurance that the data center can remain operational during interruption of power delivery from the utility.

Most of the worldwide grid energy storage is performed using pumped hydroelectricity because of its large energy capacity [1]. Many pumped hydroelectricity energy-storage plants have been constructed, each with a capacity of over 1 GW. Examples are the 3 GW Bath County Pumped Storage Station in the United States, the 2.4 GW Huizhou Pumped Storage Power Station in China, the 1.9 GW Okutataragi Pumped Storage Power Station in Japan, the 1.8 GW Tumut-3 Station in Australia, and the 1.8 GW Grand'Maison Dam in France. It is estimated that 96% of the global power storage capacity is pumped hydrostations [2]. Most of these power-storage plants rely upon rivers with hydroelectric generation to deliver the energy into the power grid.

Energy storage systems other than pumped hydro have also been implemented around the world [3]. Examples are the 3.0 GW Ouarzazate Solar Power Station with thermal storage using molten salt in Morocco, the 2.9 GW McIntosh CAES Plant in the United Stated based on compressed air storage, the 1.9 GW Cerro Dominador Solar Thermal Plant in Chile with molten salt, the 1.7 GW Solana Generating Station in the United States based on molten salt, and the 1.1 GW Estrosol Solar Power Station in Spain using molten salt.

The energy-storage Asia-Pacific market is expected to grow from $258 million in 2017 to more than $500 million by 2025 [4]. The battery storage capacity in the United States is expected to grow from 1 GW in 2019 to more than 2.5 GW by 2023. The global cumulative energy storage is expected to grow from 10 GW in 2020 to more than 900 GW by 2040. Most of the energy storage is thermal energy (molten salt), followed by batteries and compressed air.

The IGBT Device. https://doi.org/10.1016/B978-0-323-99912-0.00018-0

Data centers have proliferated due to a major transition to cloud computing and information storage. Estimates of electrical energy consumed by data centers worldwide range up to 500 TWh, which must be supplied by 100 coal-fired power plants. This amounts to about 5% of the electricity produced around the world. The typical power requirement for a data center has reached 100 MW. Any interruption of power delivered by utilities to a datacenter can produce huge losses in revenue. The average cost for interruption of operations ranges up to $500,000 per hour. Consequently, it has become commonplace to deploy energy storage at data centers to supply power whenever power delivery from the utility is compromised.

In the case of data centers, there is a push to replace back-up power using diesel generation with battery storage [5]. This is driven by social pressure to reduce carbon dioxide emissions from the increasing amount of power consumed by data centers. There is a concomitant effort to migrate to renewable energy sources, particularly wind and solar power, which makes battery energy storage essential.

The implementation of all the energy-storage systems requires power converters to both deliver power from the energy storage units to the grid and power converters to feed power back to the energy storage units to replenish the energy. The IGBT is an essential component in all these converters.

18.1 Pumped Hydro Energy Storage

Pumped hydro energy storage is performed at a massive scale using water flow from rivers. It comprises 96% of the world's energy-storage capability [6]. The basic principles for pumped hydro can be understood from the illustration in Fig. 18.1 [7].

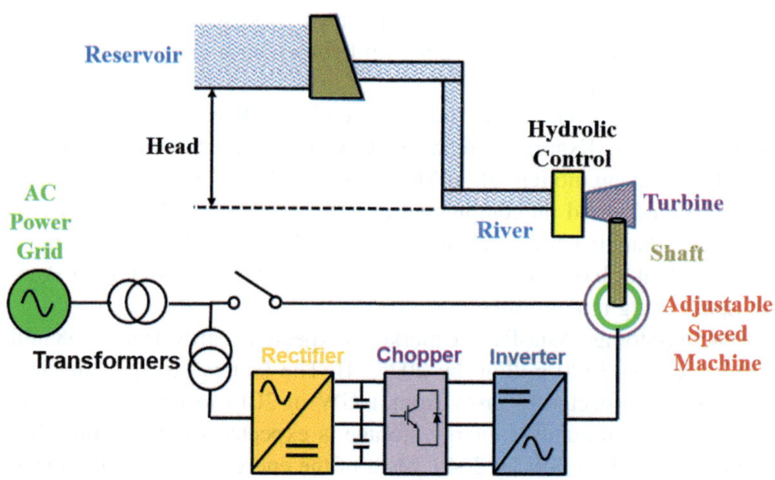

FIGURE 18.1

Pumped hydro energy storage concept.

Power is derived from the grid during periods of low electricity pricing and used by the IGBT-based rectifier, chopper, and inverter to drive an adjustable-speed machine (motor). The motor is used to drive a turbine to pump water from a river to a large reservoir located behind a dam. The reservoir and dam are built at the top of a mountain to achieve a large "head" that gives the water potential energy. Power can be derived from the water stored in the reservoir by using the hydrolic controls by running it through the turbine, similar to the process used for generating hydroelectric power. The turbine is connected to the shaft of the motor, which now behaves as a generator. The electric power generated is then fed back to the grid by closing the toggle switch. This action is performed during periods of high electricity prices. The difference in pricing for electricity as defined by market conditions forms the basis (called arbitrage) for earning income from the pumped hydro energy-storage plant. It is essential to achieve a high efficiency for pumping the water to derive profits from this venture.

An adjustable (or variable) speed machine is essential to deliver high efficiency, especially at low loads as discussed in Chapter 10. The use of IGBTs for power electronics allows obtaining efficiencies of greater than 97%. The overall efficiency for pumped hydro is in the 80% range, making this an attractive approach. Pumped hydro plants have a rapid response time (idle to full output) ranging from 20 s to a few minutes. They also have an attractive black start (ability to self-start without external electric power) capability. However, capital investment costs can be high. The capital cost for a 1 GW pumped hydro system is in the range of $1 billion for a head of 500 m. One example is the Tumut-3 station in Australia, which is shown in Fig. 18.2.

Pumped hydro energy storage can also be implemented as a closed-loop off-river system [6]. This approach makes use of two separated reservoirs (away from any river) that are located at different heights. The reservoirs can be specially constructed (referred to as "greenfield") or can make use of old mining sites, quarries, or natural reservoirs. The principles of operation in this case are the same as for the pumped hydro plant located at a river. One example is the Presenzano plant shown in

FIGURE 18.2

Tumut-3 pumped hydro energy storage in Australia.

Presenzano, Italy, Closed-Loop Pumped Hydro Energy Storage

FIGURE 18.3

Presenzano closed-loop pumped hydro energy storage in Italy.

Fig. 18.3, which is located in Italy. By using a pair of 100 Ha reservoirs with a head of 600 m, a potential energy of 24 GWh can be created after accounting for utilizing 90% of the stored water with 80% round-trip pumping efficiency. This allows delivery of 1 GW of power for a period of 24 h.

The highest per capita pumped hydro energy of 170 W is available in Japan, followed by 110 W in Taiwan, 90 W in Korea, and 55 W in the United States [6]. It is about 15 W per person worldwide.

The Australian National University has done a global search for attractive sites for pumped hydro energy-storage plants. Each site was selected based on the colocation of two closely spaced reservoirs with energy storage potential of 2–150 GWh. A total of 616,000 sites were identified around the world that are located away from major urban or protected lands. The combined energy that could become available is 23,000 TWh, which is more than sufficient to supplement 100% renewable energy generation in the future. The global population is expected to reach 10 billion by 2050. A total electricity demand of 200,000 TWh is projected if it is assumed that the entire population will demand the same level of lifestyle as the advanced economies today. The deployment of 81 TW of solar and 17 TW of wind power will be required to achieve 100% independence from fossil fuels. If energy storage for providing demand for 24 h is required, then energy storage of 500 TWh and power of 50 TW will be needed. This is well within the available energy-storage capacity of pumped hydro systems.

18.1.1 Variable Speed Pumped Storage Plant

The basic pumped hydro storage plant illustrated in Fig. 18.1 has been constructed by General Electric (GE) Renewable Energy at Linthal, Switzerland [7]. It contains four units with individual ratings of 250 MW. The storage system operates using an average head of 700 m. The adjustable speed motor operates at +/− 6% of a nominal speed of 500 rpm.

18.1.2 Voltage Sag Compensation

A pumped hydro energy-storage system can be utilized to stabilize the power fed from the grid and from wind power generation. The basic topology for this type of power delivery to critical loads, such as data centers where power interruptions are costly, is illustrated in Fig. 18.4. The pumped hydro station is based on closed-loop water circulation between a lower and an upper reservoir. The water is pumped from the lower reservoir to the upper reservoir using IGBT-based adjustable-speed motor drives. Energy from the water stored in the upper reservoir is recovered by running it through a turbine connected to a generator of electricity. The power to the critical load is then available from three resources, namely, the power grid, the wind power generation, and the pumped hydro station. The pumped hydro system is able to overcome voltage sags in the main grid or whenever the wind power output degrades due to changes in wind velocity.

The modeling of a critical load served by 9 MW of wind power supplemented by 2 MW of pumped hydro energy has been reported [8]. The pumped hydro unit supplies the reactive power needed to prevent voltage sag in the power line connected to the load.

18.1.3 Stabilizing Wind and Solar Renewable Energy Sources

A pumped hydro energy-storage system can be used to stabilize power grids that are reliant upon renewable energy sources such as wind and solar power. Both wind and solar power are prone to fluctuations in output power, depending upon weather conditions. These fluctuations can be compensated by deployment of the pumped hydro energy-storage capability [9]. The authors state that the first pumped hydro system was built in India in 1970 at Nagarjunasagar in India with a 700 MW capability.

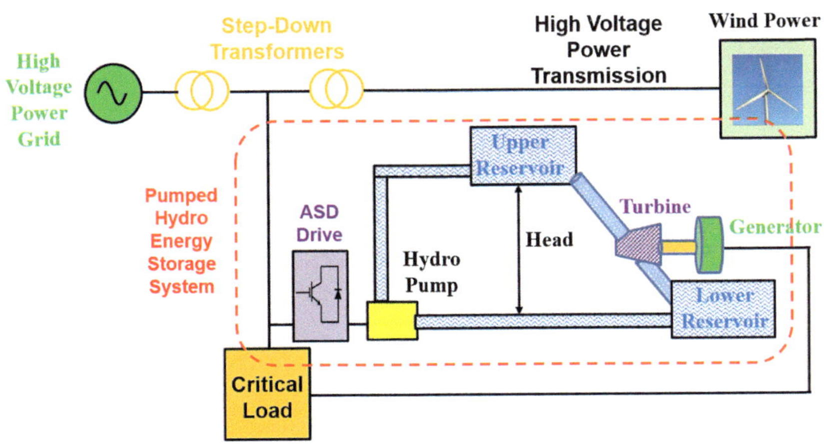

FIGURE 18.4

Closed-loop pumped hydro energy storage used for voltage sag compensation.

Eleven additional pumped hydro storage systems have been deployed since then with 3600 MW capability. Water is pumped from a lower reservoir into an upper reservoir whenever excess power is available from renewable energy sources. Power is generated from the stored energy during periods when the output of the renewable energy sources declines. Back-to-back power electronic converters are used for supplying the power. The authors state: "*The IGBT is now the main component for power electronics and also for wind turbine applications.*" The pumped hydro approach is considered important to the penetration of wind and solar power energy generation.

18.2 Other Energy-Storage Technologies

Although pumped hydro storage systems are the dominant technology, many alternate methods to store energy have become available. These approaches are compared in Fig. 18.5 on the basis of specific energy and specific power, that is, energy density and power density. Combustion engines and gas turbines offer both high energy and power density but lead to carbon emissions from the use of fossil fuels. The most commonly used, more environmentally favorable approach is the lead-acid battery-based energy-storage approach, which has a relatively high energy density but low power density. Capacitors and supercapacitors provide high power density at low energy density. Compressed air, fuel cells, and flywheels are used in niche applications.

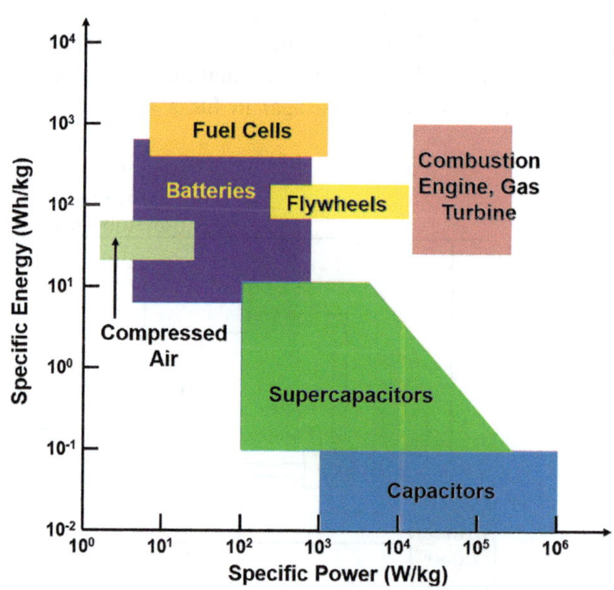

FIGURE 18.5

Comparison of nonhydro energy storage technologies.

18.2.1 Mitigating Wind Power Fault Ride Through Using Supercapacitor Storage

Wind power utilizes induction generators because they are lightweight, reliable, and low maintenance [10]. The voltage on the line delivering power from a wind farm can dip. Fault ride through (FRT) requires maintaining this dip below a certain threshold. This can be achieved by injecting reactive power into the line. The basic architecture for an FRT design with supercapacitor-based energy storage is illustrated in Fig. 18.6 [10]. This is similar to the static synchronous compensator (STATCOM) concept described in Chapter 16. The US Federal Energy Regulatory Commission (FERC) guidelines require mitigation of line voltage dips down to 15% of the nominal value. Injection of both real and reactive power generated from the energy stored in the supercapacitor using the IGBT-based DC/DC converter was found to mitigate the voltage dip, resulting in the desired FRT capability.

18.2.2 Mitigating Solar Power Voltage Fluctuations With Battery Energy Storage

The output from solar power generation can vary rapidly due to passing clouds. This has been found to produce voltage fluctuations in low-voltage (LV) feeder lines located far from the main power station. A rapid variation of more than 3% in the voltage is considered unacceptable [11]. Analysis for the case of a typical Australian LV feeder supplying 30 customers using 10 power bus lines has been performed. It serves one customer per phase per bus with each customer using a solar panel with a 4 kVA inverter. The voltage fluctuations must be kept below 7 V (3%) for the 230 V rms AC power used in Australia. The solar radiation was reduced by 400 W/m^2 in 1 s. A 12 kWh-rated lead-acid battery was used as the energy-storage element to

FIGURE 18.6

Wind power line voltage fluctuation mitigation using supercapacitor energy storage.

mitigate the voltage fluctuation. It was found that a ramp control method was able to reduce the voltage fluctuations from 3% to less than 0.5%.

18.2.3 Battery Energy Storage System

A large-scale battery energy storage system (BESS) has been successfully created using IGBTs for handling the fluctuations in output power from solar and wind generators [12]. An outdoor 1.5 MW unit has been developed using sixth-generation IGBT modules by Mitsubishi Electric. The basic topology for this large-scale BESS unit is shown in Fig. 18.7. It employs batteries operating at DC voltages ranging from 750 to 1000 V to store energy. Four 375 kW converters are used in parallel to generate the 1.5 MW of power. The outdoor unit has a size of 2.3 m height × 5.0 m width × 1.1 m depth. The converter is based on a three-level neutral-point clamped (NPC) topology to obtain a low harmonic distortion of 2.2% with a small harmonic filter. An efficiency of 98% is achieved over a broad range of output power. A second design with 2.5 MW has also been demonstrated with the same footprint by using improved seventh-generation IGBT modules.

18.2.4 Fuzzy Logic Controlled STATCOM

It is important to optimize the delivery of power from the battery energy-storage system (BESS) to mitigate fluctuations in power from a wind generation capability. A STATCOM has been analyzed with fuzzy logic control to improve the power quality in terms of the voltages on the source and load side [13]. This implementation makes use of voltage source inverters with IGBTs.

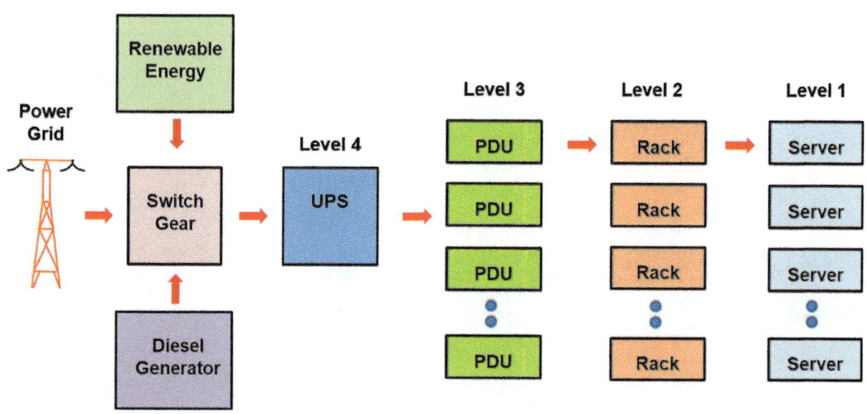

FIGURE 18.7

Large-scale battery energy storage system.

18.3 Data Center Energy Storage

Data centers are consuming increasing amounts of energy, as discussed in the introduction, due to the popularity of cloud storage and cloud computing. An example of a Google data center with energy storage and wind power generation is shown in Fig. 18.8. High-density data centers are operated by Google with server-level uninterruptible power supply (UPS) units; and by Facebook and Microsoft with rack-level UPS units. Multiple stages of AC-to-DC-to-AC power conversion within the data center have been recognized to degrade overall power delivery efficiency. Poor power delivery efficiency generates heat that must be removed with more expensive air conditioning of the data center buildings.

18.3.1 Energy-Storage Options

A comprehensive analysis of various energy-storage options has been compiled for the case of data centers and their placement at different levels of the power delivery infrastructure has been explored [14]. The typical power delivery hierarchy is illustrated in Fig. 18.9. The switch gear reduces the voltage supplied by the power grid and passes the power to a UPS unit. It has options of obtaining power from renewable sources, such as wind and solar power, and from a diesel generator when these options fail to provide enough energy. Diesel generators are being replaced with pumped hydro energy, as discussed in previous sections. The UPS units contain batteries as back-up energy storage for shorter durations of power interruption. Power distribution units deliver the input power to racks containing the servers. The servers are connected to the racks in stacks for achieving a compact footprint. Additional power conversion occurs inside each server to convert the high-voltage AC power to a low (~1 V) DC voltage required to operate the microprocessors and memory chips.

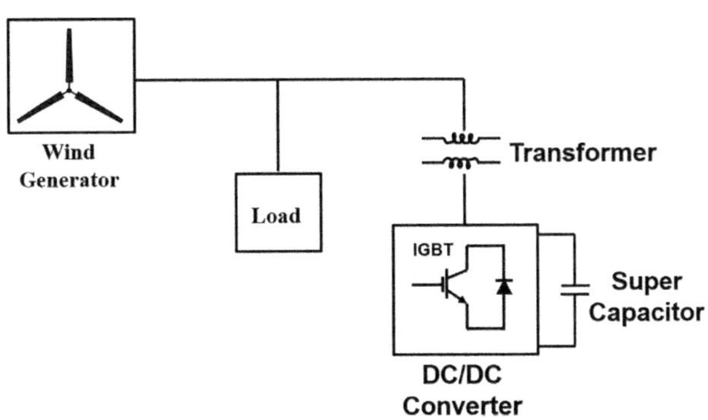

FIGURE 18.8

Google data center in the Netherlands with wind power and energy storage.

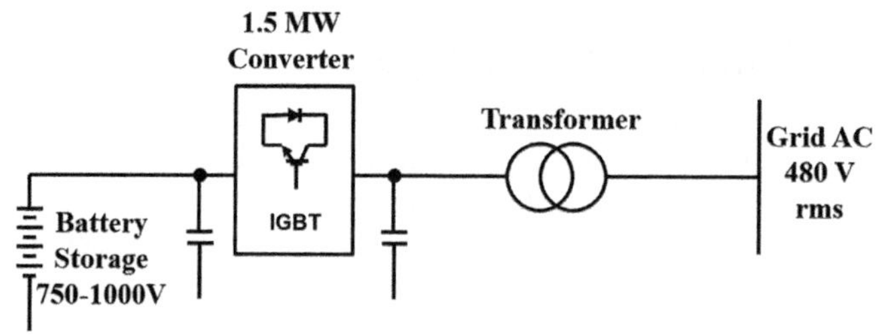

FIGURE 18.9

Data center power distribution hierarchy.

Table 18.1 Comparison of Energy Storage Options

Energy Storage Device	Lead-Acid Battery	Lithium Ion Battery	Ultra Capacitor	Flywheel	Compressed Air
Energy Cost ($/kWh)	200	525	10,000	5000	50
Power Cost ($/kW)	125	175	100	250	600
Energy Density (Wh/L)	80	150	30	80	6
Power Density (W/L)	128	450	3000	1600	0.5
Energy Efficiency (%)	75	85	95	95	68

The cost of the infrastructure for providing power at the data center is estimated to be $10–20 per watt. Reducing the wasted power is therefore economically justifiable. In addition, reducing power acquired from the grid during peak demand periods can save cost. This can be achieved by employing energy-storage capability that stores power during low demand (and cost) periods and then provides power to the data center during high cost peak demand intervals.

The relative performances of various energy storage devices (ESDs) can be compared using Table 18.1 [14]. Compressed air is the least expense technology while supercapacitors are the most expensive. However, the opposite holds true when it comes to energy and power density, which is important due to the high cost of space within a data center. In addition, compressed air storage has a notably low efficiency. Supercapacitors are useful for providing narrow bursts of power for a period of seconds and up to a minute. Compressed air can provide power for long durations at a lower power level.

18.3.2 Distributed Energy Storage

The technologies discussed above can be positioned at various levels in the data center infrastructure to optimize overall performance. Most data centers use ESDs on top of level 3 [14]. Facebook [15] and Microsoft [16] have reported ESDs placed at the rack level 2. Google [17] has reported using ESDs at the server level 1. A 4 MW data center with 8192 servers using 32 servers per rack was modeled [14]. Using power demand for a Google data center, it was determined that a single-level ESD using lead-acid batteries at the server level gave a savings of 25% compared to operation without an ESD. Using a multilevel ESD approach with a flywheel and compressed air at the data center level on top of level 4 with the lead-acid batteries at the server level gave additional savings of up to 30%.

18.3.3 DC Voltage Distribution

Power distribution in data centers has been done using AC voltage lines. However, recent advances in DC-to-DC power conversion due to the availability of high-performance IGBTs are making DC power distribution attractive [18]. A 400 V DC distribution bus is commonly used for data centers. The DC distribution approach eliminates a bulky, heavy, and expensive step-down transformer. Much smaller high-frequency transformers can be used for the DC-to-DC conversion performed at a frequency in the 100 kHz to 1 MHz range. The high operating frequency produces a faster response time for faults. The DC bus is easier to support from energy-storage devices such as ultracapacitors and batteries that operate with DC voltages.

The use of distributed UPS for a data center has also been discussed using ultracapacitors for energy storage [19]. The power in a datacenter is distributed at 400 V DC first, converted to a 48 V DC bus, and further converted to a 12 V DC bus, followed by delivery to microprocessors and memory chips at 5 V, 3.3 V, and 1 V. Google and Microsoft have used distributed UPS down to the 12 V DC bus to supply power to the microprocessor and memory chips. Facebook has implemented distributed UPS at the 48 V DC bus level.

18.4 Summary

Energy storage has become an increasingly important technology. It will play a major role in the ability to deploy renewable energy generation by solar and wind power. The output power from these sources can fluctuate intermittently due to changes in weather conditions, and solar panels stop generating power during the nighttime. The energy storage provided by pumped hydro storage can compensate for large changes in output from renewable energy sources. Short fluctuations in power can be mitigated using battery storage capability.

The proliferation of data centers is occurring due to the expansion of cloud data storage and cloud computing. The continuous operation of data centers despite interruption of power delivered by the grid is essential to avoid huge losses in revenue. Such interruptions can be avoided by using pumped hydro or battery energy storage at the data center level. In addition, energy storage at various other power delivery levels, including at the servers, can be employed for handling short dips in the power delivery to the racks.

References

[1] List of Pumped-Storage Hydroelectric Power Stations. https://en.wikipedia.org/wiki/List_of_pumped-storage_hydroelectric_power_stations.

[2] A. Blakers, et al., A review of pumped hydro energy storage, Prog. Energy 3 (2021) 1–18.

[3] List of Energy Storage Power Plants. https://en.wikipedia.org/wiki/List_of_energy_storage_power_plants.

[4] Data Center Energy Storage Market Size, Share & Trends Analysis Report By Data Center Type (Tier 1, Tier 2, Tier 3, Tier 4), By Region (North America, Europe, APAC, Latin America, MEA), And Segment Forecasts, 2020–2027. https://www.grandviewresearch.com/industry-analysis/data-center-energy-storage-market.

[5] Datacenters of the Future Will Run on Batteries. https://techwireasia.com/2021/04/the-data-centers-of-the-future-will-run-on-batteries/.

[6] A. Blakers, et al., A review of pumped hydro energy storage, Prog. Energy 3 (2021) 1–18.

[7] E. Nobile, G. Sari, A. Schwery, Variable speed hydro pumped storage as flexible enabler of intermittent renewable penetration, in: IEEE Power and Energy Society General Meeting, 2018, pp. P1–P6.

[8] N. Aghanoori, M. Masoum, Voltage Sag Compensation in Renewable Plant using HydroPump Storage, 2015, pp. 1–5.

[9] G. Salimath, N. Singh, S. Badge, Coordination and performance analysis of pumped hydro storage system integrated with solar, wind hybrid system, in: IEEE Int. Conf. on Energy, Communications, Data Analysis, and Soft Computing, 2017, pp. 160–164.

[10] M. Alam, A. Rahim, M. Abido, Supercapacitor based energy storage system for effective fault ride through of wind generation system, in: IEEE Int. Symp. Industrial Electronics, 2010, pp. 2481–2486.

[11] M. Alam, K. Muttaqi, D. Suntano, Battery energy storage to mitigate rapid voltage/power fluctuations in power grids due to fast variations of solar/wind outputs, IEEE Access 9 (2021) 12191–12202.

[12] H. Komatsu, T. Katayama, N. Kawakami, Development of large-capacity converter for battery energy storage systems, in: IEEE Int. Power Electronics Conf, 2018, pp. 1346–1350.

[13] J. Kumar, et al., Power quality improvement of a grid connected wind energy conversion system using fuzzy logic controlled STATCOM, in: International Conference on Power Electronics & IoT Applications in Renewable Energy and Its Control, 2022, pp. 1–6.

[14] D. Wang, et al., Energy Storage in Datacenters: What, Where, and How much?, Sigmetrix, 2012, pp. P1–P14. June 11–15.

[15] Opencompute, Facebook Open Compute Project, opencompute.org, 2011.

[16] Microsoft Reveals its Specialty Servers, Racks, 2011. http://www.datacenterknowledge.com/archives/.

[17] Google Server-level UPS for Improved Efficiency, 2009. http://news.cnet.com/8301-1001_3-10209580-92.html.

[18] M. Salato, U. Ghisla, Optimal power electronic architectures for DC distribution in datacenters, in: IEEE Int. Conf. on DC Microgrids, 2015, pp. 245–250.

[19] S. Zhao, et al., Lithium-ion-capacitor-based distributed UPS architecture for reactive power mitigation and phase balancing in datacenters, IEEE Trans. Power Electron. (2019) 7381–7396.

CHAPTER

IGBT Applications: Other 19

The insulated gate bipolar transistor (IGBT) has been utilized for many applications that are not covered in the previous chapters. This chapter provides a brief description of some of these applications. This includes smart homes, food processing, oil extraction, airport baggage scanners, etc. Even the power supply for the Hadron Collider that allowed discovery of the Higgs boson, a fundamental particle required for completion of the standard model, utilized IGBTs. These application examples provide further insight into the breadth of utilization of this important technology.

19.1 Smart Home

The smart home has been defined [1] as "*a dwelling incorporating a communication network that connects the key electrical appliances and services, and allows them to be remotely controlled, monitored, or accessed.*" The smart home contains an internal communications network, an intelligent control system, and home automation. Power electronics containing IGBTs are an essential element of the home automation portion of the smart home. The smart home architecture [2] is illustrated in Fig. 19.1. It contains an occupancy sensor to alert the smart home system to the presence of a person who may be the homeowner or an intruder. The smart home control system must interact with the security control modules distributed throughout the home, especially at points of egress to detect forcible entry of the premises. The infrared remote control allows owners to manipulate loads located within the home to configure them to perform appropriate functions. This includes lighting of various rooms, setting the temperature via air conditioning and heating systems, and even turning on appliances such as an oven for preheating before cooking the food. The device binding system provides the connections between various points inside the home. Performing the functions with wireless technology and secure protocols provides the desired flexibility and reliability.

19.1.1 Smart Socket and Smart Switch

The smart socket and smart switch technology [3] has been proposed for home automation systems until all the appliances for homes are configured by manufacturers with the ability to communicate with a central control unit and be remotely programmed. An example of an Efergy smart socket is shown in Fig. 19.2 with its

The IGBT Device. https://doi.org/10.1016/B978-0-323-99912-0.00025-8

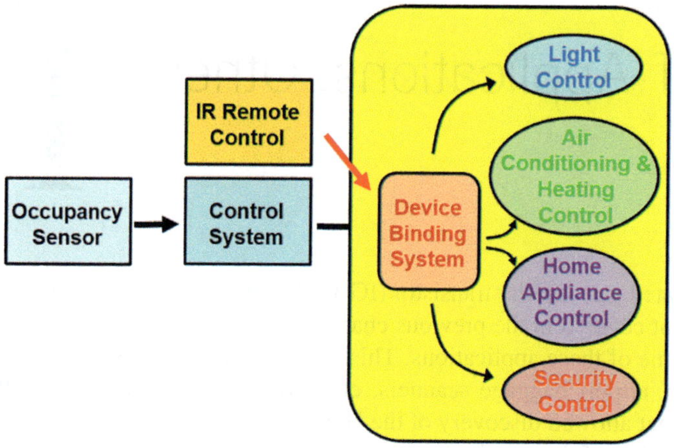

FIGURE 19.1

Smart home architecture. *IR*, infrared.

FIGURE 19.2

Efergy smart socket.

remote control and monitoring unit. The smart socket is intended for control of large power appliances such as air conditioning units, ovens, etc. The smart switch is intended for control of home lighting products. Both devices have wireless communication capability.

The block diagram for a smart socket/switch module is shown in Fig. 19.3. The smart socket/switch must be capable of receiving commands from the control module and turning on or off power being delivered to appliances (loads). It must be capable of measuring the power being consumed by the appliance and reporting this back to the control module. It should be capable of monitoring the health of the appliance and providing protection to the home power delivery lines in case of abnormal

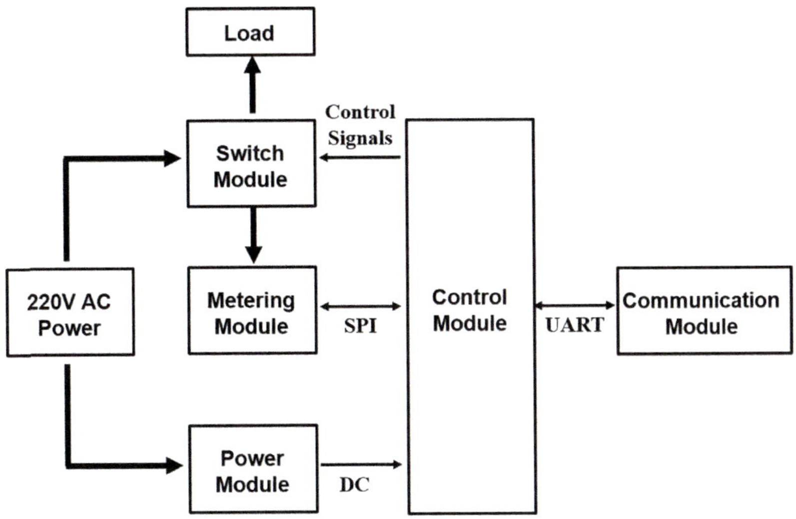

FIGURE 19.3

Smart socket/switch block diagram. *SPI*, serial peripheral interface; *UART*, universal asynchronous receiver/transmitter.

operation. The smart power module can be designed using midget relays or IGBTs as control switches.

19.1.2 Smart Power Module

Fairchild Semiconductor has reported the development of a smart home power module specifically targeted for home appliance control [4]. The targeted appliances are air conditioners, washing machines, and refrigerators. A cross-section of the μ Mini-DIP package is shown in Fig. 19.4. The IGBT and fly-back diodes are mounted on a

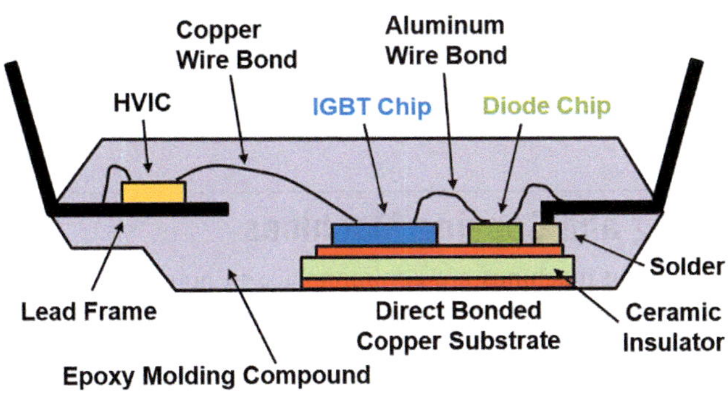

FIGURE 19.4

μMini-DIP home appliance control module. *HVIC*, high-voltage integrated circuit; *IGBT*, insulated gate bipolar transistor.

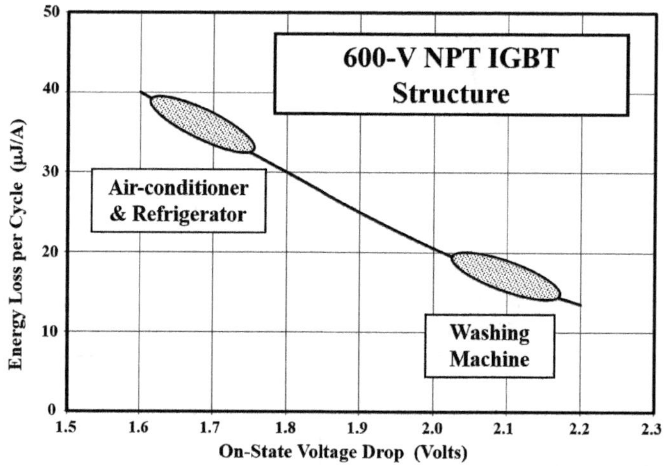

FIGURE 19.5

IGBT trade-off curve for home appliance applications. *IGBT*, insulated gate bipolar transistor.

direct bonded copper substrate to provide electrical isolation. A high-voltage integrated circuit (IC) is used for providing the gate drive with level-shifting capability. A thermistor is built into the package to allow temperature sensing and overtemperature protection. The fly-back diodes have been optimized to obtain soft reverse recovery behavior. The control circuit is designed to minimize electromagnetic interference (EMI) noise. The pin out for this package is similar to that shown in Fig. 6.12.

An optimized non-punch-through (NPT) IGBT design is used to improve the trade-off curve between the on-state voltage drop and switching losses, as shown in Fig. 19.5. The devices with a lower on-state voltage drop are suitable for air conditioning and refrigerator applications where a switching frequency of 5 kHz is used. The devices with a lower energy loss per cycle are more suitable for washing machines where a switching frequency of 15 kHz is used. The power loss in each IGBT is only 2 W when it is handling 10 A of current from a 300 V DC bus, that is, when the output load power is 3000 W [4].

19.2 Printing and Copying Machines

Printing and copying machines are commonplace in homes and offices to prepare letters, reports, and documents that are not processed in digital form. A typical printing/copying machine is shown in Fig. 19.6. The printing is performed using a process named *xerography* due to the affiliation of this technology with the Xerox company [5]. Xerox introduced the first xerographic copier in 1949 and has maintained a dominant position for this technology since that time. The advantages of this approach are

FIGURE 19.6

Printing/copying machine.

the ability to print and copy on untreated paper, the option for two-sided printing, and allowing sorting and stapling of the documents. The basic xerography process makes uses of electrostatic charge to create the image, to then use light from the image to remove charge selectively, and subsequently apply a toner to the remnant charge. The toner must finally be fixed to the paper using a thermal process.

The principles behind the xerography process are illustrated in Fig. 19.7. A corona wire is used with a high-voltage power supply to electrostatically charge the drum, as shown in Fig. 19.7(a). The photoconductor on the drum is made of semiconductor material that can respond to incident light. During exposure with the image to be printed or copied, the areas exposed to bright light become photoconductive, discharging these regions, as shown in Fig. 19.7(b). A positively charged toner dispensed on the roller adheres to the negatively charged surfaces, as shown in Fig. 19.7(c). The toner is then transferred to the paper using a larger negative charge than on the drum, as shown in Fig. 19.7(d). The toner is bonded to the paper using heat to form the final image to be printed or copied.

Fixing the toner was previously performed using halogen lamps. This process consumes 90% of the energy used in the printing/copying process [6,7]. An induction heating approach was developed to improve the energy efficiency. The inductive heating is performed using the circuit topology shown in Fig. 19.8 with two IGBTs in a push–pull configuration [6,7]. It relies on noncontact, eddy current-induced induction heating of the roller drum, which is more efficient and reliable than using halogen lamps. The induction heating process is faster and more precise, resulting in improved image quality. The circuit shown in the figure operates in a zero current switching (ZCS), pulse density modulation mode using the IGBTs and their fly-back

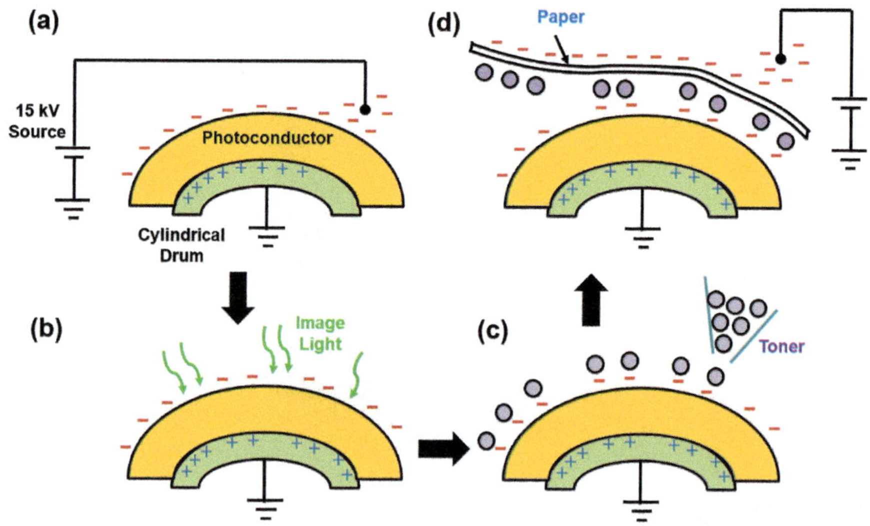

FIGURE 19.7

The xerography process. (a) electrostatic drum charging; (b) photoconductive discharging of drum; (c) toner dispensing; (d) toner transfer to paper.

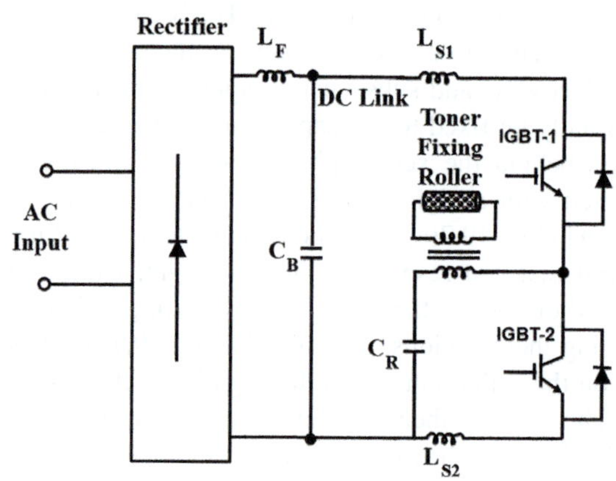

FIGURE 19.8

Inductive heating circuit for fixing roller.

diodes. The resonant frequency of the circuit is determined by capacitor C_R and the inductance of the transformer-coupled drum roller. Inductances L_{S1} and L_{S2} serve as auxiliary ZCS-assisted losses snubbers. The IGBT is operated at a frequency below the resonant frequency. The IGBTs have a rating of 600 V and 75 A. The peak switching voltage for the IGBTs is 350 V with a maximum dI/dt of 12.5 A/μs. A high efficiency of greater than 94% has been achieved with this inductive heating method.

19.3 Inductive Power Transfer

Inductive power transfer (IPT) allows power transfer over an air gap without physical contact between the primary and secondary sides. This is attractive for applications such as ultraclean environments and biomedical implants. It is also an approach for charging electric vehicles. IPT requires generating high frequency (10–100 kHz) AC power on the input side that is magnetically coupled to the load on the secondary side.

19.3.1 Stage Lighting

One interesting application for IPT is in stage lighting [8]. The traditional method for stage lighting makes use of fixed direct wiring between the power source and each of the lights. The lighting technician must set up hundreds of meters of power cables to control the intensity, direction, color, and ambiance created by the lights. These arrangements consisting of a large number of wires are not easy to reconfigure when changing the sets used for a show. The IPT configuration shown in Fig. 19.9 provides much greater flexibility. The primary high-frequency current power track needs to be set up only once. The lights can be positioned anywhere on the track rather than at previously decided plug locations. By operating the power track at a high frequency (20 kHz), the audible hum is eliminated.

The IPT pickup can be configured as a series compensation or parallel compensation circuit, as shown in Fig. 19.10. The pairs of IGBTs (IGBT-1, IGBT-2 or IGBT-3, IGBT-4) are configured as bidirectional switches with a common ground. They are rated with a blocking voltage of 1200 V and a current rating of 60 A. The IGBTs are switched at a frequency of 20 kHz to reduce audible noise. The IPT method has been found to provide a total harmonic distortion of 20–40 dB less than that observed with

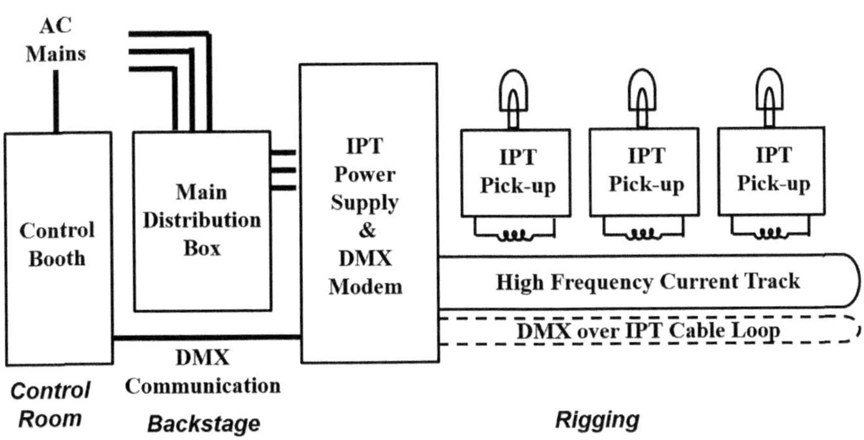

FIGURE 19.9

Inductive pick-up stage lighting system. *IPT*, inductive power transfer; *DMX*, digital multiplex.

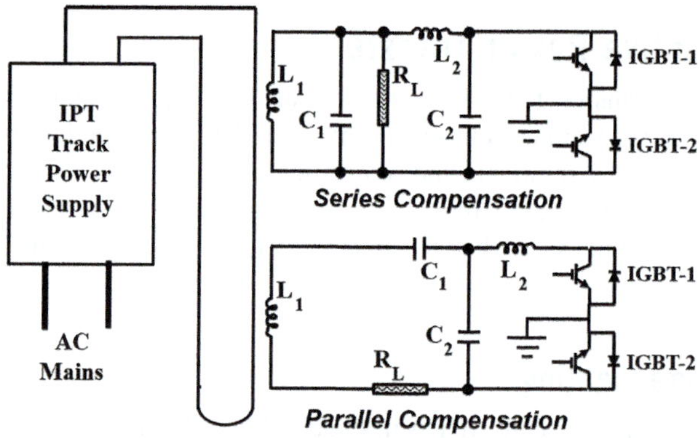

FIGURE 19.10

Series and parallel compensation circuits for IPT stage lighting.

triac-controlled lights. In addition, the efficiency for the IPT system is 30% better at light load and 10% better at full load when compared with the triac control system. The authors note that the parallel compensation circuit is easier to set up and performs better than the series compensation circuit.

19.3.2 Embedded Electric Vehicle Chargers

An effective method to charge the battery in electric vehicles is essential for the deployment of large numbers of vehicles on the road. Wireless inductively coupled chargers are considered an attractive option. A typical arrangement [9] for an inductively coupled charger is illustrated in Fig. 19.11. It consists of a high-frequency power source embedded on the road. The pickup coil located on the vehicle is positioned above the coil in the source to achieve inductive coupling. A good lateral tolerance between the coils is required to assure effective power transfer. The

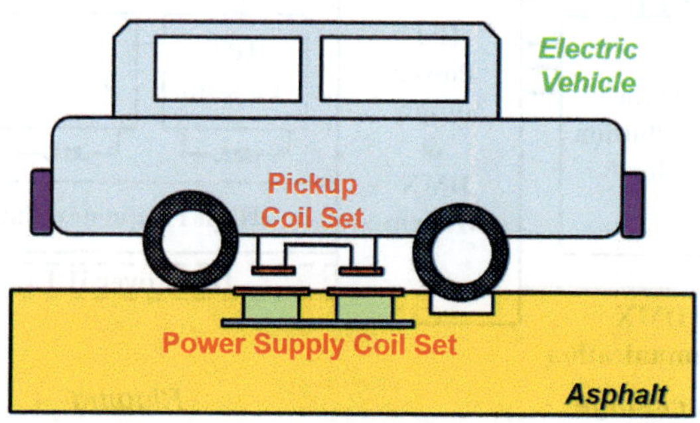

FIGURE 19.11

Electric vehicle charger using inductive pickup system.

Table 19.1 Electric Vehicle Charger Specifications

Requirement	Specification
Air-gap (Z-axis)	15 cm
Longitudinal tolerance (X-axis)	±20 cm
Lateral tolerance (Y-axis)	±40 cm
Height tolerance (Z-axis)	±5 cm

specifications for the IPT system are given in Table 19.1. This would allow creating roadway-powered electric vehicles.

The circuit topology for a roadway-powered battery charger using inductive coupling [10] is shown in Fig. 19.12. The AC input power is rectified to create a DC bus voltage of 480 V. Four IGBTs are used in an H-bridge configuration to generate the high-frequency waveform supplied to the primary side of the transformer embedded in the asphalt. The power is transferred to the pickup coil on the vehicle, which acts as the secondary winding of the transformer.

Wireless charging of electric vehicles can be accomplished with the vehicle stationary at a charging station. Alternately, the wireless power transfer could be performed with a moving vehicle with the power coils embedded in the roadway [11]. This dynamic wireless power transfer option has been investigated using a hybrid power switch comprising an Si IGBT in parallel with an SiC power MOSFET. The Si IGBT provides the low on-state voltage drop during high current operation while the SiC MOSFET provides fast turn-off to reduce switching losses. Frequent charging with this approach can allow making the batteries smaller in electric vehicles, reducing their cost and weight. A demonstration of this concept was performed using 50 coils embedded on a road, each coil driven with a dedicated 22 kW DC-to-AC converter with an efficiency of 98%.

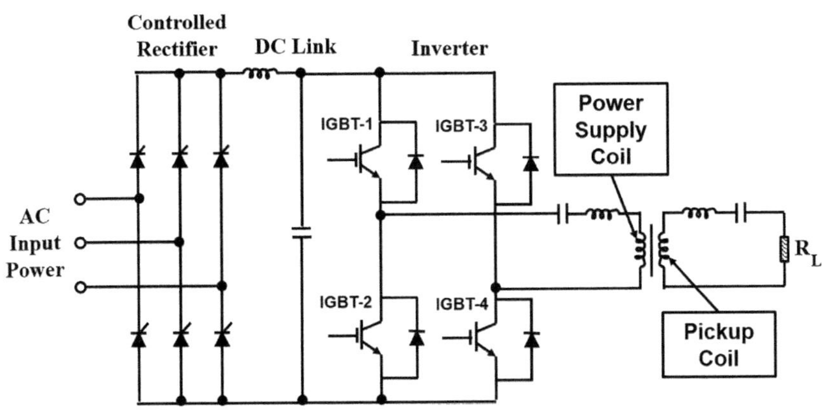

FIGURE 19.12

Inductive power transfer system using IGBTs. *IGBT*, insulated gate bipolar transistor.

19.4 Airport Security X-Ray Scanners

Due to recent acts of terrorism, travelers around the world are accustomed to going through a security check before boarding commercial aircraft. The security check includes putting carry-on baggage through an x-ray scanner like the one shown in Fig. 19.13. The airport scanner consists of an x-ray source generator, a conveyer for moving the luggage through the scanner, and an x-ray imager to view the contents of the bags [12].

The circuit diagram for the power supply for the x-ray generator in a security scanner [13] is shown in Fig. 19.14. Because a stable voltage is important for the x-ray tube in the scanner and the input AC power can vary in voltage from country to country and between different areas within a country, a rectifier stage is used followed by a DC–DC converter based on IGBTs. The DC–DC converter regulates the DC-link voltage to a stable value of 560 V, even when the output of the rectifier stage ranges from 450 to 750 V. The IGBT-based inverter stage works with a resonant circuit consisting of capacitor C_R (~1 μF) and inductor L_R (~5.2 μH) whose resonant frequency is 70 kHz. The maximum output power of the high-frequency power supply is 120 kW.

The x-ray tube operates at a voltage of 150 kV and requires up to 1 A of current. The high voltage is generated on the secondary side of the high-frequency transformer using a network of high-voltage rectifiers/capacitors, as shown in the figure. The power supply is able to maintain the x-ray anode voltage at 150 kV within 7%. Because the x-ray source is repeatedly turned on and off as it scans the bags at airports, the response time has been tested. The time to start x-ray exposure has been found to be 3 ms and the time to stop exposure is 2.4 ms.

FIGURE 19.13

Airport security x-ray scanner.

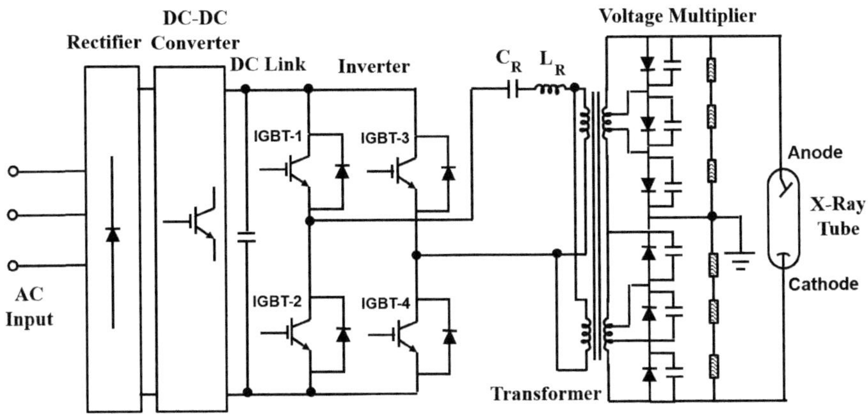

FIGURE 19.14

High-voltage IGBT-based power supply for x-ray airport scanner. *IGBT*, insulated gate bipolar transistor.

Dual-energy x-ray scanning has been introduced to discriminate between various materials inside bags and packages during security screening [14]. This approach makes use of the integrated effect of Compton scattering and photoelectric absorption of signals measured at a high x-ray energy of 140 keV and at a low x-ray energy of 80 keV. It enables discrimination between objects of different effective atomic numbers: organic material with numbers below 10, inorganic material with numbers between 10 and 20, and metallic material with numbers above 20. These materials are then displayed on the x-ray image with orange, green, and blue colors for the three ranges of atomic numbers. Contraband materials such as cocaine and heroin have an effective atomic number of between 8 and 9. Explosive materials, such as SEMTEX, C4, and TNT, have atomic numbers between 7 and 8. Contraband and dangerous materials can therefore be identified in the scanner images.

19.5 Pulse Power

Pulse power refers to the delivery of a very high power pulse for a short duration using energy stored in capacitors or inductors [15]. As an example, 1 J of stored energy can create a 1 MW power pulse with a duration of 1 μs. The stored energy can be produced by charging the capacitor or inductor over a long time duration. Pulse power has been applied to radars, particle accelerators, lasers, ion implantation equipment, and cancer treatment.

19.5.1 Marx High-Voltage Pulse Generator

The power supplies for pulse power generators often make use of the Marx circuit [16], named after Erwin Otto Marx who created it in 1924. The basic Marx circuit, shown in Fig. 19.15, contains numerous capacitors (C) that can be charged in parallel

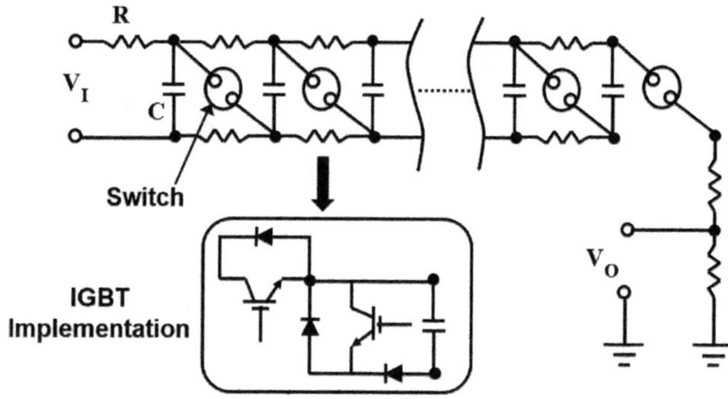

FIGURE 19.15

Marx pulse power circuit.

through resistors (R) and then suddenly connected in series by using the switches. In the circuit shown in the figure, each capacitor will eventually charge up to the input supply voltage (V_I) via the resistors. The switches are all open during this phase with the last switch isolating the network from the load. When the switches are closed, the voltage across the capacitors appears in series across the resistors R_1 plus R_2. This voltage is equal to ($n * V_I$) where n is the number of capacitors in the generator. Extremely high voltages (e.g., 600 kV) can be produced with this concept. Spark gaps were originally employed as switches in the Marx generators but have now been replaced with IGBTs.

A review of the solid-state Marx generator (MG) based on using IGBTs has been recently published [11]. Solid-state device-based Marx generators with pulse rates of 10–100,000 Hz have been enabled by IGBTs. The implementation with IGBTs is illustrated in Fig. 10.15 as an inset. The two IGBTs are configured in a similar manner to devices in the commonly used half-bridges for PWM applications allowing the use of these commercialized modules. The authors state: "*IGBTs have become the most popular device in the development of SSMGs due to their good controllability and high-power capacity.*"

19.5.2 Ion Implantation

Planar P–N junctions are the cornerstone of semiconductor technology. During the early phase of semiconductor device development, junctions were formed in silicon wafers by the deposition of oxides containing the dopant in windows cut out of a silicon dioxide layer. The dopant was then driven into the silicon by a high-temperature thermal process [17]. This process was supplanted by ion implantation technology because of its greater precision and control over the doping concentration, the doping profile, and the depth of the junction [17]. The ion implantation process consists of accelerating ions of the semiconductor dopants to an energy in the range of 50–300 keV, and bombarding the semiconductor using

FIGURE 19.16

Ion implanter.

these ions. The typical dose of the ion implant is between 10^{13} and $10^{15}/cm^2$. The semiconductor wafer surface is coated with a photoresist layer with windows created in it where the dopants are to be introduced by using a photolithographic process.

A typical ion implanter is shown in Fig. 19.16. They are classified as medium-current (10 μA–2 mA) and high-current (up to 30 mA) machines that can be used for low and high dosages of doping, respectively [18]. In addition, high-energy ion implanters are available with acceleration voltages up to 10 MeV, and high dose ion implanters are available for producing dosages of 10^{16} /cm^2.

The power supply for plasma ion implantation requires meeting the following specifications: (1) rise and fall times of less than 1 μs, (2) pulse widths of 1–100 μs, (3) voltage amplitude of 10–200 kV, and (4) 0.1–1 kHz repetition frequency. An IGBT-based modulator has been designed for a Marx generator that can satisfy these requirements [19]. The IGBT modulator is shown in Fig. 19.17 within a Marx generator. Here, the resistance R in the Marx generator shown in Fig. 19.15 is replaced by a diode and the spark gap switches are replaced by the IGBT-1 and IGBT-2. The authors state: "*IGBTs provide the capacitive gate control of MOSFETs with the lower saturation voltage drop of bipolar junction transistors and as such have become the switches of choice in solid-state pulsed power systems. As opposed to traditional vacuum tube switches, IGBTs are very compact, require negligible auxiliary power, feature simple driver circuitry, and do not emit x-rays.*" Many other publications [20–22] are available about IGBT-based Marx generators for ion implantation.

An interesting example of ion implantation pertinent to the discussion of IGBT devices is the VIISTA-1500H machine for high-energy hydrogen [23]. Deep implantation of hydrogen ions can be used to create n-type doping for the buffer layer in asymmetric (or field-stop) IGBT structures. This requires a machine with ion implantation energy of 1.5 MeV to place the hydrogen at a depth of about 30 μm.

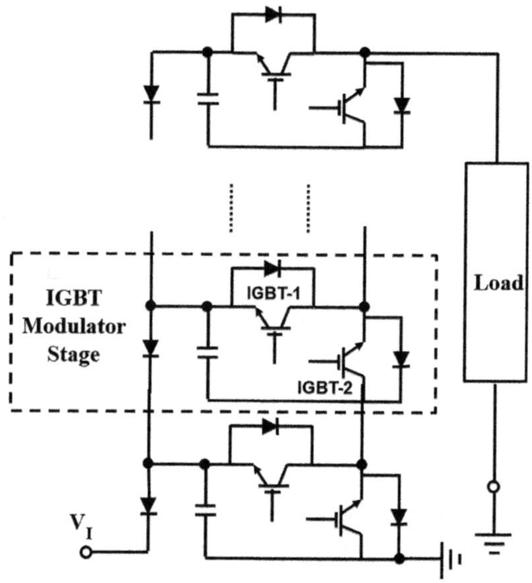

FIGURE 19.17

IGBT modulator for Marx generator. *IGBT*, insulated gate bipolar transistor.

19.5.3 Cancer Treatment Pulse Generator

A strong pulse of an electric field of a nanosecond duration can be used to burst cancer cells. Apoptosis is the normal healthy natural death of cells in organisms during growth. Cancerous growth of tumors can occur when apoptosis is interrupted. High-frequency pulsed electromagnetic waves can be used to induce apoptosis in cancer cells due to their strong electric fields. A few hundred pulses are sufficient to cure melanoma. A short pulse duration is desirable to keep the imparted energy low and avoid the generation of heat. A special resolution of cm in size is possible by using sub-ns pulse durations. Voltage pulses of 160 kV with an 800 ps width have been developed with a rise time of 200 ps for this application [24]. IGBTs with ratings of 1200 V, 10 A were employed by the authors.

19.6 Particle Physics

The basic understanding of nature is pursued under particle physics where the constituents of matter and radiation are being determined [25]. All particles and their interactions can be described by the Standard Model, a quantum field theory that contains 16 fundamental particles and the Higgs boson. The verification of the theory has required the building of particle accelerators with increasing energy over the years. Particle accelerators are used to create a beam of charged particles with increasing energy [26]. In a linear accelerator, the particles are propelled in a straight line until

FIGURE 19.18

Fermilab Tevatron particle accelerator.

they collide with a target to generate the desired interaction. The Stanford linear accelerator (SLAC) is the world's longest (1.9 miles) linear accelerator. It has been upgraded to achieve acceleration of electrons and positrons to an energy of 50 GeV [27].

Even larger energies can be generated using synchrotrons in which the particles move along a circular track of constant radius [26]. One example is the Fermilab's Tevatron, whose interior tube and exterior ring with a diameter of 4.26 miles is shown in Fig. 19.18 [28]. It can accelerate protons and antiprotons to an energy of 980 GeV. It was superseded by the Large Hadron Collider (LHC) at the European Council for Nuclear Research (Conseil Europèen pour la Recherche Nucléaire, CERN), Switzerland, with a diameter of 17 miles [29]. It could accelerate protons and antiprotons to an energy of 4 TeV, allowing discovery of the Higgs boson in 2012.

As the IGBT ratings increased, they have become the devices of choice for power supplies required for particle accelerators [30]. Writing in 2000, the authors state: "*Advances in IGBT technology are leading to rapid improvements in the capabilities of solid-state accelerator drivers. NSRC has shown that ion linacs can easily be powered by solid-state units today.*" They utilized 1700-V Toshiba IGBT modules operating with a peak current of 1200 A to create a 180 kW average power klystron modulator.

19.6.1 Stanford Linear Accelerator

The SLAC, shown in Fig. 19.19, is the world's longest linear accelerator, capable of propelling electrons and positrons to an energy of 50 GeV [26]. The SPEAR III injection kicker system is used to deflect an incoming beam of 3.3 GeV electrons using three kicker magnets K-1, K-2, and K-3 [31]. The deflection angle for the magnets K-1 and K-3 is 2.5 mrad while that for the K-2 magnet is 1.0 mrad. The magnets are designed with a magnetic gain of 8.4 mT/A. The solid-state modulator for driving the magnets is designed with IGBTs.

FIGURE 19.19

Stanford linear accelerator.

The design for the K-2 magnet is shown in Fig. 19.20. Powerex IGBT modules with a rating of 4500 V and 600 A are used. Four IGBTs and cores are used to generate a pulse of 2381 A into the magnet. The IGBT circuits are coupled to the magnet using a transformer. Each primary of the transformer has a single turn and the

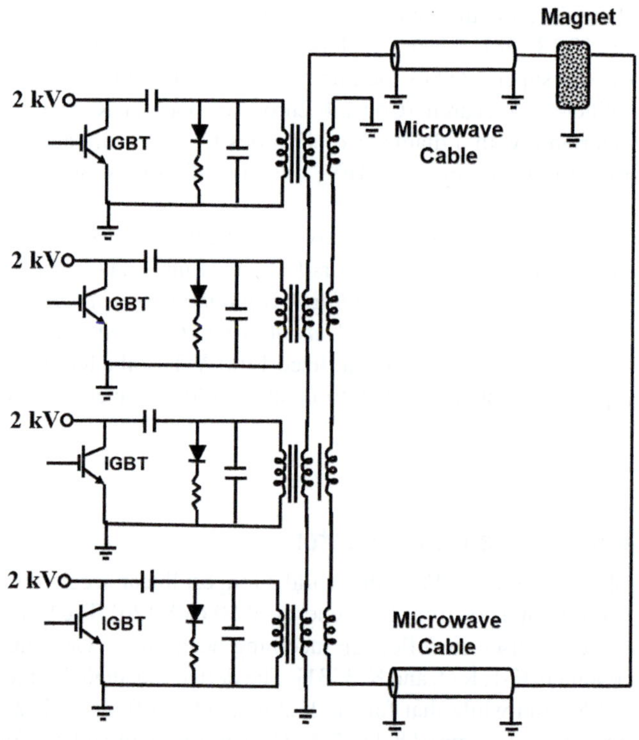

FIGURE 19.20

IGBT modulator for Stanford linear accelerator. *IGBT*, insulated gate bipolar transistor.

secondary side consists of a single turn through a toroidal core. The design makes use of a microwave coaxial cable. When the IGBT is turned on, a current pulse is generated on the secondary side of the transformer, which travels down the microwave cable into the magnet. The current magnitude is doubled to 2500 A due to reflection at the shorted magnet. Four microwave cables are used per bus in the case of magnets K-1 and K-3.

19.6.2 International Linear Collider

The International Linear Collider (ILC) is a proposed accelerator for electrons/positrons. Linear accelerators are preferred to circular accelerators, which emit synchrotron radiation due to a bent path. They are not practical for heavy particles but suitable for electron–proton interactions, which are simpler to analyze [32]. It is anticipated that physics beyond that in the Standard Model will be discovered with this new accelerator. Its construction was expected to begin in 2015–2016 and be completed by 2026, but has been put in limbo by lack of international funding commitments [33]. Some objectives of the ILC are the measurement of the mass, spin, and interaction strength of the Higgs boson, and investigation of supersymmetric particles that are dark matter candidates. The ILC has a proposed length of 19–31 miles, more than 10 times as large as the existing SLAC machine. It is being designed to propel the particles to an energy of 500–1000 GeV.

The electric field gradient in the ILC has been chosen at 31.5 mV/cm generated using superconducting radio frequency cavities operating at 1.3 GHz. Three cryomodule units that accelerate particles to 850 MeV in a distance of 1 m require 7.65 MW of power. The power rating for 1 RF unit is set at 10 MW to provide a performance margin. 560 RF units will be necessary to accelerate the electrons/positrons to an energy of 250 GeV. The RF power is generated using a klystron, operated using 120kV, 140 A pulses at a repetition frequency of 5 Hz.

A Marx generator using IGBTs is considered the most suitable option for driving the klystron [34]. It is desirable to maintain the voltage at the klystron within ±0.5% during the pulse duration of 1.7 ms. To avoid a droop in the pulse shape of the conventional Marx generator circuit, a nested droop correction circuit is implemented using IGBTs, as shown in Fig. 19.21. In this circuit, IGBT-1 and IGBT-2 are used to charge and discharge the main capacitor (C_M) in the Marx generator. IGBT-3 and IGBT-4 are parts of the nested droop correction circuit. They are used to rapidly switch energy from the droop capacitor (C_D) to regulate the height of the voltage pulse. The main capacitor is charged to 4000 V while the droop capacitor is charged to 1000 V.

To attain reliable operation of IGBTs, it is necessary to take into account cosmic ray-induced failures. IGBTs with a blocking voltage rating of 6.5 kV must be chosen to obtain reliable operation at 4000 V in the Marx modulator circuit. A 32-cell design is chosen to obtain 4000 V operation per cell for the 120 kV power supply, including a redundancy of two. For the nested droop correction portion, an IGBT blocking voltage rating of 1.7 kV is sufficient because its operating voltage is only 1050 V. A switching frequency of 40 kHz is used for IGBT-3 and IGBT-4 based on thermal

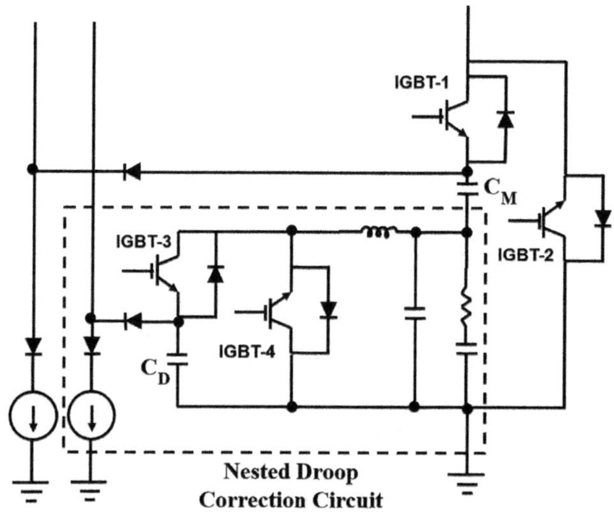

FIGURE 19.21

IGBT cell for a Marx generator driving the Klystron. *IGBT*, insulated gate bipolar transistor.

considerations. The total power loss in the semiconductor devices (IGBTs and diodes) is 360 W per cell with a total power output of 4462 W, leading to an efficiency of 92.5%. The IGBTs must be protected when the klystron undergoes arcing. This is performed using overvoltage and overcurrent protection circuits in the gate driver for the IGBT [34].

19.6.3 Fermilab Main Injector

A sextupole correction system is used in the Fermilab main injector to obtain a stable high intensity beam of protons and antiprotons with an energy range of 8.9–150 GeV [35]. Horizontal and vertical rings are used with 54 sextupole magnets connected in series. An IGBT-based 250 kW switch-mode power supply for driving these magnets was developed and implemented as far back as 1999. The sextupole magnets require drive currents ranging up to 350 A in steps of 40 A within 10 ms to achieve changes of 20 GeV. The current must be stable within ±25 mA over an 8 h period.

The circuit for the switch-mode power supply used to drive the sextupole magnets is shown in Fig. 19.22 [35]. The input AC power is rectified to create a DC bus. Four IGBTs in an H-bridge configuration are used to generate the output voltage using pulse width modulation (PWM) control. The output voltage is produced across the filters. The output voltage is given by $[(2 * \delta - 1) * V_{DC}]$ where δ is the duty cycle for IGBT-1 if the duty cycle for IGBT-3 is maintained at $(1 - \delta)$. The output voltage is zero when the duty cycle becomes 50% ($\delta = 0.5$).

Eupec IGBTs with a blocking voltage rating of 1.6 kV and a current rating of 1.2 kA were used at a PWM frequency of 10 kHz. The output of the H-bridge is filtered

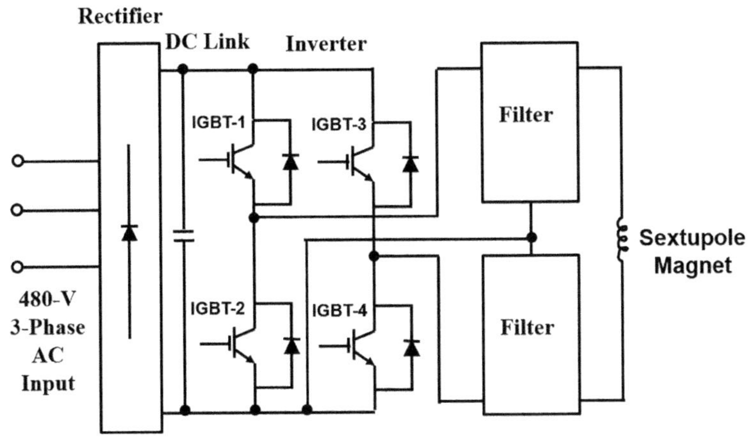

FIGURE 19.22

Switch-mode power supply for Fermilab main injector. *IGBT*, insulated gate bipolar transistor.

to reduce the peak-to-peak ripple voltage to 2.5 V. This provided current regulation in the magnet at ±78 mA at a current of 350 A. The current slew rate was limited to 4000 A/s. This power supply has been in operation at Fermilab for the main injector since October 1988 for 150 GeV particle energy.

19.6.4 Japan Hadron Facility

The Japan Hadron Facility (JHF) consists of a 50 GeV main synchrotron, a 3 GeV rapid cycling booster synchrotron, and a 200 MeV linear accelerator (linac) [36]. It is designed to propel a 10 μA beam of protons at a repetition rate of 0.3 Hz in the 50 GeV synchrotron. The 3 GeV synchrotron accelerates a 200 mA beam of protons with a pulse length of 900 ns and a repletion rate of 25 Hz. The protons are accelerated to 200 MeV in the linac with a peak current of 30 mA before injection into the 3 GeV ring. In 1998, the author stated: "*The magnet power supplies of the 50 GeV synchrotron will use IGBTs, the gating time of which is so fast and flexible as to avoid a harmful reactive (wattles) power. This is also a long-awaited device, but it is only recently that high-power devices (3.3 kV, 1200 A) have been successfully developed.*"

The power supply for the magnets in the 50 GeV synchrotron of the JHF must generate a trapezoidal pulse with a peak power of 1 MW with a maximum current of 3000 A. The block diagram for the IGBT-based power supply designed in 1999 is shown in Fig. 19.23 [37]. The specifications for the power supply are quite stringent: current ripple during the flat top of 1 μA with a very small passive filter; current stability of better than 50 μA; and tracking error of better than 300 μA. The authors state: "*By using IGBT, it becomes possible to construct a power supply basically not generating a reactive power; in spite of that, the power supply is operated*

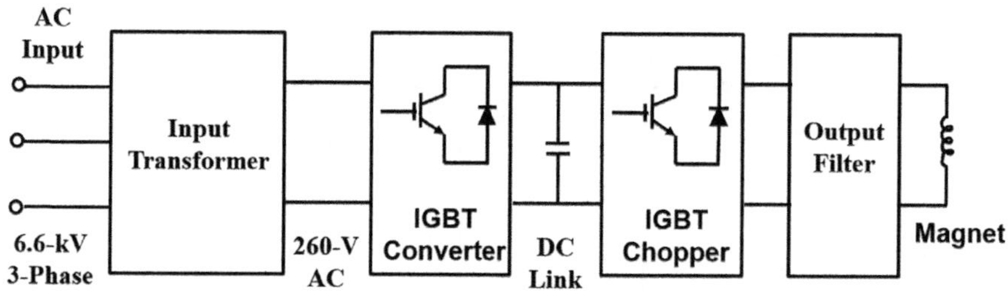

FIGURE 19.23

Block diagram of the power supply for the magnets in the 50 GeV Japan Hadron facility. *IGBT*, insulated gate bipolar transistor.

with a trapezoidal output current waveform. Furthermore, the very high switching frequency makes it easy to control the output current shape with very small tracking error and to regulate the current ripple to be very low level without a large-scale passive filter or an active filter."

The converter and chopper blocks are fabricated using IGBT modules rated with a blocking voltage of 1200 V and a current rating of 300 A. Each IGBT module contains two IGBTs and two fly-back diodes. Nine IGBT modules are required for the converter stage and 16 modules are needed for the chopper stage. The IGBTs in the converter are switched at 5 kHz to produce a stable DC link of 410 V while achieving unity input power factor. The IGBTs in the chopper stage are switched at 8 kHz, resulting in a ripple at 128 kHz because of using 16 modules with staggered timing. The high frequency of the ripple current allows making the output filter very small in size.

The IGBT-based power supply was able to deliver 1 MW of power to generate a 3000 A trapezoidal current in the 50 GeV synchrotron magnet with a ramp rate of 9700 A/s. The input power factor was close to unity when operating at an output power of 150 kW. The current ripple was measured to be less than 3 μA for a 3000 A trapezoidal waveform at a repetition frequency up to 350 Hz. The tracking error was confirmed to be below 300 μA.

19.6.5 CERN Large Hadron Collider

The Large Hadron Collider (LHC) at CERN is designed to propel protons to an energy of 4 TeV in the main ring. A Low Energy Ring (LER) is used to inject 1.5 TeV proton beams into the main ring. The LER accelerator is based on using superferric magnets. The superferric magnet requires 100,000 A of current that must be ramped up and down at the rate of 1000 A/s [38]. The power supply is designed with 10 units, each capable of delivering 10,000 A at an output voltage of 1.5 V. Each of these power supplies contains two 112:1 transformers, as shown in Fig. 19.24. The secondary side of the transformer contains 32 windings with Schottky rectifiers whose output is paralleled to deliver the current to the superferric magnet.

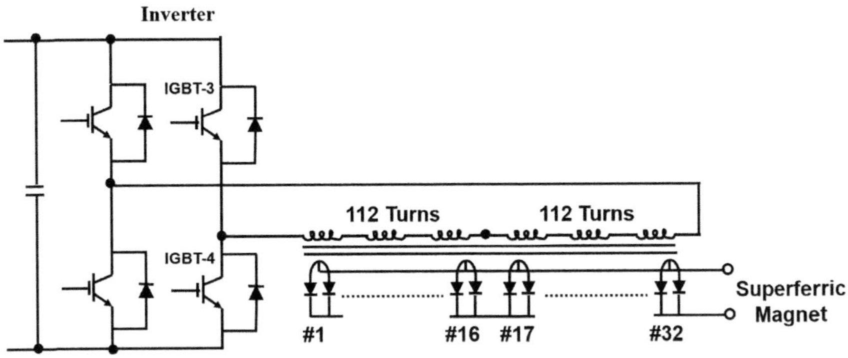

FIGURE 19.24

Power supply for the superferric magnets in the 4 TeV CERN Large Hadron Collider. *IGBT*, insulated gate bipolar transistor; *CERN*, Conseil Européen pour la Recherche Nucléaire.

The primary side of the transformer is connected to an IGBT H-bridge circuit that inverts the 400 V/600 A DC bus voltage. IGBTs with a blocking voltage of 1200 V and a current rating of 300 A are used for the H-bridge. They are switched at 2 kHz to generate the AC voltage on the primary side of the transformer. The peak power requirement is 40 MVA with an rms power rating of 500 kW. With this topology, it was possible to meet the specifications for driving the superferric magnets.

Another application of IGBTs in the LHC at CERN is to drive the current in four bumper magnets in the Low Energy Ion Ring (LEIR) [39]. The bumper magnets require a peak current of 1200 A that decreases linearly to zero in a time period ranging from 120 to 300 μs. The linearity of the current ramp must be within $\pm 2\%$ and the tracking of current magnitudes for the four magnets must be better than $\pm 1\%$. The LEIR machine cycles every 3.6 s with a burst of eight pulses at 5 Hz.

A power supply design that meets the above specifications has been designed by using IGBTs as switches. The circuit is shown in Fig. 19.25 where a resonance between the capacitor C_P and the magnet inductance L_1 is used to generate the current waveform. When IGBT-1 is turned on, a sinusoidal rising current is generated by the resonance. When the peak current amplitude is reached, IGBT-1 is turned off, allowing the current to free-wheel through diode D_1, resistor R_1, and capacitor C_1. This forms another resonant circuit, allowing the current to ramp down to zero in the magnet. The linearity of the current slope can be kept within $\pm 1.5\%$ by the choice of the components. In the circuit shown in Fig. 19.25, diode D_2 is needed to suppress oscillations of the magnet current and resistor R_2 is needed to discharge the capacitance of the transmission line. In addition, IGBT-2 and resistor R_3 are used to discharge capacitor C_1 after each pulse cycle. Diode D_1 is used to protect the power supply V_B from high voltages developed across capacitor C_1. The circuit allows continuous adjustment of the slope of the current pulse without using different capacitors that must be manually changed.

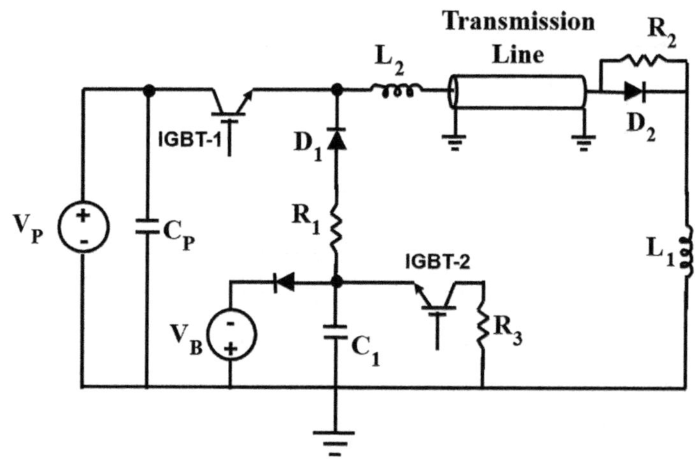

FIGURE 19.25

Power supply for the bumper magnets in the CERN Low Energy Ion Ring. *IGBT*, insulated gate bipolar transistor; *CERN*, Conseil Européen pour la Recherche Nucléaire.

IGBTs are also used in the tune measurement system for the LHC at CERN. During all phases of running the LHC from commissioning to ultimate performance luminosity runs, reliable measurements of the Betatron tune are required [40]. The tune measurements utilize the introduction of a disturbance to the proton beam with a short kick pulse of a few microseconds to a magnet. Four such magnets are installed in the LHC with one for each beam and each plane. The kick generator must produce a half-sine wave pulse superposed with a third harmonic to make the pulse closer to a rectangular shape. The duration of the pulse is 5.1 μs. During tuning for 7 TeV operation, the maximum magnet current is 1200 A. The generator must produce this pulse with a maximum voltage of 3.3 kV.

The circuit diagram for the tune generator is shown in Fig. 19.26. Ten coaxial transmission cables with lengths of 20 m are used. Together with an resistor capacitor (RC) filter and the magnet inductance, they form a harmonic circuit. When the IGBT is turned on, a 5.1 μs pulse is created with an amplitude of 1200 A. The IGBT must be capable of handling this peak current at a turn-on dI/dt of 2.5 kA/μs. IGBTs with a blocking voltage of 5.2 kV and a current rating of 1180 A made in a press-pack by Westcode were used to meet the specifications. The authors state: "*This new family of IGBTs is very rugged and possesses an exceptionally large safe operating area (SOA) and for all it is suitable for snubberless switching*." The IGBTs use a soft-punch through drift layer design. Note that an auxiliary 800 V DC supply is connected directly to the collector terminal of the IGBT. This ensures a constant turn-on delay time for the IGBT. This application is unusual because it makes use of the excellent turn-on characteristics of the IGBT.

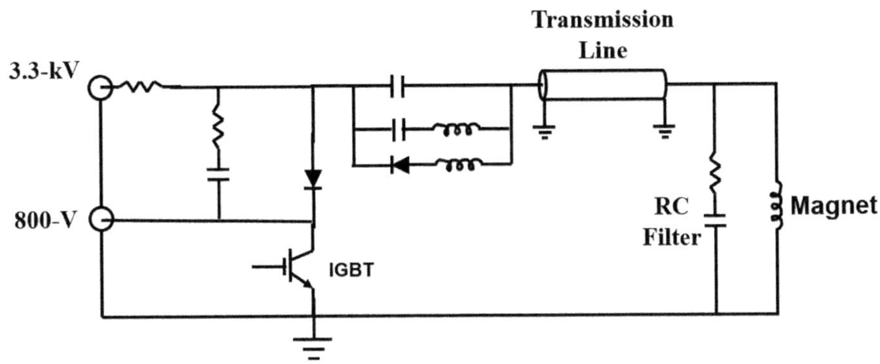

FIGURE 19.26

Tune measurement generator circuit in the CERN Large Hadron Collider. CERN, Conseil Européen pour la Recherche Nucléaire.

19.6.6 Spallation Neutron Source

Oak Ridge National Laboratory operates a spallation neutron source to allow scientists to study the molecular and magnetic structure of materials. These studies impact our understanding of fundamental physics and biological materials. The source produces short millisecond long pulses of neutrons that can interact with materials. 60 pulses per second of hydrogen ions are accelerated to 88% of the speed of light and then strike a swirling 50 ton liquid mercury target. This generates neutrons by a process called spallation when the protons interact with the mercury nucleus. The neutrons are slowed down to room temperature by passing through water.

A power supply for the extraction of the protons at a beam energy of 1.3 GeV was designed using IGBTs [41]. A switch forward resonant charging power supply was created using two IGBTs with ratings of 3.3 kV 500 A in series.

19.7 Pulsed Lasers

Pulsed lasers are used for soft tissue surgery [42]. Carbon dioxide lasers have the appropriate wavelength to create tissue ablation with coagulation to suppress bleeding [43]. Pulse dye lasers are used for destroying blood vessels without damaging the surrounding skin [44]. This can be used to treat skin stains, red scars, hemangiomas, and rosacea. They can be used to treat varicose or spider veins as well [45].

19.7.1 Power Supply

Pulsed laser and power microwave sources require the application of a high-voltage pulse of energy that is stored in a capacitor. IGBTs are used in the power supplies for charging these capacitors. The basic circuit configuration is shown in Fig. 19.27 [46]. The input 650 V DC voltage is created using a three-phase rectifier (not shown)

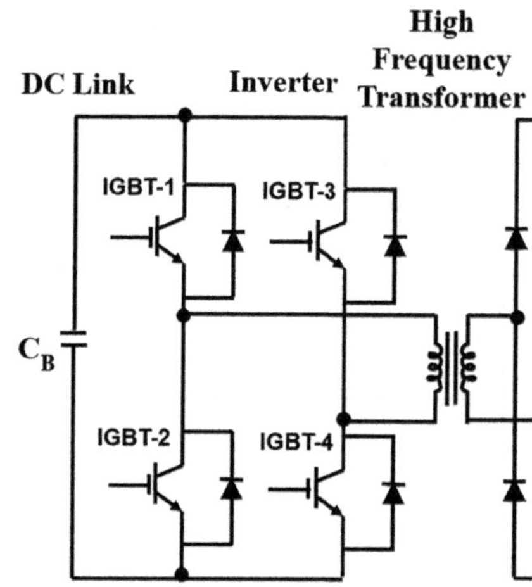

FIGURE 19.27

Pulsed laser capacitor charging circuit. *IGBT*, insulated gate bipolar transistor.

from a 480 V AC line. The circuit uses an H-bridge consisting of four IGBTs that serve as an inverter stage. The H-bridge produces a high frequency (typically 10 kHz) rectangular voltage across the primary side of the high-frequency transformer. The IGBT modules made by Semikron have a blocking voltage rating of 1200 V and a current handling capability of 1200 A. This topology can charge a 7.2 μF capacitor (C_P) to 30 kV in 40 ms. The authors state: "*From the operational experience with the previous charger we concluded that the weight and volume of the inverter, which is the core of the system, can be reduced to half of the previous size. This can be achieved by using advancements in IGBT modules as well as state-of-the-art capacitors and optimum packaging. Today, the performance of rapid capacitor chargers like the one described here is directly tied to advancements in high power semiconductor devices, namely IGBTs.*"

19.7.2 IGBT Modules

The use of IGBT modules connected in series for providing power to nitrogen lasers has been investigated [47]. The nitrogen tube requires a 25 mJ pulse of energy delivered in 20 ns, corresponding to a power of 1 MW. The power is delivered from a charged capacitor. This requires a device capable of blocking 12 kV and delivering a peak current of 500 A. IGBT modules with 1200 V and 1700 V devices connected in series and parallel were found to be capable of performing this function. The current had a rise time of 30 A/ns.

19.8 Food Sterilization

The preservation of food [48] is of great importance to society because it enables food storage for delayed consumption and the distribution of food from manufacturing locations in rural areas to where it is consumed in urban areas. Refrigeration is one very successful technology to achieve this goal by suppressing the growth of bacteria and fungi. Another approach is the sterilization of food to destroy bacteria and other microorganisms that may be prevalent in the food. One unusual method for food preservation is the application of pulsed electric fields (PEFs). The electric field enlarges the pores of the cell membranes in the food, which can kill the bacteria. The process of increasing the permeability of cell membranes using a strong electric field is referred to as *electroporation* [49]. An electric field of 30 kV/cm is required to kill cells of bacteria and fungi. PEF has the advantage of being a low-temperature process for pasteurization of liquids such as fruit juices to preserve their flavor. This process can also be used to make juice extraction process more effective by enlarging the pores in the cells of fruits and vegetables [50].

Several companies make commercial PEF sterilization equipment. The Diversified Technologies PEF System shown in Fig. 19.28A allows up to 10,000 L/h and produces >5 log bacteria kill (5 log is equivalent to 10 bacteria after a 99.999% reduction) [51]. According to the manufacturer, the PEF system does not alter the flavor of foods or juices and requires low energy consumption. They state [52]: "*A key element in this design was (the company's) recent efforts to develop IGBTs specifically for pulsed power applications (with Powerex).*" The SteriBeam XeMaticA-2LFA PUV System is shown in Fig. 19.28B. The manufacturer states [53] that the PEF system is aimed at the "*extraction of nutrients, natural colors, and juices from roots, grass, and the skin of fruits and vegetables.*"

The US Food and Drug Administration has an extensive report on the viability of food preservation by using PEFs [54]. Their report states that PEF has been applied to preserve the quality of food such as bread, milk, orange juice, and apple juice. The effectiveness of PEF depends on the intensity of the electric field and the duration of the electrical pulse. The effectiveness is also dependent on the pulse shape with a

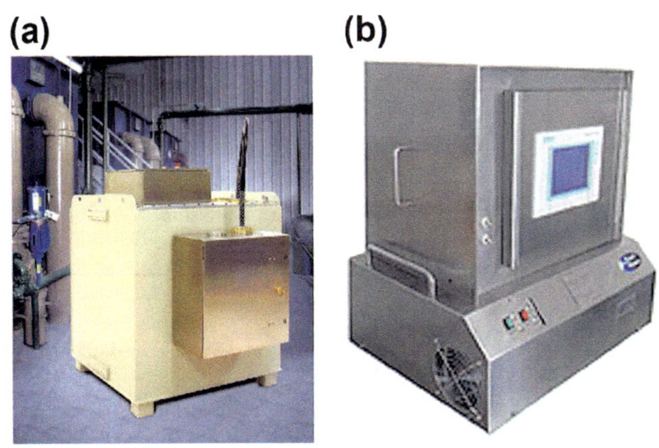

FIGURE 19.28

Pulsed electric field sterilization systems. (a) Diversified Technologies; (b) Steribeam Systems.

square wave being more effective than an exponentially decaying waveform. In addition, bipolar pulses are found to be more lethal to bacteria and fungi, and require less energy.

An IGBT-based circuit topology that can deliver the bipolar voltage pulse required for electroporation of cell membranes is shown in Fig. 19.29 [55]. With this design, it is possible to adjust the pulses in both directions (positive and negative) with an amplitude of up to 10 kV. The peak repetitive current is 200 A, leading to a peak power of 2 MW. The width of the positive or negative pulses can be adjusted from 0.8 to 10 μs and the time delay between them can be adjusted from 0 to 99 μs. The rise and fall time for the pulse is 200 ns. Any DC current applied to the stainless steel electrodes of the treatment chamber has been found to produce electrochemical erosion. This problem is circumvented by adding an inductance (L) across the chamber, as shown in the figure. This bipolar power supply has been found to be effective for sterilization. Experiments conducted using *Escherichia coli* bacteria have demonstrated an inactivation rate of 4.46 log.

To achieve these specifications, stacks of IGBTs are used for IGBT-1 and IGBT-2. Each stack makes use of 1200 V, 220 A devices. Fifteen IGBTs are used in series for each stack to support the high voltage. The voltage must be shared equally between the IGBTs in series within a stack. A capacitive voltage distribution technique is used, as shown in Fig. 19.30 for one of the IGBTs in the stack. In addition, a small (~15 mA) "leakage current" is introduced using the variable current source in parallel with the capacitor. The magnitude of the current source can be tweaked to achieve good voltage sharing between the IGBTs within a stack.

Another pulse power supply design using IGBTs is the Marx generator discussed in Section 19.5.1. A 6 kV pulse source has been reported using a two-stage Marx topology with IGBTs as switches [56]. This approach allows delivery of a 5 kV voltage pulse across a liquid chamber with electrode spacing of 1.25 mm. The resulting electric field of 4 kV/mm is sufficient to sterilize liquids containing *E. coli* bacteria. Due to the finite conductivity of the liquid, ranging from 0.1 to 1.5 S/m for apple

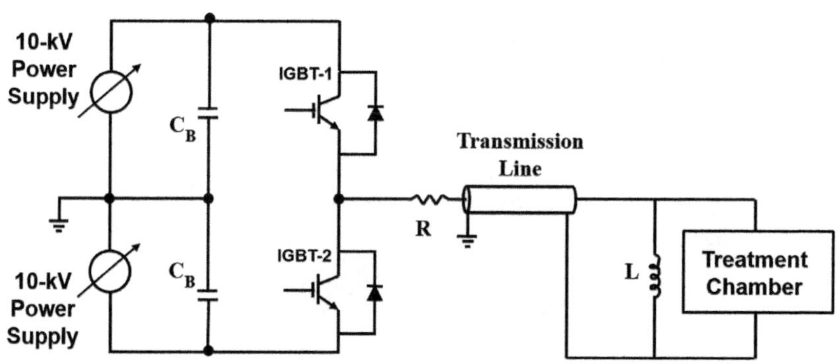

FIGURE 19.29

Bipolar pulse generator for electroporation sterilization. *IGBT*, insulated gate bipolar transistor.

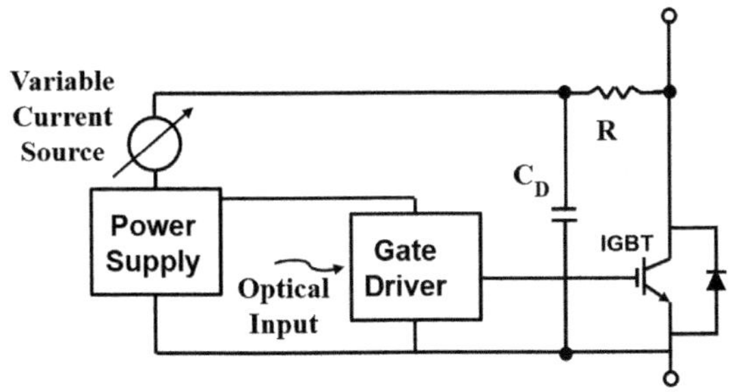

FIGURE 19.30

IGBT stack voltage sharing circuit. *IGBT*, insulated gate bipolar transistor.

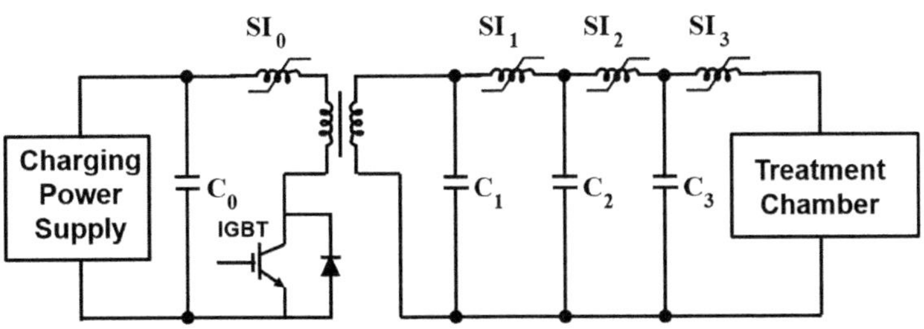

FIGURE 19.31

Three-stage pulse power generator using magnetic pulse compression. *IGBT*, insulated gate bipolar transistor.

cider to tomato juice, a current of 800 A in magnitude flows during the pulse. The capacitors must be chosen to deliver an energy of 200 kJ/kg to the liquid. The pulse power supply was designed to provide an energy of 54 J. The IGBTs used had a blocking voltage of 3.3 kV and a peak current capability of 800 A for this application.

The PEF can be generated with a power supply using magnetic pulse compression (MPC). The MPC topology is shown in Fig. 19.31 with three stages [57]. In this circuit, capacitor C_0 is first charged to 850 V using the charging power supply. The IGBT is in the off-state during this time. The IGBT is then turned on to generate the voltage pulse across the pulse transformer with a turns ratio of 1:30. This transfers energy from capacitor C_0 to capacitor C_1 whose voltage rises to 20 kV. The energy in C_1 is then transferred to capacitor C_2 after the inductor SI_1 saturates. The same process transfers energy to capacitor C_3 and then to the load, that is, the treatment chamber. The peak current in the load is 3 kA with a rise time of 7 ns.

19.9 Water Treatment

Water is an essential commodity for human existence. The evolution of life forms on Earth is attributable to the existence of water. Oceans cover 72% of the surface of the Earth [58]. Unfortunately, oceans contain saline water that is not fit for human

FIGURE 19.32

(a) Water treatment plant; (b) water desalination plant.

consumption. Only 3% of the water on Earth is classified as "fresh water," with two-thirds of this frozen in glaciers and in the polar ice caps [59]. Of the available fresh water, 87% is located in lakes, 11% in swamps, and 2% in rivers. One-fifth of the world's population lives in areas with water scarcity with insufficient fresh water to meet all the demands for drinking, washing, cooking, etc. One-third of the world's population lacks clean drinking water, leading to the prevalence of diseases.

The production of clean drinking water can greatly improve the quality of life for billions of people on Earth. This section discusses IGBT-based technologies for disinfection of water to prevent the spread of diseases using water treatment plants like that shown in Fig. 19.32(a); for the desalination of water to make this resource available in areas of scarcity using plants like that shown in Fig. 19.32(b); for the treatment of sewage for recirculation of water into the ground; and the removal of organic and inorganic material that can cause fouling of pipes in industrial plants and ships. Performing these functions with high efficiency is critical to reducing the cost of water processing and the management of electrical energy resources.

19.9.1 Disinfection

Water that is suitable for human consumption without the harmful effects of bacteria is referred to as "potable water." Disinfection refers to the killing of bacteria in water to avoid the spread of harmful diseases [60]. Chlorination is commonly used by urban municipalities to stop the growth of bacteria in the water supply. The PEF method for killing bacteria in food and juices to enhance their shelf life has also been applied for the disinfection of water. The process requires generating an electric field of about 50 kV/cm in the water.

A switch-mode power supply using IGBTs, shown in Fig. 19.33, has been demonstrated for disinfection of water [61]. This approach greatly reduces the energy required for the process and makes the equipment much smaller in size and weight when compared with directly using 50 Hz power sources. The power supply

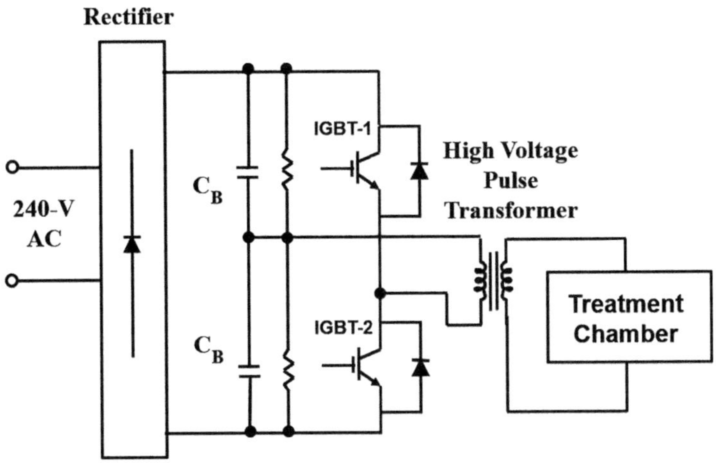

FIGURE 19.33

IGBT-based high-voltage, switch-mode power supply for disinfection of water. *IGBT*, insulated gate bipolar transistor.

generates 3 kV square-wave voltage pulses across the water treatment chamber at a frequency of 17 kHz. A 320 V DC bus is created from the 240 V AC input power source across two capacitors by using a bridge rectifier. The two IGBTs chop the DC bus voltage at 17 kHz to generate a square-wave voltage at the primary side of the pulse transformer. The transformer increases the voltage by a factor of 9.4 times to generate the 3 kV bipolar pulses across the water treatment chamber. The electrodes in the chamber have a spacing of 0.65 mm to create an electric field of 50 kV/cm.

The survival ratio (fraction of bacteria surviving out of the total population) for *E. coli* and *Serratia marcescens* bacteria after being subjected to the 40–50 kV/cm electric field has been studied as a function of exposure time. For the case of the square-wave pulses generated by the IGBT-based switch-mode power supply, the survival ratio decreases to about 3%, as shown in Fig. 19.34, after an exposure time of 15 ms. The switch-mode power supply was found to reduce the power used per unit volume of water by 92% when compared with a 50 Hz power source.

19.9.2 Desalination

The removal of salt from saline water sources, such as oceans and seas, is referred to as desalination. It is often cheaper to transport water over a large distance than to operate a desalination plant. Consequently, desalination is a relatively expensive technology that is only used in dry countries and seagoing vessels, such as submarines [62]. The largest desalination plants in the world are the Ras Al Khair, Saudi Arabia, producing 1,036,000 m^3 per day of water; the Tahweela, United Arab Emirates, producing 909,000 m^3 per day of water; and the Shuaiba 3, Saudi Arabia,

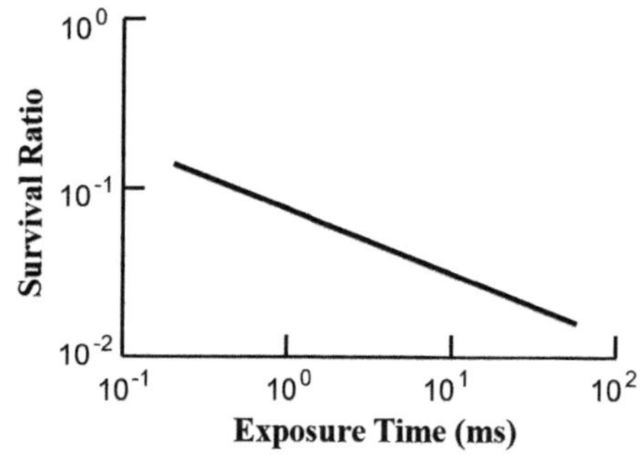

FIGURE 19.34

Survival ratio for *Escherichia coli* and *Serratia marcescens* in pulsed electric field.

producing 880,000 m^3 per day of water [63]. Israel produces 40% of its domestic water by desalination of seawater. There are at least 15 desalination plants operating in Algeria. Other countries with desalination plants include Spain and the United States. The worldwide desalination capacity grew to 120 million m^3 per day by 2020, with nearly 20,000 plants in operation [64]. Reverse osmosis has emerged as the dominant technology due to its simplicity and lowest energy requirements.

Many approaches for desalination of seawater have been considered over the years. An analysis of the economic viability of desalination was published in 1966 [65]. The author recognizes that the large amount of energy required for desalination requires colocation of the desalination plant with a power plant. Furthermore, a major portion of the cost for desalination is associated with the energy required for the process. The costs in 1966 and projection to 1980 are shown in Table 19.2. Despite the high energy requirements for the evaporation process, it was selected for development at that time because energy generation and its environmental impact were not considered as major problems. This situation is no longer true. Consequently, the most popular method for desalination of water today is the reverse osmosis process [66]. In the reverse osmosis process, a solvent (e.g., water) is forced through a semipermeable membrane from the side containing a high concentration of solute (e.g., salt) to the other side by the application of high pressure. A typical membrane is made from a polymer matrix. For the desalination of seawater, a pressure of 600–1200 psi is required.

Table 19.2 1966 Analysis of Energy Requirements for Desalination Processes

Desalination Process	Energy Requirements (kJ/m^3)	
	1963	Projected 1980
Evaporation	284,500	170,000
Freezing	170,000	100,400
Electrodialysis	142,000	86,000
Reverse osmosis	70,000	45,000

The pressure across the membrane for desalination is supplied using pumps that can consume a considerable amount of energy. Adjustable-speed drives using IGBTs are essential to maintain a high efficiency and reduce energy usage for desalination. Desalination requires a high energy consumption of 3 kWh/m^3, which raises the concern of increasing the carbon footprint if the electricity is generated using coal-fired power plants. A better option is to perform the reverse osmosis process by using energy derived from wind and solar energy sources, as shown in Fig. 19.35 [67]. The system is heavily reliant on efficient energy conversion from AC to DC, DC to DC, and DC to AC. These functions are all performed using IGBTs to achieve high efficiency with a compact, cost-effective technology.

The dearth of potable water has been identified as a national resource problem in China [68] because of 100 billion cubic meters of water shortage in 2010. To address this concern, 115 desalination plants became operational in China by 2019 with a total capacity of 1.5 million m^3 per day [64]. The energy consumption for desalination in China is estimated at 4 kWh/m^3. It is recognized that the desalination plant must have a colocated energy source. An attractive option is a non-grid-connected wind power generation plant. Grid-connected wind power generators operate at a fixed speed. A non-grid-connected wind generator can operate at a variable speed, providing efficiency improvements of up to 15%.

An innovative approach proposed for desalination of seawater in China is to make use of off-shore wind power generation colocated with the desalination process [69]. In this approach, the generated wind power is locally converted to potable water that is transmitted to the shore. The desalination plant is integrated into the towers for the

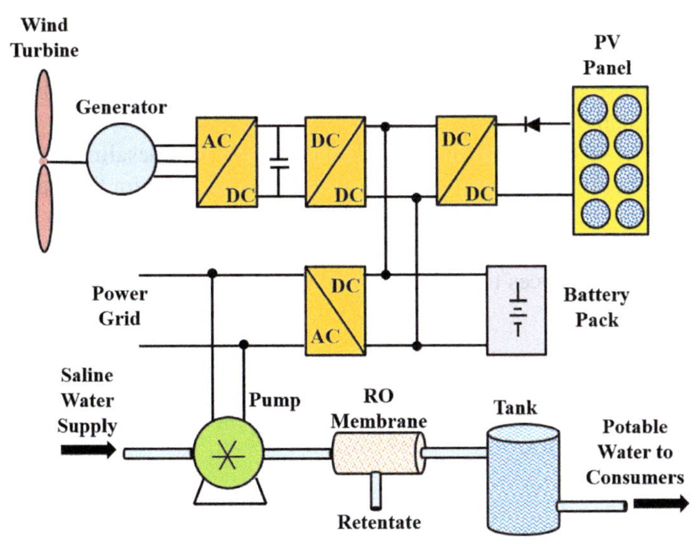

FIGURE 19.35

Reverse osmosis desalination plant with integrated wind and solar power. *PV*, photovoltaic; *RO*, reverse osmosis.

wind turbines. The advantage of this proposal is the elimination of power electronics and cables for transmission of power to the shore. The anticipated cost savings are about 30% while increasing efficiency by 15%.

The value of the integration of wind power and desalination plants has also been recognized in India [70]. There are 175 installed desalination plants in India. The authors state: "*Desalination plants have proved to be an elixir in places like Lakshadweep, Andaman, and some places in Gujarat, Tamil Nadu, and Andra Pradesh.*" The analysis of cost for small-scale desalination in India [71] indicates that a hybrid system (like that shown in Fig. 19.35) comprising wind power, solar power, and battery storage provides the most economically viable solution for desalination.

Desalination is of importance in Middle Eastern countries due to the shortage of potable freshwater sources. One example of a desalination project is the Brackish Water Reserve Osmosis desalination plant in Tunisia [72]. The proposed system utilizes a hybrid system with wind and solar power to provide energy for running the pumps in the reverse osmosis process.

A recent analysis (see Table 19.3) of energy requirements for various desalination processes for use in Morocco has been reported [73]. Multistage flash distillation is a thermal desalination process that is most widely used in the world with 50% of installed capacity for desalination. It consists of evaporation of the input saline water in successive chambers. The multieffect distillation process vaporizes the saline water and condenses it in a separate chamber using a heat exchanger to produce potable water. The vapor compression process condenses evaporated saline water by using pressure. The electrodialysis process is the oldest process for desalination using a membrane. The salts in the water are ionized and removed using ion-sensitive membranes. After taking maintenance and portability into consideration, the authors concluded that the reverse osmosis process was most suitable. A reverse osmosis desalination plant using wind power energy was found to be the most cost-effective option. A reverse osmosis desalination plant using wave power has been deemed practical for Sicily [74].

Most (76%) of the electrical energy consumption in a desalination plant is the reverse osmosis process. Energy consumption is also incurred for the seawater intake pumping station, pretreatment process, and posttreatment process [75]. The authors argue that the desalination plant can be used to balance the energy supplied by the renewable energy sources to create a more-stable electric grid.

Table 19.3 2013 Analysis of Energy Requirements for Desalination Processes

Desalination Process	Energy Requirements (kWh/m^3)
Multistage flash distillation	6–9
Multieffect distillation	10–14.5
Vapor compression	7–15
Electrodialysis	0.7–2.2
Reverse osmosis	3–13

It is estimated that 2.8 billion people around the world have water scarcity [76]. It is anticipated that 60% of the world's population may suffer from water depravation by 2025. The levelized cost of water production by the reverse osmosis desalination process has been computed for a 100,000 m^3 per day production plant. In this case, the high-pressure pumps for reverse osmosis consumed 66% of the energy, the booster pumps consumed 15% of the energy, and the seawater pumps consumed 5%. The total generation cost was found to be about $1 million per day.

19.9.3 Sewage Treatment

The treatment of leachate from sewage-filled ground is mandated by environmental laws. The leachate is treated with chemicals containing benzene compounds. These benzene compounds must be broken down to prevent environmental pollution [77]. The application of high-voltage electric pulses to the sewage has been found to reduce the concentration of benzene compounds. The pulse power supply has an output voltage of up to 60 kV with a peak current of 300 A. Its average output power is 15 kJ/s. The pulse widths range from 2 to 50 μs at a pulse repetition rate of 1–3000 Hz. The IGBT-based pulse power supply was able to reduce the benzene content in sewage from 150 ppm before treatment to 8 ppm after treatment for 10 min at a pulse repetition frequency of 1 kHz.

19.9.4 Fouling of Water Piping

Biofouling is a problem for machinery and piping that utilizes water from seawater, lakes, and rivers for their cooling system. Two examples are ships used by the US Navy and electric power plants used by utilities. Biofouling is the accumulation of organic (aquatic organisms) and inorganic (sediments) materials inside machinery and pipes. Biofouling can be mitigated using chemicals such as chlorine for water treatment. An alternative process to alleviate biofouling is with pulsed acoustic shock waves [78].

An electrohydraulic shock wave can be created by using a current pulse through an electrical arc in the water. Adiabatic expansion at the arc releases a shock wave that propagates through the pipe walls and in the water. The shock wave has been found to inhibit the settling and colonization of organic material. It has also been found to explosively remove sediments due to cavitation. The shock wave has an amplitude of 1–10 MPa with a rise time of 0.5–5 μs.

The pulse power supply used to generate the arc in the water must generate a voltage of 5–25 kV. The peak current in the arc ranges from 1 to 10 kA. Spark gaps were originally used to generate the arc but were found to be unreliable with a low lifetime. They were replaced by thyratrons, but these components are expensive. The availability of IGBTs with sufficient voltage and current ratings has made them good replacements for thyratrons in this application. In 1999, an EUPEC IGBT module with a blocking voltage of 3.3 kV and a peak repetitive current rating of 1.6–3 kA was available with a copackaged fly-back diode. This allowed designing a pulse power supply with an energy recovery circuit to improve the efficiency. The authors state: "*Finally, provided that the IGBT modules are operated within their rated safe operating area, their lifetime is virtually unlimited (at least by gas switching standards).*" This is valuable for achieving a reliable system with low maintenance costs.

19.9.5 Industrial and Pharmaceutical Pollution

Industrial pollutants such as pharmaceutical residues are difficult to remove from water. Advanced oxidation processes that employ plasma discharge can be used to remove pharmaceutical residues [79]. A corona discharge is achieved in the water by using a 10 kV pulse with a duration of 1–100 μs. The Marx generator topology shown in Fig. 19.15 is used with 12 stages of IGBT switches to obtain the 10 kV pulse.

19.10 Oil/Petroleum Extraction

As the petroleum industry matures, it must locate and extract oil from deeper and deeper reservoirs under the Earth's surface. The extraction of oil from deep underground, using oil wells like the one shown in Fig. 19.36, is difficult due to the high viscosity and inclusion of asphalt. The viscosity can be reduced by irrigating with hot water, adding chemicals, or by electric heating of the pipelines. Electric heating

FIGURE 19.36

Oil/petroleum extraction well.

technology is widely used by the oil industry because of its low cost [80]. IGBTs are used in the inverter circuits for this application. Ultrasonic power supplies based on IGBTs are also utilized in oil extraction operations. In addition, adjustable speed drives enabled by IGBTs are essential in large oil fields to reduce energy consumption. This is especially true for undersea oil extraction platforms [81].

19.10.1 Oil Pipe Heating

Heavy oil is difficult to extract from deep wells due to its high viscosity and the presence of asphalt and wax. Electrical heating technology has been widely employed by the oil industry for heavy oil extraction due to its low cost [80]. The hollow pole method with a single phase 50Hz AC source has been found to disturb the electric grid balance. The authors state: "*The inverse system, in which IGBT acts as a main inverse element, solved the nonbalanced problem of the electric power transmission line system, and wax coagulation on the pipe wall, by three-phase and low frequency supply.*"

The oil pipe heating system is schematically illustrated in Fig. 19.37. Heating the oil is achieved by passing current through the oil pipe with the well wall as the earth return path. Resistive (Joule) heating of the oil pipe increases the temperature of the oil and prevents wax buildup on the oil pipe. The power supply requirements are an output frequency ranging from 40 to 100 Hz and output power of 100 kW.

The AC–DC–AC circuit topology for heating the oil pipe is shown in Fig. 19.38. The 50Hz, 380V, three-phase AC power is converted to a DC bus by the rectifier stage and then inverted using the IGBTs to a variable frequency ranging from 40

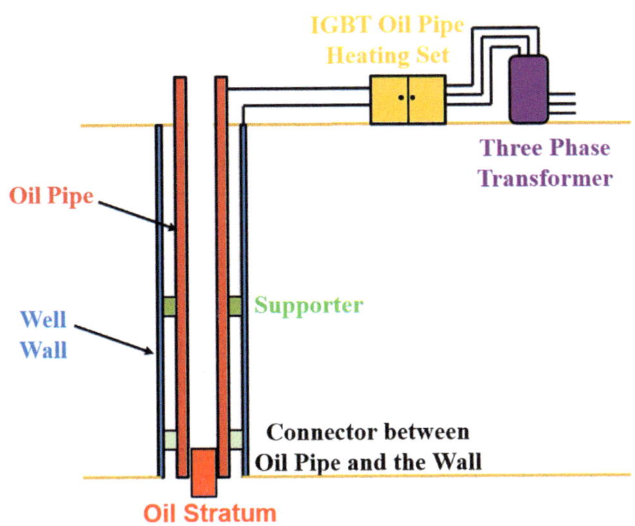

FIGURE 19.37

IGBT oil pipe heating system. *IGBT*, insulated gate bipolar transistor.

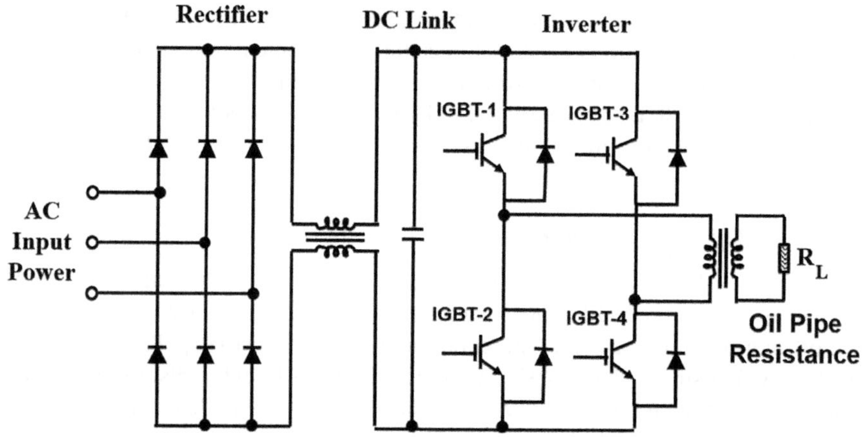

FIGURE 19.38

AC-DC-AC inverter using IGBTs for oil pipe heating. *IGBT*, insulated gate bipolar transistor.

to 100 Hz. The heating efficiency is increased by 50% with this approach. The authors state: "*The set has advantages of small volume and light weight because of (the) IGBT in the inverter system… The system has been used to coagulate oil extraction in an oil extraction factory, Liaohe Oil Fields in China, from December 2000 and has been running well.*" Their economic analysis indicates a cost savings of 21% in electrical energy by using this system.

To improve the efficiency for heating the oil in the pipe, a resonant heating method has been developed [82]. In this approach, a heating cable is introduced into the oil pipe. The magnetic field between the cable and the hollow pipe produces heat due to eddy currents and hysteresis loses. The IGBT-based resonant circuit is shown in Fig. 19.39. A capacitor C_R is used between the H-bridge with IGBTs and the cable

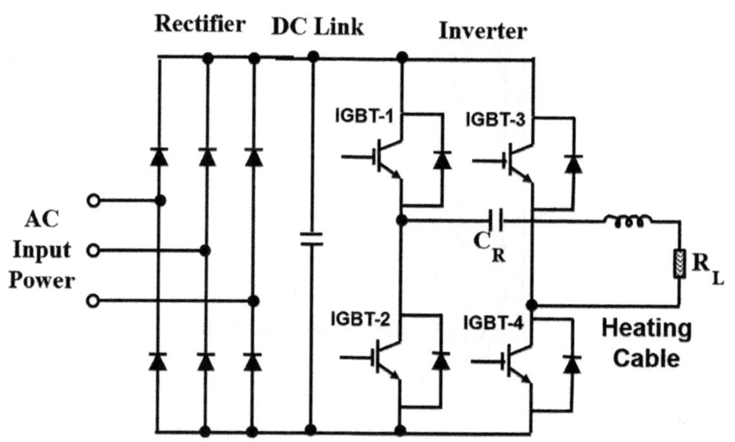

FIGURE 19.39

Resonant circuit using IGBTs for oil pipe heating. *IGBT*, insulated gate bipolar transistor.

to create the resonant circuit. It was found that the optimum frequency for heating the oil was 1 kHz. The oil temperature was increased by 69°C in 30 min.

19.10.2 Subsea Oil Extraction

The extraction of oil and gas from distant off-shore locations has become increasingly important due to the scarcity of land-based extraction sites. Major oil and gas companies have been deploying off-shore extraction in all parts of world. It is challenging to provide power to submersible pumps located at a long distance of 10 miles from the power source and at depths of 10,000 ft. The challenges of off-shore oil and gas extraction were discussed in 1998 by Conoco. [83].

The pumps required for delivering oil and gas from the bottom of the seafloor to land are rated at 2500 hp. One challenge is voltage regulation of the power transfer from shore to the pump. As an example, a 1000 hp. motor rated at 2300 V requires a full rated current of 250 A. The impedance of a typical 10,000 ft subsea cable is 2.5 Ω [84], resulting in an unacceptably large voltage drop of 625 V. Subsea substations that can instantaneously maintain the voltage at the pump are needed. The authors state: "*The use of today's most sophisticated medium-voltage variable-frequency drive technology is an absolute imperative for this application to keep the size, weight, and thus total installed cost, as low as possible*⋯ *This drive also uses pulse width modulation drive technology on the output side or inverter end, and uses well-proven 1200 V or 1700 V rated IGBTs that are stacked in a staircase fashion, in multiples of 480 V steps to get the desired voltage rating(s) on the output side of the drive.*"

In the case of phase I undersea oil fields, the oil can be extracted using the internal pressure of the reservoir. The undersea module requires regulation of the pumping speed. The electrical power needs for the undersea control module are more modest at 200–300 W per well [84]. This type of power can be supplied using conventional IGBT-based AC-DC and DC-AC converters operating at 230–600 V. However, the oilfield contains numerous oil wells that must be powered in parallel. To achieve the large cumulative power, it is necessary to use step-up and step-down transformers at the shore and the point of use at the seabed. The power transmission voltage that is the most suitable is 3.3 kV to keep the voltage drop in the power line below 10%.

19.10.3 Athabasca Oil Sands

The Athabasca oil sands in Northern Alberta in Canada hold the second-largest reservoir of oil after Saudi Arabia [85]. The Athabasca oil sands, as shown in Fig. 19.40, contain 1.7 trillion barrels of bitumen with about 10% accessible in a cost-effective manner [86]. This resource is considered favorably by the United States due its close proximity together with a stable and secure partner nation.

The cheapest method for oil extraction is by open pit mining of bitumen deposits located 80–100 m below the surface in the Fort McMurray area, although this constitutes about 6% of the reserve [86]. The other option is steam-assisted gravity

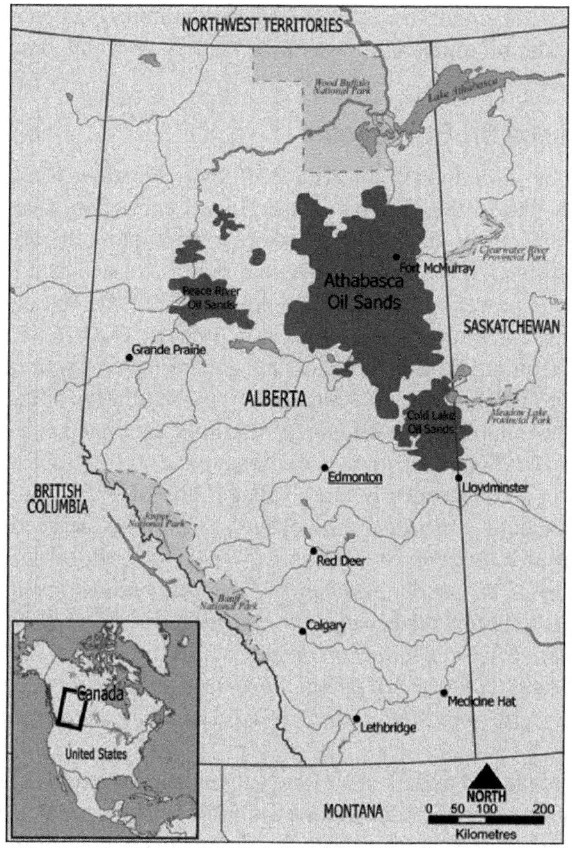

FIGURE 19.40

Athabasca oil sands in Canada.

drainage where steam is injected near the top of the McMurray formation and removed as water with bitumen. The Fort McMurray area has the largest concentration of medium-voltage adjustable-speed drives in the world. They are used to operate the conveyer belts for moving the bitumen, rotary breakers for pulverizing material, and apron feeders that utilize metal belts. Other major loads for the adjustable-speed drives are numerous pumps and fans in the oil fields. Pumps consume 59% of the energy in a typical petrochemical plant. Slurry pumps are used for transporting the oil sands over distances of several miles to containment ponds. It has been demonstrated that energy costs account for 52% of the total cost of ownership of the pumping system. The efficiency gains obtained by using IGBT-based adjustable speed drives, previously discussed in Chapter 10, provide substantial operational cost savings in the oil fields while reducing carbon dioxide emissions.

19.11 Petrochemical Plant

Petrochemicals, produced from petroleum, can be classified as olefins and aromatics [87]. Olefins include ethylene, propylene, and butadiene, which are raw materials for making plastics. Aromatics include benzene, toluene, and xylene, which are raw materials for making dyes and synthetic detergents. These products are used by consumers as engine coolants, polyester fibers, rubbing alcohols, synthetic rubbers, and epoxies.

Petrochemical plants, like the one shown in Fig. 19.41, utilize a large number of pumps, fans, and compressors. The motor drives for these application have a wide range of ratings from 7.5–300 kW [88]. The use of adjustable-speed drives in a petrochemical plant offers many advantages: (1) elimination of the control valve providing a reduction of fugitive emissions of 430 kg per year; (2) elimination of control valve piping with reduced space; (3) increased pumping capacity by the pumps; (4) continuous control of flow; (5) improved reliability; and (6) reduction of power consumption leading to cost savings. An energy savings of 30% has been reported by using adjustable-speed drives consistent with data from industrial applications (see Chapter 10). These adjustable-speed drives all utilize IGBTs in the inverter stage.

19.12 Gas Liquefaction

Gas liquefaction is commonly used by the chemical industry to reduce the size of the product for transportation. One example is liquefied natural gas and others are high-purity gasses used by the semiconductor industry. Gas liquefaction is performed by rapid reduction of the gas pressure. The Joule–Thompson valve has been replaced by a turbine generator to achieve greater efficiency.

FIGURE 19.41

Petrochemical plant.

The liquefaction of methane gas at a facility in the Oman has been reported with cryogenic turbine generators [89]. The operation of squirrel cage motors in a cryogenic ambient has been verified since the 1960s. The motor operates from a 3.3 kV supply using variable-frequency drives with IGBTs. The high switching transients (*dV/dt*) of this IGBT-based drive were found to be acceptable in this application.

19.13 Superconducting Magnetic Storage

Superconducting magnetic storage (SMES) is an energy-storage technology that takes advantage of circulating current in a superconducting coil [90]. The coil is comprised of superconducting material such as Nb/Ti/Cu whose resistance disappears at below the critical temperature. The stored energy can be recovered and delivered to the power grid when desired. The efficiency for SMES systems can be better than 95% as long as the power electronics have low losses.

The SMES system contains the superconducting magnet, the cryostat system to main its low temperature, and a power conversion module for controlling current flow through the magnet. A 1-MJ SMES system requires control of 1500 A of current by the power conversion module. The circuit topology is shown in Fig. 19.42 [91]. IGBTs are used to convert the input AC power to a DC link that feeds the SMES magnet coil. The output from the SMES is converted to the desired AC output power using the IGBT-based DC-to-AC converter. The measured online efficiency for this power storage method is 96%. The input current harmonics are less than 2.5% and the output AC voltage harmonics are less than 3% for this power supply. The same group has extended the SMES capability to 3 MJ [92].

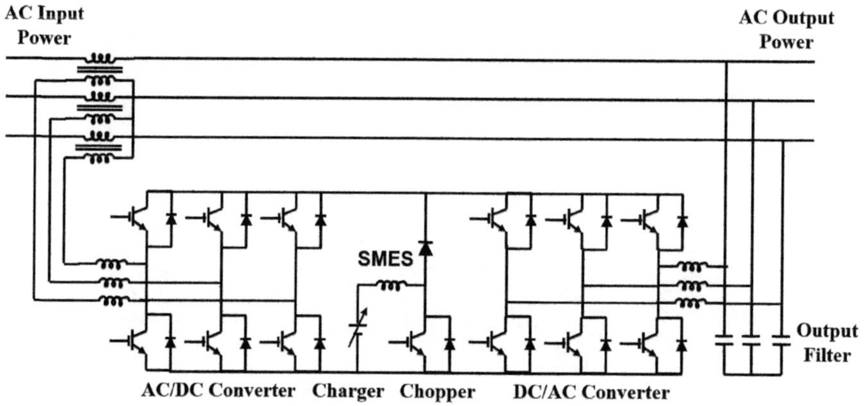

FIGURE 19.42

Power conversion circuit for superconducting magnetic storage.

One option is to cool the power electronics to the cryogenic temperatures used for the SMES energy storage or an intermediate temperature. The electrical characteristics of the IGBT were evaluated as a function of temperature down to 77°K (liquid nitrogen) [93]. Experimental evaluation of the asymmetric IGBT device demonstrated significant improvement in the trade-off curve for on-state and switching losses due to a reduction of both components. This is very promising for increasing the power-handling capability of the IGBTs and reducing the size of the power electronics due to smaller cooling requirements.

The best power conditioning approach for SMES used in conjunction with wind energy has been analyzed [94]. The voltage source converter (VSC) was compared with the current source converter (CSC). The VSC approach was implemented using asymmetric IGBTs with fly-back diodes. The CSC approach was implemented using asymmetric IGBTs with series diodes to provide reverse blocking capability. For a 500 kW system, the CSC approach was found to have 25% lower losses in the devices than for the VSC approach.

19.14 Fusion Power

A large amount of energy can be generated by the fusion reaction, which fuels stars like our Sun. Two light atoms combine by the fusion process to form a heavier nucleus while releasing the binding energy due to the strong nuclear force [95]. The coulombic repulsion between the positively charged nuclei must be overcome for the fusion process to occur. Fusion is favored for nuclei with atomic numbers below that of iron, 56. Lighter atoms, such as helium and deuterium, are usually selected in fusion experiments. The heat generated from the fusion process can be used to produce steam that can run turbines for the generation of electrical power. Fusion power has yet to be commercialized after decades of work.

The energy required to bring the nuclei together to produce the fusion reaction can be generated by increasing the temperature of the atoms beyond the ionization energy to separate the electrons from the nucleus. The temperature of the resulting gas plasma is extremely large, requiring containment by using either a strong magnetic field or lasers. The Tokamak design, shown in Fig. 19.43, allows confinement of the gas plasma within a toroidal shape [96]. The hot plasma moves around the toroid within a magnetic confinement ring. Tokamaks are the most popular designs for fusion power with 177 units in operation around the world [96].

Disruptions can develop when the current in the Tokamak plasma exceeds certain values. To stabilize the plasma, the installation of four pairs of saddle coils was reported inside the Joint European Torus in 1993 [97]. A disruption feedback amplifier was constructed using IGBTs for driving the saddle coils used to produce magnetic fields that prevent disruptions in the gas plasma. The specifications for the power amplifier are an output current of 3 kA and an output voltage of 1500 V with a peak power of 4500 kVA. A block diagram of the amplifier is shown in Fig. 19.44. The input voltage from a 36 kV AC power line is reduced using a transformer and

FIGURE 19.43

Tokamak fusion reactor.

then rectified using a thyristor bridge to create two 650 V DC buses. Twelve parallel inverters using IGBTs are used for each DC bus to generate a high-frequency (up to 10 kHz) output. The IGBTs, with a blocking voltage of 1200 V and a current rating of 200 A, are switched at 4 kHz for high current levels (200–3000-A) and at 10 kHz for low (<200 A) current levels in the saddle coils. Each inverter contains inductor capacitor (LC) filters. The output of the two sets of IGBT inverters drives the saddle coils of the magnets.

A compact power supply design using IGBTs has been reported for the Experimental Advanced Superconducting Tokamak (EAST) [98]. The Tokamak requires numerous power supplies: power supply for ion source filament, ion source arc, ion source gas, ion source Langmuir probe, extracting beam, and bending magnet. The power supplies require high-voltage (100 kV) isolation from ground. The

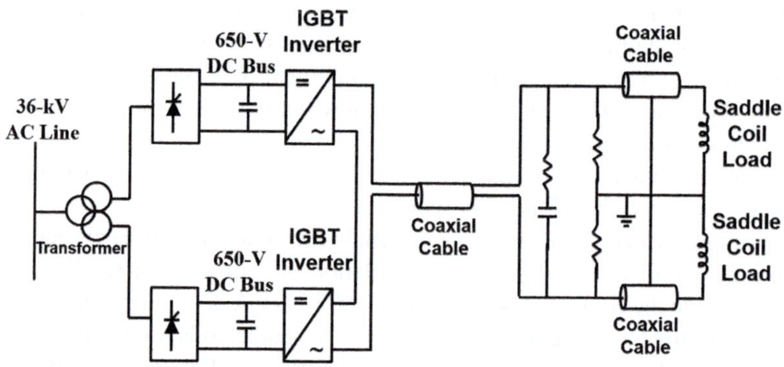

FIGURE 19.44

Disruption feedback amplifier used in the Joint European Torus Tokamak fusion reactor. *IGBT*, insulated gate bipolar transistor.

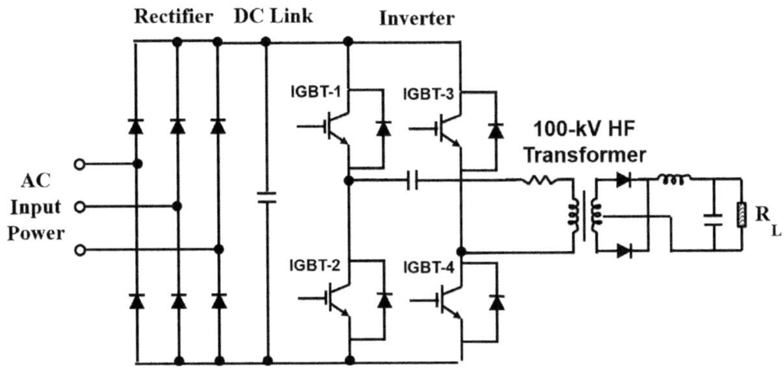

FIGURE 19.45

Compact power supply for Experimental Advanced Superconducting Tokamak reactor. *IGBT*, insulated gate bipolar transistor.

transformer coupling coefficient between the primary and secondary winding is low due to the thick insulation. The size of the power supplies can be made compact by moving up the transformer frequency from 50/60 Hz to 20 kHz by using IGBTs in a resonant circuit to generate the voltage on the primary side. The H-bridge topology for the high-frequency power supplies is shown in Fig. 19.45. Zero-current switching is used to reduce power losses in the IGBTs.

The layout for the EAST power supply is shown in Fig. 19.46. The high-frequency power supply shown in Fig. 19.45 generates the 10–100 kHz power on the primary side of various transformers. These high-voltage transformers provide power to the various loads listed in the previous paragraph. The control of the inner vessel coil to suppress fast time frame plasma disturbances is performed using IGBT-based power supplies. A three-level half-bridge topology is used [99]. Many heating systems are also required to generate the very high plasma temperatures. The IGBT-based power supplies for heating are the 6 MW low hybrid current drive (LHCD), the 2 MW LCRD drive, the 4 MW ion cyclotron range of frequencies (ICRF) drive, the 4 MW electron cyclotron resonance heating (ECRH) drive, and the 4 MW neutral beam injection (NRI) drive.

The advantages of using an IGBT-based LCL converter over the thyristor-based AC/DC converter have been shown for a Chinese fusion engineering test reactor designed for 200 MW operation [100]. It was found by simulations that a PWM-based IGBT converter could be used for 66 kV AC to 2 kV DC with low reactive power compared to the thyristor-based approach.

PEGASUS-III is a new United States-based Tokamak platform. It is a 175 MVA system with 7 MJ of stored energy and a toroidal magnetic field of 0.15–0.6 T. The poloidal field coils, the helicity injector arc currents, and the toroidal field magnets are controlled using buck converters. Thirty-six 4000 A, 900 V IGBTs are employed here [101]. The helicity injector bias voltage and current are controlled by four 10.8

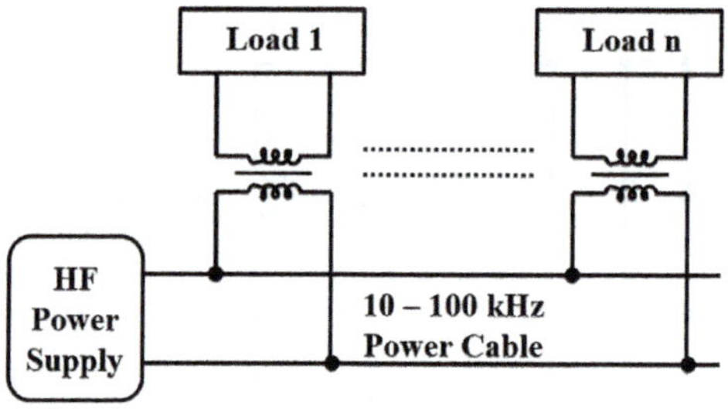

FIGURE 19.46

Tokamak power supply layout.

MVA multilevel buck converters. They use an 1800 V integrated gate commutated thyristor (IGCT) stage in series with a stage containing 900 V IGBTs to provide an injection current of up to 4000 A.

19.15 Standby Power Generators

Power blackouts can occur after severe storms, hurricanes, and earthquakes. Blackouts have also occurred in many countries due to faults on the power transmission lines that have a cascading effect on the power grid. Examples are a 2003 blackout in the United States and Canada that impacted 55 million customers and a blackout in India in 2012 that affected 670 million people [102]. Homeowners often install standby or back-up generators, such as the Generac unit shown in Fig. 19.47, to provide electrical power to the residence until the utility power is restored. The standby or backup generator can be powered by natural gas or propane. It can provide 120/240 V AC power to the house with power levels ranging from 7 to 100 kW.

FIGURE 19.47

Home back-up power generator.

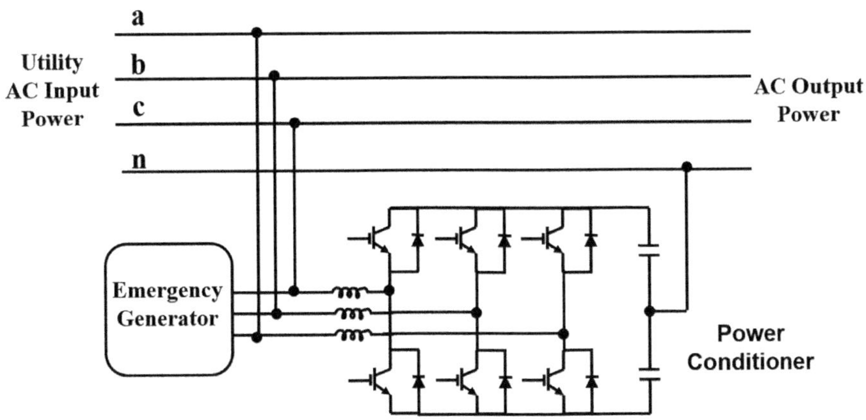

FIGURE 19.48

Insulated gate bipolar transistor-based power conditioner for emergency power generator.

A 100 kVA Korean emergency power generator system has been described [103]. This generator is loaded on a truck to provide power during emergencies. This is a three-phase, four-wire system with a 220 V line to neutral and a 380 V line-to-line voltage for a variety of loads. A major problem occurs when the generator is connected to nonlinear unbalanced loads. Problems with overheating and mechanical stress require operation of the generator at only 30% of its capacity. To overcome this problem, an IGBT-based power conditioner is included in the design [104], as shown in Fig. 19.48. The power conditioner makes the generator current sinusoidal in shape and puts it in phase with the generator voltage. The IGBTs have a voltage rating of 1200 V and a current rating of 600 A. They are operated at a PWM frequency of 12 kHz.

19.16 Roller Coasters

Linear motors are commonly used for people movers or moving walkways, conveyers for luggage delivery at airports, and conveyers for transporting ores and coal. One interesting application for linear motors is for roller coasters. Roller coasters are popular in amusement parks. These thrill rides are found in all parts of the world. The annual attendance for amusement parks is about 300 million people [104]. The advantages of linear induction motors are an absence of chains and belts, which reduces maintenance costs. They also provide the desired fast acceleration for roller coasters.

Cegelec Projects developed the "Superman—The Escape" roller coaster (shown in Fig. 19.49), and provided the specifications for the ride [105], as shown in Fig. 19.50. The linear motor enables achieving the tremendous 2G acceleration at the start of the ride. The passenger vehicle contains permanent magnets that serve

FIGURE 19.49

Superman roller coaster.

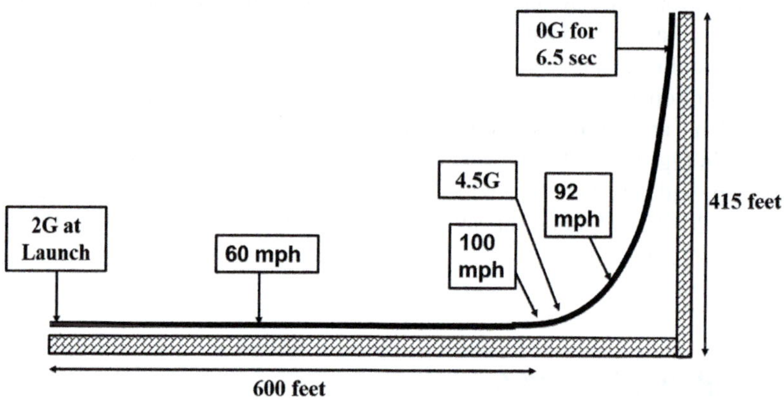

FIGURE 19.50

Acceleration profile for "Superman–The Escape" roller coaster ride.

as the rotor of a conventional motor and the tracks contain windings that serve as the stator in a conventional motor. The torque for accelerating the vehicle is provided by the field generated by the permanent magnets on the vehicle and the field generated by current in the windings on the track. The ride accelerates passengers to 100 mph within 5 s. They experience a force of 4.5 G when rounding a bend and 6.5 s of weightlessness when lifted to a height of 415 ft. The entire ride lasts only 60 s.

The linear motors are driven using a 1.8 MW power supply. Its duty cycle is 7 s at 1.8 MW; 16 s at 0 MW; 5 s at 1.3 MW, and 32 s at 0 MW. The power supply, shown

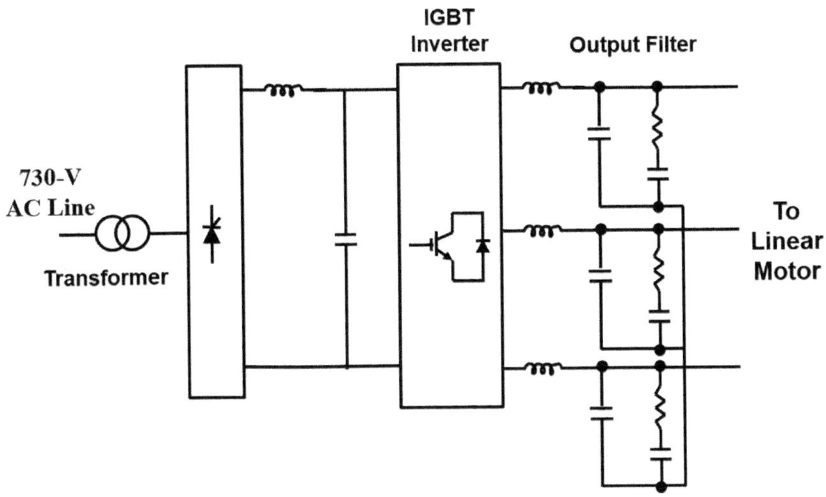

FIGURE 19.51

Linear motor drive for roller coaster. *IGBT*, insulated gate bipolar transistor.

in Fig. 19.51, derives input power from a 730 V AC line. A 780 V DC bus is created using a six-pulse thyristor converter. An inductor is included between the thyristor converter stage and the DC bus capacitor to limit the current ripple to 10%. The Delta IGBT modules used for the inverter stage are self-protected for overcurrent. The stator blocks in the roller coaster tracks have windings spaced at 20 cm. With this spacing, a speed of 100 mph for the vehicle corresponds to a frequency of 230 Hz. The IGBTs are switched at 6 kHz to achieve a 20:1 ratio between the PWM frequency and the motor fundamental frequency. The RC filters at the output of the IGBT inverter keep the dV/dt below 400 V/μs and the peak output voltage less than 1000 V.

19.17 National Aeronautics and Space Administration

The space shuttle program was created by the National Aeronautics and Space Administration (NASA) to develop a reusable low Earth orbit manned spacecraft with large payloads [106]. The first operational flights commenced in 1982 with a total of 135 missions performed until 2011. During the 1980s and 1990s, the space shuttle program was responsible for the launch of numerous satellites, interplanetary probes, and the Hubble Space Telescope. From the late 1990s, the space shuttle focused on the construction and servicing of the International Space Station (ISS).

The space shuttle was launched using its three main engines and its two solid rocket boosters operated in parallel, as shown in Fig. 19.52(a). The three main engines are started at T-minus 6.6 s and reach their 100% thrust level within 3 s. The solid rocket boosters are ignited at T-minus 0 s to produce liftoff from the launch

FIGURE 19.52

(a) Space shuttle launch; (b) space shuttle docked at International Space Station.

pad. At T-plus 126 s, the solid rocket boosters are released by firing pyrotechnic fasteners. The external tank is released at T-plus 350 s. The orbital maneuvering engines are then fired to raise the space shuttle to the upper atmosphere. The orbital maneuvering engines are also used to deorbit the space shuttle when initiating its return to Earth.

The ISS, shown in Fig. 19.52(b) with the space shuttle servicing it, is an artificial satellite designed to allow human habitation for long periods of time [107]. It is the largest satellite in orbit around the earth. The US/ESA (European Space Agency) space station Freedom and the Russian Mir-2 space station were merged in 1993 to form the ISS. It has been continuously occupied since November 2000 to allow crew members to conduct scientific experiments.

The electrical power system is critical for operation of the life support systems in the ISS and for conducting experiments [108]. The electrical power is derived by using an array of solar cells to create a primary peak DC bus voltage of 160 V. Four wing pairs of solar cells generate 138 kW of total power. The loads within the ISS are served using a 124.5 V DC secondary bus. The DC voltage of the secondary bus was chosen taking into consideration the weight of the conductors and the safety of the crew.

19.17.1 Space Shuttle Main Engine Thrust Control

Control of the thrust from the three main engines is crucial during the launch of the space shuttle. A pair of linear actuators is positioned to allow rotation of the engine or the nozzle on its bearings along orthogonal planes to move the resulting thrust vector. The inertia of the engine or nozzle must be overcome using high power level thrust vector control (TVC) systems to achieve sufficient slew rates for stable control of the vehicle during its entire flight trajectory. The hierarchical control system for the

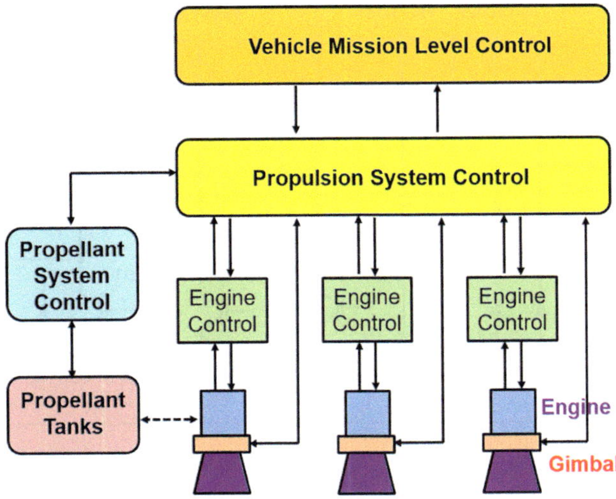

FIGURE 19.53

Space shuttle main engine propulsion system.

main engines is shown in Fig. 19.53 [109]. The vehicle mission level control is used for overall mission planning and adaptive response to payload and crew status. The propulsion system control maintains overall vehicle thrust levels, propellant mixture and usage, and propellant tank pressure. The engine control unit for each of the three main engines sends propulsion commands to its engine. The propulsion system control contains a thrust vector coordinator that determines the thrust for each of the engines and the gimbal angles.

The replacement of hydraulic TVC actuation systems with electromechanical actuators for the space shuttle control and gimbaling systems was reported in 1994 as a lighter, cleaner, safer, and more easily maintained option [110]. The electromechanical actuator design consists of two three-phase brushless DC motors, a two-pass gear reduction system, and a roller screw that translates rotational input to linear output. The motor is driven using an IGBT-based controller. The electronic controller must deliver 27 kW of power (270 V, 100 A) to the TVC actuator. A three-phase, H-bridge circuit with six IGBTs is used to drive the motor with two additional IGBTs used for regenerative braking. The IGBTs have a blocking voltage of 500 V and a current rating of 200 A. The controller met all the specifications for linearity and precision with less error than the legacy hydraulic system. The electronic system eliminates the need for a large number of maintenance personnel to inspect the hydraulic system for leaks, and reduced flight readiness time by 10 days.

TVC for the space shuttle main engines was also reported in 1999 using permanent magnet motors [111]. These motors have extremely high torque and power per unit volume and weight. The permanent magnet motor was designed with balanced three-phase windings. It is a six-pole motor with 13.6 hp. continuous and 40 hp. peak

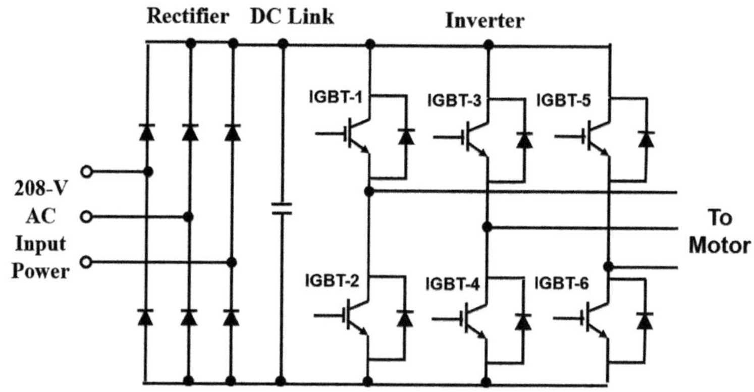

FIGURE 19.54

Space shuttle main engine thrust vector control motor drive.

rating. The IGBT-based motor drive circuit is shown in Fig. 19.54. The IGBTs are controlled using PWM signals. Each IGBT has a blocking voltage of 600 V and a current handling capability of 400 A. It enables position control without substantial ripple.

By 2012, electric thrust vector control (ETVC) systems were recognized to reduce launch costs and maintenance complexity [112]. Application of this approach to large manned vehicle projects was considered viable due to advances in several enabling technologies. The authors state: "*Among these maturing technologies, high power switching electronics such as the IGBT in particular have made it possible to further simplify actuator mechanisms and eliminate certain mechanical failure modes.*" The extensive experience with using IGBTs for motor control and regenerative braking in the electric vehicle industry was considered to be a strong motivation to apply IGBTs to TVC actuators. The authors state: "*High-power solid-state switches such as the IGBT make it possible to handle high levels of excess kinetic energy electronically through motor torque.*"

19.17.2 Space Shuttle Orbital Maneuvering System

The ETVC system was also found to be suitable for the space shuttle orbital maneuvering system (OMS) [113]. The OMS provides the thrust required for the space shuttle to perform many important "in-space" functions: (1) orbit insertion; (2) orbit circularization; (3) orbit transfer; (4) rendezvous; and (5) deorbit.

19.17.3 Space Shuttle Power Distribution

The space shuttle had an electrical power distribution system with three main bus lines, each with a fuel cell delivering DC power at 28 V [113]. Nine inverters were used in the space shuttle to convert the DC voltage to 117 V AC power at 400 Hz. These inverters were built using IGBTs.

19.17.4 International Space Station Power Distribution

The ISS, shown in Fig. 19.52B, evolved from the US space station Freedom. The design of the power distribution network on Freedom was reported in 1991 [114]. The primary power source for the space station is a 160 V DC from its solar array. The secondary power distribution is a 120 V DC bus that was specially adopted for this case due to the higher power levels compared with previous missions that used 28 and 50 V DC buses. The choice of 120 V DC was a compromise between reducing the mass of the power lines by reducing the currents and other considerations such as corona discharge and crew safety. The power delivery was anticipated to be 75 kW, which is an order of magnitude larger than previous spacecraft.

A block diagram of the power system for the space station is shown in Fig. 19.55. The power is generated from photovoltaic panels to create a 160 V DC bus. The energy from the solar cells is stored in batteries for use when the space station is blocked from the sun due to the shadow of the Earth during its regular orbits. DC–DC converters using IGBTs are needed to form the 120 V DC secondary power distribution bus. Power is fed to a variety of loads from this 120 V DC bus. Multiple secondary distribution buses are utilized in parallel. Each secondary bus is rated at 12 kW to keep the DC bus current at 100 A. The remote power controller (RPC) and remote power interrupter (RPI) units are needed for fault protection and disconnection from the power bus. The RPCs and RPIs can be implemented using vacuum relays or solid-state switches such as IGBTs (see Section 14.2.4 for solid-state breakers based on IGBTs). The power conditioners, load controllers, and motor controllers are

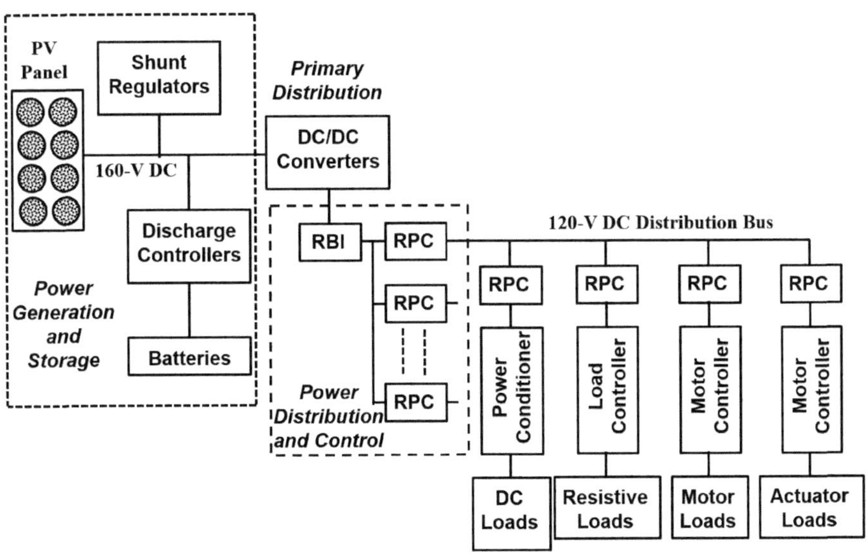

FIGURE 19.55

Space station power system. *RPC*, remote power controller; *RBI*, remote power interrupter.

IGBT-based inverters discussed in previous chapters that make use of PWM signals or resonant topologies.

An updated description of the low-voltage DC microgrid employed for the ISS has been published [115]. In the ISS, the US portion operates from a 120 V DC grid while the smaller Russian portion runs off a 28 V DC grid. The solar array wings (SAW) deliver 31 kW of power at 160 V. Batteries provide power at 150 V during solar eclipse intervals. Zero-voltage switching (ZVS) was implemented for the IGBTs to reduce switching losses.

19.17.5 Manned Interplanetary Missions

Manned missions to Mars have been proposed and analyzed many times over the years [116]. Most recently, NASA's Project Constellation in 2004 included the Orion Mars Mission for manned space exploration. It is estimated that the travel time to Mars from Earth will be 180 days and the shelf-life for the food supply would have to exceed 3 years [117]. The space shuttle used forced-convection ovens to heat food to edible temperatures with a low efficiency. Ohmic heating provides more uniform heating with better efficiency. A lightweight and compact ohmic food warming unit has been reported for replacing the convection ovens [93].

It has been determined that electrochemical reactions can be minimized by using a square wave at a frequency of 10 kHz to heat the food contained in a pouch designed with electrodes. A block diagram of the ohmic food warmer is shown in Fig. 19.56. The input power is boosted to create a high-voltage DC bus. IGBT modules were used to generate the 10 kHz square wave power for heating the food. Experiments demonstrated the ability to heat food to serving temperatures in 25 min at a power level of 200 W.

19.17.6 Cryogenic Power Electronics

NASA has been interested in the development of power electronics that can operate at cryogenic temperatures for potential applications in missions to the outer planets in the solar system. A comprehensive characterization of a variety of power devices was undertaken in the 1990s at the Power Semiconductor Research Center at temperatures down to liquid nitrogen (77 K) [93]. This range was

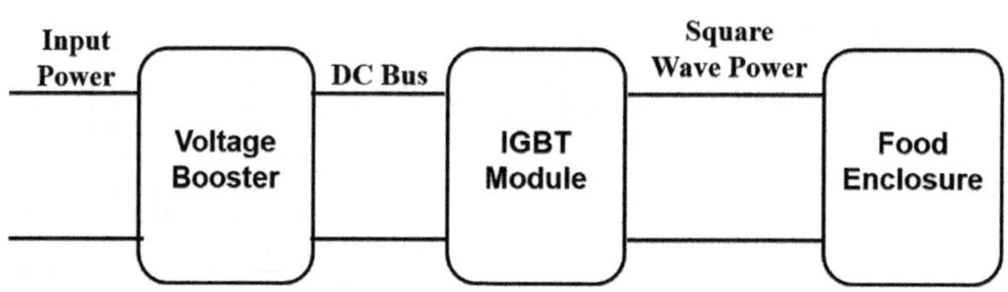

FIGURE 19.56

IGBT-based ohmic food warmer. *IGBT*, insulated gate bipolar transistor.

selected for potential use of the power electronics with high-temperature superconductors. In this work, it was demonstrated that both the on-state voltage drop and the turn-off switching time for IGBTs reduce with decreasing temperature. This makes the trade-off curve between on-state voltage drop and the turn-off energy much better at 77 K when compared with room temperature. Consequently, it was established that IGBTs are very good candidates for power electronic applications at cryogenic temperatures.

The characterization of IGBTs at even lower temperatures down to the liquid helium level (4 K) was reported in 2004 [118]. It was found that new effects become important when the temperature is reduced below 77 K. First, carrier freeze out becomes an issue. Second, the carrier mobilities begin to decrease below 77 K rather than increasing with temperature reduction. These phenomena produce a significant increase in the on-state voltage drop, creating difficulties with turning on the IGBT at below 30 K. The switching losses also begin to increase at below 60 K. The authors found that NPT (or symmetric blocking) IGBTs performed better at very low temperatures when compared with punch-through (asymmetric blocking) IGBTs.

19.17.7 IGBT Failure Analysis

NASA has supported internal and external studies on the failures of IGBTs under various electrical stresses to understand the reliability of the devices for space applications [119]. One of the first devices that was examined in these studies is the IGBT because of the relevance to health management and prognostic systems for power electronics in avionics. These systems allow anticipation of failures and the prediction of remaining life. The authors state: "*The development and advancement of this capability is also relevant to multiple NASA Exploration Systems Mission Directorate (ESMD) vehicles, including the Orion, Ares, and future vehicles, such as the Lunar Surface Access Module (LSAM). In addition, there is relevance to long-endurance robotic space missions from ESMD and Space Mission Directorate.*"

The development of an accelerated aging test capability at NASA has been reported for evaluation of IGBTs [120]. This test system can examine dielectric breakdown and failure due to thermal cycling. Dielectric breakdown refers to damage to the gate oxide of IGBTs due to electrostatic discharge. In addition, time-dependent dielectric breakdown can occur due to the generation of hot electrons in the gate oxide. Thermal cycling failures are associated with liftoff of wire bonds on the IGBT chips and desoldering or voids in the solder. These reliability problems for IGBTs were previously discussed in Section 6.7. The damage produced by a variety of ESD stresses to IGBTs has been collected using this measurement system for analysis [120].

Failure precursors are useful for prognosticating the life of IGBTs [121]. The authors state: "*These components form the backbone of avionics systems that play an ever-increasing critical role in onboard autonomous functions for vehicle controls, communications, navigation, and radar systems.*" Thermal overstress accelerated aging tests on IGBTs were performed at about 300°C. It was observed that the device approaches a latch-up condition after 210 min of switching time. Further tests performed on IGBTs reported increased charge in the gate oxide and a decrease in the on-state voltage drop of the IGBT after the stress [122]. An increase in on-state voltage drop eventually occurred, correlated with degradation of the die attach.

Predicting the remaining useful life (RUL) of IGBTs is critical to the missions sponsored by NASA. Accelerated aging tests of IGBTs have been conducted by NASA while monitoring various device parameters, such as the threshold voltage, on-state voltage drop, switching times, etc. Using this data, the back propagation neural network (BPNN) method was found to provide the best performance for obtaining the RUL for IGBTs [123].

19.18 Summary

Many applications of the IGBT have been described in this chapter that do not naturally fit into any of the different sectors of the economy covered in previous chapters. This chapter provides further validation of the breadth of the impact of this important power device technology in our society. Even more applications of the IGBT are being continually reported, indicating a bright future for this technology in spite of having been invented more than 40 years ago and continually enhanced and utilized since that time.

References

[1] L. Jiang, D.-Y. Liu, B. Yang, Smart home research, in: IEEE International Conference on Machine Learning and Cybernetics, vol. 2, 2004, pp. 659–663.

[2] D.-M. Han, J.-H. Lim, Smart home energy management system using IEEE 802.15.4 and ZigBee, IEEE Trans. Consum. Electron. 56 (2010) 1403–1410.

[3] Y. Zhao, et al., Research and thinking of friendly smart home energy system based on smart power, in: IEEE International Conference on Electrical and Control Engineering, 2011, pp. 4649–4654.

[4] T.-S. Kwon, et al., Development of new smart power module for home appliance motor drive applications, in: IEEE International Electric Machines and Drives Conference, 2011, pp. 95–100.

[5] Photocopier. http://en.wikipedia.org/wiki/Photocopier.

[6] L. Gamage, Series resonant high frequency inverter with zero current switching pulse density modulation for induction heated load, in: IEEE Ind. Electron. Soc. Meet, 2003, pp. 1739–1744.

[7] H. Sugimura, et al., Series load resonant tank high frequency inverter with ZCS-PDM control scheme for induction-heated fixing roller, in: IEEE International Conference on Industrial Technology, 2005, pp. 756–761.

[8] J.E. James, et al., Improved AC pickups for IPT systems, IEEE Trans. Power Electron. 29 (2014) 6361–6374.

[9] S.Y. Choi, et al., Asymmetrical coil sets for wireless stationary EV chargers with large lateral tolerance by dominant field analysis, IEEE Trans. Power Electron. 29 (2014) 6406–6420.

[10] S. Rosa, et al., A dynamic wireless charging system for electric vehicles based on DC/AC converters with SiC MOSFET-IGBT switches and resonant gate-drive, in: IEEE Meeting of the Industrial Electronics Society, 2016, pp. 4465–4470.

[11] Z. Zhong, et al., Review on solid-state-based Marx generators, IEEE Trans. Plasma Sci. (2021) 3625–3643.

[12] X-Ray Generator. http://en.wikipedia.org/wiki/X-ray_generator.

[13] W. Luo, Z. Wang, Development of electric control high power medical use X-ray generator, in: IEEE International Conference on Information Engineering and Computer Science, 2009, pp. 1–5.

[14] J. Chen, P. Evans, X. Wang, Enhanced color coding scheme for kinetic depth effect X-ray (KDEX) imaging, in: IEEE International Carahan Conference on Security Technology, 2010, pp. 155–160.

[15] Pulsed Power. http://en.wikipedia.org/wiki/Pulsed_power.

[16] Marx Generator. http://en.wikipedia.org/wiki/Marx_generator.

[17] S.K. Ghandhi, VLSI Fabrication Principles, John Wiley, New York, 1994.

[18] Ion Implantation. http://en.wikipedia.org/wiki/Ion_implantation.

[19] C.J.T. Casper, M.P. Bradley, Active charge/discharge IGBT modulator for Marx generator and plasma applications, IEEE Trans. Plasma Sci. 35 (2007) 473–478.

[20] J.-H. Kim, et al., High voltage pulsed power supply using IGBT stacks, IEEE Trans. Dielectr. Electr. Insul. 14 (2007) 921–926.

[21] K. Liu, et al., An all solid-state pulsed power generator based on Marx generator, in: IEEE Pulsed Power Conference, 1, 2007, pp. 720–723.

[22] L.M. Redondo, et al., Solid-state Marx type modulator for plasma ion implantation applications, in: IEEE Pulsed Power Conference, 2011, pp. 1326–1329.

[23] M. Schmeide, M. Kokot, Application of a PEM hydrogen gas generator as reliable gas source for hydrogen implantation, in: IEEE Int. Conf. on Ion Implantation Technology, 2018, pp. 243–246.

[24] K.N. Kumari, et al., High voltage high frequency pulse generator for cancer cell treatment application, in: IEEE Int. Conf. Electrical Energy Systems, 2021, pp. 654–658.

[25] Particle Physics. http://en.wikipedia.org/wiki/Particle_physics.

[26] Particle Accelerator. http://en.wikipedia.org/wiki/Particle_accelerator.

[27] List of Accelerators in Particle Physics. http://en.wikipedia.org/wiki/List_of_accelerators_in_particle_physics.

[28] Tevatron. http://en.wikipedia.org/wiki/Tevatron.

[29] Large Hadron Collider. http://en.wikipedia.org/wiki/Large_Hadron_Collider.

[30] R.J. Richter-Sand, R.J. Adler, High voltage high power klystron drivers using flexible solid state IGBT modules, in: IEEE International Power Modulator Symposium, 2000, pp. 165–170.

[31] C. Pappas, R. Cassel, An IGBT driven slotted beam pipe kicker for SPEAR III injection, in: IEEE International Particle Accelerator Conference, 2001, pp. 3747–3749.

[32] International Linear Collider. http://en.wikipedia.org/wiki/International_Linear_Collider.

[33] Japan Inches Forward with Plans to Host Next Big Particle Collider. https://www.scientificamerican.com/article/japan-inches-forward-with-plans-to-host-next-big-particle-collider/.
[34] K.J.P. Macken, et al., IGBT PEBB technology for future high energy physics machine operation applications, in: IEEE Applied Power Electronics Conference, 2011, pp. 1319–1337.
[35] S. Fang, et al., Four quadrant 250 kW switchmode power supply for Fermilab main injector, in: IEEE International Particle Accelerator Conference, 1999, pp. 3761–3763.
[36] Y. Yamazaki, The Japan Hadron Facility Accelerator, 2022. http://www.cern.ch/accelconf/a98/APAC98/5A002.pdf.
[37] M. Muto, et al., Highly-performed power supply using IGBT for synchrotron magnets, in: IEEE Particle Accelerator Conference, 5, 1999, pp. 3770–3772.
[38] S.L. Hays, B. Claypool, G.W. Foster, The 100000 amp DC power supply for a stages hadron collider superferric magnet, IEEE Trans. Appl. Supercond. 16 (2006) 1626–1629.
[39] A. Fowler, The injection bumper system for LEIR, in: IEEE Power Modulator Symposium, 2004, pp. 433–436.
[40] E. Cartier, L. Ducimetiere, E. Vossenberg, A kicker pulse generator for measurement of the tune and dynamic aperture in the LHC, in: IEEE Power Modulator Symposium, 2006, pp. 463–466.
[41] V. Peplov, et al., Progress in development of pulsed resonant charging power supply for the Spallation Neutron Source extraction kicker PFN system, in: IEEE Pulse Power Conf, 2021, pp. P1–P6.
[42] Pulsed Laser. https://en.wikipedia.org/wiki/Pulsed_laser.
[43] Laser-tissue interaction. https://www.l.com/educatiolightscalpen/surgical-co2-laser-tissue-interaction/.
[44] Pulsed Dye Lasers. https://www.bcm.edu/healthcare/specialties/dermatology/laser-surgery/laser-treatment-with-pulsed-dye-laser.
[45] Treatment of Spider Veins with the 595 nm Pulsed-Dye Laser. https://pubmed.ncbi.nlm.nih.gov/9810891/.
[46] M. Giesselmann, E. Kristiansen, Compact design of a 30 kV rapid capacitor charger, in: IEEE Pulsed Power Plasma Science Conference, 2001, pp. 640–643.
[47] A. Kluge, et al., IGBT-based switching modules for laser applications, IEEE Trans. Dielectr. Electr. Insul. 22 (2015) 1954–1962.
[48] Food Preservation. http://en.wikipedia.org/wiki/Food_preservation.
[49] Electroporation. http://en.wikipedia.org/wiki/Electroporation.
[50] User:Openman/Pulsed Electric Field Processing. http://en.wikipedia.org/wiki/User:Openman/Pulsed_Electric_Field_Processing.
[51] Sterlization System Uses Pulsed Electric Field Technology. http://news.thomasnet.com/fullstory/Sterlization-System-uses-pulsed-electric-field-technology.
[52] M. Gaudreau, et al., Solid-state power systems for pulsed power electric field (PEF) processing, in: IEEE Pulsed Power Conference, 2005, pp. 1278–1281.
[53] SteriBeam Systems—PUV and PEV Sterilization and Nutrient Extraction Enrichment Systems. http://www.Pharmaceutical-business-review.com/steriBeam_systems_puv_and_pev_sterilization_and_nutrient_extraction_enrichment_systems.
[54] Kinetics of Microbial Inactivation for Alternative Food Processing Technologies—Pulsed Electric Fields. http://www.fda.gov/Food/FoodScienceResearch/SafePracticesforFoodProcesses/ucm101662.htm.

[55] H.A. Prins, et al., Solid state pulsed plasma power source for pulsed electric field and plasma treatment of food products, in: IEEE Pulsed Power Plasma Science Conference, vol. 2, 2001, pp. 1294–1297.

[56] M.S. Moonesan, J.F. Zhang, S.H. Jayaram, IGBT based HV pulse generator for high conductivity liquid food treatment, in: IEEE Pulse Power Conference, 2011, pp. 1160–1164.

[57] T. Sakugawa, et al., Fast rise time pulsed power generator using IGBTs and coaxial magnetic pulse compression circuit, in: IEEE Pulse Power Conference, 2011, pp. 140–145.

[58] Ocean. http://en.wikipedia.org/wiki/Ocean.

[59] Water Resources. http://en.wikipedia.org/wiki/Water_resources.

[60] Water Treatment. http://en.wikipedia.org/wiki/Water_treatment.

[61] P.T. Johnstone, P.S. Bodger, Applications of a high-voltage SMPS in water disinfection, IEE Proc. Sci. Meas. Technol. 148 (2001) 41–45.

[62] Desalination. http://en.wikipedia.org/wiki/Desalination.

[63] Meet 10 of the World's Largest Desalination Plants. https:/www.aquatechtrade.com/news/desalination/worlds-largest-desalination-plants/.

[64] G. Ruan, et al., Progress and perspectives of desalination in China, Membranes (2021) P1–P18.

[65] S. Baron, The economics of desalination, IEEE Spectr. 3 (1966) 63–70.

[66] Reverse Osmosis. http://en.wikipedia.org/wiki/Reverse_osmosis.

[67] W. Qi, P.D. Christofides, Supervisory predictive control for long term scheduling of an integrated wind/solar energy generation and water desalination system, IEEE Trans. Control Syst. Technol. 20 (2012) 504–512.

[68] W. Gu, X. He, Development of an innovative seawater desalination system using non-grid-connected wind power, in: IEEE World Non-grid-connected Wind Power and Energy Conference, 2009, pp. 1–4.

[69] X. He, C. Li, W. Gu, Research on an innovative large-scale offshore wind power seawater desalination system, in: IEEE World Non-grid-connected Wind Power and Energy Conference, 2010, pp. 1–4.

[70] B.S. Saravana, K. Srinivasan, R.B. Jeyapradha, Future prospects of non-grid connected wind power in India, in: IEEE India Conference, 2011, pp. 1–4.

[71] R. Nagaraj, Renewable energy based small hybrid power system for desalination applications in remote locations, in: IEEE India Conference on Power Electronics, 2012, pp. 1–5.

[72] I.B. Ali, et al., Energy management of a reverse osmosis desalination process powered by renewable energy sources, in: IEEE Electrotechnical Conference, 2012, pp. 800–805.

[73] Y. Dahioui, K. Loudiya, Wind powered water desalination, in: IEEE Renewable and Sustainable Energy Conference, 2013, pp. 257–262.

[74] A. Viola, et al., Sea water desalination and energy consumption: a case study of wave energy converters (WEC) to desalination applications in sicily, in: IEEE Oceans Conf, 2016, pp. P1–P4.

[75] Z. Haidar, et al., Role of RO desalination plants in renewable energy integration in electric systems, in: IEEE Int. Renewable Energy Congress, 2020, pp. P1–P4.

[76] F. Mohammadi, et al., Can desalination be an economically viable solution for water scarcity? in: IEEE Global Humanitarian Technology Conference, 2019, pp. P1–P4.

[77] S.R. Jang, et al., Application of pulsed power system for water treatment of the leachate, in: IEEE Pulse Power Conference, 2009, pp. 980–983.

[78] M.S. Mazzola, A solid-state modulator for environmental applications of pulsed acoustics, in: IEEE Pulsed Power Conference, vol. 1, 1999, pp. 458–459.
[79] J. Rao, J. Kolb, All solid-state rectangular pulse generator for corona discharge in water, in: IEEE Int. Conf. on Pulsed Power, 2017, pp. P1–P4.
[80] S. Bai, et al., The application of AC-DC-AC invert system in the oil extracting under the well of oil Fields, in: IEEE International Conference on Electrical Machines and Systems, 2003, pp. 383–385.
[81] J.M. de Bedout, An industrial perspective on next generation power conversion, in: IEEE Industrial Electronics Conference, 2014, p. 67.
[82] S. He, Y. Zhang, A novel heating system for heavy oil extraction, in: IEEE International Conference on Consumer Electronics, Communications, and Networks, 2012, pp. 2292–2295.
[83] M.F. Taylor, Conceptual design for sub-sea power supplies for extremely long motor lead applications, in: IEEE Petroleum and Chemical Industry Conference, 1998, pp. 119–128.
[84] S. Xiao, et al., Design of sub-sea long distance electric power supply system, in: IEEE International Conference on Electric Utility Deregulation and Restructuring and Power Technologies, 2011, pp. 1760–1763.
[85] M. Throckmorton, R. Paes, W.C. Livoti, Adjustable speed drives as related to the oil sands process, in: IEEE Petroleum and Chemical Industries Technical Conference, 2008, pp. 1–9.
[86] Athabasca Oil Sands. http://en.wikipedia.org/wiki/Athabasca_oil_sands.
[87] Petrochemical. http://en.wikipedia.org/wiki/Petrochemical.
[88] K.W.L. Burbridge, Users view of variable speed drives on petrochemical plant, in: IEEE International Conference on Electrical Machines and Drives, 1993, pp. 582–587.
[89] R. Shively, H. Miller, Development of a submerged winding induction generator for cryogenic applications, in: IEEE International Symposium on Electrical Insulation, 2000, pp. 243–246.
[90] Superconducting Magnetic Energy Storage. http://en.wikipedia.org/wiki/Superconducting_magnetic_energy_storage.
[91] K.C. Seong, et al., Design and test of a 1-MJ SMES system, IEEE Trans. Appl. Supercond. 12 (2002) 391–394.
[92] H.J. Kim, et al., 3 MJ/750 kVA SMES system for improving power quality, IEEE Trans. Appl. Supercond. 16 (2006) 574–577.
[93] R. Singh, B.J. Baliga, Cryogenic Operation of Silicon Power Devices, Kluwer Academic Publishers, 1998.
[94] A. Rodriguez, et al., Analysis and performance comparison of different power conditioning systems for smes-based energy systems in wind turbines, in: IEEE Int. Symp. Power Electronics for Distributed Generation Systems, 2012, pp. 527–533.
[95] Fusion Power. http://en.wikipedia.org/wiki/Fusion_power.
[96] Tokamac. http://en.wikipedia.org/wiki/Tokamac.
[97] S.M. Tenconi, et al., High power, wide bandwidth linear switching amplifier using IGBTs, in: IEEE Ind. Appl. Soc. Meet, vol. 2, 1993, pp. 778–784.
[98] G. Li, Compact power supplies for tokamak heating, IEEE Trans. Dielectr. Electr. Insul. 19 (2012) 233–238.
[99] P. Fu, et al., Power supply system of EAST superconducting Tokamak, in: IEEE Conf. Industrial Electronics and Applications, 2010, pp. 457–462.

[100] H. Li, G. Li, A LCL converter for Tokamak, in: IEEE Int. Power Modulator and High Voltage Conf, 2014, pp. 615–618.

[101] M.W. Bongard, et al., Digital control and power systems for the PEGASUS-III experiment, IEEE Trans. Plasma Sci. (2022), https://doi.org/10.1109/TPS.2022.3165694. Early Access. Digital Object Identifier.

[102] K.A. Al-Salim, I. Andonovic, C. Michie, A cyclic blackout mitigation system, in: IEEE International Energy Conference, 2014, pp. 520–527.

[103] C.Y. Jeong, et al., A 100 kVA power conditioner for three-phase four-wire emergency generators, in: IEEE Power Electronics Specialists Conference, vol. 2, 1998, pp. 1906–1911.

[104] S. Lansky, Steep thrills, Time Mag. (2014) 43–44.

[105] N.J. Elliott, Novel application of a linear synchronous motor drive, in: IEE Colloquium New Power Electron. Techn, 1997, pp. 8/1–8/5.

[106] Space Shuttle. http://en.wikipedia.org/wiki/SpaceShuttle.

[107] International Space Station. http://en.wikipedia.org/wiki/InternationalSpaceStation.

[108] Electrical System of the International Space Station. http://en.wikipedia.org/wiki/Electrical_system_of_the_International_Space_Station.

[109] K. Redmill, et al., The design of intelligent hierarchical controllers for a space shuttle vehicle, in: IEEE International Symposium on Intelligent Control, 1993, pp. 64–69.

[110] J.R. Cowan, R.A. Weir, Design and test of electromechanical actuators for thrust vector control, NASA Rep. (1994) N94–N29650.

[111] T.A. Haskew, D.E. Schinstock, E.M. Waldrep, Two-phase on drive operation in a permanent magnet synchronous machine electromechanical actuator, IEEE Trans. Energy Convers. 14 (1999) 153–158.

[112] L.B. Bates, D.T. Young, Developmental testing of electric thrust vector control systems for manned launch vehicle applications, in: 41st Aerospace Mechanisms Symposium, Jet Propulsion Laboratory, 16–18 May, 2012.

[113] F. Berry, Steady state mathematical model for the DC-AC inverters on the space shuttle, in: IEEE Southeastcon Proceedings, vol. 2, 1989, pp. 455–458.

[114] S. Krauthamer, M. Gangal, R. Das, State-of-the-art of DC components for secondary power distribution of space station freedom, IEEE Trans. Power Electron. 6 (1991) 548–561.

[115] W. Chen, et al., Real-time ML-assisted hardware-in-the-loop electro-thermal emulation of LVDC microgrid on the international space station, IEEE Open J. Power Electron. (2022) 168–181.

[116] Manned Mission to Mars. http://en.wikipedia.org/wiki/Manned_mission_to_Mars.

[117] R.B. Pandit, et al., Development of a light weight Ohmic food warming unit for a Mars exploration vehicle, World Food Sci. 2 (2007) 1–15.

[118] A. Caiafa, et al., IGBT operation at cryogenic temperatures: non-punch-through and punch-through comparison, in: IEEE Power Electronics Specialists Conference, vol. 4, 2004, pp. 2960–2966.

[119] P. Wysocki, et al., Effect of electrostatic discharge on electrical characteristics of discrete electronic components, in: Annual Conference on Prognostics and Health Management Society, 2009, pp. 1–10.

[120] G. Sonnenfield, et al., An agile accelerated aging characterization and scenario simulation system for gate controlled power transistors, in: IEEE Autotestcon, 2008, pp. 208–215.

[121] B. Saha, et al., Towards prognostics for electronics components, in: IEEE Aerospace Conference, 2009, pp. 1–7.

[122] N. Patil, et al., Precursor parameter identification for insulated gate bipolar transistor (IGBT) prognostics, IEEE Trans. Reliab. 58 (2009) 271–276.

[123] C. He, et al., Machine learning based prognostics for predicting remaining useful life of IGBT—NASA IGBT accelerated ageing case study, in: IEEE Information Technology, Networking, Electronic and Automation Control Conference, 2021, pp. 1357–1361.

CHAPTER

IGBT Social Impact

20

The United Nations Intergovernmental Panel on Climate Change (IPCC) issued a report in 2021 defining the impact of human activity on the climate. Some key conclusions from the report are: (a) greenhouse gases (GHGs, primarily carbon dioxide and methane) produced by human activity are responsible for a rise in global temperatures; (b) the carbon dioxide and methane concentrations rose to 410 ppm and 1866 ppm, respectively, by 2019; (c) global surface temperatures have already risen by about 1°C due to these pollutants; (d) human activity is the main driver for the global retreat of glaciers; (e) global ocean temperatures have risen due to GHGs; (f) the mean global sea level has risen by 3.7 mm per year from 2006 to 2018 due to global warming; (g) atmospheric CO_2 concentrations in 2019 reached a higher value than in the past 2 million years; (h) heat waves, leading to more intense and widespread forest fires, are correlated with global warming; and (i) the global occurrence of intense (category 3–5) cyclones or hurricanes has increased during the past four decades due to global warming. The report concludes that global average temperatures will rise by 1.5–2°C during this century unless strong measures are undertaken to reduce and eliminate GHG emissions. Mitigating this increase in global temperature is an urgent priority for all nations to prevent severe degradation of our living environment, food production, health, and safety. This chapter documents IGBT-enabled innovations in technology that have already contributed to a reduction of GHG emissions during the past four decades. IGBT-based technologies will also play a major role in our ability to bring GHG emissions to net zero, the aspirational goal of the report. These technologies can be broadly classified as: (1) more efficient utilization of energy, (2) generation of electricity with zero emissions, and (3) transportation with zero emissions. The IGBT plays a critical role in achieving the global aspiration for net zero emissions by 2050.

In 2009, Time Magazine published an article [1] on the impact of energy efficiency on the US economy. The article states: "*This may sound too good to be true, but the (United States) has a renewable energy source that is perfectly clean, remarkably cheap, surprisingly abundant, and immediately available ⋯ This miracle juice goes by the distinctly boring name of energy efficiency*." In this article, Steven Chu, Energy Secretary at that time, states: "*I cannot impress upon you how important energy efficiency is*." The article concludes that energy efficiency is the only cost-effective "energy source" that addresses global warming, energy independence, and volatile prices. A strong argument for improving energy efficiency is also made by Lovins and Cohen in their book [2] *Climate Capitalism*. Chapter 2 in the book is

The IGBT Device. https://doi.org/10.1016/B978-0-323-99912-0.00016-7

titled "Energy Efficiency: Low-Hanging Fruit that Grows Back." They quote $130 billion annually from energy savings according to a 2007 McKinsey study. Small businesses consume half the electricity in the United States and can save 20–30% on their energy bills by using energy-efficient appliances.

The global carbon footprint at present is about four tons or 8000 pounds per capita. This corresponds to a global carbon dioxide emission of 56 trillion pounds due to human activity, assuming a population of 7 billion people. The carbon footprint for selected nations around the world is shown in Fig. 20.1 circa 2014. Developed nations, such as the United States, Australia, and Canada, have a relatively large carbon footprint of about 20 tons per year. The large carbon footprint is produced by the consumption of significant amounts of energy resources, such as gasoline and electricity, to provide convenient transportation using cars and trucks as well as to provide comforts such as air conditioning and refrigeration. In contrast, underdeveloped nations such as India, Brazil, and China have a relatively low carbon footprint of about 2 tons per year. The population of the underdeveloped nations far outstrips the population of the developed countries. It can be expected that all the citizens of the underdeveloped nations will strive to achieve the same quality of life as the developed nations. In fact, the per capita emissions from China have now doubled from 2.5 to 5 tons since 2014 [3]. Unless our energy delivery is performed with greater efficiency, it can be anticipated that the global carbon footprint will rise to even more than 20 tons per year in the future.

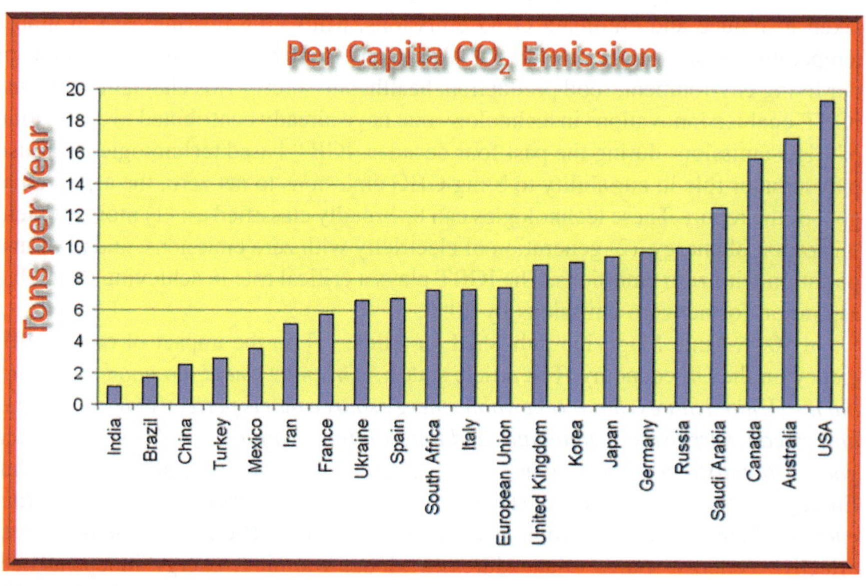

FIGURE 20.1

Annual per capita carbon footprint of selected nations.

Since the 1980s, insulated gate bipolar transistors (IGBTs) have played a critical role in the development of products with improved energy efficiency. This section highlights specific examples of energy savings derived by using IGBT-enabled products in the transportation, consumer, industrial, and lighting sectors of the economy. Based upon the energy savings derived with IGBT-enabled appliances, it is possible to compute the cost savings derived by consumers. In addition, reduced energy consumption eliminates the need for the generation of electricity from fossil-fuel power plants. This in turn reduces investments in generation capacity by the utility industry and eliminates a very large amount of carbon dioxide emissions.

The adverse impact of global warming on the climate has raised alarm bells around the world with a strong public outcry for action. It has been established that global warming is caused by the man-made generation of carbon dioxide. A consensus has now developed that all fossil fuel-based vehicles must be replaced by electric vehicles and that all fossil fuel-based electricity generation should be replaced by renewable energy generation. The IGBT plays an essential role in the implementation of both strategies, as discussed in this chapter.

20.1 Electronic Ignition System

Most cars and trucks around the globe, such as the ones shown in Fig. 20.2, rely upon the gasoline-powered internal combustion engine for powering the vehicles. According to the US Energy Information Administration (EIA), gasoline and diesel comprise 79% of all the fuel used for transportation [4]: 32% of the fuel is used by automobiles, 28% is used by light trucks, and 16% is used by other trucks for shipments of merchandise. Until the late 1980s, the Kettering ignition system for the internal combustion engine used in gasoline-powered vehicles was based upon using a mechanical distributor for controlling the timing for the spark plugs. This method was prone to poor control and inaccuracy due to contact erosion.

The availability of the IGBT enabled the introduction of reliable distributorless electronic ignition systems in the late 1980s with no moving parts [5]. The article states: "*A primary constraint to the adoption of electronics innovations was the*

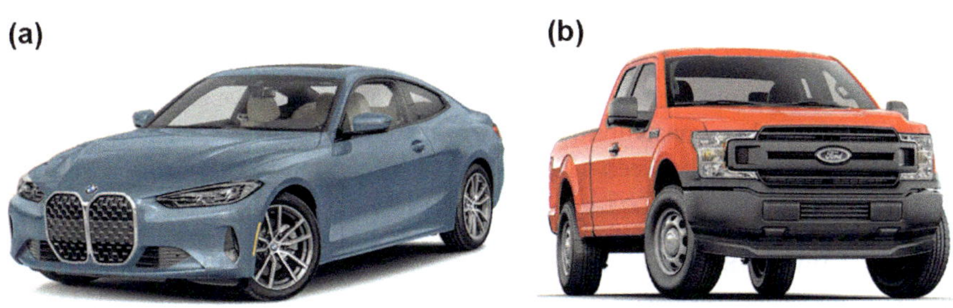

FIGURE 20.2

Gasoline powered: (a) car; (b) truck.

ability to make components durable enough to withstand the heat and vibration factors of an automobile in daily use." The IGBT was the first power switch with sufficient high-temperature operation and ruggedness attributes to allow the cost-effective and reliable implementation of the distributorless electronic ignition system. Billions of IGBTs have been sold by several power semiconductor companies for this market alone.

20.1.1 Fuel Savings

The operation of the electronic ignition system was previously discussed in detail in Section 9.1.2. It is generally acknowledged [6,7] that the distributorless electronic ignition system reduces maintenance costs while allowing more precise control of the timing for the spark plugs, which improves fuel efficiency, reduces emissions, and increases the overall power of a car. Electronic ignition systems have the advantage of operating at a higher voltage than possible with mechanical points. The larger voltage can be applied across a wider spark gap, leading to a longer and broader spark. The larger spark volume allows a leaner fuel mixture that leads to better fuel efficiency and a smooth running engine [8]. The improvements in fuel efficiency derived from using the electronic ignition systems have been documented [9,10] to range from 2 to 4 miles per gallon of fuel. A United Nations Development Program (UNDP) study on the impact of electronic ignition systems has established a 10% improvement in fuel economy while reducing environmental pollution [11]. Based upon the available data, it is reasonable to assume that the introduction of the electronic ignition system using IGBTs has produced fuel savings of at least 10% [12].

Fuel consumption in the United States and around the world has been growing over the years. The gasoline fuel used by cars and trucks in the United States and around the world is documented by the EIA and other sources [13]. The gasoline consumption data are plotted in Fig. 20.3 up to 2020. The United States consumes about one-quarter of the total gasoline consumption around the world. The gasoline consumption per year in the United States has ranged from 125 billion gallons in 1990 to a peak of 180 billion in 2005 followed by a slight decline to about 140 billion gallons in 2020. In contrast, the gasoline consumption around the world has grown steadily during this period from 480 billion gallons in 1990 to 660 billion gallons in 2009. It has remained at this level until 2020.

The gasoline savings from using the electronic ignition system enabled by IGBTs can be computed from the data in Fig. 20.3 using a conservative estimate of a 10% reduction of gasoline consumption. The worldwide cumulative savings in gasoline for the 30-year period between 1990 and 2020 is found to be 1.86 trillion gallons.

20.1.2 Consumer Cost Savings

A reduction in the fuel used by consumers for their daily transportation needs has the obvious benefit of decreasing their expenditures for this vital part of their lives. In order to quantify the economic benefit to consumers from gasoline savings, it is

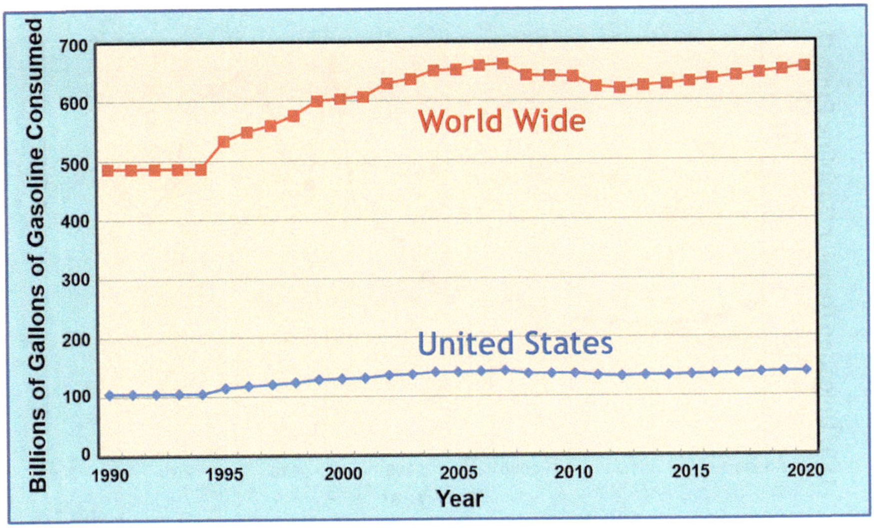

FIGURE 20.3

Gasoline consumption in the United States and around the world.

necessary to use the price for gasoline, which has varied over a large range during the 30-year period from 1990 to 2020. The price for regular unleaded gasoline in the United States has been documented by the EIA [14]. Although the price for gasoline varies across the country and is dependent on the seasons, the EIA has published the national average price for regular unleaded gasoline between 1949 and 2021 normalized to year 2000 dollars. These data are graphically represented in Fig. 20.4. It can be observed that the price for gasoline remained steady at about \$1.50 from 1990 to 2002. The price then started to steadily increase to \$2.90 in 2010. There was a sharp spike in the price of gasoline in the United States to nearly \$4.00 in 2008 and it remained near this level until 2014. It then came down and stayed at about \$2.50 until 2020. According to the EIA, the price for gasoline averaged \$3.5 between 2011 and 2015 and then averaged \$2.50 between 2016 and 2021 [15].

The price of gasoline in Europe is 3–4 times the price of gasoline in the United States [16]. For example, the prices for gasoline have averaged \$8.10 per gallon over the period from 2000 to 2005. For purposes of the analysis here, it will be assumed that the price of gasoline in the rest of the world (i.e., not US regions) is three times the average price for unleaded regular gasoline in the United States, shown in Fig. 20.4.

The gasoline cost savings for US consumers from the introduction of the IGBT-enabled electronic ignition systems is plotted in Fig. 20.5 for the period from 1990 to 2020. The cost savings have grown from the range of \$15 billion per year to \$50 billion per year as the consumption and price for gasoline have increased over the years. The cumulative cost savings for US consumers adds up to \$865 billion. Because much of the gasoline in the United States is imported, these savings also help reduce the balance of trade for the nation.

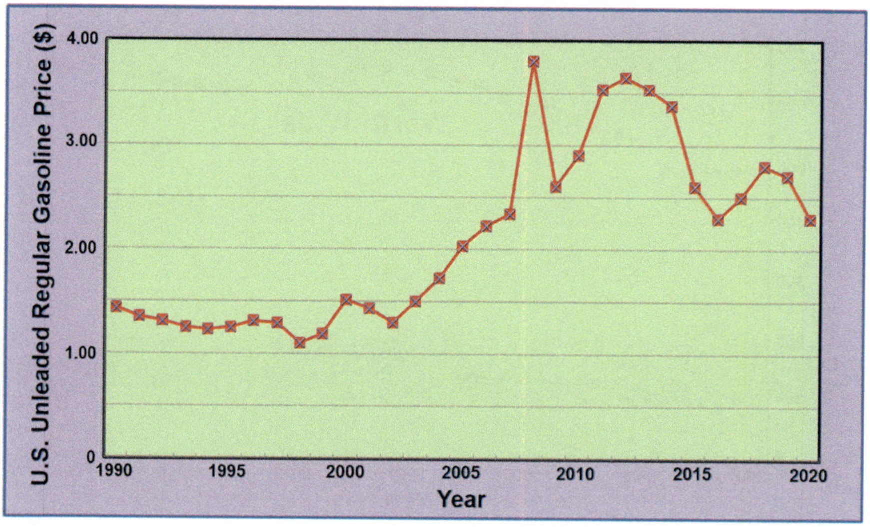

FIGURE 20.4

Prices for regular unleaded gasoline in the United States from 1990 to 2010.

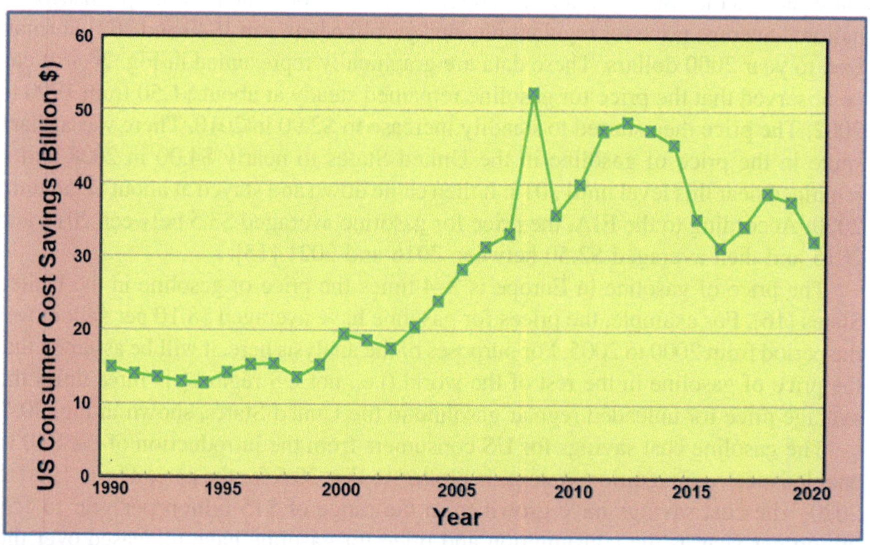

FIGURE 20.5

Gasoline cost savings due to the insulated gate bipolar transistor-enabled electronic ignition system for consumers in the United States.

The reduced gasoline consumption obtained by using IGBT-enabled electronic ignition systems also benefits consumers from around the world. The gasoline cost savings for worldwide consumers from the introduction of the IGBT-enabled electronic ignition systems is plotted in Fig. 20.6. The cost savings have grown from the range of \$200 billion per year to \$650 billion per year as the consumption and price for gasoline has increased over the years. The cumulative cost savings for worldwide consumers adds up to \$12 trillion from 1990 to 2020, which has had an important impact on the economies of nations around the world.

20.1.3 Carbon Dioxide Emission Reduction

The emission of greenhouse gases occurs from all sectors of our economy. The distribution [17], shown in Fig. 20.7, indicates that one major part (14%) of the carbon dioxide emissions emanates from the transportation sector due to gasoline consumption. In addition, this activity generates carcinogenic pollutants in the atmosphere, such as nitrous oxide (NOX) and sulfur dioxide (SOX). The IGBT-enabled improvements in efficiency in all the sectors of the economy have produced a major benefit to society in terms of reducing carbon dioxide emission.

Among the various atmospheric pollutants, the increased concentration of carbon dioxide has garnered the most attention during recent years due to its impact on global warming. Carbon dioxide is considered a greenhouse gas that allows trapping of solar infrared radiation in the atmosphere, producing global warming. The

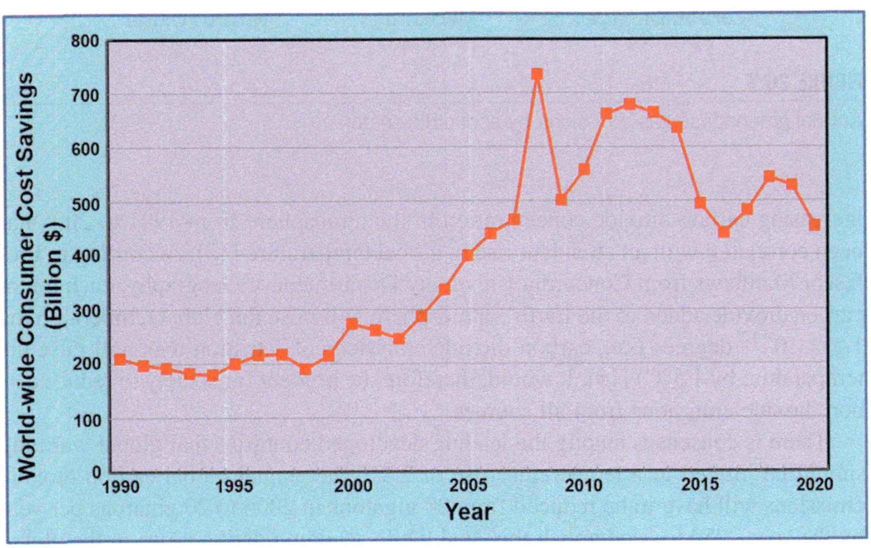

FIGURE 20.6

Gasoline cost savings due to the insulated gate bipolar transistor-enabled electronic ignition system for worldwide consumers.

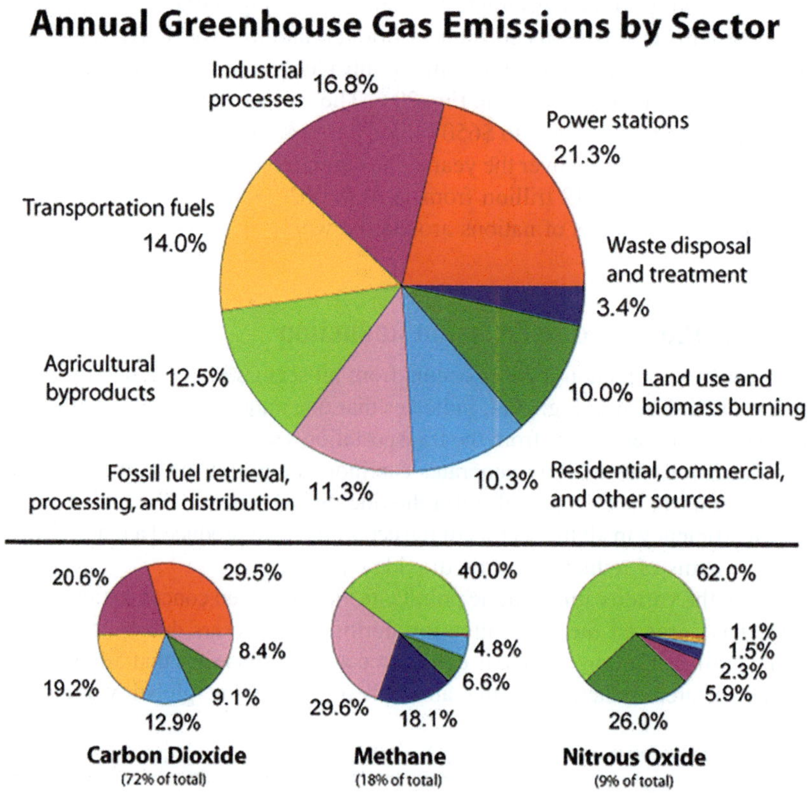

FIGURE 20.7

Annual greenhouse gas emissions by economic sector.

increasing carbon dioxide concentration in the atmosphere from 1991 to 2005 has been correlated with an attendant rise in global temperature [18]. According to Professor Matthews from Concordia University Department of Geography, each ton of carbon dioxide added to the Earth's atmosphere will raise the global temperature by 1.5×10^{-12} degrees (i.e., carbon dioxide emissions of 1 trillion tons will raise the temperature by 1.5°C) [19]. It would, therefore, be prudent for society to reduce carbon dioxide emissions from all sources.

There is consensus among the leading developed countries that global warming should be limited to a temperature rise of 2°C. The annual global carbon dioxide emissions will have to be reduced from 28 gigatons in 2006 to 20 gigatons per year by the year 2050 to accomplish this goal while accommodating a rise in the global population from 6.4 billion in 2007 to about 9.5 billion in 2050 [18]. One solution to this problem is the improvement in the efficiency of automobiles through the use of the IGBT-enabled electronic ignition system and the deployment of IGBT-enabled hybrid-electric cars. The impact due to the IGBT-enabled electronic ignition system from 1990 to 2010 is described in this section. The impact of electric and hybrid-electric cars will be felt in the future.

The emission of pollutants from gasoline-powered vehicles has been well documented [20]. In addition to carbon dioxide, the emissions from gasoline-powered vehicles include toxins that are known to be carcinogens. The US Environmental Protection Agency (EPA) states [21]: "*Motor vehicles emit several pollutants that EPA classified as known or probable human carcinogens. Benzene, for instance, is a known human carcinogen while formaldehyde, acetaldehyde, 1,3-butadiene, and diesel particulate matter are probable human carcinogens.*" In addition to these toxins, gasoline consumption produces the emission of carbon dioxide at the rate of 19.4 pounds per gallon [22].

In the previous section, it was demonstrated that the IGBT-enabled electronic ignition system has improved the fuel mileage for cars and trucks by at least 10%, resulting in a substantial reduction of gasoline consumption in the United States and around the world. The reduction of carbon dioxide emissions in pounds per year due to this reduced gasoline consumption can be computed by multiplying the gasoline savings by a factor of 19.4. The resulting annual reduction of carbon dioxide emissions in the United States and around the world is shown in Fig. 20.8 for the period from 1990 to 2020. It can be seen from this figure that there has a been a reduction of carbon dioxide emissions produced in the United States ranging from 200 billion pounds in 1990 to a maximum of 275 billion pounds in 2006. The

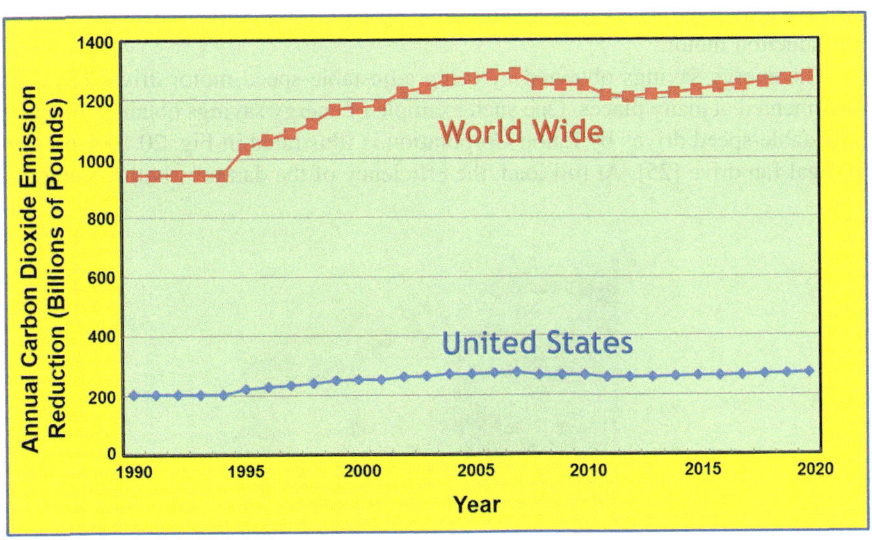

FIGURE 20.8

Reduction of annual carbon dioxide emissions due to the insulated gate bipolar transistor-enabled electronic ignition system in the United States and the world.

cumulative reduction of carbon dioxide emissions produced in the United States due to the availability of IGBT-based electronic ignition systems is 7.8 trillion pounds over the 30-year period from 1990 to 2020.

Similarly, it can be seen from the figure that there has a been a reduction of carbon dioxide emissions produced by the world, ranging from 940 billion pounds in 1990 to a maximum of 1270 billion pounds in 2020. The cumulative reduction of carbon dioxide emissions produced in the world due to the availability of IGBT-based electronic ignition systems is 36 trillion pounds over the 30-year period from 1990 to 2020. From this analysis, it is apparent that the IGBT-enabled electronic ignition system has played an important role in mitigating global warming.

20.2 Adjustable-Speed Motor Drives

According to the US Department of Energy and the Electric Power Research Institute, two-thirds of the electricity in the United States is used to run motors in consumer and industrial applications [23,24]. Motors are used in industrial and consumer sectors for (1) food production; (2) machining; (3) packaging; (4) milling; (5) oil and gas production; (6) water pumping; (7) heating and cooling of buildings and residences; (8) refrigeration; (9) irrigation; etc. As discussed in Section 10.2, induction motors were commonly used with dampers in the past, resulting in poor efficiency. A major improvement in efficiency can be derived by replacing this approach with an adjustable- or variable-speed drive for motors enabled by using IGBT-based inverters. A typical adjustable-speed drive is shown in Fig. 20.9 with an induction motor.

The energy savings obtained by using adjustable-speed motor drives has been documented at many places. One such example of energy savings obtained by using adjustable-speed drives by Eaton Corporation is illustrated in Fig. 20.10 for a centrifugal fan drive [25]. At full load, the efficiency of the damper control method is

FIGURE 20.9

Adjustable-speed motor drive with induction motor.

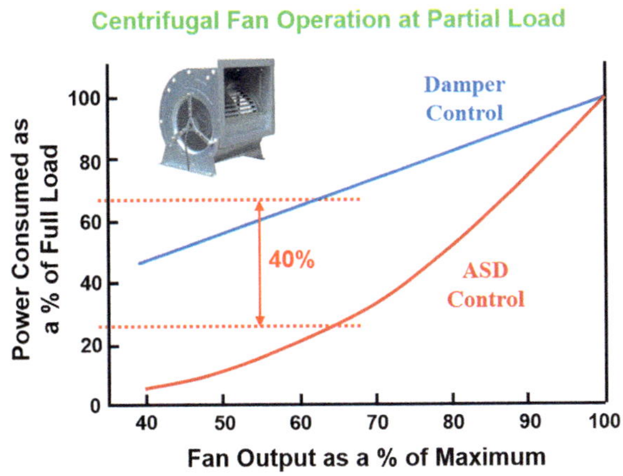

FIGURE 20.10

Example of power savings from using the adjustable-speed drive for a centrifugal fan.

satisfactory because no power is diverted to the dampers. However, for typical conditions below full load, the power consumption by the damper control method is much greater than for the adjustable-speed drive. In the range of 40–70% of full fan output, the adjustable-speed drive provides an improvement in efficiency of at least 40%.

Mayflower Vehicle Systems, manufacturer of body panels for Aston Martin and MGF cars, states that running costs have been reduced by 87.9% by using ABB variable-speed drives [26]. Washington State University concludes [27] that almost a 70% reduction in pumping energy can be obtained by using adjustable-speed drives. A very detailed report [28] documenting energy savings from adjustable-speed drives for motors has also been prepared by Arthur D. Little, Inc., for the US Department of Energy. Based upon these studies and reports, an efficiency improvement of 40% by using adjustable-speed drives can be conservatively assumed for the analysis in this section.

20.2.1 Electrical Energy Savings

The electrical energy consumed in the world has been monotonically growing since World War II. It reached 20,000 terrawatt hours by 2020 and is expected to continue to grow according to the EIA, as shown in Fig. 20.11. The total reduction of electricity power consumption in the world due to adjustable-speed drives between 1990 and 2020 can be computed using this electrical energy consumption on the basis of two-thirds of the power being utilized to drive motors and an energy savings of 40% accruing from adjustable-speed drives. The annual power savings shown in Fig. 20.12 in gigawatts is obtained by dividing two-thirds of the energy per year in terawatt hours provided in Fig. 20.11 by a factor of ($365 \times 24 = 8760$ h),

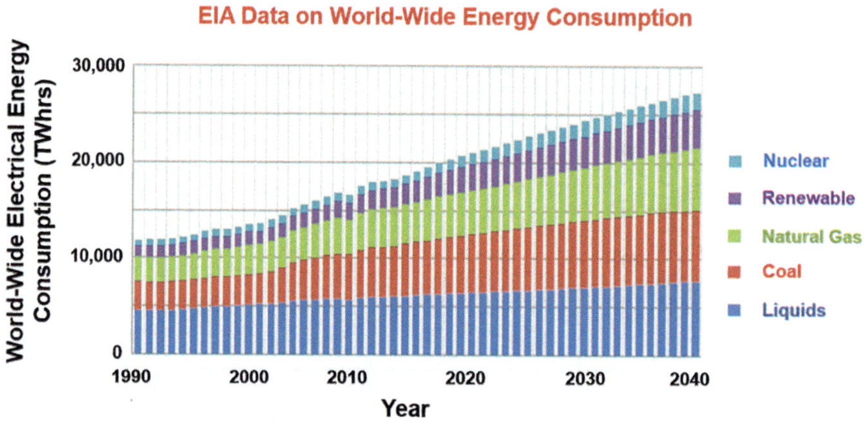

FIGURE 20.11

EIA data on electrical energy consumption in the world.

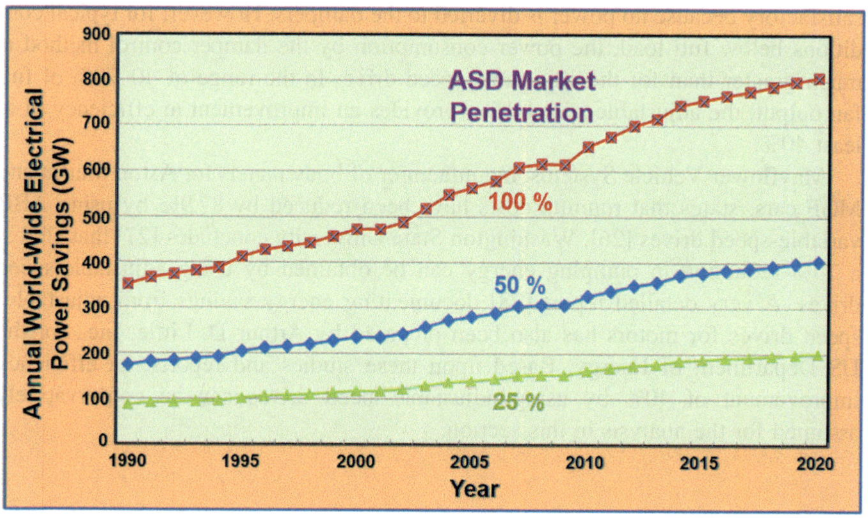

FIGURE 20.12

Annual worldwide electrical power savings due to insulated gate bipolar transistor-based adjustable-speed drives for three cases of penetration rates.

multiplying this by a factor of 1000 to convert terawatts to gigawatts, and then taking 40% of this value due to efficiency improvement with adjustable-speed drives.

It is not appropriate to assume that all the motors in the world were utilizing adjustable-speed drives during the period from 1990 to 2020 because of their higher cost than the older technology with dampers. The market penetration by adjustable-speed drives was high in Europe and Japan due to the high cost of electricity and the desire to reduce fossil fuels consumption. The market penetration was lower in developing nations but their electricity consumption was also smaller than in the developed nations. Taking these factors into account, it is reasonable to use a market penetration of 50% for this period. For this case, it can be seen from Fig. 20.12 that the savings in power consumption due to adjustable-speed drives has grown from 170 GW in 1990 to 400 GW in 2020. The cumulative global energy savings due to these power savings amounts to 77,400 TWh. The analysis indicates huge room for further savings in power when more IGBT-based adjustable-speed motor drives are used in the future to increase the penetration rate closer to 100%.

20.2.2 Electricity Cost Savings

The reduced electricity consumption by using adjustable-speed drives is not only beneficial to the end user of the electricity but also to the electricity suppliers. The construction of a 1 GW coal-fired power plant costs utilities close to $3 billion. Consequently, the 400 GW of electrical energy savings from IGBT-enabled adjustable-speed drives has saved utilities worldwide $1.2 trillion in power plant investments.

Consumers of the electricity have also accrued savings by using adjustable-speed drives to reduce their electricity bills. The cost of electricity for customers varies widely around the world [29]. The average electricity cost for consumers across the United States has been published by the EIA [30]. Their average electricity cost is 11.09 cents per kWh for residential customers, 6.72 cents per kWh for industrial customers, and 10.68 cents per kWh for transportation customers. The average electricity cost across all the sectors in the United States is 10 cents per kWh. It is generally much higher in European countries where the citizens of Denmark, Germany, and Italy pay $0.43, $0.31, and $0.37 per kWh, respectively. The price for electricity is lower in Russia ($0.13 per kWh) and Finland ($0.07 per kWh). The prices also vary widely in Asian countries where the citizens of the Philippines, Singapore, and Malaysia pay $0.16, $0.21, and $0.074 per kWh, respectively. For the analysis in this section, a worldwide average price for electricity of $0.20 per kWh will therefore be assumed.

The electricity cost savings for worldwide consumers due to smaller energy consumption resulting from IGBT-enabled adjustable-speed motor drives is shown in Fig. 20.13. These values were obtained by using the power savings shown in Fig. 20.12 for the three cases of penetration of adjustable-speed motor drives and a cost of $0.20 per kWh for electricity. It can be observed from Fig. 20.13 that customers worldwide have derived a cost savings ranging from $300 billion in 1990 to $700 billion in 2020 if 50% penetration is assumed. The cumulative electricity cost savings for worldwide customers adds up to $15.5 trillion over the period from 1990 to 2020 for the 50% penetration case. Consequently, it can be concluded that people around the world have derived an enormous cost benefit due to the IGBT-enabled adjustable-speed motor drives. The analysis indicates huge room for further savings

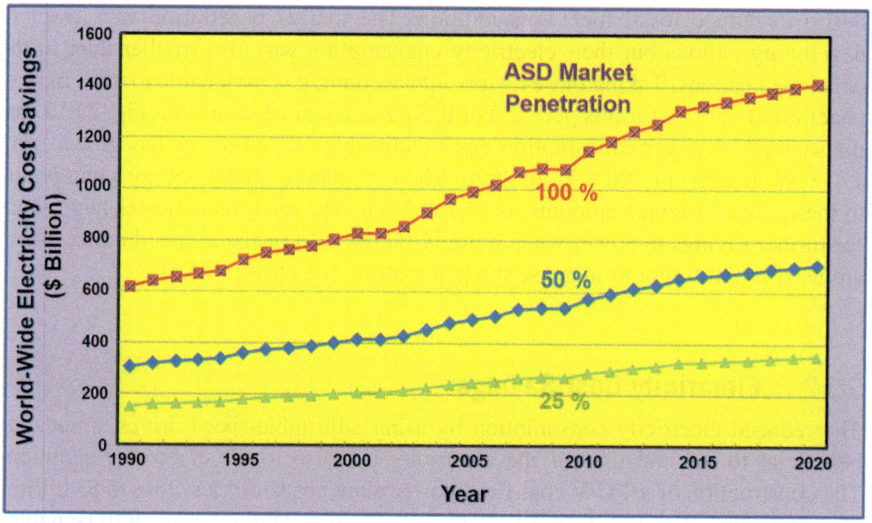

FIGURE 20.13

Electrical cost savings to consumers due to insulated gate bipolar transistor-based adjustable-speed drives.

in cost as more IGBT-based adjustable-speed motor drives are used in the future to increase the penetration rate closer to 100%.

20.2.3 Carbon Dioxide Emission Reduction

The reduction of carbon dioxide emissions due to the reduced electricity power consumption resulting from IGBT-based adjustable-speed drives can be computed by multiplying the data in Fig. 20.12 by the rate of carbon dioxide emission per kilowatt hour of electricity generated by typical power plants in the world. The EPA has analyzed the carbon dioxide emissions from various types of power plants in the United States [31]. Electricity is generated mostly (51%) from coal-fired power plants in the United States. Unfortunately, the carbon dioxide emission from coal-fired power plants is the highest among the power-generation options. The average carbon dioxide emission per kilowatt hour generated in the United States based upon all fuel sources is 1.35 pounds, according to the EPA.

Worldwide electricity generation [32] is performed using a variety of fuels: coal is used for generating 41% of the world's electricity demands and natural gas is used for generating another 21% of the world's electricity demands. The average carbon dioxide emission from the world at this time is reported to be 1.10 pounds per kWh. However, according to the above reference, coal is used to generate 79% of the electricity in China and 69% of the electricity in India. Because these two countries are the most aggressive in deploying new electricity-generation capability due to their rapid economic growth, their electricity generation in the future is likely to produce much higher levels of carbon dioxide emissions.

The reduction of carbon dioxide emissions in pounds per year due to the reduced electricity power consumption from IGBT-based adjustable-speed drives can be computed by multiplying the data in Fig. 20.12 by the rate of carbon dioxide emission per kilowatt hour of electricity generated by typical power plants. The reduction of carbon dioxide emissions due to the availability of IGBT-based adjustable-speed motor drives is shown in Fig. 20.14. These data were calculated by multiplying the energy savings per year in kilowatt hours by 1.10 pounds per kWh. It can be seen from Fig. 20.14 that there has a been a reduction of carbon dioxide emissions produced in the world, ranging from 1.65 trillion pounds per year in 1990 to 3.9 trillion pounds per year in 2020 for a 50% penetration rate for adjustable-speed drives. In this case, the cumulative reduction of carbon dioxide emissions produced in the world due to the availability of IGBT-based adjustable-speed motor drives is 85.2 trillion pounds over the 30-year period from 1990 to 2020. As the penetration rate for adjustable-speed drives rises closer to 100%, an even greater reduction of carbon dioxide emissions in the world can be anticipated in the future.

Recent recognition of the importance of abating carbon dioxide emissions by most countries worldwide that participated in the Paris climate conference in 2015 has led to a climate treaty. This treaty expects all countries to reduce their carbon dioxide emissions in the future. This impetus can be expected to increase the penetration of IGBT-based adjustable-speed motor drives well above 50%, producing the desired mitigation of carbon dioxide emissions.

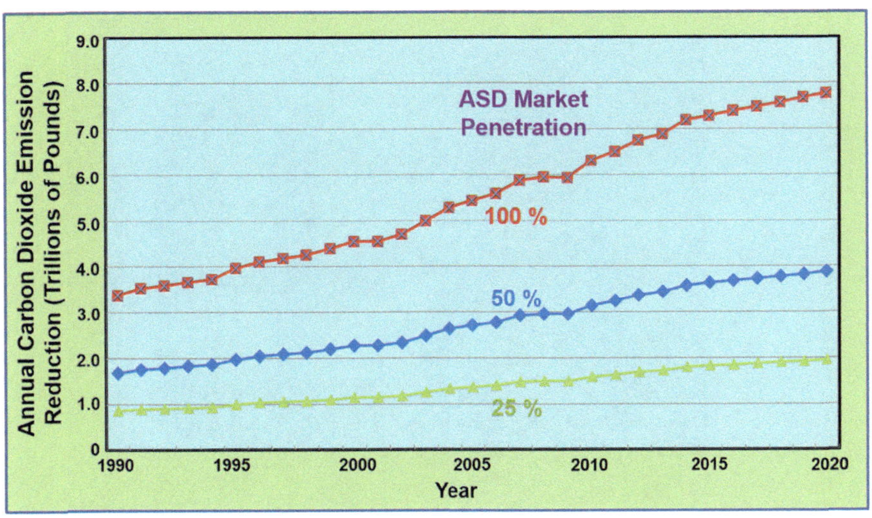

FIGURE 20.14

Annual reduction in carbon dioxide emissions due to insulated gate bipolar transistor-based adjustable-speed drives.

20.3 Compact Fluorescent Lamps

Lighting consumes about 22% of the electricity in the United States with similar levels in the rest of the world [33]. In Section 11.2, it was demonstrated that the IGBT enabled the introduction of cost-effective and reliable compact fluorescent lamps (CFLs). Today, CFLs are available to the consumer in a variety of shapes and sizes, as shown in Fig. 20.15.

There has been a worldwide movement toward replacing incandescent bulbs due to their very poor efficiency (4%) for the conversion of electric power into usable light. The most cost-effective option for replacement of the incandescent lamp in the 1990s was the CFL. This held true up to 2015 when the price of LED lamps for residential use became closer to that of CFLs. A typical CFL consumes only 15 W of power for producing the same amount of light when replacing a 60 W incandescent bulb. After a slow start in the 1990s, there has been a rapid growth in the sales of CFLs around the world since 2000.

The sales of CFLs in the United States have been reported by various sources [34]. Since 1999, the annual CFL sales grew rapidly in the United States until 2007. Walmart has promoted the sale of CFLs by dropping prices and in November 2006 announced a goal of selling 100 million CFLs by the end of 2007, a target that was met by October 2007. Unfortunately, the sales for CFLs in the US dropped by 25% from 2007 levels in 2008 and by 49% from 2007 levels in 2009 [35]. This is partly due to hysteria created by the media regarding the mercury content in the CFL lamp, even though the amount of

FIGURE 20.15

Examples of compact fluorescent lamps available to consumers.

mercury in the CFL bulb is typically 100 times less than in the home thermometer. The mercury content in CFLs has been reduced from 20 mg to less than 2 mg between 1990 and 2000 [36]. Walmart, Home Depot, and Lowes have programs to accept CFLs that have burned-out for safe disposal.

In contrast with the United States, the adoption of CFLs in the rest of the world has been progressing monotonically [37]. Worldwatch Institute states [34]: "*In the United States, CFLs accounted for more than 20% of sales in 2007. But other wealthy nations have shown much higher CFL use rates for quite some time, including 80% of households in Japan and 50% in Germany by 1996. Many developing countries have shown strong CFL market share in recent years as well: 14% of sales in China in 2003 for instance and 17% in Brazil in 2002.*" Australia became the first country to ban the sale of incandescent bulbs in 2007. The European Union, Ireland, and Canada have passed similar legislation. Worldwatch states: "*In total, more than 40 countries have announced plans to follow suit.*" Even India has launched a program [38], called Bachat Lamp Yojana (BLY), to replace the 400 million incandescent bulbs in use in the country with CFLs. In 2008, 199 million CFLs were sold in India. The government of Bangladesh has distributed 80 million CFLs to its citizens to reduce electricity demands with additional lamps available on the open market [39].

The global sales of CFLs have been monotonically growing at an annual rate of 20% [40] since the early 1990s. According to the Chinese GFL Lighting Technology Group, the production of CFLs is dominated by China with a market share of more than 80%. This conclusion is affirmed by the data published at the Second International Forum on CFL Harmonization [41]. The Worldwatch Institute states [34]: "*Between 2001 and 2006, production of compact fluorescent lamps in China, which account for roughly 85% of global output, tripled from 750 million to 2.4 billion units. The total number of CFLs in use globally nearly doubled between 2001 and 2003 alone, growing from an estimated 1.8 billion to 3.5 billion units.*" The global sales of CFL bulbs grew monotonically until 2015, reaching 4 billion before leveling off at about 4.5 billion in 2020, as shown in Fig. 20.16 [36]. This may be due to a reduction in cost of LED lamps by 2015, as these lamps provide a similar improvement in lighting efficiency.

For determining the power savings derived from the global use of CFLs, it is necessary to use the total installed CFLs for each year. A good assumption that can be made is that all the purchased CFLs have been used to replace incandescent bulbs by consumers. However, the life of a CFL is about 10,000 h, which translates to a use period of 5 years based upon being turned on for 6 h per day. It must therefore be assumed that the CFLs purchased during any particular year will be first used to replace those purchased 5 years ago with the balance being used to replace incandescent lamps. The cumulative CFLs in use globally based upon this methodology are shown in Fig. 20.17. More than 20 billion CFLs were in use worldwide by 2020. The number of CFLs in use globally shown in the figure is consistent with published estimates [34].

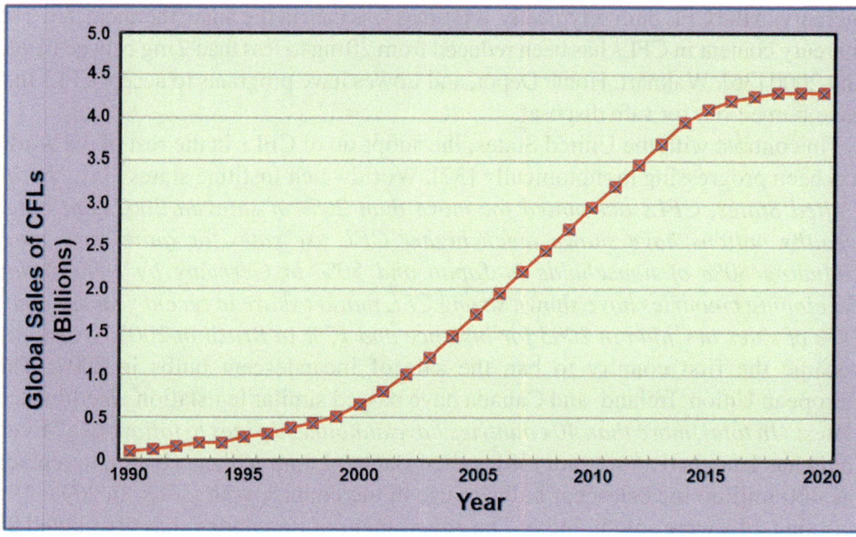

FIGURE 20.16

Annual sales of compact fluorescent lamps around the world.

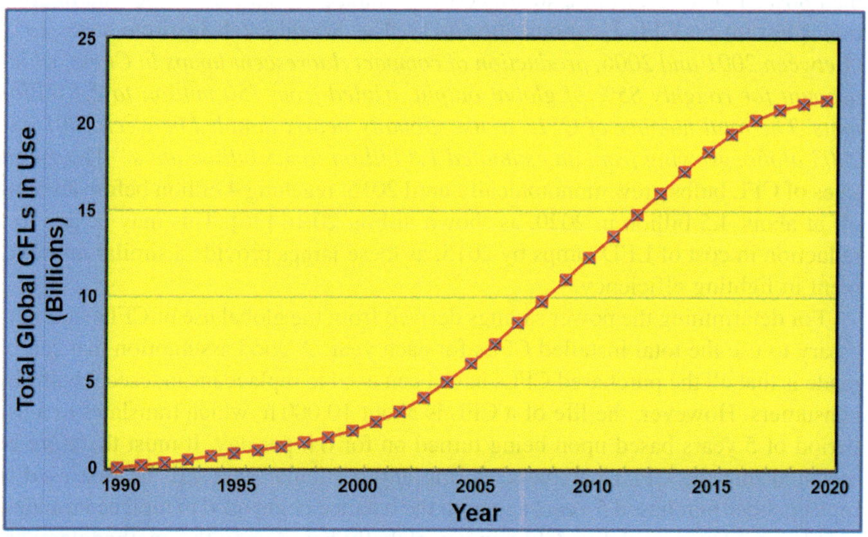

FIGURE 20.17

Total compact fluorescent lamps in use around the world.

20.3.1 Electrical Energy Savings

The power savings derived by replacing 60 W incandescent bulbs with CFLs can be computed by using the 45 W power savings per 60 W incandescent bulb. The power savings obtained from the worldwide installed CFLs shown in Fig. 20.17 is provided in Fig. 20.18. The annual power savings have grown from 4.5 GW in 1990 to 960 GW in 2020. The global cumulative energy savings with the use of CFLs is 26,000 TWh, assuming they are used for 6 h per day.

20.3.2 Electricity Cost Savings

The reduced electricity consumption by using CFLs not only benefits the end user of the electricity but also the electricity suppliers. The construction of a 1 GW coal-fired power plant costs utilities close to \$3 billion. Consequently, the 960 GW of electrical energy savings from IGBT-enabled CFLs has provided a cost savings of \$2.88 trillion to the US utility industry due to avoiding the building of power plants.

The cost savings for global customers by using CFLs can be obtained by taking the electrical power savings shown in Fig. 20.18 and multiplying by a factor of 10^6 to convert the data to kilowatts and by (365*6) to obtain the annual energy savings in kilowatt hours. This analysis assumes that each CFL is turned on for 6 h per day. The energy savings in kilowatt hours was then multiplied by the global average cost of electricity (\$0.20 per kWh) to obtain the annual cost savings. The cost savings for global customers is plotted in Fig. 20.19. The cumulative electricity cost savings

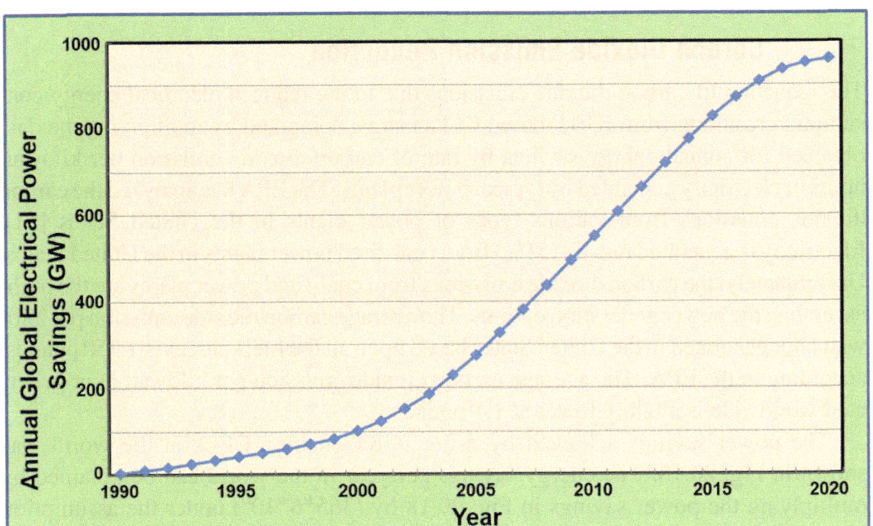

FIGURE 20.18

Global electrical power savings due to the replacement of incandescent lamps with insulated gate bipolar transistor-enabled compact fluorescent lamps.

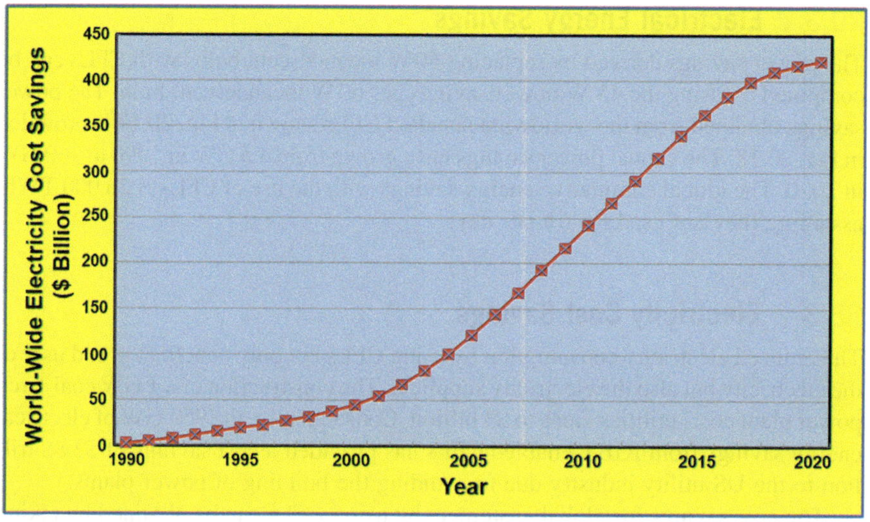

FIGURE 20.19

Global cost savings for consumers due to the replacement of incandescent lamps with insulated gate bipolar transistor-enabled compact fluorescent lamps.

for global customers from 1990 to 2020 due to smaller energy consumption resulting from IGBT-enabled CFLs is $5.2 trillion.

20.3.3 Carbon Dioxide Emission Reduction

The reduction of carbon dioxide emissions due to the reduced electrical energy consumption resulting from IGBT-based CFLs can be computed by multiplying the data obtained for annual energy savings by rate of carbon dioxide emission per kilowatt hours of electricity generated by typical power plants. The EPA has analyzed the carbon dioxide emissions from various types of power plants in the United States [31]. Electricity is generated mostly (51%) from coal-fired power plants in the United States. Unfortunately, the carbon dioxide emissions from coal-fired power plants are the highest among the power generation options. The average carbon dioxide emission per kilowatt hour generated in the United States based upon all the fuel sources is 1.350 pounds, according to the EPA. The average carbon dioxide emission per kilowatt hour generated worldwide is slightly lower at 1.1 pounds.

The power savings achieved by using IGBT-enabled CFLs in the world was shown in Fig. 20.18. The energy savings per year in the world can be obtained by multiplying the power savings in Fig. 20.18 by $(365*6*10^6)$ under the assumption that each CFL is used for 6 h per day. The reduction of carbon dioxide emissions due to the availability of IGBT-based CFLs obtained by multiplying the energy savings per year in kilowatt hour by 1.10 pounds per kWh is provided in Fig. 20.20. There has been a reduction of carbon dioxide emissions produced in the world

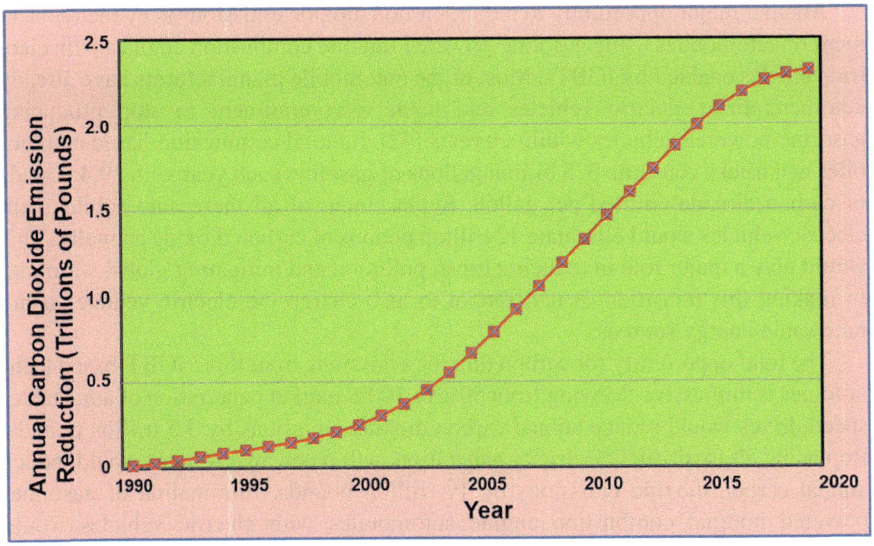

FIGURE 20.20

Global reduction in carbon dioxide emissions due to the replacement of incandescent lamps with insulated gate bipolar transistor-based compact fluorescent lamps.

ranging from 0.01 trillion pounds per year in 1990 to 2.31 trillion pounds per year in 2020. The cumulative reduction of carbon dioxide emissions produced in the world due to the availability of IGBT-based CFLs is 28.6 trillion pounds over the 30-year period from 1990 to 2020.

20.4 Future Social Impact

The previous sections have documented gasoline and electrical energy savings derived from the IGBT-enabled electronic ignition system, adjustable-speed motor drives, and CFL lamps. These benefits are expected to continue unto the future.

Additional benefits can be anticipated by the replacement of fossil fuel-based electricity generation with IGBT-based wind and solar power generation. These renewable technologies have already been deployed in the United States, China, Germany, and other countries at an increasing pace. The price per kWh for renewable power generation has now fallen even below the price of power generated with fossil fuels. The total electrical energy produced worldwide in 2020 reached 26,000 TWh. Fossil fuel-based power plants produced 68% of this electrical energy, that is, 17,680 TWh, with an average carbon dioxide emission of 1.1 pounds per kWh. Using these values, the replacement of all fossil fuel-based electrical energy generation around the world with renewable energy generation would reduce carbon dioxide emissions by 17 trillion pounds per year. This would play a major role in the mitigation of global warming due to carbon dioxide in the atmosphere.

Another major opportunity to reduce carbon dioxide emissions is by the replacement of automobiles using gasoline-powered internal combustion engines with electric vehicles enabled by IGBTs. Most of the automobile manufacturers have already commercialized electric vehicles and made a commitment to stop producing gasoline-powered vehicles within 10 years [42]. Internal combustion-based automobiles and trucks consume 625 billion gallons of gasoline each year with 19.4 pounds of carbon dioxide emitted per gallon. Replacement of all these automobiles with electric vehicles would eliminate 12 trillion pounds of carbon dioxide annually. This would play a major role in reducing urban pollution and mitigating global warming. In making this transition, it is essential to also charge the electric vehicles using renewable energy sources.

The total opportunity for carbon dioxide emissions from three IGBT-based technologies is impressive. Moving from 50% to 100% market penetration of adjustable-speed drives would reduce annual carbon dioxide emissions by 3.5 trillion pounds. Replacing all fossil-fuel electricity generation with renewable sources would reduce annual carbon dioxide emissions by 19 trillion pounds. Elimination of gasoline-powered internal combustion engine automobiles with electric vehicles would reduce annual carbon dioxide emissions by 12 trillion pounds.

The United Nations Environment Program (UNEP) published an emission gap report in 2021 [43]. The authors state that carbon dioxide emissions must be reduced by 60, 44, and 30 trillion pounds per year to keep global temperature rise below 1.5, 1.8, and 2.0°C, respectively. Taking advantage of all the above opportunities enabled by IGBTs would potentially allow reducing annual carbon dioxide emissions by 34 trillion pounds, making the UNEP targets achievable.

20.5 Summary

The IGBT has enabled improvement in the efficiency of power delivery and energy management for a wide variety of products in all sectors of the economy. In this chapter, the social benefits brought about by the availability of IGBTs have been quantified by using three case studies: (1) the gasoline savings obtained by IGBT-enabled electronic ignition systems, (2) the electricity savings obtained by IGBT-enabled adjustable-speed motor drives, and (3) the electricity savings obtained by IGBT-enabled CFLs. The resulting enormous energy and cost savings for consumers around the world is summarized in Table 20.1 during the period from 1990 to 2020. The case of a 50% penetration rate for adjustable-speed motor drives around the world was used in preparing Table 20.1.

The power savings enabled by the IGBT-based applications have also produced huge savings for the utility industry by avoiding the construction of expensive electricity-generating power plants. The cumulative cost savings for worldwide consumers over the 1990 to 2020 time frame adds up to $33.6 trillion, with additional savings for worldwide utilities of $5.4 trillion from not having to build power plants.

Table 20.1 Summary of Gasoline and Electrical Power Savings From IGBT-Enabled Technologies and the Resulting Economic Benefits to Society

IGBT-Enabled Application	Worldwide Cumulative Gasoline or Energy Savings	Worldwide Consumer Cost Savings
Electronic Ignition System	1.86 trillion gallons	$12.0 trillion
Adjustable-Speed Motor Drive	77,400 TWh	$15.5 trillion
Compact Fluorescent Lamp	26,000 TWh	$5.3 trillion
TOTAL		$32.8 trillion

The reduction of gasoline and electricity consumption due to IGBT-enabled electronic ignition systems, adjustable-speed motor drives, and CFLs has resulted in an enormous reduction in the emission of carbon dioxide. The cumulative reduction of carbon dioxide emissions over the period from 1990 to 2020 from these three application examples is provided in Table 20.2. The case of a 50% penetration rate for adjustable-speed motor drives around the world was used here. The cumulative carbon dioxide emissions reduction in the world over the 1990 to 2020 time frame adds up to 150 trillion pounds (75 Gt).

The opportunity for the reduction of carbon dioxide emissions from various sources has been recently analyzed [44]. The authors evaluated options for a reduction of annual worldwide fossil-fuel carbon dioxide emissions from today's 7.5 billion metric tons (or 15 trillion pounds) per year. This value is consistent with the global fossil-fuel carbon emissions estimate by the Carbon Dioxide Information Analysis Center at Oak Ridge National Laboratory [45]. Using this value, it can be concluded that the IGBT has enabled the elimination of worldwide carbon dioxide emissions equivalent to that produced by fossil fuels in 10 years.

Table 20.2 Summary of Carbon Dioxide Emission Reduction From IGBT-Enabled Technologies

IGB-Enabled Application	Worldwide Cumulative Gasoline or Energy Savings	Worldwide Cumulative Carbon Dioxide Emission Reduction
Electronic Ignition System	1.86 trillion gallons	36 trillion pounds
Adjustable-Speed Motor Drive	77,400 TWh	85.2 trillion pounds
Compact Fluorescent Lamp	26,000 TWh	28.6 trillion pounds
TOTAL		150 trillion pounds

In the future, further reductions in carbon dioxide emissions will be achieved by the increased generation of electricity from renewable energy sources, such as wind power, solar power, geothermal, etc., and by the replacement of gasoline-powered automobiles with electric vehicles. Time magazine has recently examined the renewable energy options [46]. The article states: "*President Obama's latest target is to have 80% of US energy needs supplied by clean sources by 2035⋯Meanwhile, the European Union is working to double the share of energy derived from wind and other renewable sources, such as solar, geothermal, and biomass, to 20% by 2020.*" All these renewable energy generation options require conversion of the generated power into a well-regulated 50 or 60 Hz AC power that can be distributed in the power grid. This is accomplished by using IGBT-based inverters, as discussed in Chapter 15. There is a future opportunity to reduce carbon dioxide emissions by 34 trillion pounds annually using IGBT-enabled technologies [47]. Consequently, IGBTs will play an ever-increasing role in creating a sustainable global society with a high standard of living while reducing our impact on the environment and mitigating global warming.

References

[1] M. Grunwald, Wasting Our Watts, Time Magazine, 2009. January 12.
[2] L.H. Lovins, B. Cohen, Climate Capitalism, Hill and Wang Publishers, New York, 2011.
[3] https://www.researchgate.net/figure/CO2-emissions-per-capita-in-selected-countries-and-world-regions-5_fig 3_251589639.
[4] Monthly Energy Review, U.S. Energy Information Administration, 2011. June.
[5] R.G. Amey, Automotive component innovation: development and diffusion of engine management technologies, Technovation 15 (1995) 211–223.
[6] K. Nice, How Automobile Ignition Systems Work. http://auto.howstuffworks.com/ignition-system5.htm.
[7] A. Faiz, C.S. Weaver, M.P. Walsh, Air Pollution from Motor Vehicles, The World Bank, Washington, DC, 1996.
[8] C. Ofria, A Short Course on Ignition Systems. http://www.familycar.com/classroom/ignition.htm.
[9] C. Fulvia. http://www.viva-lancia.com/fulvia/qanda/electrical/el.php.
[10] J.D. Hanks, G. Cunningham, M. Mallory, Electronic versus Point Distributor. http://www.hotforhotfours.com/evsd.htm.
[11] GEF Small Grants Programme, UNDP. http://sgp.undp.org/downloads/fs-krgyzstan.pdf.
[12] https://www.hot-spark.com/Electronic-Ignition-Conversion-Fuel-Savings-Calculator.htm.
[13] 2011. http://www.bts.gov/publications/national_transportation_statistics/excel/table_04_01.xls.
[14] Annual Energy Review, Table 5.24: Retail Motor Gasoline and on-Highway Diesel Fuel Prices, Energy Information Administration, 2008. http://www.eia.gov/oil_gas/petroleum/data_publications/wrgp/mogas_history.html.
[15] https://www.eia.gov/dnav/pet/hist/LeafHandler.ashx?n=pet&s=emm_epm0_pte_nus_dpg&f=m.

[16] Gasoline and Diesel Usage and Pricing, Wikipedia. http://en.wikipedia.org/wiki/Gasoline_and_diesel_usage_and_pricing.
[17] Greenhouse Gases: Why Are We Focusing on Carbon Dioxide? GreenCheck Technologies, Inc. http://gchk.biz/2010/06/16/greenhouse-gases-why-are-we-focussing-oncarbon-dioxide/.
[18] CO_2—The Major Cause of Global Warming, Time for Change Organization. http://timeforchange.org/CO2-cause-of-global-warming.
[19] Carbon Emissions Linked to Global Warming in Simple Linear Relationship, Science Daily News, 2009. June 11 http://www.sciencedaily.com/releases/2009/090610154453.htm.
[20] Report 90–1 Total Emissions from Gasoline-Powered Light-Duty Vehicles, California Air Sources Board, 1990. January http://www.arb.ca.gov/research/resnotes/notes/90-1.htm.
[21] Environmental Fact Sheet EPA400-F-92-004 Air Toxins from Motor Vehicles, U.S. Environmental Protection Agency, 1994. August http://www.epa.gov/otaq/f02004.pdf.
[22] Environmental Fact Sheet EPA420-F-05–001 Emission Facts: Average Carbon Dioxide Emissions Resulting from Gasoline and Diesel Fuel, U.S. Environmental Protection Agency, 2005. February http://www.epa.gov/oms/climate/420f05001.pdf.
[23] Energy Use Basics. http://www.fypower.org/bpg/module.html.
[24] J. Douglas, Advanced motors promise top performance, EPRI J. (1992) 24–31.
[25] Eaton Consultants Report # 00–054 Variable Frequency Drives, 2000. June www.nwalliance.org.
[26] Staggering Energy Savings Prove Value to Professional Audit. http://www.joliettech.com/cs_staggering-energy-savings.htm.
[27] Energy Efficiency Fact Sheet: Adjustable Speed Motor Drives, Cooperative Extension, Washington State University Energy Program. http://www.energy.wsu.edu/ftp-ep/pubs/engineering/motors/MotorDrvs.pdf.
[28] Opportunities for Energy Savings in the Residential and Commercial Sectors with High-Efficiency Electric Motors, Arthur D. Little, Inc., 1999. December 1. Final Report Contract No. DE-AC01-90CE23821 http://www.eere.energy.gov/buildings/info/documents/pdf/doemotor2_2_00.pdf.
[29] Electricity Pricing. en.wikipedia.org/wiki/Electricity_pricing.
[30] Average Retail Price of Electricity to Ultimate Customers by End-Use Sector, by State. U.S. Energy Information Administration. http://www.eia.gov/cneaf/electricity/epm/table5_6_b.html.
[31] Environmental Protection Agency Report Carbon Dioxide Emissions from the Generation of Electric Power in the United States, U.S. Department of Energy, 2000. July http://www.eia.gov/cneaf/electricity/page/co2_report/co2emiss.pdf.
[32] Coal and Electricity. http://www.worldcoal.org/coal/uses-of-coal/coal-electricty/.
[33] Report prepared for the U.S. Department of Energy U.S. Lighting Market Characterization, Navigant Consulting, Inc., 2002. September.
[34] Strong Growth in Compact Fluorescent Bulbs Reduces Electricity Demand, Worldwatch Institute, 2011. June 15 http://www.worldwatch.org/node/5920.
[35] L.B. Vestel, C.F.L. As. sales fall, more incentives urged, N.Y. Times, 2009. http://green.blogs.nytimes.com/2009/09/28/as-cfl-sales-fall-more-incentives-urged/.
[36] www.elcled.org.

[37] C.F.L. Widespread, Use Could Slash Global Lighting Demand for Electricity in Half, 2008, October http://www.facilitiesnet.com/lighting/article/Widespread-CFL-Use-Could-Slash-Globa-Lighting-Demand-For-Electricity-By-Half.htm.

[38] India Lights Future with Efficient Lamps and Carbon Credits, Worldwatch Institute, 2011. June 15 http://www.worldwatch.org/node/6438.

[39] M. Mollah, R. Islam, S. Islam, The forecasting effects of using compact flourescent lamps on human being, environment and electrical power quality in Bangladesh, in: IEEE International Conference on Computer Communication and Informatics, 2012, pp. P1–P6.

[40] G. Fumin, Achievement and development of Chinese CFL industry, Right Light (2002) 303–306.

[41] P. du Pont, M. Kumpengsath, International review of CFL markets in 7 Asia-Pacific countries, in: 2nd International Forum on CFL Harmonization, 2005. November 1.

[42] https://www.gm.com/commitments/electrification.html.

[43] The Heat is On, United Nations Emission Gap Report, 2021, ISBN: 978-92807-3890-2.

[44] D. Divan, F. Kreikebaum, Organic (but not green), IEEE Spectr. (2009) 49–53.

[45] Global Fossil-Fuel CO_2 Emissions, Carbon Dioxide Information Analysis Center, Oak Ridge National Laboratory. http://cdiac.ornl.gov/trends/emis/tre_glob.html.

[46] E. Heinrich, Breezing in, Time Magazine, June 20, 2011.

[47] B.J. Baliga, Enabling Green and Renewable Energy Solutions with Power Semiconductor Technology, 2021. Global Renewable Energy Researchers Meet.

CHAPTER

Synopsis

21

Previous chapters in the book have provided a detailed description of the structure and operation of the insulated gate bipolar transistor (IGBT), the design of the cell structure and the chip, and methods for controlling the device. During the last 40 years after its conception and commercialization, the IGBT has become an essential device for power management in all sectors of the economy. The various applications of IGBTs within each sector of the economy have been provided with the circuit topologies and the optimization of the device structure to maximize its performance. The enormous social impact of the IGBT has been quantified as well.

In this chapter, the state of the art in silicon IGBT performance is provided for devices with blocking voltage capability ranging from 600 to 6500 V. During the past 5 years, silicon carbide devices have become commercially available that could replace the silicon IGBT in some applications. The characteristics of the silicon carbide devices are compared with those of the silicon IGBT to understand the benefits of this emerging technology. However, its penetration of the IGBT market is constrained by the high cost of manufacturing the silicon carbide devices.

21.1 State-of-the-Art IGBT Products

The basic IGBT structure was described in Chapter 2. This structure has been extensively optimized over a period of 40 years. Devices with blocking voltage ratings ranging from 600 to 6500 V are available from many manufacturers today due to the large variety of high-volume applications described in previous chapters of the book. A representative set of devices available from Mitsubishi Semiconductor is listed in Table 21.1 with their current ratings and on-state voltage drop at 25°C as stated in the datasheets. The switching speed of these devices is also provided in the table by using the fall time for the collector current.

The datasheets for IGBT products include the on-state characteristics of the devices. To compare IGBTs with different blocking voltages and later to the silicon carbide power MOSFETs, it is necessary to make use of the on-state current density rather than the absolute current, which scales with the number of IGBT die operating in parallel inside each module. The on-state voltage drop for the IGBT increases when the blocking voltage is increased due to the thicker drift layer required to support the voltage, as discussed in Chapter 3. The on-state current density is eventually determined by thermal considerations such as maximum junction temperature and

The IGBT Device. https://doi.org/10.1016/B978-0-323-99912-0.00003-9

Table 21.1 Typical Commercially Available Insulated Gate Bipolar Transistor Products From Mitsubishi Electric

IGBT Product	Blocking Voltage (V)	Current Rating (A)	On-State Voltage (V)	Current Fall Time (μs)	On-State Current Density (A/cm²)
CM100DU-12F device A	600	100	1.6	0.25	100
CM100DU-24F device B	1200	100	1.8	0.30	85
CM100DY-34A device C	1700	100	2.2	0.35	75
CM1000HC-66R device D	3300	1000	2.45	0.30	50
CM1200HC-90R device E	4500	1200	3.50	0.35	40
CM600HG-130H device F	6500	750	3.80	0.40	30

thermal resistance of the package. The typical on-state current densities for IGBT products with different blocking voltage ratings have been published [1] and are included in Table 21.1. These current densities can be assumed to be applicable at the rated on-state current for each of the IGBT products. Using these values, it is possible to plot the on-state characteristics of the IGBT products using the on-state current density, as shown in Fig. 21.1. From this figure, it is apparent that the on-state characteristics degrade with increasing blocking voltage rating for the IGBT, that is, the on-state voltage drop increases with increasing blocking voltage rating. This holds true in general for all power devices.

The on-state operating point published on the manufacturer datasheet is shown in the figure by the star data points. In addition, the locus for a constant power dissipation density is shown in the figure by the dashed lines for three cases of 100, 125, and 150 W/cm^2 [2]. The on-state operating point for the lower blocking voltage devices A, B, and C lies on the 150 W/cm^2 [2] power dissipation curve. The on-state operating point for the larger blocking voltage devices D, E, and F lies on the 125 W/cm [2] power dissipation curve. This is due to differences in the packaging used for the various devices with a larger stand-off voltage being necessary in modules for the larger blocking voltages.

The on-state current density of state-of-the-art IGBTs is plotted in Fig. 21.2 as a function of the blocking voltage rating by using the information from Fig. 21.1. The reduction in the on-state current density for IGBTs with larger blocking voltages increases the challenges in building modules with high current ratings because more chips have to be packaged in parallel to reach a desired current rating when compared with lower blocking voltage devices. This trend is applicable to all power devices, including silicon carbide power MOSFETs, as discussed later in the chapter.

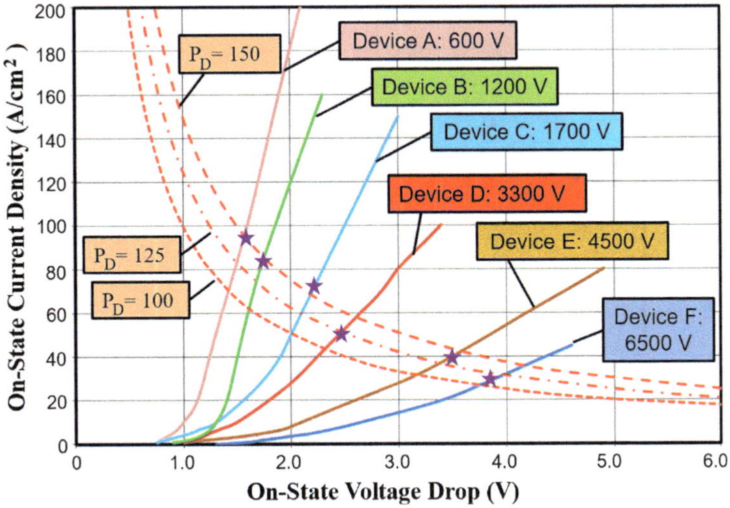

FIGURE 21.1

On-state characteristics of typical commercially available insulated gate bipolar transistor products.

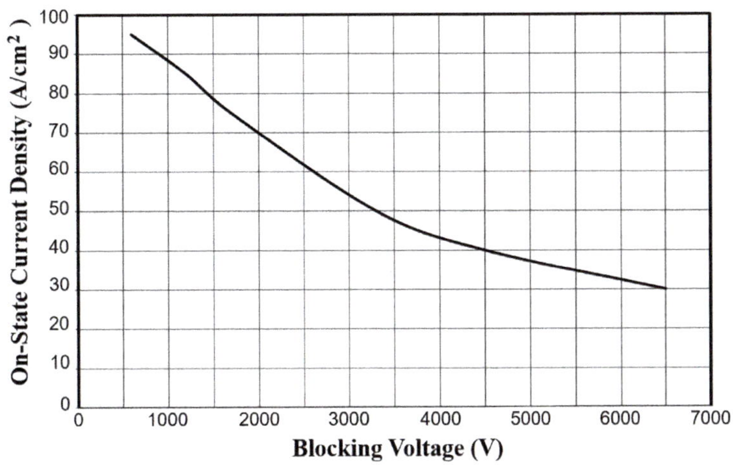

FIGURE 21.2

On-state current density for insulated gate bipolar transistors with increasing voltage rating.

The turn-off time for IGBTs is characterized by the manufacturers in their datasheets. The turnoff under inductive load switching is of most interest to the users of the devices in motor drive applications. The turn-off time consists of a delay tie and a current fall time. The current fall time is pertinent to the determination of the power losses. The current fall time for the IGBT products listed in Table 21.1 is plotted in

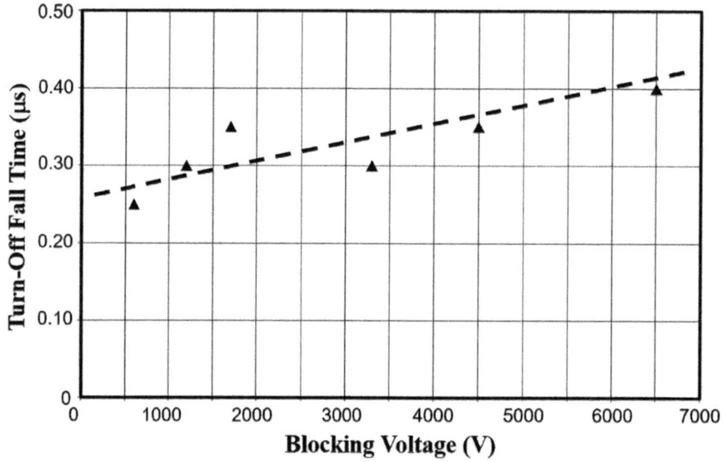

FIGURE 21.3

Turn-off fall time for insulated gate bipolar transistors with increasing voltage rating.

Fig. 21.3 as a function of the blocking voltage. There is an increase in the fall time when the blocking voltage is scaled up because of more time required to remove the stored charge from the thicker drift regions required to support the larger voltages.

21.2 Wide Bandgap Semiconductor Power Devices

Unipolar power devices, such as power MOSFETs, rely on current transport using a single carrier. Electron transport is preferable to hole transport because of the larger mobility for these carriers in semiconductors. For power MOSFETs designed to support high blocking voltages, the on-state current is constrained by the resistance within the device [2]. One of the major components of the on-resistance is from the drift region, which is designed with a low doping concentration and a large thickness to support the blocking voltage. The power MOSFET device structure must be optimized to reduce the contribution to the resistance from components such as the channel region and the contacts until the specific on-resistance (on-resistance per unit area) becomes as close as possible to that of the drift region. The optimization is constrained by the device structure (planar gate vs. trench gate) and process lithography resolution.

The specific on-resistance of the drift region can be related to the fundamental properties of the semiconductor material [3].

$$R_{on\ sp,Ideal} = \frac{4BV^2}{\varepsilon_S \mu_n E_C^3} \tag{21.1}$$

where BV is the breakdown voltage, ε_S is the dielectric constant of the semiconductor, μ_n is the mobility for electrons, and E_C is the critical electric field for breakdown.

The denominator for this equation is commonly referred to as Baliga's figure of merit (BFOM) for power devices. From the equation, it can be concluded that semiconductors with large dielectric constants, high electron mobilities, and especially high critical electric field for breakdown can be used to reduce the specific on-resistance of the drift region. Semiconductors with larger bandgap energy have a larger critical electric field for breakdown. This has motived the development of power devices from wide bandgap semiconductors ever since the BFOM was postulated 40 years ago.

The first wide bandgap semiconductor power devices were made from gallium arsenide in the 1980s because the BFOM was found to be 13.6 times larger than that for silicon [4]. In 1989, it was projected [5] based upon the available data that silicon carbide power devices would have a BFOM of more than 100. The first high-voltage silicon carbide Schottky rectifiers were reported [6] in 1992, validating the theoretical predictions. The first high-performance silicon carbide power MOSFET was reported subsequently in 1997 [7]. In 1994, it was projected that power losses in variable-speed motor drives can be reduced by an order of magnitude by replacing silicon P-i-N rectifiers with silicon carbide Schottky rectifiers and silicon IGBTs with silicon carbide power MOSFETs [8]. This provided strong motivation for worldwide investment in wide bandgap semiconductor material and then device technology. These research results from North Carolina State University and the work conducted at many other laboratories around the world eventually resulted in the announcement of silicon carbide power device products by many companies after 2005.

21.2.1 State-of-the-Art SiC Power MOSFETs

For silicon carbide power MOSFETs, the shielded planar gate device structure is preferable to the trench gate structure to prevent a high electric field in the gate oxide [9,10]. The specific on-resistance for the shielded planar gate silicon carbide power MOSFET is determined by the sum of the channel, the accumulation layer, the junction field effect transistor (JFET), and drift region resistances. An analytical model for these resistances is provided in the references. Among these resistances, the channel and drift region components are dominant. The specific on-resistance of the shielded planar silicon carbide power MOSFETs is plotted in Fig. 21.4 based on using an inversion channel mobility of 15 cm^2/V-s and a channel length of 0.5 μm with a gate drive voltage of 15 V. For blocking voltages below 2000 V, the channel resistance is dominant. For blocking voltages above 3000 V, the specific on-resistance begins to approach the drift region resistance.

The performance of state-of-the-art commercially developed silicon carbide power MOSFETs in 2014 has been reported [11]. The authors state: "*We have achieved new breakthrough performance for voltage ratings from 900 V up to 15 kV with a* $R_{ON,SP}$ *as low as 2.3 mΩ-cm*^{2} *for a breakdown voltage (BV) of 1230 V and 900 V rating, 2.7 mΩ-cm*^{2} *for a BV of 1620 V and 1200 V rating, 3.38 mΩ-cm*^{2} *for a BV of 1830 V and 1700 V rating, 10.6 mΩ-cm*^{2} *for a BV of 4160 V and*

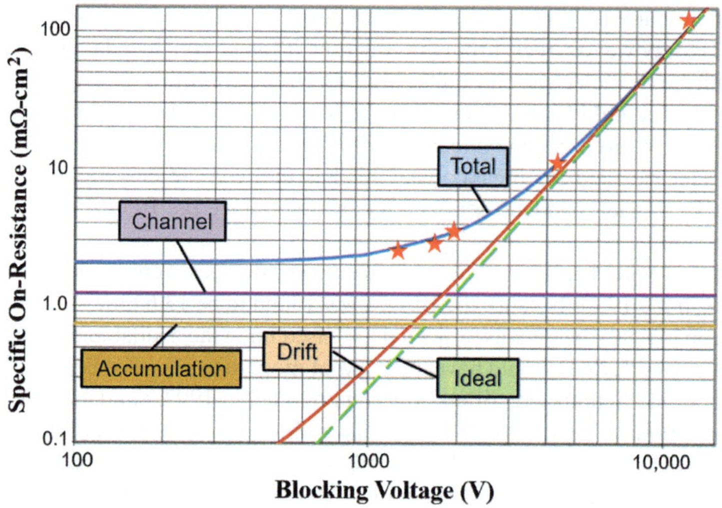

FIGURE 21.4

Specific on-resistance of state-of-the-art silicon carbide power MOSFETs.

3300 V rating, and 123 mΩ-cm² for a BV of 12 kV and 10 kV rating." These results are shown in the figure with stars as data points using the breakdown voltages. These data indicate that the specific on-resistance for devices with ratings of below 2000 V are limited by the channel resistance consistent with the analytical model. For devices with blocking voltages above 3000 V, the specific on-resistance is close to the ideal limit, as expected from the analytical model. The specific on-resistance for SiC power MOSFETs reported at the end of 2018 [12] is similar to that shown in Fig. 21.4, indicating limited progress in enhancement of this important device parameter.

The on-state characteristic of the shielded planar gate silicon carbide power MOSFETs with various blocking voltage ratings is shown in Fig. 21.5. The specific on-resistance chosen for these characteristics corresponds to the values for various devices reported in the reference [11]. The constant power dissipation curves are also shown in this figure for the cases of 100, 125, and 150 W/cm^2 [2]. Based on the analysis shown in the previous section for silicon IGBT products, it will be assumed here that the case of a power dissipation of 150 W/cm^2 is applicable for the silicon carbide devices. The on-state operating point for the shielded planar gate silicon carbide power MOSFETs is then given by the stars in the figure.

The on-state current density obtained for the shielded planar gate silicon carbide power MOSFETs is shown in Fig. 21.6 as a function of the blocking voltage rating. The on-state current density reduces with increasing blocking voltage rating, as in the case of the silicon IGBT. For purposes of comparison, the on-state current density for the silicon IGBT is included in this figure. The ratio of these on-state current levels for the devices is given in Table 21.2. It can be seen that the difference in current density is about a factor of 2.75 for the devices with blocking voltages below

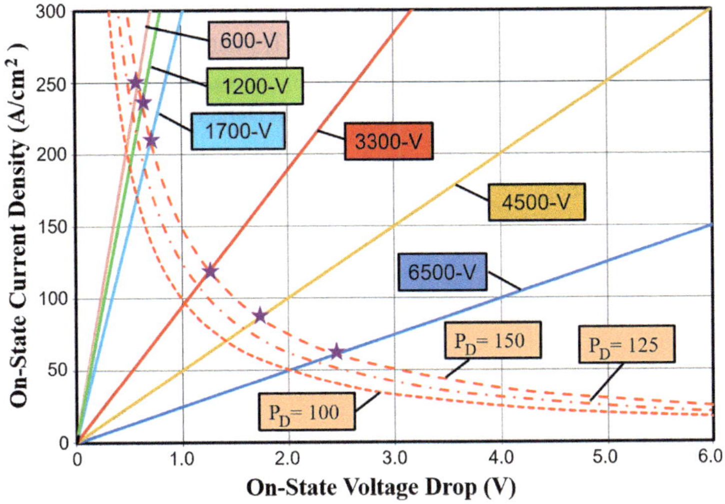

FIGURE 21.5

On-state characteristics of silicon carbide power MOSFETs with various blocking voltages.

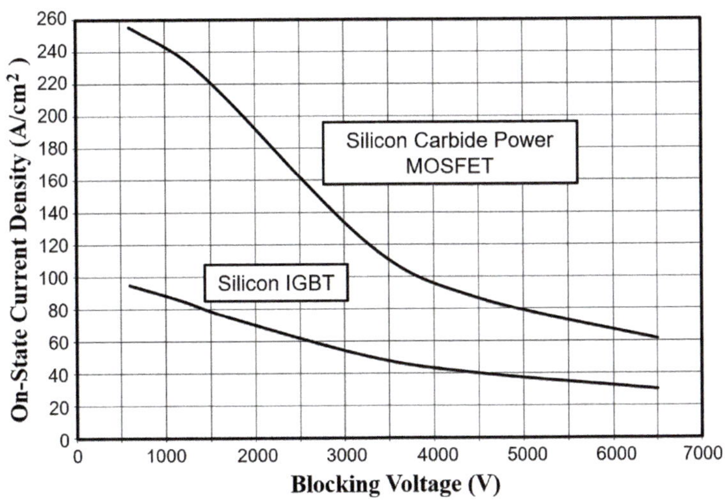

FIGURE 21.6

On-state current density for silicon carbide power MOSFETs compared with silicon insulated gate bipolar transistors.

2000 V. The difference is smaller for devices with larger blocking voltages due to the lack of maturity of the silicon carbide devices.

The turn-off time for the silicon carbide power MOSFETs has been measured under various switching conditions. It has been found that the turn-off time for the silicon carbide power MOSFETs is typically 0.1 μs [13,14]. This is three times faster

Table 21.2 Comparison of On-State Current Density for SiC Power MOSFET With Si Insulated Gate Bipolar Transistor

Blocking Voltage (V)	Si IGBT Current Density (A/cm^2)	SiC Power MOSFET Current Density (A/cm^2)	Ratio
600	95	255	2.68
1200	85	235	2.76
1700	75	210	2.80
3300	50	120	2.40
4500	40	85	2.13
6500	30	60	2.00

than the turn-off fall times for the silicon IGBT. The resulting reduction in the switching losses allows operation of converters using the silicon carbide devices at higher frequencies. This reduces the size and weight of the passive elements, such as capacitors, inductors, and filters. An improvement in the efficiency by about 0.5–1% has been reported by replacing silicon IGBTs with silicon carbide power MOSFETs for some applications [15–17]. A volume reduction by a factor of 10 with an efficiency improvement of 6% has been reported for a plug-in hybrid electric car charger [18].

21.2.2 Cost analysis

The adoption of silicon carbide power devices into applications is strongly dependent on the cost of the devices in comparison with silicon IGBTs with the same rated voltage and current. A methodology for computation of costs has been reported [18]. For a dedicated foundry running 100 mm diameter silicon carbide wafers in low volumes, the processed wafer cost is estimated to be $3000 based on the initial wafer (with epitaxial layer) cost of $1200 and the processing cost of $1800. Using this information, the cost for manufacturing a 20 A silicon carbide 1200 V power MOSFET product can be computed as shown in Table 21.3. A device yield of 60% was assumed here with an edge termination plus dicing space of 0.027 cm [19]. The price per ampere of current for the silicon carbide power MOSFET is found to be $0.65 for a gross margin of 50%. In comparison, the price per ampere of current for 1200 V silicon IGBTs is $0.11 [18]. Consequently, the cost of the silicon carbide power MOSFET is 5.9 times larger than that for the silicon IGBT. Some end users of the technology have argued that they will not replace silicon IGBTs with silicon carbide power MOSFETs unless the cost is nearly the same [20]. Major effort is underway around the world to reduce the manufacturing cost for the silicon carbide devices. However, the silicon IGBT is not likely to be rapidly displaced by silicon carbide power MOSFETs during the next decade. It may make an impact on high-performance electric vehicles in the near term due to adoption by Tesla [21]. The Tesla Model S made use of multiple IGBTs in TO-247 packages while the newer Model 3 makes use of SiC power MOSFETs.

Table 21.3 Cost Analysis for 1200 V SiC Power MOSFETs

Parameter	Units
On-state current	20 A
On-state current density	235 A/cm^2
Active area	0.085 cm^2
Die area	0.102 cm^2
Wafer size	100 mm
Yield	60%
Number of good die per wafer	463
Cost per chip	$6.47
Price per chip	$12.93
Cost per ampere	$0.65

21.3 Summary

The global IGBT market reached more than $8 billion by 2021. It is expected to grow at a cumulative annual growth rate) of 12.5% for the foreseeable future. The largest fraction of this market is the Asia/Pacific region, followed by North America and Europe. About $4 billion of this market is for electric and hybrid electric vehicles, $1.5 billion for solar inverters, $1 billion for industrial motor drives, and $0.5 billion each for electric trains, uninterruptible power supplies, and wind power. The IGBT products have a sales price of about $0.5 per ampere, making them difficult to displace with other power device technologies in the future unless they offer compelling overall system cost and performance savings.

The IGBT made a large impact on many sectors of the economy almost immediately after it was conceived and commercialized at the General Electric Company in the early 1980s. Its impact has steady grown to include every sector of the economy until society would cease to function effectively today without this device. The efficiency improvements enabled by the device have saved huge amounts of electrical energy and gasoline, leading to enormous economic and environmental benefits to billions of people in all parts of the world. The penetration of the technology has still further room to grow, indicating an even larger impact in the future. This impact will be most importantly felt in the widespread deployment of renewable energy generation and electric vehicles. A successful commitment by the worldwide community to replace power generation using fossil fuels with renewable sources and the replacement of all gasoline-powered vehicles with electrical counterparts could overcome the threat of global warming by more than 2°C [22].

References

[1] M. Rahimo, Future trends in high power bipolar metal-oxide semiconductor controlled power semiconductors, IET Circuits Devices Syst. 8 (2014) 155–167.

[2] B.J. Baliga, Fundamentals of Power Semiconductor Devices, second ed., Springer-Science, New York, 2019.

[3] B.J. Baliga, Semiconductors for high voltage vertical channel field effect transistors, J. Appl. Phys. 53 (1982) 1759–1764.

[4] B.J. Baliga, et al., Gallium arsenide Schottky power rectifiers, IEEE Trans. Electron Devices ED-32 (1985) 1130–1134.

[5] B.J. Baliga, Power semiconductor device figure of merit for high frequency applications, IEEE Electron Device Lett. 10 (1989) 455–457.

[6] M. Bhatnagar, P.K. McLarty, B.J. Baliga, Silicon-carbide high-voltage (400 V) schottky barrier diodes, IEEE Electron Device Lett. 13 (1992) 501–503.

[7] P.M. Shenoy, B.J. Baliga, The planar 6H-SiC ACCUFET, IEEE Electron Device Lett. 18 (1997) 589–591.

[8] B.J. Baliga, Power semiconductor devices for variable-frequency drives, Proc. IEEE 82 (1994) 1112–1122.

[9] B.J. Baliga, Gallium Nitride and Silicon Carbide Power Devices, World Scientific Publishing Company, Singapore, 2017.

[10] B.J. Baliga, Advanced Power MOSFET Concepts, Springer-Science, New York, 2010.

[11] J.W. Palmour, et al., Silicon carbide power MOSFETs: breakthrough performance from 900 V up to 15 kV, in: IEEE International Symposium on Power Semiconductor Devices and ICs, 2014, pp. 79–82.

[12] A. Mihaila, et al., The current status and future prospects of SiC high voltage technology, in: IEEE International Electron Devices Meeting, 2018, pp. 19.2.1–19.2.4.

[13] J. McBryde, et al., Performance comparison of 1200-V silicon and SiC devices for UPS application, in: IEEE Industrial Electronics Society Annual Conference, 2010, pp. 2657–2662.

[14] Y. Matsumoto, et al., Characteristics of the power electronics equipments applying the SiC power devices, in: IEEE Conference on Power Engineering and Renewable Energy, 2012, pp. G1–G6.

[15] O. Stalter, B. Burger, S. Lehrman, Silicon carbide D-MOS for grid-feeding solar-inverters, in: IEEE European Power Electronics and Applications Conference, 2007, pp. 1–10.

[16] K. Mino, et al., Power electronics equipments applying novel SiC power semiconductor modules, in: IEEE Applied Power Electronics Conference, 2009, pp. 1920–1924.

[17] S. Buschhorn, K. Vogel, Saving money: SiC in UPS applications, in: IEEE Power Electronics, Intelligent Motion, Renewable Energy and Energy Management Conference, 2014, pp. 765–771.

[18] A. Agarwal, WBG revolution in power electronics, in: IEEE Workshop on Wide Bandgap Power Devices & Applications, 2014.

[19] W. Sung, et al., A new edge termination for high voltage devices in 4H-SiC – multiple floating zone junction termination extension, IEEE Electron Device Lett. 32 (2011) 880–882.

[20] C. Chen, M. Su, The opportunities and challenges of wide-band-gap technologies for automotive applications, in: IEEE Workshop on Wide Bandgap Power Devices & Applications, 2014.
[21] https://www.pntpower.com/tesla-model-3-powered-by-st-microelectronics-sic-mosfets/.
[22] B.J. Baliga, Enabling Green and Renewable Energy Solutions with Power Semiconductor Technology, Global Renewable Energy Researchers Meet, 2021.

Index

Note: Page numbers followed by *f* indicate figures and *t* indicate tables.

D

F

U

V

W

9780323999120